Probability and Mathematical Statistics
Theory, Applications, and Practice in R

Probability and Mathematical Statistics
Theory, Applications, and Practice in $\mathbb{R}$

Mary C. Meyer
Colorado State University
Fort Collins, Colorado

Society for Industrial and Applied Mathematics
Philadelphia

Publications Director	Kivmars H. Bowling
Acquisitions Editor	Paula Callaghan
Developmental Editor	Gina Rinelli Harris
Managing Editor	Kelly Thomas
Production Editor	Louis R. Primus
Copy Editor	Gary Davidoff
Production Manager	Donna Witzleben
Production Coordinator	Cally A. Shrader
Compositor	Cheryl Hufnagle
Graphic Designer	Doug Smock

Library of Congress Cataloging-in-Publication Data
Names: Meyer, Mary (Mary Catherine), author.
Title: Probability and mathematical statistics : theory, applications, and
  practice in R / Mary Meyer (Colorado State University, Fort Collins,
  Colorado).
Description: Philadelphia : Society for Industrial and Applied Mathematics,
  [2019] | Includes bibliographical references and index.
Identifiers: LCCN 2019005289 (print) | LCCN 2019014686 (ebook) | ISBN
  9781611975789 | ISBN 9781611975772 (print)
Subjects: LCSH: Probabilities--Problems, exercises, etc. | Mathematical
  statistics--Problems, exercises, etc. | R (Computer program language)
Classification: LCC QA273.25 (ebook) | LCC QA273.25 .M5384 2019 (print) |
  DDC
  519.5--dc23
LC record available at *https://lccn.loc.gov/2019005289*

To the past and present students in the Masters of Applied Statistics program at Colorado State University. Many thanks for your valuable feedback while this work was under construction and for all your help with finding typos!

Contents

Preface

Probability is a fascinating field in its own right, but from a statistician's point of view, one studies probability in order to build models to do statistical inference. We'll keep the goal of data analysis and statistical inference in mind throughout the development of probability theory.

Probability and statistics are in a sense inverses of one another. For exercises in probability, a model is given with all the information (i.e., parameter values), and the task is to compute a probability of an event. For exercises in statistics, the task is to estimate the model information (parameter values) given that an event has happened. For example, a probability exercise could be the following:

> We know that 60.2% of voters support proposal Q; if we take a random sample of 20 voters, what's the probability that more than half support the proposal?

An exercise in statistics will rather say:

> It's three weeks before voting day, and we need to estimate p, the unknown proportion of voters who support proposal Q. If we take a random sample of n voters, how do we get an estimate and error bounds for p?

Clearly the second exercise is more "real world" but we have to know how to do the first before we can attempt the second. The following is another example of a probability exercise:

> The amount of fluoride (in mg) in one-liter samples of tap water in a certain city follows a normal distribution with mean 2 and standard deviation .4. If we take a random sample of five liters of tap water, what is the probability that the mean is more than 2.5 mg?

A corresponding statistics exercise, however, would look like this:

> A random sample of $n = 5$ liters of tap water has a mean of 2.5 mg/l, and the sample standard deviation is .56. Is this strong evidence that the true mean is higher than the EPA-mandated 2mg/l?

Statistics is all about doing inference in the presence of uncertainty, *and quantifying that uncertainty*. It's not enough to say, "we estimate the quantity of interest to be 2.56"—the next question will be "plus or minus what?" The person interested in the estimate will also be interested in the precision of the estimate. When we make a decision or conclusion based on data, we want to quantify our level of confidence. Are we really sure it's the right conclusion, or just somewhat sure? Statisticians can put a number on the "sureness" and interpret it. Remember, statistics means never having to say you're certain, but you do have to quantify the degree of uncertainty.

Roughly the first half of this text (through Chapter 33) is concerned with developing ideas about probability, and the second half is statistics. However, we will sneak in statistical ideas quite early. In Chapter 6, we introduce hypothesis testing terminology, so that for each of the families of probability distributions we subsequently encounter, statistics exercises make the material more interesting and more relevant to the student interested in applications. In this way, important ideas in hypothesis testing become second nature through repetition in many different contexts.

We introduce computer simulations of probability distributions in Chapter 2. Throughout the text, the R package is used to compute probabilities, to check analytically computed answers, to simulate probability distributions, to illustrate our answers with appropriate graphics, and generally to develop intuition about probability and statistics. Simulated null distributions of test statistics, and other numerical methods such as bootstrap confidence intervals, are common in the real world, where models are often too complicated to admit a nice closed-form analytical solution.

Some annotated R code and any future supplementary files or errata will be made available on the book's webpage,

www.siam.org/books/ot162

Chapter 1

Probability Basics: Sample Spaces and Probability Functions

A fundamental goal of statistical inference is to make predictions and conclusions based on data in the presence of randomness or measurement error. Probability models help us quantify the uncertainty in our results. We'll start our studies with basic probability concepts: sample spaces, some simple set theory, and probability functions.

Some definitions and examples

Data are obtained by conducting a **random experiment**, and in this context, an experiment is simply something you do to get information. An **outcome** is a possible result of the experiment.

The set of all possible outcomes of a particular experiment is called a **sample space.** Each outcome must be a distinct, smallest unit of the sample space.

An **event** is a collection of outcomes. Here are some simple examples:

1. The experiment is to roll a six-sided die. A sample space of outcomes is $S = \{1, 2, 3, 4, 5, 6\}$. An event might be $A = \{2, 4, 6\}$, which could be described as "roll is an even number."

2. The experiment is to choose a voter from the list of registered voters in Smalltown. The possible outcomes might be Y, "voter supports Proposition Q," and N, "voter does not support Proposition Q." The sample space could be represented as $S = \{Y, N\}$.

3. If the experiment is to choose eight voters from the list of registered voters in Smalltown, we could write the possible outcomes as sequences of letters such as $YYNYNYYN$, representing the opinions of the selected voters. An event might be "at least four voters support Proposition Q."

4. The experiment is to choose a card from a standard 52-card deck. The sample space can be written as

$$S = \left\{ \begin{array}{ccccccccccccc} A\heartsuit & 2\heartsuit & 3\heartsuit & 4\heartsuit & 5\heartsuit & 6\heartsuit & 7\heartsuit & 8\heartsuit & 9\heartsuit & 10\heartsuit & J\heartsuit & Q\heartsuit & K\heartsuit \\ A\diamondsuit & 2\diamondsuit & 3\diamondsuit & 4\diamondsuit & 5\diamondsuit & 6\diamondsuit & 7\diamondsuit & 8\diamondsuit & 9\diamondsuit & 10\diamondsuit & J\diamondsuit & Q\diamondsuit & K\diamondsuit \\ A\clubsuit & 2\clubsuit & 3\clubsuit & 4\clubsuit & 5\clubsuit & 6\clubsuit & 7\clubsuit & 8\clubsuit & 9\clubsuit & 10\clubsuit & J\clubsuit & Q\clubsuit & K\clubsuit \\ A\spadesuit & 2\spadesuit & 3\spadesuit & 4\spadesuit & 5\spadesuit & 6\spadesuit & 7\spadesuit & 8\spadesuit & 9\spadesuit & 10\spadesuit & J\spadesuit & Q\spadesuit & K\spadesuit \end{array} \right\}.$$

An event might be "a two is drawn," or $A = \{2\heartsuit, 2\diamondsuit, 2\spadesuit, 2\clubsuit\}$.

5. The experiment is to grow a pea plant from a randomly chosen seed and observe the phenotype color characteristics. If the flowers can be purple or white and the peas can be yellow or green, then the sample space of possible color combinations for the pea plant is

$$S = \{PY, PG, WY, WG\},$$

where P indicates that the flower is purple, W indicates that the flower is white, Y indicates that the pea is yellow, and G indicates that the pea is green. The event "flower is purple" is the subset $\{PY, PG\}$.

6. The experiment is to roll two six-sided dice. A sample space of outcomes is

$$S_1 = \{2, 3, 4, 5, 6, 7, 8, 9, 10, 11, 12\},$$

representing the sum on the dice. This is a valid sample space, because everything that *can* happen is listed, and no two outcomes can happen simultaneously. An alternative way to write the sample space, which provides more detail, is

$$S_2 = \left\{ \begin{array}{cccccc} (1,1) & (1,2) & (1,3) & (1,4) & (1,5) & (1,6) \\ (2,1) & (2,2) & (2,3) & (2,4) & (2,5) & (2,6) \\ (3,1) & (3,2) & (3,3) & (3,4) & (3,5) & (3,6) \\ (4,1) & (4,2) & (4,3) & (4,4) & (4,5) & (4,6) \\ (5,1) & (5,2) & (5,3) & (5,4) & (5,5) & (5,6) \\ (6,1) & (6,2) & (6,3) & (6,4) & (6,5) & (6,6) \end{array} \right\}.$$

Using S_2 we can define events such as "doubles are rolled," or

$$A = \{(1,1), (2,2), (3,3), (4,4), (5,5), (6,6)\}.$$

It's traditional for the sample space to be called S, and events are usually denoted by capital letters like A and B. Next we review some set notation; here the terms "set" and "event" are used interchangeably.

We say $A \subseteq B$, or "A is a *subset* of B," if each outcome in A is also an outcome of B. That is, if event A occurs, we know that event B has occurred. We use the notation $A \subset B$ if A is a *proper* subset of B, i.e., there is an outcome of B that is not an outcome of A.

The event $A \cap B$, "A intersect B," is the set of outcomes in *both* A and B.

The event $A \cup B$, "A union B," is the set of outcomes in *either* A or B, or in both.

The event A^c or "A complement" is the set of outcomes of S that *aren't* in A.

The **empty set** is the set with nothing in it; the symbol $\emptyset$ is traditionally used to represent it. We say that two events are **disjoint** or **mutually exclusive** if their intersection is empty.

By thinking about the definitions of intersection and union, you can convince yourself that $A \cap \emptyset = \emptyset$ and $A \cup \emptyset = A$ for any event A.

For finite sample spaces, a **probability function** is a rule for assigning probabilities to outcomes in a sample space. The probability function is a mapping from the sample space to real numbers in $[0, 1]$, so that all the assigned probabilities are nonnegative, and they sum to one.

A simple example is a sample space with "equally probable" outcomes, so that if there are n outcomes, each has probability $1/n$. If the experiment is to roll a fair, six-sided die, each of the six outcomes in the sample space $\{1, 2, 3, 4, 5, 6\}$ has probability $1/6$.

For example 5, the four possible phenotypes might not be equally probable; instead, Mendelian laws might indicate that the probability of phenotype PY is $9/16$, the probability of phenotype PG is $3/16$, the probability of phenotype WY is also $3/16$, and the probability of phenotype WG is $1/16$.

If the experiment is to roll two fair, six-sided dice, then if we write the sample space as having 36 outcomes as in S_2 of example 6 above, we can assign the probability $1/36$ to each of the outcomes. We can "calculate" the probability of the event A of rolling a five, for example, by adding up the probabilities of the equally likely outcomes, $A = \{(1,4), (2,3), (3,2), (4,1)\}$, to get that the probability of A is $4/36 = 1/9$. Similarly, the probability of the event "rolling a 12" is $1/36$. If we write the sample space as $S_1 = \{2, 3, 4, 5, 6, 7, 8, 9, 10, 11, 12\}$, these outcomes are *not* equally likely. (We could theoretically assign probabilities such as $1/11$ to each of the outcomes in S_1, but this would not reflect the probabilities for real dice.) We can use the probabilities for the equally likely outcomes in S_2 to compute realistic probability assignments for the outcomes in S_1.

Mathematically, we can assign any probability function to a sample space of outcomes, as long as the function follows the rules of assigning nonnegative values, and the values add to one. However, if we want the probabilities to model what happens in the real world, the probability of an outcome must represent the proportion of times the outcome would occur, if the same random experiment were repeated an unlimited number of times. Intuitively, this proportion is what we mean by the probability of an outcome or of an event. These ideas are the building blocks for statistical models where the goal is estimation and inference in the presence of randomness.

"Let's Make a Deal"

A famous probability exercise is based on a scene from a popular television game show from the 1960s. A contestant is presented with three doors and is told that behind one of the doors is a prize. She is asked to choose one of the doors—if she chooses the door with the prize, she wins it! However, after she chooses one of the three doors, the game show host (who knows where the prize is) then opens a door that does not contain the prize, and is also not the door she has chosen. The host then asks her if she wants to keep her initial choice, or does she want to choose the remaining door. What should she do, and why?

Because there are two doors remaining, it might seem that the prize is equally likely to be behind either. However, the solution is that she should switch doors, and the probability of winning the prize with this strategy is 2/3. To understand this, it is important to think of the probability of an event (winning the prize) as a proportion of times the event would occur, over repetitions of the same experiment. Suppose that the game is repeated many times, and the contestant always uses the strategy to switch doors. Clearly, she will lose only when the original choice was correct! She will switch to the door with the prize 2/3 of the time.

Infinite sample spaces

For some situations a finite sample space (as in the above examples) is insufficient. Here are two examples with infinite sample spaces:

1. Suppose the experiment is to roll a six-sided die until we see a six. Then if the outcomes in the sample space represent the number of times the die is rolled,

we can write $S = \{1, 2, 3, \ldots\}$, that is, the sample space consists of all positive integers.

2. Suppose that the experiment is to power a flashlight with a AA battery and measure the time until the battery wears out. The sample space of possible times might be $S = \{x : x > 0\}$, that is, all positive real numbers.

For the first example, it is possible to assign positive probabilities to all positive integers, and have the sum of these probabilities equal to one. We know from calculus that an infinite series can have a finite sum. Then we can figure out the probabilities for all possible events, or subsets of the sample space, simply by adding the individual probabilities.

For the second example, we can't assign probabilities to individual outcomes, because we can't make a sum over all the positive real numbers. We still want to be able to assign probabilities to subsets of real numbers, but it turns out that we can't consider *all* subsets of real numbers; we have to limit our set of subsets in order to get a proper probability function. In more theoretical treatments of probability theory, a collection of sets called a *sigma algebra* is defined, so that the domain of the function that assigns probabilities consists of the elements of the sigma algebra. However, for most practical problems (and in this book) we need only consider the sigma algebra that is based on collections of open intervals of real numbers, so that subsets of the real numbers to which we will assign probabilities can be written as intersections, unions, and complements of open intervals.

We say that a sample space is **discrete** if the number of outcomes is finite or **countably infinite**, which means that there is a one-to-one mapping between the outcomes and the positive integers. (See Appendix B.1 for a discussion of countable and uncountable infinity.) We can assign probabilities to each of the outcomes in a discrete sample space, and any subset of outcomes in a discrete sample space can be an event. In this book, sample spaces that are **uncountable** will consist of intervals on the real number line, or contiguous subsets of $\mathbb{R}^k$ for some positive integer k. We consider only events that can be written as intersections, unions, and complements of open sets, so that by assigning probabilities to these open sets, we can compute probabilities of any such events.

Probability functions

The assignment of probabilities to events that are subsets of the sample space (or elements of the sigma algebra) can be thought of as a function P whose domain is the set of all such events, and the range is real numbers between zero and one. Of course not every such function can be a proper probability function, but there are rules for probability functions that are very intuitive, given the motivation and context of a probability function. The first three rules are called the **Axioms of Probability**:

1. For any event $A \subseteq S$, we must have $P(A) \geq 0$.

2. $P(S) = 1$.

3. If $A_1, A_2, \ldots,$ is a sequence of disjoint events in S, then

$$P\left(\overset{\infty}{\underset{i=1}{\cup}} A_i\right) = \sum_{i=1}^{\infty} P(A_i).$$

If a function satisfies these three axioms, we can call it a probability function.

Axioms are a minimal set of assumptions that can be used to prove other statements. Starting with these axioms, we can derive the following additional rules:

4. $P(\emptyset) = 0$.

5. If $A_1, \ldots, A_n$ are disjoint events in S, then

$$P\left(\bigcup_{i=1}^{n} A_i\right) = \sum_{i=1}^{n} P(A_i).$$

6. $P(A^c) = 1 - P(A)$.

7. If $A \subseteq B$, then $P(A) \leq P(B)$.

8. $P(A \cup B) = P(A) + P(B) - P(A \cap B)$.

These rules might seem "obvious" intuitively, but you can prove them formally, starting with the axioms (see Exercise 1.5). **DeMorgan's laws** are not so intuitively obvious: For events A and B,

$$(A \cup B)^c = A^c \cap B^c \ \text{ and } \ (A \cap B)^c = A^c \cup B^c.$$

For any events A, B, and C, the **distributive laws** are

$$A \cap (B \cup C) = (A \cap B) \cup (A \cap C) \ \text{ and } \ A \cup (B \cap C) = (A \cup B) \cap (A \cup C).$$

Venn diagrams are useful for depicting sets. Traditionally, the sample space is shown as a rectangle and events are shown as circles or ovals. For example, the intersection of events A and B in the sample space S can be seen as

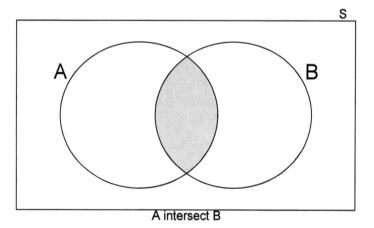

A intersect B

Venn diagrams can be used for "proving" rules in set theory. For example, Exercise 1.3 asks you to make a Venn diagram to convince yourself of the validity of DeMorgan's laws, and Exercise 1.4 asks you to make Venn diagrams to verify the distributive laws.

Venn diagrams can be used to organize probability assignments and to answer questions about probabilities. For example, suppose the "experiment" is to randomly choose a tree from the Arboles National Park. Let event A be that the tree has substantial beetle damage, and let B be the event that tree has been drought stressed in the last three years. Suppose $P(A) = 0.35$ (35% of trees have substantial beetle damage), $P(B) = 0.60$ (60% of trees have been drought stressed), and $P(A \cap B) = 0.30$ (30% of trees have been drought stressed *and* have substantial beetle damage.

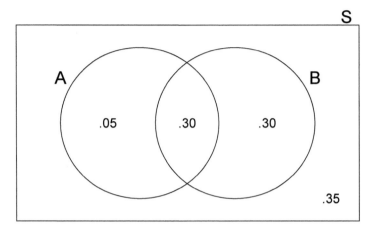

Using the information given, each section of the Venn diagram is filled in with a probability, and all the probabilities must add to one. Now we can answer questions such as, "what is the proportion of trees that have substantial beetle damage but have not been drought stressed?" The answer is $P(A \cap B^c) = .05$. Similarly, the number .35 is the proportion of trees that were not drought stressed and don't have substantial beetle damage. We can also see that half of drought-stressed trees have substantial beetle damage, while only 1/7 of trees that were not drought stressed have substantial beetle damage.

Chapter Highlights

1. A sample space is the set of all possible outcomes of an experiment. An event is a collection of outcomes.

2. If S is the sample space and A and B are events, then

    (a) $A \subseteq B$ means that outcomes in A are also in B;

    (b) $A \cap B$ is the set of outcomes in both A and B;

    (c) $A \cup B$ is the set of outcomes in A or in B or in both A and B;

    (d) A^c is the set of outcomes in S that are not in A.

3. If S is the sample space and A, B, and C are events, then

    (a) $P(A \cup B) = P(A) + P(B) - P(A \cap B)$;

    (b) DeMorgan's laws: $(A \cup B)^c = A^c \cap B^c$ and $(A \cap B)^c = A^c \cup B^c$;

(c) distributive laws:

    i. $A \cap (B \cup C) = (A \cap B) \cup (A \cap C)$ and

    ii. $A \cup (B \cap C) = (A \cup B) \cap (A \cup C)$.

4. Sample spaces can be countable (finite or countably infinite) or uncountable. For countable sample spaces, we can assign a probability to each outcome. For uncountable sample spaces, we can assign probabilities only to elements of a sigma algebra, which is a collection of subsets of the sample space. In this book the only uncountable sample spaces will be contiguous subsets of $\mathbb{R}^k$, and the only sigma algebra we will consider is based on open sets. Because it's challenging to find a set in $\mathbb{R}^k$ that is *not* a member of this sigma algebra, we can summarize by saying that for uncountable sample spaces we will find a way to assign probabilities to all open sets, and sets that are made from intersections, unions, and complements of open sets.

5. Probability functions map elements of the sample space (in the countable case) or elements of the sigma algebra (in the uncountable case) to $[0, 1]$. These functions must satisfy the axioms of probability:

  (a) For any event $A \subseteq S$, we must have $P(A) \geq 0$.

  (b) $P(S) = 1$.

  (c) If $A_1, A_2, \ldots,$ is a sequence of disjoint events in S, then

$$P \left(\mathop{\cup}_{i=1}^{\infty} A_i \right) = \sum_{i=1}^{\infty} P(A_i).$$

❧❧

Exercises

1.1 Make a Venn diagram, and use shading to simplify the expression $(A \cap B) \cup (A^c \cap B)$.

1.2 Make a Venn diagram to "explain" $P(A \cup B) = P(A) + P(B) - P(A \cap B)$.

1.3 Make Venn diagrams to convince yourself of the validity of De Morgan's laws: (a) $(A \cup B)^c = A^c \cap B^c$, and (b) $(A \cap B)^c = A^c \cup B^c$.

1.4 Make Venn diagrams with three sets to convince yourself of the validity of the distributive laws:

  (a) $A \cap (B \cup C) = (A \cap B) \cup (A \cap C)$.

  (b) $A \cup (B \cap C) = (A \cup B) \cap (A \cup C)$.

1.5 Prove the rules 4–8 from the axioms of probability, in any order. Once you have proved a rule, you can use it in another proof.

1.6 A four-sided die (as shown below) is "fair" if landing on any of the four sides is equally likely. (The result of the roll is the unseen side that it lands on.) Suppose

we roll two of these dice. Write out a sample space of equally likely outcomes.
Determine the probability of "doubles"—both dice land on the same number.

1.7 A spinner toy shown below has a movable arrow—if the arrow is spun vigorously,
it is reasonable to assume that it is equally likely to land on any of the three colors
(red, green, and blue). Both you and your friend spin the arrow.

    (a) What is the probability that the two spins
land on the same color?

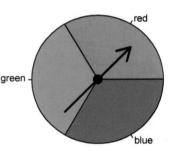

    (b) What is the probability that the two
spins land on different colors?

    (c) What is the probability that neither color is
green?

1.8 A dermatology clinic is participating in a study concerning treatments for eczema.
Patients are randomly assigned to treatment A, B, or C, and after one month of
treatment, the result is recorded as I (improved) or N (not improved). Write out a
sample space of all possible combinations for the next patient arriving at the clinic.

1.9 Suppose you are eating at a pizza parlor with two friends. You have agreed to the
following rule to decide who will pay the bill: Each person will toss a coin and the
person who gets a result different from the others will pay the bill. If all three tosses
are the same, the bill will be shared by all. Make a sample space for the random
experiment and assign reasonable probabilities. Find the probability that

    (a) you will pay for everyone;

    (b) the bill will be shared;

    (c) you get a free lunch.

1.10 Angela, Booker, Clayton, Dwanna, and Elizabeth work as office staff for a small
law firm. Two of them will be chosen at random to attend a conference in Hawaii.
(Note that "at random" means any subset of two people is equally likely to be
chosen.)

    (a) What is the probability that Angela will be one of the two chosen?

    (b) Angela really hopes that she and Dwanna are the ones who get to go. What is
the probability of this happening?

1.11 In Smalltown, 40% of households have at least one dog and 60% of households have at least one cat, while 20% of households have neither dogs nor cats. If a household is chosen at random, what is the probability that there is at least one cat *and* at least one dog?

1.12 Last year in Bucolic County, there were 1297 children ages 1–5. Of these, 115 had not had the pertussis (whooping cough) vaccination. Last year, 27 of children ages 1–5 contracted whooping cough, and two of these children had been vaccinated. (a) How many children had neither the vaccination nor the disease? (b) If we randomly select one of these 1297 children (so that each child has the same probability of being selected), what is the probability that s/he was not vaccinated and got the disease?

1.13 The experiment is to select a person from the U.S. at random. Event A is "person resides in New York," and Event B is "person is an immigrant." Suppose $P(A) = 0.066$, $P(B) = 0.072$, and $P(A \cap B) = 0.02$.

   (a) Interpret the event $A \cap B$.

   (b) Interpret the event $A \cup B$.

   (c) What is $P(A \cup B)$?

1.14 Thirty percent of students at County College have a major in STEM disciplines, and 60% of students tested out of precalculus mathematics. Further, 25% of students *both* major in a STEM discipline and tested out of precalculus mathematics.

   (a) If we choose at random a student from County College, what is the probability that the student majors in a STEM discipline but did not test out of precalculus mathematics?

   (b) What proportion of students majoring in a STEM discipline tested out of precalculus mathematics?

   (c) What proportion of students *not* majoring in a STEM discipline tested out of precalculus mathematics?

1.15 Suppose that in a certain country 10% of the elderly people have diabetes. It is also known that 30% of the elderly people are living below poverty level, and 35% of the elderly population falls into at least one of these categories.

   (a) What proportion of elderly people in this country have both diabetes and are living below poverty level?

   (b) Suppose we choose an elderly person in this country "at random."[1] What is the probability that the person will neither have diabetes nor be living at the poverty level?

1.16 Seventy percent of 35-year-olds in Urbania are married, and twenty percent of 35-year-olds have a sports car. Twenty percent of 35-year-olds are neither married nor have sports cars.

---

[1]For now, the words "at random" and "uniform" are in quotes because we are relying on intuition to understand these terms. More formal definitions will come later.

(a) What proportion of 35-year-olds in Urbania are both married and have a sports car?

(b) Suppose we choose a 35-year-old in Urbania "at random." What is the probability that the person will have a sports car and will not be married?

1.17 A biologist has determined that the proportion of honey bees with Phenotype A is .42, while the proportion of bees with Phenotype B is .28. If half of the honey bees have neither phenotype, what proportion of bees have both phenotypes?

1.18 A biologist has determined that the proportion of honey bees with Phenotype A is .42, the proportion of bees with Phenotype B is .28, and proportion of bees with Phenotype C is .24. She also knows that honey bees cannot have both Phenotype A and Phenotype C, while the proportion with both Phenotype B and Phenotype C is .06. Sixteen percent of the honey bees have none of the phenotypes.

(a) What proportion of bees have phenotype A, but neither B nor C?

(b) What proportion of bees have only one of the phenotypes?

(c) What proportion of bees have exactly two of the phenotypes?

1.19 Fruit fly phenotypes have been studied extensively by geneticists. Suppose that 30% of fruit flies have the genetic marker Cy, which results in a slight curl of the wings, and 70% have the e genetic marker, which results in a black body and wings. Further, suppose 21% of fruit flies have both markers. Now a geneticist chooses a fruit fly "at random" and defines event A to be that the fruit fly has marker Cy and event B to be that the fruit fly has marker e.

(a) Interpret the event $A \cap B$.

(b) Compute and interpret $P(B \cap A^c)$.

(c) Compute and interpret $P(B \cup A^c)$.

(d) Are the events A and B disjoint?

1.20 You are rolling five fair, six-sided dice with the goal of getting a "straight," which means 1, 2, 3, 4, 5 or 2, 3, 4, 5, 6. Your first roll is 2, 3, 5, 6, and 6. You can pick up some of the dice for one more roll. Is it better to pick up one of the sixes and try for a four, or to pick up both sixes and roll them again? If the latter, you "win" if the roll is 1 and 4 or 4 and 6.

Chapter 2

Using R to Simulate Events

In the next chapters we will learn many pencil-and-paper methods for computing probabilities of events. However, we can also write computer code to approximate probabilities of events (to as much precision as desired), either because they are too hard to calculate, or simply to verify that our calculation is correct. Computer simulations are extremely useful in the "real world," where problems are seldom as "nice" as in statistics textbooks.

The function `sample` in R can be used to choose randomly from a finite set. Suppose we want to simulate ten rolls of a fair, six-sided die. The following line of code will produce ten random numbers with integer values from 1 to 6, with equal probabilities (1/6) for each value.

```
x=sample(1:6,10,replace=TRUE)
```

The option `replace=TRUE` means that values can be repeated. In contrast, `sample(1:10,4)` will choose, at random, a subset of four *distinct* numbers from one to ten. We don't need to specify `replace=FALSE`, because sampling without replacement is the default.

This R function can be used to simulate or mimic tosses of a coin, rolls of a die, or drawing samples of voters, or to simulate selections with equal probabilities, from a finite set, with or without replacement.

Example: Suppose we want to know the probability of getting at least five sixes when we roll a die ten times. We will learn how to calculate this exactly when we study binomial probabilities in Chapter 8, but in the meantime we can use R to simulate a large number of experiments in which the die is rolled ten times, and then simply count up the number of these experiments in which five or more sixes appear:

```
nloop=1000000  ## this is the number of experiments
numsix=1:nloop   ## vector to record the number of sixes in each experiment
for(iloop in 1:nloop){   ## perform the experiment nloop times
   x=sample(1:6,10,replace=TRUE)   ## roll the six-sided die ten times
   numsix[iloop]=sum(x==6)    ## count the number of sixes
}
sum(numsix>=5)/nloop    ## find the desired proportion
```

The proportion of times that at least five sixes appear is about .0155. We know intuitively that a larger `nloop` will result in a more precise approximation of the true
probability. We can also look at a histogram of the simulated distribution of the number
of sixes in 10 rolls of a fair die:

```
br=0:11-1/2
hist(numsix,breaks=br,main="",freq=FALSE,xaxt="n",ylim=c(0,.35),col='beige')
for(i in 0:10){text(i,sum(numsix==i)/nloop+.02,i)}
```

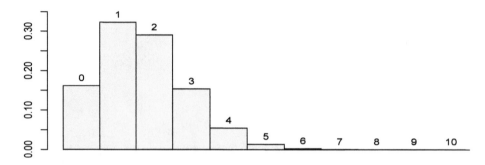

We specified the histogram bins with the option `breaks=`. If we leave out this
option, R will choose the bins for us, and these might not show the integer values. For
more information about histograms in R, we can type `help(hist)` or `?hist` at the R
prompt.

Example: Suppose an ornithologist is curious about whether a certain species of bird in
New Zealand prefers reddish rocks to other colors when building its nest. She chooses
a spot near a nesting area, and sets out many rocks that are either reddish, bluish, or
gray, in approximately equal numbers, then watches until ten different birds have come
to choose rocks. If the birds are actually equally likely to choose any of the three colors,
what is the probability that at least eight of the ten chose reddish rocks?

We can compute this probability through simulations, where, for instance, the colors
are labeled 1=reddish, 2=bluish, and 3=gray. Each loop in the code mimics an experiment with ten birds. The vector `numred` stores the numbers of reddish rocks chosen
by the birds in each of the loops, and the assumption that each of the colors is equally
likely is used to simulate the rock colors chosen. In this way we can answer the following question: *if* the rock colors are all equally likely, what's the probability that at least
eight of the ten are reddish?

```
nloop=1000000
numred=1:nloop
for(iloop in 1:nloop){
    x=sample(1:3,10,replace=TRUE)     ## choose 10 rocks, three colors
    numred[iloop]=sum(x==1)       ## how many of the 10 are red?
}
sum(numred>=8)/nloop
```

This will produce a proportion close to .0034. If the birds were choosing randomly,
the probability of at least eight of ten choices of reddish rocks is quite unlikely, so if
this event is observed by the ornithologist, she might become confident that the choice
was *not* random; that is, the birds prefer reddish rocks. (This somewhat convoluted
reasoning is the basis for formal statistical hypothesis testing, which will be introduced
in Chapter 6.)

Example: Suppose twenty percent of households in Bucolic County have no children, 25% have one child, 30% have two children, 15% have three children, and 10% have at least four children. If we randomly choose 100 households in Bucolic county, what is the probability that the number of households with at least four children is greater than the number with no children?

We can simulate the sample of 100 households using the `sample` command with the `prob` option to specify the probabilities:

```
x=sample(0:4,100,replace=TRUE,prob=c(.2,.25,.3,.15,.10))
```

The vector `x` will contain 100 values between zero and four, inclusive. On average, the vector `x` will contain twenty zeros, but some simulated vectors will have a higher number, and some will have a lower number, due to the random variation. We can do this many times and count the proportion of samples for which the number of households with at least four children is greater than the number of households with no children:

```
nloop=100000
count=0
for(iloop in 1:nloop){
   x=sample(0:4,100,replace=TRUE,prob=c(.2,.25,.3,.15,.1))
   if(sum(x==4)>sum(x==0)){count=count+1}
}
count/nloop
```

We find that about 2.5% of the samples have more households with at least four children than with no children. This approximates the probability that, in a single sample, there will be more households with at least four children than with no children.

The function `runif(n)` will choose, "at random," n real numbers in the interval $(0, 1)$. The numbers are distributed "uniformly" in $(0, 1)$; this concept will be developed more formally in Chapter 13. For now, if $0 \leq a < b \leq 1$, then the probability that number generated by `runif(1)` lands in the interval (a, b) is $b - a$. Further, the randomly generated numbers are "independent," meaning that if one number lands in (a, b), this does not affect the probability that another lands in (a, b). The concept of independence of events will be more formally developed in Chapter 4, but for now we will rely on our intuition.

This function can be used to sample in any interval: to sample 40 random numbers uniformly in the interval $(1, 5)$, the command is `runif(40,1,5)`. Without the second and third arguments, the default interval $(0, 1)$ is used.

Example: The `runif` function might be used to model an arrival time. Suppose an airport shuttle makes round trips from the parking lot to the terminal, and each trip takes 20 minutes. Now suppose a traveler arrives "at random," so that the time he or she waits for the shuttle, in minutes, is a real number between 0 and 20, which might be simulated using `runif`:

```
x=runif(1,0,20)
```

The value x will be a realization of the waiting time in the random experiment. Suppose on a particular day two company executives arrive separately (independently) at the terminal at random times; what is the probability that they both have to wait more than ten minutes? To answer this, we can simulate many pairs of random wait times, and count how many are both more than ten (do you have a guess?).

```
nloop=1000000
numwait=0
for(iloop in 1:nloop){
    x=runif(2,0,20)
    if(x[1]>10&x[2]>10){numwait=numwait+1}
}
numwait/nloop
```

This will give about 25% for the proportion of pairs of waiting times that are both more than 10 minutes.

Now suppose there are three shuttles operating at the airport, where the round trips were started "independently" so that if we know the arrival time of one shuttle, that doesn't give us any information about the arrival time of the other two shuttles. They all take 20 minutes for each trip. A traveler will get on the first shuttle that arrives, so that the wait time can be modeled as the *minimum* of the three random numbers between zero and twenty. What is the probability that a traveler arriving at random will have to wait more than 5 minutes?

The following R code mimics experiments in which a traveler arrives and gets on the first shuttle that appears. The vector `waittime` stores the wait times, i.e., the minimum of three numbers chosen at random in the interval $(0, 20)$.

```
nloop=1000000
waittime=1:nloop
for(iloop in 1:nloop){
    x=runif(3)*20
    waittime[iloop]=min(x)
}
sum(waittime>5)/nloop
```

We find that about 42.2% of wait times are more than five minutes.

One million simulated experiments gives a fairly precise estimate of the probabilities. We can also get an idea of the *distribution* of the wait times. When only one shuttle was operating, the wait times were "evenly distributed" in the interval $(0, 20)$, but for three shuttles we imagine that the distribution of weight times is "piled up" at the left end of the interval. The idea of a distribution of a random quantity is very important and will be explored formally in future chapters, but here we can get an intuitive feel for the distribution of the wait times, simply by making a histogram of the simulated wait times:

```
hist(waittime,breaks=40,main="Waiting times",col='beige')
```

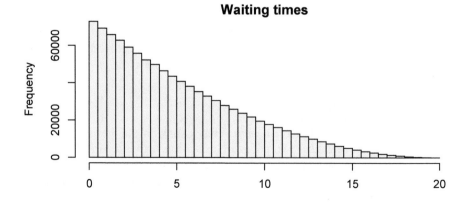

We see that our intuition that "wait times are likely to be smaller when there are three shuttles instead of one" is correct. The proportion of observations greater than 5 (which we found to be .422) roughly corresponds to the proportion of the area of the histogram that is to the right of the number 5.

We will continue to use simulations throughout the book, to compute probabilities and distributions that are still too hard to solve analytically, and also to verify our analytical calculations. Further, simulations will be used to build intuition about concepts in probability and statistics, and to solidify our understanding of these concepts.

Chapter Highlights

- The R function `sample` will choose values "at random" from a finite set of values.

  - The values will be sampled "without replacement" (at most one of each value chosen), unless the option `replace=TRUE` is used.

  - The values will be sampled with equal probabilities unless the probabilities are specified with the option `prob`.

  - When `replace=TRUE`, the values are "independent," which means that each does not depend on the other values.

- The R function `runif(n)` will choose n values "at random" from the interval $(0, 1)$.

- The R function `runif(n,a,b)` will choose n values "at random" from the interval (a, b).

Exercises

2.1 Suppose you roll a fair, six-sided die six times. Write R code to approximate the probability that each number comes up exactly once.

2.2 What is the probability of a "full house" when five dice are rolled? This is defined (in the game *Yahtzee*) as three of one number and two of another (different) number. Write R code to approximate this probability.

2.3 What is the probability of a "large straight" when five dice are rolled? This is defined (in the game *Yahtzee*) as the numbers 1,2,3,4,5 or 2,3,4,5,6. Write R code to approximate this probability.

2.4 Refer to Exercise 1.8. Suppose the true probability of improvement, if treatment A is given, is .8, but if either treatment B or treatment C is given, the probability of improvement is only .5. Also suppose that as patients come in, the treatments are assigned at random, so that each treatment has probability 1/3 of being assigned, and each assignment is "independent," that is, the assignments don't depend on previous assignments. If five patients enter the study, what is the probability that the result is "improved" for all five of them? Do simulations in R to approximate this probability.

2.5 Refer to Exercise 1.8. Suppose 40 patients arrive at the clinic and each is randomly assigned to one of the three treatments. Suppose that the probability of improving is .2, for all three treatments. Do simulations in R to approximate the probability that the proportion of treatment A patients who improve is at least twice the proportion of treatment B patient who improve.

2.6 Refer to Exercise 1.17. Suppose a biologist chooses two bees "at random." What is the probability that one has only Phenotype A and the other has only Phenotype B? Write R code to approximate this probability.

2.7 In China, 48% of the population have blood type O, 28% have blood type A, 19% have type B, and 5% have type AB. Suppose four people in China are chosen "at random." Write R code to simulate the probability that there is one of each blood type.

2.8 Three numbers are chosen "at random" from the interval $(0, 2)$. Write R code to determine the probability that the sum is less than 1.

2.9 Four children from a neighborhood in Smalltown attend a holiday party. There are 90 children total at the party, and 50 prizes are to be distributed "at random" to the children. That is, each prize is equally likely to go to each of the 90 children, regardless of whether a child already has prizes.

    (a) Write R code to determine the probability that each of the four children gets exactly one prize.

    (b) Write R code to determine the probability that each of the four children gets at least one prize.

2.10 Lengths of filaments are produced at a factory, and some have flaws that can be determined only through extensive testing. The proportion of filaments that don't have any flaws is .55, while 25% have exactly one flaw, 15% have exactly two flaws, and the remaining 5% have at least three flaws. Four filaments will be used in a device. If two or more of the filaments have at least three flaws, then the device will certainly fail within the first week. Otherwise, if one or more of the filaments have at least two flaws, the device with fail within the first week, with probability 1/2. If none of the filaments has more than one flaw, then the device will certainly not fail within the first week. If the four filaments are chosen "at random," what is the probability that the device fails within one week? Write R code to approximate this probability.

2.11 A corporation has 6 executives in Chicago, 8 in Los Angeles, and 12 in Tampa. Three executives are chosen "at random" to represent the corporation at a retreat. Use simulations in R to approximate the probability that none is from Tampa.

2.12 Fifteen percent of employees at a large company are managers, twenty percent are support staff, and the remaining 65% are workers in the factory. Suppose 20% of managers are women, 85% of support staff are women, and 40% of the factory workers are women. If the CEO forms a committee having two managers, two support staff, and four factory workers, choosing "at random" from each group, what is the probability that there are no women on the committee?

    (a) Write R code to approximate this probability, with the assumption that the numbers of employees are large enough in each group so that we can sample with replacement and the probability of choosing the same person twice is very small.

(b) Now suppose there are 100 employees at the company, with the above percentages applying, and do the sampling without replacement. Compare your answer to that in part (a).

2.13 A drawer contains 12 pairs of socks, where each pair is a different color. Sam draws four socks "at random" from the drawer (without replacement). Write R code to determine the probability that there is no pair among the four socks.

2.14 Suppose we have a necklace with 16 beads, 8 of which are red and 8 are blue. The 16 beads will be mixed, then randomly placed on the string and made into a circle. What is the probability that there are no instances of 5 adjacent beads of the same color? Write R code to simulate this probability.

Chapter 3

The Basic Principle of Counting, Permutations, and Combinations

Assigning probabilities often involves counting outcomes in a sample space. For example, suppose we have four components in an electrical device, and each can be installed right-side up or upside down. The device works if at least two are right-side up. Suppose the person who installs the components does so "at random." (We are still using our intuitive understanding of randomness, but we will later develop more formal ideas of probability distributions and independence.) If we want to find the probability of the device working, we need to count the number of ways the four components can be installed, and we also need to count the number of ways that at least two are right-side up. The proportion of the ways for which at least two are right-side up is the desired probability.

Because the number of components is small, we can write out the entire sample space by being methodical:

$$
S = \left\{
\begin{array}{lllll}
RRRR, \\
RRRU, & RRUR, & RURR, & URRR, \\
RRUU, & RURU, & RUUR, & URRU, & URUR, & UURR, \\
UUUR, & UURU, & URUU, & RUUU, \\
UUUU
\end{array}
\right\},
$$

where, for example, $RRUU$ represents the first two components being right-side up, and the last two being upside down. There are 16 outcomes and for all but 5 of them, at least two components are right-side up. If each of the outcomes is equally likely (i.e., if the components were installed "at random"), then the probability that the device works is 11/16.

If there are five or more components, we might not want to write out the entire sample space. Instead, we use ideas from "combinatorics," or the mathematics of counting.

We can start with the **Basic Principle of Counting**, or the mn **rule**:

> If Experiment 1 has m possible outcomes, and Experiment 2 has n possible outcomes, then the number of possible combinations of outcomes when the experiments are performed in succession is mn.

This is readily generalized to k experiments, where the ith experiment has m_i outcomes, for $i = 1, \ldots, k$. If the experiments are performed in succession, the total number of combinations of outcomes is $\prod_{i=1}^{k} m_i$.

The basic principle is used to determine that the number of outcomes in the electrical components example is 16: Each device has two possible orientations, so the total number of orientations is $2 \times 2 \times 2 \times 2 = 16$.

Example: The experiment is to toss a coin, then roll a six-sided die. Because there are two possible outcomes for the coin toss and six for the die roll, there are 12 possible outcomes for the experiment, and the sample space can be written as

$$\left\{ \begin{array}{cccccc} H1, & H2, & H3, & H4, & H5, & H6, \\ T1, & T2, & T3, & T4, & T5, & T6 \end{array} \right\},$$

where T2 means "toss is tails and roll is a 2."

Example: A mathematical sciences committee is to be formed with one professor each from the math, applied math, and statistics departments. If there are 25 math professors, 30 applied math professors, and 15 statistics professors, then there are $25 \times 30 \times 15 = 11{,}250$ possible committees.

Permutations and **combinations** are ways to count the number of subsets of a finite set of objects. For permutations, the order is important, and for combinations, the order is not important.

Let's start with **permutations,** where the order is important. How many ways are there to arrange the letters ABCDE (with no repeated letters)? Think about filling in the spaces

$$\underline{\quad} \quad \underline{\quad} \quad \underline{\quad} \quad \underline{\quad} \quad \underline{\quad}$$

one at a time. There are five ways to fill the first space. Subsequently, there are four ways to fill the next space (because you've already used one of the letters). There are three letters left for the second space, etc. Therefore,

> The number of ways to order n distinct objects is $n!$,

where $n!$ reads "n factorial" and is the product of all the integers from one to n. (We also define $0! = 1$. To see why this actually makes sense, see Appendix B.5.) We could also say, "the number of ways to *permute* n distinct objects is $n!$."

Now suppose that we're interested in the number of four-letter substrings from the six letters ABCDEF, for example DEAC or ABED, where order is important (so the string ABCD is different from DABC), and no letter is repeated. We want to fill in only four spaces, and there are six possibilities for the first space, five for the second, etc., so the answer is $6 \times 5 \times 4 \times 3 = 360$ substrings. We can write this as $6!/2!$ substrings. The generalization of this reasoning is:

> The number of ordered subsets of size m, from a set of $n \geq m$ distinct objects, is $n!/(n-m)!$.

Example: A club has 25 members, from which we choose a president, vice president, secretary, and treasurer. How many choices are there? We have 25 choices for president, then we have to choose one of the remaining 24 for vice president, etc., so there are

$$25!/(25-4)! = 25 \times 24 \times 23 \times 22 = 303{,}600 \text{ choices.}$$

For **combinations**, we want to count the number of unordered subsets of size m, of a set of size $n \geq m$. To illustrate, let's count the subsets of size four of the set of six letters ABCDEF, where order is *not* important, that is, ABCD and DACB are the same subset. For small sets, we can simply enumerate in a methodical way,

$$\{ABCD, ABCE, ABCF, ABDE, ABDF, ABEF, ACDE, ACDF, ACEF,$$

$$ADEF, BCDE, BCDF, BCEF, BDEF, CDEF\},$$

but let's figure out a formula. We know that there are $6!/2!$ *ordered* subsets, and for each subset such as ABCD, there are $4!$ orderings. So, we should divide the number of ordered subsets by the number of orderings within the subset, to get the number of unordered subsets:

$$\frac{6!}{2!\,4!} = 15$$

unordered subsets. More generally:

> Given a set of size $n > 0$, the number of unordered subsets of size k, $0 \leq k \leq n$, is
> $$\binom{n}{k} = \frac{n!}{k!(n-k)!}.$$

(Remember that $0! = 1$.) The n and k, vertically arranged within parentheses, is a standard representation for a combination, and reads "n choose k." Another common notation is $_nC_k$—this is how a combination key is usually marked on a calculator.

Example: A club has 25 members, from which we choose a committee of size 4. The number of possible committees is "twenty-five choose four," that is,

$$\binom{25}{4} = \frac{25!}{4!\,21!} = 12{,}650.$$

Example: Let's go back to the electrical device from the beginning of this chapter. We know that there are 16 equally likely ways for the electrician to install the components. To count the number of ways for *exactly* two to be right-side up, we choose two out of four spaces, to get $\binom{4}{2} = 6$ ways. The device will also work if there are exactly three components right-side up, and there are $\binom{4}{3} = 4$ ways to do this. (Of course, choosing three out of four to be right-side up is equivalent to choosing one out of four to be upside down, which is clearly four ways.) Finally, there is $\binom{4}{4} = 1$ way to install all the components right-side up, so there are $6 + 4 + 1 = 11$ elements of the sample space for which the device works, and the probability of the device working (under the equally likely probability function) is $11/16$.

Example: Now we can compute probabilities of poker hands, an important application of our methods! First, we need to determine how many standard poker hands are possible, In other words, how many subsets of size five are there of a set of 52 cards? The answer is

$$\binom{52}{5} = 2{,}598{,}960$$

possible poker hands. Let's suppose that if the deck is well shuffled and five cards
are dealt "at random," then each of these hands is equally likely. An example out-
come would be $(4\diamondsuit, 2\clubsuit, 2\heartsuit, Q\heartsuit, 9\clubsuit)$. Of course, this is the same outcome or hand as
$(Q\heartsuit, 9\clubsuit, 4\diamondsuit, 2\heartsuit, 2\clubsuit)$, because order is not important.

The number of these hands that are all hearts is the number of subsets of size five
from the set of 13 hearts, that is,

$$\binom{13}{5} = 1287.$$

An example of such an outcome is $(Q\heartsuit, 9\heartsuit, 5\heartsuit, 3\heartsuit, 2\heartsuit)$.

So, the probability of getting all hearts in a draw of five cards from a well-shuffled
deck is

$$\frac{\binom{13}{5}}{\binom{52}{5}} = \frac{1{,}287}{2{,}598{,}960} = .0004952,$$

showing exactly how unlikely it is to be dealt five hearts. The probability of a "flush,"
meaning all same suit, must be four times this, because there are four suits. In other
words, the probability of being dealt four clubs, or four spades, or four diamonds, is also
.0004952, and we can add these probabilities because the events are disjoint. Therefore,
the probability of a flush is .00198. We can expect to be dealt a flush in about two out
of a thousand poker hands.

The binomial theorem states that

$$(x + y)^n = \sum_{k=0}^{n} \binom{n}{k} x^k y^{n-k}.$$

This can be proved by mathematical induction, or with a combinatorial argument. For
the latter, consider "multiplying out" the product

$$(x_1 + y_1)(x_2 + y_2) \cdots (x_n + y_n).$$

We get a sum of 2^n terms, where for each term we choose either x_i or y_i from the ith
term, systematically going from $i = 1, \ldots, n$. In this way the first term is $x_1 x_2 \cdots x_n$,
and the last term is $y_1 y_2 \cdots y_n$. How many terms have one x factor and $n - 1$ y factors?
These are $x_1 y_2 \cdots y_n$, $y_1 x_2 y_3 \cdots y_n$, etc; there are n of these terms. How many have
k of the x_i's and $n - k$ of the y_i's? The answer is $\binom{n}{k}$. Now, if all of the x_i's have
the same value x, and all of the y_i's have the same value y, these $\binom{n}{k}$ terms can be
combined, and the binomial theorem is the result.

Example: To expand $(x + 1)^5$, we use the formula

$$(x+1)^5 = \sum_{k=0}^{5} \binom{5}{k} x^k 1^{n-k} = \sum_{k=0}^{5} \binom{5}{k} x^k = x^5 + 5x^4 + 10x^3 + 10x^2 + 5x + 1.$$

Chapter Highlights

1. The basic principle of counting: For experiments $1, 2, \ldots, k$ where experiment i has m_i outcomes, the total number of outcomes when all the experiments are performed is $\prod_{i=1}^{k} m_i$.

2. The number of ways to choose m items from n items when order is important:

$$\frac{n!}{(n-m)!}$$

3. The number of ways to choose m items from n items when order is *not* important:

$$\binom{n}{m} = \frac{n!}{m!(n-m)!}$$

4. The binomial theorem is

$$(x+y)^n = \sum_{k=0}^{n} \binom{n}{k} x^k y^{n-k}.$$

Exercises

3.1 ID numbers for a health club are made with two letters followed by three numbers. How many different IDs can be made with this scheme?

3.2 How many distinct license plates can made if all the plates have two letters followed by four numbers?

3.3 Given an area code, how many different (seven digit) phone numbers are possible if none of the numbers can start with zero?

3.4 Suppose an electrical device has seven components, each of which can be installed right-side up or upside down. The device will work if at least five components are right-side up. If an electrician installs the components "at random," what is the probability that the device works? Write R code to check your answer.

3.5 A corporation has 6 executives in Chicago, 8 in Los Angeles, and 12 in Tampa. If three executives are chosen "at random" to represent the corporation at a retreat, what is the probability that none is from Tampa? (This answer was approximated through simulations in Exercise 2.11.)

3.6 Suppose there are 24 kids in a sports competition.

  (a) How many different sets of winners of gold, silver, and bronze medals are there?

  (b) If three of them are to get (unranked) medals, how many different sets of winners are there?

3.7 A shop makes deluxe ice cream sundaes with three scoops of ice cream. If there are 12 ice cream flavors, how many different sundaes can be made if each of the scoops is a different flavor? (You can decide whether or not order is important, and explain why.)

3.8 The numbers 1 through 20 are printed on a lottery ticket. A player circles three numbers. At the official lottery drawing, balls with the numbers 1 through 20 are randomized in an urn. Three balls are drawn at random, without replacement. The player wins if the circled numbers match the three numbers drawn from the urn. What is the probability that the player wins? Compute the probability, then write code in R to simulate draws and check your answer.

3.9 In a lottery game, a player writes three different numbers between 1 and 20 on three lines of a lottery ticket. At the official lottery drawing, balls with the numbers 1 through 20 are randomized in an urn. Three balls are drawn at random, one at a time without replacement. The player wins if the written numbers match the three numbers drawn from the urn, *in the same order*. What is the probability that the player wins? Write code in R to simulate draws and confirm the probability of winning.

3.10 Ten kids at a summer tennis camp are to be ranked by a tournament scheme.

    (a) How many different rankings are there?

    (b) What if the six girls and four boys are to be ranked among themselves? How many different ways are there to order six girls and four boys?

3.11 For a random draw of 5 cards from a standard 52-card deck, what is the probability of a full house, i.e., two of one value and three of another value (such as two queens and three sevens)?

3.12 A deck of 40 cards contains ten of each of the following colors: red, blue, green, and yellow. If the deck is well shuffled, and a hand of four cards is randomly chosen (without replacement), what is the probability that all four colors are in the hand? Check your answer using simulations in R.

3.13 (a) How many ways are there to give 10 candies to three kids, where each kid gets at least one candy? (Suppose the kids are Samantha, Joey, and Allison. You can give four to Samantha, and three each to the other two, or you can give eight to Allison, and one each to the others, etc.)

    (b) Generalize the problem in part (a) to n candies and m kids, where $n \geq m$.

3.14 Suppose there are n components of which m are defective, where $m < n/2$. How many ways are there to order the n components, where there are no two consecutive defective components?

3.15 A child has 12 blocks, of which 6 are blue, 4 are red, 1 is yellow, and 1 is green. If the child puts the blocks in a line, how many color arrangements are there? (The reasoning behind this solution will be used again in the chapter on multinomial distributions.)

3.16 Scientists at State University want to test the conjecture that higher levels of omega-3 fatty acids in the diet are beneficial to brain development in mammals. They have 12 baby rats; 6 will be randomly chosen to be fed a diet high in the omega-3 fatty acids, and the others will be fed a "regular" diet. Two months later all the rats will be tested with a maze, with the idea that smarter rats will run the maze faster. Suppose the diet actually has no effect on rat intelligence. What is the probability that the fastest four rats were all in the omega-3 group? (These ideas are used in a statistical procedure called a *rank-sum test*.) Check your answer using simulations in R.

3.17 You are rolling 5 dice with the goal of getting a least three of the same number. The numbers from your first roll of five dice are 1, 1, 4, 4, 5. You can either pick up the 5 and try to get a 1 or 4, or you can pick up three dice, such as the 1, 1, and 5, and try to get one or more 4's. Which alternatives gives a higher probability of getting at least three of the same number?

3.18 Show that
$$\binom{n}{k} = \binom{n}{n-k}.$$

3.19 Expand $(x + 2y)^5$.

3.20 Expand $(x^2 + 1)^4$.

Chapter 4

Conditional Probability, Independence, and Tree Diagrams

The **conditional probability** of an event is computed when we have information about another event in the sample space. Suppose we want to compute the probability of event B, given that event A has happened. In the figure below, we know that the outcome is in the set A (total shaded area), and we're asking what is the probability that the event is also in B (dark shaded area).

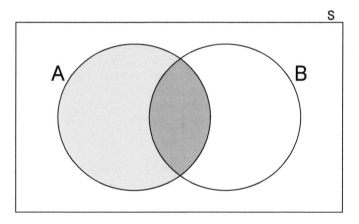

We define the conditional probability of event B given event A as

$$P(B|A) = \frac{P(B \cap A)}{P(A)}.$$

The intuition behind this is that we know that A has happened. For what proportion of times that A happens does B also happen? Or, how does knowing that A has happened change our guess about B?

Example: Suppose we are interested in the prevalence of diabetes in a certain community. If an experiment is to select "at random" (with equal probabilities) a person in the community, and B is the event that the person is found to have diabetes, then $P(B)$ is the prevalence, or the proportion of people in the community with diabetes. Now suppose we are further interested in the prevalence among a subgroup, such as people over 60 years of age. If A is the event that a randomly selected person in the community is over 60, then $P(B|A)$ is simply the prevalence of diabetes in the over-60 subgroup. The question, "what proportion of over-60 people in the community have diabetes?" has the same answer as "suppose we randomly select a person in the community and discover s/he is over 60; what is the probability that the person also has diabetes?"

Example: The "experiment" is to choose a tree "at random" from the Arboles National Park. Let event A be that the tree has substantial beetle damage, and let B be the event that the tree has been drought stressed in the last three years. Suppose $P(A) = 0.35$ (35% of trees have substantial beetle damage), $P(B) = 0.60$ (60% of trees have been drought stressed), and $P(A \cap B) = 0.30$ (30% of trees have been drought stressed *and* have substantial beetle damage). Given that the randomly selected tree has been drought stressed, what is the probability that it has substantial beetle damage? We simply compute

$$P(A|B) = \frac{P(A \cap B)}{P(B)} = \frac{.30}{.60} = .50.$$

In other words, half of the drought-stressed trees have substantial beetle damage.

The result $P(A|B) + P(A^c|B) = 1$ is intuitive, but let's prove it anyway, starting with the definition

$$P(A|B) + P(A^c|B) = \frac{P(A \cap B)}{P(B)} + \frac{P(A^c \cap B)}{P(B)} = \frac{P(A \cap B) + P(A^c \cap B)}{P(B)} = \frac{P(B)}{P(B)} = 1.$$

We used the identity $P(A \cap B) + P(A^c \cap B) = P(B)$, as illustrated in Exercise 1.1.

Definition: Events A and B are **independent** if $P(B|A) = P(B)$.

Intuitively, if events A and B are independent, then the occurrence of event A doesn't affect whether or not B occurs, and vice versa. Under independence, the probability of event B, given that event A has happened, is the same as the probability of event B, given that event A has *not* happened. Knowing that A has happened does not change your guess about B. You are asked in Exercise 4.19 to show that if $P(B|A) = P(B)$, we must have $P(A|B) = P(A)$.

Example: In the previous example about Arboles National Park, are events A and B independent? No, because $P(A|B) \neq P(A)$. Trees that were drought stressed are *more* likely to have substantial beetle damage, compared to trees that were not drought stressed.

The **multiplication rule for independent events:** If Events A and B are independent, then $P(A \cap B) = P(A)P(B)$. This is easy to show from the definitions of conditional probability and independence. If A and B are independent, we have

$$P(A) = P(A|B) = \frac{P(A \cap B)}{P(B)},$$

and multiplying through by $P(B)$ gives the result.

We can use the multiplication rule for checking independence, which is sometimes more convenient than using the definition of independence. Furthermore, we can use the multiplication rule for independent events to assign probabilities.

Example: Suppose $P(A) = .4$ and $P(B) = .5$, and events A and B are independent. We can fill out the Venn diagram because we know the intersection must be $P(A \cap B) = P(A)P(B) = .2$:

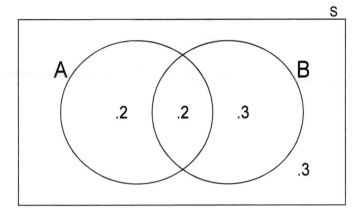

Tree diagrams

The Venn diagram is a useful way to organize and keep track of probabilities of events. Another useful drawing is a **tree diagram**; this is especially handy when the problem involves conditional probabilities, or when the experiment consists of two or more parts, performed in sequence.

Here is a classical example that is useful for illustration. Suppose we have two urns, A and B, and Urn A contains three red balls and one green ball, while Urn B contains two red balls and four green balls. If we first select an urn at random (i.e., with equal probability), then select a ball at random from the urn, what is the probability that the ball is red?

The following tree diagram is drawn, branches are labeled, and probabilities are assigned in a very natural way. Here, the first step is "choose urn," and the two initial branches are labeled with the urn names. The two urns are equally probable, so we put "1/2" on each of the branches.

For the second set of branches, containing the "choose ball" alternatives, it's natural to write in conditional probabilities, given the event on the first branch. Let R be the event that a red ball is chosen, and let G be the event that a green ball is chosen. Given that Urn A was chosen, what is the probability of choosing a red ball? This probability, $P(R|A) = 3/4$, is added to the tree. On each set of branches on the tree, probabilities should add to one.

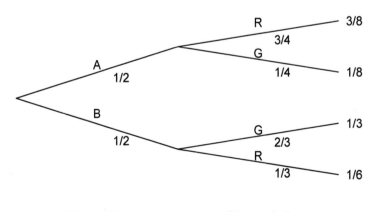

Choose Urn   Choose Ball

The last step is to "multiply through" the tree, to get the numbers on the right-hand side. These are probabilities of intersections of events. The top number, 3/8, is $P(A \cap R)$, i.e., the probability that a red ball is drawn from Urn A. We have $P(G \cap A) = P(G|A)P(A) = 1/8$, the probability that a green ball is drawn from Urn A, etc. All of the numbers at the far right have to add up to one, of course; this is a good way to check your work.

To answer the question "what is the probability that the ball is red?" we add the two branches that end in R:

$$P(R) = P(R \cap A) + P(R \cap B) = 3/8 + 1/6 = 13/24.$$

Example: Suppose we have an urn with two red balls and one green ball. The experiment is to draw balls "at random" from the urn (without replacement) until we get the green ball. We could write the sample space as $\{G, RG, RRG\}$, where, for example, RRG means two red balls were drawn before the green ball, or we could write the sample space as $\{1, 2, 3\}$, indicating the draw on which a green ball is obtained. To assign probabilities, we can draw a tree diagram as shown below. The tree looks lopsided because we stop when we get a green ball. Here the "end" branch probabilities have to add to one. We can see that the three elements of the sample space are actually equally probable.

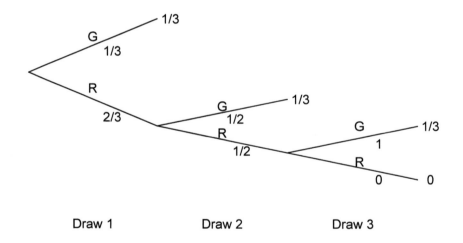

Draw 1   Draw 2   Draw 3

Imperfect testing examples

An important application of these conditional probability ideas involves medical testing for a rare disease or for illegal drug use. This example also illustrates the difference between $P(A|B)$ and $P(B|A)$, and the importance of not confusing these conditional probabilities.

Suppose a pharmaceutical company finds a cheap way to test for a genetic predisposition for a certain type of cancer. It may seem like a good idea to have a policy to test everyone, regardless of symptoms or risk factors. Let's investigate this.

Suppose the test is very good but not perfect. If a person has the predisposition, then the test is positive (correct in identifying the predisposition) 99% of the time. We can write this as $P(+|D) = 0.99$, where the "+" symbolizes the event that the test is positive, and "D" is the event that the person actually has the predisposition. Further, suppose that if a person doesn't have the predisposition, the probability of a negative result (i.e., the test is correct) is 98%. We write this as $P(-|N) = 0.98$, where the "$-$" symbolizes the event that the test is negative, and "N" is the event that the person does not have the predisposition for the disease. We can easily calculate $P(-|D) = 0.01$ and $P(+|N) = 0.02$, but we cannot calculate $P(D|+)$ from the given information.

Assume further that only one tenth of one percent of people actually have this predisposition; that is, $P(D) = 0.001$, and $P(N) = 0.999$. Suppose that a randomly selected person is tested for the predisposition, and the test is positive. The probability that the person actually has the predisposition, given that the test is positive, is $P(D|+)$.

To calculate this conditional probability, we can make a tree diagram. The first step is to select a person at random. He or she either has the predisposition or not, and we put the proportions on the first set of branches (see figure below). The second step is to test. We put the conditional probabilities on the second set of branches as shown.

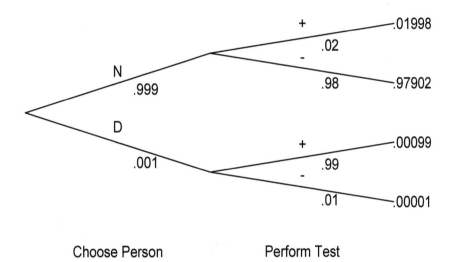

We see that the majority of cases (97.902%) are negative results for people without the predisposition. There are two ways to get a positive result: correct results for people with the predisposition and incorrect results for people without the predisposition. Of the people who test positive, there are more who do not have the predisposition, simply because the vast majority of the people in the population do not have the predisposition. We can calculate $P(D|+)$ as follows. According to the definition of conditional probability,

$$P(D|+) = \frac{P(D \cap +)}{P(+)}$$

$$= \frac{0.00099}{0.00099 + 0.01998}$$

$$= \frac{0.00099}{0.2097} = 0.0472.$$

We see that less than five percent of the people who test positive actually have the predisposition. The *false positive rate* is .9528. It might not be such a good idea to do widespread testing if a positive result leads to an invasive procedure or excessive worrying.

In the medical-testing world, the word "sensitivity" is used for $P(+|D)$ and the word "specificity" means $P(-|N)$. The "prevalence" is the proportion of the population with the condition being tested for. The above example shows a test can have a high specificity and a high sensitivity, and yet have a high false positive rate. This happens when the prevalence is low; see Exercise 4.34 for the same example with a higher prevalence.

Writing R code to simulate this experiment is a useful exercise to ensure understanding of these concepts (and we can check our answer). In the following bit of code, we "sample" a person from a population for which .1% have the predisposition. Then we use if statements to simulate the testing procedure. We increment the denominator if the test is positive, but the numerator is incremented if the test is positive and the person has the disease. Doing this in a loop will simulate the conditional probability, and increasing `nloop` will give more precision in the simulated probability.

```
nloop=100000
numerator=0;denominator=0
for(iloop in 1:nloop){
    disease=sample(0:1,1,prob=c(.999,.001))  ## sample a person
    if(disease==0){
        test=sample(0:1,1,prob=c(.98,.02))    ## if no disease
    }else{
        test=sample(0:1,1,prob=c(.01,.99))   ## if disease
    }
    if(test==1){
        denominator=denominator+1
        if(disease==1){ numerator=numerator+1}
    }
}
numerator/denominator
```

Example: Suppose a mother is researching whether her daughter should be vaccinated against whooping cough. This vaccination is not always effective, and there may be side effects, so she wants more information before making the decision. On the website of the Center for Disease Control for her state, she finds there were 64 cases of children's whooping cough in the state in the previous year. Of these, 32 of the children had been vaccinated, and 32 had not. She concludes, "obviously it doesn't make any difference whether or not you have the vaccination." We can see that she has her conditional probabilities backwards!

Let W be the event that a randomly selected child gets whooping cough, and let V be the event that the child is vaccinated. We will use NW for the event that the child did not get whooping cough, and NV for "not vaccinated." The mother is comparing $P(V|W)$ with $P(NV|W)$. The important comparison, though, is $P(W|V)$ and $P(W|NV)$, that is, are children less likely to get whooping cough given that they had the vaccine.

For purposes of illustration, let's suppose that 95% of children in the state were vaccinated last year, and see if we can compare $P(W|V)$ and $P(W|NV)$ with this information. Suppose $P(V|W)$ and $P(NV|W)$ are both 1/2, the proportions from the previous year. The tree diagram is

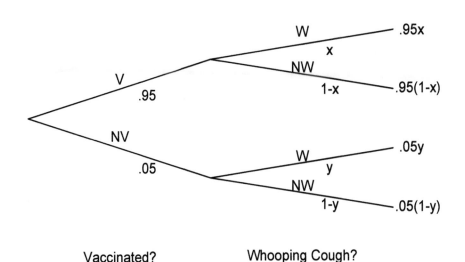

Vaccinated?   Whooping Cough?

Here x represents $P(W|V)$ and y represents $P(W|NV)$, the quantities we would like to compare. We are given that $P(V|W) = 0.5$, and we know from the definition of conditional probability that

$$P(V|W) = \frac{P(V \cap W)}{P(W)} = \frac{0.95x}{0.95x + 0.05y},$$

so

$$0.5 = \frac{0.95x}{0.95x + 0.05y} \Rightarrow 0.95x = 0.475x + 0.025y.$$

We can solve for y to get $y = 19x$. We see that if a child is vaccinated, he or she is 19 times less likely to get whooping cough than a child who is not vaccinated.

To determine the effectiveness of the vaccine, we must compare the probability of getting the disease, given that the vaccine was given, with the probability of getting the disease, given that the vaccine was *not* given. In this case, we find that the vaccine does drastically lower the probability of getting the disease, even if it is not completely effective.

Definition: Suppose we can specify events $A_1, A_2, \ldots, A_n$ so that each outcome of S is in an event A_i for exactly one $i = 1, \ldots, n$. Formally, we say that events $A_1, A_2, \ldots, A_n$ form a **partition** of the sample space S if

1. for $i \neq j$, $A_i \cap A_j = \emptyset$, and

2. $S = A_1 \cup A_2 \cup \cdots \cup A_n$.

The Law of Total Probability

Suppose we have a partition $A_1, A_2, \ldots, A_n$ and an event B. In the following figure, the oval represents the event B.

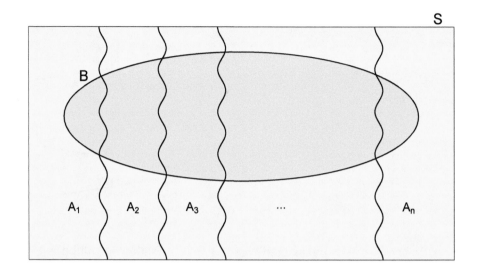

We have

$$P(B) = P(B \cap A_1) + P(B \cap A_2) + \cdots + P(B \cap A_n),$$

because all the intersections on the right are disjoint. Using the definition of conditional probability from the beginning of this chapter, we have

$$P(B \cap A_i) = P(B|A_i)P(A_i)$$

for each $i = 1, \ldots, n$, and the law of total probability is

$$P(B) = P(B|A_1)P(A_1) + P(B|A_2)P(A_2) + \cdots + P(B|A_n)P(A_n).$$

Example: Suppose we have three urns with red and green balls. Urn A has one red and one green ball, Urn B has two red and one green ball, and Urn C has nine red and one green ball. We choose an urn at random, then choose a ball at random from that urn. What is the probability of getting a green ball? Let G be the event that the ball is green, and let A be the event that Urn A is chosen, and similarly for events B and C. Then we can write

$$P(G) = P(G|A)P(A) + P(G|B)P(B) + P(G|C)P(C)$$

$$= \left(\frac{1}{2}\right)\left(\frac{1}{3}\right) + \left(\frac{1}{3}\right)\left(\frac{1}{3}\right) + \left(\frac{1}{10}\right)\left(\frac{1}{3}\right) = \frac{14}{45}.$$

We could have gotten the same thing from a tree diagram:

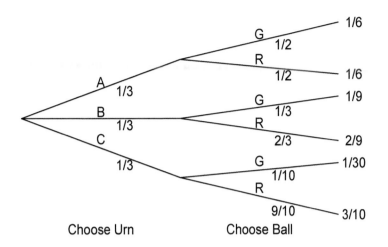

Now we add up the probabilities after the "G" branches to get the answer. In most problems, you can use either the formula or a tree diagram, whichever you feel more comfortable with.

Bayes' rule

Bayes' rule is built on the law of total probability. It's used when you want to find a conditional probability, and you have information about the "opposite" conditional probability. In other words, you want to find $P(A_1|B)$ and you know $P(B|A_1)$. The formula is derived from the definition of conditional probability and the law of total probability:

$$P(A_1|B) = \frac{P(A_1 \cap B)}{P(B)} = \frac{P(B|A_1)P(A_1)}{P(B|A_1)P(A_1) + P(B|A_2)P(A_2) + \cdots + P(B|A_n)P(A_n)},$$

where $A_1, A_2, \ldots, A_n$ form a partition of the sample space.

To continue the previous example, suppose you have drawn a green ball. What is the probability that it came from Urn A? In other words, what is $P(A|G)$? Bayes' rule says

$$P(A|G) = \frac{P(G|A)P(A)}{P(G|A)P(A) + P(G|B)P(B) + P(G|C)P(C)},$$

and we have already computed the denominator. The numerator is $(1/2)(1/3) = 1/6$, so $P(A|G) = 15/28$.

Example: Suppose a computer company receives 60% of chips from Supplier A, 30% from Supplier B, and 10% from Supplier C. Further, chips from Supplier A are 2% defective, chips from Supplier B are 1% defective, and chips from Supplier C are 5% defective. If a chip is randomly chosen and found to be defective, what is the probability that it came from supplier C?

Let's define some events: Let event A be that the randomly selected chip is from Supplier A, and similarly define events B and C. Let event D be that the randomly selected chip is defective. The problem is to find $P(C|D)$; we are given $P(D|C) = .05$

but of course this is not the same as $P(C|D)$. All the elements needed for using Bayes' rule are given. We plug in

$$P(C|D) = \frac{P(D|C)P(C)}{P(D|A)P(A) + P(D|B)P(B) + P(D|C)P(C)}$$

$$= \frac{(.05)(.1)}{(.02)(.6) + (.01)(.3) + (.05)(.1)} = .25.$$

In other words, 1/4 of defective chips that the company receives are from Supplier C, although the company gets only 10% of chips from this supplier.

Note that we have already used Bayes' rule and the law of total probability when we did the vaccination and disease testing examples. With these explicit rules, you don't have to draw a tree diagram. For each problem, you can decide whether you prefer drawing a diagram or using the formulas.

Chapter Highlights

1. Definition of conditional probability: The probability of event B given that event A has happened is
$$P(B|A) = \frac{P(B \cap A)}{P(A)}.$$

2. Events A and B are **independent** if $P(B|A) = P(B)$.

3. If events A and B are independent, then $P(A \cap B) = P(A)P(B)$.

4. The events $A_1, A_2, \ldots, A_n$ form a **partition** of the sample space S if

    (a) for $i \neq j$, $A_i \cap A_j = \emptyset$, and

    (b) $S = A_1 \cup A_2 \cup \cdots \cup A_n$.

5. Law of total probability: If $A_1, A_2, \ldots, A_n$ form a partition of the sample space, then for any event B in the sample space,

    $$P(B) = P(B|A_1)P(A_1) + P(B|A_2)P(A_2) + \cdots + P(B|A_n)P(A_n).$$

6. Bayes' rule: If $A_1, A_2, \ldots, A_n$ form a partition of the sample space, then for any event B in the sample space,

$$P(A_1|B) = \frac{P(A_1 \cap B)}{P(B)} = \frac{P(B|A_1)P(A_1)}{P(B|A_1)P(A_1) + P(B|A_2)P(A_2) + \cdots + P(B|A_n)P(A_n)}.$$

Exercises

4.1 Sixteen percent of students at State College have been to Europe, and five percent have been to South America. Eighty-three percent of the students have been neither to Europe nor to South America. If a student is chosen at random and is found to have been to South America, what is the probability that the student has also been to Europe?

4.2 An educator has determined that 40% of statistics majors do their homework very neatly, that 45% of statistics majors graduate with a GPA over 3.5, and that 55% of statistics majors do their homework very neatly *or* graduate with a GPA over 3.5. Determine the probability that a "randomly chosen" student will graduate with a GPA over 3.5, given that s/he does homework very neatly.

4.3 In Smalltown, 40% of households have at least one dog and 60% of households have at least one cat, while 20% of households have neither dogs nor cats. If a household is chosen at random and found to have at least one dog, what is the probability that it also has at least one cat?

4.4 Suppose that 65% of circuit breakers last at least 10 years, and 50% last at least 15 years. Suppose a circuit breaker has lasted 10 years; what is the probability that it lasts at least 15 years?

4.5 Suppose 70% of sales people at Sellmore, Incorporated are women. Further, 20% of sales people got a promotion last year, and 24% of sales people are men who were not promoted. Suppose we randomly select a sales person. Are the events $A =$ "sales person is a woman" and $B =$ "sales person got promoted" independent?

4.6 Suppose that in a certain country 10% of the elderly people have diabetes. It is also known that 30% of the elderly people are living below the poverty level, and 35% of the elderly population falls into at least one of these categories.

   (a) Given that a randomly selected elderly person is living below the poverty level, what is the probability that she/he has diabetes?

   (b) Are the events "has diabetes" and "living below the poverty level" disjoint in this elderly population? Explain.

   (c) Are the events "has diabetes" and "living below the poverty level" independent in this elderly population? Explain.

4.7 In an industrialized country, 20% of the adults are aged 65 and older. Furthermore, 12% of the adults have at least one symptom of heart disease, but 25% of people aged 65 and older have at least one symptom of heart disease. If an adult from this country is selected at random, what is the probability that the adult is under age 65 and has no symptoms of heart disease?

4.8 The "Mathematical Recreations of Lewis Carroll" included this problem: A bag contains a counter that is either black or white (with equal probabilities). A white counter is added to the bag, and the bag is shaken. A counter is drawn from the bag at random. If the counter is white, what is the probability that the remaining counter is white?

4.9 In 60% of households in a retirement community, there was at least one upper respiratory infection last winter. However, in the subset of households with at least one cat, 80% of these households saw at least one upper respiratory infection last winter. Further, 45% of households have at least one cat. If a household is chosen at random and found not to have a cat, what is the probability that there was at least one upper respiratory infection last winter?

4.10 Referring to Exercise 4.5, consider the experiment of randomly selecting a sales person. Define the events A = "sales person is a woman" and B = "sales person got promoted." Suppose we know only that the proportion of men who got promoted is the same as the proportion of women who got promoted. Show that events A and B are independent.

4.11 Suppose we have three numbers chosen "at random" in the interval $(0, 1)$, obtained for example by the R command `runif(3)`. What is the probability that all three numbers are greater than 1/2, given that the sum is less than 2. Use simulations in R to compute this conditional probability.

4.12 The experiment is to roll four fair, six-sided dice. Given that all dice show numbers that are three or greater, what is the probability that all the dice show the same number? Use simulations in R to compute this conditional probability.

4.13 Suppose a couple decides on the following scheme for planning their family: They will have children until they have at least one boy and at least one girl, or until they have three children. Assume that for each child, the probability of its being a girl is 0.5, that the genders of successive children are independently determined, and they don't have twins. Write out the sample space for this experiment and assign probabilities.

4.14 The "one child rule" in some rural parts of China had been amended to the following. All couples are allowed one baby. If the baby is a girl, they are allowed to have exactly one more. If this rule is exactly followed, and ignoring possibilities of twins, infertility, etc., what will be the resulting proportion of boys to girls in this community? Assume that for each child the probability of its being a girl is 0.5.

4.15 Charley Whiney and Sam Stoic are best friends and do a lot of things together. When they disagree about what to do next, they often flip a (fair) coin to see who gets to decide. If Charley wins the coin toss, they do what he wants to do. However, if Charley loses, he always says, "Oh, come on! Two out of three!" and they extend the contest (so that they toss at most twice more). What proportion of times do they end up doing what Charley Whiney wants to do?

4.16 There are three baskets with balls:

- Basket 1: two reds.
- Basket 2: one red and one blue.
- Basket 3: two blues.

A basket is chosen "at random." We draw a ball and it is blue. What is the probability that the other ball in that basket is red?

4.17 Sixty percent of adults over 60 have received a vaccine for shingles. The vaccine is not perfect, and sometimes vaccinated people get the disease. Suppose 20% of adults over 60 have had the disease, while 25% of adults over 60 have had neither the vaccine nor the disease. An adult over 60 is randomly selected.

   (a) If the adult is found to have had the disease, what is the probability that the adult had been vaccinated?

   (b) Are the events "the adult is found to have had the disease" and "the adult had been vaccinated" independent?

4.18 Refer to Exercise 1.16.

    (a) Given that a randomly selected person is not married, what is the probability that s/he has a sports car?

    (b) What percent of unmarried people have sports cars?

4.19 Show that if $P(B|A) = P(B)$, then we must have $P(A|B) = P(A)$.

4.20 In a region of the U.S., 15.14% of the adult population are smokers, 0.86% are smokers with emphysema, and 0.24% are nonsmokers with emphysema.

    (a) What is the probability that a person, selected at random, has emphysema?

    (b) Given that the person is a smoker, what is the probability that the person has emphysema?

    (c) Given that the person is a nonsmoker, what is the probability that the person has emphysema?

    (d) Are the events "person is a smoker" and "person has emphysema" independent? Explain.

4.21 Thirty percent of people over age 65 have condition A, and forty percent of people over age 65 have condition B. If the occurrences of the conditions are independent, what proportion of people over age 65 have neither condition?

4.22 Sam has a drawer with six blue socks and four white socks. He takes two socks from the drawer at random. What is the probability that Sam has a matching pair of socks (i.e., the same color)?

4.23 Sam has a drawer with five blue socks and three white socks. He draws socks at random, one at a time and without replacement, until he has a matching pair (two of the same color). What is the probability that the pair is blue?

4.24 Refer to Exercises 1.8 and 2.4. Suppose one person arrives at the clinic and participates in the study. Using the probabilities given in Exercise 2.4, compute the probability that the person improves. If five patients independently enter the study, what is the probability that the result is "improved" for all five of them?

4.25 Maxine and Susan are glass blowers and make colored vases for a boutique. Maxine is very careful and 60% of her vases have no flaws, 30% have one flaw, and 10% have two flaws. Susan is faster but less careful and only 30% of her vases have no flaws, 50% have one flaw, and 20% have two flaws. However, 70% of the vases in the boutique were made by Susan and only 30% of the vases were made by Maxine.

    (a) If we select a vase at random and find that it has no flaws, what is the probability that Susan made it?

    (b) What proportion of the vases in the boutique have two flaws?

4.26 Suppose that 20% of employees at Watchdog, Inc. are guilty of embezzlement. The company wants to give everyone lie detector tests, but are not allowed to force employees to take the test. Suppose 90% of guilty employees refuse to take the test, and 10% of guilty employees agree to take the test. Of the employees not guilty of embezzlement, 80% agree to take the test and 20% refuse. What proportion of employees who take the test are guilty?

4.27 Only one fifth of one percent of circuit breakers are flawed, but if a flawed circuit breaker is used in a power plant, there is a 90% chance of an explosion. If a nonflawed circuit breaker is used, the probability of an explosion is only 5%.

   (a) If we select a circuit breaker at random and use it in the power plant, what is the probability of an explosion?

   (b) Suppose we select a circuit breaker at random and use it in the power plant, and an explosion occurs. What is the probability that the circuit breaker was flawed?

4.28 Eighty percent of students in a difficult statistics course attended lectures regularly. Ninety percent of those who attended regularly got at least a C in the course, while only forty percent of those who did not attend regularly got at least a C. If we randomly choose a student in the course and find that the student got at least a C, what is the probability that the student attended regularly?

4.29 The immunization for the flu is not completely effective, and 2% of folks who get immunized in the fall still get the flu the following winter. However, 5% of folks who do not get immunized will get the flu. Suppose that 80% of Smalltown residents get a flu immunization. What percent of Smalltown residents will get the flu?

4.30 Suppose it has been determined that 5% of the U.S. population carries a gene mutation that increases the risk of diabetes by a factor of ten. Specifically, only 0.2% of people without the gene mutation develop diabetes, but 2% of people with the gene mutation will develop it. If we randomly select a person who has developed diabetes, what is the probability that the person has the gene mutation?

4.31 A factory has two machines producing computer boards. Ten percent of the boards that Machine A makes are defective, while only two percent of the boards made by Machine B are defective. Both machines made the same number of boards last week. One of the boards is randomly selected, and it proves to be defective. What is the probability that it was made by Machine A?

4.32 Suppose the CIA is considering mandatory lie detector tests for all its agents, in an attempt to ferret out spies. They get the best lie detector machine available. If a person is guilty, there is a 99% chance of detecting guilt with the machine. If the person is innocent, there is a 7% of the machine reading guilty. Further, one half of one percent of agents are really spies, and the rest are not. Suppose a randomly selected agent takes the test, and it reads guilty. What is the probability that the agent is a spy?

4.33 Suppose that 5% of people who are 65 or older will have a heart attack in a given year, compared with 1% of people who are younger than 65. Suppose that a certain population has 10% of its members over 65. What is the overall heart attack rate?

4.34 Refer to the first example on imperfect testing, where the sensitivity of the test is $P(+|D) = .99$ and the specificity is $P(-|N) = .98$. Although the prevalence is low in the general population ($P(D) = .001$), there is a subgroup of the population (perhaps with a family history) in which the prevalence is $P(D) = .08$. Suppose a clinic decides to give the test *only* to the people in this group. Now, if the result is positive, what is the probability that the person has the condition?

4.35 The small country of Kazbecstan is known for its poppy fields and opium production. Thirty percent of the residents in Kazbecstan are first or second generation immigrants from Bacmenia, and seventy percent are native Kazbecstanis. Of the native Kazbecstanis, 25% carry opium for sale, while only 10% of Bacmenians carry opium for sale. Because of the large numbers of drug dealers, police have adopted a stop-and-frisk policy. But because the police are all native Kazbecstanis, they are more likely to target Bacmenians for stop-and-frisk. Suppose 15% of Bacmenians have been subjected to stop-and-frisk, while only 2% of Kazbecstanis have been subjected to stop-and-frisk. The stopped person is jailed if opium is found. What proportion of people being jailed for carrying opium are Bacmenians?

Chapter 5

Discrete Random Variables and Expected Value

Random variables are the building blocks of statistical models. They are functions of outcomes of a random experiment, and because we like numbers so much, we insist that they take on numerical values only. Hence, we have the following definition. A **random variable** is a mapping of outcomes in the sample space to the real numbers.

For example, if the experiment is to toss a coin repeatedly until a head is observed, we might write the sample space as $\{H, TH, TTH, TTTH, \ldots\}$. The random variable X associated with this experiment might be the number of tosses. The values that the random variable can take are 1, 2, 3, etc. The set of possible values is called the **support** of the random variable.

A random variable is traditionally denoted by a capital letter from the end of the alphabet, such as X or Y. The **probability distribution** of the random variable is inherited from the probabilities associated with outcomes or events in the sample space. The probability distribution assigns probabilities to values or sets of values in the support of the random variable.

A **discrete** random variable takes on a finite or countably infinite number of values. In this chapter we will consider probability distributions for discrete random variables, called **probability mass functions** (or just "mass functions" for short). Mass functions assign probabilities to the possible values of the random variable—these are determined by the probability function that assigns probabilities to the outcomes in the sample space. This chapter will be followed by several chapters about various important families of discrete distributions, such as geometric, binomial, Poisson, and others.

A **continuous** random variable takes values in an interval on the real line, or in a union of intervals. In this case the set of values for the random variable is uncountable, so we cannot assign probabilities to the individual values. Instead, the distribution of a continuous random variable is described by a **probability density function**; we will tackle these density functions after the chapters on discrete distributions.

Example: The experiment is to roll two fair dice, where the sample space of equally probable outcomes is shown below (and was considered in Chapter 1):

$$
S = \left\{ \begin{array}{cccccc}
(1,1) & (1,2) & (1,3) & (1,4) & (1,5) & (1,6) \\
(2,1) & (2,2) & (2,3) & (2,4) & (2,5) & (2,6) \\
(3,1) & (3,2) & (3,3) & (3,4) & (3,5) & (3,6) \\
(4,1) & (4,2) & (4,3) & (4,4) & (4,5) & (4,6) \\
(5,1) & (5,2) & (5,3) & (5,4) & (5,5) & (5,6) \\
(6,1) & (6,2) & (6,3) & (6,4) & (6,5) & (6,6)
\end{array} \right\}.
$$

There are many mappings from this set to the real numbers; let's consider the random variable X defined as the number of sixes. What is the probability mass function for X? The random variable takes only three possible values: $X = 0$ if no sixes are rolled, $X = 1$ for one six, and $X = 2$ for "double sixes." Given that the probability for each outcome is $1/36$, we can compute $P(X = 0) = 25/36$, $P(X = 1) = 10/36$, and $P(X = 2) = 1/36$.

Probability mass functions are often displayed as "bar charts" where there is one bar per value and the heights of the bars correspond to the probabilities. For this example we have the following chart:

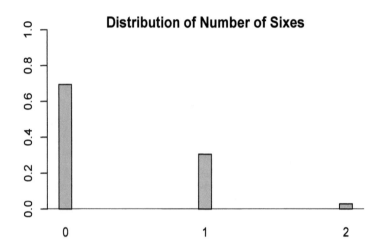

A probability mass function for any discrete random variable X must have the following properties.

1. $P(X = x_i) \geq 0$ for all x_i in the support of X.

2. $\sum_i P(X = x_i) = 1$, where the sum is over the support.

These two properties are inherited from the probability functions defined on the sample space, given that the random variable is a mapping from the sample space.

The **expected value** of a discrete random variable is defined as

$$E(X) = \sum_i x_i P(X = x_i),$$

where the sum is over the support (all possibly values of the discrete random variable). If we repeat the experiment "many times" and record the value of the random variable each time, then the average of the recorded values is the expected value of the random variable. The expected value of a random variable is also referred to as the **mean** or **average** value of the random variable. It's also the "balance point" of the bar chart, if you imagine that the bars are rods, all of the same diameter and composition.

For the first example, the expected number of sixes on the dice is $(0)(25/36) + (1)(10/36) + (2)(1/36) = 12/36 = 1/3$. If we mark this point on the horizontal line of the bar chart, this is where the chart would balance.

We can define a new random variable Y as a function of a discrete random variable X, say $Y = g(X)$ for some function g. The distribution of Y is inherited from the distribution of X, and so we can compute the expected value:

$$E(Y) = E(g(X)) = \sum_i g(x_i)P(X = x_i).$$

Example: Suppose we have a box of 10 unlabeled resistors, two of which are 4 ohms, three are 10 ohms, and five are 20 ohms. We select a resistor "at random" (each with probability 1/10) and put it in the circuit with a 2-volt battery. The power (in watts) for a circuit is V^2/R, where R is the resistance (in ohms) and V is the voltage (in volts). What is the expected power? Let X be the resistance and Y the power; then $Y = 4/X$. We compute

$$E(Y) = \sum_i \frac{4}{x_i}P(X = x_i)$$

$$= \frac{4}{4}P(X = 4) + \frac{4}{10}P(X = 10) + \frac{4}{20}P(X = 20)$$

$$= .2 + (.4)(.3) + (.2)(.5) = .42.$$

The expected power is .42 watts.

If g is linear, for example, $g(x) = a + bx$, we could use the following simple result:

$$E(a + bX) = a + bE(X).$$

This is easy to prove, using the definition of expected value and properties of summation:

$$E(a + bX) = \sum_i (a + bx_i)P(X = x_i)$$

$$= \sum_i [aP(X = x_i) + bx_iP(X = x_i)]$$

$$= a\sum_i P(X = x_i) + b\sum_i x_iP(X = x_i)$$

$$= a + bE(X).$$

Using the same ideas for the proof, we can get the following more general formula for several functions $g_1, \ldots, g_k$ and constant values $a_1, \ldots, a_k$:

$$E\left(\sum_{j=1}^k a_j g_j(X)\right) = \sum_{j=1}^k a_j E(g_j(X)).$$

Definition: The **variance** of a random variable is a measure of the "spread" of its distribution and is defined as the expected squared distance of the random variable from its mean:

$$V(X) = E[(X - E(X))^2].$$

The square root of the variance is the **standard deviation**.

We have a formula for computing variances that is typically easier to use, compared to the definition. If $\mu = E(X)$, then $V(X) = E(X^2) - \mu^2$. To prove this, we use our properties of expectation:

$$\begin{aligned} V(X) = E[(X - \mu)^2] &= E[X^2 - 2X\mu + \mu^2] \\ &= E(X^2) - 2E(X)\mu + \mu^2 \\ &= E(X^2) - 2\mu^2 + \mu^2 \\ &= E(X^2) - \mu^2. \end{aligned}$$

To compute the variance of the random variable X representing the number of sixes in the dice example, we first compute the expected value of the square of the number of sixes. This is

$$E(X^2) = (0^2)(25/36) + (1^2)(10/36) + (2^2)(1/36) = 7/18,$$

so

$$V(X) = 7/18 - (1/3)^2 = 5/18.$$

The standard deviation of X is $\sqrt{5/18}$, or about .527.

The distribution of a discrete random variable can be specified as a list or a table of values, but when the number of values of the random variable is large, or the values depend on one or more unspecified parameters, it will be more convenient to use a formula to specify the probabilities.

Example: Suppose $0 < a < 1$ and k is a positive integer. A random variable X might be defined with the following probability mass function:

$$P(X = i) = \frac{a^{i-1}(1 - a)}{1 - a^k} \quad \text{for } i = 1, 2, \ldots, k.$$

It can be seen that this is a valid probability mass function because the probabilities are positive and

$$\sum_{i=1}^{k} a^{i-1} = \frac{1 - a^k}{1 - a}.$$

(See Appendix B.2 for the derivation of this summation formula.) Let's find the expected value of this random variable:

$$E(X) = \frac{(1 - a)}{a(1 - a^k)} \sum_{i=1}^{k} ia^i = \frac{(1 - a)}{a(1 - a^k)}[a + 2a^2 + 3a^3 + \cdots + ka^k].$$

We can simplify this by letting $S = a + 2a^2 + 3a^3 + \cdots + ka^k$; then $aS = a^2 + 2a^3 + \cdots + ka^{k+1}$, and subtracting the first equation from the second gives

$$(1 - a)S = a + a^2 + a^3 + \cdots + a^k - ka^{k+1} = \frac{a - a^{k+1}}{1 - a} - ka^{k+1}.$$

Putting all of this together, we have

$$E(X) = \frac{(1-a)}{a(1-a^k)} \left[\frac{a-a^{k+1}}{1-a} - ka^{k+1} \right] / (1-a) = \frac{1}{1-a} - \frac{ka^k}{1-a^k}.$$

For example, if $a = .6$ and $k = 5$, the bar chart for the distribution is

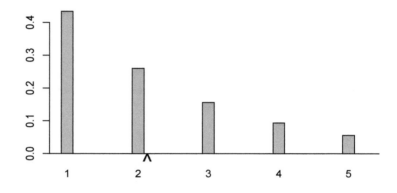

The expected value (about 2.078) is marked on the bar chart as the balance point.

Random variables and their probability distributions are used in statistical models for estimation and prediction of real-world phenomena. A data set could be assumed to be a **random sample** from a distribution, that is, a set of independent realizations of the random variable. For example, suppose $X_1, \ldots, X_n$ are independent random variables all with the same distribution as the random variable X. The average of the random sample, $\bar{X} = (\sum_{i=1}^n X_i)/n$, is the **sample mean**, not to be confused with the population mean $E(X)$.

Similarly, **sample variance** is $\sum_{i=1}^n (X_i - \bar{X})^2/(n-1)$. The sample mean and the sample variance are random variables, because different samples produce different values for sample mean and sample variance, even though the distribution of the random variable used to generate the sample is the same. In contrast, the population mean and the population variance are not random, they are constants.

Suppose the values of the sample are $x_1, \ldots, x_n$, stored in the R vector x. The R function mean(x) will return the average of the values of the vector x and the R function var with return the sample variance. When the sample size is large, we expect the sample mean to be close to the population mean, and the sample variance to be close to the population variance; we will formalize these notions in later chapters. For now, we can use them to check our calculations of the population mean and variance, using R.

A random sample from a distribution contains *independent* realizations of a random variable. We have not yet formally discussed independent random variables (only independent events), but your intuition will serve—if two random variables are independent, then the probability that one takes on a value does not depend on the value of the other.

Because the sample is random, the sample mean is a random variable. Let's go back to the simple example where X is the number of sixes when two fair dice are rolled. A sample of size 20 can be obtained by rolling a pair of dice 20 times and recording the numbers of sixes, or through simulations in R. The command

```
x=sample(0:2,20,prob=c(25/36,10/36,1/36),replace=TRUE)
```

will make a vector mimicking 20 rolls. The command `mean(x)` will provide the average number of sixes in 20 rolls of a pair of dice. Of course, repeating the two commands will likely result in a different sample mean; if might be .15 for one sample and .4 for another.

As n gets larger, the mean of a random sample will get closer to the distribution mean and, similarly, the sample variance will get closer to the variance of the distribution. For example, one million rolls of two dice can easily be simulated in R:

```
x=sample(0:2,1000000,prob=c(25/36,10/36,1/36),replace=TRUE)
```

The vector `x` will have a sample mean very close to .334, and the sample variance will be close to .278. Computing the mean of large samples in this way can be used to check your calculations. Later we will be interested in the distribution of the sample mean itself, which can be approximated with many simulated samples.

Example: Let's write some R code to sample from the more complicated distribution in the previous example. Setting values for a and k, and the sample size n, we first compute the vector of the k probabilities, then generate a random sample and compute the mean.

```
a=.6;k=5;n=10
pr_a=1:k
for(i in 1:k){
    pr_a[i] = a^(i-1)*(1-a)/(1-a^k)
}
x=sample(1:k,n,prob=pr_a,replace=TRUE)
mean(x)
```

Due to randomness, the sample mean can be anything from 1 to 5. Because the sample mean is itself a random variable, if we run the exact same code again, we will likely get a different sample mean, but for large sample sizes, we expect the sample mean to be close to the distribution mean.

Results about the sample mean will be important when we want to "make an inference" about the distribution. In this example, suppose we know k but not a. Then the sample mean will give us some information about a that will enable us to make an educated guess—perhaps a "best" guess—and we can also quantify the uncertainty of our guess. Larger samples will give more precise information.

This is an illustration of the relationship between probability and statistics that we discussed in the preface. A probability problem might be "given $a = .6$ and $k = 5$, find the probability that the sample mean is at least 4," while a statistics problem might be "if $k = 5$ and the sample mean is 4, is this strong evidence that $a > .3$?"

Right now, the important thing is to recognize the difference between the mean of the distribution and the mean of a sample from the distribution. The mean of a distribution is a fixed number, and the mean of a sample is a random variable that will take a different value for a new sample.

For now, we can check our calculations for the mean and variance of distributions, using simulations in R. We can use the idea that if a sample from a distribution is really large, then the sample mean and the sample variance ought to be close to the true mean and the true variance of the distribution. We will prove this formally in future chapters, but for now our intuition allows us to believe it without formal proof.

Chapter Highlights

1. A **random variable** maps the outcomes in a sample space to the real numbers.

2. The **support** of a random variable is the set of values it can take.

3. A **discrete random variable** has a finite or countably infinite support.

4. A probability mass function is used to assign probabilities to values that the random variable X can take. If the mass function is denoted by P, we must have

   (a) $P(X = x_i) \geq 0$ for all x_i in the support of X, and

   (b) $\sum_i P(X = x_i) = 1$, where the sum is over the support of X.

5. If the discrete random variable X can take values $x_1, x_2, \ldots$, then the expected value is defined as
$$E(X) = \sum_i x_i P(X = x_i),$$
where the sum is over the support of X.

6. If the discrete random variable X can take values $x_1, x_2, \ldots$, then the expected value of the random variable $g(X)$ is defined as
$$E(g(X)) = \sum_i g(x_i) P(X = x_i),$$
where the sum is over the support of X.

7. The variance of a random variable X is defined as
$$V(X) = E\left[(X - E(X))^2\right] = E(X^2) - E(X)^2,$$
and the standard deviation is the square root of the variance.

8. If X is a random variable and a and b are real numbers, then
$$E(a + bX) = a + bE(X).$$

9. Although we have not formally proved this yet, if we have a large random sample from the distribution of the random variable, then under mild conditions (including that the mean and variance of the distribution have to be finite), the sample mean will be close to the mean of the random variable, and the sample variance will be close to the variance of the random variable.

Exercises

5.1 Suppose 70% of the members of a large labor union are women. If a committee consists of two members chosen "at random," then the number of women on the committee is a random variable, say Y.

(a) Make a mass function for Y and draw the bar chart representing the distribution.

(b) Find the expected value of Y and mark it on the bar chart.

(c) Compute the variance of Y, and write code in R (using the distribution from (a)) to check your answer.

5.2 A quiz consists of two multiple choice questions with choices (a), (b), and (c) for each. If an unprepared student marks answers at random, what are the probabilities of the following events:

(a) Both answers are correct.

(b) Exactly one correct answer.

(c) Both answers wrong.

(d) Let Y be the random variable representing the number of correct answers on the quiz. What is $E(Y)$?

5.3 Last summer there were 500 campers at Camp Nearahive. Some of the campers had bad luck with bees, and the distribution of bee stings per camper is summarized in the following table:

# y of stings	0	1	2	3	4
proportion $p(y)$	0.6	0.2	0.1	0.08	0.02

This table also gives the distribution of the random variable Y, which represents the number of bee stings for a randomly selected camper.

(a) Graph the bar chart.

(b) What proportion of campers got at least one bee sting?

(c) *How many* campers got exactly 3 bee stings?

(d) What is the average number of bee stings per camper?

(e) Check your answer to part (d) using simulations in R.

(f) Compute the standard deviation of the number of bee stings per camper.

(g) Use simulations in R to check your answer to part (f).

5.4 The experiment is to toss three fair coins.

(a) Write out a sample space of equally likely outcomes.

Suppose the random variable Y is the number of heads.

(b) Write out a mass function for Y.

(c) What is the expected value of Y?

(d) What is the variance of Y?

5.5 A large jar of chocolates contains 20% caramel fudge. Suppose you decide to select candies at random from the jar, until you get a caramel fudge or until you have four candies, whichever comes first. The variable of interest is the number of candies that you get. (Note that "large" means you can compute probabilities as if you were sampling with replacement.)

(a) Write out the distribution for the possible values:

$y = \text{\# candies}$	1	2	3	4
probability $f(y)$				

(b) What is the expected number of candies chosen?

(c) What is the variance of number of candies chosen?

(d) Write code in R, using your answer to (a), to check your answers to (b) and (c).

5.6 A box has five switches, three of which are defective. An engineer draws switches at random from the box, one by one and without replacement, until he gets a switch that is not defective. What is the expected number of switches that he draws from the box?

5.7 The support for a discrete random variable Y consists of three values $\{-a, 0, a\}$ for a constant $a > 0$. The mass function is defined as

$$P(Y = 0) = 1/2, P(Y = -a) = P(Y = a) = 1/4.$$

Find the expected value and variance of Y.

5.8 A random variable X has the following distribution: For $p \in (0, 1)$,

- $P(X = -2) = P(X = 2) = p/6$,
- $P(X = -1) = P(X = 1) = p/3$, and
- $P(X = 0) = 1 - p$.

Find the expected value and variance of X.

5.9 A random variable X has the following distribution: For a constant value $p \in (0, 2)$,

- $P(X = 2) = p/6$,
- $P(X = 1) = p/3$, and
- $P(X = 0) = 1 - p/2$.

Find the variance of X.

5.10 In the American version of the casino game roulette, a wheel with 38 slots is spun, and a marble drops (at random) into one of the slots. There are 18 red slots, 18 black slots, and 2 green slots. If you bet $100 on red, you will win $100 if the marble drops in a red slot; otherwise you lose $100. Calculate your expected winnings.

5.11 Bag A has two gold balls and six brown balls. Bag B has two gold balls and two brown balls. A game consists of two stages. First, you toss a fair coin; if heads, you get bag A; if tails, you get bag B. You (randomly) choose two balls from the bag (without replacement). You win $1000 for each gold ball you draw.

(a) Let Y be the number of gold balls, and find the probability mass function for Y.

(b) What is the expected amount of money that you win?

5.12 An urn has three blue balls and one gold ball. A player draws balls from the urn, one at a time and without replacement, until the gold ball is drawn. A player will receive $100 when the gold ball is drawn, but has to pay $N per draw. What is the value of N that makes this game "fair," i.e., the expected value of the amount won is zero?

5.13 Refer to Exercise 1.17. Let the random variable Y be the number of the phenotypes exhibited by a randomly selected honeybee. (a) Find the probability mass function for Y. (b) Find E(Y).

5.14 An electrical circuit has components that are chosen "at random." The resistance (R) can be 4, 8, or 16 ohms, and the voltage (V) can be 1.5, 2, 2.5, or 3 volts. If the given values for resistance and for voltage are equally likely and chosen independently, what is the expected value and variance of the resulting current, which is $I = V/R$ amperes. Although this can be computed "by hand," it's faster to do simulations in R.

5.15 For $b > 0$ and $\mu > 0$, define a discrete random variable Y by specifying

$$P(Y = y) = \frac{e^{-1/2}}{y!\, 2^y}, \quad y = 0, 1, 2, \ldots.$$

Find $E(Y)$. (*Hint:* Use a change of variable, and the fact that the probabilities must add up to 1.)

5.16 Harry will play a lottery game repeatedly until he wins or until he runs out of money. It costs $20 to play the game, and he has only $60. The probability of winning for each game is $p = .15$, and this probability does not depend on outcomes of other games. Let Y be the number of times he plays.

(a) Find the probability mass function for Y.

(b) Find E(Y).

5.17 Bag H has 4 gold and 16 brown marbles, while Bag T has 2 gold and 4 brown marbles. A game is proposed in which you flip a fair coin, and if it is heads, you choose a marble at random from Bag H; if it is tails, you choose a marble at random from Bag T. If you get a gold marble, you win $100. However, if you get a brown marble, you have to pay $20. What is the expected number of dollars that you win if you play this game?

5.18 Suppose X is a random variable with mean μ and variance v. Find an expression for the variance of $Y = a + bX$.

Chapter 6

Hypothesis Testing Terminology and Examples

The hypothesis test is one of the most common statistical procedures. A conjecture is made, data are obtained, and a conclusion is formed that is based on the data and an appropriate model. The statistical techniques used to analyze the data come from probability theory, and interpretations of the results are based on imaginary repeated tests and probabilities of outcomes. In the next chapters, we combine probability ideas with hypothesis testing ideas to provide examples of how probability theory is used by statisticians. Here we introduce all the necessary language, which is standard for statistics theory and practice.

The hypotheses: There are two hypotheses in a formal statistical test. The **null hypothesis** is the status quo, the default, the "nothing's happening," or the "no difference" hypothesis. This is labeled as H_0. The **alternative hypothesis** is the claim that we are testing. It is the "new thing" or "there *is* a difference" claim, and is labeled H_a, or sometimes H_1. For now, we will consider only situations in which the null hypothesis is "simple." This means that under H_0 the model is completely specified, and we can compute probabilities of various events under the assumption that H_0 is true.

The decision: We take some data, do some calculations, and come up with a decision. The decision is traditionally phrased in terms of the null hypothesis, so that we can either *reject* the null hypothesis, or *accept* the null hypothesis. Some statisticians prefer the phrase "there is not enough evidence to reject the null hypothesis" in lieu of "accept the null hypothesis." This is because the hypothesis testing procedure is designed to be "conservative"—the decision is based on evidence *against* the null hypothesis rather than the amount of support *for* the null hypothesis. It's as if the null hypothesis is believed ahead of time, and we have to be thoroughly convinced otherwise, before we will reject it. If there is not much evidence either way, we will accept H_0.

The decision rule: This rule determines the decision (reject or accept the null hypothesis), and is based on the data; usually on a summary of the data such as a sample mean. The decision rule is usually phrased "we will reject H_0 if . . . ."

Errors: If there are two possible decisions, then there are two possible mistakes we can make. A **Type I Error** is rejecting the null hypothesis when it is really true. A **Type II Error** is accepting the null hypothesis when it is really false.

The test size: The size of the test is the probability of making a Type I Error and is represented by the Greek letter α. In the case that H_0 is true, α is the probability that we have the bad luck to reject it. Given a decision rule and a model for the data, we can calculate this probability α. Alternatively, we can formulate the decision rule based on a desired test size. In this case, a common choice of test size is $\alpha = .05$; then the probability of rejecting H_0 when it is true is 5%. The test size can be thought of as our "stake" in H_0, with smaller test sizes being more "conservative." If $\alpha = .01$ instead of .05, then the probability of making a Type I Error is smaller, and the null hypothesis is "harder" to reject, so that the status quo is more likely to be accepted. To guard against mistakenly rejecting H_0, we formulate the decision rule to make the test size small. However, as we will see, decreasing the test size has a tendency to increase the probability of a Type II Error—we are then more likely to accept H_0 when H_a is true.

Statistically significant: Test results are called "statistically significant" if the null hypothesis is rejected.

p-value: The p-value is the probability of seeing as much or more evidence for the alternative hypothesis than we saw in our data, when the null hypothesis is true. The p-value quantifies the amount of evidence *against* the null hypothesis. The smaller the p-value, the more evidence against H_0, in favor of H_a. Once we decide on the test size α, the decision rule can be stated as "reject H_0 if the p-value is less than α."

Power: The power of a test is the probability that the null hypothesis is rejected, given that the alternative is true. This is one minus the probability of a Type II Error. Of course we would like the test size to be small and the power to be large, so that we feel confident about our conclusion.

This is a lot of terminology to absorb if you haven't seen it before, but the ideas will be used repeatedly in the remainder of the book—you will have plenty of opportunity to get a good intuitive understanding of all the aspects of statistical hypothesis testing.

Example: A farmer claims that organic tomatoes are higher in vitamin A than conventional tomatoes. Suppose we know that the average amount of vitamin A in a conventional tomato is 1500 IU (international units), but the amount of vitamin A in a tomato varies according to a distribution, so that some tomatoes have more vitamin A than others. Therefore, having one or two organic tomatoes with high levels of vitamin A might not provide strong evidence that, *on average*, levels of vitamin A are higher in organic tomatoes. Statistical modeling allows us to quantify the amount of evidence supporting the claim.

Let's outline a formal statistical hypothesis test of the farmer's claim. We will choose, "at random," four organic tomatoes and determine the amount of vitamin A in each. The hypotheses are

$$H_0 : \mu = 1500 \quad \text{versus} \quad H_a : \mu > 1500,$$

where μ is the mean amount of vitamin A (in IU) in organically grown tomatoes. In this example, the null hypothesis is simple (only one value of the mean is allowed), while the alternative is *composite*, which means that a range of values is allowed.

The decision rule might be something like "reject H_0 if all four of the organic tomatoes have greater than 1500 IU of vitamin A," or perhaps "reject H_0 if the *average* vitamin A of the four of the organic tomatoes is greater than 1600 IU." To calculate the test size and power for these decision rules, we have to know (or assume) something about the distribution of vitamin A units in the populations of conventional and organic tomatoes. See Exercises 13.23, 18.16, and 19.17 for various models in this context. Exercise 6.4 of this chapter asks you to calculate test size and power using the first decision rule, when the distribution is assumed to be symmetric.

A Type I Error occurs if the grower concludes that the organic tomatoes have more vitamin A, when the true mean vitamin A is really the same as for conventional tomatoes. This might happen by chance if the sample mean is unusually large compared to the true mean.

A Type II Error occurs if the grower concludes that the organic tomatoes have the same amount of vitamin A as the conventionally grown tomatoes, when the mean vitamin A in the organic tomatoes really is higher.

After the data have been collected, a p-value can be calculated using the assumptions about the distribution of amounts of vitamin A in tomatoes. The p-value is a summary of the evidence against H_0. Suppose the grower obtains the measurements, performs the statistical test, and reports a p-value of 0.15. This means that if the null hypothesis was really true, and the procedure was repeated many times (that is, many samples of four organic tomatoes were collected and the measurements obtained), then for about 15% of these samples there would be as much or more evidence that the organic tomatoes have higher vitamin A. In other words, the probability of seeing that much evidence for H_a when H_0 is true is 15%, and perhaps this would not be considered a compelling amount of evidence that the true average level for organic tomatoes is higher.

On the other hand, if the p-value is reported to be .0015, then the probability of seeing this much evidence for H_a when H_0 is true is quite small, and the amount of evidence for the alternative hypotheses *would* be compelling. Either H_0 is true and something very unusual happened, or H_a is true. In this way the p-value quantifies the amount of evidence against H_0—the smaller the p-value, the more evidence that H_a is true. The test size α, which is determined before the data are collected, is a "cutoff" p-value. We reject H_0 if the p-value is less than α, so α is a measure of how much evidence against H_0 will cause you to reject it.

It's important to remember that you can understand and interpret a reported p-value even if you don't understand the statistical methodology that was used to calculate it! You need only know what the null and alternative hypotheses are. For example, suppose you are reading an article comparing cancer treatments, where subjects with a certain type of cancer are randomly assigned to the standard treatment or the new treatment. The null hypothesis is that the survival times for the two treatments are the same, and the alternative is that the subjects given the new treatment have longer survival times. If you read that that p-value for the test is .003, then you know this is strong evidence that the new treatment has longer survival times. For, if the null hypothesis were really true, and the study was repeated "many" times, we would see evidence this strong or stronger in only .3% of the studies. Either the null hypothesis is true and the results from this particular test were a "fluke," or the alternative is true.

In the course of this book, we will learn how to compute test sizes and p-values for many different kinds of tests, but the next example is constructed so that we can already do the calculations.

Example: Suppose a six-sided die is used in a gambling game, and you suspect the die is not fair. For a fair die, the probability of attaining any of the six faces on a given roll is 1/6. Suppose you think that the die might be weighted so that the probability of rolling a one is actually *less* than 1/6. You want to test $H_0 : p = 1/6$ versus $H_a : p < 1/6$, where p is the true probability of getting a one.

The null hypothesis corresponds to the die being fair. A Type I Error occurs if we reject H_0 and claim that the die is not fair, when the die really is fair. The consequences of a Type I Error might be to make a false accusation—we want the test size to be small to avoid this.

A Type II Error occurs if we accept H_0 and decide that the die is fair when it is not. The consequence of a Type II Error might be to let someone get away with cheating, which is not desirable either.

Now we need to design the experiment, which can be as simple as rolling the die 10 times and counting the number of ones. Our decision rule might be "reject H_0 if we don't get any ones."

The test size is the probability that we get no ones when the die is really fair. We can compute this using our probability ideas from Chapter 3. It's a reasonable assumption that the rolls are independent. For each roll, the probability of *not* getting a one is 5/6 if H_0 is true, so the probability of getting 10 rolls that are not ones is $\alpha = (5/6)^{10} = .1615$.

Typically we want the test size to be small. For the suggested decision rule, we have a 16.15% probability of concluding that the die is not fair when it really is. If rejecting H_0 means making an accusation, then we might want the test size to be smaller, so that it is less likely to accuse an innocent person! Suppose we want the test size to be close to $\alpha = .01$, and we still want the decision rule to be "reject H_0 if there are no ones." How many rolls of the die should we use? We want $(5/6)^n = .01$, where n is the number of rolls. We can solve this for n by observing that $n \log(5/6) = \log(.01)$, so $n = \log(.01)/\log(5/6) = 25.26$. To be conservative we can take $n = 26$ rolls. Then the test size is $\alpha = (5/6)^{26} = .0087$.

This is a nice small test size, and we are happy that there is little probability of making a false accusation, i.e., making a Type I Error. But we ought to check our Type II Error probability. That is, suppose someone is cheating; what is the probability that they are *not* caught? To calculate the probability of a Type II Error, we assume that H_a is true, but the alternative hypothesis does not specify a value for p. We need to make a guess about the true probability p of rolling a six, and report the Type II Error probability in terms of the guess.

Suppose $p = .08$, less than half of what it should be. When we roll the die 26 times, the probability of the event "no sixes" is $(.92)^{26} = .114$, so the probability of at least one six (accepting H_0) is .886. We see that Type II Error probability is large. Equivalently, the power of the test, .114, is small. (Conventionally, power is reported instead of the Type II Error probability, though these quantities contain the same information).

Our test is unsatisfying because, although the test size is small, the power is also small. An alternative decision rule might be "reject H_0 if we get at most three ones when the die is rolled 50 times." We'll be able to calculate the test size and power for this decision rule when we study the binomial distribution in Chapter 8.

Example: Suppose 75% of the members of a large labor union are women. Four committees of two members each will be chosen by the president of the union, and two of the members, chatting over coffee, conjecture that men are more likely than women to be chosen for the committees. They want to test the null hypothesis that men and women are equally likely to be chosen versus the alternative that men are more likely to be chosen. They decide to reject the null hypothesis if all four of the committees have at least one man. What is the test size?

If the members are chosen "at random," then the probability of selecting two women for any of the committees is about $(.75)(.75) = .5625$, so the probability of at least one man is $1 - .5625 = .4375$. The probability that all four committees have at least one man, given that the committees are chosen "at random" and independently, is $\alpha = .4375^4 = .0366$.

Example: Historically, 35% of eighteen-year-olds on the east side of Largetown have been arrested at least once. The historical distribution of number of arrests in this population is compiled in the table below:

Number of arrests	0	1	2	3
Proportion	.65	.18	.10	.07

Five years ago, youth programs were established in these neighborhoods, and sociologists would like to test the null hypothesis that the arrest rates for eighteen-year-olds have remained the same versus the alternative that the situation has improved. They will choose six eighteen-year-olds "at random" and reject the null hypothesis if none has been arrested. The test size is calculated to be $\alpha = .65^6 = .0754$. If the distribution has remained the same, there is a 7.54% chance of erroneously rejecting the null hypothesis.

Suppose the new distribution of arrests is shown in the table below. What is the power of the test?

Number of arrests	0	1	2	3
Proportion	.82	.10	.05	.03

The power is the probability that H_0 is rejected (all six have not been arrested) under the new distribution—this is $.82^6 = .304$. This power is rather low—even if this new distribution is true, we are more likely to accept H_0 than to reject it.

Part of the challenge of hypothesis testing is choosing a decision rule for which the test size is small and the power is large. There may be several choices for the decision rule, and if they all have the same test size, the "best" one has the largest power. This is complicated by the fact that the alternative is often stated in terms of the mean or a parameter being "greater than" or "less than" the null mean or parameter, so that we have to make a "guess" before the power is computed. Generally, increasing the sample size (number of observations) improves the power, so power calculations are performed before collecting data, in order to determine an appropriate sample size.

In Chapter 57 we will discuss most powerful tests and learn how to find a decision rule that gives the largest power for a desired test size.

Chapter Highlights

1. The null hypothesis H_0 is the "status quo" and the alternative is H_a. We take data and compute a data summary that is used to make a decision: We either reject H_0 or we accept H_0.

2. The decision rule specifies the conditions under which we reject H_0.

3. If we reject H_0 when it is true, we make a Type I Error. If we accept H_0 when it is false, we make a Type II Error.

4. The test size α is the probability that we reject H_0 when it is true (i.e., the probability of a Type I Error).

5. The p-value is the probability that we see "our data" (the computed summary) or "more extreme" (supports H_a more) when H_0 is true.

6. The power of the test is the probability that H_0 is rejected, given that H_a is true.

🙂🙂

Exercises

6.1 A sociologist makes a claim that girls read more than boys. A skeptical colleague wants to test this claim for sixth graders.

 (a) Let μ_G be the average number of hours of reading per week for the population of sixth grade girls, and let μ_B be the average number of hours of reading per week for the population of sixth grade boys. State the hypotheses to be tested in terms of μ_G and μ_B.

 (b) Suppose random samples of boys and girls are chosen, and the sample average hours per week of reading is observed for both boys and girls. A statistical test is performed, and the p-value is reported as 0.07. Interpret this p-value in terms of repeated samples.

6.2 Education researchers are concerned with the amount of coffee that students at the local high schools consume on a daily basis. They decide to do a study to determine the effect of coffee consumption on concentration ability, and they recruit 28 students who are willing to participate in the study. They are randomly divided into two groups. The first group consumes three cups of coffee in one hour, and the second group does not have any coffee. All of the students take a memory test in which concentration is important, and the scores are compared. Let μ_C be the mean score for the coffee group, and μ_N be the mean score for the no-coffee group.

 (a) State the hypotheses to be tested in terms of μ_C and μ_N.

 (b) The data analysis results in a p-value of .012. Interpret the p-value in terms of imaginary repeated studies.

6.3 There is a new ultrasound system that will be marketed to hospitals. The company that makes them claims that it detects more Type Q tumors in early stages than the old technology. A hospital would like to test this claim by using one new machine and one old machine, and recording the proportion of (known) tumors detected with each machine.

(a) State the hypotheses.

(b) What are Type I and Type II Errors in the context of the problem?

(c) What are the possible consequences of these errors?

(d) Discuss the merits of a high or low test size, from the hospital's point of view, based on a comparison of the consequences of the Type I and Type II Errors.

(e) Would the CEO of the company building the new machine prefer a lower or a higher test size for this study? Explain.

6.4 Continue the example in which the farmer claims that organic tomatoes are higher in vitamin A than conventional tomatoes, and the hypotheses to be tested are

$$H_0 : \mu = 1500 \quad \text{versus} \quad H_a : \mu > 1500,$$

where μ is the average amount of vitamin A (in IU) in organically grown tomatoes. Let's assume that the distribution of vitamin A is symmetric, so that the *median* amount of vitamin A in conventional tomatoes is also 1500 IU. In other words, half of the tomatoes have more than 1500 IU of vitamin A, and half have less. Suppose the farmer will select four organic tomatoes "at random" and reject H_0 if all of the tomatoes have greater than 1500 IU of vitamin A.

(a) What is the test size α for this decision rule?

(b) Suppose the alternative is true and, in particular, suppose 80% of organic tomatoes have more than 1500 IU of vitamin A. What is the power of the test?

6.5 Interest is in testing whether a coin is "fair," that is, $H_0 : p = .5$ versus $H_a : p \neq .5$, where p is the probability that the coin toss is heads. The coin is tossed eight times. We reject the null hypothesis if we get all heads or all tails.

(a) What is the test size α for this decision rule?

(b) Suppose the alternative is true and, in particular, $p = .8$. What is the power of the test?

6.6 Household incomes in Smalltown were found to have the following distribution in the last census:

Income	$\leq \$20,000$	$\$20,001-\$40,000$	$\$40,001-\$80,000$	$\geq \$80,001$
Proportion	.12	.43	.36	.09

A city council member is worried that in the 8 years since the census, incomes in Smalltown have dropped. He wants to test the null hypothesis that the income distribution is the same, against the (rather vague) alternative that households in Smalltown have gotten poorer. He will randomly select four households in

Smalltown and reject the null hypothesis if all four have incomes that are $40,000 or less.

    (a) What is the test size α for this decision rule?

    (b) Suppose that the city council member's fears are correct, and that households have gotten poorer. In fact, the proportion of households with incomes less than $40,000 is now .7 instead of .55. In this case, what is the power for this test?

6.7 A candy shop near an elementary school has a promotional gimmick in which, for each purchase of candy, the purchaser gets a scratch-off ticket. The ticket is rubbed with a coin to reveal whether or not a prize is won. The owner of the shop claims that 1/20th of the tickets lead to prizes. The third grade class is skeptical of this claim. They want to test the null hypothesis that the owner's claim is correct, against the alternative that the probability of winning a prize is *less* than 1/20. All 50 of the children purchase candy at the store, and none wins a prize. What is the p-value for the test?

6.8 Refer to Exercise 3.16. Suppose the scientists want to test formally the null hypothesis that the omega-3 fatty acids have no effect on brain development versus the alternative that rats fed high levels of the fatty acids will be smarter and hence run the maze faster. If the decision rule is to reject the null hypothesis if the four fastest rats are all in the omega-3 group, what is the test size?

6.9 There tends to be a problem with kids getting bee stings at Camp Nearahive. If we randomly select a child that has been to a summer session at the camp, and Y is the number of bee stings received by the child, then the distribution of Y has been determined to be that in the following table:

y	0	1	2	3	4
$P(Y = y)$	0.6	0.2	0.1	0.08	0.02

Suppose a counselor thinks that this year more kids are getting stung than in the past. She decides to test the null hypothesis that the traditional (above) distribution is still correct versus the alternative hypothesis that the bee sting problem is *worse*. Her study design is to select two children at random and observe the number of bee stings. Her decision rule is, "reject the null hypothesis if the children get at least two stings each."

    (a) Find the test size α.

    (b) Suppose the alternative is true, and the new distribution of Y is shown below. What is the power of the test?

y	0	1	2	3	4
$P(Y = y)$	0.2	0.4	0.26	0.10	0.04

6.10 Researchers at a psychology clinic are interested in identifying possible causes of attention deficit disorder (ADD). They will find a large sample of children, some diagnosed with ADD and some not, and collect information about possible causes, such as whether the birth was premature, number of siblings, whether or not there is a stay-at-home parent, etc. For 12 possible causes, they perform 12 hypothesis tests, e.g., H_0 : probability of ADD is the same for premature births as for full term births versus H_a : probability of ADD is *higher* for premature births.

(a) Suppose the null hypothesis is true for all 12 hypothesis tests. What is the probability of rejecting at least one, with $\alpha = .05$ for each test, if we can assume that the tests are independent?

(b) What should α be for each of the 12 tests if we want the probability of rejecting at least one to be .05 when the null hypothesis is true for all 12? (This is called a *Bonferroni adjustment* of test size.)

(c) Repeat part (a) for $\alpha = .01$.

(d) Repeat part (b), supposing instead that we want the probability of rejecting at least one to be .01 when the null hypothesis is true for all 12.

Chapter 7

Simulating Distributions for Hypothesis Testing

The last example in the previous chapter involved testing a null hypothesis distribution of numbers of arrests of eighteen year olds in Largetown. For six eighteen-year-olds sampled at random, the decision rule was "reject H_0 if none had been arrested." The test size was satisfactory, but the decision rule did not provide good power even if the new distribution for numbers of arrests showed substantial improvement over the old (null) distribution.

Let's consider a new test design, using the *sample average* number of arrests. The sample average is a **test statistic**, or a summary of the collected data. Recall that the null hypothesis distribution is

Number of arrests	0	1	2	3
Proportion	.65	.18	.10	.07

If this distribution still holds, the expected number of arrests for a randomly selected eighteen-year-old (using ideas from Chapter 5) is

$$(0)(.65) + (1)(.18) + (2)(.10) + (3)(.07) = .59,$$

and we would expect a sample average to be close to this average, especially for larger samples. If the new distribution is

Number of arrests	0	1	2	3
Proportion	.82	.10	.05	.03

then we would expect the sample average to be smaller than .59, so that smaller sample averages support the alternative more.

Suppose we decide to reject the null hypothesis if the average number of arrests in a sample of size 30 is less than .3. What is the test size? We don't have the analytical tools to compute this probability, because we don't know how to compute the distribution of the sample mean. We can, however, write some R code to do the job!

Using ideas from Chapter 2, we can simulate data from the null distribution. In the following piece of code, we generate one million random samples of size 30 from the historical (null hypothesis) distribution, and for each sample we calculate the mean.

```
nloop=1000000
sampmean=1:nloop
```

```
for(iloop in 1:nloop){
   x=sample(0:3,30,replace=TRUE,prob=c(.65,.18,.10,.07))
   sampmean[iloop]= mean(x)
}
hist(sampmean,br=150)
sum(sampmean<.3)/nloop
```

The histogram of the one million sample means is shown below. The distribution is discrete, because there are a finite number of possible values that the sample mean can take. The simulated sample mean distribution shown in the histogram approximates the distribution of a mean of a sample of size 30, chosen "at random." Because we simulated such a large number of samples, the distribution is very close to the true distribution.

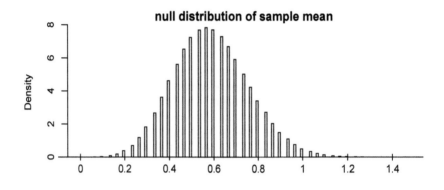

We find 2.6% of the sample means are below .3, so the test size for this decision rule is $\alpha = .026$.

Suppose we want our test size to be the traditional $\alpha = .05$. What should the decision rule be? We can use our million sample means to find the value c for which (approximately) 5% are less than c. The following R code accomplishes this, given the previously simulated distribution:

```
ss=sort(sampmean)
ss[.05*nloop]
```

We find the fifth percentile of the distribution is $c = .333$. In other words, if we reject H_0 if the sample average number of arrests is .333 or smaller, then our test size is close to $\alpha = .05$. (Because there are only a finite number of possible test sizes, we can't get exactly $\alpha = .05$ in this case.)

What is the power of this test if the true distribution is given as in the second table? We can again use simulations to find this, but this time we sample from the distribution associated with the alternative hypothesis:

```
nloop=1000000
sampmean=1:nloop
for(iloop in 1:nloop){
   x=sample(0:3,30,replace=TRUE,prob=c(.82,.10,.05,.03))
   sampmean[iloop]= mean(x)
}
hist(sampmean, br=100)
sum(sampmean<=.333)/nloop
```

We find that the power is about .611.

If we would like the power to be at least .8 with $\alpha \approx .05$, we could accomplish this using the same type of decision rule (using the sample average), with a larger sample size. We could repeat the simulations with increasing sample sizes, changing the decision rule with each sample size, until we find the smallest sample size that gives both the desired test size and the desired power. If we do this, we will find that a sample of size 50 will suffice. In practice, these types of simulated power calculations are quite common!

Next, suppose we have collected our sample of size 30, and we find that the sample mean is .20. What is the p-value? Recall that the p-value is the probability of our data (the sample mean is .20) or more extreme (the sample mean is less than .20), when H_0 is true. This is found using the simulated null distribution and the command

```
sum(sampmean<=.20)/nloop
```

We find that the p-value is .007, which provides strong evidence that the distribution has changed, and that arrests are *less* likely than for the old distribution.

In the above example we could use the sample mean because the observations were *ordinal*. There are a finite number of possible values, but the values are numerical. A discrete random variable can alternatively be *nominal*, so that there is no natural ordering to the values. For example, suppose we have a six-sided die that will be used in a gambling game. Three sides are red, two sides are blue, and one side is gold. We want to check that the die is fair, that is, the probability of rolling red is 1/2, blue has probability 1/3, and gold has probability 1/6.

The null hypothesis is that the die is fair and the probabilities of the three colors are as listed above, and the alternative is simply that the probabilities are something else. We decide to roll the die 60 times and record the colors. We need to determine a test statistic that summarizes the amount of evidence for the alternative.

The expected number of red rolls is 30, and we expect 20 blue and 10 gold rolls. The test statistic should measure the deviation from these expectations. In Chapter 59 we will formally derive a test procedure based on some optimality principles, but here we will just use our intuition and do something that makes sense. A straightforward idea for a test statistic might be

$$T = |n_R - 30| + |n_B - 20| + |n_G - 10|,$$

where n_R is the number of observed red rolls, n_B is the number of observed blue rolls, n_G is the number of observed gold rolls. Clearly, larger values will support the alternative hypothesis. Because we have no idea how to derive the distribution of this test statistic when the null hypothesis is true, we resort to simulations.

The following code simulates the distribution of our test statistic when the null hypothesis is true (the die is fair). Within the loop, the die is rolled 60 times, and the test statistic is computed. The values of the T statistic are stored in tstat.

```
n=60;nloop=1000000
tstat=1:nloop
for(iloop in 1:nloop){
    rolls=sample(1:3,60,replace=TRUE,prob=c(1/2,1/3,1/6))
    nr=sum(rolls==1)
    nb=sum(rolls==2)
    ng=sum(rolls==3)
    tstat[iloop]=abs(nr-30)+abs(nb-20)+abs(ng-10)
```

```
}
hist(tstat,br=100,main='null distribution',freq=FALSE,ylab='Probability')
```

One million simulated tests when the null hypothesis is true gives a very precise distribution of the test statistic, shown in the histogram below. If the decision rule is to reject the null hypothesis when T is at least 17, we can find the test size using

```
sum(tstat>=17)/nloop
```

to find a test size of about .042.

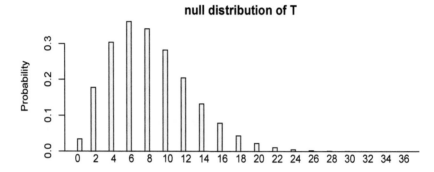

If we require $\alpha \leq .01$, we find that the decision rule must be "reject H_0 if ≥ 22," which provides a test size of about .0093.

Simulations also will provide the power of the test under various scenarios. Suppose the following are the true probabilities for the colors: the probability of red is .65, the probability of blue is .3, and the probability of gold is .05. This is substantially different from a fair die, and we would like to have a high probability of detecting the difference and rejecting the null hypothesis.

We repeat the above simulations with prob=c(.65,.3,.05)) and produce the histogram of the distribution of the test statistic when the die has these color probabilities:

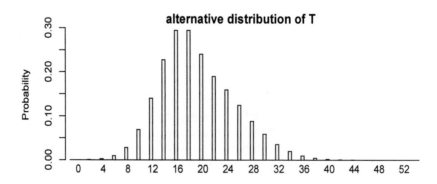

To find the power for the decision rule with the smaller test size, we use

```
sum(tstat>=22)/nloop
```

and the power is about .347. Perhaps we need to change the design of our experiment and decide to roll the die a larger number of times so that we can have both a small test size and large power.

Although we will, in the rest of the book, learn of many standard test statistics and how to compute the null distribution, in the real world the "standard" situation is rarer than in textbooks, and it is not uncommon for statisticians to resort to simulations to provide test sizes, p-values, and power.

Chapter Highlights

1. A **test statistic** is a summary of the data that are used in hypothesis testing.

2. When the null hypothesis model is completely specified, we can simulate data using the model. The simulated data sets can be used to determine the distribution of the test statistic under the null hypothesis. From this we can compute a test size and a p-value.

3. For a specified alternative distribution, we can use simulations to approximate the distribution of the test statistic under the alternative, in order to compute the power of the test.

Exercises

7.1 For Exercise 6.9, we considered a distribution of bee stings for kids at Camp Near-ahive. We wanted to test the null hypothesis that the historical distribution is correct versus the alternative that kids are more likely to get bee stings. Suppose that there are 30 kids at the camp, and we consider that their numbers of bee stings are a random sample from the true distribution. For Exercise 5.3, we computed the expected number of bee stings, for a randomly selected camper, to be 0.72 if the null distribution is true. Suppose we decide to compute the *sample* average number of bee stings for the 30 kids, and reject H_0 if the sample average is greater than 1.0.

   (a) Find the test size for this decision rule using simulations in R.

   (b) Suppose we want to set the test size to be $\alpha = .01$ (or as close to this as we can get). Use your estimated distribution to find the new decision rule: "reject H_0 if the sample mean is greater than _____."

   (c) Given the decision rule in (b), do simulations to compute the power of the test if the true distribution is the alternative given in Exercise 6.9.

7.2 A weaving machine in a garment factory produces lengths of linen cloth. In a randomly selected yard of cloth, the number of flaws follows the following distribution:

# flaws	0	1	2	3
Probability	.6	.2	.15	.05

The manager of the factory has the opportunity to buy a new machine and wants to test the null hypothesis—that the distribution of numbers of flaws for the new machine is the same as for the old machine—against the alternative that the new machine makes, on average, fewer flaws. The manager will make 8 yards of linen cloth on the new machine. Assume that the numbers of flaws in each yard are independent random variables.

(a) The manager uses this decision rule: "reject the null hypothesis if there are no flaws for all 8 yards of linen." What is the test size for this decision rule?

(b) Suppose the true distribution for flaws with the new machine is

# flaws	0	1	2	3
Probability	.75	.15	.1	0

What is the power for the test in (a)?

(c) You think that the manager really ought to make a decision rule based on the average number of flaws in the 8 yards. Write code to simulate the distribution of T, the average number of flaws in the 8 yards, when the null hypothesis is true. What is the decision rule that gives a test size as close as possible to .05?

(d) What is the power for your decision rule in (c) when the true distribution is that in (b)?

7.3 The historical distribution of the number of cats per household in Gatoburg is in the table

# cats	0	1	2	3	4
Proportion of households	.3	.3	.2	.1	.1

The Humane Society believes that currently households are more likely to have more cats. They decide to randomly sample 20 households and find the number of cats in each household. They will reject the null hypothesis that the cat distribution has not changed, in favor of the alternative that folks are more likely to have more cats if the average number of cats per household in the sample is at least two.

(a) Find the test size for this decision rule using simulations in R.

(b) Suppose that the average number of cats per household in the sample is 1.85. What is the p-value?

(c) Suppose that the true current distribution of number of cats per household in Gatoburg is

# cats	0	1	2	3	4
Proportion of households	.15	.3	.25	.2	.1

Find the power of the test.

7.4 A botanist is studying color patterns of a species of flowering plant. If a certain genetic rule holds, 50% of the plants have pink flowers, 25% have white flowers, and 25% have red flowers. The botanist randomly selects 100 seeds to test the null hypothesis that the genetic rule holds. The alternative is that the rule does not hold, but there is no specific distribution given for the alternative. The botanist will construct a test statistic, $T = |X_1 - 50| + |X_2 - 25| + |X_3 - 25|$, where X_1 is the number of pink flowers, X_2 is the number of white flowers, and X_3 is the number of red flowers. Clearly, larger values of T support the alternative hypothesis.

(a) Through simulations, find the "critical value" c of T that gives a test size of $\alpha = .05$. That is, find c so that the decision rule is "reject H_0 when T is greater than c" and the probability of rejecting H_0, when H_0 is true, is .05.

(b) Suppose that the botanist's seeds result in 42 pink flowering plants, 28 white, and 30 red. Do we reject the null hypothesis that the genetic rule holds?

7.5 Refer to Exercise 2.7. Suppose researchers in a remote area of China are interested in determining whether the inhabitants share the same distribution of blood types as in China as a whole. The null hypothesis is that they do, and the alternative is that the blood distribution is different. Suppose 100 people are selected "at random" from this remote area, and their blood types are determined. The test statistic is

$$T_1 = |X_O - 48| + |X_A - 28| + |X_B - 19| + |X_{AB} - 5|,$$

where X_O is the number of people in the sample who have type O blood, etc.

(a) Through simulations, find the "critical value" c of T_1 that gives a test size of $\alpha = .05$. That is, find c so that the decision rule is "reject H_0 when T_1 is greater than c" and the probability of rejecting H_0, when H_0 is true, is .05.

(b) Suppose that 54 people in the sample have type O, 32 have type A, 8 have type B, and 6 have type AB. Is the null hypothesis rejected?

7.6 Refer to Exercise 7.5. Suppose a different researcher decides to test the same hypothesis, only with a different test statistic. She suggests (for a sample of 100 people) the statistic

$$T_2 = \frac{(X_O - 48)^2}{48} + \frac{(X_A - 28)^2}{28} + \frac{(X_B - 19)^2}{19} + \frac{(X_{AB} - 5)^2}{5}.$$

(a) Through simulations, find the "critical value" c of T_2 that gives a test size of $\alpha = .05$. That is, find c so that the decision rule is "reject H_0 when T_2 is greater than c" and the probability of rejecting H_0, when H_0 is true, is .05.

(b) Suppose that 54 people in the sample have type O, 32 have type A, 8 have type B, and 6 have type AB. Is the null hypothesis rejected?

7.7 Refer to Exercises 7.5 and 7.6. Suppose the real distribution of blood types in the remote region is 40% have type O, 40% have type A, 12% have type B, and 8% have type AB. Which of the two test statistics, T_1 or T_2, will provide more power?

Chapter 8

Bernoulli and Binomial Random Variables

The sample space of a **Bernoulli experiment** has only two possible outcomes; these are often denoted as "success" and "failure":

$$S = \{\text{success, failure}\}.$$

If the probability of a success is $p \in (0, 1)$, then the probability of a failure is $1 - p$.

The **Bernoulli random variable** has value 1 if the outcome is "success" and 0 if the outcome is "failure."

We can use the formulas from Chapter 5 to find the expected value and variance of a Bernoulli random variable. Suppose X is a Bernoulli random variable where the probability of success is p; for a convenient shorthand we can write $X \sim \text{Bernoulli}(p)$. Then

$$\text{E}(X) = (1)(p) + (0)(1 - p) = p.$$

The variance is $\text{E}(X^2) - p^2$, so we calculate

$$\text{E}(X^2) = (1^2)(p) + (0^2)(1 - p) = p$$

and compute the variance $\text{V}(X) = p - p^2 = p(1 - p)$.

The **binomial experiment** or **binomial trials** and the **binomial random variable** are generalizations of the Bernoulli ideas.

We define a **binomial experiment** as repeated independent Bernoulli experiments or "trials," so that each trial has two possible outcomes, which are generically called "success" and "failure." The probability of a success is p for each trial, and a success or failure in one trial does not affect the probabilities in another trial. A **binomial random variable** is the number of successes in a binomial experiment.

For example, suppose there are $n = 3$ trials. The sample space can be written as

$$\{SSS, SSF, SFS, FSS, SFF, FSF, FFS, FFF\},$$

where S stands for "success" and F for "failure." The binomial random variable Y denoting the number of successes can take four values, from zero to three. For a convenient shorthand, we write $Y \sim \text{Binom}(n, p)$, when Y is a random variable denoting the number of successes in a binomial experiment with n trials with probability p of success.

We can find the probability of each outcome in the sample space using ideas from Chapter 3. For example, if $n = 3$ and $p = .75$, we can compute $P(SSF) = (.75)^2(1 - .75) = .1406$, because the trials are independent. In this way, we can compute $P(Y = 2)$ as $P(SSF) + P(SFS) + P(FSS) = (3)(.1406) = .4218$.

To motivate the general formula for the binomial probabilities, let's consider the problem of rolling a fair die five times, where the random variable Y is the number of sixes. You can see that Y is a binomial random variable, because the outcomes for the rolls are independent, and the probability of a six is 1/6 for each roll.

We can figure out the probability of *all* sixes in five tosses using our multiplication rule for independent events:

$$P(Y = 5) = \left(\frac{1}{6}\right)^5 \approx 0.0001286.$$

We can also figure out the probability of no sixes:

$$P(Y = 0) = \left(\frac{5}{6}\right)^5 \approx 0.4019.$$

What about the probability of exactly one six? It's tempting to try

$$P(Y = 1) \stackrel{?}{=} \left(\frac{1}{6}\right)\left(\frac{5}{6}\right)^4 \approx 0.08038,$$

since this is one six and four nonsixes. But this is not right, because there are five outcomes associated with the event $Y = 1$. This event is shown as the following subset of the sample space, where an "F" is a not-six and an "S" represents a six.

$$\{SFFFF, FSFFF, FFSFF, FFFSF, FFFFS\}.$$

Each of these five outcomes has probability $(5/6)^4(1/6)$. Therefore, the probability of getting exactly one six in five rolls is

$$P(Y = 1) = 5\left(\frac{5}{6}\right)^4\left(\frac{1}{6}\right) \approx 0.4019.$$

What about exactly two sixes? There are 10 different ways to get two sixes in five rolls:

$$\{SSFFF, SFSFF, SFFSF, SFFFS, FSSFF,$$
$$FSFSF, FSFFS, FFSSF, FFSFS, FFFSS\},$$

so

$$P(Y = 2) = 10\left(\frac{5}{6}\right)^3\left(\frac{1}{6}\right)^2 \approx 0.1608.$$

In fact, the number of ways to get exactly two sixes in five rolls is $\binom{5}{2}$—the number of ways we can choose the two positions for the S symbols, in a string of two S and three F symbols.

Generalizing, using the ideas from Chapter 3, the number of ways to get k sixes in n rolls is $\binom{n}{k}$. The probability of getting exactly three sixes is then

$$P(Y = 3) = \binom{5}{3}\left(\frac{5}{6}\right)^2\left(\frac{1}{6}\right)^3 \approx 0.0322.$$

Continuing with this reasoning, we can make the table below. The probabilities add to one, within roundoff error.

y	$\text{Prob}(Y = y)$
0	0.4019
1	0.4019
2	0.1608
3	0.0322
4	0.0032
5	0.0001

Now we have "derived" the probability mass function for the binomial random variable: Suppose we have n independent trials, where each trial has outcomes "success" and "failure," and the probability of a success on each trial is p. If Y is a random variable representing the number of successes, then the probability of getting exactly k successes in n trials is

$$P(Y = k) = \binom{n}{k} p^k (1 - p)^{n-k}$$

for $k = 0, \ldots, n$.

Examples of binomial mass functions are shown in the bar charts below:

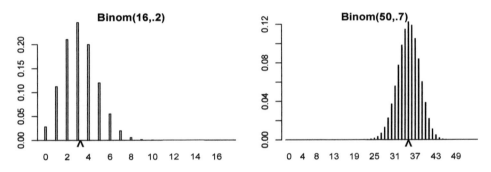

Example: Sixty percent of voters in Largetown support Proposal Q. Take a random sample of ten voters (with replacement, to ensure independence); what is the probability that exactly eight of the ten support the proposal? We plug into the binomial formula $n = 10, r = 8, p = 0.60$. We get

$$\binom{10}{8} (.6)^8 (.4)^2 = 0.121.$$

We can interpret this probability in terms of repeated samples: If we repeat the experiment (taking a random sample of size 10 from the population) many times, about 12.1% of the samples will have exactly eight voters supporting Proposal Q.

What is the probability that *at least* eight of the ten support the proposal? We add together the probabilities of exactly eight out of ten, exactly nine out of ten, and exactly ten out of ten, to get

$$\binom{10}{8} (.6)^8 (.4)^2 + \binom{10}{9} (.6)^9 (.4)^1 + \binom{10}{10} (.6)^{10} (.4)^0 = .121 + .040 + .006 = .167.$$

The probability of getting at least eight out of ten voters in the sample who support the proposal is about 16.7%.

We should show that our binomial probabilities make a proper mass function. It's clear that the probabilities are all nonnegative, but let's confirm that they add to one. Suppose that the random variable Y is distributed as a binomial random variable with n trials and probability p; that is, $Y \sim \text{Binom}(n, p)$. Then there are $n + 1$ possible values for Y. Recall from Chapter 3 that the binomial theorem says

$$(p + q)^n = \sum_{k=0}^{n} \binom{n}{k} p^k q^{n-k},$$

and let $q = 1 - p$. Then the left-hand side is 1, and the right-hand side is $\sum_{k=0}^{n} P(Y = k)$.

Next we will compute the expected value and variance of a binomial random variable. For $Y \sim \text{Binom}(n, p)$, the expected value of Y is computed using the formula from Chapter 5:

$$\sum_{k=0}^{n} kP(Y = k) = \sum_{k=1}^{n} kP(Y = k) \quad (\text{don't need } k = 0 \text{ term})$$

$$= \sum_{k=1}^{n} k \binom{n}{k} p^k (1 - p)^{n-k}$$

$$= np \sum_{k=1}^{n} \frac{(n-1)!}{(k-1)!(n-k)!} p^{k-1} (1 - p)^{n-k},$$

where in the last step we canceled the k with the first term in the factorial in the denominator. We can simplify this using a change of variables: $j = k - 1$ and $m = n - 1$, and noticing that $n - k = m - j$. Then

$$E(Y) = np \sum_{j=0}^{m} \binom{m}{j} p^j (1 - p)^{m-j}$$

$$= np,$$

because the elements of the sum are the $m + 1$ probabilities for a $\text{Binom}(m, p)$ random variable. The mean values for the binomial distributions shown in the above bar charts are marked with "⌃".

How do we compute the variance of Y? If we want to use the formula $V(Y) = E(Y^2) - E(Y)^2$, we will need to compute $E(Y^2)$. If we try to do the same change-of-variable trick as with the expected value, we will get stuck when trying to cancel terms in the factorial part. However, it's not hard to find $E[Y(Y - 1)]$:

$$E[Y(Y - 1)] = \sum_{k=0}^{n} k(k - 1)P(Y = k)$$

$$= \sum_{k=2}^{n} k(k - 1) \binom{n}{k} p^k (1 - p)^{n-k}$$

$$= n(n-1)p^2 \sum_{k=2}^{n} \frac{(n-2)!}{(k-2)!(n-k)!} p^{k-2} (1 - p)^{n-k},$$

where now we have canceled $k(k - 1)$ with the first two terms of the factorial in the denominator. We can simplify this using a change of variables: $j = k - 2$ and $m = n - 2$, and noticing that $n - k = m - j$. Then

$$E[Y(Y-1)] = n(n-1)p^2 \sum_{j=0}^{m} \binom{m}{j} p^j (1-p)^{m-j}$$
$$= n(n-1)p^2,$$

because the elements of the sum are the $m + 1$ probabilities for a Binom(m, p) random variable. Finally we have the variance of $Y \sim$ Binom(n, p):

$$V(Y) = E(Y^2) - E(Y)^2 = E[Y(Y-1)] + E(Y) - E(Y)^2$$
$$= n(n-1)p^2 + np - n^2 p^2 = np(1-p).$$

Example: It is hoped that a new cancer treatment is more effective than the old treatment. With the old treatment, it is known that 60% of patients go into remission for at least one year. We want to test the new treatment with $n = 20$ cancer patients. If p is the proportion of patients who would go into remission for at least a year with the new treatment, we set up the hypotheses

$$H_0 : p = .6 \quad \text{versus} \quad H_a : p > .6.$$

The independence assumption seems reasonable, because whether or not one patient goes into remission should not depend on the status of other patients. Suppose we decide to reject the null hypothesis if at least 16 of the 20 patients go into remission for at least a year. What is the test size α?

Let Y be the number of patients that go into remission, then $Y \sim$ Binom(p). The test size α is the probability of rejecting H_0 (i.e., $Y \geq 16$), given H_0 (i.e., $p = .6$). We calculate

$$\alpha = \binom{20}{16}(.6)^{16}(.4)^4 + \binom{20}{17}(.6)^{17}(.4)^3 + \binom{20}{18}(.6)^{18}(.4)^2$$
$$+ \binom{20}{19}(.6)^{19}(.4) + (.6)^{20} \approx .051.$$

That is, if the null hypothesis is true and the new treatment is not better than the old treatment, we have a 5.1% probability of making a Type I Error.

If the true proportion of patients who go into remission for at least a year with the new treatment is 0.8, what is the power for this test? Recall that the power is the probability that the null hypothesis is rejected when the alternative is true. This is the probability that at least 16 of the 20 patients go into remission for at least a year, when the probability of remission is .8. This is

$$\binom{20}{16}(.8)^{16}(.2)^4 + \binom{20}{17}(.8)^{17}(.2)^3 + \binom{20}{18}(.8)^{18}(.2)^2 + \binom{20}{19}(.8)^{19}(.2) + (.8)^{20}$$
$$\approx .630.$$

Instead of doing all that calculating, we can use R to get binomial probabilities. The function pbinom(q,n,p) will return the probability that Y is less than or equal to q, where $Y \sim$ Binom(n, p). Therefore, to get the probability of at least 16 out of 20 patients going into remission when $p = .8$, we simply type

```
1-pbinom(15,20,.8)
```

at the R prompt, to get .630.

Suppose that we want the power to be at least 80%, and we also want the test size α to be at most .01. We need to increase the sample size!

Let's try $n = 40$. The first step is to find the decision rule, and for now we will do this by trial and error, using pbinom. The command

```
1-pbinom(31,40,.6)
```

returns .0061. That is, if $Y \sim \text{Binom}(40, .6)$, then $P(Y \geq 32) = .0061$. In other words, if our decision rule is to reject H_0 if at least 32 out of 40 patients go into remission, then our test size is $\alpha = .0061$. (If our decision rule is to reject H_0 if at least 31 out of 40 patients go into remission, then our test size will be greater than .01. For a discrete random variable, we can't always get the exact test size that we want.)

The corresponding power for $p = .8$ is 1-pbinom(31,40,.8), which is about .593. If we want the power to be at least .8 (a common target for these kinds of tests), we'll need an even larger sample size! We can repeat this procedure: for the choice of n, find the decision rule (using $p = .6$) that gives α close to but not exceeding .01; then using that decision rule, find the power (using $p = .8$). The goal is to find the smallest n for which the resulting power is at least .8. After trying various values of n, we conclude that for $n = 55$ we can reject H_0 when at least 42 patients go into remission, and when $p = .8$ this test has power .803.

Other useful R functions related to the binomial distribution include dbinom, rbinom, and qbinom. The command dbinom(x,n,p) will return $P(X = x)$ when $X \sim \text{Binom}(n, p)$. We can use rbinom to generate random values from a binomial distribution, which can be useful for checking calculations. For example, suppose we want to verify that $E[X(X - 1)] = n(n - 1)p^2$. For $p = .4$ and $n = 10$, $n(n - 1)p^2 = 14.4$; we can randomly generate a large number of values from this distribution and find the average of the value times one minus the value:

```
x=rbinom(100000,10,.4)
mean(x*(x-1))
```

This code will return a number very close to 14.4, with a little "random error."

The "quantile function" qbinom(r,n,p) will return the smallest value x such that $P(X \leq x) \geq r$. This qbinom command can be used to find the decision rule for the above example, instead of using trial and error. Quantiles will be discussed more formally in Chapter 15.

Chapter Highlights

1. A Bernoulli trial results in a "success" or a "failure." A Bernoulli random variable has the value 1 for a success and the value 0 for a failure.

2. If Y is the number of successes in n independent trials where, for each trial, the probability of success is p, then we say $Y \sim \text{Binom}(n, p)$, and

$$P(Y = y) = \binom{n}{y} p^y (1 - p)^{n-y} \text{ for } y = 0, 1, \ldots, n.$$

3. If $Y \sim \text{Binom}(n, p)$, then $E(Y) = np$ and $V(Y) = np(1 - p)$.

---

4. In R, the command

- dbinom(y,n,p) will return the probability that Y is equal to y, where $Y \sim \text{Binom}(n, p)$;

- pbinom(y,n,p) will return the probability that Y is less than or equal to y, where $Y \sim \text{Binom}(n, p)$;

- rbinom(N,n,p) will return N randomly generated values of a $\text{Binom}(n, p)$ random variable;

- qbinom(r,n,p) will return the smallest value y such that $P(Y \leq y) \geq r$, where $Y \sim \text{Binom}(n, p)$.

Exercises

8.1 Suppose $Y \sim \text{Binom}(20, .2)$.

(a) Use R to simulate many values from this distribution and verify that $E(Y) = 4$ and $V(Y) = 3.2$.

(b) Approximate the value of $E(Y^3)$ using simulations in R.

8.2 If a fair six-sided die is rolled 20 times, what is the probability of getting at most 2 sixes?

8.3 Researchers want to determine if a magician has ESP. They set up a test that consists of eight trials. In each trial, a card is randomly selected (with replacement) from a standard deck. The magician guesses the suit of the card. The null hypothesis is that she does not have ESP, so she is just guessing randomly, and the alternative is that she is *more* likely to guess the suit. Suppose that she is successful for 5 out the 8 trials. What is the p-value for this experiment? (Note: There are equal numbers of four suits in a standard deck of 52 cards.)

8.4 Which of the following are binomial experiments? For those that are not, state the assumption that is not met.

(a) An urn has twelve green balls and eight blue balls. The experiment is to randomly choose five balls, one at a time, *with* replacement. The random variable Y is the number of blue balls.

(b) An urn has twelve green balls and eight blue balls. The experiment is to randomly choose five balls, one at a time, *without* replacement. The random variable Y is the number of blue balls.

(c) It is known that 40% of registered voters in Largetown support Proposition Q, a referendum on school prayer. A city council member randomly selects 12 people from the list of registered voters. The random variable Y is number that support Proposition Q.

(d) It is known that 40% of registered voters in Largetown support Proposition Q, a referendum on school prayer. A city council member selects 12 people from Largetown. The first six were approached at Largetown College, and the next six at church on Sunday. The random variable Y is number that support Proposition Q.

8.5 Marie can hit 80% of free throws in basketball. Suppose she enters a competition in which each player has ten free-throw attempts. What is the probability of her hitting at least nine out of ten attempts? What assumptions have to be made to solve this problem using the binomial ideas? Are they reasonable assumptions?

8.6 Suppose 10.9% of the U.S. population reside in the state of California. Suppose we select a random sample of size 20 of U.S. residents. What is the probability that at least two reside in California?

8.7 A pizza parlor is running a promotion. Everyone who comes to the restaurant gets to spin a wheel that gives them a 10% chance of getting a free pizza. If you go with five friends, and each of you spins the wheel, what is the chance of winning at least one pizza for your group?

8.8 The current shingles vaccine is known to reduce the incidence of shingles, but is not completely effective. Suppose that 2% of subjects who receive the shingles vaccine contract shingles within five years (compared to 4% who do not receive the vaccine). The makers of new shingles vaccine claim it is more effective; in fact they claim that only 1% of subjects receiving the vaccine will contract shingles within five years. Let p be the proportion of subjects with the new vaccine who contract shingles within five years. The FDA wants to test the null hypothesis that $p = .02$ against the alternative $p = .01$. Suppose they vaccinate $n = 300$ subjects, and use this decision rule: reject H_0 if at most one subject contracts shingles within five years.

(a) Find the test size.

(b) Find the power.

8.9 Brand A switches are ordered by the manager of a factory that builds gizmos. He finds that 20% of Brand A switches fail within one year of use, which is the warranty period for the gizmos. The manager buys 16 of Brand B switches to see if they will last longer. He wants to test the null hypothesis $H_0 : p = .2$ versus $H_a : p < .2$, where p is the proportion of Brand B switches that fail within one year. He decides to reject H_0 (and switch to Brand B) if at most one of the 16 switches fails within a year.

(a) What is the test size α for this decision rule?

(b) If the proportion of Brand B switches that fail within one year is really .1, what is the power of the test?

(c) Would the owner of the Brand B company prefer that the test size be larger or smaller? Explain.

8.10 A department store promotion consists of giving lottery tickets to customers who spend $10 or more on purchases. The prize is a $50 gift certificate for the store. The management claims that the probability of winning on each store purchase is four percent, that is, the proportion of winning tickets is 0.04. A consumer group wants to test the store's claim, that is,

$$H_0: \theta = 0.04 \quad \text{versus} \quad H_a: \theta < 0.04.$$

The test consists of making 100 purchases at the store and counting the number of winning tickets out of the 100 tickets received.

(a) What is the probability of fewer than three winning tickets if the store's claim is valid?

(b) Suppose we get two winning tickets out of our 100 tickets received. What is the p-value for the data?

(c) If we had set $\alpha = 0.05$, would we reject or accept H_0 if we found two winning tickets?

8.11 The proportion of patients with high blood pressure in nursing homes in a state is known to be 0.42. A social worker suspects that this proportion is higher in the inner city. He takes a random sample of 12 nursing home residents from the inner city and tests

$$H_0: \theta = 0.42 \quad \text{versus} \quad H_a: \theta > 0.42,$$

where θ is the proportion of nursing home residents in the inner city who have high blood pressure. He decides to reject the null hypothesis if the number of residents in the sample with high blood pressure is at least eight.

(a) Calculate the size of this test.

(b) If the true proportion of nursing home residents in the inner city who have high blood pressure is .54, what is the power of this test?

8.12 Refer to Exercise 8.11. The social worker decides a larger sample size is required, so he takes a random sample of 20 nursing home residents from the inner city. He decides to reject H_0 if the number of residents in the sample with high blood pressure is at least 12.

(a) What is the size of the test?

(b) Given that the true proportion of nursing home residents in the inner city who have high blood pressure is .54, what is the power of this test?

8.13 A suspicious casino guest is wondering if a die used in a game of chance is fair; he thinks that the probability that a six is rolled might be larger than 1/6. He will roll the die ten times to test $H_0 : p = 1/6$ versus $H_a : p > 1/6$, where p is the probability of rolling a six. His decision rule is "reject H_0 if four or more sixes are rolled." What is the test size α?

8.14 Your friend makes a claim that, when people are asked to choose a number (integer) from one to ten, they are more likely to choose three or seven, compared to the other numbers. Suppose you want to test this claim. Let p be the probability that a "randomly selected" person will say three or seven when asked to choose a number from one to ten. You test $H_0 : p = .2$ versus $H_a : p > .2$, by asking 12 such people to choose a number. Suppose you get four threes, two sevens, and six other numbers. What is the p-value for the test?

8.15 Consider the die testing example from Chapter 6. The null hypothesis was that the probability p of rolling a one is 1/6 (die is fair) and the alternative is that p is less than 1/6. We will test the hypothesis by rolling the die n times and counting the number of ones.

   (a) Suppose we roll the die 40 times, and reject if the number of ones is three or fewer. What is the test size?

   (b) What is the power of the test in (a) when the true probability of rolling a one is .08?

   (c) Suppose we roll the die 40 times, and we want the test size to be at most $\alpha = .03$. What is the decision rule?

   (d) What is the power for the test in (c) when the true probability of rolling a one is .08?

   (e) We'd like the test size to be not greater than .03, and the power to be at least .8 when the true probability of rolling a one is .08. What is the smallest sample size needed to attain the desired test size and power?

8.16 The manager of a computer repair shop orders a certain type of chip from Company A, and he knows that 5% of these chips are defective. He finds that Company B also offers these chips, at the same price. He decides to buy 100 chips from Company B to test the null hypothesis $H_0 : \mu = .05$ versus $H_a : \mu < .05$, where μ is the proportion of Company B chips that are defective.

   (a) He will reject H_0 if three or fewer of the new chips are defective. What is the test size?

   (b) Suppose that in fact only 2% of Company B chips are defective. Using the decision rule of part (a), what is the power of the test?

8.17 Company A and Company B both make the same kind of computer chips, but 20% of the chips that Company A makes are defective, and only 5% of Company B's chips are defective. Suppose a computer repair shop has one box each of Company A and Company B chips, but the labels have fallen off and the boxes and chips look identical. An employee chooses a box at random and tests twelve of the chips. If at most one chip is defective, what is the probability that the box was from Company A?

8.18 Suppose $X_1, \ldots, X_{10}$ are independent Bernoulli random variables, each with probability p of success.

   (a) Suppose $p = .5$. You are given that $\sum_{i=1}^{10} X_i = 5$. What is the (conditional) probability that $X_1 = \cdots = X_5 = 0$ and $X_6 = \cdots = X_{10} = 1$?

   (b) Suppose $p = .1$. You are given that $\sum_{i=1}^{10} X_i = 5$. What is the (conditional) probability that $X_1 = \cdots = X_5 = 0$ and $X_6 = \cdots = X_{10} = 1$?

8.19 Eddie proposes the following game: You roll five fair six-sided dice. For every six, he pays you $100, but for every nonsix, you pay him $25.

   (a) Calculate your expected profit. Do you want to play?

   (b) Suppose you want to propose a fair game, where your expected profit is exactly zero. That is, for every six, Eddie will pay you $100, but for every nonsix, how much will you pay him?

8.20 Eddie and Mary each roll ten fair six-sided dice.

(a) Using simulations in R, approximate the probability that Eddie rolls at least twice as many sixes as Mary.

(b) Eddie proposes this game: He gives Mary $10, then they each roll ten fair six-sided dice. If Eddie rolls at least twice as many sixes as Mary, she has to give him d dollars. Find the value of d that makes this game fair, that is, the expected net exchange of money is zero.

Chapter 9

The Geometric Random Variable

The **geometric random variable** is also defined using binomial trials, but without specifying a fixed number of trials. We again have independent trials, each with only two possible outcomes labeled "success" and "failure," and for each trial the probability of success is p. Now our random variable Y is defined as *the number of failures before the first success occurs.* The notation we will use is

$$Y \sim \text{Geom}(p).$$

The support for the random variable is the nonnegative integers; $Y = 0$ if the first trial is a success. The support is an infinite set, but since it is only countably infinite, the random variable is discrete.

Let's derive a formula for the distribution. Clearly, $P(Y = 0) = p$; this is the probability that you get a success on the first attempt. The event $Y = 1$ occurs when there is one failure and then one success, so $P(Y = 1) = (1 - p)p$.

The event $Y = k$ occurs when there are k failures in a row, then one success, so

$$P(Y = k) = (1 - p)^k p \ \text{ for } \ k = 0, 1, 2, 3, \ldots.$$

Let's check to see if this is a proper probability mass function. It's clear that all the probabilities are nonnegative, so we need only check that they sum to one.

We use the formula (see Appendix B.2) for a "geometric series"

$$\sum_{k=0}^{\infty} a^k = \frac{1}{1 - a} \ \text{ for } \ 0 < a < 1$$

to get

$$\sum_{k=0}^{\infty} P(Y = k) = \sum_{k=0}^{\infty} (1 - p)^k p = p \sum_{k=0}^{\infty} (1 - p)^k = p/p = 1.$$

Examples of geometric mass functions are shown in the following bar charts. It is not hard to see from the formula that the probabilities $P(Y = k)$ must be decreasing in k for every p.

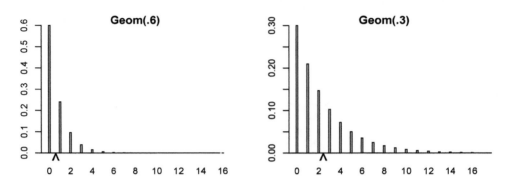

To find the expected value of a geometric random variable, we need another summation formula (see Appendix B.2): For $a \in (0,1)$,

$$\sum_{k=1}^{\infty} ka^k = \frac{a}{(1-a)^2}.$$

Now the expected value of $Y \sim \text{Geom}(p)$ is computed:

$$\text{E}(Y) = \sum_{k=0}^{\infty} k\text{P}(Y = k) = \sum_{k=1}^{\infty} k(1-p)^k p = p \sum_{k=1}^{\infty} k(1-p)^k = \frac{p(1-p)}{p^2} = \frac{1-p}{p}.$$

Intuitively, it makes sense that the number of trials before a first success is typically larger when the probability of a success is smaller.

Now let's consider finding $\text{V}(Y) = \text{E}(Y^2) - \text{E}(Y)^2$. We need yet another summation formula from Appendix B.2: For $a \in (0,1)$,

$$\sum_{k=1}^{\infty} k^2 a^k = \frac{a + a^2}{(1-a)^3}.$$

Using this, we have

$$\text{E}[Y^2] = \sum_{k=0}^{\infty} k^2 \text{P}(Y = k)$$

$$= p \sum_{k=1}^{\infty} k^2 (1-p)^k$$

$$= p \frac{(1-p) + (1-p)^2}{p^3}$$

$$= \frac{(1-p)(2-p)}{p^2}.$$

Finally,

$$\text{V}(Y) = \text{E}[Y^2] - \text{E}(Y)^2 = \frac{(1-p)(2-p)}{p^2} - \frac{(1-p)^2}{p^2} = \frac{1-p}{p^2}.$$

Example: Suppose that you play a game that involves rolling a fair die. You will win $100 if you roll a six, but you have to pay $10 each time you roll. If you come to the game with only $30, what is the probability that you lose all your money before winning?

If X is the number of (losing) rolls before winning, then $X \sim$ Geom$(1/6)$, and we want to compute P$(X \geq 3) = 1 - [P(X = 0) + P(X = 1) + P(X = 2)]$. This is $1 - (p + (1-p)p + (1-p)^2 p) = .5787$. You're likely to lose all your money!

Now suppose you come to the game with \$300. What is the probability that you lose all your money before winning? Instead of doing a lengthy sum, we'll use the R command pgeom(x,p), which returns P$(X \leq x)$ when $X \sim$Geom(p). In this case, 1-pgeom(29,1/6) returns .0042.

Now suppose you have as much money as you need, but you have to stop when you roll a six. What is the expected winnings for the game? If Y is the number of dollars that you win, then $Y = 100 - 10(X + 1)$. Then E$(Y) = 90 - 10$E$(X) = 100 - 10(1-p)/p = 40$. It's definitely worth playing!

Hypothesis testing example: A die is being used in a gambling game, but you suspect it is not a fair die and, in particular, you think that the probability of rolling a one is less than 1/6. You decide to test this by rolling the die until a one appears and counting the rolls. If p is the true probability of rolling a one, we want to test

$$H_0 : p = 1/6 \quad \text{versus} \quad H_a : p < 1/6.$$

Suppose you decide to reject the null hypothesis if you count at least 15 rolls before a one appears (in other words, reject if there are no ones in the first 15 rolls). The test size is the probability of rejecting when $p = 1/6$, which can be computed as P$(X \geq 15)$ when $X \sim$ Geom$(1/6)$. Using 1-pgeom(14,1/6), we see this is $\alpha = .0649$.

If the true probability of rolling a one is only 2%, we can compute the power of the test, P$(X \geq 15)$, when $p = .02$. The command 1-pgeom(14,.02) returns .739.

Chapter Highlights

1. A sequence of binomial trials is performed. If Y is the number of failures before the first success occurs, then $Y \sim$ Geom(p).

2. For $Y \sim$ Geom(p),

$$P(Y = y) = p(1-p)^y \text{ for } y = 0, 1, 2, \ldots.$$

3. For $Y \sim$ Geom(p), E$(Y) = (1-p)/p$ and V$(Y) = (1-p)/p^2$.

4. In R, the command

   - pgeom(y,p) returns P$(Y \leq y)$, when $Y \sim$Geom(p);
   - rgeom(N,p) returns N randomly generated values from a Geom(p) distribution;
   - dgeom(N,p) returns P$(Y = y)$ when $Y \sim$Geom(p);
   - qgeom(r,p) will return the smallest value y such that P$(Y \leq y) \geq r$, where $Y \sim$Geom(p).

Exercises

9.1 Suppose $Y \sim \text{Geom}(.4)$.

  (a) Using simulations in R, verify that $E(Y) = 1.5$ and $V(Y) = 3.75$.

  (b) Using simulations in R, approximate $E(Y^3)$.

9.2 If $Y \sim \text{Geom}(p)$, find a formula for $P(Y \geq k)$ for some nonnegative integer k.

9.3 If $Y \sim \text{Geom}(p)$, and $0 < k_1 < k_2$ are integers, show that $P(Y \geq k_2 | Y \geq k_1) = P(Y \geq k_2 - k_1)$.

9.4 Eddie suggests the following game: You toss three coins, repeatedly, until you get all heads. For each repetition, you have to pay him \$10. When you get all heads, he pays you \$50.

  (a) Find your expected winnings from this game.

  (b) Verify your answer using simulations in R.

9.5 A surgical procedure has an 80% chance of being successful each time it is performed. Assuming independence,

  (a) what is the probability that at least five of the next seven procedures are successful?

  (b) what is the expected number of successful surgeries before a failure occurs?

9.6 Referring to Exercise 9.5, suppose a team of surgeons develops a new technique that they think will be successful more often. They will test the procedure by performing it sequentially on the patients that show up at the clinic. The null hypothesis is that the success rate for the new surgical technique is the same as the old (80%), and the alternative is that the new technique has a *higher* success rate. They decide that they will reject H_0 if the number of successful surgeries before a failure occurs is at least 12.

  (a) Find the test size α for this decision rule.

  (b) If the new technique really has a 90% success rate, find the power for the test.

9.7 An urn has four green balls and two red balls.

  (a) Cindy will draw balls from the urn, one by one and *without* replacement, until she gets a green ball. Let X be the number of balls drawn. Find the distribution of X and the expected value of X.

  (b) Cindy will draw balls from the urn, one by one and *with* replacement, until she gets a green ball. Let X be the number of balls drawn. Find the distribution of X and the expected value of X.

9.8 A company buys devices from providers A and B. Each time a device is used, it has a probability of breaking, which does not change with successive uses, so that independence can be assumed. Devices from provider A have a probability .1 of breaking with each use, and devices from provider B break with probability .2 with each use. Seventy-five percent of the company's devices were made by provider A.

  If a device at the company is chosen at random, what is the probability that it can be used successfully at most twice before it breaks?

9.9 A company buys devices from provider B. Each time a device is used, it has probability $p_B = .2$ of breaking, which does not change with successive uses, so that independence can be assumed. Provider A makes similar devices, and a sales representative tells the manager at the company that their devices have a lower probability of breaking. The manager decides to test this claim, and the sales representative gives him three devices from provider A. The null hypothesis is $p_A = .2$ and the alternative is $p_A < .2$, where p_A is the failure probability for provider A devices. We will assume that, as with provider B devices, this probability does not change with successive uses, and independence can be assumed. The manager will reject H_0 if the total number of times the three devices can be used successfully is at least 20. Write some R code to determine the test size. If the true probability of breakage with each use is $p_A = .1$, what is the power of this test?

9.10 (a) A lab has five identical-looking white rabbits, and it is known that two of them have a blood cancer. A scientist will sacrifice the rabbits one by one (choosing randomly), testing for the cancer, stopping when the test is positive for the cancer. Let the random variable Y be the number of rabbits tested. Find E(Y).

(b) A lab supply company sells white mice of a specially bred variety, 40% of which have cancerous cells in their bones. A scientist will purchase mice one by one, and test for the cancer, stopping when a cancerous mouse is found. If Y is the number of mice tested, what is the expected value of Y? (Assume independence.)

9.11 Suppose a Brand A nail gun has a probability of .01 of jamming every time it is used, but a Brand B nail gun has a probability of .02 of jamming every time it is used. Let X be the number of times the Brand A nail gun can be used successfully before it jams, and let Y be the number of times the Brand B nail gun can be used successfully before it jams. Assume that X and Y can be reasonably modeled as independent geometric random variables, and use simulations in R to compute $P(Y \geq X)$.

9.12 In a game of craps, a player will roll two fair six-sided dice until a two *or* a seven appears.

(a) Compute analytically the probability that the two appears first. (*Hint:* Let A_k be the event that there was no two or seven on the first $k - 1$ rolls, then a two on the kth roll. Then you need to sum all the $P(A_k)$.)

(b) Verify your answer using simulation in R.

9.13 Eddie proposes the following game: You will each roll a fair six-sided die until a six appears. If the sum of the number of rolls is greater than 12, he pays you $10. Otherwise, you pay him $10. Using simulations in R, determine your expected profit from the game.

Chapter 10

The Poisson Random Variable

The Poisson distribution is useful for modeling counts. The number of gamma ray emissions per unit time, from a radioactive source such as cesium, follows a Poisson model, and counts of natural events such as number of hurricanes or earthquakes in a given time span are often modeled with Poisson distributions.

For $\lambda > 0$, we say $Y \sim \text{Pois}(\lambda)$ if the possible values for Y are the nonnegative integers, and

$$P(Y = k) = \frac{e^{-\lambda}\lambda^k}{k!} \quad \text{for } k = 0, 1, 2, \ldots.$$

The bar charts of some example distributions show that, although the support is the same as for the geometric random variable, the shape of the distribution can be quite different:

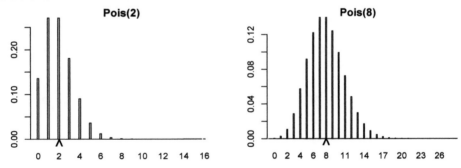

The first order of business is to check to see that we have a proper probability mass function. We see that all of the probabilities are nonnegative, so we need only show that they sum to one. For this, we use the Taylor expansion of $f(x) = e^x$, about $x = 0$. Recall that if a function f is "infinitely differentiable" on an interval containing x_0, then for x in that interval,

$$f(x) = f(x_0) + f'(x_0)(x - x_0) + \frac{1}{2}f''(x_0)(x - x_0)^2 + \frac{1}{3!}f''(x_0)(x - x_0)^3 + \cdots$$

$$+ \frac{1}{k!}f^{(k)}(x_0)(x - x_0)^k + \cdots.$$

Now we simply plug in $x_0 = 0$ and $f(x) = e^x$ to get

$$e^x = 1 + x + \frac{x^2}{2} + \frac{x^3}{3!} + \cdots + \frac{x^k}{k!} + \cdots,$$

and we divide both sides by e^x to get

$$1 = e^{-x} + e^{-x}x + \frac{e^{-x}x^2}{2} + \frac{e^{-x}x^3}{3!} + \cdots + \frac{e^{-x}x^k}{k!} + \cdots$$

for any $x \in \mathbb{R}$. If we plug in λ for x, we get the Poisson probabilities on the right.

Next, let's find the expected value of $Y \sim \text{Pois}(\lambda)$. Using the formula we get

$$\text{E}(Y) = \sum_{k=0}^{\infty} k \frac{e^{-\lambda}\lambda^k}{k!}.$$

To simplify this, we use the trick of manipulating the terms in the summation, until the sum equals one and we can cross it out. We can start the sum at $k = 1$ instead of at $k = 0$; then the obvious thing to try is canceling the k with the first term in the factorial, to get

$$\text{E}(Y) = \sum_{k=1}^{\infty} k \frac{e^{-\lambda}\lambda^k}{k!} = \sum_{k=1}^{\infty} \frac{e^{-\lambda}\lambda^k}{(k-1)!} = \lambda \sum_{k=1}^{\infty} \frac{e^{-\lambda}\lambda^{k-1}}{(k-1)!}.$$

Doing a quick change of variable $j = k - 1$, we get

$$\text{E}(Y) = \lambda \sum_{j=0}^{\infty} \frac{e^{-\lambda}\lambda^j}{j!}.$$

We recognize that the terms in the sum are the probabilities for a $\text{Pois}(\lambda)$ random variable, so the sum is one, and $\text{E}(Y) = \lambda$.

Next we find the variance $\text{V}(Y) = \text{E}(Y^2) - \text{E}(Y)^2$. Because we again want to use the cancel-with-terms-in-the-factorial trick, we modify the formula

$$\text{V}(Y) = \text{E}(Y^2) - \text{E}(Y^2) = \text{E}[Y(Y-1)] + \text{E}(Y) - \text{E}(Y^2).$$

To evaluate the first term in the expression for the variance, we use the definition of expected value and simplify:

$$\begin{aligned}
\text{E}[Y(Y-1)] &= \sum_{k=0}^{\infty} k(k-1)\text{P}(Y=k) = \sum_{k=2}^{\infty} k(k-1)\text{P}(Y=k) \\
&= \sum_{k=2}^{\infty} k(k-1)\frac{e^{-\lambda}\lambda^k}{k!} \\
&= \lambda^2 \sum_{k=2}^{\infty} \frac{e^{-\lambda}\lambda^{k-2}}{(k-2)!} \\
&= \lambda^2 \sum_{j=0}^{\infty} \frac{e^{-\lambda}\lambda^j}{j!} \quad \text{(substitute } j = k-2\text{)} \\
&= \lambda^2 \quad \text{(recognize the summation is 1).}
\end{aligned}$$

Finally, $\text{V}(Y) = \lambda^2 + \lambda - \lambda^2 = \lambda$. For a Poisson random variable, the variance is equal to the mean.

Example: Suppose the number of people killed in automobile crashes per day in Colorado can be modeled as a Poisson random variable with mean 1.47. On about how many days per year are no people killed?

Let Y be the number of people killed in Colorado automobile crashes on a randomly selected day; then $Y \sim \text{Pois}(1.47)$:

$$P(Y = 0) = e^{-1.47} = .23.$$

So, on average there are $(365)(.23) \approx 84$ days in the year, on which no one is killed in Colorado automobile crashes.

On a randomly selected day, what is the probability that *at least* three people die in crashes, given that the number of people killed is at least one?

Here $P(Y \geq 1) = 1 - P(Y = 0) = .770$ and

$$P(Y \geq 3) = 1 - P(Y = 0) - P(Y = 1) - P(Y = 2) = .184.$$

Then

$$P(Y \geq 3 | Y \geq 1) = \frac{P(Y \geq 3 \& Y \geq 1)}{P(Y \geq 1)} = \frac{P(Y \geq 3)}{P(Y \geq 1)} = .239.$$

Finally, let's find the probability that fewer than six people are killed in automobile crashes in Colorado, for a randomly selected day. Instead of doing "by hand" calculations, we can use the R command:

```
ppois(k,lambda)
```

will return $P(Y \leq k)$, when $Y \sim \text{Pois}(\texttt{lambda})$. Therefore, we use `ppois(5,1.47)` to get about .996 for the proportion of days in Colorado that fewer than six people are killed in automobile crashes. On only about .4% of days do six or more people die in crashes in Colorado.

Of course, there are reasons to question the assumption that the distribution is Poisson. If crashes are more likely in winter or on holidays, then we need a more complicated model. In *Poisson regression* (beyond the scope of this book), the number of deaths can be modeled as Poisson, but with the mean value depending on predictor values such as weather and holidays. This type of count model is one of the most important applications of the Poisson random variable.

Hypothesis testing example

A machine produces bolts of cloth, with occasional flaws. The number of flaws per yard of cloth follows a Poisson random variable with mean 0.6. The manager of the company is considering getting a new machine and decides to test the null hypothesis—that the new machine produces flaws at the same rate as the old machine—against the alternative hypothesis that the new machine produces fewer flaws, on average. He decides to produce five yards of cloth on the new machine, and reject the null hypothesis if none of the yards contains flaws. Assuming independence, what is the test size α?

The test size α is the probability of rejecting H_0 (all of the lengths of cloth have no flaws) given that H_0 is true (the flaws per yard are distributed as Poisson with mean 0.6). For each yard, the probability of no flaws is $e^{-.6} = .5488$. The probability of all five having no flaws (using the independence assumption) is $\alpha = (.5488)^5 = .0498$.

Poisson-binomial connection

Suppose we want to model the number of flaws in a three-meter section of string, or the number of misprints on a page of newsprint, or the number of people injured monthly on the job in a large corporation. In each case, we could model this number as a binomial random variable with large n and small p. For the last example, there is a large number n of people, and for each person, a small probability p of injury. For the string example, we could divide up the string into many tiny pieces, and for each piece, there is a small probability p of having a flaw at that piece. For the newsprint example, there are many characters, and each one has a small probability of being misprinted.

In general, assume we have a large n and a small p so that $\lambda = np$ is "moderate" (neither small nor large). Recall that np is the expected value for the binomial random variable.

We want to show that this large-n–small-p binomial random variable is approximately Poisson. Starting with the binomial probabilities then substituting $p = \lambda/n$, we have for $k = 1, \ldots, n$

$$P(Y = k) = \frac{n!}{k!(n-k)!} p^k (1-p)^{n-k}$$

$$= \frac{n!}{k!(n-k)!} \left(\frac{\lambda}{n}\right)^k \left(1 - \frac{\lambda}{n}\right)^{n-k}$$

$$= \frac{n(n-1)\cdots(n-k+1)}{n^k} \frac{\lambda^k}{k!} \left(1 - \frac{\lambda}{n}\right)^{-k} \left(1 - \frac{\lambda}{n}\right)^n.$$

Notice that there are k terms in the numerator of the first fraction. Now we think of fixing k and letting n be "large," or you can think of k as being small compared with n. This makes sense for the large-n–small-p, because if p is very small, we expect the values Y takes to be much smaller than n. With this in mind, we can see that the first fraction is approximately one, and approaches one as k is fixed and n goes to infinity. The second fraction is part of the Poisson probability formula, so we want to keep it. The third term goes to one as n increases. The last term goes to $e^{-\lambda}$ as n goes to infinity by l'Hôpital's rule (see Appendix B.4). Therefore, in the large-n–small-p binomial case, we have

$$P(Y = k) \approx \frac{e^{-\lambda} \lambda^k}{k!},$$

where $\lambda = np$.

This relationship between the binomial and Poisson distributions provides some intuition for determining when the Poisson model is appropriate. The bar charts shown below show how closely these distributions are related: Even for $n = 50$, the binomial distribution with mean equal to 6 is similar to the Pois(6) distribution, and when $n = 500$, the probabilities are so similar that the bar charts look identical:

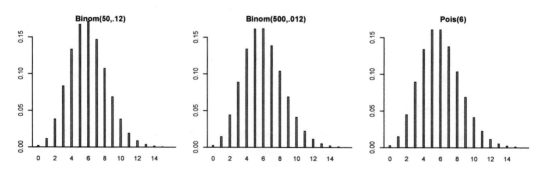

Chapter Highlights

1. If $Y \sim \text{Pois}(\lambda)$, then the support of Y is the nonnegative integers, and

$$P(Y = y) = \frac{e^{-\lambda}\lambda^y}{y!}.$$

2. If $Y \sim \text{Pois}(\lambda)$, then $E(Y) = \lambda$ and $V(Y) = \lambda$.

3. If n is "large" and p is "small," then $Y \sim \text{Binom}(n, p)$ has a distribution that is similar to $Y \sim \text{Pois}(\lambda)$, where $\lambda = np$.

4. In R, the command

   (a) `ppois(y,lambda)` will return $P(Y \leq y)$, where $Y \sim \text{Pois(lambda)}$;

   (b) `dpois(y,lambda)` will return $P(Y = y)$, where $Y \sim \text{Pois(lambda)}$;

   (c) `rpois(N,y,lambda)` will return N randomly generated values from a Pois(lambda) distribution;

   (d) `qpois(r,p)` will return the smallest value y such that $P(Y \leq y) \geq r$, where $Y \sim \text{Pois(lambda)}$.

Exercises

10.1 If $Y \sim \text{Pois}(\lambda)$, simplify the ratio of probabilities

$$\frac{P(Y = k + 1)}{P(Y = k)}.$$

10.2 Show that the Poisson mass function is decreasing in $k = 0, 1, \ldots$ if and only if $\lambda < 1$.

10.3 For $\lambda \geq 1$, determine the mode of the Poisson distribution.

10.4 For Exercise 9.3, we proved that if $Y \sim \text{Geom}(p)$, then $P(Y \geq k_2 | Y \geq k_1) = P(Y \geq k_2 - k_1)$. Is this true if $Y \sim \text{Pois}(\lambda)$? Prove it or find a counterexample.

10.5 The number of times a geyser erupts in one day follows a Poisson distribution with mean 2.5. On a randomly selected day, what is the probability that the geyser erupts at least three times?

10.6 Suppose $Y \sim \text{Pois}(\lambda)$.

   (a) Find $E[1/(Y + 1)]$.

   (b) Check your answer to part (a) using simulations in R for $\lambda = 2$. (You can simulate Poisson random variables with the R command `rpois`. For example, `rpois(10000,2)` will return a vector of 10,000 replicates of a Poisson random variable with mean 2.)

10.7 The number of daily on-the-job accidents at ABC, Inc., follows a Poisson distribution with mean .3. If we randomly select two days from last year, what is the probability that on both days there were no accidents? (Assume independence.)

10.8 The number of bee stings for a randomly selected camper at Camp Nearahive follows a Poisson distribution with mean 0.8. If a newspaper randomly chooses three campers to be interviewed, what is the probability that the three have had at least two bee stings each? (Assume independence, so that the number of stings for one camper doesn't depend on the number of stings for another camper.)

10.9 Traditionally, the number of bee stings for a randomly selected camper at Camp Nearahive follows a Poisson distribution with mean 0.8. However, a camp counselor conjectures that the bee problem is worse this year. She wants to test $H_0 : \lambda = .8$ versus $H_a : \lambda > .8$, where λ is the true expected number of bee stings per camper this year. She decides to randomly select five campers, and reject H_0 if they have all had at least one bee sting.

   (a) What is the test size (assuming independence)?

   (b) If the true expected number of bee stings per camper is $\lambda = 1.2$, what is the power of the test?

10.10 The number of times a flowering plant blooms in a season is assumed to follow a Poisson distribution with mean λ. A botanist is interested in testing $H_0 : \lambda = 3$ versus $H_a : \lambda > 3$. She has five plants and will observe the number of times each blooms in the coming season. She decides to reject H_0 if the numbers of blooms for at least four of the plants is three or more.

   (a) Find the test size α (assume independence for your calculations).

   (b) If the true mean number of blooms is $\lambda = 4.5$, find the power of the test.

10.11 The number of sports injuries per week at elementary schools in Largetown can be modeled as a Poisson random variable. Historically, the average number of sports injuries per week is 3.7. The school board, thinking this too high, imposed a mandatory training course for coaches and gym instructors. Now they would like to test the null hypothesis that the distribution of the number of sports injuries per week has not changed, with the alternative that the average number of sports injuries per week is *smaller* (still distributed as Poisson). The decision rule is "if there are two or fewer injuries in each of the next three weeks, reject H_0." Assuming independence,

   (a) what is the test size?

   (b) if the new distribution has mean 2.1, what is the power of the test?

10.12 A fabric manufacturing company has two looms to make madras cloth. Loom A makes cloth with number of flaws per yard following a Poisson random variable with mean 1/2. Loom B makes cloth with number of flaws per yard following a Poisson random variable with mean 2. Eighty percent of the company's madras cloth is made using Loom A. If a yard of cloth made by the company is randomly chosen and found to have no flaws, what is the probability that is came from Loom A?

10.13 Maxine and Susan are glass blowers and make colored vases for a boutique. The number of flaws in a randomly selected vase made by Maxine follows a Poisson distribution with mean .5, and the number of flaws in a randomly selected vase made by Susan follows a Poisson distribution with mean 1. Further, suppose 70% of the vases in the boutique were made by Susan.

   (a) If we randomly select one of Susan's vases, what is the expected number of flaws?

   (b) If we select a vase at random and find that it has no flaws, what is the probability that Susan made it?

   (c) What proportion of the vases made for the boutique have two flaws?

10.14 The average number of hurricanes to hit the U.S. mainland per year has historically been about 1.8. Someone concerned about climate change wonders if the expected number of hurricanes has increased. Suppose we model the number of hurricanes to hit the U.S. mainland next year to be a Poisson random variable with mean λ. We want to test $H_0 : \lambda = 1.8$ against the alternative $H_a : \lambda > 1.8$. If 7 hurricanes hit the U.S. mainland next year, what is the p-value for our test?

10.15 Suppose the annual number of tropical storms in the Gulf of Mexico can be modeled as a Poisson random variable. Historically, the average is about $\lambda = 12.1$. If next year the number is 20 tropical storms, is the statement "the frequency of tropical storms has increased" justified? Explain.

10.16 Suppose $Y \sim \text{Binom}(1000, .003)$. Calculate $P(Y = k)$ for $k = 0, 1, \ldots, 10$ and compare to the corresponding $P(X = k)$ for $X \sim \text{Poisson}(3)$. (You can use R if you like!)

10.17 The plant pauciflora has beautiful blossoms that are popular in cut flower arrangements. However, the number of blossoms for a randomly selected stalk follows a Poisson distribution with mean only 1.5. A new hybrid is developed to have more blossoms, and the botanists want to test $H_0 : \lambda = 1.5$ versus the alternative $H_a : \lambda > 1.5$, where λ is the average number of blossoms for a randomly selected stalk of the new hybrid. Assume that the number of blossoms for a randomly selected new hybrid stalk follows a Poisson distribution.

   (a) The experiment will be to randomly select four hybrid stalks and count the number of blossoms per stalk. The decision rule is to reject H_0 if all four stalks have at least two blossoms. What is the test size?

   (b) For the experiment and decision rule in (a), what is the power of the test if the true mean number of blossoms for a randomly selected new hybrid stalk is $\lambda = 3$?

   (c) The developers of the new hybrid think that the power computed in part (b) is too low, so they decide to select seven stalks, and will reject H_0 if at least six stalks out of the seven have at least two blossoms. What is the test size?

   (d) For the experiment and decision rule in (c), what is the power of the test if the true mean number of blossoms for a randomly selected new hybrid stalk is $\lambda = 3$?

10.18 Books from a certain publisher have typos with a frequency that follows a Poisson distribution with an average of 0.3 typos per page. A new publisher claims to have a better rate. You want to test $H_0 : \lambda = .3$ versus $H_a : \lambda < .3$. You can obtain a random sample of 10 pages, and you will reject the null hypothesis if all of the pages have no typos.

   (a) Find the test size for this decision rule.

   (b) Suppose the new publisher's typo rate really follows a Poisson distribution with an average of 0.05 typos per page. Find the power of your test.

10.19 Typists from School A make mistakes according to a Poisson distribution with an average of two mistakes per page of typing. Typists from School B make mistakes according to a Poisson distribution with an average of four mistakes per page of typing. A company has eight typists, two from School A and six from School B. A typist is randomly selected and given one page to type. If no mistakes are made, what is the probability that the typist went to School A?

10.20 Suppose the ABC company of Exercise 10.7 has a new plant in a state with tighter safety regulations, so that the managers and workers go to semiannual safety training. The CEO is interested to see whether the training actually results in fewer accidents. Suppose that for a randomly selected week of the new plant's operation, there were no accidents. Test $H_0 : \lambda = .3$ versus $H_a : \lambda < .3$, where λ is the expected number of daily accidents at the ABC company. Assume that the number of accidents per day follows a Poisson distribution, and the numbers of accidents in sequential days are independent random variables. Report a p-value for the test.

10.21 The number of homicides per week in a major city historically follows a Poisson distribution with mean $\lambda_0 = 1.23$. Suppose a political candidate claims that the homicide rate has increased, so that the rate is now larger. A fact-checking group wishes to test $H_0 : \lambda = 1.23$ versus $H_a : \lambda > 1.23$, where λ is the current homicide rate. They will use the decision rule "reject H_0 if each of the next three weeks has at least two homicides."

   (a) Find the test size.

   (b) If the new homicide rate is $\lambda = 1.85$, find the power of the test.

Chapter 11

The Hypergeometric Random Variable

The hypergeometric distribution is *not* a generalization of the geometric distribution, in spite of the name. You can think of it as a "sampling without replacement" version of the binomial. The classical example is a good illustration: Suppose we have an urn with 6 red balls and 7 green balls. We choose 5 at random, *without replacement*. The number of balls that are red follows a hypergeometric distribution. (Note that if we chose balls *with* replacement, we would use the binomial with $p = 6/13$.)

We can find the hypergeometric probabilities using ideas developed in Chapter 3. First, the number of ways to choose 5 balls from 13 is the denominator. We are assuming that all these ways are equally likely, which is a reasonable assumption for draws "at random." The numerator is the number of ways to choose 2 red balls and 3 green balls from 6 red balls and 7 green balls. By our mn rule, this is the number of ways to choose 2 red balls from 6 red balls *times* the number of ways to choose 3 green balls from 7 green balls. Therefore,

$$P(\text{exactly two red}) = \frac{\binom{6}{2}\binom{7}{3}}{\binom{13}{5}} \approx .408.$$

To generalize this idea, suppose there are M red balls and N green balls in the urn, and we choose n balls without replacement. If n is smaller than either M or N, then the number of red balls in the selection can range from zero to n. However, if n is larger than M but smaller than N, the number of red balls can range only from zero to M; there must be at least $n - M$ green balls. Also, if n is larger than N, there has to be at least $n - N$ red balls. Therefore, if Y is the number of red balls from such a random draw, the values Y can take range from $\min(0, n - N)$ to $\min(M, n)$.

Definition: The random variable Y is a hypergeometric random variable with parameters M, N, and n, $n \le M + N$ if the values Y can take are $\min(0, n - N), \ldots,$ $\min(M, n)$, and

$$P(Y = k) = \frac{\binom{M}{k}\binom{N}{n-k}}{\binom{M+N}{n}}.$$

We can write $Y \sim HG(M, N, n)$. In the "urn language," this is the distribution of the number of red balls in a sample of size n, taken without replacement from an urn with M red balls and N not-red balls. The denominator is the number of samples of size n from the urn, and the numerator is the number of these samples that have k red and $n - k$ not-red. When you are computing these probabilities, a quick spot check is to make sure the top numbers in the numerator add to the top number in the denominator, and similarly the bottom numbers in the numerator must add to the bottom number in the denominator.

Some example bar charts show the probabilities and the mean values:

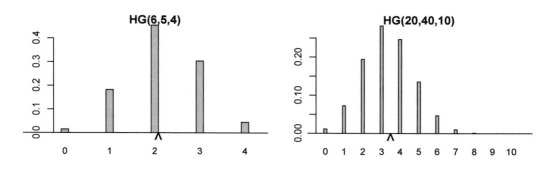

To prove that these probabilities add to one, we need a nice little identity concerning combinations. Start by noticing that

$$(a + 1)^M (a + 1)^N = (a + 1)^{M+N}.$$

If we use the binomial theorem on both sides, we get

$$\left[\sum_{i=0}^{M} \binom{M}{i} a^i \right] \left[\sum_{j=0}^{N} \binom{N}{j} a^j \right] = \left[\sum_{k=0}^{M+N} \binom{M + N}{k} a^k \right].$$

Now we match up the coefficients of a^k on both sides. For example, on the right, one of the terms is

$$\binom{M + N}{2} a^2$$

and on the left we have the corresponding three terms

$$\binom{M}{0} \binom{N}{2} a^2 + \binom{M}{1} \binom{N}{1} a^2 + \binom{M}{2} \binom{N}{0} a^2,$$

which means

$$\binom{M + N}{2} = \binom{M}{0} \binom{N}{2} + \binom{M}{1} \binom{N}{1} + \binom{M}{2} \binom{N}{0}.$$

Now we take the coefficients of a^n, where $0 \le n \le \min(M, N)$, to get

$$\binom{M + N}{n} = \binom{M}{0} \binom{N}{n} + \binom{M}{1} \binom{N}{n - 1} + \cdots + \binom{M}{n} \binom{N}{0},$$

or

$$\binom{M+N}{n} = \sum_{k=0}^{n} \binom{M}{k}\binom{N}{n-k}.$$

If we divide both sides by $\binom{M+N}{n}$, on the right are the probabilities of the hypergeometric, and on the left is one.

To compute the expected value of $Y \sim \mathrm{HG}(M,N,n)$, for $n \le \min(M,N)$, we start with the definition

$$E(Y) = \sum_{k=0}^{n} \left[k \frac{\binom{M}{k}\binom{N}{n-k}}{\binom{M+N}{n}} \right] = \frac{\sum_{k=0}^{n} \left[k \binom{M}{k}\binom{N}{n-k} \right]}{\binom{M+N}{n}}.$$

Working with the numerator only, we can start the sum from one and write

$$\sum_{k=1}^{n} \left[k \binom{M}{k}\binom{N}{n-k} \right] = M \sum_{k=1}^{n} \left[\frac{(M-1)!}{(k-1)!(M-k)!} \binom{N}{n-k} \right]$$

$$= M \sum_{k=1}^{n} \left[\binom{M-1}{k-1}\binom{N}{n-k} \right]$$

$$= M \sum_{j=0}^{m} \left[\binom{L}{j}\binom{N}{m-j} \right],$$

making the change of variables $j = k-1$, $L = M-1$, and $m = n-1$. We notice that the summation consists of the numerators of the probabilities for all the values of an $\mathrm{HG}(L,N,m)$ random variable, so the last term becomes

$$M \binom{L+N}{m} = M \binom{M-1+N}{n-1},$$

switching back to the original variables. Going back to the expected value, we have

$$E(Y) = \frac{M \binom{M-1+N}{n-1}}{\binom{M+N}{n}} = \frac{Mn}{M+N}.$$

The variance is computed similarly, but we will skip the derivation:

$$V(Y) = n \frac{MN}{M+N} \frac{M+N-n}{M+N-1}.$$

Example: Electronic parts come in boxes of 25. A purchaser inspects the box by randomly choosing five (without replacement), and rejecting the box if he finds at least one defective. If a box actually has four defective parts, what is the probability the purchaser finds exactly two defective parts? We choose two of the four defective parts, and three of the 21 nondefective parts, so the probability is

$$\frac{\binom{4}{2}\binom{21}{3}}{\binom{25}{5}} = .150.$$

What is the probability that the box is rejected by the purchaser (i.e., he finds at least one of the defectives)? The probability of no defectives in the sample of size 5 is

$$\frac{\binom{4}{0}\binom{21}{5}}{\binom{25}{5}} = .383,$$

so the probability of rejecting is $1 - .383 = .617$.

We can check these answers using the `sample` function in R (see Chapter 2) to simulate draws from the box. Suppose we label the parts 1 through 25, with parts 22–25 being defective. The following bit of code draws five parts from the box, 100,000 times, and counts the draws for which there are exactly two defective parts.

```
cnt2def=0;nloop=100000
for(iloop in 1:nloop){
    draw=sample(1:25,5)
    if(sum(draw>21)==2){cnt2def=cnt2def+1}
}
cnt2def/nloop
```

Math class example

Samantha is complaining to her aunt that her math teacher asks boys to solve problems at the board more than he asks girls. Her aunt is a statistician, so she suggests that Samantha collect data to test the null hypothesis that girls and boys are equally likely to be asked versus the alternative that boys are more likely to be asked. They decide that Samantha will keep track of the first 10 kids asked to solve problems at the board the next day. Suppose there are 14 boys and 17 girls in the math class, and 7 of the 10 kids called to solve problems are boys. (Suppose that a child is not called more than once in the same day, so that the sampling is without replacement.) What is the p-value for the test?

The p-value is the probability of "our data" (7 out of 10 are boys) or "more extreme" (8, 9, or 10 out of 10 are boys), given that the null hypothesis is true (boys and girls are equally likely to be called). Under H_0, these probabilities can be calculated using the hypergeometric formula: We add up the probabilities of 7, 8, 9, and 10 boys being chosen, when there are 14 boys and 17 girls and children are chosen "at random."

$$p = \frac{\binom{14}{7}\binom{17}{3}}{\binom{31}{10}} + \frac{\binom{14}{8}\binom{17}{2}}{\binom{31}{10}} + \frac{\binom{14}{9}\binom{17}{1}}{\binom{31}{10}} + \frac{\binom{14}{10}\binom{17}{0}}{\binom{31}{10}} = .0626.$$

There is some evidence against H_0, but it is not decisive. If the null hypothesis were true, and each day there were 10 children chosen, then on about 6.26% of the days, there would be at least 7 boys chosen.

The hypergeometric random variable in R

The dhyper function in R returns probabilities for a hypergeometric random variable. The command

```
dhyper(k,M,N,n)
```

will return $P(Y = k)$, where $Y \sim HG(M, N, n)$.

The phyper function is useful for hypothesis testing, as it returns *sums* of probabilities for a hypergeometric random variable. Specifically,

```
phyper(k,M,N,n)
```

will return $P(Y \leq k)$, where $Y \sim HG(M, N, n)$. The probability for the math class example would be obtained with

```
1-phyper(6,14,17,10)
```

that is, $P(Y \geq 7) = P(Y > 6) = 1 - P(Y \leq 6)$.

Capture-and-release example

Ecologists are often interested in estimating the number B of some species of animal in some region. They capture a number M of these animals and tag them. The animals are released, and the ecologists wait for some time, long enough for "random mixing" of the animals. Then they capture n animals and count the number x of these animals that have tags.

The estimate of B follows this reasoning. If X is the number of tagged animals in the second sample, then under some assumptions about the animal populations, we could model $X \sim HG(M, B - M, n)$, so,

$$P(X = x) = \frac{\binom{M}{x}\binom{B - M}{n - x}}{\binom{B}{n}}.$$

Here we know M, n, and x, but we don't know B. We want to find the value of B that makes this probability the largest possible for the value of x that we have observed. (In a future chapter, we will call this the *maximum likelihood estimate* of B.) That is, different values of B will give different bar charts of the distribution. We want to choose the value of B that puts the tallest bar at the observed value.

To illustrate, suppose we captured and tagged $M = 20$ animals, and recaptured $n = 30$. The distribution of X, with $B = 100$, is shown in the top-left plot below.

The probability of seeing $X = 4$ is about .127 in this case. The top-right plot shows the distribution of X, with $B = 140$; here $P(X = 4) = .231$. Finally, in the bottom plot, if $B = 180$, $P(X = 4) = .214$, a little smaller.

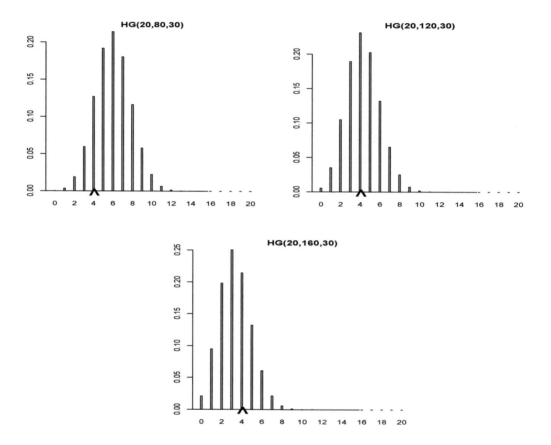

Our goal to estimate B is to choose out of all possible values the one that maximizes $P(X = 4)$. To do this analytically, let $\psi_x(B) = P(X = x)$ if $X \sim HG(M, B - M, n)$, where M and n are given. Then for $B \geq x - 1$,

$$\frac{\psi_x(B)}{\psi_x(B-1)} = \frac{(B-M)(B-n)}{B(B-M-n+x)},$$

after a lot of canceling. Multiplying out the numerator and denominator,

$$\frac{\psi_x(B)}{\psi_x(B-1)} = \frac{B^2 - nB - MB + nM}{B^2 - nB - MB + Bx}.$$

This ratio is larger than one if $B < Mn/x$, and less than one if $B > Mn/x$. This says that the probabilities are increasing in B until $B \approx Mn/x$, then they are decreasing. Thus, the value of B that maximizes the probability is Mn/x. For our example with $M = 20$, $n = 30$, and $x = 4$, B is estimated to be 150.

Chapter Highlights

1. Suppose we select "at random" a subset of n objects from $M + N > n$ objects, where M is the number of objects with attribute A and N is the number of objects that do not have attribute A. If Y is the number of objects with attribute A in our subset, then $Y \sim \mathrm{HG}(M, N, n)$.

2. For $Y \sim \mathrm{HG}(M, N, n)$,

$$P(Y = y) = \frac{\binom{M}{y}\binom{N}{n-y}}{\binom{M+N}{n}} \text{ for } y = 0, \ldots, \min(M, n).$$

3. For $Y \sim \mathrm{HG}(M, N, n)$,

$$\mathrm{E}(Y) = \frac{Mn}{M+N} \text{ and } \mathrm{V}(Y) = n\frac{MN}{M+N}\frac{M+N-n}{M+N-1}.$$

4. In R the function

- `phyper(k,M,N,n)` will return $P(Y \leq k)$, where $Y \sim \mathrm{HG}(M, N, n)$;
- `dhyper(k,M,N,n)` will return $P(Y = k)$, where $Y \sim \mathrm{HG}(M, N, n)$;
- `rhyper(S,M,N,n)` will return S randomly generated values from an $\mathrm{HG}(M, N, n)$ distribution;
- `qhyper(r,p)` will return the smallest value y such that $P(Y \leq y) \geq r$, where $Y \sim \mathrm{HG}(M, N, n)$.

Exercises

11.1 Researchers want to determine if a magician has ESP. They take ten red cards and four black cards, shuffle them, and place them face down on the table. They ask the magician to turn over the black cards (she knows there are four black cards). The null hypothesis is that she is just turning cards over "at random," and the alternative is that she is more likely to turn over black cards. Suppose she turns over three black cards and one red card. Summarize the evidence that the magician has ESP by computing a p-value for the test.

11.2 An urn contains 20 red balls and five black balls. Suppose a game consists of choosing three balls at random, without replacement, from the urn. For each black ball, the player gets \$100, and for each red ball, the player gets \$10. What is the expected number of dollars that the player will win?

11.3 The following two questions are similar but not the same:

(a) A box of 40 ears of corn from Sunshine Farms has four ears with at least one worm beneath the husk. If Annie randomly chooses eight of these ears of corn for dinner, what is the probability that there are at least six ears without worms?

(b) It is known that 10% of ears of corn from Sunshine Farms have at least one worm beneath the husk. If Annie randomly chooses eight of these ears of corn for dinner, what is the probability that there are at least six ears without worms?

11.4 A fifth grade class has eight boys and twelve girls.

(a) The teacher will choose, at random, a group of four children to attend an exhibit next Monday. What is the probability that at least three girls are in the group?

(b) The teacher assigns numbers 1–20 to the children, then calls them to the board one by one to do math problems by rolling a fair 20-sided die and calling out the number. If four children are called to the board in this manner, what is the probability that at least three are girls?

11.5 A box of parts from Manufacturer A contains 10 defective parts and 10 good parts. A box from Manufacturer B also contains 20 parts, only 2 of which are defective. Suppose you choose a box at random, and choose 3 parts without replacement. If at least 2 of the 3 are defective, what is the probability that the box was from Manufacturer A?

11.6 In a class of 20 kindergartners, 10 are wearing blue shirts, 5 are wearing green shirts, and 5 are wearing red shirts. A group of 10 kindergartners is chosen at random to go outside for play time.

(a) What is the probability that exactly two of the group are wearing red shirts?

(b) What is the probability that none of the group is wearing a red shirt?

11.7 Consider again the math class example of this chapter. Samantha is disappointed that the null hypothesis cannot be rejected at $\alpha = .05$. Suppose that Samantha and her aunt had instead decided to observe the next 15 students who were called to the board to solve problems, and 10 were boys (note that this is a smaller percentage than in the original example). Compute the p-value.

11.8 In a certain community live three ethnic groups: A, B, and C. Ethnic group A has historically been socially and economically advantaged, and groups B and C are referred to as "minorities." Suppose there is a school with 40 kids from group A, and 30 from each of the other two groups. Five children are to be chosen to represent the school at a community event. Some parents think "they always choose three kids from group A and token kids from the other two groups." If the kids are chosen "at random," what is the proportion of times that there are three from A and one each from B and C?

11.9 A purchaser of electrical components buys them in lots of size 10. It is his policy to inspect three components randomly from a lot (without replacement) and to accept the lot only if all three are nondefective. If 30% of the lots have four defective components and 70% of the lots have only one defective component, what proportion of the lots does the purchaser reject?

11.10 A naturalist conjectures that the Australian Bowerbird prefers blue stones for nest building. He sets out 10 blue stones and 10 red stones near a Bowerbird habitat and watches until 8 stones have been taken. Suppose 7 of the 8 stones are blue. What is the p-value for testing the null hypothesis that the birds are equally likely to take either color against the alternative that they prefer the blue stones?

11.11 Refer to Exercise 3.16. Suppose the scientists want to test formally the null hypothesis that the omega-3 fatty acids have no effect on brain development versus the alternative that rats fed high levels of the fatty acids will be smarter and hence run the maze faster. If the decision rule is to reject the null hypothesis if at least five of the six fastest rats are in the omega-3 group, what is the test size?

11.12 A basketball club has eight tall players and twelve short players. If a team of five players is chosen "at random" to play against a different club, what is the probability that the team has at most one short player? ("At random" means that all 20 players have equal probability to be on the chosen team.)

11.13 Eddie has a box with four gold and eight brown balls. Your box has three gold and six brown balls.

(a) You each choose two balls from your box, at random and without replacement. Who has the larger probability that both balls are gold?

(b) Use simulations in R to determine the probability that the sum of the number of gold balls (yours plus Eddie's) is greater than two.

11.14 Suppose 13 cards are dealt from a well-shuffled deck. (The deck has 52 cards, 13 cards of each suit.)

(a) What is the probability that at least eight are spades?

(b) What is the probability that at least eight are of a single suit?

11.15 In the card game bridge, a well-shuffled deck of standard playing cards is dealt, 13 cards each to 4 players. The player opposite you is your partner. (The deck has 52 cards, 13 cards of each suit.)

(a) What is the probability that the number of spades in your hand and in your partner's hand totals exactly ten?

(b) What is the probability that the number of spades in your hand and in your partner's hand totals at least ten?

11.16 In the card game bridge, a well-shuffled deck of standard playing cards is dealt, 13 cards each to 4 players. The player opposite to you is your partner. Suppose you and your partner have eight spades between you, so that your opponents must have the remaining five spades. The number of tricks you can take often depends on the "split" of the five cards between the opponents.

(a) What is the probability that the split is 3 and 2, that is, that one of your opponents has three spades and the other has two spades.

(b) What is the probability that the split is 4 and 1, that is, that one of your opponents has four spades and the other has one spade.

(c) What is the probability that the split is 5 and 0, that is, that one of your opponents has five spades and the other has no spades.

Chapter 12

The Negative Binomial Random Variable

The negative binomial distribution can be thought of as a generalization of the geometric distribution. It is another random variable related to binomial trials, but instead of counting the number of failures before the *first* success occurs (as for the geometric random variable), we count the number of failures before the rth success occurs.

The classical example is, suppose you have a coin where the probability of heads is p, and you toss it repeatedly. The number Y of tails that occur before the rth head appears is a negative binomial random variable with parameters r and p, written $Y \sim \text{NB}(r, p)$. Of course, if $r = 1$, then Y could be said to be a geometric random variable with parameter p.

The negative binomial random variable takes nonnegative integer values. It's clear that $P(Y = 0) = p^r$, because in this case (no failures) we must have r consecutive successes. For the general formula, we consider that $Y = k$ if the rth success lands on the $(k+r)$th trial. Then we must have had $r - 1$ successes on the first $k + r - 1$ trials, as well as k failures. The order of the previous successes and failures is not important, so we consider the number of ways to fill in the blank spaces below with $r - 1$ successes and k failures:

$$\underbrace{\rule{1cm}{0.4pt} \ \rule{1cm}{0.4pt} \ \rule{1cm}{0.4pt} \ \rule{1cm}{0.4pt} \ \rule{1cm}{0.4pt} \ \rule{1cm}{0.4pt} \ \cdots \ \rule{1cm}{0.4pt}}_{r+k-1 \ \text{trials}} \ \underset{}{S}$$

There are $\binom{k + r - 1}{k}$ arrangements, and each arrangement has k failures and r successes (including the last success). We get the following formula. If $Y \sim \text{NB}(r, p)$ with $0 < p < 1$ and r a positive integer, then

$$P(Y = k) = \binom{k + r - 1}{k} p^r (1 - p)^k \quad \text{for} \ k = 0, 1, 2, \ldots.$$

The bar charts show the flexibility of the shape of the negative binomial distribution.

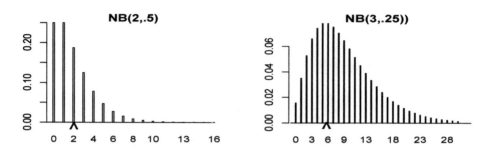

Instead of going through a lengthy calculation to compute the mean and variance of the negative binomial random variable, we will cheat and use some results that will be proved in later chapters, but are very intuitive.

We can think of a negative binomial random variable Y as the sum of independent geometric random variables. If Y_1 is the number of failures before the first success occurs, and Y_2 is the number of *subsequent* failures before the second success, Y_3 is the number of failures after the second success and before the third success, etc., then $Y_1 + Y_2 + \cdots + Y_r = Y$. We will see (in Chapter 24) that the expected value of the sum is the sum of the expected values, and, when random variables are independent, the same is true for the variances. Hence,

$$\mathrm{E}(Y) = \frac{r(1-p)}{p} \ \text{ and } \ \mathrm{V}(Y) = \frac{r(1-p)}{p^2}.$$

Example: Eddie wants to play this game with you: You repeatedly toss a fair die, and he will give you \$100 on the fourth time that you roll a six. However, you have to give him \$5 every time that you roll. What is your expected profit? Do you want to play?

Let Y be the number of nonsixes before the fourth six appears. The profit will be $W = 100 - 5(Y + 4)$ and $\mathrm{E}(W) = 80 - 5\mathrm{E}(Y)$. The random variable Y is negative binomial, with $p = 1/6$ and $r = 4$. Using the formula, we get $\mathrm{E}(Y) = 20$, so your expected profit is -20. Tell him you don't want to play!

Chapter Highlights

1. For repeated binomial trials, if Y is the number of failures before the rth success occurs, then $Y \sim \mathrm{NB}(r, p)$.

2. If $Y \sim \mathrm{NB}(r, p)$, then

$$P(Y = y) = \binom{y + r - 1}{y} p^r (1 - p)^y \ \text{ for } \ y = 0, 1, 2, \ldots.$$

3. If $Y \sim \mathrm{NB}(r, p)$, then

$$\mathrm{E}(Y) = \frac{r(1-p)}{p} \ \text{ and } \ \mathrm{V}(Y) = \frac{r(1-p)}{p^2}.$$

4. The R command `dnbinom(k,r,p)` R command will return $P(Y = k)$ when $Y \sim \mathrm{NB}(r, p)$, and the command `pnbinom(k,r,p)` will return $P(Y \le k)$.

Exercises

12.1 Eddie wants to play this game with you: You will repeatedly roll a pair of fair six-sided dice, and he will give you $100 the second time you roll "doubles." However, you have to pay him $8 before each roll of the pair of dice.

    (a) What are your expected winnings?

    (b) What is the probability of making a profit?

12.2 A certain type of cancer is fairly deadly. The probability that a patient with this cancer goes into remission after treatment is only .2. Suppose a team of clinicians wants to study the genetics of the patients going into remission. They will treat patients entering the clinic sequentially. What is the probability that they have to treat at least 20 patients before they find the 5th patient for the study?

12.3 A clinician is screening patients at a nursing home for a certain mild disease, in order to find three patients with the disease to participate in a study. Suppose 25% of the patients have the disease, and the clinician can screen two patients every hour. Assuming independence, what is the probability that the third patient with the disease is found in the fourth hour of testing?

12.4 Exosso clinic is conducting an experiment to compare treatments of bone spurs. When patients arrive for treatment, they are randomly assigned by rolling a fair six-sided die. If the die shows one or two, they get Treatment A. If the die shows three or four, they get Treatment B. If the die shows five or six, they get Treatment C.

    (a) What is the probability that at least three of the next five patients get assigned to Treatment A?

    (b) Suppose the patients come to the clinic sequentially. Dr. Osto decides to stop the experiment when there are five patients assigned to Treatment C. What is the expected (total) number of patients to get assigned to treatments in this experiment?

12.5 An institution has intermittent internet service, so that each time someone attempts to connect, the probability of actually connecting is only .25. The institution charges $2 per attempt, even if the attempt is unsuccessful. Suppose we can assume independence of attempts, that is, the probability of connecting does not depend on the results of other attempts.

    (a) Suppose you must get one connection. What is your expected expense?

    (b) You and your colleague both need a connection. What it the probability of spending more than $20 for the two connections?

12.6 Refer to Exercise 9.9. Can you solve this problem with negative binomial ideas, rather than simulations in R? If yes, solve the problem; if no, explain.

12.7 Refer to Exercise 9.13. Can you solve this problem with negative binomial ideas, rather than simulations in R? If yes, solve the problem; if no, explain.

Chapter 13

Continuous Random Variables and Density Functions

A **continuous random variable** takes values in an interval or set of intervals on the real line. Because this is an uncountable set, the distribution may not be specified by providing probabilities for the individual values; in fact, the probability that a continuous random variable takes on a specific value is always zero. Recall from the discussion of sigma algebras in Chapter 1 that we need only worry about specifying the probability that the random variable takes values in intervals of the real line. Integral calculus provides a nifty way to do this. For a continuous random variable Y, we specify a nonnegative **density function** $f_Y(y)$ whose domain is the real numbers; then the probability that the random variable takes on a value in an interval is the area under the density function, in that interval.

Specifically, a density function $f_Y(y)$ for the continuous random variable Y must have the following properties:

1. $0 \leq f_Y(y) < \infty$ for $y \in (-\infty, \infty)$;

2. $\int_{-\infty}^{\infty} f_Y(y)dy = 1$; and

3. $P(Y \in [a, b]) = \int_a^b f_Y(y)dy$ for $-\infty \leq a \leq b \leq \infty$.

Property 1 ensures that the density, and hence the probabilities, are nonnegative, and property 2 says that the total area under the density curve is 1. Property 3 is the formula for computing probabilities in intervals; these are the areas under the density curve in the interval. Another property of the probability density function that we will encounter in Chapter 15 ensures that

$$P(Y \in (a, b)) = P(Y \in (a, b]) = P(Y \in [a, b)) = P(Y \in [a, b]),$$

and we must have $P(Y = a) = P(Y \in [a, a]) = \int_a^a f_Y(y)dy = 0$.

The **support** of a continuous random variable is the set of points $y \in \mathbb{R}$ for which $f(y) > 0$.

For the density function below, the total area under the curve has to be one, and the probability that the random variable has values between 3 and 5 is shown as the shaded area. If we knew the functional form, we could integrate it between 3 and 5, to obtain the probability.

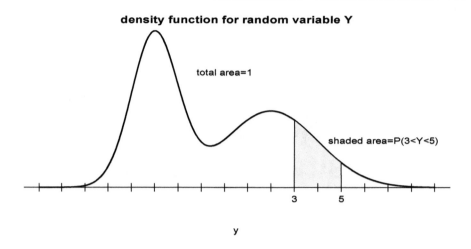

density function for random variable Y

Example: Suppose it is known by mechanical engineering principles that the density for the braking time in seconds for a certain model of car traveling at 60 miles per hour, under dry highway conditions, is given by the sketch below:

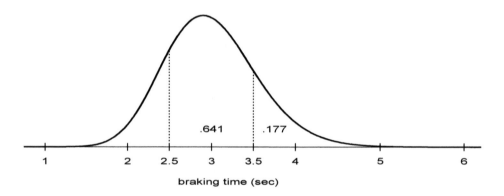

braking time (sec)

The proportion of braking times that are between 2.5 and 3.5 seconds is .641, and $.641 + .177 = .818$ is the proportion of braking times that are greater than 2.5 seconds. The proportion of braking times that are less than 2.5 seconds is found by ensuring that all of the proportions add to one: this is .182.

In general, if we know the functional form of the distribution, we use calculus to determine the area under the curve between any two values. There are a few types of densities for which we can compute areas without calculus, the simplest being the **uniform** density.

The density function for a **uniform random variable** is just a horizontal line between two values a and b, where $-\infty < a < b < \infty$, as shown in the figure below. The height must be $1/(b-a)$, because the area under the line has to be one. We say that a random variable is "uniformly distributed between the values a and b" if its density is constant on (a, b) and zero elsewhere. If a random variable Y has such a distribution, we use the notation $Y \sim \text{Unif}(a, b)$. To find the probability that Y is in a certain interval that is a subset of $[a, b]$, we use the formula for the area of a rectangle.

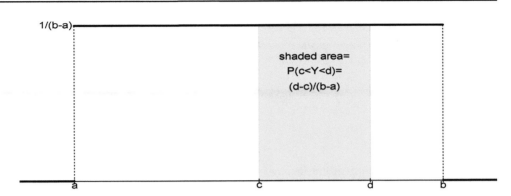

We learned in Chapter 2 that the R command `runif(n)` will return n independent replicates of a uniform random variable in the interval $(0, 1)$. More generally, the command `runif(n,a,b)` will return n independent replicates of a uniform random variable in the interval (a, b). Further, `punif(y,a,b)` will return $P(Y \leq y)$, for $Y \sim \text{Unif}(a, b)$. Of course, this is zero if $y \leq a$, and one if $y \geq b$.

To find areas under a **triangular density**, we can integrate a linear function, or we can use the formulas for the areas of triangles or trapezoids. For example, suppose a vending machine dispenses 8-ounce cups of coffee. The actual amount of coffee varies randomly, and we model the dispensed amount of coffee, in ounces, as following the density shown below:

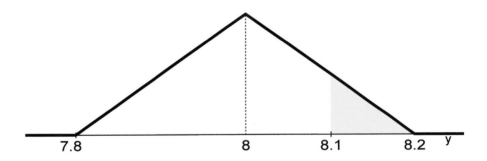

The proportion of cups of coffee that contain more than 8.1 ounces is shown by the shaded area. We'll look at two ways to compute this. First, we could determine the functional form of the density and integrate it between 8.1 and 8.2. To determine the density function, we first observe that the height of the triangle must be 5, to ensure that the area is one. The density function uses the equations of the lines forming the top of the triangle. Using some algebra we can calculate that

$$f(y) = \begin{cases} 25y - 195 & \text{for } y \in (7.8, 8), \\ 205 - 25y & \text{for } y \in (8, 8.2), \\ 0 & \text{otherwise.} \end{cases}$$

Then the shaded area is

$$\int_{8.1}^{8.2} f(y)dy = \int_{8.1}^{8.2} [205 - 25y]dy$$

$$= 205(8.2 - 8.1) - \frac{25}{2}(8.2^2 - 8.1^2) = .125.$$

An alternative way to do the computation is to use the formula for the area of a triangle. The height of the shaded triangle is half the total height, or 2.5 (using similar triangles). Then the proportion of cups of coffee that contain more than 8.1 ounces is

$$\frac{1}{2} \times \text{base} \times \text{height} = \frac{1}{2}(.1)(2.5) = .125,$$

much easier!

Example: Consider the density function

$$f_Y(y) = \begin{cases} ce^{-2y} & \text{for } y > 0 \\ 0, & \text{otherwise,} \end{cases}$$

where $c > 0$ is some constant. We will use the shorthand notation

$$f_Y(y) = ce^{-2y} \text{ for } y > 0,$$

where it is assumed that $f_Y(y) = 0$ for the unspecified ranges of y. We can find the value of c that makes f_Y a density function. We must have

$$1 = \int_0^\infty ce^{-2y} dy = -\frac{1}{2}ce^{-2y} \Big|_0^\infty = \frac{c}{2},$$

so $c = 2$. Suppose we model the length of time between serious accidents at a mining site, in years, as a random variable Y with density $f_Y(y)$. Now we can calculate quantities like the probability that the time between accidents is greater than one year:

$$P(Y > 1) = 2 \int_1^\infty e^{-2y} dy = e^{-2} \approx .1353.$$

The probability that the time between accidents is less than half a year is

$$P(Y < 1/2) = 2 \int_0^{1/2} e^{-2y} dy = 1 - e^{-1} \approx .6321;$$

these probabilities are shown as areas under the density in the following figure:

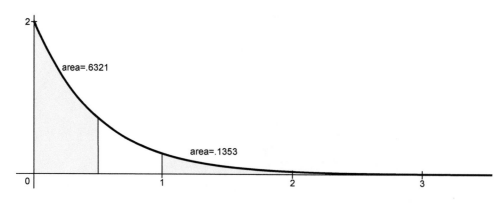

This density belongs to the family of exponential densities, which will be encountered again in Chapter 16.

Expected value of a continuous random variable

Let Y be a continuous random variable with density $f_Y(y)$. The expected value of Y is defined to be

$$E(Y) = \int_{-\infty}^{\infty} y f_Y(y) dy.$$

As with discrete random variables, the expected value is an average. If we imagine sampling from the distribution a "very large" number of times, the average of all these values will be the expected value.

Example: Let's find the expected value for the time between accidents using the density in the previous example about time between accidents:

$$E(Y) = \int_{-\infty}^{\infty} y f_Y(y) dy$$

$$= 2 \int_{0}^{\infty} y e^{-2y} dy$$

$$= \frac{1}{2} \int_{0}^{\infty} u e^{-u} du \qquad \text{(letting } u = 2y\text{)}$$

$$= \frac{1}{2},$$

using integration by parts or the gamma function (see Appendix B.5). The average length of time between accidents is half a year.

There is also a "balance point" interpretation for the mean of a continuous random variable. If we imagine the density cut out of sheet metal, the point of balance is the mean of the density. This is nice because if a density is symmetric, the mean is in the middle and we don't have to do any calculations. For skewed densities, we can check our calculation using our intuition about balance. A skewed density is shown below, with the mean marked; we can imagine the density balancing on this point.

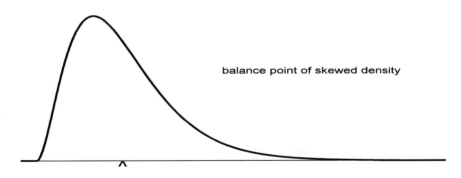

balance point of skewed density

Example: Suppose the density shown below represents the distribution of waiting times (in minutes) for a tram.

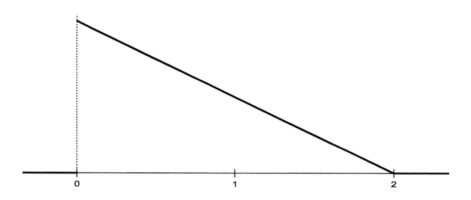

Let's find the expected value of the random variable. First, we find the equation of the line. The height of the density is one unit (to make the area equal to one), so the density is

$$f_X(x) = \begin{cases} 1 - x/2 & \text{for } x \in (0, 2), \\ 0 & \text{otherwise.} \end{cases}$$

(To simplify notation, we could write $f_X(x) = 1 - x/2$ for $x \in (0, 2)$, and it is understood that $f(x) = 0$ for $x \notin (0, 2)$.) Then if the random variable X follows the density $f_X(x)$, i.e., $X \sim f_X$, we have

$$\mathrm{E}(X) = \int_0^2 x(1 - x/2)dx = 2/3.$$

The average waiting time is 2/3 of a minute, or 40 seconds. If we mark the point $x = 2/3$ on the plot of the density, it seems reasonable that the density balances at this point.

Expected value of a function of a continuous random variable

Let Y be a continuous random variable with density $f_Y(y)$. The expected value of $g(Y)$ is defined to be

$$\mathrm{E}(g(Y)) = \int_{-\infty}^{\infty} g(y)f_Y(y)dy.$$

If the function is linear, we get a nice result analogous to that for discrete random variables:

$$\begin{aligned} \mathrm{E}(a + bY) &= \int_{-\infty}^{\infty} (a + by)f_Y(y)dy \\ &= a \int_{-\infty}^{\infty} f_Y(y)dy + b \int_{-\infty}^{\infty} y f_Y(y)dy \\ &= a + b\mathrm{E}(Y). \end{aligned}$$

Variance of a continuous random variable

If Y is a continuous random variable, its variance is

$$\mathrm{V}(Y) = \mathrm{E}((Y - \mu)^2) = \int_{-\infty}^{\infty} (y - \mu)^2 f_Y(y)dy,$$

where $\mu = E(Y)$. This has the same "measure of spread" interpretation as for the discrete case: The variance of a random variable is its expected squared distance from its mean.

To compute variances, we typically prefer to use the formula $V(Y) = E(Y^2) - \mu^2$, which is easily derived. In fact, the derivation is the same as for the discrete case (Chapter 5).

Example: Let's find the mean and variance of $Y \sim \text{Unif}(a, b)$. We can write the density as

$$f_Y(y) = \frac{1}{b-a} \text{ if } y \in (a, b)$$

(recalling our convention that $f_Y(y) = 0$ for values of y that are not in (a, b)). Then

$$E(Y) = \int_{-\infty}^{\infty} y f_Y(y) dy = \int_a^b \frac{y}{b-a} dy = \frac{1}{2}\frac{b^2 - a^2}{b-a} = \frac{a+b}{2},$$

which is what you probably guessed, given the balance point interpretation of the mean. To get the variance of Y, we use the formula $V(Y) = E(Y^2) - E(Y)^2$.

$$E(Y^2) = \int_{-\infty}^{\infty} y^2 f_Y(y) dy = \int_a^b y^2 \frac{1}{b-a} dy = \frac{1}{3}\frac{b^3 - a^3}{b-a} = \frac{a^2 + ab + b^2}{3},$$

so

$$V(Y) = \frac{a^2 + ab + b^2}{3} - \frac{(a+b)^2}{4} = \frac{(b-a)^2}{12}.$$

Example: Let's find the variance of the triangular density $f(x) = 1 - x/2$ for $x \in (0, 2)$, as shown in the above example with tram waiting times. First,

$$E(Y^2) = \int_0^2 y^2(1 - y/2) dy = 2/3,$$

and we have already calculated $E(Y) = 2/3$; so $V(Y) = 2/3 - (2/3)^2 = 2/9$.

Example: A hydrologist in Monsville models the maximum one-day rainfall Y (inches) in a randomly selected year as, following the density,

$$f_Y(y) = \frac{3}{2}\left(\frac{y}{2} + 1\right)^{-4} \text{ for } y \geq 0,$$

shown in the plot below.

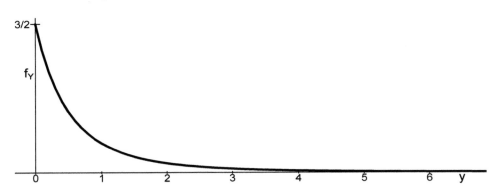

This is a special case of the Pareto density, which is used to model phenomena where the values taken tend to be mostly "small" but occasionally "really big." The probability that the maximum one-day rainfall in a randomly selected year is less than one inch is

$$P(Y < 1) = \int_0^1 \frac{3}{2} \left(\frac{y}{2} + 1 \right)^{-4} dy = 1 - \left(\frac{2}{3} \right)^3 \approx .704,$$

so in about 70.4% of years, the maximum rainfall in a day will not exceed one inch. Suppose the river running through Monsville will overflow its banks if the daily rainfall is four inches or more. What is the proportion of years that the river overflows its banks? This is

$$P(Y > 4) = \int_4^\infty \frac{3}{2} \left(\frac{y}{2} + 1 \right)^{-4} dy = \frac{1}{27} \approx .037,$$

so the river overflows its banks in about 3.7% of years.

Suppose there is severe flooding in Monsville if there is more than seven inches of rain in one day. What is the proportion of years with severe flooding?

$$P(Y > 7) = \int_7^\infty \frac{3}{2} \left(\frac{y}{2} + 1 \right)^{-4} dy = \frac{8}{729} \approx .011,$$

so this is approximately the 100-year flood level.

The average maximum one-day rainfall in Monsville is

$$E(Y) = \frac{3}{2} \int_0^\infty y \left(\frac{y}{2} + 1 \right)^{-4} dy$$

$$= 6 \int_1^\infty (u - 1) u^{-4} du \qquad \text{(transforming } u = y/2 + 1)$$

$$= 6 \int_1^\infty (u^{-3} - u^{-4}) du = 1.$$

The value $y = 1$ might not look like the balance point, but this density is "heavy-tailed"—that is, the tail goes to zero "slowly" (more slowly than exponentially), and the "mass" in the tail results in the balance point being pulled to the right.

Let's compute the variance next. First, we'll find $E(Y^2)$:

$$E(Y^2) = \frac{3}{2} \int_0^\infty y^2 \left(\frac{y}{2} + 1 \right)^{-4} dy$$

$$= 12 \int_1^\infty (u - 1)^2 u^{-4} du \qquad \text{(transforming } u = y/2 + 1)$$

$$= 12 \int_1^\infty (u^2 - 2u + 1) u^{-4} du = 4.$$

Then $V(Y) = 4 - 1 = 3$.

In Chapters 16 through 18, important families of densities commonly used in statistical modeling will be introduced.

Chapter Highlights

1. A continuous random variable Y takes values in an interval or set of intervals. Its distribution is described by a nonnegative density function $f_Y(y)$, where

$$P(Y \in (a, b)) = \int_a^b f_Y(y) dy$$

and $P(Y \in (-\infty, \infty)) = 1$.

2. The **support** of a continuous random variable is the set of points $y \in \mathbb{R}$ for which $f(y) > 0$.

3. A continuous random variable Y with density $f_Y(y)$ has expected value

$$E(Y) = \int_{-\infty}^{\infty} y f_Y(y) dy,$$

and for a function g of the random variable,

$$E[g(Y)] = \int_{-\infty}^{\infty} g(y) f_Y(y) dy.$$

4. A continuous random variable Y with density $f_Y(y)$ and expected value μ has variance
$$V(Y) = E[(Y - \mu)^2] = E(Y^2) - \mu^2.$$

5. A **uniform** random variable Y can have support $[a, b]$ for any $-\infty < a < b < \infty$. The density is
$$f_Y(y) = \frac{1}{b - a} \text{ for } a \leq y \leq b.$$

Also,

$$E(Y) = \frac{a + b}{2} \text{ and } V(Y) = \frac{(b - a)^2}{12}.$$

❧❧❧

Exercises

13.1 Let Y be a random variable representing waiting times (in minutes) for a tram, and suppose Y has the density shown below.

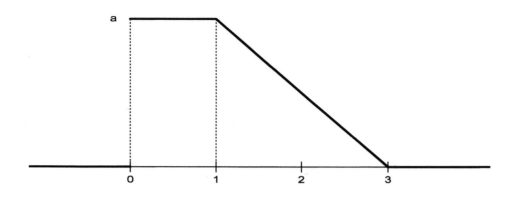

(a) Find the value of a.

(b) What proportion of waiting times are greater than 2 minutes?

(c) What is the median waiting time, that is, for what time t is the probability of waiting more than t minutes equal to 1/2?

(d) What is the mean waiting time?

(e) Suppose two people arrive at the tram stop, independently and "at random." What is the probability that they both have to wait more than 2 minutes?

13.2 Let Y be a random variable representing failure time of a switch in years. Suppose we can model the density of Y as in the plot below.

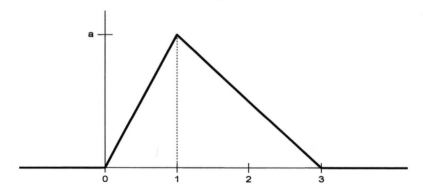

(a) Find the height a of the density.

(b) The probability that a randomly selected switch lasts longer than _____ is .1.

(c) Find the expected failure time.

13.3 The time to wait Y, in minutes, for service at a counter follows the triangular density below.

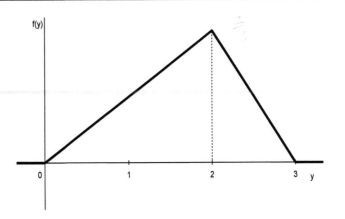

(a) Find $P(Y > 1)$.

(b) Find b so that $P(Y < b) = 1/2$.

(c) Find $E(Y)$.

(d) Only 5% of customers wait more than _____ minutes.

13.4 Suppose $Y \sim \text{Unif}(0, 2)$. Calculate $E(Y^3)$ and check your answer using R simulations.

13.5 Suppose $Y \sim \text{Unif}(a, b)$ for some $0 < a < b < \infty$. Calculate $E(1/Y)$ and check your answer using simulations in R, when $a = 1$ and $b = 4$.

13.6 A machine at a factory makes steel rods whose lengths are always between two known quantities a and b, which are set by the operator. However, the actual length varies randomly between a and b, and "middle" values are more likely.

The factory manager wants to model the distribution of these lengths as quadratic between a and b as shown in the plot on the right. Help him by providing the density function.

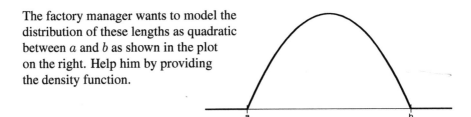

13.7 Let a and b be constants such that $0 < a < b < \infty$. The random variable Y has probability density

$$f(y) = ce^{-y} \text{ for } y \in (a, b).$$

(a) What is c?

(b) Sketch the density for $a = 1$ and $b = 3$.

(c) Is $E(Y)$ smaller than, greater than, or equal to $(a + b)/2$?

13.8 The random variable Y has probability density

$$f(y) = cye^{-y} \text{ for } y > 0.$$

(a) What is c?

(b) Sketch the density.

(c) What is E(Y)? (*Hint:* Remember your gamma function (Appendix B.5).)

(d) What is the mode of the density (i.e., the value of y for which $f(y)$ is largest)?

13.9 The random variable Y has probability density

$$f(y) = cye^{-y^2} \quad \text{for} \ y > 0.$$

(a) What is c?

(b) What is P($Y > 2$)?

(c) What is E(Y)? (*Hint:* Do a change of variable, and remember your gamma function (Appendix B.5).)

(d) Find V(Y).

13.10 Consider the function

$$f(x) = \frac{cx^\theta}{\sqrt{1 + x^{\theta+1}}} \quad \text{for } x \in (0, 1),$$

where $\theta > 0$ is an unknown parameter. Find c (in terms of θ) so that $f(x)$ is a density function.

13.11 The random variable Y has the density

$$f(y) = \frac{3}{(y+1)^4} \quad \text{for} \ 0 < y < \infty.$$

(a) Find the expected value of Y.

(b) Find the variance of Y.

13.12 The random variable Y has the density

$$f(y) = \frac{2}{(y+1)^3} \quad \text{for} \ 0 < y < \infty.$$

(a) Find the expected value of Y.

(b) Try to find the variance of Y and show that it does not exist.

13.13 The random variable Y has the density

$$f(y) = \frac{a}{(y+1)^{a+1}} \quad \text{for} \ 0 < y < \infty$$

for some $a > 2$.

(a) Verify that this is a valid density function.

(b) Find the mean of Y in terms of a.

(c) Find the variance of Y in terms of a.

(d) Find $E((Y+1)^2)$ in terms of a.

(e) Find $E(\log(Y))$ (*Hint:* You will need integration by parts and l'Hôpital's rule.)

13.14 For $a \in [-1, 1]$, the random variable Y has the density

$$f_Y(y) = \frac{3}{2}(ay+1), \quad y \in [-1, 1].$$

Find $E(Y)$ and $V(Y)$ (in terms of a).

13.15 Suppose Y is a continuous random variable having the density function

$$f_Y(y) = \frac{1}{\pi(1+y^2)} \quad \text{for} \quad y \in \mathbb{R}.$$

(a) Verify that this is a valid density function. (*Hint:* Use symmetry, and remember trig substitution!)

(b) Try to find the expected value of Y and show that it does not exist.

13.16 Suppose $\theta > 0$ and $Y_1, \ldots, Y_n$ is a random sample from the density

$$f_Y(y) = \frac{cy}{\theta} e^{-y^2/\theta} \quad \text{for} \quad y > 0.$$

Hint: Use substitution and the gamma function to do the following.

(a) Find the constant c to make this a proper density.

(b) Find $E(Y)$.

(c) Find $E(Y^2)$.

(d) Find $V(Y)$.

13.17 Suppose $\theta > 0$ and $Y_1, \ldots, Y_n$ is a random sample from the density

$$f(y) = cy^{\theta-1} \quad \text{for} \quad y \in (0, 1).$$

(a) Find the constant c (as a function of θ) to make this a proper density.

(b) Find $E(Y)$.

(c) Find $E(\log(Y))$. (*Hint:* Use integration by parts.)

13.18 The random variable Y has a "double-exponential" distribution

$$f_Y(y) = \frac{a}{2} e^{-a|y|}$$

for $y \in (-\infty, \infty)$. Find the expected value and the variance of Y (in terms of a).

13.19 The response time (in seconds) of a mechanism is modeled as a random variable Y with density

$$f_Y(y) = 3(1-y)^2 \quad \text{for} \quad y \in (0, 1).$$

(a) What is the probability that the response time is less than .2 seconds?

(b) What is the expected response time of the mechanism?

13.20 The profit from a purchase of one share of stock in a specific portfolio can be modeled as a random variable Y with density

$$f_Y(y) = c[4 - (y - 1)^2] \text{ for } y \in (-1, 3).$$

(a) Find c to make f_Y a proper density function.

(b) What is the probability that the profit is more than 2 units?

(c) What is $E(Y - 1)$?

13.21 Suppose $\theta > 0$ and $Y_1, \ldots, Y_n$ is a random sample from the density

$$f(y) = \frac{c\theta^4}{(y + \theta)^5} \text{ for } y > 0.$$

(a) Find the constant c to make this a proper density.

(b) Find $E(Y + \theta)$.

(c) Find $E(Y)$.

13.22 When airplanes land on aircraft carriers, the pilots have to be very precise about the landing place. Suppose that for a particular pilot the actual touchdown point is a (positive or negative) distance Y from a target line, in meters, where Y follows the density

$$f_Y(y) = \frac{ce^{-y}}{(1 + e^{-y})^2}, \quad y \in \mathbb{R},$$

where a negative distance means that the pilot touches down before the target line, and a positive distance means after the target line.

(a) Find the constant c that makes this function a proper density.

(b) Show that the density is symmetric.

(c) What is the probability that the pilot touches down more than two meters from the target (in either direction)?

(d) Find a distance a so that the probability that the pilot touches down within a meters from the target is 99%.

13.23 A farmer claims that organic tomatoes are higher in vitamin A than conventional tomatoes. Suppose we know that the vitamin A in a conventional tomato is distributed according to the plot below on the left.

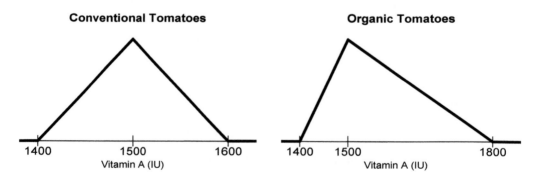

The farmer will randomly choose 4 organic tomatoes and determine the amount of vitamin A in each. He wants to test

H_0 : the distribution for vitamin A in organic tomatoes is the same as for conventional tomatoes

versus

H_a: organic tomatoes have a distribution of vitamin A that is, on average, higher than for conventional tomatoes.

Suppose the farmer's decision rule is "reject H_0 if all four tomatoes have higher than 1500 IU of vitamin A."

(a) What is the test size?

(b) Suppose the true distribution of vitamin A in organic tomatoes is distributed as in the right plot of the above figure. What is the power for the farmer's test?

(c) Suppose the farmer wanted the test size to be $\alpha = .02$. Then the decision rule would be "reject H_0 if all four tomatoes had more than _____ IU of vitamin A."

(d) What is the power for the decision rule in part (c) if the true distribution of vitamin A in organic tomatoes is distributed as in the right plot?

13.24 An ecotourism company buys batteries from Powerbright, Inc., for powering flashlights and other equipment. The lifetime Y, in days, of a randomly selected battery can be modeled using the following density:

$$f_Y(y) = \frac{1}{8}e^{-y/8} \text{ for } y > 0.$$

(a) Find the expected lifetime of a battery.

(b) If the company buys four (randomly selected) batteries, what is the probability that they all last at least seven days?

(c) Suppose Everlast, Inc., another battery manufacturer, claims that their batteries last longer, on average, than Powerbright's batteries. The company decides to test this claim, using a statistical hypothesis test. The null hypothesis is that the Everlast batteries have the same distribution as the Powerbright batteries, and the alternative is that the lifetimes of the Everlast batteries are, on average, longer.

The company obtains four (randomly selected) Everlast batteries, and measures their lifetimes. They will reject the null hypothesis if all four batteries last at least seven days. What is the size of this test?

(d) Suppose the true lifetimes of Everlast batteries follow the distribution

$$g_Y(y) = \frac{1}{12}e^{-y/12} \text{ for } y > 0.$$

What is the power of the test in part (c)?

Chapter 14

Moments and the Moment Generating Function

A **moment** is a fancy name for the expectation of the random variable raised to a positive integer. The mean of a random variable is its first moment, and for each positive integer n, the **nth moment** of a random variable Y is

$$\mu'_n = \mathrm{E}(Y^n).$$

(The prime notation is standard for moments, and does not mean derivative in this case.) If Y is discrete, the nth moment is the sum over the values in the support:

$$\mu'_n = \sum_{all\ y} y^n \mathrm{P}(Y = y).$$

If Y is continuous,

$$\mu'_n = \int_{-\infty}^{\infty} y^n f_Y(y) dy.$$

The **nth central moment** of Y is

$$\mu_n = \mathrm{E}[(Y - \mu)^n],$$

where $\mu = \mathrm{E}(Y)$, the first moment. The variance is the second central moment, and of course the first central moment is zero.

The **moment generating function**, or **MGF**, of a random variable Y is defined as

$$M_Y(t) = \mathrm{E}(e^{tY}).$$

The MGF is going to be very useful for deriving the distributions of transformations of random variables, for some important applications in statistics. For now we can just notice that the MGF generates moments (hence the name!).

First, we can notice that $M_Y(0) = 1$ for any random variable Y, because $M_Y(0)$ is just the expected value of one.

Next, take the first derivative $M'_Y(t)$. For discrete random variables, this is

$$M'_Y(t) = \frac{d}{dt} \sum_{all\ y} e^{ty}P(Y = y)$$

$$= \sum_{all\ y} \frac{d}{dt} e^{ty}P(Y = y)$$

$$= \sum_{all\ y} y e^{ty}P(Y = y)$$

for values of t for which the sum converges. Then $M'_Y(0) = E(Y)$.

Taking the second derivative,

$$M''_Y(t) = \frac{d^2}{dt^2} \sum_{all\ y} e^{ty}P(Y = y)$$

$$= \sum_{all\ y} \frac{d^2}{dt^2} e^{ty}P(Y = y)$$

$$= \sum_{all\ y} y^2 e^{ty}P(Y = y).$$

Then $M''_Y(0) = E(Y^2)$. In general,

$$M_Y^{(k)}(t) = \frac{d^k}{dt^k} \sum_{all\ y} e^{ty}P(Y = y)$$

$$= \sum_{all\ y} \frac{d^k}{dt^k} e^{ty}P(Y = y)$$

$$= \sum_{all\ y} y^k e^{ty}P(Y = y),$$

and $M_Y^{(k)}(0) = E(Y^k)$.

Example: Let's compute the MGF for a geometric random variable $Y \sim \text{Geom}(p)$:

$$M_Y(t) = E(e^{tY}) = \sum_{k=0}^{\infty} e^{tk}p(1 - p)^k$$

$$= p \sum_{k=0}^{\infty} [e^t(1 - p)]^k.$$

If t is "small" so that $0 < e^t(1 - p) < 1$ or $t < -\log(1 - p)$, then the sum is $(1 - e^t(1 - p))^{-1}$, and

$$M_Y(t) = \frac{p}{1 - e^t(1 - p)}.$$

Let's check to see that this MGF really does generate the moments for the geometric distribution. Using the chain rule,

$$M'_Y(t) = \frac{p(1 - p)e^t}{[1 - e^t(1 - p)]^2}.$$

Then we plug in zero:

$$M_Y'(0) = \frac{p(1-p)}{[1-(1-p)]^2} = \frac{1-p}{p} = \mathrm{E}(Y),$$

as advertised. Taking the second derivative using the quotient rule, we get (after some cancellation)

$$M_Y''(t) = \frac{p(1-p)e^t[1 + e^t(1-p)]}{[1 - e^t(1-p)]^3}.$$

Plugging in zero for t, we get

$$M_Y''(0) = \frac{(1-p)(1-2p)}{p^2} = \mathrm{E}(Y^2).$$

We can verify that this gives $\mathrm{V}(Y) = (1-p)/p^2$.

Example: Next, let's find the MGF for the binomial distribution. If $Y \sim \mathrm{Binom}(n,p)$, then

$$M_Y(t) = \mathrm{E}(e^{tY}) = \sum_{k=0}^{n} e^{tk} \binom{n}{k} p^k(1-p)^{n-k}$$

$$= \sum_{k=0}^{n} \binom{n}{k} (pe^t)^k (1-p)^{n-k}$$

$$= (pe^t + 1 - p)^n,$$

by the binomial theorem.

Let's take the first derivative of the MGF, to confirm that we get the mean. Using the chain rule,

$$M_Y'(t) = n(pe^t + 1 - p)^{n-1}(pe^t)$$

and

$$M_Y'(0) = np.$$

The MGF of a continuous random variable Y has the same definition as for the discrete case:

$$M_Y(t) = \mathrm{E}(e^{tY}) = \int_{-\infty}^{\infty} e^{ty} f_Y(y)dy.$$

The derivative is

$$M_Y'(t) = \frac{d}{dt} \int_{-\infty}^{\infty} e^{ty} f_Y(y)dy$$

$$= \int_{-\infty}^{\infty} \left[\frac{d}{dt} e^{ty}\right] f_Y(y)dy$$

$$= \int_{-\infty}^{\infty} ye^{ty} f_Y(y)dy,$$

so $M_Y'(0) = \mathrm{E}(Y)$, as is the case for a discrete random variable Y. We can similarly show that the kth moment of Y can be obtained by $M^{(k)}(0)$, showing that the name "moment generating function" is appropriate for the continuous random variable.

Example: Let's find the MGF of a uniform random variable. If $X \sim \text{Unif}(a, b)$ for $a < b$, then

$$M_X(t) = \frac{1}{b-a} \int_a^b e^{tx} dx$$
$$= \frac{e^{tb} - e^{ta}}{t(b-a)}.$$

Let's obtain the expected value of X for $a = 0$ and $b = 1$. The derivative is

$$M_X'(t) = \frac{d}{dt} \frac{e^t - 1}{t}$$
$$= \frac{te^t - e^t + 1}{t^2}.$$

When we plug in zero, the ratio is undefined; we have to take the limit as t goes to zero, using l'Hôpital's rule. Then

$$\lim_{t \to 0} \frac{te^t - e^t + 1}{t^2} = \lim_{t \to 0} \frac{te^t}{2t} = \frac{1}{2},$$

as expected.

Uniqueness property

An interesting and useful fact about the MGF is that it essentially defines a random variable. If two random variables have the same MGF, then they have the same distribution. This fact will be helpful later on when we are transforming random variables in order to make test statistics. One limitation of the usefulness is that not all random variables have closed-form MGFs, and for some random variables, the MGF is undefined.

Chapter Highlights

1. The **nth moment** of a random variable Y is

$$\mu_n' = \text{E}(Y^n).$$

2. The **nth central moment** of Y is

$$\mu_n = \text{E}[(Y - \mu)^n].$$

3. For a random variable Y, its MGF is

$$M_Y(t) = E(e^{ty}).$$

4. We have $M_Y(0) = 1$, and if the kth moment of the random variable is defined, then

$$M^{(k)}(0) = E(Y^k).$$

☙❧

Exercises

14.1 Show how to compute the moment generating function for the Poisson random variable.

14.2 Go through the steps to show that $M_Y''(0) = E(Y^2)$ if Y is a continuous random variable.

14.3 Find the moment generating function for a continuous random variable Y with density

$$f(y) = \frac{1}{\theta} e^{-y/\theta} \text{ for } y > 0,$$

where $\theta > 0$.

14.4 Find the moment generating function for a continuous random variable Y with density

$$f(y) = \frac{\theta}{2} e^{-\theta|y|},$$

where $\theta > 0$ and the support is the entire real line.

14.5 Let Y be a random variable with density $f_Y(y) = y/2$ for $y \in (0, 2)$. Find the moment generating function for Y.

14.6 Find the moment generating function for an $NB(2, p)$ random variable.

14.7 For $a > 0$, let Y be a random variable with density

$$f_Y(y) = e^{a-y} \text{ for } y \in (a, \infty)$$

and find the moment generating function for Y.

14.8 For $a > 0$, let Y be a random variable with density

$$f_Y(y) = \frac{e^{-y}}{1 - e^{-a}} \text{ for } y \in (0, a)$$

and find the moment generating function for Y.

Chapter 15

The Cumulative Distribution Function and Quantiles of Random Variables

Any random variable Y (e.g., continuous or discrete) has a cumulative distribution function (CDF) $F_Y(y)$ defined as

$$F_Y(y) = P(Y \le y).$$

For continuous random variables,

$$F_Y(y) = \int_{-\infty}^{y} f_Y(t)dt,$$

and for discrete random variables,

$$F_Y(y) = \sum_{all\ values\ t \le y} P(Y = t).$$

A continuous random variable can be defined as having a continuous CDF, while a discrete random variable can be defined as having a CDF that is a step function. Traditionally, capital letters are used for the CDF while the lower case is used for the density or mass function.

Example: The bee sting distribution of Exercise 5.3 has the mass function

y	0	1	2	3	4
$P(Y = y)$	0.6	0.2	0.1	0.08	0.02

The CDF is the step function shown below:

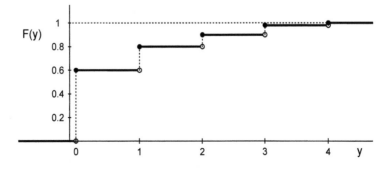

Note that, for example, $P(Y < 2) = .8$, but $P(Y \leq 2) = .9$.

Example: The CDF for a continuous random variable Y with uniform density on $(0, 1)$ is

$$F_Y(y) = \begin{cases} 0 & \text{for } y < 0, \\ y & \text{for } y \in [0, 1], \\ 1 & \text{for } y > 1. \end{cases}$$

The density and the CDF are compared in the plot—note that the density for a continuous random variable does not have to be continuous, but the CDF is continuous.

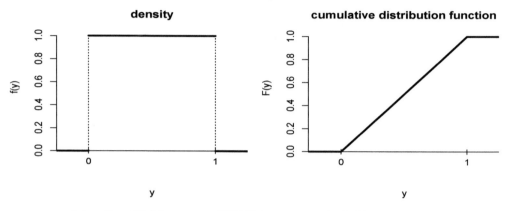

A function $F(y)$ is a proper CDF if these properties hold:

1. $F(y)$ is defined for all $y \in \mathbb{R}$;

2. $0 \leq F(y) \leq 1$ for all $y \in \mathbb{R}$;

3. F is nondecreasing; and

4. F is right continuous, that is,

$$\lim_{y \searrow y_0} F(y) = F(y_0).$$

Example: Suppose we want to find the CDF for the triangular density shown below:

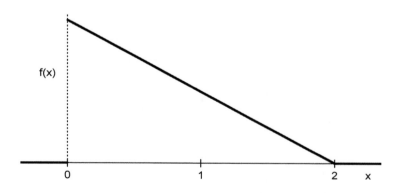

The density function is $f(x) = 1 - x/2$ on $(0, 2)$, so for $x \in (0, 2)$,

$$F(x) = \int_{-\infty}^{x} f(t)dt = \int_{0}^{x} (1 - t/2)dt = x - x^2/4.$$

We also have $F(x) = 0$ for $x \leq 0$ and $F(x) = 1$ for $x \geq 2$. The plot of the CDF is shown:

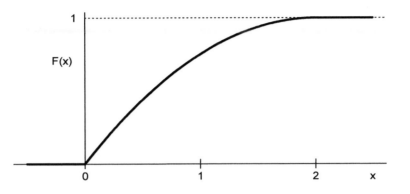

Quantiles and percentiles

For $q \in (0, 1)$, the qth quantile of a distribution with CDF F is the value of x such that $F(x) = q$. For $a \in (0, 100)$, the ath percentile is simply the $(a/100)$th quantile.

If we want to find the 50th percentile of the distribution in the above triangular density example, we can solve $F(x) = 1/2$; this involves a quadratic equation. We find the 50th percentile is about .5858. Notice that this is different from the expected value, which we previously found to be 2/3.

The 50th percentile of a random variable is also called the **median**. The 25th percentile is called the first quartile and the 75th percentile is the third quartile.

Example: If we want to find the median of a symmetric density such as the trapezoidal density shown below, we can simply choose the "middle" point—that is, the median is 1.5.

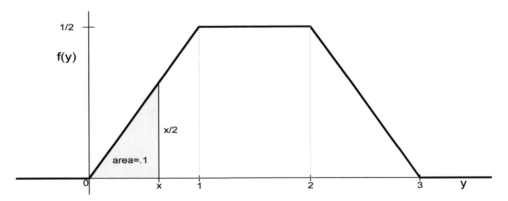

The quartiles are also easy using notions of areas of triangles and rectangles: The first quartile is 1 and the third is 2. The 10th percentile takes some calculation, however. The 25th percentile is at 1, so the tenth percentile must be in $(0, 1)$. Call this x, as shown on the plot. We need to find x so that the area to the left under the density (shaded in the plot) is equal to one tenth. Because the slope of the line segment forming the left part of the density is 1/2, the height of the shaded triangle must be $x/2$. Then the area (base times height divided by two) is $x^2/4$. Setting $x^2/4 = .1$ gives $x \approx .632$.

Chapter Highlights

1. The cumulative distribution function (CDF) of a random variable Y is defined as

$$F_Y(y) = P(Y \leq y).$$

2. A continuous random variable has a continuous CDF, and a discrete random variable has a CDF that is a step function.

3. The qth quantile of a random variable Y is the value c for which $F_Y(c) = q$ for $q \in (0,1)$.

Exercises

15.1 Find a formula for the CDF for $Y \sim \text{Geom}(p)$, and sketch it for your choice of p.

15.2 Suppose $a > 0$, and Y is a random variable such that $P(Y = 0) = 1/2$, and $P(Y = a) = P(Y = -a) = 1/4$. Determine the CDF for Y, and sketch it for your choice of a.

15.3 Let Y be a random variable representing waiting time for service, in minutes. Suppose we can model the density of Y as in the plot.

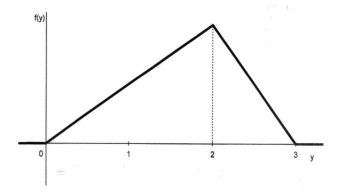

(a) Find the CDF $F_Y(y)$ and sketch it.

(b) What is the 10th percentile of waiting times?

15.4 Suppose $\theta > 0$ and the random variable Y is uniform on $(0, \theta)$. Find the CDF for Y and sketch it.

15.5 Suppose $\theta > 0$ and the random variable Y is uniform on $(\theta, 2\theta)$. Find the CDF for Y and sketch it.

15.6 Let Y be a random variable representing failure time of a switch in years. Suppose we can model the density of Y as in the plot.

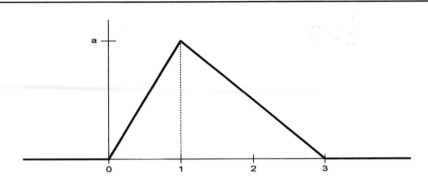

(a) Find the CDF of $F_Y(y)$ and sketch it.

(b) What is the 90th percentile of failure times?

15.7 Find the CDF for the density shown, and sketch it:

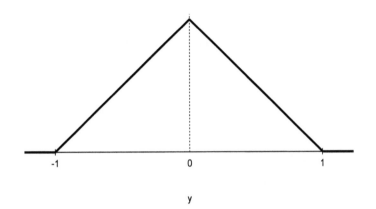

y

15.8 Let Y be a random variable representing waiting times (in minutes) for a tram, and suppose Y has the density shown below.

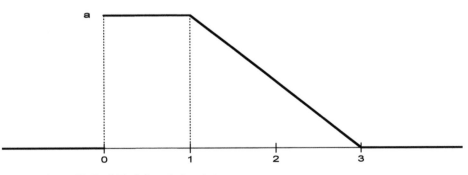

Find the CDF of $F_Y(y)$ and sketch it.

15.9 The random variable Y has the density

$$f_Y(y) = 2e^{-2y} \quad \text{for} \ y > 0.$$

(a) Sketch the density and its CDF.

(b) What is the 90th percentile of the density?

15.10 Suppose Y is a continuous random variable with density

$$f_Y(y) = ye^{-y^2/2} \ \text{ for } \ y > 0.$$

(a) Find the CDF for Y.

(b) Find the median (50th percentile) of the distribution.

15.11 Suppose Y is a random variable with density function

$$f_Y(y) = \frac{e^{-y}}{1 - e^{-3}} \ \text{ for } \ y \in [0, 3].$$

(a) Find the CDF for Y.

(b) Find the median (50th percentile) of the distribution.

15.12 The random variable X has a density that is a quadratic on $(0, 2)$ as shown below.

(a) Find the CDF for X.

(b) Find the median (50th percentile) of the distribution.

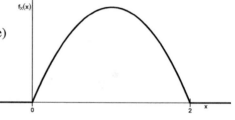

15.13 Find the CDF of the density given in Exercise 13.13.

15.14 Find the CDF of the density given in Exercise 13.22.

Chapter 16

The Exponential, Gamma, and Inverse Gamma Random Variables

An exponential random variable Y has nonnegative support (takes values $y \geq 0$), and has density function

$$f_Y(y) = \beta e^{-\beta y}, \quad y \geq 0,$$

for some parameter $\beta > 0$. We use the notation $Y \sim \text{Exp}(\beta)$ to indicate that the random variable Y has such a density. It is straightforward to verify that the density function integrates to one. The cumulative distribution function (CDF) is

$$F_Y(y) = \int_{-\infty}^{y} f_Y(t)dt = \beta \int_{0}^{y} e^{-\beta t}dt = 1 - e^{-\beta y} \quad \text{for } y > 0,$$

and $F_Y(y) = 0$ for $y \leq 0$. Further, $F_Y(y)$ increases to one as y gets larger. The density and CDF for an exponential random variable with $\beta = 2$ are shown:

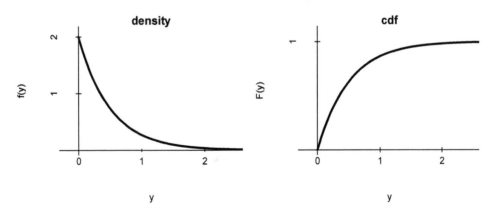

The exponential is a good model for a random variable that takes positive values that are likely to be "small" but occasionally are "large." We will see that for sequences of events where the counts of events in a given amount of time can be modeled as Poisson, the time between events can be modeled as exponential. For example, the time between sequential emissions of gamma rays from a radioactive source fits an exponential distribution quite well, while the number of emissions in a given length of time follows a Poisson distribution.

Let's find the expected value of $Y \sim \text{Exp}(\beta)$:

$$E(Y) = \int_0^\infty y\beta e^{-\beta y} dy$$

$$= \frac{1}{\beta} \int_0^\infty x e^{-x} dx$$

$$= \frac{1}{\beta}\Gamma(2) = \frac{1}{\beta},$$

because $\Gamma(2) = 1! = 1$. In the second step, we used the change of variable $x = \beta y$. (For a review of the gamma function, see Appendix B.5.)

To compute the variance, we use $V(Y) = E(Y^2) - E(Y)^2$, and compute

$$E(Y^2) = \int_0^\infty y^2 \beta e^{-\beta y} dy$$

$$= \frac{1}{\beta^2} \int_0^\infty x^2 e^{-x} dx$$

$$= \frac{1}{\beta^2}\Gamma(3) = \frac{2}{\beta^2}.$$

Then $V(Y) = \frac{2}{\beta^2} - \frac{1}{\beta^2} = \frac{1}{\beta^2}$. For the exponential random variable, the standard deviation is the same as the mean.

Next, we consider the moment generating function (MGF) for $Y \sim \text{Exp}(\beta)$:

$$M_Y(t) = E(e^{tY}) = \int_0^\infty e^{ty}\beta e^{-\beta y} dy$$

$$= \beta \int_0^\infty e^{-y(\beta - t)} dy$$

$$= \frac{\beta}{\beta - t} \int_0^\infty e^{-x} dx$$

$$= \frac{\beta}{\beta - t}.$$

We have used the change of variable $x = y(\beta - t)$ and assumed that $t < \beta$. For an $\text{Exp}(\beta)$ random variable, the MGF is defined only for $t < \beta$, but we need the MGF (and its derivatives) only on an open interval containing zero.

Example: A computing software company has determined that the lengths of calls (in minutes) to the customer service department follow an exponential distribution with mean 20. To find the proportion of calls that last for more than one hour, we integrate the density from 60 to infinity. That is, if $X \sim \text{Exp}(1/20)$, then

$$P(X > 60) = \int_{60}^\infty \left[\frac{1}{20}e^{x/20}\right] dx = \int_3^\infty e^{-u} du = e^{-3}.$$

Approximately 5% of calls are one hour or longer. Another solution is to use the CDF:

$$P(X > 60) = 1 - P(X \le 60) = 1 - F(60) = 1 - (1 - e^{-60/20}) = e^{-3}.$$

An exponential distribution is *memoryless*, which means that if $Y \sim \text{Exp}(\beta)$, then

$$P(Y > y + t | Y > t) = P(Y > y)$$

for all nonnegative y and t. To prove this, we can use our definition of conditional probability to show

$$P(Y > y + t | Y > t) = \frac{P(Y > y + t \text{ and } Y > t)}{P(Y > t)}$$

$$= \frac{P(Y > y + t)}{P(Y > t)}$$

$$= \frac{e^{-\beta(y+t)}}{e^{-\beta t}}$$

$$= e^{-\beta y} = P(Y > y).$$

This memoryless property might imply that this distribution is *not* a good model for lifetimes, in cases where the object or organism "wears out." If a lifetime of a switch, say, has an exponential distribution, then the probability of it lasting for more than 50 units of time is the same as the probability of it lasting more than 150 units of time, given that it is already 100 units old. If the switch is more likely to fail as it ages, then another distribution should be used to model the lifetimes.

Example: The lifetimes of a certain brand of electric switch are distributed as exponential with mean .32 years (the lifetimes are "memoryless"). A sales representative for a different brand claims that their switches last longer. A consumer buys two of the new brand to test the null hypothesis that $1/\beta = .32$ versus $1/\beta > .32$, where $1/\beta$ is the average lifetime of the new switch. He decides to reject the null hypothesis if both switches last more than .5 years (and he assumes that the lifetimes of the new switches also are exponential random variables). What is the test size α?

We must compute the probability of rejecting H_0 (both switches last more than .5 years) when H_0 is true ($\beta = .32$). If the lifetimes are independent, then we can multiply the individual probabilities of the switches lasting more than .5 years, to get

$$\alpha = (e^{-.5/.32})^2 = .0439.$$

The parameter β is called the **rate** parameter of the distribution. The **scale** parameter is $\theta = 1/\beta$, which is also the mean of the distribution. In R, the default way to specify an exponential distribution is with the rate parameter. That is, `pexp(3,1/4)` will return the area to the left of 3, under an exponential density with rate $\beta = 1/4$ and mean $\theta = 4$.

A random variable Y has a **gamma distribution** with parameters $\alpha > 0$ and $\beta > 0$ if Y has nonnegative support and density

$$f_Y(y) = \frac{\beta^\alpha}{\Gamma(\alpha)} y^{\alpha-1} e^{-\beta y}, \quad y \geq 0.$$

It's clear that $f_Y(y) \geq 0$ for all y in the support, but we need to show that the density integrates to one. In the calculations below, we make the change of variable $x = \beta y$:

$$\int_{-\infty}^{\infty} f_Y(y) dy = \frac{\beta^\alpha}{\Gamma(\alpha)} \int_0^\infty y^{\alpha-1} e^{-\beta y} dy$$

$$= \frac{\beta^\alpha}{\Gamma(\alpha)} \int_0^\infty \left(\frac{x}{\beta}\right)^{\alpha-1} e^{-x} \frac{dx}{\beta}$$

$$= \frac{1}{\Gamma(\alpha)} \int_0^\infty x^{\alpha-1} e^{-x} dx$$

$$= 1,$$

by the definition of the gamma function (see Appendix B.5 for review).

The family of gamma densities is quite flexible because it has two parameters. The gamma density has a vertical asymptote at zero if $0 < \alpha < 1$, has a y-intercept of β when $\alpha = 1$, and has a "bump" when $\alpha > 1$. For each of the gamma densities below, you can point to the mean α/β and see that this is also the balance point for the density.

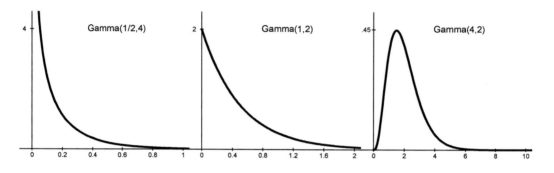

The parameter α is called the *shape* parameter, and β is the *rate* parameter. Of course, if $\alpha = 1$, we get an exponential density with rate β, so the exponential distribution is a special case of the gamma distribution. Let's find the expected value for other values of α.

For $Y \sim \text{Gamma}(\alpha, \beta)$, starting with the definition of expected value, we have

$$
\begin{aligned}
E(Y) &= \int_{-\infty}^{\infty} y f_Y(y) dy = \frac{\beta^\alpha}{\Gamma(\alpha)} \int_0^\infty y^\alpha e^{-\beta y} dy \\
&= \frac{\Gamma(\alpha+1)\beta^\alpha}{\Gamma(\alpha)\beta^{\alpha+1}} \left[\frac{1}{\Gamma(\alpha+1)\beta^\alpha} \int_0^\infty y^\alpha e^{-\beta y} dy \right] \\
&= \frac{\Gamma(\alpha+1)\beta^\alpha}{\Gamma(\alpha)\beta^{\alpha+1}} \\
&= \alpha/\beta,
\end{aligned}
$$

because the term in the brackets is the integral of a Gamma($\alpha + 1, \beta$) density over its support. We also used an identity from Appendix B.5: $\Gamma(\alpha + 1) = \alpha\Gamma(\alpha)$.

To get the variance, we first get $E(Y^2)$:

$$
\begin{aligned}
E(Y^2) &= \frac{\beta^\alpha}{\Gamma(\alpha)} \int_0^\infty y^{\alpha+1} e^{-\beta y} dy \\
&= \frac{\Gamma(\alpha+2)\beta^\alpha}{\Gamma(\alpha)\beta^{\alpha+2}} \left[\frac{\beta^{\alpha+2}}{\Gamma(\alpha+2)} \int_0^\infty y^{\alpha+1} e^{-\beta y} dy \right] \\
&= \frac{\Gamma(\alpha+2)\beta^\alpha}{\Gamma(\alpha)\beta^{\alpha+2}} \\
&= \frac{(\alpha+1)\alpha}{\beta^2},
\end{aligned}
$$

noting that the term in the brackets is the integral of a Gamma($\alpha + 2, \beta$) density. Then

$$
V(Y) = \frac{(\alpha+1)\alpha}{\beta^2} - \left(\frac{\alpha}{\beta}\right)^2 = \frac{\alpha}{\beta^2}.
$$

Next we find the MGF. If $Y \sim \text{Gamma}(\alpha, \beta)$, then

$$M_Y(t) = \text{E}(e^{tY}) = \frac{\beta^\alpha}{\Gamma(\alpha)} \int_0^\infty e^{ty} y^{\alpha-1} e^{-\beta y} dy$$

$$= \frac{\beta^\alpha}{\Gamma(\alpha)} \int_0^\infty y^{\alpha-1} e^{-y(\beta-t)} dy.$$

To simplify, let $b = \beta - t$ (and again assume $t < \beta$), so

$$M_Y(t) = \frac{\beta^\alpha}{b^\alpha} \left[\frac{b^\alpha}{\Gamma(\alpha)} \int_0^\infty y^{\alpha-1} e^{-by} dy \right]$$

$$= \left(\frac{\beta}{\beta - t} \right)^\alpha,$$

because the term in the brackets is the integral of a $\text{Gamma}(\alpha, b)$ density over its support.

Example: If $Y \sim \text{Gamma}(4, 1/2)$, find $\text{P}(Y > 10)$. We can write

$$\text{P}(Y > 10) = \frac{1}{96} \int_{10}^\infty y^3 e^{-y/2} dy,$$

but we can't do the integral by finding an antiderivative. Instead, we have to do repeated integration by parts, or numerical integration, or we could use the pgamma function in R, which finds areas under gamma densities. The command

pgamma(10,4,1/2)

returns the area to the *left* of 10, under a $\text{Gamma}(4, 1/2)$ density. So we want

1-pgamma(10,4,1/2)

which gives about .265.

An important subfamily of the gamma family of random variables are the **chi-squared** random variables. If $X \sim \text{Gamma}(\nu/2, 1/2)$, then we say $X \sim \chi^2(\nu)$, for integers $\nu = 1, 2, \ldots$. We can see that for $X \sim \chi^2(\nu)$, $\text{E}(X) = \nu$ and $\text{V}(X) = 2\nu$. This distribution is used in many statistics applications, and we will encounter it again in future chapters.

A random variable Y has an **inverse gamma distribution** with parameters $\alpha > 0$ and $\beta > 0$ if Y has nonnegative support and density

$$f_Y(y) = \frac{\beta^\alpha}{\Gamma(\alpha)} y^{-\alpha-1} e^{-\beta/y}, \quad y > 0.$$

If Y has such a distribution, we write $Y \sim \text{InvGamma}(\alpha, \beta)$.

To show that the density integrates to one, we use the gamma function ideas again. In the calculations below, we make the change of variable $x = \beta/y$, so $dx = -(\beta/y^2)dy$ and $dy = -(\beta/x^2)dx$:

$$\int_{-\infty}^{\infty} f_Y(y)dy = \frac{\beta^\alpha}{\Gamma(\alpha)} \int_0^\infty y^{-\alpha-1} e^{-\beta/y} dy$$

$$= -\frac{\beta^\alpha}{\Gamma(\alpha)} \int_\infty^0 \left(\frac{x}{\beta}\right)^{\alpha+1} e^{-x} \frac{\beta}{x^2} dx$$

$$= \frac{1}{\Gamma(\alpha)} \int_0^\infty x^{\alpha-1} e^{-x} dx$$

$$= 1.$$

Note that the negative sign got "canceled" by reversing the integration limits.

The family of inverse gamma densities is also quite flexible, several examples are shown below.

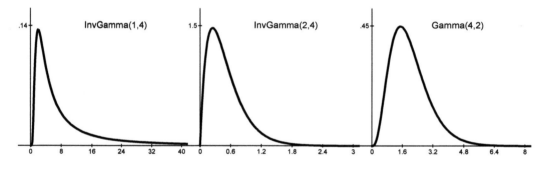

Exercise 16.15 asks you to find the expected value and the variance of an inverse gamma random variable. The areas under the curve can be found using the `pigamma` function in the R package `pscl`. The statement `pigamma(c,a,b)` returns the area to the left of c, under an InvGamma(a,b) density.

Neither the CDF nor the MGF exist in closed form for the inverse gamma distribution. Quantiles can be found using the R function `qigamma(q,a,b)`, which returns the value c for which the area to the left of c, under an InvGamma(a,b) density, is q.

Chapter Highlights

1. An **exponential random variable** Y with rate $\beta > 0$ has density

$$f_Y(y) = \beta e^{-\beta y} \text{ for } y > 0,$$

(and zero otherwise) and CDF

$$F_Y(y) = 1 - e^{-\beta y} \text{ for } y > 0.$$

The expected value is $1/\beta$ and the variance is $1/\beta^2$.

2. The exponential random variable is alternatively specified with mean $\theta > 0$. Then the density can be written

$$f_Y(y) = \frac{1}{\theta} e^{-y/\theta} \text{ for } y > 0,$$

(and zero otherwise) and CDF

$$F_Y(y) = 1 - e^{-y/\theta} \text{ for } y > 0.$$

The expected value is θ and the variance is θ^2. (This specification is more popular for textbooks, but we use the inverse mean as the default parameter, to match the default specification in R.)

3. The MGF for the exponential random variable with rate β is

$$M_Y(t) = \frac{\beta}{\beta - t}.$$

4. A **gamma random variable** Y with parameters $\alpha > 0$ and $\beta > 0$ has density

$$f_Y(y) = \frac{\beta^\alpha}{\Gamma(\alpha)} y^{\alpha-1} e^{-\beta y}, \ \ y \geq 0$$

(and zero otherwise). There is no closed form for the CDF, but the mean and variance are

$$E(Y) = \frac{\alpha}{\beta} \text{ and } V(Y) = \frac{\alpha}{\beta^2}.$$

5. The density function gives us this useful integration formula:

$$\int_0^\infty y^{\alpha-1} e^{-\beta y} dy = \frac{\Gamma(\alpha)}{\beta^\alpha}.$$

6. The MGF for the gamma random variable is

$$M_Y(t) = \left(\frac{\beta}{\beta - t} \right)^\alpha.$$

7. A special subfamily of the gamma random variables is the family of chi-squared random variables. We say $X \sim \chi^2(\nu)$ if $X \sim \text{Gamma}(\nu/2, 1/2)$.

8. An **inverse gamma random variable** Y with parameters $\alpha > 0$ and $\beta > 0$ has density

$$f_Y(y) = \frac{\beta^\alpha}{\Gamma(\alpha)} y^{-\alpha-1} e^{-\beta/y}, \ \ y \geq 0$$

(and zero otherwise). There is no closed form for the CDF, but the mean (if $\alpha > 1$) and variance (if $\alpha > 2$) are

$$E(Y) = \frac{\beta}{\alpha - 1} \text{ and } V(Y) = \frac{\beta^2}{(\alpha - 1)^2 (\alpha - 2)}.$$

9. These are the corresponding R functions:

- The function dexp(y,beta) will return $f_Y(y)$, where $Y\sim$Exp(beta).

- The function pexp(y,beta) will return $P(Y \le y)$, where $Y\sim$Exp(beta).

- The function qexp(q,beta) will return the value of c for which $P(Y \le c) = q$, where $Y\sim$Exp(beta).

- The function rexp(n,beta) will return n independent replicates from an Exp(beta) distribution.

- The function dgamma(y,alpha,beta) will return $f_Y(y)$, where $Y \sim$ Gamma(alpha,beta).

- The function pgamma(y,alpha,beta) will return $P(Y \le y)$, where $Y \sim$ Gamma(alpha,beta).

- The function qgamma(c,alpha,beta) will return the value of c for which $P(Y \le c) = q$, where $Y\sim$Gamma(alpha,beta).

- The function rgamma(n,alpha,beta) will return n independent replicates from a Gamma(alpha,beta) distribution.

- The function dchisq(y,k) will return $f_Y(y)$, where $Y\sim \chi^2(k)$.

- The functionpchisq(y,k) will return $P(Y \le y)$, where $Y\sim \chi^2(k)$.

- The function qchisq(q,k) will return the value of c for which $P(Y \le c) = q$, where $Y\sim \chi^2(k)$.

- The function rchisq(n,k) will return n independent replicates from a $\chi^2(k)$ distribution.

If the pscl library is installed, we have the following:

- The function densigamma(y,alpha,beta) will return $f_Y(y)$, where $Y\sim$InvGamma(alpha,beta). (The function digamma does something entirely different.)

- The function pigamma(y,alpha,beta) will return $P(Y \le y)$, where $Y\sim$InvGamma(alpha,beta).

- The function qigamma(q,alpha,beta) will return the value of c for which $P(Y \le c) = q$, where $Y\sim$InvGamma(alpha,beta).

- The function rigamma(n,alpha,beta) will return n independent replicates from an InvGamma(alpha,beta) distribution.

❧❧❧

Exercises

16.1 Find the median of a random variable having an exponential density with mean θ.

16.2 Suppose Y is an exponential random variable with mean 1.5. What is $E(Y^4)$?

16.3 Suppose Y is an exponential random variable with mean θ. Find an expression for $E(Y^n)$. Verify your answer using R simulations, for $n = 4$ and $\theta = 2$.

16.4 Suppose Y is a Gamma(α, β) random variable. Find an expression for $E(Y^n)$. Verify your answer using R simulations for $n = 4$, $\alpha = 3$, and $\beta = 1$.

16.5 In R, plot some gamma densities and mark the mean on the plot. For example, to plot a Gamma$(2, 1/3)$ random variable, try

```
xpl=0:2500/100
d1=dgamma(xpl,2,1/3)
plot(xpl,d1,type="l")
points(6,0,pch=20,cex=2)
```

Now change the parameters to see other shapes of the gamma density. You might have to change the range of the `xpl` argument to get a plot that captures the features of the distribution.

16.6 Compute the mode of $Y \sim$ Gamma(α, β), where $\alpha > 0$ and $\beta > 0$. (The mode is the value of y that maximizes the density function.) *Hint:* Be careful about the value of α.

16.7 Compute the integral

$$\int_0^\infty x^8 e^{-4x} dx.$$

16.8 The weights Y, in milligrams, of experimental E. coli populations one day after feeding follow a density

$$f_Y(y) = cy^2 e^{-6y} \text{ for } y > 0.$$

Find the constant c to make this function a proper density.

16.9 Supplier A makes light bulbs whose lifetimes are known to follow an exponential density with mean 56 days. Supplier B claims their light bulbs will last longer. The manager of a building decides to test this claim by using one light bulb from Supplier B and measuring its lifetime. The null hypothesis is that $\theta = 56$ and the alternative is $\theta > 56$, where θ is the true mean for light bulbs from Supplier B. Suppose the Supplier B bulb lifetimes also follow an exponential density. The manager wants to use the decision rule "reject H_0 if the lifetime is greater than c days." If the test size is $\alpha = .05$, what is c? If $\theta = 72$, what is the power of this test?

16.10 (Continuation of Exercise 16.9) The power for the test is too small, so the manager decided to get a larger sample size. This time, he will obtain five light bulbs from Supplier B, and reject H_0 if the mean lifetime is greater than 100 days. We don't know yet how to find the distribution of the average of exponential lifetimes, but you can find the test size using simulated data in R. Also find the power of this test if $\theta = 72$, and compare to your answer for Exercise 16.9.

16.11 A Gamma$(\nu/2, 1/2)$ density is called a "chi-squared density with ν degrees of freedom." Take derivatives of the MGF of a Gamma$(\nu/2, 1/2)$ density to find the mean and variance of the chi-squared density with ν degrees of freedom.

16.12 Suppose $Y \sim$ Gamma(α, β) for some $\alpha > 0$ and $\beta > 1$. Find $E(e^Y)$.

16.13 Suppose Y is an exponential random variable with mean θ. Find $V(Y^2)$.

16.14 Suppose $Y \sim \text{Gamma}(n, \beta)$ for an integer $n > 2$ and $\beta > 0$. Find $V(1/Y)$.

16.15 Suppose $Y \sim \text{InvGamma}(\alpha, \beta)$.

   (a) Assume $\alpha > 1$ and derive $E(Y)$.

   (b) Assume $\alpha > 2$ and derive $V(Y)$.

Chapter 17

The Beta Random Variable

The beta distribution has two parameters and its support is the unit interval. We say $Y \sim \text{Beta}(\alpha, \beta)$, where $\alpha > 0$ and $\beta > 0$ if Y takes values in $(0,1)$ and has density

$$f_Y(y) = \frac{\Gamma(\alpha + \beta)}{\Gamma(\alpha)\Gamma(\beta)} y^{\alpha-1}(1-y)^{\beta-1} \text{ for } y \in (0,1).$$

It's a little tricky to show that this density integrates to one. We have to do a transformation of variables in multiple integration using a Jacobian, but we'll need this for Chapter 30 so we may as well review it here. A more detailed review can be found in Appendix B.9.

We start with the definition of the gamma function:

$$\Gamma(\beta) = \int_0^\infty y^{\beta-1} e^{-y} dy,$$

so

$$\Gamma(\alpha)\Gamma(\beta) = \int_0^\infty \int_0^\infty x^{\alpha-1} y^{\beta-1} e^{-(x+y)} dx dy.$$

Now we define a change of variables $x = uv$ and $y = u(1-v)$, so that the inverse transformation is $u = x + y$ and $v = x/(x+y)$. We notice that when (x, y) range over the positive quadrant of the reals, the corresponding (u, v) values are in a strip where u can be any positive number but $v \in (0, 1)$. The Jacobian is

$$\begin{vmatrix} \frac{\partial x}{\partial u} & \frac{\partial x}{\partial v} \\ \frac{\partial y}{\partial u} & \frac{\partial y}{\partial v} \end{vmatrix} = \begin{vmatrix} v & u \\ 1-v & -u \end{vmatrix} = -u,$$

and the absolute value of the Jacobian is u.

Continuing with the change of variables,

$$\Gamma(\alpha)\Gamma(\beta) = \int_0^\infty \int_0^1 (uv)^{\alpha-1} \left[u(1-v)\right]^{\beta-1} e^{-u} u \, du \, dv$$

$$= \int_0^\infty \int_0^1 u^{\alpha+\beta-1} v^{\alpha-1}(1-v)^{\beta-1} e^{-u} du \, dv$$

$$= \int_0^\infty u^{\alpha+\beta-1} e^{-u} du \int_0^1 v^{\alpha-1}(1-v)^{\beta-1} dv$$

$$= \Gamma(\alpha + \beta) \int_0^1 v^{\alpha-1}(1-v)^{\beta-1} dv,$$

so finally,

$$\int_0^1 v^{\alpha-1}(1-v)^{\beta-1}dv = \frac{\Gamma(\alpha)\Gamma(\beta)}{\Gamma(\alpha+\beta)}.$$

This shows that the beta density integrates to one. It is also a nice integration formula that we can use in the derivation of the moments of the beta random variable.

Let's find the mean of $Y \sim \text{Beta}(\alpha, \beta)$. By definition, we have

$$\text{E}(Y) = \frac{\Gamma(\alpha+\beta)}{\Gamma(\alpha)\Gamma(\beta)} \int_0^1 y^\alpha (1-y)^{\beta-1}dy.$$

Using the nice integration formula,

$$\begin{aligned}
\text{E}(Y) &= \frac{\Gamma(\alpha+\beta)}{\Gamma(\alpha)\Gamma(\beta)} \frac{\Gamma(\alpha+1)\Gamma(\beta)}{\Gamma(\alpha+\beta+1)} \\
&= \frac{\Gamma(\alpha+\beta)}{\Gamma(\alpha+\beta+1)} \frac{\Gamma(\alpha+1)}{\Gamma(\alpha)} \\
&= \frac{\alpha}{\alpha+\beta},
\end{aligned}$$

where we used the identity $\Gamma(a+1) = a\Gamma(a)$ twice. Similarly,

$$\begin{aligned}
\text{E}(Y^2) &= \frac{\Gamma(\alpha+\beta)}{\Gamma(\alpha)\Gamma(\beta)} \int_0^1 y^{\alpha+1}(1-y)^{\beta-1}dy \\
&= \frac{\Gamma(\alpha+\beta)}{\Gamma(\alpha+\beta+2)} \frac{\Gamma(\alpha+2)}{\Gamma(\alpha)} = \frac{(\alpha+1)\alpha}{(\alpha+\beta+1)(\alpha+\beta)}.
\end{aligned}$$

Now we can find the variance:

$$\text{V}(Y) = \frac{(\alpha+1)\alpha}{(\alpha+\beta+1)(\alpha+\beta)} - \left(\frac{\alpha}{\alpha+\beta}\right)^2 = \frac{\alpha\beta}{(\alpha+\beta+1)(\alpha+\beta)^2}.$$

There's no closed form for the cumulative distribution function or for the moment generating function, for the beta random variable.

It is easy to verify that the $\text{Beta}(1,1)$ distribution coincides with the $\text{Unif}(0,1)$ distribution. If $\alpha > 1$, then the value of the density at $y = 0$ is zero, and if $\alpha < 1$, then there is an asymptote at zero. Similarly, if $\beta > 1$, then the value of the density at $y = 1$ is zero, and there is an asymptote if $\beta < 1$. Some examples of beta densities are given in the plots below.

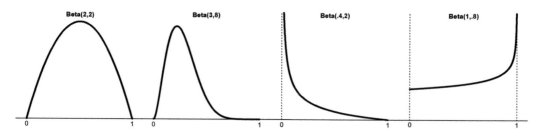

You can explore the flexibility of the family of beta distributions in R by plotting the densities using the dbeta function. For example, you can draw a Beta$(2, 2)$ density function using

```
xpl=0:1000/1000
plot(xpl,dbeta(xpl,2,2),type="l")
```

The pbeta function will let you find areas under the density. pbeta(.3,4.2), for example, will return the area to the left of .3, under a Beta$(4, 2)$ density. Finally, qbeta(c,alpha,beta) returns the number between zero and one, for which the area to the left is c. For example, qbeta(.25,4,2) returns .5458, so the area to the left of .5458, under a Beta$(4, 2)$ density, is .25.

Beta distributions are useful to model random variables that take values on the unit interval. We will see them again in Chapter 33 when we look at order statistics for a sample of uniform random variables. Further, beta random variables are often used in Bayesian analysis, as we will see Chapters 45 and 49.

Chapter Highlights

1. The random variable Y has a beta density with parameters α and β if Y has support $(0, 1)$ and density

$$f_Y(y) = \frac{\Gamma(\alpha + \beta)}{\Gamma(\alpha)\Gamma(\beta)} y^{\alpha-1}(1 - y)^{\beta-1} \text{ for } y \in (0, 1).$$

2. There is no closed form for the cumulative distribution function.

3. The mean is
$$\mathrm{E}(Y) = \frac{\alpha}{\alpha + \beta},$$

and the variance is
$$\mathrm{V}(Y) = \frac{\alpha\beta}{(\alpha + \beta + 1)(\alpha + \beta)^2}.$$

4. There is no closed form for the moment generating function.

5. The function dbeta(y,alpha,beta) will return $f_Y(y)$, where $Y \sim$ Beta(alpha,beta).

6. The function pbeta(y,alpha,beta) will return $\mathrm{P}(Y \leq y)$, where $Y \sim$ Beta(alpha,beta).

7. The function qbeta(q,alpha,beta) will return the value of c for which $\mathrm{P}(Y \leq c) = q$, where $Y \sim$ Beta(alpha,beta).

8. The function rbeta(n,alpha,beta) will return n independent replicates from a Beta(alpha,beta) distribution.

Exercises

17.1 Let Y have a Beta(α, β) density, where $\alpha > 0$ and $\beta > 0$. Find $E(Y(1 - Y)^2)$. Verify your answer using simulation in R for $\alpha = 2$ and $\beta = 1$.

17.2 Let $Y \sim$ Beta(α, β), where $\alpha > 0$ and $\beta > 0$, and let $X = 4Y^2(1 - Y)$. Find $E(X)$. Verify your answer using simulation in R for $\alpha = 2$ and $\beta = 3$.

17.3 Solve the integral

$$\int_0^1 x^3(1 - x)^4 dx.$$

17.4 The random variable Y has probability density

$$f(y) = cy^2(1 - y)^4 \quad \text{for } 0 < y < 1.$$

What is c?

17.5 Suppose Y is a random variable with density

$$f_Y(y) = 6y(1 - y) \quad \text{for } y \in [0, 1].$$

Find the value for $E(Y^2(1 - Y)^4)$.

17.6 Consider a Beta(α, β) density, where $\alpha > 1$ and $\beta > 1$. Show that this distribution is "unimodal" (increasing and then decreasing) and find the mode as a function of the parameters. What is the mode if $\alpha = \beta$?

17.7 Let $Y \sim$ Beta$(\alpha, 1)$, where $\alpha > 0$. Find $E[\log(Y)]$. (*Hint:* Use integration by parts.)

17.8 Suppose the waiting time for an airport tram belonging to Grand Hotel is supposed to be uniform on $(0, 1)$ hour. This would be reasonable if the round trip from the airport to the hotel takes 60 minutes, and travelers arrive "at random" at the airport tram stop. A disgruntled traveler thinks that the wait is longer than it should be and wants to do a hypothesis test. Suppose the true waiting time for a randomly arriving traveler can be modeled as Beta$(\alpha, 1)$, and consider testing $H_0 : \alpha = 1$ versus $H_a : \alpha > 1$. He will take a random sample of 12 travelers and find out how long they waited. He will reject the null hypothesis if nine or more waited for more than half an hour.

  (a) What is the test size?

  (b) If the true distribution of waiting times is Beta$(2, 1)$, what is the power of the test?

17.9 For an experiment in particle physics, a filament in a bubble chamber must be evenly charged along its length. When the settings in the chamber are incorrect, the charge tends to be greater in the middle. To test the settings, a number n of particles are released in the chamber to settle on the filament. If the settings are correct and the filament is evenly charged, the positions of the particles (independently) follow a uniform distribution on the length. Based on symmetry considerations, the physicists believe that the charge can be modeled by a Beta(θ, θ) density, and they want to test $H_0 : \theta = 1$ versus $H_a : \theta > 1$. They will count the number X of particles that settle in the middle half of the filament, in the $(1/4, 3/4)$ interval where the length is one unit.

(a) If $n = 40$ particles, what is the decision rule for test size $\alpha = .10$? State the decision rule as "reject H_0 when the observed X is greater than _____."

(b) Suppose the settings are messed up, and $\theta = 2$. What is the power of the test?

17.10 Suppose the number of customers who come into a computer repair shop on a randomly selected day is a Poisson random variable with mean 20. The times that these customers arrive are distributed uniformly in the eight-hour period that the shop is open. Finally, the amount of time (in hours) each customer requires can be modeled as a Beta$(1, 3)$ random variable. The store owner wants to have enough employees in the shop so that on 90% of days, no one has to wait for service. How many employees is this? Solve this difficult problem using simulations in R.

Chapter 18

The Normal Random Variable

The famous "bell-shaped" density is the most important of the probability distributions, largely because of the central limit theorem, which is the topic for a future chapter. There are two parameters, called μ and σ^2, which we will show are the mean and the variance of the distribution. We write $Y \sim N(\mu, \sigma^2)$, and its density function is

$$f_Y(y) = \frac{1}{\sqrt{2\pi\sigma^2}} e^{-\frac{1}{2}\frac{(y-\mu)^2}{\sigma^2}}.$$

Let's sketch this density using notions from calculus. First, we notice that the density is symmetric about μ; that is, when y is some distance to the right of μ, the density has the same value as when y is the same distance to the left of μ. The density is positive over the entire real line, and the limit as y gets large in either direction is zero.

Taking the first derivative,

$$f_Y'(y) = \frac{1}{\sqrt{2\pi\sigma^2}} e^{-\frac{1}{2}\frac{(y-\mu)^2}{\sigma^2}} \left[-\frac{(y-\mu)}{\sigma^2} \right],$$

so we know that f_Y is increasing to the left of μ and decreasing to the right, while $f_Y'(\mu) = 0$.

Taking another derivative,

$$f_Y''(y) = -\frac{1}{\sqrt{2\pi\sigma^2}} \left[e^{-\frac{1}{2}\frac{(y-\mu)^2}{\sigma^2}} - \frac{(y-\mu)^2}{\sigma^2} e^{-\frac{1}{2}\frac{(y-\mu)^2}{\sigma^2}} \right]$$

$$= \frac{1}{\sqrt{2\pi\sigma^2}} e^{-\frac{1}{2}\frac{(y-\mu)^2}{\sigma^2}} \left[\frac{(y-\mu)^2}{\sigma^2} - 1 \right].$$

This is zero when $(y-\mu)^2/\sigma^2 = 1$, negative when $(y-\mu)^2/\sigma^2 < 1$, and positive when $(y-\mu)^2/\sigma^2 > 1$. We conclude that there are two points of inflection, where $y = \mu \pm \sigma$. The density is concave on $(\mu - \sigma, \mu + \sigma)$ and convex otherwise. This is useful when we make a sketch of the density "by hand."

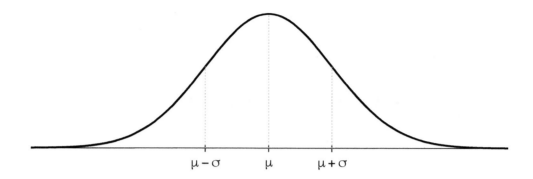

We can verify that the area under the density curve is one, using our gamma function (see Appendix B.5). First consider the case where $\mu = 0$ and $\sigma = 1$; then

$$\int_{-\infty}^{\infty} f_Y(y)dy = \frac{1}{\sqrt{2\pi}} \int_{-\infty}^{\infty} e^{-y^2/2}dy = \sqrt{\frac{2}{\pi}} \int_0^{\infty} e^{-y^2/2}dy,$$

by symmetry. We can do a change of variables: Let $u = y^2/2$, then $du = ydy$, so $dy = (2u)^{-1/2}du$. Then the integral becomes

$$\int_{-\infty}^{\infty} f_Y(y)dy = \frac{1}{\sqrt{\pi}} \int_0^{\infty} u^{-1/2}e^{-u}du = \frac{1}{\sqrt{\pi}}\Gamma\left(\frac{1}{2}\right) = 1,$$

because the gamma function evaluated at $1/2$ is $\sqrt{\pi}$.

For the general case,

$$\frac{1}{\sqrt{2\pi\sigma^2}} \int_{-\infty}^{\infty} e^{-\frac{1}{2}\frac{(y-\mu)^2}{\sigma^2}}dy;$$

let $x = (y - \mu)/\sigma$, $dx = 1/\sigma$, and the integral becomes

$$\frac{1}{\sqrt{2\pi}} \int_{-\infty}^{\infty} e^{-\frac{1}{2}x^2}dx = 1.$$

Computing the mean is easy with a change of variable $x = (y - \mu)/\sigma$:

$$
\begin{aligned}
\mathrm{E}(Y) &= \frac{1}{\sqrt{2\pi\sigma^2}} \int_{-\infty}^{\infty} ye^{-\frac{1}{2}\frac{(y-\mu)^2}{\sigma^2}}dy \\
&= \frac{1}{\sqrt{2\pi}} \int_{-\infty}^{\infty} (\sigma x + \mu)e^{-\frac{1}{2}x^2}dx \\
&= \frac{\sigma}{\sqrt{2\pi}} \int_{-\infty}^{\infty} xe^{-\frac{1}{2}x^2}dx + \frac{\mu}{\sqrt{2\pi}} \int_{-\infty}^{\infty} e^{-\frac{1}{2}x^2}dx \\
&= \mu,
\end{aligned}
$$

because in the first term, the integrand is an odd function, and the second term is μ times the integral of a density function. Of course, we knew ahead of time that the mean was μ, because of the symmetry.

For the variance, we use the original definition instead of the expression that usually makes the calculations easier. With the same change of variable, we get

$$V(Y) = E[(Y - \mu)^2] = \frac{1}{\sqrt{2\pi\sigma^2}} \int_{-\infty}^{\infty} (y - \mu)^2 e^{-\frac{1}{2}\frac{(y-\mu)^2}{\sigma^2}} dy$$

$$= \frac{1}{\sqrt{2\pi}} \int_{-\infty}^{\infty} (\sigma x)^2 e^{-\frac{1}{2}x^2} dx$$

$$= \frac{2\sigma^2}{\sqrt{2\pi}} \int_{0}^{\infty} x^2 e^{-\frac{1}{2}x^2} dx,$$

by symmetry. Now we must make another change of variable $u = x^2/2$, $du = xdx$. Then

$$V(Y) = \frac{2\sigma^2}{\sqrt{2\pi}} \int_{0}^{\infty} (2u)^{1/2} e^{-u} du$$

$$= \frac{2\sigma^2}{\sqrt{\pi}} \Gamma(3/2) = \sigma^2,$$

using $\Gamma(3/2) = \Gamma(1/2)/2 = \sqrt{\pi}/2$.

There is no closed-form expression for the normal CDF, because there is no anti-derivative for e^{-x^2}. Calculations such as $P(a < Y < b)$ have to be done numerically except for a few special cases. For instance, we know $P(Y > \mu) = 1/2$, because of the symmetry. Extensive tables of areas under the normal density are found in many statistics textbooks, and we can use the R function pnorm(1.3) to get the area to the left of 1.3, under a *standard* normal density, i.e., the mean is zero and the variance is one. The command pnorm(1.3,1,.5) will return the area to the left of 1.3, under a normal density with mean $\mu = 1$ and standard deviation $\sigma = .5$. It's important to remember that R uses standard deviation and not variance.

The **68-95-99.7 rule** is useful for estimating probabilities without looking them up in a table or using R. If $Y \sim N(\mu, \sigma^2)$,

- $P(\mu - \sigma < Y < \mu + \sigma) \approx .68$,

- $P(\mu - 2\sigma < Y < \mu + 2\sigma) \approx .95$,

- $P(\mu - 3\sigma < Y < \mu + 3\sigma) \approx .997$.

Using this we can get, for example, that if $Y \sim N(3, 4)$, then $P(Y > 5) \approx (1 - .68)/2 = .16$.

Our next task is to find the MGF. For now we'll consider only $\mu = 0$ and $\sigma^2 = 1$; we'll do the more general case later when we have a useful result about linear combinations of random variables.

If $Y \sim N(0, 1)$, then

$$M_Y(t) = E(e^{tY}) = \frac{1}{\sqrt{2\pi}} \int_{-\infty}^{\infty} e^{ty} e^{-\frac{1}{2}y^2} dy$$

$$= e^{t^2/2} \frac{1}{\sqrt{2\pi}} \int_{-\infty}^{\infty} e^{-\frac{1}{2}(y-t)^2} dy$$

$$= e^{t^2/2}.$$

Normality is a common assumption in statistical theory, and a wealth of statistical procedures are related to this distribution. In later chapters, we will see how the t, F, and χ^2 distributions are related to the normal distribution, and we will learn many hypothesis testing methods and methods for constructing confidence intervals, based on the normality assumption. Let's go through a hypothesis example, using what we have learned so far.

Example: A machine produces rods for building a device. When the machine is "in spec" the lengths of the rods (in mm) follow a normal distribution with mean 120 mm and standard deviation $\sigma = 5$ mm. Suppose that to check the machine, we take a random sample of size $n = 4$ rods, and measure the lengths. We want to test the null hypothesis that the machine is in spec, i.e., has the correct mean length, versus the alternative that the machine is not in spec. We can state these hypotheses more formally as

$$H_0 : \mu = 120 \quad \text{versus} \quad H_a : \mu \neq 120,$$

where μ is the true average rod length.

Suppose we have this decision rule: "reject the null hypothesis unless *all four* of the rods are between 110 and 130 mm in length." What is the test size? Under the null hypothesis, the probability that a randomly selected rod has a length in this interval (plus or minus two standard deviations from the mean) is about .95 by our 68-95-99.7 rule. Therefore, the probability that all randomly chosen rods have lengths in this interval is about $(.95)^4 = .8145$. Therefore, the probability of rejecting H_0, when it is true, is $\alpha = 1 - .8145 = .1855$.

If the decision rule were "reject the null hypothesis unless all four of the rods are between 105 and 135 mm in length," then our test size would be $\alpha = 1 - (.997)^4 = .0119$. In practice, a test size is chosen based on the consequences of the Type I and Type II Errors. If we reject H_0 when it's really true (Type I Error), the consequence is that we unnecessarily shut down production to fix the machine. If we accept H_0 when it's false, the consequence is that we might get too many out-of-spec parts. The costs of these two types of errors are balanced to come up with a reasonable test size.

If it's easy and quick to adjust the machine, while it's slow and expensive to make the rods, we would probably choose a larger test size, because Type I Errors would not be that bad. On the other hand, if it's time-consuming and expensive to shut down the machine and make repairs, but rods that are two big or too small can be easily recycled, then a Type I Error is expensive compared to a Type II Error, and we'd prefer a smaller test size.

Let's be traditional and choose $\alpha = .05$, and find the decision rule that states, "reject H_0 unless all four rods have lengths that are within D mm of 120," with D chosen so that the test size is .05. The value D is chosen so that the probability that a single rod has a length that is within D mm of 120 is $(.95)^{1/4} = .9873$. The density for the rod lengths is shown below, with the desired area shaded.

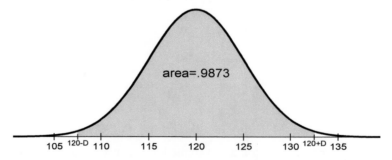

The small areas to either side of the shaded area are then $(1 - .9873)/2 = .006371$, and the command qnorm(.006371,120,5) returns 107.5 as the left boundary of the shaded area. Computing $120 - D = 107.5$ gives $D = 12.5$. Therefore the decision rule "reject H_0 unless all four rods are within 12.5 mm of 120 mm" has test size $\alpha = .05$.

Let's find the power of this test if the alternative is true and $\mu = 125$; assume that the standard deviation is still $\sigma = 5$. Recall that the power is the probability of correctly rejecting H_0, when the alternative is true. In our context, this is the probability that *not* all four rods have lengths that are within the interval $(107.5, 132.5)$, which is $120 \pm D$. The probability that a single rod has a length in this interval is the area shaded in the next plot, under the density assumed for the alternative rod length distribution:

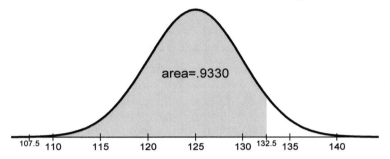

The probability that all four rods have length in this interval is $.9330^4 = .7578$, so the power for the test is $1 - .7578 = .2422$, when the true rod length average is 125 instead of 120. The power for the test is not very high; perhaps we should choose a larger sample size.

Chapter Highlights

1. If $Y \sim \mathrm{N}(\mu, \sigma^2)$, we say that Y is a normal (or Gaussian) random variable with mean μ and variance σ^2. Its density is

$$f_Y(y) = \frac{1}{\sqrt{2\pi\sigma^2}} e^{-\frac{1}{2}\frac{(y-\mu)^2}{\sigma^2}}.$$

2. There is no closed form for the CDF.

3. If $Y \sim \mathrm{N}(0, 1)$, then Y has moment generating function

$$M_Y(t) = e^{t^2/2}.$$

4. The **68-95-99.7 rule** for $Y \sim \mathrm{N}(\mu, \sigma^2)$:

   - $\mathrm{P}(\mu - \sigma < Y < \mu + \sigma) \approx .68;$
   - $\mathrm{P}(\mu - 2\sigma < Y < \mu + 2\sigma) \approx .95;$
   - $\mathrm{P}(\mu - 3\sigma < Y < \mu + 3\sigma) \approx .997.$

5. The R command `pnorm(c,mu,sig)` returns the area to the left of c, under a normal density with mean mu and standard deviation sig.

6. The R command `qnorm(q,mu,sig)` returns the value c so that the area to the left of c, under a normal density with mean mu and standard deviation sig, is q.

Exercises

18.1 Using tables or R, find the following areas:

    (a) The area to the left of 3.45, under a normal density with mean 2 and variance 1.

    (b) The area to the right of 4.8, under a normal density with mean 6 and variance 2.

18.2 Using tables or R, find the following probabilities:

    (a) If Y is a normal random variable with mean 24 and variance 2, find $P(Y < 22)$.

    (b) If Y is a normal random variable with mean 100 and variance 4, find $P(\frac{Y-102}{2} < 1.2)$.

18.3 If Y has a normal density with mean $\mu = -1$ and standard deviation $\sigma = 2$, find $P(|Y| > 1)$.

18.4 Suppose Y is a normally distributed random variable with mean 100 and variance 4. Find two *different* intervals (a, b) so that $P(Y \in (a, b)) = .6$.

18.5 Suppose the distribution of petal lengths of the Gorzola flower can be modeled as normally distributed with mean 2.25 inches and standard deviation 0.06 inches.

    (a) Sketch the density for petal lengths, marking the mean, and one standard deviation away from the mean on either side. Remember to get the points of inflection over these values.

    (b) What proportion of flowers have petal lengths of less than 2.13 inches? Shade this area in your sketch.

18.6 Consider a density of the form

$$f(y) = de^{-(ay^2 - 2by + c)/2},$$

where a, b, c, and d are constants. What are the mean and variance (in terms of a, b) for a random variable with this density?

18.7 Compute the following integral in terms of the constants a, b, and c:

$$\int_{-\infty}^{\infty} e^{-(ay^2 - 2by + c)/2} dy.$$

18.8 NuttyKorn is a snack consisting of caramel popcorn and peanuts. They advertise that the weight of peanuts (in ounces) in a 26-ounce box of snack follows a normal distribution with mean 5 and standard deviation .2.

A consumer group has grave fears that NuttyKorn is skimping on the peanuts. They want to test $H_0 : \mu = 5$ versus $H_a : \mu < 5$, where μ is the *true* average weight of peanuts in a 26-ounce box of snack. (Assume that the standard deviation is known to be 0.2.) They will randomly select four boxes and reject H_0 if at least three of the boxes have less than 4.8 ounces of peanuts.

    (a) What is the test size for this decision rule? (Use the 68-95-99.7 rule.)

    (b) Suppose that the true average weight of peanuts is only $\mu = 4.6$ ounces (and the standard deviation is again .2). What is the power of the test?

18.9 The travel time on the Metrorail from Point A to Point B follows a normal distribution having an average of 20 minutes, with a standard deviation of 2.5 minutes. Mike gets on the Metrorail at Point A, at 6pm, and he is scheduled to meet a friend at Point B at 6:25pm. What is the probability that Mike will be late meeting his friend; that is, that his travel time will be longer than 25 minutes?

18.10 Filaments made at Gloglobe factory are supposed to contain 2.75 mg of chromelite. Because of randomness in the manufacturing process, the amount of chromelite in a filament is actually a random variable. If a filament has more than 2.77 mg or less than 2.73 mg of chromelite, it must be thrown out. Suppose Machine A makes filaments with a chromelite distribution which is normally distributed with mean $\mu_A = 2.75$ mg and standard deviation $\sigma_A = 0.01$ mg of chromelite, and Machine B makes filaments with a chromelite distribution which is normally distributed with mean $\mu_A = 2.76$ mg and standard deviation $\sigma_A = 0.005$ mg of chromelite. Which machine is better, in terms of a smaller proportion thrown out?

18.11 The amount of vitamin C in conventionally grown oranges follows a normal distribution with mean 60 mg and standard deviation 8 mg. The amount of vitamin C in organically grown oranges follows a normal distribution with mean 70 mg and standard deviation 5 mg. A basket of oranges contains thirty randomly selected conventional oranges and ten randomly selected organic oranges. If an orange is selected at random from the basket and found to have less than 60 mg of vitamin C, what is the probability that the orange is organic?

18.12 A biological research supply company breeds mice that have higher levels of a specific growth hormone than standard laboratory mice. In these specially bred mice, the hormone levels are normally distributed with mean 120 units and standard deviation 8 units. For the standard mice, the mean is 100 units with standard deviation 10 units.

A research lab buys 100 of these expensive specially bred mice, but then a careless lab assistant allows them to mix with 300 of the cheaper, standard mice. The mice look identical. If a mouse is randomly selected, what is the probability that this growth hormone level is greater than 112 units?

18.13 A telecommunications company buys 80% of its filaments from Supplier A and 20% from Supplier B. The tensile strength of a randomly selected filament from Supplier A follows a normal distribution with mean $\mu_A = 108$ and standard deviation $\sigma_A = 4$, while the tensile strength of a randomly selected filament from Supplier B follows a normal distribution with mean $\mu_B = 106$ and standard deviation $\sigma_B = 6$.

A filament will break when used in a device if its tensile strength is less than 100. Suppose a filament bought by the telecommunications company is randomly selected and put in a device. If the filament breaks, what is the probability that it was obtained from Supplier B?

18.14 A company purchases optical fibers in boxes of 100 units. One third of the boxes in the company's warehouse are from Supplier A, while two thirds are from

Supplier B. The tensile strength of the Supplier A fibers are known to follow a normal density with mean 100 and standard deviation 10, while the tensile strength of the Supplier B fibers are known to follow a normal density with mean 80 and standard deviation 20. The boxes in the warehouse are not labeled and have gotten all mixed up. Suppose an engineer chooses a box at random and tests the tensile strength of two fibers in the box. Both fibers have tensile strength greater than 100. What is the probability that the box was from Supplier A?

18.15 Children are tested for a condition related to diabetes by measuring the level of a certain blood sugar. The level for children without the condition is approximately normally distributed with mean 100 mg and standard deviation 15 mg. The level for children with the condition is also approximately normal, with mean 140 mg and standard deviation 20 mg. A child is to be tested for the condition. We set up hypotheses:

H_0: child does not have condition   versus H_a: child has condition.

The null hypothesis will be rejected if the child's level is above 130.

(a) What are the probabilities of Type I and Type II Errors?

(b) What are the potential practical consequences of making a Type I Error? of making a Type II Error?

(c) Now change the decision rule to "reject H_0 if the level is above 120." What are the probabilities of Type I and Type II Errors?

(d) Note that, for the same amount of data, decreasing one type of error probability will increase the other. Given your consequences of each type of error in this situation, which decision rule do you think is better? Explain. (There is no right answer, you just have to be consistent with your answer for (b) and show that you understand what these errors mean.)

18.16 A farmer claims that organic tomatoes are higher in vitamin A than conventional tomatoes. Suppose we know that the distribution of vitamin A in a conventional tomato is normal with mean 1500 IU (international units) and standard deviation 100 IU. The grower will randomly choose 4 organic tomatoes and determine the amount of vitamin A in each. He wants to test

$$H_0 : \mu = 1500 \quad \text{versus} \quad H_a : \mu > 1500,$$

where μ is the mean amount of vitamin A (in IU) in organically grown tomatoes. The decision rule is "reject H_0 if all four tomatoes have greater than 1500 IU of vitamin A."

(a) What is the test size?

(b) Suppose the true distribution for organic tomatoes is normal with mean 1600 and standard deviation 100. What is the power for the test?

(c) Suppose the farmer wants the test size to be $\alpha = .05$. The decision rule is "reject H_0 if all four tomatoes have greater than c IU of vitamin A." What is c?

(d) What is the power for the test in (c) if the true distribution is that described in (b)?

18.17 Repeat the testing example in this chapter for a larger sample size. That is, suppose we want to test $H_0 : \mu = 120$ versus $H_a : \mu \neq 120$ for $n = 8$ rods selected "at random" from the distribution of rod lengths. Assume that the standard deviation of rod lengths is $\sigma = 5$ mm.

(a) Find a value D so that the decision rule "reject H_0 unless all eight rods are within D mm of 120" has test size $\alpha = .05$.

(b) Find the power of the test in part (a) when $\mu = 125$.

Chapter 19

CDF Tricks: Probability Plots and Random Number Generation

Quantiles and percentiles of a distribution can be found using the cumulative distribution function (CDF), which was defined in Chapter 15. For example, to find the 10th percentile of a distribution with CDF F, we solve $F(x) = .1$ for x. This percentile will exist and be unique for a continuous distribution. For many families of distributions, the quantiles are easy to find using R. For example, `qgamma(.25,8,2)` will return the 25th percentile of a Gamma$(8, 2)$ density, and `qnorm(.95,10,2)` will return the 95th percentile of a normal distribution with mean 10 and standard deviation 2. Further, `qnorm(1:n/(n+1),10,2)` will return n quantiles from this normal distribution, as shown for $n = 20$ in the plots below. On the left, the areas under the normal curve, between consecutive quantiles, are all equal to $1/(n + 1)$. On the right, there are n equally spaced vertical ticks in (0,1), and these are mapped to the quantiles through the CDF.

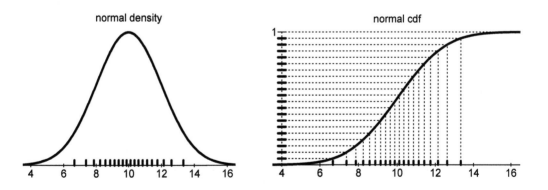

If we have a large random sample from a distribution, the 10th percentile of the sample will be close to the 10th percentile of the distribution, and the other sample percentiles will roughly line up with the population percentiles. This idea can be used to assess whether a random sample came from a particular distribution.

For many statistical procedures (i.e., the various t-tests, regression, and ANOVA models), we assume that some population has a normal density. With "real" data, this assumption can be assessed by comparing the observed quantiles with normal density quantiles. A visual comparison is accomplished plotting the sample quantiles against the quantiles of the density in question, using a **quantile-quantile plot**, also known as a **Q-Q plot** or a **probability plot**.

Because we expect approximately ten percent of the sample to be less than the 10th percentile, 20 percent of the sample to be less than the 20th percentile, etc., if we plot the sorted sample of size n against the n population quantiles, the points should be close to the diagonal line if the sample is indeed from that population. In the figure below, a sorted sample from a standard normal population is plotted against the normal quantiles for $n = 25$, $n = 50$, and $n = 200$. The specific commands for the plot are

```
n=25
plot(qnorm(1:n/(n+1)),sort(rnorm(n)),xlab="normal quantiles",
                        ylab="sorted sample")
lines(c(-2,2),c(-2,2))
```

The points get closer to the line as the sample size grows, especially for the middle range of points.

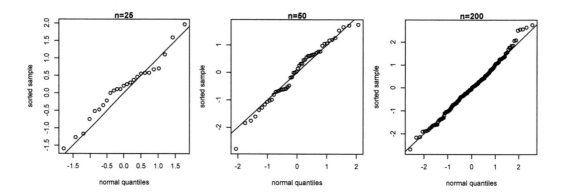

If the population from which the sample was drawn is normal but not *standard* normal, the points will lie near a straight line that is not the diagonal line. This is because a standard normal can be obtained by a linear transformation of the normal random variable (we will show this in Chapter 21), so the quantiles are transformed linearly as well.

On the other hand, if the population from which the sample was drawn is *not* normal, then the points will likely not lie near a straight line. Examples of plots of sorted samples against normal quantiles are shown when the samples of size $n = 200$ are drawn from nonnormal populations. In the first plot, a sorted sample from a bell-shaped but heavier-tailed density is plotted against the normal quantiles. The t-density will be defined in Chapter 34 and looks like a standard normal but with a larger "spread." The middle section looks "roughly linear," but for a sample from the $t(3)$ distribution the maximum and minimum points tend to be farther from zero than for a sample from a normal density. This is shown by the points on the right deviating upwards from the line, and the points on the left deviating downwards. If a density is skewed to the right, as for a Gamma$(3, 2)$, the Q-Q plot using normal quantiles will be convex, as shown in the middle plot below. Finally, if the true population has a finite support such as for a Beta$(3, 2)$, the Q-Q plot using normal quantiles can be S-shaped, as in the third plot.

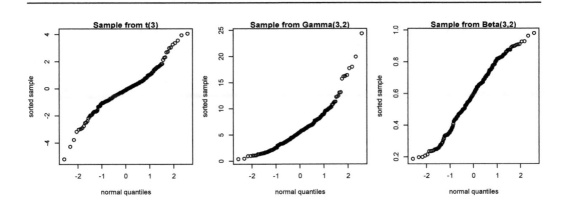

We can similarly check to see if a sample came from a chi-squared distribution, or any other target distribution (recall that a $\chi^2(m)$ random variable is a special case of the gamma random variable). Suppose we have a sample from a population that is supposed to have a $\chi^2(1)$ distribution. We can assess this by plotting the sorted sample against the appropriate quantiles. In the plots below, the sample on the left is from a $\chi^2(1)$ population, but the sample on the right is from a Gamma(2,3) population, both with $n = 200$. When the sample is indeed from the target population, we see that the points lie roughly on the diagonal line, with a few stragglers. However, there is systematic deviation from the line when the sample comes from a distribution other than the target distribution.

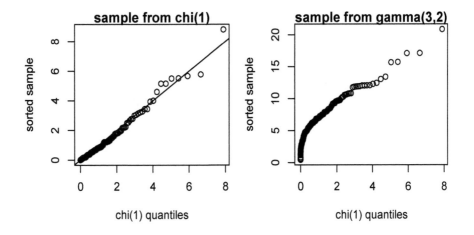

The CDF can also be used to generate a sample "at random" from the density function. Recall that we can use R to generate uniform random variables using the `runif` function. Given these uniform random variable values, we can easily generate values of a random variable X, if we can compute the inverse function for the CDF F_X^{-1}. We simply generate $U \sim \text{Unif}(0,1)$, and compute $X = F_X^{-1}(U)$. To show that this generates $X \sim F_X$, notice that

$$P(X \leq x) = P(F_X^{-1}(U) \leq x) = P(U \leq F_X(x)) = F_X(x)$$

because $P(U \leq u) = u$.

Example: Suppose we want to generate values of a triangular random variable, shown below on the left. We previously computed the CDF $F(x) = x - x^2/4$ on $(0, 2)$.

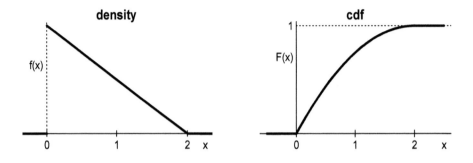

Let $u = F(x)$ and solve for x to get the inverse CDF: $x - x^2/4 = u$ becomes $x^2 - 4x + 4u = 0$, and solving for x gives $x = 2 - 2\sqrt{1 - u}$ for $x \in (0, 2)$. That is, $F^{-1}(u) = 2 - 2\sqrt{1 - u}$ for $u \in [0, 1]$.

Here is some R code for generating a sample of size 10,000:

```
u=runif(10000)
x=2-2*sqrt(1-u)
```

We can check our calculations by making a histogram and superimposing our density function:

```
hist(x,freq=FALSE,br=40)
lines(c(0,2),c(1,0),lwd=3)
```

Given densities f_1 and f_2, and a number $\alpha \in (0, 1)$, we can define a density f that is a **mixture** of these densities as

$$f(x) = (1 - \alpha)f_1(x) + \alpha f_2(x).$$

It's easy to show that f is a proper density: It must be nonnegative, and it integrates to one. If we can write code to sample both from f_1 and f_2, we can sample from f, using a uniform density and the following steps:

1. Sample $U \sim \text{Unif}(0, 1)$.

2. If $U < \alpha$, sample X from f_2; otherwise, sample X from f_1.

Then, for any $x \in \mathbb{R}$,

$$F_X(x) = \text{P}(X \le x) = \text{P}(X \le x | U < \alpha)\text{P}(U < \alpha) + \text{P}(X \le x | U \ge \alpha)\text{P}(U \ge \alpha)$$

using the law of total probability from Chapter 4. Now, $P(X \leq x|U < \alpha) = F_2(x)$ and $P(X \leq x|U \geq \alpha) = F_1(x)$, by the rules of the algorithm. Therefore, because $P(U < \alpha) = \alpha$ and $P(U \geq \alpha) = 1 - \alpha$, we have

$$F_X(x) = (1 - \alpha)F_1(x) + \alpha F_2(x),$$

and the density for X (take derivatives with respect to x) is the mixture density given above.

The following code samples from a mixture of two normals using the algorithm, makes a histogram, and draws the density function on the histogram. The mixture is 80% from a standard normal, and 20% from a normal with mean $\mu = 3$ and variance $\sigma^2 = .25$.

```
bign=1000000
alpha=.2
u=runif(bign)
x=rnorm(bign)
nu=sum(u<alpha)
x[u<alpha]=rnorm(nu,3,.5)
hist(x,br=100,freq=FALSE)
xpl=c(-50:60/10)
lines(xpl,alpha*dnorm(xpl,3,.5)+(1-alpha)*dnorm(xpl),lwd=3)
```

The original densities are shown as dotted curves:

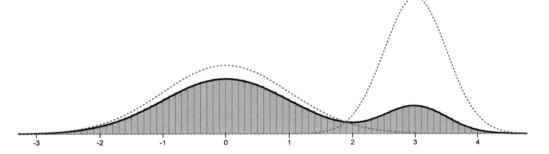

Chapter Highlights

1. If $X_1, \ldots X_n$ is a random sample from a distribution with CDF $F(x)$, then the plot of the sorted sample against the quantiles $F^{-1}(i/(n+1))$, $i = 1, \ldots, n$, will lie roughly on a diagonal line. This type of plot is often used for checking to see if the sample came from the distribution.

2. If we can simulate a random variable U from a Unif$(0, 1)$ distribution, we can use the transformation $X = F_X^{-1}(U)$ to simulate a random variable with CDF F_X.

3. Consider the mixture of densities $\alpha \in (0, 1)$ and $f(y) = (1-\alpha)f_1(y) + \alpha f_2(y)$. We can sample from f if we can sample from both f_1 and f_2: Sample $U \sim$ Unif$(0, 1)$; if $U < \alpha$, simulate from f_2; otherwise simulate from f_1.

Exercises

19.1 The data set `case1602` in the R library `Sleuth2` contains a variable `Baseline`, representing a before-treatment cholesterol level for 20 patients. Make a probability plot to see if the measurements are approximately normally distributed.

19.2 The data set `case0301` in the R library `Sleuth2` contains a variable `Rainfall`, representing volume of rainfall in a target area, for seeded and unseeded cloud systems. Make a probability plot to see if the rainfall measurements for the unseeded clouds are approximately normally distributed. Then do the same for the logarithm of the measurements.

19.3 Someone tells you that the square of a standard normal random variable is a random variables with a $\chi^2(1)$ distribution. (We will prove this in Chapter 30 and again in Chapter 31.) Being a skeptic, you want to verify their claim using a probability plot. Using R, generate a large number of values from a standard normal distribution, and plot the sorted squares of the values against the quantiles of a $\chi^2(1)$ distribution. Do the points lie roughly on a straight line?

19.4 Someone tells you that the sum of two independent $Exp(\beta)$ random variables is a random variable with a $Gamma(2, \beta)$ distribution. (We will prove this in Chapter 30 and again in Chapter 31.) Being a skeptic, you want to verify their claim using a probability plot. Using R, generate a large number of pairs of values from independent $Exp(\beta)$ random variables, and plot the sorted sum against the quantiles of a $Gamma(2, \beta)$ distribution, for your choice of β. Do the points lie roughly on a straight line?

19.5 Write R code to simulate random variables from the density

$$f(x) = \frac{2}{(x+1)^3} \quad \text{for } x > 0.$$

Make a histogram of 10,000 simulated values and superimpose the density function to check your work.

19.6 Write R code to simulate random variables from the density

$$f(x) = 2\theta^2 x^{-3} \quad \text{for } x > \theta.$$

Make a histogram of 10,000 simulated values using $\theta = 2$, and superimpose the density function to check your work.

19.7 Write R code to simulate random variables from the density

$$f(x) = \frac{1}{80} \left[\frac{x}{240} + 1 \right]^{-4} \quad \text{for } x > 0.$$

Make a histogram of 10,000 simulated values and superimpose the density function to check your work.

19.8 Write R code to simulate random variates from the density shown in the plot for any value of $\theta > 0$. Check your answer for $\theta = 3$ by simulating a large number of values, and superimposing the density function over a histogram.

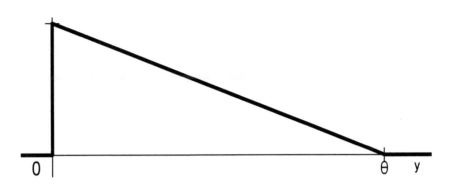

19.9 Write R code to simulate random variates from the density shown in the plot.

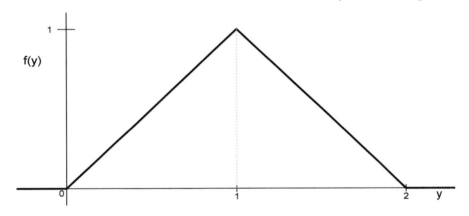

(a) Make a histogram of 10,000 values and superimpose the density function to check your work.

(b) Get a random sample of 100 values from this triangular density, and plot them against the appropriate quantiles from a standard normal density. Do you think that you could tell, from this plot, that your sample is not from a normal density? Repeat for a random sample of 1000 values from the triangular density.

19.10 Consider the family of densities

$$f(y) = \frac{\theta}{2} \left(\frac{y}{2} + 1 \right)^{-(\theta+1)} \quad \text{for } y > 0,$$

where $\theta > 0$ is a parameter. Write R code to simulate random variates from this density, and check your work for $\theta = 10$ by getting a histogram of a large number of values and superimposing the density.

19.11 Consider the density
$$f(x) = \frac{x}{3} e^{-x^2/6} \quad \text{for } x > 0.$$

(a) Find the CDF; sketch both the density and the CDF.

(b) Find the mean and median of a random variable having the density.

(c) Write R code to simulate random variates from the density. Get a histogram and superimpose the density function to check your work.

(d) Plot your sorted sample against the quantiles from a Gamma$(2, 6)$ density. Comment on the shape of the probability plot.

19.12 Write R code to simulate from the density

$$f(y) = \frac{2\theta^2}{y^3} \text{ for } y > \theta.$$

Make a histogram of 10,000 values and superimpose the density function to check your work.

19.13 Refer to Exercises 13.22 and 15.14. Write code to simulate from the given density.

(a) Use this code to approximate the variance of a random variable with the given density. That is, generate a large sample from the density, then use var(x) where the vector x contains the sample.

(b) Get a random sample of 100 values from this density and plot them against the appropriate quantiles from a standard normal density. Do you think that you could tell, from this plot, that your sample is not from a normal density? Repeat for a random sample of 1000 values from the density.

19.14 Consider the density

$$f(y) = \frac{\theta}{(y + 1)^{\theta+1}} \text{ for } y > 0$$

for some $\theta > 0$. Write R code to sample from this density. Get a histogram using $\theta = 4$, and superimpose the density function to check your work.

19.15 Consider the density

$$f(y) = 3\theta y^2 e^{-\theta y^3} \text{ for } y > 0$$

for some $\theta > 0$. Write R code to sample from this density. Get a histogram using $\theta = 5$, and superimpose the density function to check your work.

19.16 Consider the density

$$f(x) = \frac{\lambda}{2\sqrt{x}} e^{-\lambda\sqrt{x}} \text{ for } x > 0$$

for some $\lambda > 0$. Write R code to sample from this density. Get a histogram using $\lambda = .5$, and superimpose the density function to check your work. (*Hint:* This is a very "heavy-tailed" density so you can get a histogram only of the values in $(0, 20)$.)

19.17 Consider again the distributions in Exercise 13.23, representing vitamin A content of conventional and organic tomatoes. In that problem, a sample of size four of organic tomatoes was obtained, and the decision rule was "reject H_0 if all four tomatoes had greater than c_1 units of Vitamin A," where c_1 was calculated so that the test size was $\alpha = .05$.

Suppose instead that the decision rule was "reject H_0 if the *average* vitamin A content of four tomatoes is greater than c_2." Using simulations in R, find c_2 so that the test size is $\alpha = .05$. Then find the power of the test, given the alternative distribution of vitamin A, and compare to the power of the original test. Which test is better?

19.18 The maximum one-day rainfall Y (inches) in a randomly selected year was modeled following the density

$$f_Y(y) = \frac{3}{2} \left(\frac{y}{2} + 1 \right)^{-4} \quad \text{for } y \geq 0$$

in an example in Chapter 13. The average for this distribution was calculated to be $\mu = 1$, and the variance was $\sigma^2 = 3$.

(a) Write code to sample from this distribution, and check your answer by getting a histogram of many simulated values and superimposing the density function. Also check by comparing the mean and variance of your simulated values.

(b) Suppose a climate scientist predicts that the changing climate will result in more floods in future years, because on average the maximum one-day rainfall will be *larger* than predicted by the density. She decides to test H_0 : *the future distribution of maximum one-day rainfall will stay the same* against the alternative of larger values. Her test statistic will be the average maximum one-day rainfall values, in the next eight years. She will reject H_0 if this average is greater than 2. If these eight values can be considered to be independent, what is the size of her test? Answer by simulating many samples of size eight from the distribution and computing the average for each sample. The proportion of averages greater than 2 will be the approximate test size.

Chapter 20

Chebyshev's Inequality

Chebyshev's inequality can be used to bound probabilities for random variables, where the mean and variance are known but the distribution is unknown. It's also used frequently when proving limit theorems involving random variables. In this chapter we'll be concerned only with the former, but we'll see Chebyshev's theorem again in future chapters.

We'll start with a useful result for random variables with positive support: If Y is a random variable taking values in $[0, \infty)$, where $E(Y)$ exists, then

$$P(Y \geq 1) \leq E(Y).$$

Proof for continuous random variables:

$$
\begin{aligned}
E(Y) &= \int_0^\infty y f_Y(y) dy \\
&\geq \int_1^\infty y f_Y(y) dy \\
&\geq \int_1^\infty f_Y(y) dy \\
&= P(Y \geq 1).
\end{aligned}
$$

Proof for discrete random variables:

$$
\begin{aligned}
E(Y) &= \sum_{all\ y} y P(Y = y) \\
&\geq \sum_{all\ y \geq 1} y P(Y = y) \\
&\geq \sum_{all\ y \geq 1} P(Y = y) \\
&= P(Y \geq 1).
\end{aligned}
$$

This result is easily extended to **Markov's inequality** for random variables Y with support in $[0, \infty)$, where $E(Y)$ exists. For $c > 0$,

$$P(Y \geq c) \leq \frac{1}{c}E(Y).$$

We have already proved Markov's inequality for the special case of $c = 1$, and we use this to prove the more general result:

$$P(Y \geq c) = P\left(\frac{Y}{c} \geq 1\right)$$

$$\leq E\left(\frac{Y}{c}\right) \quad \text{(by the first result)}$$

$$= \frac{1}{c}E(Y).$$

Now we can prove **Chebyshev's inequality:** Let Y be any random variable with mean μ and variance $\sigma^2 < \infty$. For any $c > 0$,

$$P\left(|Y - \mu| \geq c\sigma\right) \leq \frac{1}{c^2}.$$

The statement of Chebyshev's inequality in words is as follows:

> The probability that a random variable takes a value that is more than c standard deviations away from the mean is no bigger than $1/c^2$.

Proof:

$$P\left(|Y - \mu| \geq c\sigma\right) = P\left(\frac{|Y - \mu|}{c\sigma} \geq 1\right)$$

$$= P\left(\frac{(Y - \mu)^2}{c^2\sigma^2} \geq 1\right)$$

$$\leq \frac{1}{c^2}E\left(\frac{(Y - \mu)^2}{\sigma^2}\right)$$

$$= \frac{1}{c^2},$$

because $E[(Y - \mu)^2] = \sigma^2$.

Note that the theorem is not very useful unless $c > 1$! Let's check it against our normal distribution. We know that the probability that a normal random variable is more than two standard deviations above its mean is about .16. According to Chebyshev, this probability has to be less than .25. Chebyshev's theorem is considered "conservative" for most distributions.

Markov's and Chebyshev's inequalities will be useful for limit theorems in probability and statistics. For example, the law of large numbers of Chapter 47 is proved using Chebyshev's inequality.

Chapter Highlights

1. Markov's inequality: For a random variable Y with positive support, and any $c > 0$,

$$P(Y > c) \leq \frac{E(Y)}{c}.$$

2. Chebyshev's inequality: For random variables with finite standard deviation σ and mean μ,

$$P\left(|Y - \mu| \geq c\sigma\right) \leq \frac{1}{c^2}.$$

Exercises

20.1 Using Chebyshev's inequality, find the bounds and fill in the first blanks with "greater" or "less."

(a) If Y is a random variable with mean 6 and standard deviation 2, then the probability that Y is between 0 and 12 is _____ than _____.

(b) If Y is a random variable with mean 0 and standard deviation 1, then the probability that Y is greater than 10 is _____ than _____.

(c) If Y is a random variable with mean 10 and standard deviation 3, then the probability that Y is between 0 and 17 is _____ than _____.

(d) If Y is a random variable with mean 0 and standard deviation 3, then the probability that Y^2 is greater than 10 is _____ than _____.

20.2 Suppose Y is an exponential random variable with mean θ; find $P(Y > 4\theta)$.

(a) Using Markov's inequality, find a bound for $P(Y > 4\theta)$, and compare this to the true probability.

(b) Using Chebyshev's inequality, find a bound for $P(Y > 4\theta)$, and compare this to the true probability.

20.3 The maximum height of a river in a "randomly selected" year is a random variable Y. Suppose it is know that the expected value of Y is 12, and if Y exceeds 24, there will be a flood in a nearby neighborhood.

(a) Using Markov's inequality, find a bound for the probability of a flood for a randomly selected year.

(b) Suppose we also know that the standard deviation of Y is 4. Using Chebyshev's inequality, find a bound for the probability of a flood for a randomly selected year.

(c) Now suppose we also know that the standard deviation of Y is only 2. Using Chebyshev's inequality, find a bound for the probability of a flood for a randomly selected year.

20.4 Technicians working for an appliance repair service know that for a randomly selected service call, the average time for the job is .62 hours.

    (a) Using Markov's inequality, find a bound for the probability that the time required for a randomly selected service call is greater than two hours.

    (b) Suppose we also know that the standard deviation of Y is .28. Using Chebyshev's inequality, find a bound for the probability that the time required for a randomly selected service call is greater than two hours.

20.5 A new mayor of a large city is considering the staffing for calls to a emergency telephone number. The average number of calls per hour is known to be 340. The mayor wants to be confident that there are enough staffers for the calls.

    (a) Use Markov's inequality to find a bound for the 90th percentile of the number of calls per hour.

    (b) Suppose the variance for the number of calls is 20,000. Use Chebyshev's inequality to find a bound for the 90th percentile of the number of calls per hour.

20.6 Consider the density $f(x) = 2(x+1)^{-3}$ for $x > 0$.

    (a) Confirm that a random variable X having this density has a finite mean but undefined variance.

    (b) Use Markov's inequality to bound $P(X > 3)$ and compare this to the true probability.

Chapter 21

Transformation of Continuous Random Variables, Using the CDF Method

Suppose Y is a continuous random variable with known density $f_Y(y)$ and cumulative distribution function (CDF) $F_Y(y)$, and g is a function whose domain contains the support of Y. We are interested in the random variable $X = g(Y)$. We know how to compute $E(X)$, but we often want more information than the first moment of the new random variable. In this chapter we will use the **CDF method** to find the *distribution* of a random variable X that is a function of a continuous random variable Y. We can do this by finding the CDF $F_X(x)$ using the known CDF $F_Y(y)$, then taking the derivative with respect to x to get the density.

Example: Suppose $Y \sim \text{Unif}(0, 1)$ and $X = Y^2$. What is the distribution of the random variable X? (You can develop your intuition by making guesses—if we sample values between zero and one, and square these values, do they get bigger or smaller?) First, we determine the support of X: it's clear that X takes values in $(0, 1)$. Using the definition of CDF, we have for $x \in (0, 1)$:

$$
\begin{aligned}
F_X(x) &= P(X \leq x) \\
&= P(Y^2 \leq x) \\
&= P(Y \leq \sqrt{x}) \qquad \text{(because } Y \text{ takes only positive values)} \\
&= F_Y(\sqrt{x}) \\
&= \sqrt{x},
\end{aligned}
$$

where we used the known CDF for Y: $F_Y(y) = y$ for $y \in (0, 1)$. At this point we can check to make sure $F_X(0) = 0$ and $F_X(1) = 1$; that is, F_X goes from zero to one over the support of X. Then

$$
f_X(x) = \frac{d}{dx} F_X(x) = \frac{1}{2\sqrt{x}} \quad \text{for } x \in (0, 1).
$$

Let's check our calculations using R. Here is a short piece of code that simulates a large number of replicates from a uniform, squares the values, makes a histogram, and superimposes our calculated density on top of the histogram. The plots of the histograms of Y and X are shown with their respective densities. It's satisfying to verify your calculations this way!

```
y=runif(100000)
x=y^2
hist(x,freq=FALSE,br=50)
xpl=0:100/100
lines(xpl,1/2/sqrt(xpl),lwd=2,col=2)
```

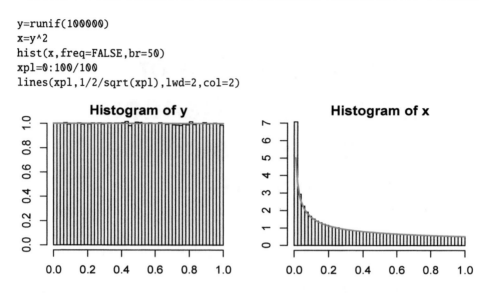

The same steps can be taken for any "reasonably smooth" increasing function g, and any continuous random variable Y. If we can compute the inverse function $g^{-1}(x)$, then the CDF for $X = g(Y)$ can be computed:

$$
\begin{aligned}
F_X(x) &= \mathrm{P}(X \le x) \\
&= \mathrm{P}\left(g(Y) \le x\right) \\
&= \mathrm{P}\left(Y \le g^{-1}(x)\right) \\
&= F_Y\left(g^{-1}(x)\right).
\end{aligned}
$$

The density for X is then

$$
f_X(x) = F_X'(x) = f_Y\left(g^{-1}(x)\right)\frac{d}{dx}g^{-1}(x).
$$

This formula is appropriate only if g is strictly increasing on the support of Y, and is defined only where g^{-1} is differentiable.

To use the formula for our initial example, we compute $g^{-1}(x) = \sqrt{x}$, then

$$
f_X(x) = F_X'(x) = f_Y\left(g^{-1}(x)\right)\frac{d}{dx}g^{-1}(x) = \frac{1}{2\sqrt{x}},
$$

because in this case $f_Y(g^{-1}(x)) = 1$.

Exercise 21.2 asks you to find a similar formula for the density of $X = g(Y)$, when g is a decreasing function and Y is a continuous random variable with density $f_Y(y)$.

Example: Suppose $Y \sim \mathrm{Unif}(-a, a)$ for $a > 0$. What is the distribution of $X = Y^2$? We cannot use the formula, because the function g is not increasing over the support of Y. We can still use the ideas, with a little care. This time X takes values in a different interval than Y; the support of X is $(0, a^2)$. We have $f_Y(y) = 1/(2a)$ on $(-a, a)$, so the CDF is

$$
F_Y(y) = \frac{y + a}{2a}.
$$

Using the definition of CDF, we have for $x \in (0, a^2)$,

$$
\begin{aligned}
F_X(x) &= \mathrm{P}(X \le x) \\
&= \mathrm{P}(Y^2 \le x) \\
&= \mathrm{P}(-\sqrt{x} \le Y \le \sqrt{x}) \\
&= \mathrm{P}(Y \le \sqrt{x}) - \mathrm{P}(Y \le -\sqrt{x}) \\
&= \frac{\sqrt{x} + a}{2a} - \frac{-\sqrt{x} + a}{2a} = \frac{\sqrt{x}}{a}.
\end{aligned}
$$

Before we take the derivative, check to see that $F_X(0) = 0$ and $F_X(a^2) = 1$. Then

$$
f_X(x) = \frac{d}{dx} F_X(x) = \frac{1}{2a\sqrt{x}} \quad \text{for } x \in (0, a^2).
$$

We can write R code to check this answer. The following code simulates many replicates of the uniform Y, then computes $X = Y^2$. A histogram of X should follow the density function if the calculations are correct. Superimposing the density function on the histogram confirms the calculations.

```
a=2
y=runif(100000,-a,a)
hist(y,freq=FALSE,br=40,main=expression(paste("histogram of  Y")))
lines(c(-2,2),c(.25,.25),col=2,lwd=2)
x=y^2
hist(x,freq=FALSE,br=40,main=expression(paste("histogram of X=",Y^2)))
xpl=1:400/100
lines(xpl,1/2/a/sqrt(xpl),col=2,lwd=2)
```

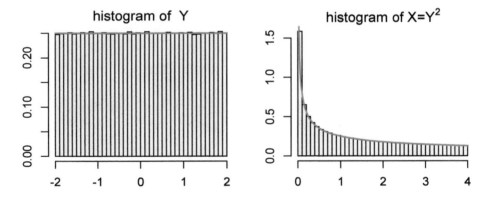

Let's look at the general case of a "shift" transformation, $X = Y + b$, where Y is a continuous random variable. Intuitively, we're just "moving" the distribution by b units, to the right if b is positive; note that the support for X is the support for Y, moved over b units. Let's get a formula:

$$
\begin{aligned}
F_X(x) &= \mathrm{P}(X \le x) \\
&= \mathrm{P}(Y + b \le x) \\
&= \mathrm{P}(Y \le x - b) \\
&= F_Y(x - b),
\end{aligned}
$$

so $f_X(x) = \frac{d}{dx} F_Y(x - b) = f_Y(x - b)$. This confirms our intuition about "moving" the density.

If we consider a more general linear transformation $X = aY + b$, where $a > 0$, then we're "moving and stretching" but the density should have the same *shape*. If the support for Y is (c, d), then the support for X is $(ac + b, ad + b)$. Let's get a formula:

$$\begin{aligned} F_X(x) &= \mathrm{P}(X \le x) \\ &= \mathrm{P}(aY + b \le x) \\ &= \mathrm{P}(Y \le (x - b)/a) \\ &= F_Y((x - b)/a), \end{aligned}$$

so

$$f_X(x) = \frac{d}{dx} F_Y\left(\frac{x - b}{a}\right) = \frac{1}{a} f_Y\left(\frac{x - b}{a}\right).$$

The sketch of f_X could be obtained from the sketch of f_Y if we erase the scale on the horizontal axis and write in a new scale.

If we don't know the CDF for the original random variable, we can still use the CDF method—we take the derivative and plug in the known density function, as shown in the next example.

Example: Let $Z \sim \mathrm{N}(0, 1)$; show $Y = \sigma Z + \mu \sim \mathrm{N}(\mu, \sigma^2)$. First, we observe that Y takes values on the real line; the support of Y is the same as the support of Z. Next we apply the CDF method (which we can do even if the CDF for Z is not known):

$$\begin{aligned} F_Y(y) &= \mathrm{P}(Y \le y) \\ &= \mathrm{P}(\sigma Z + \mu \le y) \\ &= \mathrm{P}\left(Z \le \frac{y - \mu}{\sigma}\right) \\ &= F_Z\left(\frac{y - \mu}{\sigma}\right). \end{aligned}$$

We don't have an expression for F_Z, but we can take the derivative with respect to y and use our expression for f_Z:

$$f_Y(y) = \frac{d}{dy} F_Z\left(\frac{y - \mu}{\sigma}\right) = \frac{1}{\sigma} f_Z\left(\frac{y - \mu}{\sigma}\right) = \frac{1}{\sigma\sqrt{2\pi}} e^{(y-\mu)^2/(2\sigma^2)}.$$

This shows that $Y \sim \mathrm{N}(\mu, \sigma^2)$. All normal random variables are linear transformations of the standard normal; this is why they all have the same shape.

Example: Let $Z \sim \mathrm{N}(0, 1)$; find the density for $Y = Z^2$. First we notice that the support for Y is the positive real line. Then we can use the CDF method: for $y > 0$,

$$\begin{aligned} F_Y(y) &= \mathrm{P}(Y \le y) \\ &= \mathrm{P}(Z^2 \le y) \\ &= \mathrm{P}(-\sqrt{y} \le Z \le \sqrt{y}) \\ &= \mathrm{P}(Z \le \sqrt{y}) - \mathrm{P}(Z \le -\sqrt{y}) \\ &= F_Z(\sqrt{y}) - F_Z(-\sqrt{y}). \end{aligned}$$

We don't have a closed form for F_Z, but we can still take the derivative:

$$
\begin{aligned}
f_Y(y) &= \frac{d}{dy} F_Y(y) \\
&= \frac{d}{dy} F_Z(\sqrt{y}) - \frac{d}{dy} F_Z(-\sqrt{y}) \\
&= \frac{1}{2\sqrt{y}} f_Z(\sqrt{y}) + \frac{1}{2\sqrt{y}} f_Z(-\sqrt{y}) \qquad \text{(using the chain rule)} \\
&= \frac{1}{\sqrt{y}} f_Z(\sqrt{y}),
\end{aligned}
$$

because f_Z is an even function. Now we plug this into the expression for the standard normal random variable to get

$$
f_Y(y) = \frac{1}{\sqrt{2\pi}} y^{-1/2} e^{-y/2}.
$$

This looks like a gamma density function, with $\beta = 1/2$ and $\alpha = 1/2$. (Recall that $\Gamma(1/2) = \sqrt{\pi}$.) This special gamma density has its own name; it is called a chi-squared density with one degree of freedom. We'll come back to the chi-squared densities, and define a *family* of densities, indexed by the positive integers, later in the book. It's a very important density for many types of statistical inference, including least-squares regression and models with categorical data. We use the term "degrees of freedom" for the index of the density, because there is an analogy to the dimension of a subspace (this is made explicit in a course on linear models).

Example: Suppose $Y \sim \text{Exp}(1/\theta)$ (so Y has mean θ), and interest is in finding the density for $X = e^Y$. The support for X is $(1, \infty)$. We know $F_Y(y) = 1 - e^{-y/\theta}$, and for $x > 1$,

$$
\begin{aligned}
F_X(x) &= P(X \le x) \\
&= P(e^Y \le x) \\
&= P\left(Y \le \log(x)\right) \\
&= 1 - e^{-\log(x)/\theta} \\
&= 1 - e^{\log(x^{-1/\theta})} \\
&= 1 - x^{-1/\theta}.
\end{aligned}
$$

Then the density for X is

$$
f_X(x) = F_X'(x) = \frac{1}{\theta} x^{-(1/\theta + 1)} \quad \text{for } x > 1.
$$

We can check that this is a proper density by integrating the density function over the support. We notice, though, that if $\theta \ge 1$, the mean for this density is undefined. We can get a sample from this density in R using our CDF tricks, but the sample mean will vary wildly from sample to sample. The variance is undefined if $\theta \ge 1/2$, and the third moment will be undefined if $\theta \ge 1/3$, etc.

For any θ, this density is considered "heavy-tailed." Even if θ is small, values that are many standard deviations away from the mean occur with a much higher frequency than for "light-tailed" distributions such as for a normal random variable. When $\theta = 1/3$, the expected value of X is 3, and the variance is 3/4. We can also calculate that almost 1% of the distribution is more than four standard deviations above the mean. In contrast, the probability of a standard normal random variable being larger than four is about .000032.

We can check our calculations using R code, generating a large sample from an exponential with mean 1/3, say, then squaring the elements of the sample. We get a histogram of our squared values, and then overlay our density function. The maximum of the sample is likely to be in the hundreds, so to get a histogram for which we can see the features of the density, we truncate the sample as shown below.

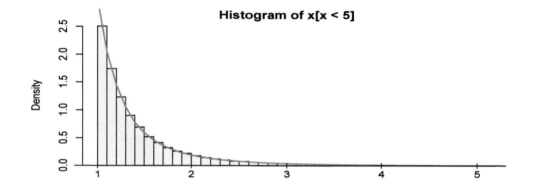

Here is the R code for the histogram:

```
y=rexp(1000000,3)
x=exp(y)
hist(x[x<5],br=50,freq=FALSE,col="beige",xaxt="n")
xpl=0:600/10
fpl=xpl^(-4)*3
lines(xpl,fpl,lwd=2,col=2)
```

Chapter Highlights

1. If Y is a continuous random variable with known density f_y, and g is a function whose domain contains the support of Y, the CDF method can often be used to determine the distribution of $X = g(Y)$. Steps for the CDF method:

   • Determine the support of X.

   • Find the CDF $F_X(x)$ as a function of the CDF $F_Y(y)$.

   • Then take the derivative of $F_X(x)$, with respect to x, to get the density.

   • Check your answer!

2. For $X = aY + b$, where $a > 0$,

$$f_X(x) = \frac{1}{a} f_Y\left(\frac{x-b}{a}\right).$$

If the support for X is (c, d), then the support for Y is $(ac + b, ad + b)$.

3. All normal random variables are linear transformations of the standard normal random variable.

4. If $Y \sim \mathrm{N}(0, 1)$, then Y^2 has a chi-squared density with one degree of freedom (an important special case of a gamma distribution).

Exercises

21.1 Let $X \sim \mathrm{Unif}(0, 1)$, find the density function for $Y = X^n$, where n is a positive integer. Verify your answer for $n = 3$ using simulations in R.

21.2 Suppose Y is a continuous random variable with density $f_Y(y)$. Using the CDF method, find a formula for the density of $X = g(Y)$ when g is a strictly decreasing function whose domain contains the support of Y.

21.3 Suppose X is a continuous random variable with support $(-a, a)$, with density $f_X(x)$ that is symmetric about the vertical axis (i.e., $f_X(x)$ is an even function). Find an expression for the density of $Y = |X|$.

21.4 Find the density for $X = e^Y$ if $Y \sim \mathrm{N}(\mu, \sigma^2)$. (This is called a log-normal density.) Verify your answer in R by generating many Y values (using your choice of μ and σ), getting a density histogram for $X = e^Y$, and adding the curve for your solution to the histogram.

21.5 Let $Y \sim \mathrm{Beta}(\alpha, \beta)$ and find the density of $X = 1 - Y$.

21.6 Let $Y \sim \mathrm{Gamma}(\alpha, \beta)$ and find the density of $X = 2Y + 1$.

21.7 Let $Y \sim \mathrm{Gamma}(\alpha, \beta)$ and find the density of $X = \beta Y$.

21.8 Show that if $Y \sim \mathrm{Gamma}(\alpha, \beta)$, then $X = 1/Y \sim \mathrm{InvGamma}(\alpha, \beta)$.

21.9 Suppose $Y \sim \mathrm{Exp}(\theta)$ and find the density for $X = \sqrt{Y}$. Using simulations in R, confirm that your answer is correct using $\theta = 2$.

21.10 Let θ be uniformly distributed on $(-\pi/2, \pi/2)$. Find the density function for $R = A\sin(\theta)$. Verify your answer in R by choosing $A = 2$, generating many values of θ, getting a density histogram for $2\sin(\theta)$, and adding the curve for your solution to the histogram.

21.11 Let $U \sim \text{Unif}(0, 4)$ and $Y = \sqrt{U}$. Find the density for Y, and check your answer using simulations in R.

21.12 Let $U \sim \text{Unif}(0, 1)$ and $Y = -2\log(U)$. Find the density for Y, and check your answer using simulations in R.

21.13 Let U be a random variable that is uniformly distributed on (a, b), where $0 < a < b < \infty$. Find the density of $Y = 1/U$, and check your answer for some values of a and b using simulations in R.

21.14 Let U be a random variable that is uniformly distributed on $(0, 1)$. Find the density of $Y = 1 - U$.

21.15 Let U be a random variable that is uniformly distributed on $(0, 1)$. Find the density of $Y = -2\log(1 - U)$. Do simulations in R to check your answer.

21.16 Let $Y \sim \text{Exp}(\theta)$, and $X = Y^2$. Find the density of X, and check your answer for some value of θ using simulations in R.

21.17 Suppose $X \sim \text{Gamma}(2, 2)$ and $Y = \sqrt{X}$. Find the density function for Y, and check your answer using simulations in R.

21.18 Suppose $Y \sim \text{Gamma}(\alpha, 1)$, and $X = \log(Y)$. Find the density function for X, and check your answer using simulations in R.

21.19 Suppose Y is a random variable with density function

$$f_Y(y) = \frac{2}{\theta}y e^{-y^2/\theta} \quad \text{for } y > 0.$$

(a) Using the CDF method, show that $X = Y^2$ is an exponential random variable, and determine $\text{E}(X)$ in terms of θ.

(b) Verify your answer using simulations in R by generating a large vector of realizations from an exponential variable with your choice of parameter, taking the square root of the values of this vector, making a histogram (with freq=FALSE), then superimposing the density curve for f_Y.

21.20 Let X be a random variable with density

$$f_X(x) = \frac{3}{(x+1)^4} \quad \text{for } x > 0.$$

Find constants a and b so that $Y = a + bX$ has Y of mean 12 and standard deviation 15.

21.21 For $\theta > 0$, let X be a random variable with density

$$f_X(x) = \theta x^{\theta-1} \quad \text{for } x \in (0, 1).$$

Find constants a and b so that $Y = a + bX$ has support on $(2, 4)$, and derive the density for Y.

21.22 Suppose X is a random variable that follows the triangular density shown:

Let $Y = X^2$. Find the density function for Y and sketch it.

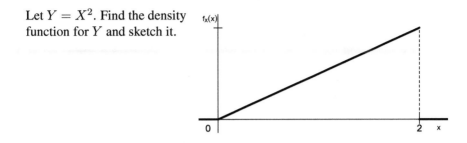

21.23 The random variable X has the density shown below.

(a) Let $Y = X^2$; find the density for Y and sketch it.

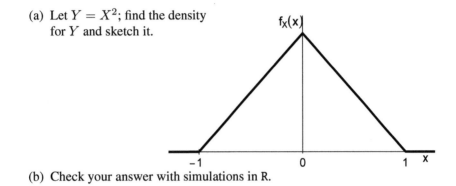

(b) Check your answer with simulations in R.

21.24 Suppose that for some $\theta > 0$, Y is a random variable with density

$$f_\theta(y) = \theta(1 - y)^{\theta - 1} \ \text{ for } \ y \in (0, 1).$$

Let $W = -\log(1 - Y)$ and find the density of W.

Chapter 22

Log-Normal, Weibull, and Pareto Random Variables

Some important families of distributions are derived through transformations, in a way that is natural to an application. A random variable has a **log-normal** distribution if its logarithm has a normal distribution. This two-parameter distribution has support on the positive real numbers, and is unimodal with a skewed tail. In biological applications, it's useful for modeling sizes—weights or lengths of organisms or plants. In economics or engineering/technology applications, it can be used to model times between events.

The **Weibull** distribution can be thought of as an extension of the gamma family. It also has positive support, and is popular for modeling survival times. The **Rayleigh** distribution is a special case of the Weibull and has its own name because of its uses in modeling magnitudes of directional phenomena, such as wind speeds.

The **Pareto** distribution is obtained through a transformation of the exponential random variable. It is useful whenever a "heavy-tailed" density is appropriate—for example, in modeling climate, or modeling probabilities of extreme events for insurance purposes.

Log-normal random variables

If X is a normal random variable with mean μ and variance σ^2, then $Y = e^X$ has a log-normal distribution. We'll write $Y \sim \text{LN}(\mu, \sigma^2)$ to indicate that the random variable Y has this distribution. From the definition, it's clear that the support for Y is the set of positive real numbers. Let's find the CDF for Y and take the derivative to get the density. For $y > 0$,

$$
\begin{aligned}
F_Y(y) &= \mathrm{P}(Y \leq y) \\
&= \mathrm{P}(e^X \leq y) \\
&= \mathrm{P}(X \leq \log(y)) \\
&= F_X(\log(y)).
\end{aligned}
$$

We don't have a closed-form expression for the CDF of a normal random variable, so we simply take the derivative:

$$
\begin{aligned}
f_Y(y) &= \frac{d}{dy} F_Y(y) = \frac{d}{dy} F_X(\log(y)) \\
&= \frac{1}{y} f_X(\log(y)) \\
&= \frac{1}{y\sqrt{2\pi\sigma^2}} \exp\left(-\frac{(\log(y) - \mu)^2}{2\sigma^2}\right).
\end{aligned}
$$

189

Simulating from a log-normal density is easy: We generate a normal random variable X, and $Y = e^X$. Let's make a histogram of many values generated in this way to check our derivation. The code on the left generates the histogram on the right, superimposing the density function that we derived. You can play with this code by changing mu and sig to see how this changes the density.

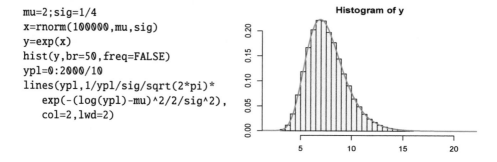

```
mu=2;sig=1/4
x=rnorm(100000,mu,sig)
y=exp(x)
hist(y,br=50,freq=FALSE)
ypl=0:2000/10
lines(ypl,1/ypl/sig/sqrt(2*pi)*
    exp(-(log(ypl)-mu)^2/2/sig^2),
    col=2,lwd=2)
```

Let's derive the mean and variance of a log-normal random variable:

$$E(Y) = \frac{1}{\sqrt{2\pi\sigma^2}} \int_0^\infty \exp\left(-\frac{(\log(y) - \mu)^2}{2\sigma^2}\right) dy.$$

Making a substitution $u = (\log(y) - \mu)/\sigma$, we have $du = dy/(y\sigma)$ and $y = \exp(\sigma u + \mu)$. When we make this substitution, the integration limits change. We have

$$E(Y) = \frac{1}{\sqrt{2\pi}} \int_{-\infty}^\infty \exp(\sigma u + \mu) \exp(-u^2/2) \, du.$$

Now we complete the square in the exponent:

$$-u^2/2 + \sigma u + \mu = \mu + \frac{1}{2}\sigma^2 - \frac{1}{2}(u^2 - 2\sigma u + \sigma^2) = \mu + \frac{1}{2}\sigma^2 - \frac{1}{2}(u - \sigma)^2,$$

so

$$\begin{aligned} E(Y) &= e^{\mu + \sigma^2/2} \left[\frac{1}{\sqrt{2\pi}} \int_{-\infty}^\infty \exp\left(-\frac{1}{2}(u - \sigma)^2\right) du\right] \\ &= e^{\mu + \sigma^2/2}, \end{aligned}$$

because the expression in the brackets is the area under a normal density with mean σ and variance 1. Exercise 22.1 asks you to compute the second moment; you can use the same variable substitution.

The median is straightforward, because the CDF for Y is

$$F_Y(y) = F_X(\log(y)),$$

and we know $F_X(\mu) = \frac{1}{2}$. Therefore, $\log(m) = \mu$, and the median is $m = e^\mu$.

The moment generating function (MGF) does not exist for the log-normal random variable. It's not that the integral is intractable, but the integral is simply undefined. If we write, for $Y \sim LN(\mu, \sigma^2)$,

$$M_Y(t) = E(e^{tY}) = \frac{1}{\sqrt{2\pi\sigma^2}} \int_0^\infty e^{ty} \frac{1}{y} \exp\left(-\frac{(\log(y) - \mu)^2}{2\sigma^2}\right) dy,$$

and again make the substitution $u = (\log(y) - \mu)^2/\sigma$, we get

$$M_Y(t) = \frac{1}{\sqrt{2\pi}} \int_{-\infty}^{\infty} e^{te^{u\sigma+\mu}} e^{-u^2/2} du,$$

which "blows up" unless $t = 0$. This nonexistence of the MGF leads to strange phenomena; for example, there exist random variables that have the same kth moments as a log-normal for $k = 1, 2, \ldots$, but have quite different distributions. However, when the MGFs exist for two random variables, and their moments are the same, then the random variables have the same distributions.

Weibull random variables

A Weibull random variable results from a power of an exponential random variable. Suppose $X \sim \text{Exp}(1/\beta)$ (so that $\text{E}(X) = \beta$)), and let $Y = \beta^{(\alpha-1)/\alpha} X^{1/\alpha}$ for $\alpha > 0$. We will write $Y \sim \text{Weib}(\alpha, \beta)$ to indicate a random variable with this distribution. We use the CDF rule to derive the density. The support of Y is the positive reals, so for $y > 0$, we have

$$\begin{aligned} F_Y(y) = \text{P}(Y \le y) &= \text{P}(\beta^{(\alpha-1)/\alpha} X^{1/\alpha} \le y) \\ &= \text{P}(X \le y^\alpha/\beta^{\alpha-1}) \\ &= F_X(y^\alpha) = 1 - e^{-(y/\beta)^\alpha}. \end{aligned}$$

Then

$$\frac{d}{dy} F_Y(y) = f_Y(y) = \frac{\alpha}{\beta^\alpha} y^{\alpha-1} e^{-(y/\beta)^\alpha} \quad \text{for } y > 0.$$

When $\alpha = 1$, we recognize the density for an exponential random variable. To further check our derivation by finding the area under the density function, we make the substitution $u = (y/\beta)^\alpha$ and $du = \alpha y^{\alpha-1}/\beta^\alpha$, so

$$\frac{\alpha}{\beta^\alpha} \int_0^\infty y^{\alpha-1} e^{-(y/\beta)^\alpha} = \int_0^\infty e^{-u} du = 1.$$

You might be wondering, why not define the Weibull to simply be $X^{1/\alpha}$ or, better yet, X^α, instead of having the complicated coefficient and power? In fact, there are other parameterizations of the Weibull random variable, which are all multiples of a power of an exponential random variable, but this particular version is coded into R. The command `rweibull(100,alpha,beta)` will generate 100 values of a Weibull random variable with the parameters `alpha` and `beta` matching the definitions given above. To check this, we can use the following code to generate $\text{Exp}(1/\beta)$ random variables in a vector `x`, and for a given α plot the sorted values of $y = \beta^{(\alpha-1)/\alpha} x^{1/\alpha}$ against the quantiles of a $\text{Weib}(\alpha, \beta)$ random variable.

```
a=2;b=4;n=1000
x=rexp(n,1/b)
y=b^((a-1)/a)*x^(1/a)
plot(qweibull(1:n/(n+1),a,b),sort(y))
lines(c(0,12),c(0,12),col=2,lwd=2)
```

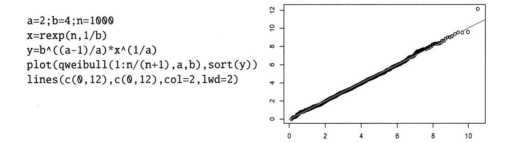

You can use `pweibull(y,alpha,beta)` to get $P(Y \leq y)$, where $Y \sim \text{Weib}(\alpha, \beta)$, and `qweibull(p,alpha,beta)` returns y for which $P(Y \leq y) = c$. Finally, `dweibull(y,alpha,beta)` returns $f(y)$, the value of the $\text{Weib}(\alpha, \beta)$ density at y. The flexibility of the Weibull family is illustrated below:

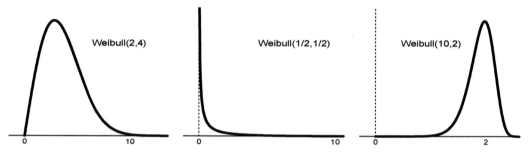

To get the moments of a Weibull random variable, we use the same substitution: For $Y \sim \text{Weib}(\alpha, \beta)$,

$$E(Y) = \frac{\alpha}{\beta^\alpha} \int_0^\infty y^\alpha e^{-(y/\beta)^\alpha} = \beta \int_0^\infty u^{1/\alpha} e^{-u} du = \beta \Gamma(1 + 1/\alpha).$$

A similar calculation leads to

$$E(Y^2) = \beta^2 \Gamma(1 + 2/\alpha),$$

so that the variance of a $\text{Weib}(\alpha, \beta)$ random variable is

$$V(Y) = \beta^2 \left[\Gamma(1 + 2/\alpha) - \Gamma(1 + 1/\alpha)^2 \right].$$

The **Rayleigh** family of distributions is a subfamily of the Weibull distributions with $\alpha = 2$, and the β parameter of the Weibull is often indicated as σ. We will use $Y \sim \text{Rayleigh}(\sigma^2)$ if $Y \sim \text{Weib}(2, \sigma)$, so that the Rayleigh density is

$$f(y) = \frac{2y}{\sigma^2} e^{-y^2/\sigma^2} \quad \text{for } y > 0.$$

Pareto random variables

Pareto distributions are commonly used to model "heavy-tailed" phenomena. This means that while most of the values of random variables having this distribution will be "small," occasionally we can see a value that is "quite large." Specifically, if X is an $\text{Exp}(\alpha)$ random variable with $\alpha > 0$ (and mean $1/\alpha$), and we define for $\beta > 0$ the transformation

$$Y = \beta(e^X - 1),$$

then Y is a Pareto random variable, written $Y \sim \text{Pareto}(\alpha, \beta)$. Let's derive the density function. The CDF for Y is

$$\begin{aligned}
F_Y(y) = P(Y \leq y) &= P\left(\beta(e^X - 1) \leq y \right) \\
&= P\left(X \leq \log(1 + y/\beta) \right) \\
&= 1 - e^{-\alpha \log(1 + y/\beta)} \\
&= 1 - \left(1 + \frac{y}{\beta} \right)^{-\alpha} = 1 - \frac{\beta^\alpha}{(y + \beta)^\alpha}.
\end{aligned}$$

Taking the derivative with respect to y gives the density function

$$f_Y(y) = \frac{\alpha \beta^\alpha}{(y + \beta)^{\alpha+1}} \quad \text{for } y > 0.$$

The parameter α is called the "shape" parameter, and β is the "scale" parameter.

Functions related to the Pareto distribution can be found in R in the package `actuar`. Let's check our formulation of the Pareto distribution against that in the package. If we generate a large number of exponential random variables with parameter α, then compute the transformation, we can plot these against the Pareto quantiles. First the package must be installed.

```
alp=8;bet=3;n=1000
x=rexp(n,alp)
y=bet*(exp(x)-1)
plot(qpareto(1:n/(n+1),alp,bet),sort(y),
    xlab="pareto quantiles",
    ylab="sorted sample")
lines(c(0,10),c(0,10),col=2,lwd=3)
```

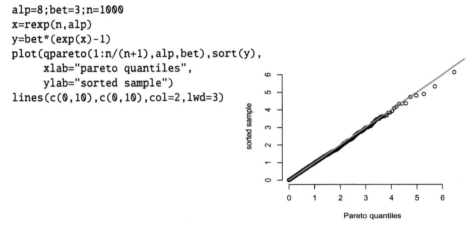

The Pareto densities $f_Y(y)$ are decreasing in y:

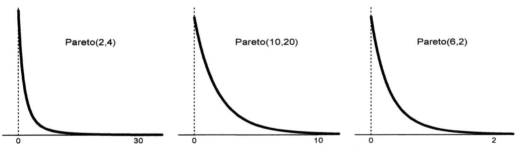

Let's find the expected value of a Pareto(α, β) random variable Y. We use the change of variable $u = y + \beta$:

$$\begin{aligned}
\mathrm{E}(Y) &= \alpha\beta^\alpha \int_0^\infty y(y + \beta)^{-\alpha-1} dy \\
&= \alpha\beta^\alpha \int_\beta^\infty (u - \beta)u^{-\alpha-1} du \\
&= \alpha\beta^\alpha \left[\int_\beta^\infty u^{-\alpha} du + \beta \int_\beta^\infty u^{-\alpha-1} du \right] \\
&= \alpha\beta^\alpha \left[\frac{1}{\alpha - 1}\beta_{-\alpha+1} - \frac{\beta}{\alpha}\beta^{-\alpha} \right] \\
&= \frac{\beta}{\alpha - 1}.
\end{aligned}$$

We see that the expected value is not defined unless $\alpha > 1$. You are asked to compute the variance (and find a further restriction on α) in Exercise 22.8.

Chapter Highlights

1. For $\sigma^2 > 0$ and $\mu \in \mathbb{R}$, the log-normal random variable $Y \sim \text{LN}(\mu, \sigma^2)$

   (a) has density

   $$f_Y(y) = \frac{1}{y\sqrt{2\pi\sigma^2}} \exp\left(-\frac{(\log(y) - \mu)^2}{2\sigma^2}\right) \quad \text{for } y > 0;$$

   (b) has mean and variance

   $$\text{E}(Y) = e^{\mu + \sigma^2/2} \quad \text{and} \quad \text{V}(Y) = e^{2\mu + \sigma^2}(e^{\sigma^2} - 1),$$

   and the median is e^μ.

   (c) If $X \sim \text{N}(\mu, \sigma^2)$ and $Y = e^X$, then $Y \sim \text{LN}(\mu, \sigma^2)$.

2. For $\alpha > 0$ and $\beta > 0$, the Weibull random variable $Y \sim \text{Weibull}(\alpha, \beta)$

   (a) has density

   $$f_Y(y) = \frac{\alpha}{\beta^\alpha} y^{\alpha - 1} e^{-(y/\beta)^\alpha} \quad \text{for } y > 0;$$

   (b) has mean and variance

   $$\text{E}(Y) = \beta\Gamma(1 + 1/\alpha) \quad \text{and} \quad \text{V}(Y) = \beta^2 \left[\Gamma(1 + 2/\alpha) - \Gamma(1 + 1/\alpha)^2\right].$$

   (c) If $X \sim \text{Exp}(1/\beta)$ and $Y = \beta^{(\alpha-1)/\alpha} X^{1/\alpha}$, then $Y \sim \text{Weibull}(\alpha, \beta)$.

3. The Rayleigh(σ^2) random variable is a Weibull$(2, \sigma)$ random variable.

4. For $\alpha > 0$ and $\beta > 0$, the Pareto random variable $Y \sim \text{Pareto}(\alpha, \beta)$

   (a) has density

   $$f_Y(y) = \frac{\alpha\beta^\alpha}{(y + \beta)^{\alpha+1}} \quad \text{for } y > 0;$$

   (b) has the following mean and variance defined when $\alpha > 1$ and $\alpha > 2$, respectively:

   $$\text{E}(Y) = \frac{\beta}{\alpha - 1} \quad \text{and} \quad \text{V}(Y) = \frac{\beta^2 \alpha}{(\alpha - 1)^2(\alpha - 2)}.$$

   (c) If X is an $\text{Exp}(\alpha)$ random variable and $Y = \beta(e^X - 1)$, then $Y \sim \text{Pareto}(\alpha, \beta)$.

Exercises

22.1 Derive the second moment and the variance of an $LN(\mu, \sigma^2)$ random variable.

22.2 Using the `qnorm` and `pnorm` functions in R, find

(a) the 95th percentile of an $LN(0, 1)$ random variable, and

(b) the probability that an $LN(2, 1/2)$ random variable is larger than 16.

22.3 Suppose the annual snowfall (in inches) in a region can be modeled as an $LN(1, 2)$ random variable.

(a) What is the probability that the snowfall is greater than 20 inches in a randomly selected year?

(b) What are the mean and variance of the annual snowfall?

22.4 Suppose the annual snowfall (in inches) in a mountainous region can be modeled as a log-normal random variable. The average annual snowfall is 20 inches, with standard deviation 20 inches.

(a) Find the parameters of a log-normal random variable that has this mean and variance.

(b) What proportion of years have snowfall greater than 60 inches?

22.5 Using the `qweibull` and `pweibull` functions in R, find

(a) the 95th percentile of a Weibull$(2, 4)$ random variable, and

(b) the probability that a Weibull$(2, 4)$ random variable is larger than 8.

22.6 Switches for a transportation system are manufactured either by Company A or by Company B. Lifetimes of Company A switches (in years) follow a Weibull$(2, 20)$ distribution, while Company B switch lifetimes (in years) follow a Weibull$(4, 15)$ distribution. Further, 80% of switches used for the system are produced by Company B.

(a) If a switch used for the system is chosen at random, what is the probability that it lasts for more than 20 years?

(b) Suppose a switch is chosen at random and found to last more than 20 years. What is the probability that it was made by Company A?

22.7 Parts vital for the life-support system on the space shuttle have lifetimes (in hours) that are distributed as Weibull$(2, 20)$ random variables. Suppose that when a part wears out, it can be replaced by a new part. Write R code to determine the minimum number of parts that should be brought on a mission that lasts 110 hours if the probability that the parts last more than 110 hours (used in succession) is greater than 99.9%.

22.8 Show how to compute the variance of a Pareto(α, β) random variable. What is the condition on α, for the variance to be defined?

22.9 Use Chebyshev's inequality to bound the probability that a Pareto$(3, 4)$ random variable is greater than 42, and compare this to the actual probability.

22.10 Suppose the annual maximum single-day rainfall (in millimeters) in a farming community in India can be modeled as a Pareto$(4, 12)$ random variable.

(a) What is the average maximum single-day rainfall?

(b) What is the 90th percentile of the maximum annual rainfall?

(c) The field behind the town hall will flood if the single-day rainfall is greater than 10 mm (1 cm). What proportion of years does the flood occur?

(d) The downtown area will flood if the single-day rainfall is greater than 20 mm (2 cm). What proportion of years does the flood occur?

22.11 Recall the definition of a chi-squared random variable from Chapter 16. Show that if $X \sim \chi^2(2)$, then $Y = X^{1/2}$ has one of the densities described in this chapter.

Chapter 23

Jointly Distributed Discrete Random Variables

Often we need to consider two or more random variables at the same time, where the random variables might vary together. The **joint distribution** of two random variables describes the probabilities associated with *pairs* of values that the two can take.

In this chapter we will consider the joint distribution of two discrete random variables (the joint probability mass function), and in the next chapter we consider the case of two continuous random variables (the joint probability density). We will briefly generalize these concepts to consider the joint distribution of three or more random variables. The joint distribution of more than two random variables will be applied in a couple of ways. First, the distribution of the multinomial random variable will be formulated as a joint distribution of several discrete random variables that are related to each other. Second, we'll consider the joint distribution of n independent random variables when we talk about sampling distributions in later chapters.

The joint probability mass function of two discrete random variables Y_1 and Y_2 is a function $f(y_1, y_2)$ that assigns probabilities to all possible pairs of values, so that $P(Y_1 = y_1, Y_2 = y_2) = f(y_1, y_2)$. Suppose S is the joint support of the random variables, that is, the ordered pairs of values in $\mathbb{R}^2$ that Y_1 and Y_2 can take. We must have $0 < f(y_1, y_2) \le 1$ for all pairs $(y_1, y_2) \in S$, and

$$\sum\sum_{(y_1, y_2) \in S} f(y_1, y_2) = 1.$$

When we describe the joint distribution of two discrete random variables, we can sometimes use a formula, but often it's more convenient to make a table of probabilities, where the values that Y_1 can take, for example, run along the top of the table and the values Y_2 can take run along the side.

Example: Suppose we have a box of five switches, two of which are defective. We test them one by one, at random and without replacement. Let Y_1 be the draw on which the first defective is found, and Y_2 is the draw on which the second is found. For example, if the first two tested are both defective, we have $Y_1 = 1$ and $Y_2 = 2$. Find the joint probability mass function for Y_1 and Y_2.

For problems where there are a small number of values the pair can take, we can compute the probability for each pair and enter the probability into the table. The random variable Y_1 can take values 1, 2, 3, and 4, while Y_2 can take values 2, 3, 4, and 5.

The probability that $Y_1 = 1$ and $Y_2 = 2$ is $(2/5)(1/4) = 1/10$. The probability that $Y_1 = 1$ and $Y_2 = 3$ is $(2/5)(3/4)(1/3) = 1/10$, etc. After using the same reasoning for all possible pairs, we get the following mass function:

		Y_1			
		1	2	3	4
	2	1/10	0	0	0
Y_2	3	1/10	1/10	0	0
	4	1/10	1/10	1/10	0
	5	1/10	1/10	1/10	1/10

Now we can compute probabilities such as $P(Y_1 > 2) = 3/10$ simply by adding the probabilities in the table that correspond to the event $Y_1 > 2$.

Example: Suppose we toss a coin for which the probability of heads is p. Let Y_1 be the toss on which the first head occurs, and let Y_2 be the toss on which the second head occurs. This time we don't have a finite support for the joint distribution; we can have $Y_1 = y_1$ and $Y_2 = y_2$ for any $1 \le y_1 \le y_2 - 1 < \infty$. For $y_2 \ge y_1 + 1$ and $y_1 \ge 1$, we can use our ideas from the geometric distribution to determine that

$$P(Y_1 = y_1, Y_2 = y_2) = \left[p(1-p)^{y_1 - 1} \right] \left[p(1-p)^{y_2 - y_1 - 1} \right],$$

where the term in the first square bracket is the probability that it takes y_1 tosses to get to the first head, and the term in the first square bracket is the probability that it takes $y_2 - y_1$ *more* tosses to get to the second head (and all the toss outcomes are independent). This simplifies to

$$P(Y_1 = y_1, Y_2 = y_2) = p^2 (1-p)^{y_2 - 2} \quad \text{for } 1 \le y_1 < y_2 < \infty.$$

Let's make sure these probabilities add to one. Summing over the support (and noticing that it's easier to have the y_2 sum on the "outside"), we get

$$p^2 \sum_{y_2=2}^{\infty} \sum_{y_1=1}^{y_2-1} (1-p)^{y_2-2} = p^2 \sum_{y_2=2}^{\infty} (y_2-1)(1-p)^{y_2-2}.$$

Now let's make a change of variable $k = y_2 - 2$, so our expression for the sum of the probabilities is

$$p^2 \sum_{k=0}^{\infty} (k+1)(1-p)^k = p^2 \sum_{k=1}^{\infty} k(1-p)^k + p^2 \sum_{k=0}^{\infty} (1-p)^k.$$

Using our summation expressions from Appendix B.2, this simplifies to

$$p^2 \left[\frac{1-p}{p^2} \right] + p^2 \left[\frac{1}{p} \right] = 1.$$

The **marginal distribution** for Y_1 is simply the probability mass function for Y_1. If we obtain repeated random samples from the joint distribution, and observe only the Y_1 values, the proportions corresponding to the values will describe the marginal distribution.

Given the joint distribution values, we can compute the marginal for Y_1 by summing over the Y_2 values:

$$f_1(y_1) = P(Y_1 = y_1) = \sum_{all \ y_2} f(y_1, y_2).$$

Similarly, the marginal for Y_2 is

$$f_2(y_2) = \mathrm{P}(Y_2 = y_2) = \sum_{all\ y_1} f(y_1, y_2).$$

Example: For the switches example, we can find the marginal distributions for both Y_1 and Y_2 by adding up columns and rows, respectively. This is intuitive—the probability that $Y_1 = 4$ is 1/10, but $\mathrm{P}(Y_1 = 3) = 2/10$, because there are two ways Y_1 can be 3, and we have to add these probabilities. Thus, the marginal distribution of Y_1 is given in the table

y	1	2	3	4
$\mathrm{P}(Y_1 = y)$	4/10	3/10	2/10	1/10

Similarly, the marginal distribution for Y_2 for the switches example is

y	2	3	4	5
$\mathrm{P}(Y_2 = y)$	1/10	2/10	3/10	4/10

Example: Going back to the coin toss example, we know that the marginal distribution of $Y_1 - 1$ ought to be geometric with probability p. Let's verify this for Y_1 using the formula for marginal distribution:

$$f_1(y_1) = \sum_{y_2 = y_1 + 1}^{\infty} p^2 (1-p)^{y_2 - 2}.$$

If we make a change of variable to $k = y_2 - y_1 - 1$, we have

$$f_1(y_1) = p^2 \sum_{k=0}^{\infty} (1-p)^{k + y_1 - 1}$$

$$= p^2 (1-p)^{y_1 - 1} \sum_{k=0}^{\infty} (1-p)^k$$

$$= p^2 (1-p)^{y_1 - 1} \frac{1}{p} = p(1-p)^{y_1 - 1}.$$

To find the marginal distribution for Y_2, we have

$$f_2(y_2) = p^2 \sum_{y_1=1}^{y_2 - 1} (1-p)^{y_2 - 2} = p^2 (y_2 - 1)(1-p)^{y_2 - 2} \quad \text{for } y_2 = 2, 3, \ldots.$$

To verify that this adds to one, we compute

$$\sum_{y_2=2}^{\infty} f_2(y_2) = p^2 \sum_{y_2=2}^{\infty} (y_2 - 1)(1-p)^{y_2 - 2},$$

and making the change of variable $k = y_2 - 2$, this is

$$p^2 \sum_{k=0}^{\infty} (k+1)(1-p)^k = p^2 \left[\sum_{k=0}^{\infty} k(1-p)^k + \sum_{k=0}^{\infty} (1-p)^k \right] = p^2 \left[\frac{1-p}{p^2} + \frac{1}{p} \right] = 1.$$

The **conditional mass function** of Y_2, given $Y_1 = y_1$, is a probability mass function that is computed from the joint mass function and the marginal for Y_1 in a natural way, given the definition of conditional probabilities from Chapter 4. Given $Y_1 = y_1$, the probabilities for each value of y_2 are

$$f(y_2|y_1) = \frac{P(Y_1 = y_1, Y_2 = y_2)}{P(Y_1 = y_1)} = \frac{f(y_1, y_2)}{f_1(y_1)}.$$

It's important to remember two things. First, $f(y_2|y_1)$ is a function of y_2, with y_1 as a fixed number. Second, $f(y_2|y_1)$ is a probability mass function, for any y_1 in the support of Y_1. It's clear that all the values of the conditional mass function are nonnegative, and we can show that they must add to one,

$$\sum_{all\ y_2} f(y_2|y_1) = \sum_{all\ y_2} \frac{f(y_1, y_2)}{f_1(y_1)} = \frac{\sum_{all\ y_2} f(y_1, y_2)}{f_1(y_1)} = \frac{f_1(y_1)}{f_1(y_1)} = 1,$$

by definition of the marginal distribution.

Example: For the switches example, the conditional distribution of Y_2 given $Y_1 = 2$ can be specified as follows: $P(Y_2 = 1|Y_2 = 2) = 0$ and $P(Y_2 = 3|Y_1 = 2) = P(Y_2 = 4|Y_1 = 2) = P(Y_2 = 5|Y_1 = 2) = 1/3$. Note that the values of the conditional mass function add to one.

Example: For the coin toss example, let's find the conditional distribution of Y_2, given $Y_1 = y_1$. We already have the marginal for Y_1, so

$$f(y_2|y_1) = \frac{p^2(1-p)^{y_2-2}}{p(1-p)^{y_1-1}} = p(1-p)^{y_2-y_1-1} \quad \text{for} \ \ y_2 = y_1 + 1, y_1 + 2, \dots.$$

Next, let's find the conditional distribution of Y_1, given $Y_2 = y_2$ (do you have a guess?):

$$f(y_1|y_2) = \frac{p^2(1-p)^{y_2-2}}{p^2(y_2 - 1)(1-p)^{y_2-2}} = \frac{1}{y_2 - 1} \quad \text{for} \ \ y_1 = 1, \dots, y_2 - 1.$$

In other words, given a value of y_2, the distribution of Y_1 is uniform on the positive integers less than y_2.

The random variables Y_1 and Y_2 can alternatively be described by their **joint cumulative distribution**. For any $(c_1, c_2) \in \mathbb{R}^2$,

$$F(c_1, c_2) = P(Y_1 \leq c_1 \text{ and } Y_2 \leq c_2).$$

Note that c_1 and c_2 don't have to be in the support of the random variables; the joint cumulative distribution is defined over the entire real plane. For c_1 larger than all values in the support of Y_1, and c_2 larger than all values in the support of Y_2, we have $F(c_1, c_2) = 1$ and, similarly, for c_1 smaller than all values in the support of Y_1, and c_2 smaller than all values in the support of Y_2, we have $F(c_1, c_2) = 0$.

For discrete Y_1 and Y_2, the joint CDF is a two-dimensional step function, and (as we will see) for continuous Y_1 and Y_2, the joint CDF is continuous.

We can get the **marginal CDF** for each random variable from the joint CDF. For any $c_1 \in \mathbb{R}$,

$$\begin{aligned} F_1(c_1) &= P(Y_1 \leq c_1) \\ &= P(Y_1 \leq c_1 \text{ and } Y_2 \leq \infty) \\ &= F(c_1, \infty), \end{aligned}$$

and similarly for Y_2: $F_2(c_2) = F(\infty, c_2)$. The marginal CDF can alternatively be calculated from the marginal distribution: for any $c_2 \in \mathbb{R}$,

$$F_2(c_2) = P(Y_2 \le c_2) = \sum_{all\ y_2 \le c_2} f_2(y_2).$$

For the switches example, you can check that the marginal CDF for Y_2 is

$$F_2(c_2) = \begin{cases} 0 & \text{for } c_2 < 2, \\ 1/10 & \text{for } 2 \le c_2 < 3, \\ 3/10 & \text{for } 3 \le c_2 < 4, \\ 6/10 & \text{for } 4 \le c_2 < 5, \\ 1 & \text{for } c_2 \ge 5. \end{cases}$$

The joint CDF for the switches example is shown for a piece of the plane:

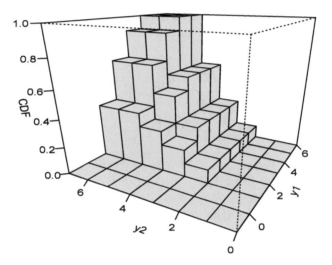

Random variables Y_1 and Y_2 are **independent** if

$$P(Y_1 \in A_1 | Y_2 \in A_2) = P(Y_1 \in A_1)$$

for any sets A_1 in the support of Y_1 and A_2 in the support of Y_2. Discrete random variables are independent if the conditional distribution of Y_1, given any value of Y_2, is equal to the marginal distribution of Y_2.

It follows that if Y_1 and Y_2 are independent discrete random variables, then

$$P(Y_1 \in A_1, Y_2 \in A_2) = P(Y_1 \in A_1)P(Y_2 \in A_2),$$

and in particular, for independent discrete Y_1 and Y_2,

$$P(Y_1 = y_1, Y_2 = y_2) = P(Y_1 = y_1)P(Y_2 = y_2).$$

This means that the joint mass function for independent Y_1 and Y_2 is simply the product of the marginals:

$$f(y_1, y_2) = f_1(y_1)f_2(y_2)$$

for all (y_1, y_2) in the support.

Given jointly distributed discrete random variables Y_1 and Y_2, the **expected value of a function** $g(Y_1, Y_2)$ of the random variables is computed using the formula

$$E\left[g(Y_1, Y_2)\right] = \sum \sum_{all(y_1, y_2)} g(y_1, y_2) f(y_1, y_2),$$

where $f(y_1, y_2) = P(Y_1 = y_1, Y_2 = y_2)$.

For the switches example, the expected number of good switches tested between the two defective switches is

$$E(Y_2 - Y_1) = \sum \sum_{all(y_1, y_2)} (y_2 - y_1 - 1) f(y_1, y_2)$$

$$= (0)\left(\frac{4}{10}\right) + (1)\left(\frac{3}{10}\right) + (2)\left(\frac{2}{10}\right) + (3)\left(\frac{1}{10}\right) = 1.$$

We can derive a nice result about the **expected value of a product of independent random variables**. If Y_1 and Y_2 are independent with mass functions $f_1(y_1)$ and $f_2(y_2)$, then

$$E(Y_1 Y_2) = \sum \sum_{(y_1, y_2)} y_1 y_2 f(y_1, y_2) = \sum \sum_{(y_1, y_2)} y_1 y_2 f_1(y_1) f_2(y_2)$$

$$= \sum_{y_1} y_1 f_1(y_1) \sum_{y_2} y_2 f_1(y_2) = E(Y_1) E(Y_2).$$

In Exercise 23.6 you are asked to verify that this formula does *not* hold for the switches example, and in Exercise 23.7 you show that this formula also does not hold for the coin toss example.

In Exercise 23.12 you are asked to prove the following: If Y_1 and Y_2 are independent random variables, where $E(Y_1) = \mu_1$, $E(Y_2) = \mu_2$, $V(Y_1) = \sigma_1^2$, and $V(Y_2) = \sigma_2^2$, then

$$V(Y_1 Y_2) = \mu_1^2 \sigma_1^2 + \mu_2^2 \sigma_2^2 + \sigma_1^2 \sigma_2^2.$$

Given jointly distributed discrete random variables Y_1 and Y_2 with mass function $f(y_1, y_2)$, functions $g_1(y_1, y_2)$ and $g_2(y_1, y_2)$, and constants a_1 and a_2,

$$E\left[a_1 g_1(Y_1, Y_2) + a_2 g_2(Y_1, Y_2)\right]$$

$$= \sum \sum_{all(y_1, y_2)} \left[a_1 g_1(y_1, y_2) + a_2 g_2(y_1, y_2)\right] f(y_1, y_2)$$

$$= a_1 \sum \sum_{all(y_1, y_2)} \left[g_1(y_1, y_2)\right] f(y_1, y_2) + a_2 \sum \sum_{all(y_1, y_2)} \left[g_2(y_1, y_2)\right] f(y_1, y_2)$$

$$= a_1 E\left[g_1(Y_1, Y_2)\right] + a_2 E\left[g_2(Y_1, Y_2)\right].$$

The above result is easily generalized to

$$E\left[a_1 g_1(Y_1, Y_2) + \cdots + a_n g_n(Y_1, Y_2)\right] = a_1 E\left[g_1(Y_1, Y_2)\right] + \cdots + a_n E\left[g_n(Y_1, Y_2)\right].$$

One useful special case is

$$E(a_1 Y_1 + a_2 Y_2) = a_1 E(Y_1) + a_2 E(Y_2).$$

Finally, we consider n jointly distributed discrete random variables, say $Y_1, Y_2, \ldots,$ Y_n. The joint mass function $f(y_1, y_2, \ldots, y_n)$ assigns probabilities to all possible combinations of values of the random variables, that is,

$$f(y_1, y_2 \ldots, y_n) = P(Y_1 = y_1, Y_2 = y_2 \ldots, Y_n = y_n).$$

We must have $0 < f(y_1, y_2, \ldots, y_n) \leq 1$ for all pairs $(y_1, y_2, \ldots, y_n) \in S$, where $S \in \mathbb{R}^n$ is the joint support, and

$$\sum_{(y_1, y_2, \ldots, y_n) \in S} \sum f(y_1, y_2, \ldots, y_n) = 1.$$

The concepts for a pair of jointly distributed random variables readily generalize to n random variables. For example, the marginal distribution for Y_1 is the probability mass function for Y_1, which we compute given the joint distribution values

$$f_1(y_1) = P(Y_1 = y_1) = \sum \sum \cdots \sum_{all \ (y_2, \ldots, y_n)} f(y_1, y_2, \ldots, y_n).$$

The marginals for random variables $Y_2, \ldots, Y_n$ are computed similarly.

If the random variables $Y_1, Y_2, \ldots, Y_n$ are **pairwise independent**, that is, any pair is independent as defined above, then

$$P(Y_1 \in A | (Y_2, \ldots, Y_n) \in B) = P(Y_1 \in A),$$

where A is in the support of the marginal for Y_1, and B is in the support of the joint distribution of $Y_2, \ldots, Y_n$.

The expected value of a function $g(Y_1, Y_2, \ldots, Y_n)$ of the random variables is computed using the formula

$$E[g(Y_1, Y_2, \ldots, Y_n)] = \sum \sum \cdots \sum_{all(y_1, y_2, \ldots, y_n)} g(y_1, y_2 \ldots, y_n) f(y_1, y_2 \ldots, y_n).$$

Using the same ideas, we can derive the formula

$$E\left[\sum_{i=1}^{n} a_i g_i(Y_1, Y_2 \ldots, Y_n)\right] = \sum_{i=1}^{n} a_i E[g_i(Y_1, Y_2 \ldots, Y_n)].$$

An important special case is

$$E(a_1 Y_1 + a_2 Y_2 + \cdots + a_n Y_n) = a_1 E(Y_1) + a_2 E(Y_2) + \cdots + a_n E(Y_n).$$

We can simulate jointly distributed random variables in R using either a formula or the probabilities from a table. Let's simulate from the switches distribution to illustrate. The following code is straightforward; first we generate $U \sim \text{Unif}(0, 1)$, then set $(Y_1, Y_2) = (1, 2)$ if $U < 1/10$. We set $(Y_1, Y_2) = (1, 3)$ if $U \in [1/10, 2/10]$, $(Y_1, Y_2) = (1, 4)$ if $U \in [2/10, 3/10]$, etc.

```
u=runif(1)
if(u<.1){
    y1=1;y2=2
}else if(u<.2){
    y1=1;y2=3
```

```
}else if(u<.3){
   y1=1;y2=4
}else if(u<.4){
   y1=1;y2=5
}else if(u<.5){
   y1=2;y2=3
}else if(u<.6){
   y1=2;y2=4
}else if(u<.7){
   y1=2;y2=5
}else if(u<.8){
   y1=3;y2=4
}else if(u<.9){
   y1=3;y2=5
}else{
   y1=4;y2=5
}
```

To get a sample of size n from the joint distribution, we can save the values of y_1 and y_2 in vectors:

```
n=1000
y1=1:n;y2=1:n
for(i in 1:n){
  u=runif(1)
  if(u<.1){
     y1[i]=1;y2[i]=2
  }else if(u<.2){
     y1[i]=1;y2[i]=3
  }else if(u<.3){
     y1[i]=1;y2[i]=4
  }else if(u<.4){
     y1[i]=1;y2[i]=5
  }else if(u<.5){
     y1[i]=2;y2[i]=3
  }else if(u<.6){
     y1[i]=2;y2[i]=4
  }else if(u<.7){
     y1[i]=2;y2[i]=5
  }else if(u<.8){
     y1[i]=3;y2[i]=4
  }else if(u<.9){
     y1[i]=3;y2[i]=5
  }else{
     y1[i]=4;y2[i]=5
  }
}
```

For a large sample, the average of the y_1 values should be near the expected value of the random variable Y_1. The proportion of samples for which $Y_1 = 2$ should be close to $3/10$, the probability from the marginal distribution of Y_1. You can check your answers to the exercises through simulations.

The code for sampling from the joint distribution in the coin toss example is shorter. Here we simulate a string of 100 coin tosses with $p = 1/2$, and record on what tosses we find the first and second heads. Although the support of this joint distribution is infinite, the probability of having to toss more than 100 times is so small that it can be set to zero for practical purposes. The following code simulates 10,000 pairs of values from the joint distribution:

```
n=10000; obs=1:100
y1=1:n;y2=1:n
for(i in 1:n){
    x1=sample(c(0,1),100,replace=TRUE)
    x2=obs[x1==1]
    y1[i]=x2[1]
    y2[i]=x2[2]
}
```

To check values of our computed marginal distribution, we can find the proportion of values for which $y_2 = 3$: sum(y2==3)/n returns a value close to $1/4$. The probability that $Y_1 = 1$, given that $Y_2 = 4$, can be computed analytically as

$$P(Y_1 = 1|Y_2 = 4) = \frac{P(Y_1 = 1 \text{ and } Y_2 = 4)}{P(Y_2 = 4)} = \frac{1/16}{3/16} = 1/3,$$

and we can check against our simulations with the R command

```
sum(y1==1&y2==4)/sum(y2==4).
```

Chapter Highlights

1. The joint distribution of two discrete random variables (i.e., the joint mass function) defines the probabilities of occurrences for all possible ordered pairs of values. The mass function can be defined using a table of all possible values, or by a formula.

2. The marginal distribution of one of the random variables is simply its distribution, ignoring the other random variable. We can get the marginal probability mass function $f_1(y_1)$ of Y_1 from the joint probability mass function $f(y_1, y_2)$ of Y_1 and Y_2 using the formula
$$f_1(y_1) = \sum_{all y_2} f(y_1, y_2).$$

3. The conditional distribution of Y_1 given $Y_2 = y_2$ is a mass function assigning probabilities to the values in the support of Y_1, if we know that $Y_2 = y_2$. The formula is
$$f(y_1|y_2) = P(Y_1 = y_1|Y_2 = y_2) = \frac{f(y_1, y_2)}{f_2(y_2)}.$$

4. The joint cumulative distribution for random variables Y_1 and Y_2:
$$F(c_1, c_2) = P(Y_1 \leq c_1 \text{ and } Y_2 \leq c_2)$$
for any $c_1, c_2 \in \mathbb{R}$.

5. Random variables Y_1 and Y_2 are **independent** if

$$P(Y_1 \in A_1 | Y_2 \in A_2) = P(Y_1 \in A_1)$$

for all sets A_1 in the support of Y_1 and all sets A_2 in the support of Y_2.

6. If random variables Y_1 and Y_2 are independent, then the joint mass function is the product of the marginals. That is,

$$f(y_1, y_2) = f_1(y_1) f_2(y_2).$$

It's important to remember that this works only if Y_1 and Y_2 are independent.

7. If random variables Y_1 and Y_2 are independent, then

$$E(Y_1 Y_2) = E(Y_1) E(Y_2).$$

It's important to remember that this works only if Y_1 and Y_2 are independent.

8. If Y_1 and Y_2 are independent random variables, where $E(Y_1) = \mu_1$, $E(Y_2) = \mu_2$, $V(Y_1) = \sigma_1^2$, and $V(Y_2) = \sigma_2^2$, then

$$V(Y_1 Y_2) = \mu_1^2 \sigma_1^2 + \mu_2^2 \sigma_2^2 + \sigma_1^2 \sigma_2^2.$$

It's important to remember that this works only if Y_1 and Y_2 are independent.

9. For discrete random variables Y_1 and Y_2, the expected value of $g(Y_1, Y_2)$ is

$$E\left[g(Y_1, Y_2)\right] = \sum_{all(y_1,y_2)} \sum g(y_1, y_2) f(y_1, y_2),$$

where $f(y_1, y_2) = P(Y_1 = y_1, Y_2 = y_2)$.

10. The expected value of a sum is the sum of the expected values:

$$E\left[\sum_{i=1}^{n} g_i(Y_1, Y_2)\right] = \sum_{i=1}^{n} E\left[g_i(Y_1, Y_2)\right]$$

(even if Y_1 and Y_2 are *not* independent).

❧❧❧

Exercises

23.1 Using the definition of independence, prove that if Y_1 and Y_2 are independent discrete random variables, then

$$P(Y_1 = y_1, Y_2 = y_2) = P(Y_1 = y_1) P(Y_2 = y_2).$$

23.2 An urn has six balls. Three are red, two are blue, and one is green. You choose two at random, with replacement. Let Y_1 be the number of red balls drawn, and let Y_2 be the number of green balls drawn.

(a) Find the joint distribution of Y_1 and Y_2.

(b) Find the marginal distribution of Y_1.

(c) Find the conditional distribution of Y_2, given that $Y_1 = 0$.

(d) Are Y_1 and Y_2 independent? Why or why not?

23.3 An urn contains one red ball, one blue ball, and one gold ball. Balls are drawn one at a time, without replacement, until a gold ball appears. Let Y_1 be the number of red balls drawn, and let Y_2 be the number of blue balls drawn.

(a) Find the joint distribution of Y_1 and Y_2.

(b) Find the marginal distribution of Y_2.

(c) Find the conditional distribution of Y_1, given $Y_2 = 0$.

(d) Are Y_1 and Y_2 independent? Why or why not?

23.4 Suppose you roll two fair six-sided dice. Let Y_1 be the number of ones and Y_2 the number of twos.

(a) What is the joint distribution of Y_1 and Y_2?

(b) What is the conditional distribution of Y_2 given $Y_1 = 1$?

23.5 Two dice are rolled. Let Y_1 be the smaller of the two values, and let Y_2 be the larger of the two values. (Y_1 and Y_2 can be equal.)

(a) Compute the joint probability mass function of Y_1 and Y_2.

(b) Find the marginal distribution of Y_1.

(c) Find the conditional distribution of Y_2 given that $Y_1 = 4$.

(d) Are Y_1 and Y_2 independent? Why or why not?

23.6 For the switches example in this chapter, do the following:

(a) Find $E(Y_1)$ using the marginal for Y_1.

(b) Find $E(Y_2)$ using the marginal for Y_2.

(c) Find $E(Y_1 Y_2)$ using the joint mass function, and show that this does not equal $E(Y_1)E(Y_2)$.

23.7 For the coin toss example in this chapter, do the following:

(a) Find $E(Y_1)$ using the marginal for Y_1.

(b) Find $E(Y_2)$ using the marginal for Y_2.

(c) Find $E(Y_1 Y_2)$ using the joint mass function, and show that this does not equal $E(Y_1)E(Y_2)$.

23.8 A drawer has two red socks and four blue socks. You draw socks one at a time, without replacement, until you have a pair (either two red or two blue). Let Y_1 be the number of red socks you have drawn, and let Y_2 be the number of blue socks.

(a) What is the joint distribution of Y_1 and Y_2?

(b) What is the marginal distribution of Y_2?

(c) What is the probability that you get a blue pair of socks?

(d) What is the distribution of Y_1, given $Y_2 = 2$?

23.9 A species of annual garden plant produces blossoms and fruit. A randomly selected plant will produce Y_1 blossoms and Y_2 fruits in a season, where the joint
distribution of Y_1 and Y_2 is described by the mass function with probabilities given
in the table:

		Y_2 (fruits)		
		0	1	2
	0	.10	0	0
Y_1	1	.12	.26	0
(blossoms)	2	.14	.30	.08

(a) Find the marginal distribution of the number of blossoms.

(b) Find the conditional distribution of the number of fruits when the number of
blossoms is 2.

(c) Are Y_1 and Y_2 independent? Why or why not?

(d) Each fruit is formed from a blossom, so that there must be at least as many
blossoms as fruits; this is reflected in the table. Find the expected number of
blossoms that do not form fruit.

(e) Write R code to simulate from the joint distribution and check your answers
to parts (a) and (b).

23.10 A species of annual garden plant produces blossoms and fruit. A randomly selected plant will produce Y_1 blossoms and Y_2 fruits in a season, where the joint
distribution of Y_1 and Y_2 is described by the formula. For $a, b \in (0, 1)$,

$$P(Y_1 = y_1, Y_2 = y_2) = (1 - a)(1 - ab)a^{y_1} b^{y_1 - y_2}$$
$$\text{for } y_2 = 0, 1, \ldots, y_1 \text{ and } y_1 = 0, 1, 2, \ldots.$$

(a) Show that this is a valid mass function (probabilities add to one).

(b) If $a = .8$ and $b = .6$, find the conditional distribution of the number of fruits
when the number of blossoms is 2.

(c) Are Y_1 and Y_2 independent? Why or why not?

(d) Each fruit is formed from a blossom, so that there must be at least as many
blossoms as fruits; this is reflected in the formula. Find the expected number
of blossoms that do not form fruit.

23.11 A factory has two machines to produce widgets. The number of failures in a
randomly selected month for machine #1 is Y_1, while the number of failures in a
randomly selected month for machine #2 is Y_2. The random variables Y_1 and Y_2
have support on the nonnegative integers, and the joint mass function is

$$P(Y_1 = y_1, Y_2 = y_2) = e^{-6} \frac{4^{y_1} 2^{y_2}}{y_1! y_2!}.$$

(a) Show that this is a valid mass function (probabilities add to one).

(b) What is the probability that neither machine fails in a randomly selected month?

(c) Are Y_1 and Y_2 independent? Why or why not?

(d) What is the expected number of machine failures in the factory in a randomly selected month?

(e) Suppose it costs $200 to repair machine #1, but only $150 to repair machine #2. What is the expected monthly repair cost?

23.12 Prove the following claim: If Y_1 and Y_2 are independent random variables, where $E(Y_1) = \mu_1$, $E(Y_2) = \mu_2$, $V(Y_1) = \sigma_1^2$, and $V(Y_2) = \sigma_2^2$, then

$$V(Y_1 Y_2) = \mu_1^2 \sigma_2^2 + \mu_2^2 \sigma_1^2 + \sigma_1^2 \sigma_2^2.$$

Chapter 24

Jointly Continuously Distributed Random Variables

Two random variables, Y_1 and Y_2, are **jointly continuous** with **joint density** $f(y_1, y_2)$ if

$$P((Y_1, Y_2) \in A) = \iint\limits_{A} f(y_1, y_2) dy_1 dy_2$$

for a set $A \subseteq \mathbb{R}^2$. The joint density is a function of two variables and can be pictured as a surface over the (y_1, y_2) plane. For any set $A \subseteq \mathbb{R}^2$, the probability that the random ordered pair (Y_1, Y_2) falls in A is the *volume* under the density, over the set A. In the plot below, the surface represents the joint density function, with support in the positive quadrant. The square in the (y_1, y_2) plane represents the set A. The volume above A and under the surface, represented by the square column, is the probability that (Y_1, Y_2) falls in A.

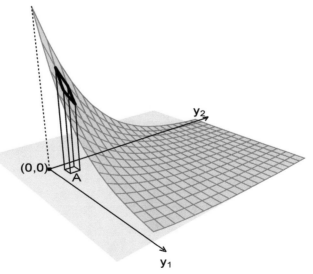

For jointly continuous Y_1 and Y_2 with density $f(y_1, y_2)$, we must have $f(y_1, y_2) \geq 0$ for all $(y_1, y_2) \in \mathbb{R}^2$, and

$$\int_{-\infty}^{\infty} \int_{-\infty}^{\infty} f(y_1, y_2) dy_1 dy_2 = 1.$$

That is, the volume under the density surface, over the (y_1, y_2) plane, must equal one.
The joint cumulative distribution function is defined as

$$F(y_1, y_2) = P(Y_1 \le y_1 \text{ and } Y_2 \le y_2) = \int_{-\infty}^{y_1} \int_{-\infty}^{y_2} f(y_1, y_2) dy_1 dy_2,$$

and we could alternatively get the density from the CDF:

$$f(y_1, y_2) = \frac{\partial^2}{\partial y_1 \partial y_2} F(y_1, y_2).$$

Example: A simple example of jointly continuous random variables is the **uniform distribution** over a rectangle. Suppose Y_1 and Y_2 are jointly uniformly distributed on the unit square $(0, 1) \times (0, 1)$ (the support of the density is the unit square). We can use ideas about volume to calculate probabilities. In order to get the volume under the density to be 1, the height of the density over the unit square must be 1, so that the density function sits over the unit square at a height of one unit. The support of the joint density is shown on the left in the figure below, and the density function (on the right) is a flat, square surface.

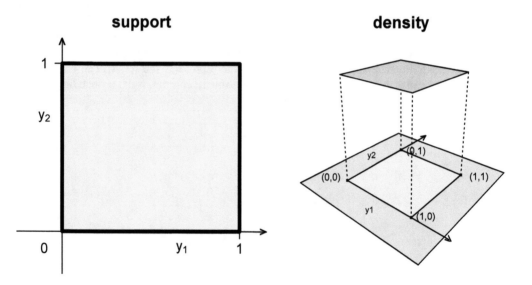

This density function can be written as

$$f(y_1, y_2) = 1 \text{ on } 0 < y_1 < 1 \text{ and } 0 < y_2 < 1,$$

with the implication that $f(y_1, y_2) = 0$ elsewhere.

To find $P(Y_2 > 1/2)$, we consider the volume that is over the area of the support corresponding to $Y_2 > 1/2$. This area is shaded in the picture of the support on the right, and is also shaded in the (y_1, y_2) plane on the left. It's easy to see that the volume under the density, over the rectangle, is 1/2.

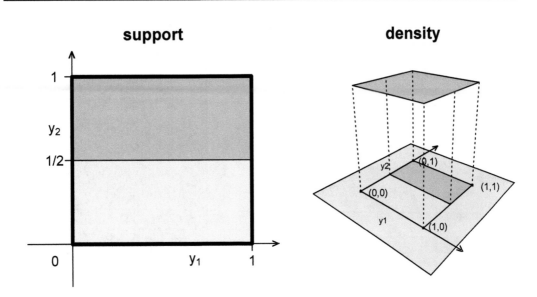

To find $P(Y_1 + 2Y_2 > 2)$, we first shade the corresponding area of the support, as shown in the plot. The volume under the density that is over this area, i.e., the double integral of the density function over this triangle, is the area of the triangle times the height of the density (this works only because the density function is constant). Then $P(Y_1 + 2Y_2 > 2) = 1/4$.

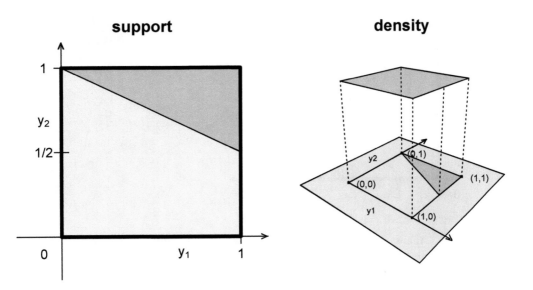

Let's think about **simulating pairs of jointly distributed random variables** using R. For the uniform distribution over the unit square, it suffices to generate two vectors of uniform $(0, 1)$ random variable values. The commands on the left below will generate the points on the right and check one of the calculations. The proportion of points that fall in the area is the estimated probability. For a more precise estimate of the probability, we can make n larger.

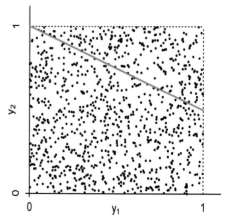

```
n=1000
y1=runif(n)
y2=runif(n)

plot(y1,y2)
lines(c(0,1),c(1,1/2),col=2)

sum(y1+2*y2>2)/n
```

Example: Let's look next at a jointly uniform density, where the support is a triangle in the (y_1, y_2) plane, rather than a square. If the support is the area shown on the left below, and the density is uniform over this triangle, then the density function on the right is a flat triangle over this area, but it must be at height $= 2$, so that the volume under the density is one.

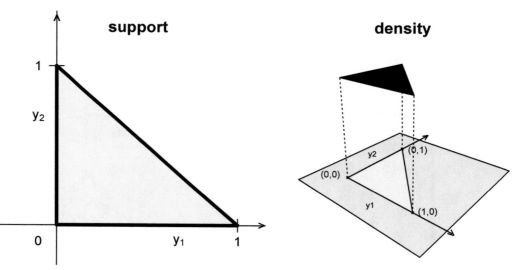

Again, calculus is not necessary for some probability calculations. For example, to compute $P(Y_1 > Y_2)$, we can find the area of the subset of the support for which $Y_1 > Y_2$, and multiply this area by 2 (the height) to get the volume and hence the probability, which is 1/2.

To simulate values from this density, we can simulate values over the unit square as we did for the previous example, then "reject" the values that are not in our new support. In this case we reject about half of the generated values, so we start with at least twice as many values as desired. Code to do this generation of values is shown on the left below; we also check the above calculation by finding the proportion of values where $Y_1 > Y_2$.

```
n=2200
y1=runif(n)
y2=runif(n)
use=y1+y2<1
y1=y1[use];y2=y2[use]

plot(y1,y2)
lines(c(0,1/2),c(0,1/2),col=2)

sum(y1>y2)/sum(use)
```

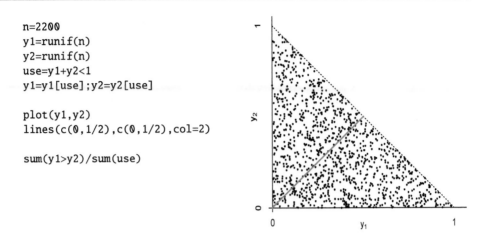

Example: Next, suppose the support of the joint density is again the unit square, but the density is not a constant function. Let

$$f(y_1, y_2) = cy_1 \text{ on } 0 < y_1 < 1 \text{ and } 0 < y_2 < 1,$$

where c is a constant. We can determine the value of c using the idea that the volume has to be one:

$$\int_{-\infty}^{\infty} \int_{-\infty}^{\infty} f(y_1, y_2) dy_1 dy_2 = \int_0^1 \int_0^1 cy_1 dy_1 dy_2 = \int_0^1 cy_1 dy_1 = \frac{c}{2}.$$

Setting this equal to one gives $c = 2$. The density is shown below, where the height of the density is 2 at $(1, 0)$ and $(1, 1)$ is 2.

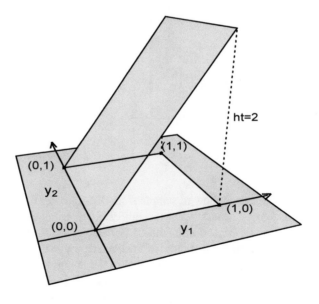

To find $P(Y_1 > 1/2)$, we integrate the joint density over the rectangle $\{(y_1, y_2) : 1/2 < y_1 < 1, 0 < y_2 < 1\}$:

$$P(Y_1 > 1/2) = 2 \int_{1/2}^1 \int_0^1 y_1 dy_2 dy_1 = 2 \int_{1/2}^1 y_1 dy_1 = 3/4.$$

Intuitively, it makes sense that this probability should be bigger than 1/2, because the density function is greater for greater values of y_1.

To find $P(2Y_2 < Y_1)$, we integrate the density over a triangular area in the (y_1, y_2) plane (see Appendix B.7 for a review of multiple integration):

$$P(2Y_2 < Y_1) = 2 \int_0^1 \int_0^{y_1/2} y_1 dy_2 dy_1$$

$$= \int_0^1 y_1^2 dy_1$$

$$= 1/3.$$

On the left is the area over which the density is integrated; when we look at the picture of the density on the right, the volume of the wedge-shaped piece, under the density and above the triangle, is 1/3.

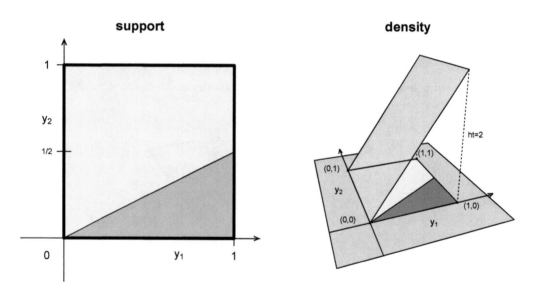

Let's think about how to sample from this joint density using R. To accomplish this, we again start by generating points uniformly distributed over the support, but we use a different kind of "rejection method." For each generated point in the support, we keep it or reject it based on the height of the density above the point. We want the probability of keeping the point to be proportional to the height of the density at that point. In this way we are more likely to keep points with large y_1 values, and the proportion of "kept" points that are over a subset A of the support is approximately the volume of the density over that subset.

The code on the left accomplishes this by generating three uniform vectors of length n: one each for the y_1 and y_2 values, and a third that is uniform on $(0, 2)$, because 2 is the maximum height of the density. It also checks the above probability calculation; about 1/3 of the kept points land in the triangle below the red line.

```
n=2200
y1=runif(n)
y2=runif(n)
u=runif(n)*2
use=2*y1>u
y1=y1[use];y2=y2[use]

plot(y1,y2)
lines(c(0,1),c(0,1/2),col=2)

sum(2*y2<y1)/sum(use)
```

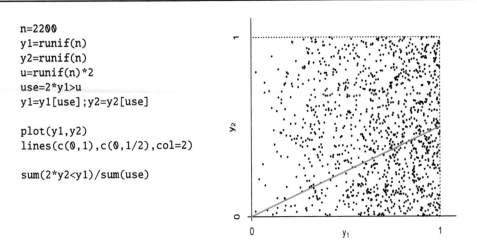

Expected value

Given jointly distributed continuous random variables Y_1 and Y_2, the expected value of a function $g(Y_1, Y_2)$ of the random variables is

$$\mathrm{E}\left[g(Y_1, Y_2)\right] = \int_{-\infty}^{\infty} \int_{-\infty}^{\infty} g(y_1, y_2) f(y_1, y_2) dy_1 dy_2,$$

where $f(y_1, y_2)$ is the joint density function.

Example: Consider the density function $f(y_1, y_2) = 2y_1$ on the unit square as in the last example. Suppose we want to find $\mathrm{E}(Y_1)$. To get an intuitive feel for what this means, picture sampling from the density. Our sampled points are more likely to have larger values of Y_1, because that's where the density is larger. We'd guess that the expected value for Y_1 must be greater than 1/2. To compute the value,

$$\mathrm{E}(Y_1) = 2 \int_0^1 \int_0^1 y_1^2 dy_2 dy_1 = 2 \int_0^1 y_1^2 dy_1 = \frac{2}{3}.$$

This answer can be checked using the simulated values from the last piece of code above: Simply add the command `mean(y1)`.

What about the expected value for Y_2? The density is constant in y_2, and looking at the density while we imagine sampling points from it, intuitively we know that the expected value for Y_2 must be 1/2. Let's compute it to be sure:

$$\mathrm{E}(Y_2) = 2 \int_0^1 \int_0^1 y_2 y_1 dy_2 dy_1 = \int_0^1 y_1 dy_1 = \frac{1}{2}.$$

We will now compute $\mathrm{E}(Y_1 Y_2)$, but first look at the density function again, and make a guess, to develop your intuition. We have

$$\mathrm{E}(Y_1 Y_2) = 2 \int_0^1 \int_0^1 y_1^2 y_2 dy_2 dy_1 = \int_0^1 y_1^2 dy_1 = \frac{1}{3}.$$

Again, these answers can be verified using the simulated values from this distribution.

Example: Suppose a device contains two components. Let random variables Y_1 and Y_2 be the lifetimes (in years) of components A and B, respectively, and they are jointly distributed with density

$$f(y_1, y_2) = ce^{-(y_1+y_2)/5} \text{ for } y_1 > 0, y_2 > 0.$$

(This density is pictured in the first plot of this chapter.) What is the value of the constant c that makes this a valid joint density function? To determine this, we integrate the density function over the (y_1, y_2) plane, set this to one, and solve for c. We have

$$\int_{-\infty}^{\infty} \int_{-\infty}^{\infty} f(y_1, y_2) dy_1 dy_2 = \int_0^{\infty} \int_0^{\infty} ce^{-(y_1+y_2)/5} dy_1 dy_2 = 25c,$$

so $c = 1/25$.

What is the probability that both components fail within 5 years? This is

$$P(Y_1 < 5, Y_2 < 5) = \frac{1}{25} \int_0^5 \int_0^5 e^{-(y_1+y_2)/5} dy_1 dy_2$$

$$= \left[\frac{1}{5} \int_0^5 e^{-y_1/5} dy_1 \right] \left[\frac{1}{5} \int_0^5 e^{-y_2/5} dy_2 \right]$$

$$= (1 - e^{-1})^2 \approx .3996.$$

What is the probability that component A fails before component B? (Make a guess!) We integrate the joint density over the set of (y_1, y_2) such that $0 < y_1 < y_2 < \infty$; this is

$$P(Y_1 < Y_2) = \frac{1}{25} \int_0^{\infty} \int_0^{y_2} e^{-(y_1+y_2)/5} dy_1 dy_2$$

$$= \frac{1}{5} \int_0^{\infty} \left[1 - e^{-y_2/5} \right] e^{-y_2/5} dy_2$$

$$= \frac{1}{5} \int_0^{\infty} e^{-y_2/5} dy_2 - \frac{1}{5} \int_0^{\infty} e^{-2y_2/5} dy_2 = 1/2.$$

By the symmetry of y_1 and y_2, we could have guessed that each component is equally likely to fail first.

Suppose a device uses the components in sequence, so that component A is used first, and when it fails, component B is substituted. The lifetime of the device is the random variable $Y_1 + Y_2$. We can find the expected lifetime of the device:

$$E(Y_1 + Y_2) = \frac{1}{25} \int_0^{\infty} \int_0^{\infty} (y_1 + y_2) e^{-(y_1+y_2)/5} dy_1 dy_2$$

$$= \frac{1}{25} \int_0^{\infty} \int_0^{\infty} y_1 e^{-y_1/5} e^{-y_2/5} dy_1 dy_2$$

$$+ \frac{1}{25} \int_0^{\infty} \int_0^{\infty} y_2 e^{-y_1/5} e^{-y_2/5} dy_1 dy_2$$

$$= \frac{1}{5} \int_0^{\infty} y_1 e^{-y_1/5} dy_1 + \frac{1}{5} \int_0^{\infty} y_2 e^{-y_2/5} dy_2$$

$$= 5 + 5 = 10.$$

In the last integrals we used the useful formula from Chapter Highlight 5 in Chapter 16, recognizing the form of a gamma density.

What is the probability that the device lasts less than 10 years? To find this we integrate the joint density over the set $\{(y_1, y_2) : y_1 > 0, y_2 > 0, y_1 + y_2 < 10\}$:

$$\begin{aligned}
P(Y_1 + Y_2 < 10) &= \frac{1}{25} \int_0^{10} \int_0^{10-y_2} e^{-y_1/5} e^{-y_2/5} dy_1 dy_2 \\
&= \frac{1}{5} \int_0^{10} \left[1 - e^{-(10-y_2)/5}\right] e^{-y_2/5} dy_2 \\
&= 1 - 3e^{-2} \approx .594.
\end{aligned}$$

We won't try to simulate from this density yet. The rejection trick for the last example will not work, because the maximum of the density does not exist. However, we will discover in Chapter 26 that Y_1 and Y_2 are independent random variables, which will allow us to easily simulate the joint values.

Example: The common ragweed plant produces several types of pollen, but type A and type B pollen tend to be allergens. Let Y_1 and Y_2 be the proportion of pollens, produced by a randomly selected ragweed plant, that are type A and type B pollen, respectively. Suppose the joint density for these proportions can be modeled as

$$f(y_1, y_2) = c(1 - (y_1 + y_2)^2) \text{ for } 0 < y_1 + y_2 < 1.$$

Let's find the constant c that makes this function a density. Integrating the density over the support (shown in the figure below), we solve

$$\begin{aligned}
1 &= c \int_0^1 \int_0^{1-y_2} \left[1 - (y_1 + y_2)^2\right] dy_1 dy_2 \\
&= \frac{c}{2} - c \int_0^1 \int_0^{1-y_2} (y_1 + y_2)^2 dy_1 dy_2 \\
&= \frac{c}{2} - \frac{c}{3} \int_0^1 (y_1 + y_2)^3 \big|_0^{1-y_2} dy_2 \\
&= \frac{c}{2} - \frac{c}{3} \int_0^1 (1 - y_2^3) dy_2 \\
&= \frac{c}{4},
\end{aligned}$$

so $c = 4$. This is shown in the plot on the right as the height of the density function.

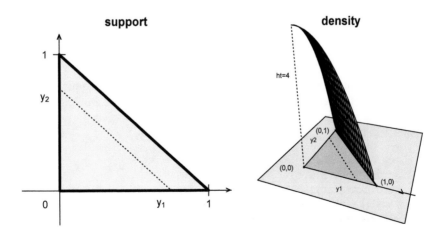

Suppose the researchers are interested in determining the probability that the ragweed pollen is more than 75% allergens. This is the volume sitting under the density function over the slice of the support above $y_1 + y_2 = 3/4$, the dotted line shown on the support in the plot on the right in the above figure. It's easier to integrate over the triangle beneath the dotted line to get the probability that the ragweed pollen is *less* than 75% allergens. Then we can subtract this result from one to get the desired probability. The probability that $Y_1 + Y_2 < .75$ (the ragweed pollen is less than 75% allergens) is

$$
4 \int_0^{3/4} \int_0^{3/4 - y_2} \left[1 - (y_1 + y_2)^2 \right] dy_1 dy_2 = \frac{9}{8} - 4 \int_0^{3/4} \int_0^{3/4 - y_2} (y_1 + y_2)^2 dy_1 dy_2
$$

$$
= \frac{9}{8} - \frac{4}{3} \int_0^{3/4} (y_1 + y_2)^3 \Big|_0^{3/4 - y_2} dy_2
$$

$$
= \frac{9}{8} - \frac{4}{3} \int_0^{3/4} \left[\left(\frac{3}{4} \right)^3 - y_2^3 \right] dy_2
$$

$$
\approx .809.
$$

The probability that the ragweed pollen is *more* than 75% allergens is $1 - .809 = .191$.
The expected proportion of allergens in ragweed pollen is

$$
\mathrm{E}(Y_1 + Y_2) = 4 \int_0^1 \int_0^{1 - y_2} (y_1 + y_2) \left[1 - (y_1 + y_2)^2 \right] dy_1 dy_2
$$

$$
= 4 \int_0^1 \int_0^{1 - y_2} (y_1 + y_2) dy_1 dy_2 - 4 \int_0^1 \int_0^{1 - y_2} (y_1 + y_2)^3 dy_1 dy_2
$$

$$
= 2 \int_0^1 (y_1 + y_2)^2 \Big|_0^{1 - y_2} dy_2 - \int_0^1 (y_1 + y_2)^4 \Big|_0^{1 - y_2} dy_2
$$

$$
= 2 \int_0^1 (1 - y_2^2) dy_2 - \int_0^1 (1 - y_2^4) dy_2
$$

$$
= 2(1 - 1/3) - (1 - 1/5) = 8/15.
$$

Let's simulate from this density to practice simulations and to check our answers. First, we simulate points in the support:

```
n=100000
y1=runif(n)
y2=runif(n)
use=y1+y2<=1
y1=y1[use]
y2=y2[use]
```

Next, we use the rejection method to simulate a random sample from the joint density. Because the height is 4, we sample from a uniform random variable on $(0, 4)$ for every simulated point (y_1, y_2) in the support, and keep the points for which the sampled value is less than the height of the joint density at (y_1, y_2):

```
new.n=sum(use)
u=runif(new.n,0,4)
keep=u<4*(1-(y1+y2)^2)
y1=y1[keep]
y2=y2[keep]
```

Finally, we have a random sample from the density, and we compute the proportion of points for which the sum is less than .75:

```
sum(y1+y2>.75)/sum(keep)
```

Densities in higher dimensions

Densities can be defined in higher dimensions; for example a uniform density in $\mathbb{R}^3$ could be defined as $f(y_1, y_2, y_3) = 1$ with support on $[0, 1] \times [0, 1] \times [0, 1]$. For higher-dimensional densities, the ideas are the same: A density function f defined in $\mathbb{R}^n$ must be nonnegative and integrate to one over $\mathbb{R}^n$. Further, the probability that $(Y_1, \ldots, Y_n) \in A \subset \mathbb{R}^n$ is

$$\iiint_A f(y_1, \ldots, y_n) dy_1 \cdots dy_n,$$

and

$$E(g(Y_1, \ldots, Y_n)) = \int_{-\infty}^{\infty} \int_{-\infty}^{\infty} \int_{-\infty}^{\infty} g(y_1, \ldots, y_n) f(y_1, y_2, y_3) dy_1 \cdots dy_n.$$

Most of the higher-dimensional joint distributions that we will encounter later in this book will be samples from a population for which we want to do estimation and inference.

Chapter Highlights

1. The joint distribution of two continuous random variables Y_1 and Y_2 can be described by a density function $f(y_1, y_2)$ that is nonnegative and integrates to one over the (y_1, y_2) plane.

2. Probabilities are calculated as

$$P((Y_1, Y_2) \in A) = \iint_A f(y_1, y_2) dy_1 dy_2.$$

3. The joint cumulative distribution function is defined for any point in the plane:

$$F(y_1, y_2) = P(Y_1 \le y_1 \text{ and } Y_2 \le y_2) = \int_{-\infty}^{y_1} \int_{-\infty}^{y_2} f(y_1, y_2) dy_1 dy_2.$$

4. The expected value of $g(Y_1, Y_2)$ is

$$E[g(Y_1, Y_2)] = \int_{-\infty}^{\infty} \int_{-\infty}^{\infty} g(y_1, y_2) f(y_1, y_2) dy_1 dy_2.$$

5. Values from joint distributions may be simulated in R if the maximum value of the density function is finite.

   (a) To generate values (y_1, y_2) from a density that is uniform over a rectangle, we simply generate y_1 and y_2 values that are uniform over the sides of the rectangle.

   (b) To generate values (y_1, y_2) from a density that is uniform over a subset of a rectangle, we generate values uniformly in the rectangle, then reject values that are not in the desired support.

   (c) To generate values (y_1, y_2) from a density that is not uniform, but has maximum value M, we first generate values uniformly in the support, using (a) or (b). Then we generate a vector of uniform values on $(0, M)$, the same length as the number of points generated (or kept) in the support. Finally, we reject points in the support for which the corresponding value of the third vector is greater than the value of the density over that point. The remaining points are a random sample from the joint density.

Exercises

24.1 Draw a picture to convince yourself that if $a_1 < b_1$ and $a_2 < b_2$,

$$P(a_1 \leq Y_1 \leq b_1 \text{ and } a_2 \leq Y_2 \leq b_2) = F(b_1, b_2) - F(a_1, b_2) - F(b_1, a_2) + F(a_1, a_2).$$

24.2 Suppose Y_1 and Y_2 are continuous random variables with joint density $f(y_1, y_2)$, and suppose $E(Y_1) = \mu_1$ and $E(Y_2) = \mu_2$. Find $E(a_1 Y_1 + a_2 Y_2)$.

24.3 The random variables Y_1 and Y_2 have a joint density that is uniform on the triangle shown below in the (y_1, y_2) plane:

   (a) Find $P(Y_2 > 1/2)$.

   (b) Find $P(Y_1 > Y_2)$.

   (c) Find $E(Y_1)$.

   (d) Find $E(Y_2)$.

   (e) Find $E(Y_2 - 4Y_1)$

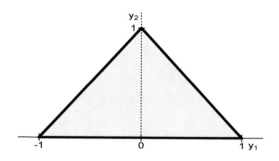

   (f) Write some code in R to simulate points from this density, and check your answers to (a)–(e).

24.4 Suppose Y_1 and Y_2 are jointly uniform on the unit square $0 < y_1 < 1$ and $0 < y_2 < 1$.

(a) For $a \in (0,1)$, find $P(Y_1 + Y_2 < a)$.

(b) For $a \in [1,2)$, find $P(Y_1 + Y_2 < a)$.

24.5 Suppose Y_1 and Y_2 are jointly continuous random variables, with joint density function $f(y_1, y_2) = cy_2$, where the support of the density is over the triangle shown.

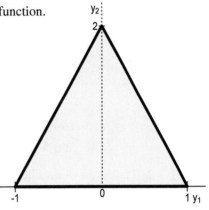

(a) Find c to make this a proper density function.

(b) Find $P(Y_1 > 0)$.

(c) Find $P(Y_1 > Y_2)$.

(d) Find $E(Y_1)$.

(e) Find $E(Y_2)$.

(f) Write some R code to simulate points from this distribution; check your answers to (b)–(e).

24.6 Let Y_1 and Y_2 be jointly continuous random variables with density

$$f(y_1, y_2) = cy_1^2 y_2^2 \text{ for } 0 < y_1 + y_2 < 1, \ y_1 > 0, \ y_2 > 0.$$

(a) Find c to make this a proper density function.

(b) Find $P(Y_2 > 1/2)$.

(c) Find $P(Y_1 > Y_2)$.

(d) Find $P(Y_2 > Y_1)$.

(e) Write some R code to simulate points from this distribution; check your answers to (b)–(d).

24.7 Suppose Y_1 and Y_2 are jointly distributed with density

$$f(y_1, y_2) = ce^{-(y_1/2 + y_2/4)} \text{ for } 0 < y_1, y_2 < \infty.$$

(a) Find c.

(b) Find $P(Y_1 > 2)$.

(c) Find $E(Y_1^2)$.

24.8 Suppose Y_1 and Y_2 are jointly distributed with density

$$f(y_1, y_2) = ce^{-(y_1/2 + y_2/4)} \text{ for } 0 < y_2 < y_1 < \infty.$$

(a) Find c.

(b) Find $P(Y_1 > 2)$.

(c) Find $E(Y_1^2)$.

24.9 Let Y_1 and Y_2 have joint density $f(y_1, y_2) = c(1 - y_2)$ on $0 \leq y_1 \leq y_2 \leq 1$.

    (a) Determine the value of c.

    (b) Find $P(Y_1 > 1/2)$.

    (c) Find $P(Y_2 > 1/2)$.

    (d) Find $P(Y_1 + Y_2 < 1)$.

    (e) Find $E(Y_1 Y_2)$.

    (f) Write R code to simulate from the joint density and check your answers to (b)–(e) above.

24.10 Let Y_1 and Y_2 be jointly continuous random variables with density

$$f(y_1, y_2) = cy_1 e^{-y_1 y_2/2} \quad \text{for } 0 < y_1 < y_2 < \infty.$$

    (a) Determine the value of c.

    (b) Find $E(Y_1)$.

24.11 The parts per billion of arsenic found in randomly selected grass stems in a downtown neighborhood of Largetown is modeled by a random variable Y_1, while the parts per billion of antimony found in randomly selected grass stems in the same neighborhood is modeled by a random variable Y_2. The joint distribution of Y_1 and Y_2 is given by

$$f(y_1, y_2) = ce^{-(y_1/8 + y_2/12)} \quad \text{for } y_1 > 0, y_2 > 0.$$

    (a) Determine the value of c.

    (b) The EPA is concerned if the sum of the parts per billion of arsenic and antimony is greater than 10. Find $P(Y_1 + Y_2 > 10)$.

    (c) Find $E(Y_1 + Y_2)$.

24.12 Two city workers in Largetown are on a ditch-digging crew, assigned to a particular area. They are supposed to take turns digging, while the other rests. If Y_1 and Y_2 represent the proportions of time that worker 1 and worker 2 spend digging, in a randomly selected hour, these random variables can be modeled as having joint density

$$f(y_1, y_2) = c(1 - y_1 - y_2) \quad \text{for } y_1 > 0, \ y_2 > 0, \ \text{and} \ y_1 + y_2 < 1.$$

    (a) Determine the value of c.

    (b) Find $P(Y_1 < 1/2)$, and interpret in the context of the problem.

    (c) Find $E(Y_1 + Y_2)$, and interpret in the context of the problem.

    (d) Write R code to simulate from the joint density and check your answers to (b)–(c) above.

24.13 The power cell for a machine in an industrial plant is known to emit a humming sound before it fails. Suppose that if a new power cell is installed, the time before it starts humming, Y_1, and the lifetime Y_2 are jointly distributed with density

$$f(y_1, y_2) = ce^{.9y_1 - y_2} \quad \text{for } 0 < y_1 < y_2 < \infty.$$

    (a) Determine the value of c.

    (b) Find the expected lifetime of the power cell.

    (c) Find the expected time that the power cell hums before it fails.

24.14 The lifetime for a circuit element is known to have an exponential distribution with mean 200 days. If two circuit elements with lifetimes Y_1 and Y_2 are installed, what is the probability that they fail within 10 days of each other? That is, what is $P(|Y_1 - Y_2| < 10)$? Assume that the lifetimes Y_1 and Y_2 are independent.

24.15 A furniture delivery service operates in Smalltown, USA. The drivers know that the time it takes for each delivery follows an exponential distribution with mean of one hour, and that the times are independent. Suppose they have two deliveries on Saturday morning, starting at 9:00. What is the probability that they will finish by noon?

Chapter 25

Marginal Distributions for Jointly Continuous Random Variables

Given jointly continuous random variables Y_1 and Y_2, the **marginal density** for Y_1 is simply the density for Y_1 if we sample from the joint distribution and ignore the value of Y_2. This can be written as

$$f_1(y_1) = \int_{-\infty}^{\infty} f(y_1, y_2) dy_2.$$

The calculation of this marginal density function is sometimes referred to as "integrating out" y_2. Similarly, the marginal density for Y_2 is obtained by integrating out y_1:

$$f_2(y_2) = \int_{-\infty}^{\infty} f(y_1, y_2) dy_1.$$

It's important to remember that the marginals are themselves density functions. It's clear they have to be positive, and we can show that they integrate to one:

$$\int_{-\infty}^{\infty} f_1(y_1) dy_1 = \int_{-\infty}^{\infty} \left[\int_{-\infty}^{\infty} f(y_1, y_2) dy_2 \right] dy_1 = \int_{-\infty}^{\infty} \int_{-\infty}^{\infty} f(y_1, y_2) dy_2 dy_1 = 1.$$

Example: Suppose the random variables Y_1 and Y_2 are jointly uniform on the triangle.

The support can be written as

$$\{(y_1, y_2) : y_2 > 0, 2y_1 + y_2 < 2, y_2 - 2y_1 < 2\},$$

and the joint density can be written as

$$f(y_1, y_2) = \frac{1}{2} I\{y_2 > 0, 2y_1 + y_2 < 2, y_2 - 2y_1 < 2\},$$

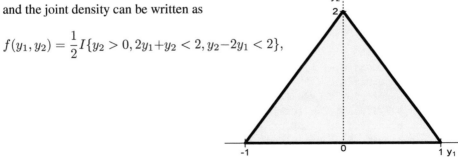

where I is the **indicator function**: $I(A) = 1$ if the expression A is true, and $I(A) = 0$ otherwise. In other words, the density sits above the triangle, at a height of $1/2$. Writing the formula for the density using an indicator function will be helpful for computing conditional densities in the next chapter.

To develop your intuition about marginal densities, you can imagine sampling from the joint density, and looking, say, only at the Y_2 values. The distribution of these values is the marginal for Y_2. The support for the marginal distribution for Y_2 is $(0, 2)$. Because the points are equally likely to be sampled all over the support triangle, and because the support is wider where the values of Y_2 are smaller, we expect the density $f_2(y_2)$ to be larger for smaller values.

To compute the marginal density for Y_2, we fix a value of $y_2 \in (0, 2)$ and integrate out y_1. In the plot on the left below, we can see the integration limits for y_1, given a value of $y_2 \in (0, 2)$.

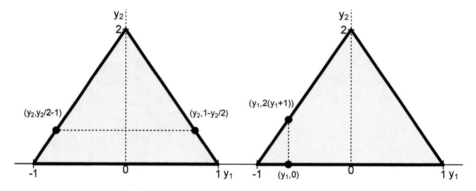

Then we compute

$$f_2(y_2) = \int_{-\infty}^{\infty} f(y_1, y_2) dy_1$$
$$= \frac{1}{2} \int_{y_2/2-1}^{1-y_2/2} dy_1$$
$$= 1 - \frac{y_2}{2}$$

on $(0, 2)$.

If we sampled from this joint density using R code, as in the last chapter, then made a histogram of just the y_2 values, this histogram would be in the shape of this density. The code on the left below makes the histogram shown on the right, with the density function superimposed.

```
n=100000
y1=runif(n,-1,1)
y2=runif(n,0,2)
use=2*y1+y2<2&y2<2+2*y1
y1=y1[use];y2=y2[use]
hist(y2,freq=FALSE,br=30)
lines(c(0,2),c(1,0),col=2)
```

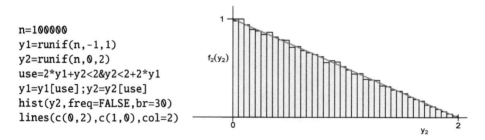

We have to compute the marginal for Y_1 in two steps. For $y_1 \in (-1, 0)$, the limits of integration for y_2 are zero and $2(y_1 + 1)$, as shown in the above plot on the right. So, for $y_1 \in (-1, 0)$,

$$
\begin{aligned}
f_1(y_1) &= \int_{-\infty}^{\infty} f(y_1, y_2) dy_2 \\
&= \frac{1}{2} \int_0^{2(y_1+1)} dy_2 \\
&= y_1 + 1.
\end{aligned}
$$

For $y_1 \in (0, 1)$, we find the limits of integration over y_2 are zero and $2(1 - y_1)$, so

$$
\begin{aligned}
f_1(y_1) &= \frac{1}{2} \int_0^{2(1-y_1)} dy_2 \\
&= 1 - y_1.
\end{aligned}
$$

We can write the marginal density function as

$$
f_1(y_1) = \begin{cases} 1 + y_1 & \text{for } y_1 \in (-1, 0], \\ 1 - y_1 & \text{for } y_1 \in (0, 1), \\ 0 & \text{otherwise.} \end{cases}
$$

We can check our work by verifying that the area under this density function is 1:

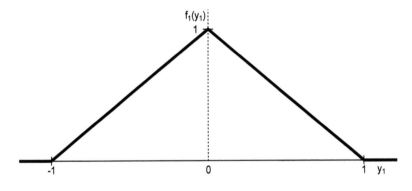

If we want to find $P(Y_1 \in A)$, we can compute either

$$
P(Y_1 \in A) = \iint_{Y_1 \in A} f(y_1, y_2) dy_1 dy_2
$$

or

$$
P(Y_1 \in A) = \int_A f_1(y_1) dy_1,
$$

and we should get the same answer.

Example: Let's look again at the ragweed example in the last chapter, where the proportions of pollen that are the type A and type B allergens are modeled as Y_1 and Y_2, respectively, with joint density

$$
f(y_1, y_2) = 4 \left[1 - (y_1 + y_2)^2 \right] \text{ for } 0 < y_1 + y_2 < 1.
$$

The random variable Y_1 can take values in $(0, 1)$, and for $y_1 \in (0, 1)$ the marginal density is found by integrating the joint density over possible values of y_2, given the value of y_1. So, for $y_1 \in (0, 1)$,

$$f_1(y_1) = 4 \int_0^{1-y_1} \left[1 - (y_1 + y_2)^2\right] dy_2$$

$$= 4 \left[1 - y_1 - \frac{1}{3}(y_1 + y_2)^3 \Big|_0^{1-y_1}\right]$$

$$= \frac{4}{3}y_1^3 - 4y_1 + \frac{8}{3}.$$

The marginal density is shown below; you can check to see if the area is one by integrating the density function over its support.

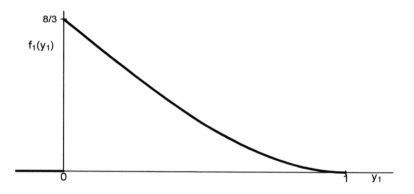

The marginal density for Y_2 has the same formula as that for Y_1.

Code to simulate from the joint density and make the histogram for y_1 values is shown on the left, with the histogram on the right, with the density function superimposed.

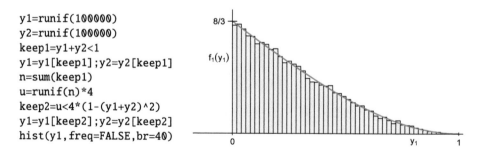

```
y1=runif(100000)
y2=runif(100000)
keep1=y1+y2<1
y1=y1[keep1];y2=y2[keep1]
n=sum(keep1)
u=runif(n)*4
keep2=u<4*(1-(y1+y2)^2)
y1=y1[keep2];y2=y2[keep2]
hist(y1,freq=FALSE,br=40)
```

The **marginal CDF** $F_1(y_1)$ is simply the CDF for the density $f_1(y_1)$. It can be computed in the usual way from f_1,

$$F_1(y_1) = \int_{-\infty}^{y_1} f_1(t) dt,$$

or from the joint CDF,

$$F_1(y_1) = F(y_1, \infty).$$

Chapter Highlights

1. Given the joint density $f(y_1, y_2)$ of continuous random variables Y_1 and Y_2, the marginal density for Y_1 is

$$f_1(y_1) = \int_{-\infty}^{\infty} f(y_1, y_2)\,dy_2.$$

2. The function $f_1(y_1)$ is a proper density function and can be thought of as the distribution of Y_1 when (Y_1, Y_2) are sampled from the joint density and the value of Y_2 is ignored.

3. Indicator functions can be used to indicate the support of the random variable as part of the density function:

$$I(\text{expression}) = \begin{cases} 1 & \text{if expression is true,} \\ 0 & \text{otherwise.} \end{cases}$$

Exercises

25.1 The random variables Y_1 and Y_2 have a joint density that is uniform on the triangle shown below in the (y_1, y_2) plane:

(a) Find the marginal distribution of Y_1 and sketch it.

(b) Find the marginal distribution of Y_2 and sketch it.

(c) Write code to simulate from the joint density to check your answers to (a) and (b).

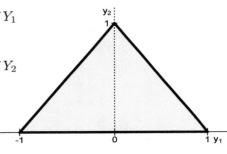

25.2 Let Y_1 and Y_2 have joint density

$$f(y_1, y_2) = \frac{6}{5}(1 - y_1 y_2^2) \quad \text{for } 0 \le y_1 \le 1, \ 0 \le y_2 \le 1.$$

(a) Find the marginal for Y_1 and sketch it.

(b) Find the marginal for Y_2 and sketch it.

(c) Write code to simulate from the joint density to check your answers to (a) and (b).

25.3 Suppose the random variables Y_1 and Y_2 are jointly continuous random variables where the support is the unit square, and

$$f(y_1, y_2) = \frac{9}{5}\left(1 - \sqrt{y_1 y_2}\right) \quad \text{for } 0 < y_1 < 1, \ 0 < y_2 < 1.$$

The density is shown in the plot:

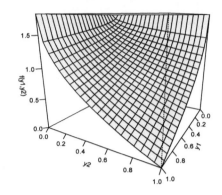

(a) Find the marginal density for Y_1.

(b) Find the marginal density for Y_2.
    (Can you use your answer to (a)?)

(c) Write R code to simulate from this
    joint density, and make a histogram
    of the y_1 values, superimposing
    your density calculated in (a).

25.4 Let Y_1 and Y_2 have joint density $f(y_1, y_2) = 6(1 - y_2)$ on $0 \leq y_1 \leq y_2 \leq 1$.

(a) Find the marginal for Y_1 and sketch it.

(b) Find the marginal for Y_2 and sketch it.

(c) Write code to simulate from the joint density to check your answers to (a)
    and (b).

25.5 Let Y_1 and Y_2 have joint density $f(y_1, y_2) = 24y_1(1 - y_2)$ on $0 \leq y_1 \leq y_2 \leq 1$.

(a) Find the marginal for Y_1 and sketch it.

(b) Find the marginal for Y_2 and sketch it.

(c) Write code to simulate from the joint density to check your answers to (a)
    and (b).

25.6 Suppose Y_1 and Y_2 are jointly distributed with density

$$f(y_1, y_2) = \frac{1}{8}e^{-(y_1/2 + y_2/4)} \text{ for } 0 < y_1, y_2 < \infty.$$

(a) Find the marginal for Y_1 and describe it in terms of one of our "named"
    densities.

(b) Find the marginal for Y_2 and describe it in terms of one of our "named"
    densities.

25.7 Suppose Y_1 and Y_2 are jointly distributed with density

$$f(y_1, y_2) = \frac{3}{8}e^{-(y_1/2 + y_2/4)} \text{ for } 0 < y_2 < y_1 < \infty.$$

(a) Find the marginal for Y_1—can you describe it in terms of one of our "named"
    densities?

(b) Find the marginal for Y_2—can you describe it in terms of one of our "named"
    densities?

25.8 Find the marginal distributions for Y_1 and Y_2, given the joint density in Exercise 24.11.

25.9 Find the marginal distributions for Y_1 and Y_2, given the joint density in Exercise 24.12. Interpret these distributions in the context of the problem.

25.10 Find the marginal distributions for Y_1 and Y_2, given the joint density in Exercise 24.13. Interpret these distributions in the context of the problem.

Chapter 26

Conditional Distributions and Independent Random Variables

For continuous random variables Y_1 and Y_2, with joint density $f(y_1, y_2)$, we define the **conditional density** for Y_1, given $Y_2 = y_2$, as

$$f(y_1|y_2) = \frac{f(y_1, y_2)}{f_2(y_2)},$$

where the denominator is the marginal density for Y_2. Even though the expression for the conditional density contains both y_1 and y_2, it's important to keep in mind that the function $f(y_1|y_2)$ has only one argument, and y_2 is a constant value. Conceptually, we fix a value of y_2 and determine how Y_1 is distributed, given that $Y_2 = y_2$.

The support of the conditional density $f(y_1|y_2)$ may depend on the value of y_2. To illustrate, let's look at the first example from the last chapter, where the joint density is uniform over the triangle shown:

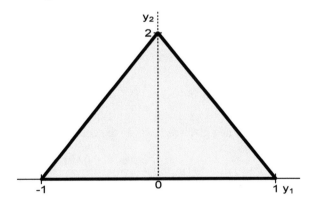

If we are given $Y_2 = 1$, say, then we know that the value of Y_1 must be in $(-1/2, 1/2)$ and, intuitively, we think the distribution of Y_1, given $Y_2 = 1$, must be uniform over this interval, because the height of the density is constant. Let's check this intuition using the definition. We found the marginal density for Y_2 to be $f_2(y_2) = 1 - y_2/2$; then

$$f(y_1|y_2) = \frac{f(y_1, y_2)}{f_2(y_2)} = \frac{\frac{1}{2}I\{y_2 > 0, 2y_1 + y_2 < 2, y_2 - 2y_1 < 2\}}{1 - \frac{y_2}{2}}.$$

This looks like a complicated function in two variables, but when you remember that y_2 is a constant and this expression is a density function in y_1, you can see that it is simply a uniform density. Plugging in $y_2 = 1$, we get

$$f(y_1|y_2 = 1) = \frac{\frac{1}{2}I\{y_1 < 1/2, y_1 > -1/2\}}{1 - \frac{1}{2}} = I\{-1/2 < y_1 < 1/2\},$$

which is an expression for the uniform density on $(-1/2, 1/2)$.

The conditional density must be a proper density; this will be something to check to ensure your calculations are correct. To continue the example for another value of Y_2,

$$f(y_1|y_2 = 1/2) = \frac{\frac{1}{2}I\{y_1 < 3/4, y_1 > -3/4\}}{1 - \frac{1}{4}} = \frac{2}{3}I\{-3/4 < y_1 < 3/4\},$$

which is an expression for the uniform density on $(-3/4, 3/4)$. For any value of y_2, the condition density $f(y_1|y_2)$ will integrate to one.

Example: Suppose Y_1 and Y_2 have joint density

$$f(y_1, y_2) = (y_1 + y_2)I\{0 < y_1 < 1, 0 < y_2 < 1\},$$

shown in the plot at the right.

To "see" the conditional distribution for Y_1, given $y_2 = 1/2$, say, visualize the cross-section of the density at $y_2 = 1/2$. The effect of the calculations is simply to scale this cross-section to be a proper density.

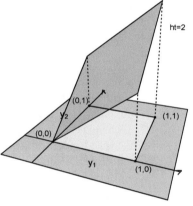

To find $f(y_1|y_2)$, we first need the marginal $f_2(y_2)$. Because the support is a rectangle, the integral is easy to set up:

$$f_2(y_2) = \int_0^1 (y_1 + y_2)dy_1 = \frac{1}{2} + y_2$$

for $0 < y_2 < 1$. Then given $y_2 \in (0, 1)$ the conditional density for y_1 is

$$f(y_1|y_2) = \frac{(y_1 + y_2)I\{0 < y_1 < 1\}}{\frac{1}{2} + y_2}.$$

To sketch this, we choose a value of y_2, say $y_2 = 1/2$, then

$$f(y_1|y_2 = 1/2) = \left(y_1 + \frac{1}{2}\right)I\{0 < y_1 < 1\}.$$

If $y_2 = 1/4$, we have

$$f(y_1|y_2 = 1/4) = \frac{4}{3}\left(y_1 + \frac{1}{4}\right)I\{0 < y_1 < 1\}.$$

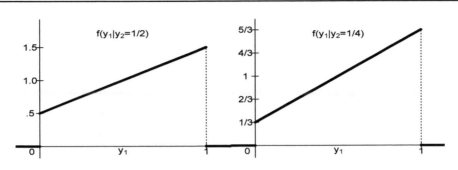

You can check that both of these functions are proper densities.

Example: Let Y_1 and Y_2 have joint density $f(y_1, y_2) = 6(1 - y_2)I\{0 \le y_1 \le y_2 \le 1\}$. Find the conditional distribution $f(y_2|y_1)$ and sketch it.

We found the marginal for Y_1 in Exercise 25.4: $f_1(y_1) = 3(1-y_1)^2 I\{0 < y_1 < 1\}$. Then for a fixed $y_1 \in (0, 1)$,

$$f(y_2|y_1) = \frac{2(1 - y_2)I\{y_1 \le y_2 \le 1\}}{(1 - y_1)^2}.$$

For example, if $y_1 = 3/4$, this is

$$f(y_2|y_1 = 3/4) = 32(1 - y_2)I\{.75 < y_2 < 1\}.$$

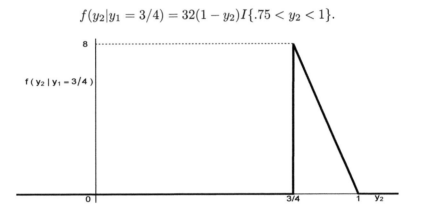

Recall the definition of independent random variables: The random variables Y_1 and Y_2 are **independent** if, for any sets A and B on the real line, $P(Y_1 \in A|Y_2 \in B) = P(Y_1 \in A)$.

For continuous random variables, this means (if $P(Y_2 \in B) \ne 0$)

$$\frac{P\left((Y_1 \in A) \cap (Y_2 \in B)\right)}{P(Y_2 \in B)} = P(Y_1 \in A)$$

or

$$\frac{\int_B \int_A f(y_1, y_2)dy_1 dy_2}{\int_B f_2(y_2)dy_2} = \int_A f_1(y_1)dy_1.$$

This means that

$$\int_B \int_A f(y_1, y_2)dy_1 dy_2 = \left(\int_A f_1(y_1)dy_1\right)\left(\int_B f_2(y_2)dy_2\right),$$

which can happen for *all* A and B only if

$$f(y_1, y_2) = f_1(y_1)f_2(y_2).$$

It follows that if Y_1 and Y_2 are independent, then the conditional density for Y_1 given $Y_2 = y_2$ is

$$f(y_1|y_2) = \frac{f(y_1, y_2)}{f_2(y_2)} = \frac{f_1(y_1)f_2(y_2)}{f_2(y_2)} = f_1(y_1)$$

for any y_1 in the support of Y_1. Of course, we would have guessed this from our intuitive notions of independence.

Now we can easily check if jointly continuous random variables are independent. For example, if $f(y_1, y_2) = e^{-(y_1+y_2)}$ on $y_1 > 0, y_2 > 0$, we can factor the joint density into two marginals: $f_1(y_1) = e^{-y_1}$ on $y_1 > 0$ and $f_2(y_2) = e^{-y_2}$ on $y_2 > 0$.

Conversely, we can write down joint densities of independent random variables given their marginals. If Y_1 and Y_2 are independent random variables with densities $f_1(y_1)$ and $f_2(y_2)$, then their joint density is the product of these marginals.

Note that if the support of the joint density is *not* rectangular, then the random variables *can't* be independent. If the support of the joint density is a triangle, as in the first example, the support of the conditional density $f(y_2|y_1)$ will always depend on the value of y_1, and vice versa.

For another example, suppose $f(y_1, y_2) = 2e^{-(y_1+y_2)}$ on $0 < y_1 < y_2 < \infty$. This cannot be factored into a product of marginals because the support of one random variable depends on the value of the other.

If Y_1 and Y_2 are independent random variables with densities f_1 and f_2, respectively, then

$$\begin{aligned}
\mathrm{E}(Y_1 Y_2) &= \int_{-\infty}^{\infty} \int_{-\infty}^{\infty} y_1 y_2 f_1(y_1) f_2(y_2) dy_1 dy_2 \\
&= \int_{-\infty}^{\infty} y_1 f_1(y_1) dy_1 \int_{-\infty}^{\infty} y_2 f_2(y_2) dy_2 \\
&= \mathrm{E}(Y_1)\mathrm{E}(Y_2).
\end{aligned}$$

It's important to remember that this formula is only for independent random variables!

Random variables $Y_1, \ldots, Y_n$ form an independent set or "are independent" if any two are independent. (This is also called *pairwise independent*.) We can extend the last formula to n independent random variables:

$$\mathrm{E}(Y_1 Y_2 \cdots Y_n) = \mathrm{E}(Y_1)\mathrm{E}(Y_2) \cdots \mathrm{E}(Y_n).$$

Now we can define some new terms that will be very useful for statistical inference about a distribution. First, we say that the random variables $Y_1, \ldots, Y_n$ are iid if they are independent and identically distributed. Next, if $Y_1, \ldots, Y_n$ are iid with common density or probability mass function f_Y, then we say that $Y_1, \ldots, Y_n$ is a **random sample** from the distribution. We can write

$$Y_1, \ldots, Y_n \overset{iid}{\sim} f_Y \ \text{ or } \ Y_1, \ldots, Y_n \overset{ind}{\sim} f_Y.$$

We have been using language such as "selecting at random" and relying on intuition to capture this concept that we have now defined formally.

Chapter Highlights

1. If Y_1 and Y_2 are jointly continuous random variables, then the conditional density of Y_1, given $Y_2 = y_2$, is

$$f(y_1|y_2) = \frac{f(y_1, y_2)}{f_2(y_2)},$$

where $f_2(y_2)$ is the marginal distribution of Y_2, and $f(y_1, y_2)$ is the joint density.

2. Random variables Y_1 and Y_2 are independent if

$$P(Y_1 \in A_1 | Y_2 \in A_2) = P(Y_1 \in A_1).$$

Equivalently, Y_1 and Y_2 are independent if and only if

$$f(y_1, y_2) = f_1(y_1) f_2(y_2),$$

where f, f_1, and f_2 are probability density functions or mass functions. Remember to be careful about the support! Include indicator functions in the expressions of the densities.

3. If Y_1 and Y_2 are independent with densities $f_1(y_1)$ and $f_2(y_2)$, respectively, then the joint density is $f(y_1, y_2) = f_1(y_1) f_2(y_2)$.

4. If Y_1 and Y_2 are independent random variables, then $E(Y_1 Y_2) = E(Y_1)E(Y_2)$.

5. If $Y_1, \ldots, Y_n$ are iid, this means that the random variables are pairwise independent (any pair is independent) and they all have the same probability distribution.

6. A **random sample** is a set of iid random variables.

Exercises

26.1 For a constant $a > 0$, the joint density of Y_1 and Y_2 is

$$f(y_1, y_2) = \frac{2}{a} e^{-y_1} e^{-y_2/a} \text{ for } 0 < y_2 < a y_1 < \infty.$$

(a) Find the conditional density of Y_2, given $Y_1 = 3$, and sketch it for some choice of a.

(b) Find the conditional density of Y_1, given $Y_2 = a$, and sketch it.

26.2 Ecologists are modeling the behavior of insects arriving at a nest. If Y_1 is the length of time (in minutes) an insect spends at the nest, and Y_2 is the time (in minutes) spent in feeding young at the nest, they stipulate that the joint density of Y_1 and Y_2 is

$$f(y_1, y_2) = \frac{1}{\theta^2} e^{-y_1/\theta} \text{ on } 0 \le y_2 \le y_1 < \infty.$$

(a) Show that f is a proper joint density.

(b) Find the marginal density for Y_1; sketch it for $\theta = 2$.

(c) Find the marginal density for Y_2; sketch it for $\theta = 2$.

(d) Find the conditional density of Y_2, given $Y_1 = 4$, and sketch it. Interpret it in the context of the problem.

(e) Are Y_1 and Y_2 independent random variables? Explain.

(f) Find the expected length of time that the insects spend feeding young at the nest.

26.3 An electronic device has two components with random lifespans (in days) Y_1 and Y_2. When one component fails, the device fails. The joint density for the random variables is

$$f(y_1, y_2) = 2e^{-y_1 - 2y_2} \text{ on } y_1 > 0, y_2 > 0.$$

(a) What is the probability that the device fails within one day?

(b) What is the marginal density for the lifetime of the first device?

(c) Are the lifespans independent? Explain.

26.4 Physicists want to model the position of a particle in a magnetic field. If the particle is at the origin at time zero, then the position after one time unit has the density

$$f(x, y) = \frac{e^{-x/y} e^{-y}}{y} \text{ on } x > 0, y > 0,$$

where x is the absolute value of the horizontal distance from the origin, and y is the absolute value of the vertical distance from the origin.

(a) Show that f is a proper density.

(b) Find the probability that $Y > 1$.

(c) Find the conditional distribution of X given that $Y = y$ (for a y value of your choice), and sketch it.

26.5 Consider again the situation in Exercise 24.12. Find the conditional density representing the proportion of an hour that worker 1 spends digging, given that worker 2 digs for half of the hour. Sketch the conditional density.

26.6 Consider again the situation in Exercise 24.13. Find the conditional density representing the lifetime of the power cell, given that it started humming 12 units of time after it was installed. Sketch the conditional density.

26.7 Suppose Y_1 and Y_2 are independent exponential random variables, and both have mean θ. Find the distribution of Y_1, given $Y_1 + Y_2 = s$.

26.8 The random variables Y_1 and Y_2 have joint density

$$f(y_1, y_2) = ce^{-y_2} \text{ for } 0 \le y_1 \le y_2 < \infty.$$

(a) Find c to make f a density function.

(b) Find the conditional distribution of Y_2 given $Y_1 = 4$, and sketch it.

26.9 Suppose X is a random variable with mean μ_X and variance σ_X^2, and Y is a random variable with mean μ_Y and variance σ_Y^2. Further, X and Y are independent. Find an expression for $V(XY)$.

Chapter 27

Covariance of Two Random Variables

If Y_1 and Y_2 are jointly distributed random variables, we can define their **covariance** to be

$$\text{cov}(Y_1, Y_2) = \text{E}\left[(Y_1 - \mu_1)(Y_2 - \mu_2)\right],$$

where $\mu_1 = \text{E}(Y_1)$ and $\mu_2 = \text{E}(Y_2)$. Common notation for covariance of Y_1 and Y_2 is σ_{12}.

The covariance of a random variable with itself is simply the variance of the random variable. Also, $\text{cov}(Y, -Y) = -\text{V}(Y)$.

The covariance can be any real number. Intuitively, if the joint distribution of Y_1 and Y_2 is such that Y_1 tends (on average) to be larger than its mean when Y_2 is also larger than *its* mean, then the covariance is positive. If Y_1 tends to be smaller than its mean when Y_2 is larger than its mean, then the covariance is negative. Sometimes we say that Y_1 and Y_2 have a "positive relationship" when the covariance is positive.

Formula for computation: The expression for covariance simplifies for easier calculation "by hand:"

$$\text{cov}(Y_1, Y_2) = \text{E}\left[(Y_1 - \mu_1)(Y_2 - \mu_2)\right]$$

$$= \text{E}(Y_1 Y_2 - \mu_1 Y_2 - \mu_2 Y_1 + \mu_1 \mu_2)$$

$$= \text{E}(Y_1 Y_2) - \text{E}(Y_1)\text{E}(Y_2).$$

Example: Let Y_1 and Y_2 have a joint density

$$f(y_1, y_2) = 2I\{y_1 > 0, y_2 > 0, y_1 + y_2 < 1\};$$

find the covariance. First make a guess as to whether the covariance is positive or negative, based on the picture of the support shown below.

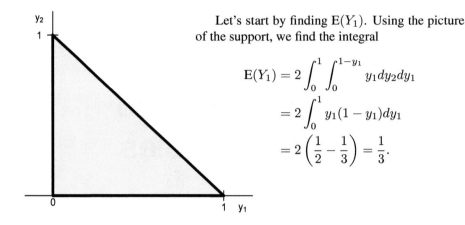

Let's start by finding $E(Y_1)$. Using the picture of the support, we find the integral

$$E(Y_1) = 2 \int_0^1 \int_0^{1-y_1} y_1 \, dy_2 \, dy_1$$

$$= 2 \int_0^1 y_1(1 - y_1) \, dy_1$$

$$= 2 \left(\frac{1}{2} - \frac{1}{3} \right) = \frac{1}{3}.$$

It is easy to see (by a symmetry argument) that $E(Y_2)$ is also $1/3$. To find the covariance,

$$E(Y_1 Y_2) = \int_0^1 \int_0^{1-y_1} y_1 y_2 \, dy_2 \, dy_1 = \frac{1}{12},$$

then $\text{cov}(Y_1, Y_2) = -1/36$. Intuitively, we understand that the covariance must be negative: When Y_1 is above 1/2, Y_2 must be below 1/2. "Larger" values of Y_1 typically occur with "smaller" values of Y_2.

Example: Exercise 23.9 concerns a species of annual garden plant that produces blossoms and fruit. A randomly selected plant will produce Y_1 blossoms and Y_2 fruits in a season, where the joint distribution of Y_1 and Y_2 is described by the mass function with probabilities given in the table below:

		Y_2 (fruits)		
		0	1	2
	0	.10	0	0
Y_1	1	.12	.26	0
(blossoms)	2	.14	.30	.08

There are a lot of zero values for the random variables, which makes the expected values easy to calculate. There are only four possible values of $Y_1 Y_2$, so the expected value of this product is

$$E(Y_1 Y_2) = .26 + 2(.30) + 4(.08) = 1.18.$$

Also, $E(Y_1) = 1.42$ and $E(Y_2) = .72$. Then $\text{cov}(Y_1, Y_2) = .1576$.

The covariance of two random variables depends on the scale. If the random variables are rescaled (to be in meters instead of feet, for example), then the covariance changes as well (see Exercise 27.1). To have a measure of joint variation that is independent of scaling, we define **correlation**.

The **correlation** of jointly distributed random variables Y_1 and Y_2 is

$$\rho_{12} = \frac{\text{cov}(Y_1, Y_2)}{\sqrt{V(Y_1)} \sqrt{V(Y_2)}} = \frac{\sigma_{12}}{\sigma_1 \sigma_2},$$

where σ_{12} is the covariance of Y_1 and Y_2, $\sigma_1^2 = V(Y_1)$, and $\sigma_2^2 = V(Y_2)$ (so σ_1 and σ_2 are standard deviations).

To show that the correlation has the claimed property of scale invariance, let $\mu_1 = E(Y_1)$ and $\mu_2 = E(Y_2)$. Suppose $X_1 = aY_1$ and $X_2 = bY_2$, where a and b are both positive. Then the correlation between X_1 and X_2 is

$$\frac{\text{cov}(X_1, X_2)}{\sqrt{V(X_1)}\sqrt{V(X_2)}} = \frac{\text{cov}(aY_1, bY_2)}{\sqrt{V(aY_1)}\sqrt{V(bY_2)}}$$
$$= \frac{E\left[(aY_1 - a\mu_1)(bY_2 - b\mu_2)\right]}{\sqrt{a^2 V(Y_1)}\sqrt{b^2 V(Y_2)}}$$
$$= \frac{E\left[(Y_1 - \mu_1)(Y_2 - \mu_2)\right]}{\sqrt{V(Y_1)}\sqrt{V(Y_2)}},$$

which is the correlation between Y_1 and Y_2.

To find the correlation for the random variables Y_1 and Y_2 in the above joint density example with triangular support, we first find the variance of Y_1. We calculate $E(Y_1^2) = 1/6$, so that $V(Y_1) = 1/18$. By symmetry, $V(Y_2) = 1/18$ as well, and the correlation is

$$\rho_{12} = \frac{1/36}{\sqrt{1/18}\sqrt{1/18}} = -1/2.$$

For the blossoms and fruit example, we calculate $V(Y_1) = E(Y_1^2) - E(Y_1)^2 = 2.46 - 1.42^2 = .4436$. Similarly, $V(Y_2) = E(Y_2^2) - E(Y_2)^2 = .88 - .72^2 = .3616$. The correlation is

$$\rho_{12} = \frac{.1576}{\sqrt{.4436}\sqrt{.3616}} = .3935.$$

We can show that the correlation has to be between -1 and $+1$ by considering for any constant c

$$V(Y_1 + cY_2) = E\left[\{(Y_1 + cY_2) - (\mu_1 + c\mu_2)\}^2\right]$$
$$= E\left[\{(Y_1 - \mu_1) + c(Y_2 - \mu_2)\}^2\right]$$
$$= E\left[(Y_1 - \mu_1)^2 + 2c(Y_1 - \mu_1)(Y_2 - \mu_2) + c^2(Y_2 - \mu_2)^2\right]$$
$$= \sigma_1^2 + 2c\rho_{12}\sigma_1\sigma_2 + c^2\sigma_2^2.$$

For the last equality, we used the fact that $\text{cov}(Y_1, Y_2) = \rho_{12}\sigma_1\sigma_2$. If we plug in $c = -\rho_{12}\sigma_1/\sigma_2$, we get

$$V\left(Y_1 - \frac{\rho_{12}\sigma_1}{\sigma_2}Y_2\right) = \sigma_1^2 - \rho_{12}^2\sigma_1^2 = \sigma_1^2(1 - \rho_{12}^2).$$

Because the variance of any random variable cannot be negative, we know that $\sigma_1^2(1 - \rho_{12}^2) \geq 0$, which means $\rho_{12}^2 \leq 1$.

The correlation is a kind of standardized measure of how two random variables vary together. If you are told "the covariance of Y_1 and Y_2 is 3," that doesn't tell you *how* closely related Y_1 and Y_2 are, unless you also know the variance of Y_1 and Y_2. On the other hand, if you know that the correlation between Y_1 and Y_2 is .8, you know the random variables are quite closely related.

If Y_1 and Y_2 are independent, then they are not correlated. We can compute

$$\text{cov}(Y_1, Y_2) = \text{E}(Y_1 Y_2) - \text{E}(Y_1)\text{E}(Y_2) = 0,$$

because for independent random variables, the expectation of the product is the product of the expectations.

However, there are cases in which the covariance is zero but the random variables are not independent. This can happen when the support of one random variable depends on the value of the other.

Example: Let Y_1 and Y_2 be jointly uniform over $\{y_2 > 0, y_1 + y_2 < 1, y_2 < y_1 + 1\}$. We know that the random variables are not independent; looking at the support we see that if $Y_2 = 3/4$, then Y_1 can't be larger than $1/4$ or smaller than $-1/4$.

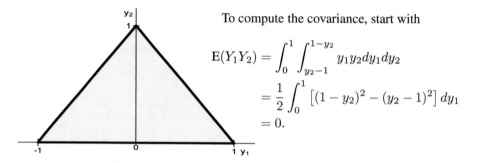

To compute the covariance, start with

$$\text{E}(Y_1 Y_2) = \int_0^1 \int_{y_2-1}^{1-y_2} y_1 y_2 dy_1 dy_2$$
$$= \frac{1}{2} \int_0^1 \left[(1 - y_2)^2 - (y_2 - 1)^2\right] dy_1$$
$$= 0.$$

We won't bother computing $\text{E}(Y_2)$, because by symmetry $\text{E}(Y_1) = 0$, and that's all we need to calculate $\text{cov}(Y_1, Y_2) = 0$. The covariance is zero, and yet the random variables are not independent.

Now we can compute a formula for variance of linear combinations of random variables. We already know from Chapters 23 and 24 that

$$\text{E}\left(a_1 Y_1 + a_2 Y_2 + \cdots + a_n Y_n\right) = a_1\text{E}(Y_1) + a_2\text{E}(Y_2) + \cdots + a_n\text{E}(Y_n).$$

Let's compute the variance of $a_1 Y_1 + a_2 Y_2$ and then generalize to more terms. Using the definition of variance,

$$\text{V}(a_1 Y_1 + a_2 Y_2) = \text{E}\left[(a_1 Y_1 + a_2 Y_2)^2\right] - \text{E}(a_1 Y_1 + a_2 Y_2)^2$$
$$= \text{E}\left[a_1^2 Y_1^2 + 2a_1 a_2 Y_1 Y_2 + a_2^2 Y_2^2\right] - \left[a_1\text{E}(Y_1) + a_2\text{E}(Y_2)\right]^2.$$

Let's use $\mu_i = \text{E}(Y_i)$ for $i = 1, \ldots, n$. Then this is

$$\text{V}(a_1 Y_1 + a_2 Y_2) = \text{E}\left[a_1^2 Y_1^2 + 2a_1 a_2 Y_1 Y_2 + a_2^2 Y_2^2\right] - \left[a_1\mu_1 + a_2\mu_2\right]^2$$
$$= a_1^2\text{E}(Y_1^2) + 2a_1 a_2\text{E}(Y_1 Y_2) + a_2^2\text{E}(Y_2^2) - a_1^2\mu_1^2 - 2a_1 a_2\mu_1\mu_2 + a_2^2\mu_2^2$$
$$= a_1^2\left[\text{E}(Y_1^2) - \mu_1^2\right] + 2a_1 a_2\left[\text{E}(Y_1, Y_2) - \mu_1\mu_2\right] + a_2^2\left[\text{E}(Y_2^2) - \mu_2^2\right]$$
$$= a_1^2\text{V}(Y_1) + 2a_1 a_2\text{cov}(Y_1, Y_2) + a_2^2\text{V}(Y_2).$$

Following the same ideas, we can derive the formula

$$\text{V}\left(\sum_{i=1}^n a_i Y_i\right) = \sum_{i=1}^n a_i^2\text{V}(Y_i) + 2\sum\sum_{i<j} a_i a_j\text{cov}(Y_i, Y_j).$$

An important special case of this formula is when the random variables $Y_1, Y_2, \ldots, Y_n$ are pairwise independent. Then all of the covariances are zero, so we get that, for independent random variables, the variance of the sum is the sum of the variances.

Example: Let's go back to the blossoms and fruit example from Exercise 23.9 and compute the variance of the number of blossoms that do not fruit. Let $X = Y_1 - Y_2$ be the number of these blossoms, and so

$$V(X) = V(Y_1) + V(Y_2) - 2\text{cov}(Y_1, Y_2).$$

Plugging in the values we have already calculated gives

$$V(X) = .4436 + .3616 - 2(.1576) = .49.$$

The R function `cov` will compute the **sample covariance** for two vectors. For example, if `x` and `y` are vectors of the same length, then `cov(x,y)` returns the sample covariance $\sum_{i=1}^{n} \left[(x_1 - \bar{x})(y_i - \bar{y}) \right] / (n-1)$, where $\bar{x}$ and $\bar{y}$ are the sample means. Similarly, `cor(x,y)` returns the **sample correlation** of the vectors. This is computed as the sample covariance divided by the two sample standard deviations. Like the sample mean and the sample variance, discussed at the end of Chapter 5, the sample covariance is random variable. As the sample size gets larger, the sample covariance will get closer to the true covariance (if this is finite). This can be used to check your covariance and correlation calculations, using simulations in R. Alternatively, R can be used to give an approximation of the covariance and correlation when the calculations would be too tedious.

Example: Suppose we roll five fair six-sided dice and define Y_1 to be the number of ones while Y_2 is the number of twos. We can see that the correlation of Y_1 and Y_2 would be negative, and we could write out the joint distribution and do the calculations by hand, but you can see that it would be a lot of work. Instead, we can use the code

```
nloop=100000
y1=1:nloop
y2=1:nloop
for(iloop in 1:nloop){
    x=sample(1:6,5,replace=TRUE)
    y1[iloop]=sum(x==1)
    y2[iloop]=sum(x==2)
}
cor(y1,y2)
```

to get that the correlation of Y_1 and Y_2 is about $-.202$.

Let's use R to check our computed covariance for the first example of this chapter, where the joint distribution is uniform over a triangle. Using the methods of Chapter 24, we can sample a large number of pairs of values from the above joint density and compute the sample covariance:

```
y1=runif(10000)
y2=runif(10000)
keep=y1+y2<1
y1=y1[keep];y2=y2[keep]
cov(y1,y2)
```

returns a value close to $-1/36$.

Chapter Highlights

1. The covariance of random variables Y_1 and Y_2 is defined as

    $$\text{cov}(Y_1, Y_2) = \text{E}\left[(Y_1 - \mu_1)(Y_2 - \mu_2)\right],$$

    where $\mu_1 = \text{E}(Y_1)$ and $\mu_2 = \text{E}(Y_2)$. This simplifies to

    $$\text{cov}(Y_1, Y_2) = \text{E}(Y_1 Y_2) - \mu_1 \mu_2.$$

2. The correlation of random variables Y_1 and Y_2 is defined as

    $$\rho_{12} = \frac{\text{cov}(Y_1, Y_2)}{\sqrt{\text{V}(Y_1)\text{V}(Y_2)}}.$$

3. If Y_1 and Y_2 are independent, then $\text{cov}(Y_1, Y_2) = 0$, but not the converse!

4. We can find the variance of a linear combination of random variables as a function of their variances and covariances:

    $$\text{V}\left(\sum_{i=1}^{n} a_i Y_i\right) = \sum_{i=1}^{n} a_i^2 \text{V}(Y_i) + 2\sum\sum_{i<j} a_i a_j \text{cov}(Y_i, Y_j).$$

5. If vectors **x** and **y** in R both have length n, and values $x_1, \ldots, x_n$ and $y_1, \ldots, y_n$, respectively, then the command `cov(x,y)` will return the sample covariance

    $$\frac{1}{n-1} \sum_{i=1}^{n} \left[(x_i - \bar{x})(y_i - \bar{y})\right],$$

    where $\bar{x}$ is the average of the **x** values and $\bar{y}$ is the average of the y values. The command `cor(x,y)` will give the sample correlation, which is the sample covariance divided by the sample standard deviations for **x** and y.

Exercises

27.1 Let μ_1 and σ_1^2 be the mean and variance of Y_1, respectively, and similarly let μ_2 and σ_2^2 be the mean and variance of Y_2. Finally, define σ_{12} as the covariance of Y_1 and Y_2.

   (a) If $X_1 = a_1 Y_1 + b_1$ and $X_2 = a_2 Y_2 + b_2$, for constants a_1, a_2, b_1, and b_2, find $\text{cov}(X_1, X_2)$.

   (b) If $X_1 = a_1 Y_1 + b_1$ and $X_2 = a_2 Y_2 + b_2$, for constants a_1, a_2, b_1, and b_2, where a_1 and a_2 are positive, find the correlation between X_1 and X_2.

   (c) If $X_1 = a_1 Y_1 + b_1$ and $X_2 = a_2 Y_2 + b_2$, for constants a_1, a_2, b_1, and b_2, where a_1 and a_2 are negative, find the correlation between X_1 and X_2.

   (d) If $X_1 = a_1 Y_1 + b_1$ and $X_2 = a_2 Y_2 + b_2$, for constants a_1, a_2, b_1, and b_2, where $a_1 > 0$ and $a_2 < 0$, find the correlation between X_1 and X_2.

27.2 Suppose Y_1 and Y_2 have the joint mass function described in the table below.

    (a) Find the correlation of Y_1 and Y_2.

    (b) Check your answer using simulations in R.

		Y_1	
		-1	1
	-1	.3	.2
Y_2			
	1	.2	.3

27.3 The joint density for the random variables Y_1 and Y_2 is

$$f(y_1, y_2) = c(1 - y_1 y_2) \text{ for } 0 \leq y_1 \leq 1, 0 \leq y_2 \leq 1.$$

    (a) Find the covariance of Y_1 and Y_2.

    (b) Are Y_1 and Y_2 independent? Explain.

    (c) Check your answer to part (a) using simulations in R.

27.4 A deck has six cards: two are red, two are blue, and two are yellow. The deck is shuffled and two cards are chosen (without replacement). Let Y_1 be the number of red cards that are chosen, and let Y_2 be the number of blue cards that are chosen.

    (a) Find the joint mass function for Y_1 and Y_2.

    (b) Find the covariance of Y_1 and Y_2.

27.5 Consider the joint density red and blue balls drawn from the urn described in Exercise 23.3.

    (a) Find the covariance of these random variables.

    (b) Are the random variables independent? Explain.

27.6 Recall that in Chapter 23 an example used repeated independent coin tosses, where p is the probability for heads on each toss, Y_1 is the number of the toss on which the first head appears, and Y_2 is the number of the toss on which the second head appears. Using the solution to Exercise 23.7, find the covariance of Y_1 and Y_2.

27.7 Consider the joint density of number of blossoms (Y_1) and number of fruit (Y_2) for a randomly selected plant described in Exercise 23.10.

    (a) Find the covariance of Y_1 and Y_2.

    (b) Are Y_1 and Y_2 independent? Explain.

27.8 Consider the joint density of number of failures for the machines that make widgets, described in Exercise 23.11.

    (a) Find the covariance of these random variables.

    (b) Are the random variables independent? Explain.

27.9 Consider the joint density of concentrations of arsenic (Y_1) and antimony (Y_2) in grass stems in Largetown, given in Exercise 24.11.

(a) Find the covariance of Y_1 and Y_2.

(b) Are Y_1 and Y_2 independent? Explain.

27.10 Consider the joint density of times spent digging for the workers in Exercise 24.12.

(a) Find the covariance of these random variables.

(b) Are the random variables independent? Explain.

27.11 Ecologists are modeling the nesting behavior of pairs of a species of sea bird. They observe each of the pairs for an hour. If Y_1 is the proportion of the hour that the male spends in nest building and Y_2 is the proportion of the hour that the female spends in nest building, they postulate that the joint density of Y_1 and Y_2 is

$$f(y_1, y_2) = cy_1 y_2 \text{ on } 0 \le y_2 + y_1 \le 1, y_1 > 0, y_2 > 0.$$

(a) Find c to make the function a proper joint density.

(b) Find the marginal density for Y_2, sketch it, and interpret in the context of the problem.

(c) Find the marginal density for Y_1, sketch it, and interpret in the context of the problem.

(d) Find the conditional density of Y_2, given $Y_1 = 1/4$, and sketch it. Interpret in the context of the problem.

(e) Find the expected proportion of the hour that the male spends building the nest.

(f) Find the covariance of Y_1 and Y_2, and interpret the sign in the context of the problem.

(g) Write R code to simulate from this joint density and check your answers to (b), (c), (e), and (f).

27.12 Suppose you roll ten fair six-sided dice, and define Y_1 to be the sum of the numbers on the dice, and Y_2 to be the number of sixes. Using simulations in R, approximate the correlation of Y_1 and Y_2.

27.13 Let $Y_1, \ldots, Y_n$ be a collection of independent random variables, all with mean zero and variance σ^2. Let $x_1, \ldots, x_n$ be a collection of real numbers such that $\sum_{i=1}^{n} x_i = 0$. Show that the sum $Y_1 + \cdots + Y_n$ of the random variables and the linear combination $\sum_{i=1}^{n} Y_i x_i$ have zero covariance.

Chapter 28

The Multinomial Distribution

The multinomial distribution is a generalization of the binomial ideas. We again have independent trials, with probabilities of the outcomes unchanging over trials, but for each trial we have more than two possible outcomes.

Example: Suppose a treatment for a certain type of disease results in three possible outcomes: complete cure (for 60% of cases), death (for 10% of cases), or partial cure (for 30% of cases). If we randomly select 10 patients for treatment, what is the probability that we get 4 complete cures, 3 partial cures, and 3 deaths?

Example: Suppose 40% of Ameraucana hens lay green eggs, 40% lay brown eggs, 15% lay pink eggs, and 5% lay blue eggs. If Elizabeth gets five Ameraucana pullets (young hens), what is the probability that she gets one blue egg layer and four green egg layers?

When there are only two possible outcomes, the results can be summarized by only one random variable: the number of "successes." However, when there are k possible outcomes, we need $k - 1$ random variables to summarize the results. For the above example, if the number of blue egg layers, the number of pink egg layers, and the number of green egg layers are specified, we know the number of brown egg layers.

Actually, it is convenient to summarize the counts of the outcomes in a k-tuple of random variables $(Y_1, \ldots, Y_k)$, where Y_i is the number of trials with outcome i, for $i = 1, \ldots, k$. Of course we must have $\sum_{i=1}^{k} Y_i = n$, so any $k - 1$ values of the random variables will determine the remaining value.

Let's go back to the first example, and try to compute the probability. If we let Y_1 be the number of patients with complete cures, Y_2 be the number of patients who die, and Y_3 be the number of patients with partial cures, we are looking for $P(Y_1 = 4, Y_2 = 3, Y_3 = 3)$. Of course, $P(Y_1 = y_1, Y_2 = y_2, Y_3 = y_3) = 0$ whenever $y_1 + y_2 + y_3 \neq 10$.

The probability that the *first* four patients get complete cures, the *next* three patients die, and the *final* three patients get partial cures is $(.6)^4(.1)^3(.3)^3 = 0.0000035$. However, there are many different trial orderings for which the values of Y_1, Y_2, and Y_3 are 4, 3, and 3, respectively. We have to count these orderings, using ideas from Chapter 3.

Counting the trial orderings is analogous to the arranging-blocks problem. How many ways are there to arrange four red blocks, three green blocks, and three blue blocks if the blocks of any given color are indistinguishable? There are 10! ways to arrange distinct blocks, and within these arrangements there are 4! ways to swap around the red blocks, 3! ways to swap around the blue blocks, and 3! ways to swap around the green blocks. Therefore, we have

$$\frac{10!}{4!3!3!} = 4200$$

ways to arrange the blocks, and 4200 trial orderings for which the values of Y_1, Y_2, and Y_3 are 4, 3, and 3, respectively.

Using this reasoning for the orderings of outcomes, the probability of 4 complete cures, 3 partial cures, and 3 deaths is $4200 \times 0.0000035 = .0147$.

In general, suppose we have n trials with k possible outcomes each, with probabilities $p_1, p_2, \ldots, p_k$, where $\sum_{i=1}^{k} p_i = 1$. If Y_i is the number of the trials resulting in the ith outcome, $i = 1, \ldots, k$, then the formula for the multinomial probabilities is

$$P(Y_1 = y_1, Y_2 = y_2, \ldots, Y_k = y_k) = \frac{n!}{y_1! y_2! \cdots y_k!} p_1^{y_1} p_2^{y_2} \cdots p_k^{y_k}$$

if $\sum_{i=1}^{k} y_i = n$. The probability is zero if $\sum_{i=1}^{k} y_i \neq n$.

To prove that the probabilities add to one, we can use the multinomial theorem, which is a generalization of the binomial theorem. We'll skip this proof, but it is similar to the proof that the binomial probabilities add to one, using the binomial theorem.

To find the expected value of each of the random variables, we borrow binomial distribution ideas. Suppose we're interested in $E(Y_1)$. We can redefine the first outcome to be a "success" and all of the other types of outcomes to be "failures." The expected number of successes is then np_1. Generalizing this idea gives

$$E(Y_i) = np_i$$

for $i = 1, \ldots, k$. Similarly, we get

$$V(Y_i) = np_i(1 - p_i)$$

for $i = 1, \ldots, k$.

Let's compute $\text{cov}(Y_r, Y_s)$ for $r \neq s$. Intuitively, we think the covariances should be negative, because if there are more of one type of outcome, then there tends to be less of another.

To compute the covariance $\text{cov}(Y_r, Y_s)$, we define sequences of indicator random variables, where $U_i = 1$ if the ith trial results in the rth outcome, and zero otherwise. Also define $W_i = 1$ if the ith trial results in the sth outcome, and zero otherwise. Then $Y_r = \sum_{i=1}^{n} U_i$ and $Y_s = \sum_{i=1}^{n} W_i$. Now,

$$E(Y_r Y_s) = E\left[\left(\sum_{i=1}^{n} U_i\right)\left(\sum_{j=1}^{n} W_j\right)\right]$$

$$= E\left[\sum_{i=1}^{n} U_i W_i + \sum\sum_{i \neq j} U_i W_j\right]$$

$$= E\left[\sum_{i=1}^{n} U_i W_i\right] + E\left[\sum\sum_{i \neq j} U_i W_j\right].$$

$$= \sum_{i=1}^{n} E\left[U_i W_i\right] + \sum\sum_{i \neq j} E\left[U_i W_j\right].$$

First, for any i we have $U_iW_i = 0$, because if $U_i = 1$, the $W_i = 0$, and vice versa. Therefore the first sum is zero. Next, because the trials are independent, $E[U_iW_j] = E(U_i)E(W_j) = p_rp_s$. So the second term is a sum of terms that are all equal to p_rp_s. Finally,

$$E(Y_rY_s) = (n^2 - n)p_rp_s,$$

because there are $n^2 - n$ pairs of i, j, where i and j run from 1 to n, but $i \neq j$. Then

$$\text{cov}(Y_r, Y_s) = (n^2 - n)p_rp_s - (np_r)(np_s) = -np_rp_s,$$

where $r \neq s$.

Example: A machine makes metal rods for a device. The lengths of the rods are normally distributed with mean 25 mm and standard deviation 1 mm. Only the rods that are between 23 and 27 mm in length can be used for the device. In a random sample of seven rods, what is the probability that exactly one is too large and exactly one is too small?

Suppose Y_1 is the number of rods that are too small, Y_2 is the number that can be used, and Y_3 is the number that are too big. Then using the 68-95-99.7 rule, we have $p_1 = .025$, $p_2 = .95$, and $p_3 = .025$. Using the formula,

$$P(Y_1 = 1, Y_2 = 5, Y_3 = 1) = \frac{7!}{1!5!1!}(.025)^1(.95)^5(.025)^1 = .020.$$

Suppose that the rods that are between 23 and 27 mm in length can be sold for \$4 each, and the rods that are greater than 27mm can be sold for \$1 each, and the rods that are less than 23mm have to be discarded. If it costs 75 cents to make a rod, what is the expected profit for a random sample of 10 rods?

The profit in dollars is

$$X = 4Y_2 + Y_3 - 7.5,$$

and the expected profit is

$$E(X) = 4E(Y_2) + E(Y_3) - 7.5 = 4(9.5) + .25 - 7.5 = 30.75.$$

The R function dmultinom(x,prob) is useful for computing multinomial probabilities, where x is a vector of counts and prob is a vector of probabilities. For example, the command dmultinom(c(1,5,1),prob=c(.025,.95,.025)) returns the probability that $Y_1 = 1$, $Y_2 = 5$, and $Y_3 = 1$, when $n = 7$, $k = 3$, and $p_1 = .025$, $p_2 = .95$, $p_3 = .025$; this probability is .020.

The R function rmultinom(N,n,prob) generates N vectors of length k from a multinomial distribution with n trials, according to probabilities in the vector prob, which has length k. The probabilities in prob must add to one. The function returns a matrix with k rows and n columns.

The rmultinom function is useful for simulations. For example, suppose that $(Y_1, \ldots, Y_k)$ is a multinomial with n trials and probabilities $(p_1, \ldots, p_k)$. We want to find $P(Y_1 > Y_2)$. To compute this by hand, we need to find all configurations of the vector for which $Y_1 > Y_2$, and add up the probabilities. Instead, we can write code to simulate many multinomial random vectors and count the proportion that have $Y_1 > Y_2$. Suppose $n = 10$, $k = 4$, and $p = (.1, .2, .3, .4)$. The following code approximates $P(Y_1 > Y_2)$:

```
n=100000
y=rmultinom(n,10,c(.1,.2,.3,.4))
sum(y[1,]>y[2,])/n
```

About 18.4% of samples have $Y_1 > Y_2$.

Chapter Highlights

1. For n independent, identical trials with k possible outcomes, suppose the outcome probabilities are $p_1, p_2, \ldots, p_k$, where $\sum_{i=1}^{k} p_i = 1$. If Y_i is the number of the trials resulting in the ith outcome, $i = 1, \ldots, k$, then if $\sum_{i=1}^{k} y_i = n$,

$$P(Y_1 = y_1, Y_2 = y_2, \ldots, Y_k = y_k) = \frac{n!}{y_1! y_2! \cdots y_k!} p_1^{y_1} p_2^{y_2} \cdots p_k^{y_k}.$$

2. $E(Y_i) = np_i$ and $V(Y_i) = np_i(1 - p_i)$.

3. If $i \neq j$, $\text{cov}(Y_i, Y_j) = -np_i p_j$.

4. The R command dmultinom(x,prob) returns the probability that a multinomial random vector has values matching the vector x, when the probabilities are those in the vector prob.

5. The R command rmultinom(N,n,prob) generates a matrix of k rows and n columns, where k is the length of prob, and the columns contain realizations of multinomial random vectors with the given probabilities.

Exercises

28.1 The lifetimes of switches of a certain brand follow an exponential distribution with mean 26 days. Suppose we get a sample of 5 switches and determine the lifetimes. What is the probability that exactly two lifetimes are less than 20 days, and exactly two lifetimes are more than 30 days?

28.2 The amount of vitamin C in a randomly selected orange, in international units, from a species of tree follows a normal distribution with mean 100 and standard deviation 8. Suppose we have a random sample of 8 of these oranges. What is the probability that 4 oranges have between 92 and 108 units of vitamin C, 2 have less than 92 units, and 2 have more than 108 units?

28.3 The number of flaws in lengths of cord produced by a machine follows a Poisson distribution with mean $\lambda = .2$.

   (a) Suppose we take a random sample of 10 lengths of cord. What is the probability that five have no flaws, four have one flaw, and one has more than one flaw?

(b) Suppose each cord costs $2 to make, and a flawless cord can be sold for $5. A cord with one flaw can be sold for $1, but a cord with more than one flaw has to be discarded. What is the expected profit for each cord?

28.4 Breeders of aquarium fish know that for a certain breed, 10% of offspring are a "fancy" type that can be sold for $50 when they reach maturity. However, 40% of offspring are unviable and will die when they reach maturity, and the remaining 50% are a healthy "regular" type that can be sold for $8 each. If the fish cost $4 each to raise to maturity, what is the expected profit for raising 100 fish?

28.5 Eight seeds for a hybrid flowering plant will be randomly selected. Botanists know that 50% of these plants have pink flowers, 25% have white flowers, and 25% have red flowers.

(a) What is the probability that, from these eight seeds, exactly two are red, exactly two are white, and the remaining four are pink?

(b) What is the probability that, from these eight seeds, all of the plants have either pink or red flowers?

(c) Use simulations in R to find the probability that the number of white flowers is greater than the number of pink flowers.

28.6 Suppose we roll five fair six-sided dice, and let Y_1 be the number of ones and Y_2 the number of twos. Find the correlation between Y_1 and Y_2, and compare it to the approximation we computed using R, in the last example of Chapter 27.

28.7 A weaving machine produces lengths of linen for making bedsheets. The machine often "catches the weft," which produces a small flaw in the cloth. One randomly selected yard of cloth from the machine has 0–3 flaws with the following probabilities:

Number of flaws	0	1	2	3
Probability	.6	.2	.1	.1

The company manager has an opportunity to buy a new machine. She wants to test the null hypothesis that the new machine produces flaws with the same probabilities as the old machine against the alternative that the new machine produces fewer flaws.

She will choose 10 yards of fabric produced by the new machine, and reject the null hypothesis if either of the following holds:

- all 10 have no flaws, or
- 9 have no flaws and one has one flaw.

(a) What is the test size?

(b) Suppose the true distribution of flaws per yard, for the new machine, has the following probabilities:

Number of flaws	0	1	2	3
Probability	.85	.1	.04	.01

What is the power of the test?

Chapter 29

Conditional Expectation and Variance

For jointly continuous random variables X and Y, the expected value of Y, given $X = x$, is defined as

$$E(Y|X = x) = \int_{-\infty}^{\infty} yf(y|x)dy.$$

Recall that the conditional density of Y, given $X = x$, is defined as

$$f(y|x) = \frac{f(x,y)}{f_X(x)},$$

where $f(x,y)$ is the joint density and $f_X(x)$ is the marginal density for X.

For jointly discrete random variables X and Y, the expected value of Y, given $X = x$, is defined as

$$E(Y|X = x) = \sum_{all\ y} yf(y|x),$$

where the conditional density of Y, given $X = x$, is defined as

$$f(y|x) = P(Y = y|X = x) = \frac{P(Y = y \text{ and } X = x)}{P(X = x)} = \frac{f(x,y)}{f_X(x)},$$

where $f(x,y)$ is the joint mass function, and $f_X(x)$ is the marginal mass function for X.

In either case, the expectation $E(Y|X = x)$ is a function of x. Let's do an example to see this more concretely.

Let X and Y be jointly uniform on the triangle shown:

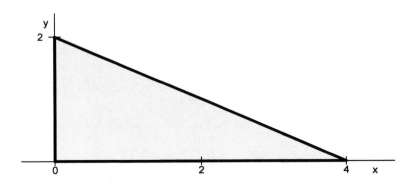

The density can be written as

$$f(x, y) = \frac{1}{4} I \{x \geq 0, y \geq 0, x + 2y \leq 4\}.$$

It's straightforward to compute $E(X) = 4/3$ and $E(Y) = 2/3$.

Using results from Chapters 24 and 25, we get the marginal for X:

$$f_X(x) = \frac{1}{4} \int_0^{2-x/2} dy = \frac{4-x}{8} \quad \text{for } x \in (0, 4).$$

Then for $x \in (0, 4)$,

$$f(y|x) = \frac{2}{4-x} I \left\{ 0 \leq y \leq \frac{4-x}{2} \right\}$$

is the conditional density for Y, given $X = x$. Note that this is a uniform density, where the support and height are determined by the value of x. Then

$$E(Y|X = x) = \int_{-\infty}^{\infty} y f(y|x) dy = \frac{2}{4-x} \int_0^{(4-x)/2} y \, dy = \frac{4-x}{4}.$$

This makes sense intuitively, and in fact we could have just guessed the answer. If $x = 3$, say, then the conditional for Y, given $X = 3$, is uniform on $(0, 1/2)$, so its expected value is $1/4$—and this is what we get by plugging into the formula.

Now let's consider the expression for the conditional expectation of Y, given X, but without plugging in $X = x$:

$$E(Y|X) = \frac{4-X}{4}.$$

This is a random variable! It's a function of the random variable X, and so it has an expected value and variance. We calculate

$$E[E(Y|X)] = E\left(\frac{4-X}{4}\right) = \frac{4 - E(X)}{4} = \frac{4 - 4/3}{4} = 2/3,$$

which is the (marginal) expected value for Y. In fact, a very useful formula for conditional expectation holds:

$$E[E(Y|X)] = E(Y).$$

To prove this for the continuous case, we start with the definition

$$
\begin{aligned}
E(Y) &= \int_{-\infty}^{\infty} \int_{-\infty}^{\infty} y f(x, y) dy dx \\
&= \int_{-\infty}^{\infty} \int_{-\infty}^{\infty} y f(y|x) f_X(x) dy dx \\
&= \int_{-\infty}^{\infty} \left[\int_{-\infty}^{\infty} y f(y|x) dy \right] f_X(x) dx \\
&= \int_{-\infty}^{\infty} [E(Y|X = x)] f_X(x) dx \\
&= E[E(Y|X)].
\end{aligned}
$$

You are asked in Exercise 29.1 to prove the result for the discrete case.

Many of the interesting conditional expectation problems involve a "mixed" or hybrid joint distribution, where X is discrete and Y is continuous, or vice versa. We haven't yet looked at these types of joint distributions, but we'll assert that the formula applies in this case, and work through some examples.

Example: A bicycle courier service in the business district of Largetown specializes in quick deliveries. Suppose the number of delivery jobs obtained daily by the service follows a Poisson distribution with mean λ. The time to delivery for a randomly selected job (in minutes) follows a gamma distribution with parameters α and β, and the times for various jobs can be considered to be independent. The owner of the service is trying to figure out how many employees to hire. Let's help her out by calculating the expected number of minutes per day necessary to make all of the deliveries!

Let X be the number of deliveries in a randomly selected day—note that we know that marginal distribution for X: this is $\text{Pois}(\lambda)$. Let Y_i be the delivery time for the ith job of the day, and we want to know the expected value of $T = \sum_{i=1}^{X} Y_i$:

$$
\begin{aligned}
\text{E}(T) &= \text{E}\left[\text{E}(T|X)\right] \\
&= \text{E}\left[\text{E}\left(\sum_{i=1}^{X} Y_i \middle| X\right)\right] \\
&= \text{E}\left[\sum_{i=1}^{X} \text{E}(Y_i|X)\right] \\
&= \text{E}\left[X\alpha/\beta\right] \\
&= \lambda\alpha/\beta.
\end{aligned}
$$

So, if $\lambda = 45$, $\alpha = 4$, and $\beta = 1/2$, the expected number of minutes per day is 360.

The owner is grateful but would like to know the standard deviation of the number of minutes per day as well. This way, she could aim for being able to cover two standard deviations over the mean number of minutes.

There is a corresponding formula for the variance that is more complicated but proved in the same way. For jointly distributed random variables X and Y,

$$
\text{V}(Y) = \text{E}\left[\text{V}(Y|X)\right] + \text{V}\left[\text{E}(Y|X)\right].
$$

We can prove that the formula holds by starting with the definition of variance and using the formula for conditional expectation twice:

$$
\begin{aligned}
\text{V}(Y) &= \text{E}(Y^2) - \text{E}(Y)^2 \\
&= \text{E}\left[\text{E}(Y^2|X)\right] - \text{E}\left[\text{E}(Y|X)\right]^2 \\
&= \text{E}\left[\text{E}(Y^2|X)\right] - \text{E}\left[\text{E}(Y|X)^2\right] + \text{E}\left[\text{E}(Y|X)^2\right] - \text{E}\left[\text{E}(Y|X)\right]^2 \\
&= \text{E}\left[\text{E}(Y^2|X) - \text{E}(Y|X)^2\right] + \text{E}\left[\text{E}(Y|X)^2\right] - \text{E}\left[\text{E}(Y|X)\right]^2 \\
&= \text{E}\left[\text{V}(Y|X)\right] + \text{V}\left[\text{E}(Y|X)\right].
\end{aligned}
$$

In the middle equation, we have added and subtracted the term we need to regroup as expectation and variance.

Applying this to the delivery problem, we get

$$\begin{aligned}
V(T) &= E\left[V(T|X)\right] + V\left[E(T|X)\right] \\
&= E\left[V\left(\sum_{i=1}^{X} Y_i|X\right)\right] + V\left[E\left(\sum_{i=1}^{X} Y_i|X\right)\right] \\
&= E\left[X\alpha/\beta^2\right] + V\left[X\alpha/\beta\right] \\
&= \lambda\alpha/\beta^2 + \lambda\alpha^2/\beta^2 = \lambda\alpha(1+\alpha)/\beta^2.
\end{aligned}$$

So, if $\lambda = 45$, $\alpha = 4$, and $\beta = 1/2$, the variance of the number of minutes per day is 3600, and the standard deviation is 60 minutes.

Simulations using R provide an easy way to check this answer. The following code returns an average number of minutes per day of about 360 with a standard deviation near 60.

```
n=10000   ## number of simulated days
minutes=1:n;lam=45;alp=4;bet=.5
## simulate days one by one:
for(i in 1:n){
    njobs=rpois(1,lam)
    times=rgamma(njobs,alp,bet)
    minutes[i]=sum(times)
}
mean(minutes)
sd(minutes)
```

Example: A machine in a factory produces computer chips. Each chip has a probability θ of being defective, and the chips are produced independently, that is, the probability that one chip is defective doesn't depend on the designation of the other chips. Suppose that the machine makes 100 chips every day. Every morning when the machine is turned on, there is a different value of θ. In fact, θ is a random variable that follows a Beta(1,4) distribution. If Y is the number of defectives produced on a randomly selected day, what is the expected value of Y?

Fortunately, we don't have to compute the probability mass function for Y. We know that $Y|\theta \sim \text{Binom}(n, \theta)$, so we can use the formula

$$E(Y) = E\left[E(Y|\theta)\right] = E\left[100\theta\right] = 20,$$

because the expected value of θ is $1/5$.

What is the standard deviation of the number of defectives produced on a randomly selected day? Using the formula for the variance, we have

$$\begin{aligned}
V(Y) &= E\left[V(Y|\theta)\right] + V\left[E(Y|\theta)\right] \\
&= E\left[100\theta(1-\theta)\right] + V\left[100\theta\right] \\
&= 100E(\theta) - 100E(\theta^2) + 10{,}000V(\theta) \\
&= 100E(\theta) - 100\left[V(\theta) + E(\theta)^2\right] + 10{,}000V(\theta) \\
&= 20 - 100\left(\frac{2}{75} + \frac{1}{25}\right) + 10{,}000\frac{2}{75} = 280,
\end{aligned}$$

so the standard deviation of the number of defectives is about 16.7.

Chapter Highlights

1. For random variables X and Y,

$$E(Y) = E\left[E(Y|X)\right]$$

and

$$V(Y) = E\left[V(Y|X)\right] + V\left[E(Y|X)\right].$$

2. These formulas are useful when the distribution of $Y|X = x$, and the marginal for X, are known but the marginal for Y is *not* known.

❧

Exercises

29.1 Derive the formula for conditional expectation in the discrete case. That is, if X and Y are jointly discrete random variables, then

$$E\left[E(Y|X)\right] = E(Y).$$

29.2 A machine in a factory produces computer chips. Each chip has a probability θ of being defective, and the chips are produced independently, that is, the probability that one chip is defective doesn't depend on the designation of the other chips. The machine is run continually until a defective chip is produced, at which time it is shut down for repairs. After each repair, the machine is started again, with a different probability θ of producing a defective item. Suppose that at each start, θ is an independent draw from a uniform density on $(1/4, 1/2)$. Let Y be the number of good chips produced after a repair and before the machine shuts down, and find the expected value of Y.

29.3 A game at a carnival costs one dollar to play. It involves randomly choosing n lottery tickets from a large barrel. Suppose ten percent of the lottery tickets are golden, and the player gets one dollar for every golden ticket chosen. The game proceeds as follows: The player rolls two fair six-sided dice to determine the number N (from two to twelve), then receives N lottery tickets. Let Y be the number of dollars won by the player, i.e., the number of golden tickets chosen. Find the mean and variance of Y. (*Hint:* If you don't feel like computing the variance of the dice roll by hand, feel free to write some R code!)

29.4 Maxine is a talented glass blower; she makes vases to sell at a craft fair. Glass vases are hard to make, and some of them contain flaws that reduce the price. Suppose that the price obtained for each vase is a random variable that depends on the number of flaws. In fact, if there are x flaws, the price (in dollars) is exponential with mean $1/[\theta(x + 1)]$. Suppose the number of flaws in a randomly selected vase follows a Poisson distribution with mean λ.

   (a) What is the average price for Maxine's vases in terms of λ and θ?

   (b) Calculate the average price for $\theta = 1/100$ and $\lambda = .8$.

(c) Check your answer to (b) using simulations! You can get a random number x of flaws from a Poisson(.8), then generate a random price from an exponential with mean $100/(x + 1)$. Do this in a loop and get your average price.

(d) Using your code for (c), fill in the blank: The probability that a randomly selected vase costs more than $\$_____$ is .05.

29.5 The number of flaws N in a length of rod is distributed as a Poisson random variable, with an average of five flaws. The tensile strength of the rod depends on the number of flaws, and is approximately normally distributed with mean $100 - N$ and variance 6.

(a) What are the expected value and variance of the tensile strength of the population of rods?

(b) Although we can calculate a formula for the mean and the variance of the distribution of tensile strengths, we don't know the distribution itself. Suppose we want to know the percentiles of the distribution. Find these through simulations! Report the median of the distribution, and the 5th percentile.

29.6 The number of flaws X in lengths of filaments in a detecting device follows a Poisson distribution with mean 4. The lifetime Y of the filament follows an exponential distribution where the mean depends on the number of flaws; given a number x of flaws, the mean lifetime is $20(x + 1)^{-1}$. Find the expected value of Y.

29.7 Lengths of cords produced by a company have tensile strengths that vary, and cords with less strength have higher probability of breaking. Let X be the breaking probability of a randomly selected cord and assume that $X \sim$ Beta$(3, 6)$. A randomly chosen cord will be used in a pulley device until it breaks. Assuming independent trials, let Y be the number of times that a randomly selected cord is used successfully (so $Y = 0$ if it breaks the first time it is used).

(a) Find E(Y).

(b) Find V(Y).

(c) Verify your answers to (a) and (b) using simulations in R.

29.8 A facility has 10 pumps that are cleaned every morning. During the day they are likely to get gummed up and stop working. The probability of getting gummed up depends on the level of impurities in the water, which varies from day to day, depending on the preceding weather conditions. Suppose that the level of impurities X on a randomly selected day follows an exponential distribution with mean 8 (using standard units). Given a level x of impurities, the probability that an individual pump gets gummed up is $1 - e^{-x/24}$. Let Y be the number of pumps that get gummed up on a randomly selected day. If the pumps operate independently, what is the expected value of Y?

29.9 Samples of water taken during the summer at random points around the shore of Lake Hatfelter have mercury concentrations (in micrograms) that follow a Gamma$(3, 1/2)$ density. The blueglass dragonfly is abundant around the lake shore, but the insects' reproduction is affected by the mercury level. Given a level

x of mercury, the number of eggs laid by a female dragonfly follows a Poisson distribution with mean $200/x^2$. Suppose a female dragonfly is selected at random from the lake shore. What is the expected number of eggs it lays? Do simulations in R to check your answer.

29.10 The amount of precipitate X in a certain chemical assay follows a normal distribution, but the mean of X depends on the amount Y of reactant used. In fact, given an amount y of reactant, the amount X of precipitate is distributed as normal with mean $100/y$ and variance one. Suppose the chemists cannot control exactly the amount of reactant, and Y follows a Gamma$(2, 1/2)$ distribution. What is $E(X)$?

29.11 An urn has 20 brown balls. Eddie replaces Y of the brown balls with gold balls, where Y is a Poisson random variable with mean 2. Then you choose three balls from the urn (without replacement). What is the expected number of gold balls that you choose? Check your answer using simulations in R. (We can ignore the possibility of Y being greater than 20, because in the very unlikely case in which that happens, we start over.)

29.12 An urn contains one gold marble. Eddie will add Y blue marbles, where Y is a Poisson random variable with mean 4. Then you draw marbles from the urn, at random and with replacement, until you get a gold marble. Let X be the number of blue marbles drawn from the urn, and compute $E(X)$.

Chapter 30

Transformations of Jointly Distributed Random Variables

We address two important types of transformations in this chapter; both are transformations of jointly distributed random variables, say Y_1 and Y_2. First, we find the distribution of $X = g(Y_1, Y_2)$. Second, we find the joint distribution of X_1 and X_2, where $X_1 = g_1(Y_1, Y_2)$ and $X_2 = g_2(Y_1, Y_2)$. The techniques in this chapter mostly apply to jointly continuous random variables; to transform jointly discrete random variables, we can often compute the probabilities directly.

Example: Let Y_1 and Y_2 be independent Poisson random variables, with means λ_1 and λ_2, respectively. What is the distribution of $X = Y_1 + Y_2$? Here, the joint mass function is the product of the individual mass functions, and we can write

$$f(y_1, y_2) = P(Y_1 = y_1, Y_2 = y_2) = \left(\frac{e^{-\lambda_1} \lambda_1^{y_1}}{y_1!} \right) \left(\frac{e^{-\lambda_2} \lambda_2^{y_2}}{y_2!} \right).$$

Then

$$P(X = x) = \sum_{u=0}^{x} P(Y_1 = u, Y_2 = x - u)$$

$$= \sum_{u=0}^{x} \left[\left(\frac{e^{-\lambda_1} \lambda_1^{u}}{u!} \right) \left(\frac{e^{-\lambda_2} \lambda_2^{x-u}}{(x - u)!} \right) \right]$$

$$= e^{-(\lambda_1 + \lambda_2)} \sum_{u=0}^{x} \left[\lambda_2^{x} \left(\frac{\lambda_1}{\lambda_2} \right)^{u} \frac{1}{u!(x - u)!} \right],$$

where we have done some rearranging, with the goal of using the binomial theorem to get rid of the summation. Now we can multiply and divide by $x!$ so that the factorials in the sum can be written as a combination. We get

$$P(X = x) = \frac{e^{-(\lambda_1 + \lambda_2)} \lambda_2^{x}}{x!} \sum_{u=0}^{x} \binom{x}{u} \left(\frac{\lambda_1}{\lambda_2} \right)^{u}$$

$$= \frac{e^{-(\lambda_1 + \lambda_2)} \lambda_2^{x}}{x!} \left(1 + \frac{\lambda_1}{\lambda_2} \right)^{x}$$

$$= \frac{e^{-(\lambda_1 + \lambda_2)} (\lambda_1 + \lambda_2)^{x}}{x!}.$$

Recognizing that this is the expression for a Poisson probability, we conclude that X is a Poisson random variable with mean $\lambda_1 + \lambda_2$.

For transformations involving continuous random variables, we can often use the CDF method to find the distribution of $X = g(Y_1, Y_2)$.

Example: Suppose U_1 and U_2 are independent Unif$(0, 1)$ random variables, and define $Y = U_1 + U_2$. What is the density for Y? We can use the same ideas as in Chapter 21, where we find the CDF for Y, then take the derivative to get the density. First note that Y takes values in the interval $(0, 2)$.

The support of the joint density for (U_1, U_2) is shown on the right; the density function is one over the support and zero elsewhere. For $y \in (0, 1)$,

$$F_Y(y) = P(Y \le y) = P(U_1 + U_2 \le y)$$
$$= \frac{y^2}{2},$$

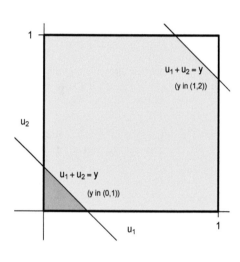

which is the volume of the density over the darkly shaded triangle formed by the support and the line $u_1 + u_2 = y$.

For $y \in (1, 2)$, the total shaded area is formed by the support and the line $u_1 + u_2 = y$. We calculate $P(Y \le y)$ as the volume of the density over the total shaded area. This is

$$F_Y(y) = P(Y \le y) = 1 - \frac{1}{2}(2 - y)^2.$$

(These calculations were also needed for Exercise 24.4.) Therefore, the density for Y is

$$f_Y(y) = \begin{cases} y & \text{for } y \in (0, 1), \\ 2 - y & \text{for } y \in [1, 2), \end{cases}$$

shown below:

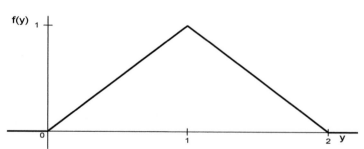

These calculations are easy to check with R simulations: Simply generate many pairs (U_1, U_2), compute $Y = U_1 + U_2$ for each pair, then make a histogram with `freq=FALSE`, and superimpose the density function.

For the second type of transformation, let $f_Y(y_1, y_2)$ be the joint density for random variables Y_1 and Y_2, and we want to find the joint distribution of X_1 and X_2, where $X_1 = g_1(Y_1, Y_2)$ and $X_2 = g_2(Y_1, Y_2)$. We can do this if the inverse transformation $Y_1 = h_1(X_1, X_2)$ and $Y_2 = h_2(X_1, X_2)$ exists.

This transformation requires the Jacobian, which is the determinant of the matrix of partial derivatives:

$$J = \det \begin{bmatrix} \frac{\partial h_1}{\partial x_1} & \frac{\partial h_1}{\partial x_2} \\ \frac{\partial h_2}{\partial x_1} & \frac{\partial h_2}{\partial x_2} \end{bmatrix}.$$

The formula for the transformation is

$$f_X(x_1, x_2) = f_Y(h_1(x_1, x_2), h_2(x_1, x_2))|J|.$$

Example: Let Y_1 and Y_2 be independent standard normal random variables, and let $X_1 = Y_1 + Y_2$ and $X_2 = Y_1 - Y_2$. The inverse transformation is straightforward: $Y_1 = (X_1 + X_2)/2$ and $Y_2 = (X_1 - X_2)/2$. Then the Jacobian is

$$J = \det \begin{bmatrix} \frac{1}{2} & \frac{1}{2} \\ \frac{1}{2} & -\frac{1}{2} \end{bmatrix} = -\frac{1}{2},$$

and the joint density for X_1, X_2 is

$$\begin{aligned} f_X(x_1, x_2) &= \frac{1}{2} f_Y\left(\frac{x_1 + x_2}{2}, \frac{x_1 - x_2}{2}\right) \\ &= \frac{1}{2} \frac{1}{\sqrt{2\pi}} \exp\left[-\frac{1}{2}\left(\frac{x_1 + x_2}{2}\right)^2\right] \frac{1}{\sqrt{2\pi}} \exp\left[-\frac{1}{2}\left(\frac{x_1 - x_2}{2}\right)^2\right] \\ &= \frac{1}{4\pi} \exp\left[-\frac{1}{4}\left(x_1^2 + x_2^2\right)\right] \\ &= \frac{1}{\sqrt{4\pi}} \exp\left[-\frac{1}{2}\left(\frac{x_1^2}{2}\right)\right] \frac{1}{\sqrt{4\pi}} \exp\left[-\frac{1}{2}\left(\frac{x_2^2}{2}\right)\right] \end{aligned}$$

for any real numbers x_1 and x_2. Because we can factor the densities, we know that X_1 and X_2 are independent normal random variables, both with mean zero and variance two. We could have figured out the mean and variance of X_1 and X_2 without finding the joint density, but the independence might be a little surprising!

Example: Let's do the same thing but with exponential random variables. Let Y_1 and Y_2 be independent exponential variables with mean θ, and let $X_1 = Y_1 + Y_2$ and $X_2 = Y_1 - Y_2$. We get the same Jacobian as in the previous example, but we have to be careful about the support of the pair (X_1, X_2). It's clear that X_2 cannot be larger than X_1, for instance. We can find the support and the density for the new random variables at the same time by using indicator functions in the specification of the density for (Y_1, Y_2).

Let's write the joint density of Y_1 and Y_2 as

$$f_Y(y_1, y_2) = \frac{1}{\theta^2} e^{-(y_1+y_2)/\theta} I\{y_1 > 0, y_2 > 0\}.$$

Then we have

$$
\begin{aligned}
f_X(x_1, x_2) &= \frac{1}{2} f_Y\left(\frac{x_1+x_2}{2}, \frac{x_1-x_2}{2}\right) \\
&= \frac{1}{2\theta^2} \exp\left[-\left(\frac{x_1+x_2}{2} + \frac{x_1-x_2}{2}\right)\Big/\theta\right] I\{x_1+x_2 > 0, x_1 - x_2 > 0\} \\
&= \frac{1}{2\theta^2} \exp\left[-x_1/\theta\right] I\{-x_1 < x_2 < x_1, x_1 > 0\}.
\end{aligned}
$$

The support of the joint density $f_X(x_1, x_2)$ is shown as the shaded region of the plot. This joint density integrates to one:

$$
\begin{aligned}
&\frac{1}{2\theta^2} \int_0^\infty \int_{-x_1}^{x_1} e^{-x_1/\theta} dx_2 dx_1 \\
&= \frac{1}{\theta^2} \int_0^\infty x_1 e^{-x_1/\theta} dx_1,
\end{aligned}
$$

which is one because the integrand is a Gamma$(2, 1/\theta)$ density function.

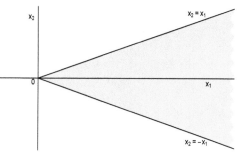

Note that the random variables X_1 and X_2 are *not* independent. The indicator function that determines the support cannot be factored into a piece involving only x_1 and a piece involving only x_2.

We can also see that the marginal density $f_{X_1}(x_1)$ is Gamma$(2, 1/\theta)$; the calculation of the integral of the joint density over its support shows us that if we had integrated over only x_2, we get that density for X_1. Hence, we have shown that the sum of two independent exponentials with mean θ is distributed as Gamma$(2, 1/\theta)$

Let's check this answer using simple simulations in R. The following code generates pairs of exponential random variables, each with mean $\theta = 2$, and sums them. The distribution of the sum is verified using a histogram and also using a probability plot; these are shown below.

```
n=2000
par(mfrow=c(1,2))
y1=rexp(n,1/2)
y2=rexp(n,1/2)
x=y1+y2
hist(x,freq=FALSE,br=50,col='beige')
xpl=0:300/10
lines(xpl,dgamma(xpl,2,1/2),lwd=2,col=2)
plot(qgamma(1:n/(n+1),2,1/2),sort(x),xlab="Gamma(2,1/2) quantiles")
lines(c(0,30),c(0,30),col=2,lwd=2)
lines(c(0,30),c(0,30),col=2,lwd=2)
```

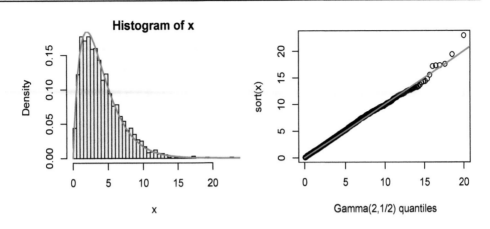

Example: Let Y_1 and Y_2 be independent exponential variables with mean θ, and now let $X_1 = Y_1/(Y_1 + Y_2)$ and $X_2 = Y_1 + Y_2$. If Y_1 and Y_2 are times to complete jobs 1 and 2, then X_2 is the total time to complete the jobs, and X_1 is the proportion of time spent doing the first job.

The inverse transformation functions are $y_1 = x_1 x_2$ and $y_2 = x_2(1 - x_1)$, so the Jacobian is

$$J = \det \begin{bmatrix} x_2 & x_1 \\ -x_2 & 1 - x_1 \end{bmatrix} = x_2,$$

and the joint density function for (X_1, X_2) is

$$
\begin{aligned}
f_X(x_1, x_2) &= \frac{1}{\theta^2} \exp\left[-\left(x_1 x_2 + x_2(1 - x_2)\right)/\theta\right] x_2 I\{x_1 x_2 > 0, x_2(1 - x_1) > 0\} \\
&= \frac{x_2}{\theta^2} \exp(-x_2/\theta) I\{0 < x_1 x_2 < x_2 < \infty\} \\
&= \frac{x_2}{\theta^2} \exp(-x_2/\theta) I\{x_2 > 0\} I\{0 < x_1 < 1\}.
\end{aligned}
$$

This tells us that $X_2 \sim \text{Gamma}(2, 1/\theta)$ and $X_1 \sim \text{Unif}(0, 1)$, and X_1 and X_2 are independent. We have again shown that the sum of two independent exponentials with mean θ has a gamma distribution with parameters $(2, 1/\theta)$, which was the result from the previous example. That the proportion of time spent on the first job is uniform seems intuitive; however, in Exercise 30.5 we see that this might not be the case if the distributions are not exponential.

Chapter Highlights

1. If the density for the jointly continuous random variables Y_1 and Y_2 is $f(y_1, y_2)$ and $X = g(Y_1, Y_2)$, then the CDF for X can be found by integrating the joint density over the set $\{(y_1, y_2) : g(y_1, y_2) \leq x)\}$, i.e.,

$$F_X(x) = P(X \leq x) = \int\int_{\{(y_1, y_2): g(y_1, y_2) \leq x)\}} f(y_1, y_2) dy_1 dy_2.$$

Then

$$f(x) = \frac{dF_X(x)}{dx}.$$

2. If the density for the jointly continuous random variables Y_1 and Y_2 is $f(y_1, y_2)$, then the joint distribution of the random variables $X_1 = g_1(Y_1, Y_2)$ and $X_2 = g_2(Y_1, Y_2)$ can be computed using the formula

$$f_X(x_1, x_2) = f_Y(h_1(x_1, x_2), h_2(x_1, x_2))|J|,$$

where $Y_1 = h_1(X_1, X_2)$ and $Y_2 = h_2(X_1, X_2)$ are the inverse transformation functions, and $|J|$ is the determinant of the Jacobian:

$$J = \det \begin{bmatrix} \frac{\partial h_1}{\partial x_1} & \frac{\partial h_1}{\partial x_2} \\ \\ \frac{\partial h_2}{\partial x_1} & \frac{\partial h_2}{\partial x_2} \end{bmatrix}.$$

Exercises

30.1 As in the fourth example of this chapter, let Y_1 and Y_2 be independent exponential variables with mean θ, and let $X_1 = Y_1 + Y_2$ and $X_2 = Y_1 - Y_2$. Finish the example by finding the marginal density for X_2. Check your work through simulations in R.

30.2 For some $\theta > 0$, suppose $Y_1 \sim \text{Unif}(0, \theta)$ and $Y_2 \sim \text{Unif}(0, \theta)$ are independent random variables. Find the density of the random variable $X = Y_1 + Y_2$ and sketch it. Check your answer using simulations in R, with $\theta = 2$.

30.3 For some $\theta > 0$, suppose $Y_1 \sim \text{Unif}(0, 2\theta)$ and $Y_2 \sim \text{Unif}(0, \theta)$ are independent random variables. Find the density of the random variable $X = Y_1 + Y_2$ and sketch it. Check your answer using simulations in R, with $\theta = 2$.

30.4 The random variables Y_1 and Y_2 have joint density that is uniform on the area bounded by the lines $y_2 > 0$, $y_1 + y_2 < 4$, and $y_2 < y_1$. Find the density for $U = Y_1 + Y_2$ and sketch it. Do simulations in R to check your answer.

30.5 The random variables U_1 and U_2 are independent random variables, both distributed as uniform over $(0, 1)$. Define $Y = U_1/(U_1 + U_2)$. Find the density for Y and sketch it. Check your answer using simulations in R.

30.6 Suppose Y_1 and Y_2 are independent random variables, both exponentially distributed, but $E(Y_1) = 1$ and $E(Y_2) = 2$. Find the density of the random variable $X = Y_1 + Y_2$. Check your answer using simulations in R.

30.7 Random variables Y_1 and Y_2 have a joint distribution that is uniform over the support shown in the figure. If $X = Y_2/Y_1$, find the density for X and sketch it. Verify your answer using simulations in R.

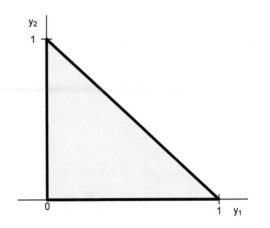

30.8 The lifetimes of switches of a certain brand follow an exponential distribution with parameter θ. Two switches are randomly chosen (assume independence). The first is used until it fails, then the second one is used until it fails. Find the distribution of the sum of the lifetimes using the CDF method.

30.9 The lifetimes Y_1 and Y_2 of two components of a device are jointly distributed as

$$f(y_1, y_2) = \frac{1}{8} y_1 e^{-(y_1+y_2)/2} \text{ on } y_1 > 0, y_2 > 0.$$

(a) Find $P(Y_1 > 1 \text{ and } Y_2 > 1)$.

(b) Are Y_1 and Y_2 independent random variables? Explain.

(c) Find the marginal for Y_1 and sketch it.

(d) The relative efficiency of the devices is $X = Y_2/Y_1$. Find the density for X and sketch it.

(e) Using R code, simulate from the marginals for Y_1 and Y_2, and get a histogram of Y_2/Y_1. Superimpose your density function computed in (d) to check your answer.

30.10 A device has two components, and the lifetimes of these components are modeled by the random variables Y_1 and Y_2. The first component cannot fail before the second component fails, and the joint distribution of Y_1 and Y_2 is determined to be

$$f(y_1, y_2) = 2e^{-y_1} e^{-y_2} \text{ for } 0 < y_2 < y_1 < \infty.$$

The random variable $X = Y_1 - Y_2$ can be interpreted as the time between failures Find the density for X and sketch it.

30.11 A device has two components, and the lifetimes of these components are modeled by the random variables Y_1 and Y_2. The first component cannot fail before the second component fails, and the joint distribution of Y_1 and Y_2 is determined to be

$$f(y_1, y_2) = 2e^{-y_1} e^{-y_2} \text{ for } 0 < y_2 < y_1 < \infty.$$

The random variable $X_1 = Y_1 - Y_2$ can be interpreted as the time between failures, and $X_2 = Y_1 + Y_2$ might be the lifetime of the device.

(a) Find the joint density for X_1 and X_2.

(b) Using (a), find the marginal for X_1 and compare it to the answer to Exercise 30.10.

(c) Using (a), find the marginal for X_2.

30.12 Let U_1 and U_2 be independent uniform random variables, both on $(0, 1)$. Let $Y_1 = U_1 U_2$ and let $Y_2 = U_2$.

(a) Find the joint density of Y_1 and Y_2.

(b) What is the marginal density for Y_1?

(c) Check your answer to part (b) using simulations. Generate U_1 and U_2 deviates using runif, and compute values for Y_1. Get the histogram of Y_1 values and superimpose your density function in part (b).

Chapter 31

Transformations Using Moment Generating Functions

Recall that the moment generating function (MGF) for a random variable Y is

$$M_Y(t) = E(e^{tY}).$$

An important property of MGFs is their uniqueness:

> Let $M_X(t)$ be the moment generating function for the random variable X, and let $M_Y(t)$ be the moment generating function for the random variable Y. If $M_X(t) = M_Y(t)$ for all t in an open interval containing zero, then X and Y have the same probability distribution.

The general proof is beyond the scope of this book, but we can prove it for discrete random variables with the same finite support, using this algebra result: If a polynomial $f(x) = c_0 + c_1 x + c_2 x^2 + \cdots + c_n x^n$ is zero for all x in some open interval of the real line, then $c_0 = c_1 = \cdots = c_n = 0$.

Suppose the random variables X and Y both have support $\{0, 1, 2, \ldots, n\}$, and have the same MGFs; that is, $M_X(t) = M_Y(t)$ for t in an open interval containing zero. Using the definition of MGF, we have

$$\sum_{k=0}^{n} e^{tk} P(X = k) = \sum_{k=0}^{n} e^{tk} P(Y = k),$$

which implies that

$$\sum_{k=0}^{n} z^k \left[P(X = k) - P(Y = k) \right] = 0,$$

where $z = e^t$. Because this is zero for all z in an open interval containing one, the coefficients $P(X = k) - P(Y = k)$ must be zero for all k, which means that the probability distribution of X is the same as that for Y.

This uniqueness property is very useful for finding the distributions of transformed random variables, especially for sums of independent random variables.

Let X and Y be independent random variables with MGFs $M_X(t)$ and $M_Y(t)$, and let $Z = X + Y$. Then

$$M_Z(t) = E(e^{tZ}) = E(e^{t(X+Y)}) = E(e^{tX} e^{tY}) = E(e^{tX}) E(e^{tY}) = M_X(t) M_Y(t),$$

where the second-to-last equality is valid because X and Y are independent.

We have a flood of nice easy results from this simple idea. First, let's revisit the sum of two independent exponential random variables with the same parameter β (we did this in the last chapter using the Jacobian transformation method and in Exercise 30.8 using the CDF method).

Example: We have already computed the moment generating function for many of our "named" distributions, and we know that if $Y \sim \text{Exp}(\beta)$, then

$$M_Y(t) = \frac{\beta}{\beta - t}.$$

Now if Y_1 and Y_2 are independent exponential random variables with rate parameter β, their sum has MGF

$$M_{Y_1 + Y_2}(t) = \left(\frac{\beta}{\beta - t}\right)^2,$$

which we recognize as the MGF for a Gamma$(2, \beta)$ random variable. Wasn't that easy? Not only that, but we can keep going: If $Y_1, Y_2, \ldots, Y_n$ are independent exponential random variables with parameter β (mean $1/\beta$), and $Z = Y_1 + Y_2 + \cdots + Y_n$, then

$$M_Z(t) = \text{E}(e^{t(Y_1 + \cdots + Y_n)}) = \text{E}(e^{tY_1})\text{E}(e^{tY_2}) \cdots \text{E}(e^{tY_n}) = \left(\frac{\beta}{\beta - t}\right)^n,$$

which is the MGF for a Gamma(n, β) random variable.

Similarly, we get that the distribution of the sum of independent $Y_1 \sim$ Gamma(α_1, β) and $Y_2 \sim$ Gamma(α_2, β) is distributed as Gamma$(\alpha_1 + \alpha_2, \beta)$.

Example: In Chapter 18, we had computed the MGF only for the standard normal random variable, and promised to do the more general case later. Our old result is, if $Z \sim \text{N}(0, 1)$, then

$$M_Z(t) = e^{t^2/2}.$$

In Chapter 21, we derived this result: If $Z \sim \text{N}(0, 1)$ and $Y = \sigma Z + \mu$, then $Y \sim \text{N}(\mu, \sigma^2)$. The moment generating function for $Y \sim \text{N}(\mu, \sigma^2)$ is then

$$M_Y(t) = \text{E}(e^{t(\sigma Z + \mu)}) = e^{t\mu}\text{E}(e^{t\sigma Z}) = e^{t\mu}M_Z(t\sigma) = e^{t\mu}e^{\sigma^2 t^2/2} = e^{t\mu + \sigma^2 t^2/2}.$$

Example: Suppose $Y_1 \sim \text{N}(\mu_1, \sigma_1^2)$ and $Y_2 \sim \text{N}(\mu_2, \sigma_2^2)$, and $X = Y_1 + Y_2$. Then

$$M_X(t) = M_{Y_1}(t)M_{Y_2}(t) = e^{t\mu_1 + \sigma_1^2 t^2/2}e^{t\mu_2 + \sigma_2^2 t^2/2} = e^{t(\mu_1 + \mu_2) + (\sigma_1^2 + \sigma_2^2)t^2/2},$$

which shows that X is normally distributed, with mean $\mu_1 + \mu_2$ and variance $\sigma_1^2 + \sigma_2^2$. In other words, the sum of two independent normal random variables is a normal random variable, and the means and variances add.

Example: We can use the MGF method to show that the square of a standard normal random variable is a chi-squared random variable. We did this example using the CDF method, but it's easier with our new method. We have $Z \sim \text{N}(0, 1)$ and $Y = Z^2$. Then

$$M_Y(t) = \text{E}(e^{tY}) = \text{E}(e^{tZ^2}) = \frac{1}{\sqrt{2\pi}} \int_{-\infty}^{\infty} e^{tz^2} e^{-z^2/2} dz$$

$$= \frac{1}{\sqrt{2\pi}} \int_{-\infty}^{\infty} e^{-\frac{z^2}{2}(1 - 2t)} dz.$$

If we let $\sigma^2 = (1 - 2t)^{-1}$, then

$$M_Y(t) = \frac{\sigma}{\sqrt{2\pi\sigma^2}} \int_{-\infty}^{\infty} e^{-\frac{z^2}{2\sigma^2}} dz$$
$$= \sigma = (1 - 2t)^{-1/2}.$$

We recognize this as the MGF for a Gamma$(1/2, 1/2)$ random variable, which is also known as $\chi^2(1)$.

Example: The last example extends nicely to a very important result. Suppose that $Z_1, Z_2, \ldots, Z_n$ are independent standard normal random variables, and let $Y = Z_1^2 + Z_2^2 + \cdots + Z_n^2$. The MGF for Y is

$$M_Y(t) = \mathrm{E}(e^{t(Z_1^2 + Z_2^2 + \cdots + Z_n^2)})$$
$$= \mathrm{E}(e^{tZ_1^2})\mathrm{E}(e^{tZ_2^2}) \cdots \mathrm{E}(e^{tZ_n^2})$$
$$= (1 - 2t)^{-n/2},$$

which is the MGF for the Gamma$(n/2, 1/2)$ or $\chi^2(n)$.

Chapter Highlights

1. Uniqueness of the MGF: Let $M_X(t)$ be the MGF for the random variable X, and let $M_Y(t)$ be the MGF for the random variable Y. If $M_X(t) = M_Y(t)$ for all t, then X and Y have the same probability distribution.

2. Let X and Y be independent random variables with MGFs $M_X(t)$ and $M_Y(t)$, and let $Z = X + Y$. Then
$$M_Z(t) = M_X(t)M_Y(t).$$

3. If $Y_1, \ldots, Y_n$ are independent Exp(β) random variables, then
$$X = Y_1 + \cdots + Y_n \sim \mathrm{Gamma}(n, \beta).$$

4. If $Y_1 \sim \mathrm{Gamma}(\alpha_1, \beta)$ and $Y_2 \sim \mathrm{Gamma}(\alpha_2, \beta)$, and Y_1 and Y_2 are independent, then $Y_1 + Y_2$ is distributed as Gamma$(\alpha_1 + \alpha_2, \beta)$.

5. The sum of two independent normal random variables is a normal random variable, and the mean and the variances add.

6. The square of a standard normal random variable has a chi-square distribution with one degree of freedom.

7. If $Y_1 \sim \chi^2(\nu_1)$ and $Y_2 \sim \chi^2(\nu_2)$ are independent, then $Y_1 + Y_2 \sim \chi^2(\nu_1 + \nu_2)$.

8. If $Z_1, \ldots, Z_n$ are independent standard normal random variables, then $Y = \sum_{i=1}^{n} Z_i^2$ has a $\chi^2(n)$-density.

9. The moment generating function for $Y \sim \mathrm{N}(\mu, \sigma^2)$ is
$$M_Y(t) = e^{t\mu + \sigma^2 t^2/2}.$$

Exercises

31.1 Using simulations in R, verify the result that the sum of squares of four independent standard normal random variables has a $\chi^2(4)$-density. You can either use our histogram method as we did in the chapters on transformations or a probability plot as in Chapter 19.

31.2 Using simulations in R, verify the result that the sum of four independent $\mathrm{Exp}(\theta)$ random variables has a $\mathrm{Gamma}(4, \theta)$ density. You can choose a value of θ, and either use our histogram method as we did in the chapters on transformations or a probability plot as in Chapter 19.

31.3 Suppose $Y_1 \sim \mathrm{Pois}(\lambda_1)$ and $Y_2 \sim \mathrm{Pois}(\lambda_2)$ are independent, and $X = Y_1 + Y_2$. Find the MGF for X and infer the distribution.

31.4 Let $U \sim \mathrm{Unif}(0, 1)$ and $Y = -\log(U)/5$.

   (a) Using the method of moment generating functions, find the density for Y.

   (b) Generalize your derivation to $Y = -\log(U)/a$.

   (c) Verify your answer to (a), using simulations in R.

31.5 Suppose $Y_1 \sim \mathrm{Binom}(n_1, p)$ and $Y_2 \sim \mathrm{Binom}(n_2, p)$ are independent, and $X = Y_1 + Y_2$. Find the MGF for X and infer the distribution.

31.6 Suppose $Y_1 \sim \mathrm{Pois}(\lambda_1)$ and $Y_2 \sim \mathrm{Pois}(\lambda_2)$. Find the conditional mass function for Y_1, given $Y_1 + Y_2 = m$.

31.7 Let $Y \sim \mathrm{Unif}(0, 1)$, and use the MGF method to find the density for $X = 1 - Y$.

31.8 Suppose $Y_1 \sim \chi^2(\nu_1)$ and $Y_2 \sim \chi^2(\nu_2)$, and Y_1 and Y_2 are independent. Find the density for $X = Y_1 + Y_2$.

31.9 Return to Exercise 16.10. Compute the test size and power without using simulations.

31.10 In Exercise 14.6, we derived the MGF for an $\mathrm{NB}(2, p)$ random variable. Show that this MGF is also that of a sum of two independent $\mathrm{Geom}(p)$ random variables. Find the MGF for the sum of r independent $\mathrm{Geom}(p)$ random variables, and argue that this is the MGF for an $\mathrm{NB}(r, p)$ random variable.

Chapter 32

The Multivariate Normal Distribution

If Y_1 and Y_2 are independent normal random variables, we can write down their joint density simply by multiplying together the marginals:

$$f_Y(y_1, y_2) = \frac{1}{2\pi} \exp\left\{-\frac{1}{2}\left[\frac{(y_1 - \mu_1)^2}{\sigma_1^2} + \frac{(y_2 - \mu_2)^2}{\sigma_2^2}\right]\right\},$$

where μ_1 and σ_1^2 are the mean and variance of Y_1, and μ_2 and σ_2^2 are the mean and variance of Y_2. The figures below show two such bivariate normal densities, the first with $\sigma_1^2 = \sigma_2^2$, the second with $\sigma_1^2 > \sigma_2^2$:

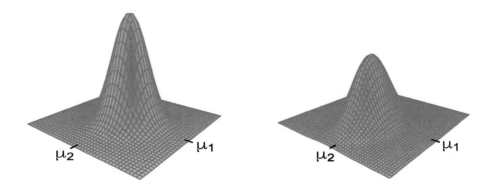

Now suppose Y_1 and Y_2 are normal random variables that are not independent. If we want to specify the covariance as well as the two means and two variances (of the marginal distributions), how do we write down the density?

To answer this question, let's consider the following way of constructing pairs of nonindependent normal random variables. Suppose Z_1 and Z_2 are independent standard normal random variables, and define

$$Y_1 = 4 + Z_1 + 2Z_2 \text{ and } Y_2 = 3Z_1 + Z_2.$$

We can easily determine that the mean of Y_1 is $\mu_1 = 4$, the mean of Y_2 is $\mu_2 = 0$. The variance of Y_1 is $V(Z_1) + 4V(Z_2) + 4\text{cov}(Z_1, Z_2) = 5$ and, similarly, $V(Y_2) = 10$.

The covariance of the two new random variables is

$$\text{cov}(Y_1, Y_2) = \text{cov}(4 + Z_1 + 2Z_2, 3Z_1 + Z_2) = 3V(Z_1) + 7\text{cov}(Z_1, Z_2) + 2V(Z_2) = 5.$$

We know that Y_1 and Y_2 are normally distributed, because we showed in Chapter 31 that sums of independent normal random variables are again normal random variables.

Some simple R code can confirm these calculations and develop our intuition:

```
z1=rnorm(1000,0,1)
z2=rnorm(1000,0,1)
y1=4+z1+2*z2
y2=3*z1+z2
cov(y1,y2)
```

Plots of the generated values of the random variables are shown below; the points for the pair of independent standard normal random variables are located roughly in a circle about the origin, but the points of the values of Y_1 and Y_2 form an ellipse centered at $(4, 0)$ and tilted to show the positive correlation between the random variables.

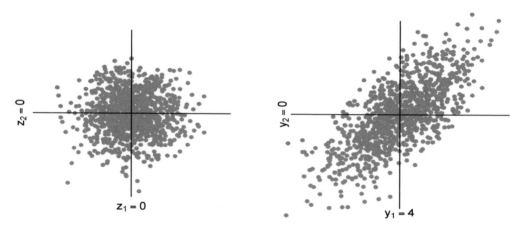

Let's find the joint density function for the pair (Y_1, Y_2), where, to be more general,

$$Y_1 = a_1 + b_1 Z_1 + c_1 Z_2 \quad \text{and} \quad Y_2 = a_2 + b_2 Z_1 + c_2 Z_2,$$

and again Z_1 and Z_2 are independent standard normal random variables. We know that

$$f_Z(z_1, z_2) = \frac{1}{2\pi} \exp\left\{ -\frac{1}{2}(z_1^2 + z_2^2) \right\}$$

for any real numbers z_1 and z_2.

As in our preliminary example, we can find $E(Y_1) = a_1$ and $E(Y_2) = a_2$. Similarly, we find $V(Y_1) = b_1^2 + c_1^2$ and $V(Y_2) = b_2^2 + c_2^2$, and the covariance is $\text{cov}(Y_1, Y_2) = b_1 b_2 + c_1 c_2$.

The inverse transformation is found solving two equations and two unknowns:

$$\begin{pmatrix} y_1 - a_1 \\ y_2 - a_2 \end{pmatrix} = \begin{pmatrix} b_1 & c_1 \\ b_2 & c_2 \end{pmatrix} \begin{pmatrix} z_1 \\ z_2 \end{pmatrix},$$

so

$$\begin{pmatrix} z_1 \\ z_2 \end{pmatrix} = \begin{pmatrix} b_1 & c_1 \\ b_2 & c_2 \end{pmatrix}^{-1} \begin{pmatrix} y_1 - a_1 \\ y_2 - a_2 \end{pmatrix}.$$

We require that the 2×2 matrix is invertible, or that $b_1 c_2 \neq c_1 b_2$. We will come back to interpret this in terms of the random variables. Letting $d = b_1 c_2 - c_1 b_2$, we have

$$z_1 = \frac{1}{d}[c_2(y_1 - a_1) - c_1(y_2 - a_2)] \quad \text{and} \quad z_2 = \frac{1}{d}[b_1(y_2 - a_2) - b_2(y_1 - a_1)].$$

The Jacobian is

$$J = \det \begin{pmatrix} c_2/d & -c_1/d \\ -b_2/d & b_1/d \end{pmatrix} = \frac{c_2 b_1 - c_1 b_2}{d^2} = \frac{1}{d},$$

and the joint density for Y_1 and Y_2 is

$$f(y_1, y_2) = \frac{1}{|d|} \exp\left\{ -\frac{1}{2d^2} \left[(c_2(y_1 - a_1) - c_1(y_2 - a_2))^2 + (b_1(y_1 - a_1) \right. \right.$$

$$\left. \left. - b_2(y_2 - a_2))^2 \right] \right\}$$

$$= \frac{1}{|d|} \exp\left\{ -\frac{1}{2d^2} \left[(c_2^2 + b_2^2)(y_1 - a_1)^2 - 2(c_1 c_2 + b_1 b_2)(y_1 - a_1)(y_2 - a_2) \right. \right.$$

$$\left. \left. + (b_1^2 + c_1^2)(y_2 - a_2)^2 \right] \right\}$$

for any real numbers y_1 and y_2. This looks complicated, but if we define $\sigma_1^2 = b_1^2 + c_1^2$ (the variance of Y_1), $\sigma_2^2 = b_2^2 + c_2^2$ (the variance of Y_2), and $\sigma_{12} = b_1 b_2 + c_1 c_2$ (the covariance), as well as $\mu_1 = a_1$ and $\mu_2 = a_2$, then

$$f(y_1, y_2) = \frac{1}{2\pi|d|} \exp\left\{ -\frac{1}{2d^2} [\sigma_2^2 (y_1 - \mu_1)^2 - 2\sigma_{12}(y_1 - \mu_1)(y_2 - \mu_2) \right.$$

$$\left. + \sigma_1^2 (y_2 - \mu_2)^2] \right\}.$$

We can simplify the notation even more by defining a **covariance matrix**

$$\Sigma = \begin{pmatrix} \sigma_1^2 & \sigma_{12} \\ \sigma_{12} & \sigma_2^2 \end{pmatrix}$$

so that $\Sigma_{ij} = \text{cov}(Y_i, Y_j)$. We also define vectors

$$y = \begin{pmatrix} y_1 \\ y_2 \end{pmatrix} \quad \text{and} \quad \mu = \begin{pmatrix} a_1 \\ a_2 \end{pmatrix},$$

and we notice that

$$f(y_1, y_2) = \frac{1}{2\pi} \det(\Sigma)^{-1/2} \exp\left\{ -\frac{1}{2}(y - \mu)^\top \Sigma^{-1}(y - \mu). \right\}$$

In this way the density function is specified in terms of the mean vector and the covariance matrix. We can write

$$Y \sim N(\mu, \Sigma).$$

The multivariate standard normal distribution is indicated by

$$\boldsymbol{Y} \sim \mathrm{N}(\boldsymbol{0}, \boldsymbol{I}),$$

where $\boldsymbol{0}$ is a vector of zeros and $\boldsymbol{I}$ is the identity matrix that has ones on the diagonal and zeros elsewhere.

Below we see two bivariate normal density functions; on the left we see positive covariance, and negative covariance is demonstrated on the right.

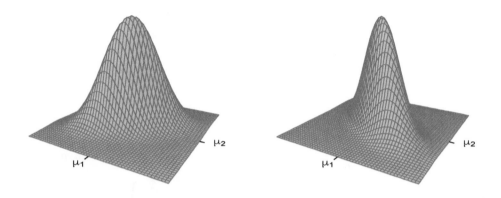

Let's go back to our prohibition of $b_1 c_2 = c_1 b_2$. If this equality holds, then

$$Y_2 = a_2 + b_2 Z_1 + \frac{c_1 b_2}{b_1} Z + 2 = a_2 + \frac{b_2}{b_1}[b_1 Z_1 + c_1 Z_2]$$

and

$$Y_2 = a_2 - \frac{a_1 b_2}{b_1} + \frac{b_2}{b_1} Y_1.$$

So Y_2 is a linear transformation of Y_1, and the correlation between Y_1 and Y_2 is one. The density is zero except for above a line in the (y_1, y_2) plane, because there is really only one random variable. In this case if we tried to make a covariance matrix, it would not be invertible.

We can similarly generate larger sets of jointly varying normal random variables. Suppose we have k independent standard normal random variables $Z_1, \ldots, Z_k$, and define

$$Y_1 = \mu_1 + a_{11} Z_1 + \cdots + a_{1k} Z_k,$$
$$Y_2 = \mu_2 + a_{21} Z_1 + \cdots + a_{2k} Z_k,$$
$$\vdots$$
$$Y_k = \mu_k + a_{k1} Z_1 + \cdots + a_{kk} Z_k,$$

which can be written in vector notation as

$$\boldsymbol{Y} = \boldsymbol{\mu} + \boldsymbol{A} \boldsymbol{Z}$$

if we define $\boldsymbol{Y} = (Y_1, \ldots, Y_K)^\top$, $\boldsymbol{\mu} = (\mu_1, \ldots, \mu_K)^\top$, and the $k \times k$ matrix $\boldsymbol{A}$ has elements $A_{ij} = a_{ij}$ for $i, j = 1, \ldots, k$.

We can define a variance-covariance matrix for the k random variables in $\boldsymbol{Y}$ as

$$\Sigma_{ij} = \sum_{\ell=1}^{k} a_{i\ell} a_{j\ell}$$

or, in matrix notation, $\boldsymbol{\Sigma} = \boldsymbol{AA}^{\top}$.

If $\boldsymbol{A}$ is nonsingular, then $\boldsymbol{\Sigma}$ is invertible, and the joint density is

$$f(y_1, \ldots, y_k) = [2\pi]^{-k/2} \det(\boldsymbol{\Sigma})^{-1/2} \exp\left\{ -\frac{1}{2}(\boldsymbol{y} - \boldsymbol{\mu})^{\top} \boldsymbol{\Sigma}^{-1} (\boldsymbol{y} - \boldsymbol{\mu}) \right\}.$$

In Chapter 27 we saw that it was possible for two random variables to have zero co-variance, but not be independent. We can see here that if two normal random variables have zero covariance, then they *must* be independent. For, in this case, $\boldsymbol{\Sigma}$ is a diagonal matrix, so we are able to factor the joint density function into the density functions for two univariate random variables.

Instead of defining the multivariate normal in terms of linear combinations of independent standard normal random variables, we could simply specify the mean vector $\boldsymbol{\mu}$ and the covariance matrix $\boldsymbol{\Sigma}$. We will get a proper density if $\boldsymbol{\Sigma}$ can be written as $\boldsymbol{AA}^{\top}$ for a nonsingular matrix $\boldsymbol{A}$. (In other words we require $\boldsymbol{\Sigma}$ to be *positive definite*.) Otherwise, we could specify a covariance matrix for which the covariance between two random variables is larger than the product of the standard deviations, which does not make sense. For the bivariate normal density, this means that $\sigma_1^2 \sigma_2^2 > \sigma_{12}^2$.

We have seen how to determine the variances and covariances of linear transformations of a set of independent standard normal random variables. Suppose next that we want to simulate from a multivariate normal density, where we specify the means, variances, and covariances. That is, we want to generate (using R) some variates of $\boldsymbol{Y}$, where $\boldsymbol{Y} \sim \mathrm{N}(\boldsymbol{\mu}, \boldsymbol{\Sigma})$, and $\boldsymbol{\mu}$ and $\boldsymbol{\Sigma}$ are given.

We must find a matrix $\boldsymbol{A}$ such that $\boldsymbol{AA}^{\top} = \boldsymbol{\Sigma}$. Then for $\boldsymbol{Z} \sim \mathrm{N}(\boldsymbol{0}, \boldsymbol{I})$, the random vector $\boldsymbol{Y} = \boldsymbol{AZ}$ has the desired multivariate normal density. Such an $\boldsymbol{A}$ can be found using the Cholesky decomposition of the matrix $\boldsymbol{\Sigma}$. Then we simply compute $\boldsymbol{Y} = \boldsymbol{\mu} + \boldsymbol{AZ}$.

In R we can use the command `amat=t(chol(sigma))` if the desired covariance matrix is in the matrix `sigma`. Suppose, for example, we want $\boldsymbol{\mu} = (2, 1)^{\top}$ and

$$\boldsymbol{\Sigma} = \begin{pmatrix} 4 & 2 \\ 2 & 9 \end{pmatrix}.$$

The following piece of code will generate 1000 pairs of normal random variables with the desired joint multivariate normal density:

```
smat=matrix(c(4,2,2,9),nrow=2)
z1=rnorm(1000)
z2=rnorm(1000)
y=t(chol(smat))%*%rbind(z1,z2)
y1=y[1,]+2
y2=y[2,]+1
plot(y1,y2,xlab=expression(y[1]),ylab=expression(y[2]),pch=20)
```

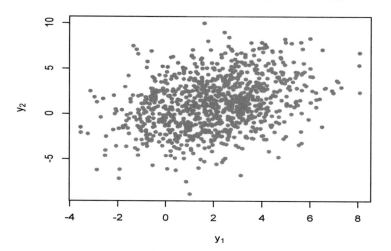

Chapter Highlights

1. Suppose we have a linear transformation of k independent standard normal random variables $Z_1, \ldots, Z_k$, defined by $Y_i = \mu_i + \sum_{j=1}^{k} a_{ij} Z_j$ for $i, j = 1, \ldots, k$. Then $Y_1, \ldots, Y_k$ are jointly distributed normal random variables having a multivariate normal density. For each $i = 1, \ldots, k$,

$$\mathrm{E}(Y_i) = \mu_i \ \text{ and } \ \mathrm{V}(Y_i) = \sum_{\ell=1}^{k} a_{i\ell}^2.$$

   For $i, j = 1, \ldots, k$,

$$\mathrm{cov}(Y_i, Y_j) = \sum_{\ell=1}^{k} a_{i\ell} a_{j\ell}.$$

2. Let $\boldsymbol{A}_{ij} = a_{i,j}$ for $i, j = 1, \ldots, k$. The variance-covariance matrix for a set of k normal random variables having elements defined as above is $\Sigma_{ij} = \mathrm{cov}(Y_i, Y_j)$, and $\boldsymbol{\Sigma} = \boldsymbol{A}\boldsymbol{A}^\top$. If $\boldsymbol{Y} = (Y_1, \ldots, Y_k)^\top$ are jointly normal random variables with mean $\boldsymbol{\mu} = (\mu_1, \ldots, \mu_k)^\top$ and $\mathrm{cov}(\boldsymbol{Y}) = \boldsymbol{\Sigma}$, we write

$$\boldsymbol{Y} \sim \mathrm{N}(\boldsymbol{\mu}, \boldsymbol{\Sigma}).$$

3. The joint density function for $\boldsymbol{Y} = (Y_1, \ldots, Y_k)^\top$ is

$$f(y_1, \ldots, y_k) = [2\pi]^{-k/2} \det(\boldsymbol{\Sigma})^{-1/2} \exp\left\{-\frac{1}{2}(\boldsymbol{y} - \boldsymbol{\mu})^\top \boldsymbol{\Sigma}^{-1}(\boldsymbol{y} - \boldsymbol{\mu})\right\}.$$

4. If two normal random variables have zero covariance, they are independent.

Exercises

32.1 Suppose X and Y are random variables with joint density function

$$f(x,y) = \frac{1}{\pi\sqrt{3}}\exp\{-(x^2+xy+y^2)/2\}.$$

(a) What is the covariance of X and Y?

(b) Simulate many pairs of X and Y in R to check your answer to (a).

32.2 Suppose X and Y are random variables with joint density function

$$f(x,y) = d\exp\{-(x^2+2axy+y^2)/2\}.$$

(a) Find the marginal density of Y by "integrating out" x.

(b) Find d (as a function of a) to make this joint density (and the marginal density found in (a)) a proper density function.

(c) What are $\text{var}(X)$ and $\text{var}(Y)$? What is the covariance of X and Y?

(d) What is the range of possible values of a, so that $\text{V}(X)\text{V}(Y) > \text{cov}(X,Y)^2$?

32.3 Suppose X and Y are normal random variables with $\text{E}(X) = 2$, $\text{E}(Y) = 1$, $\text{V}(X) = 9$, $\text{V}(Y) = 4$, and $\text{cov}(X,Y) = 2$. Write the joint density *not* in vector notation, and simplify.

32.4 Suppose Y_1 and Y_2 are normal random variables with $\text{E}(Y_1) = 2$, $\text{E}(Y_2) = 1$, $\text{V}(Y_1) = 9$, $\text{V}(Y_2) = 5$, and $\text{cov}(Y_1,Y_2) = 2$.

(a) Express Y_1 and Y_2 as linear combinations of independent standard normal random variables Z_1 and Z_2.

(b) Check your answer to (a) by simulating values of Y_1 and Y_2 in R.

Chapter 33

Order Statistics

In this chapter we are concerned with the distributions of **order statistics**. That is, if we take a random sample from a known distribution, then sort the sample, how are the sorted random variables distributed? We will be particularly interested in the distribution of the minimum and the maximum of the sample, but we can also find the distribution of the second largest, or third smallest, etc.

Example: To demonstrate the ideas, let's start by finding the distribution of the maximum of two independent uniform random variables. Let U_1 and U_2 be independent Unif$(0, 1)$, and let $Y = \max(U_1, U_2)$. We want to find the density function for Y.

We note that Y can take values in $(0, 1)$. Our old friend the CDF method will give us the density. First, the CDF for $y \in (0, 1)$ is

$$
\begin{aligned}
F_Y(y) = \mathrm{P}(Y \le y) &= \mathrm{P}(\max(U_1, U_2) \le y) \\
&= \mathrm{P}(U_1 \le y \ \text{ and } \ U_2 \le y) \\
&= \mathrm{P}(U_1 \le y)\mathrm{P}(U_2 \le y) \qquad \text{(using independence)} \\
&= y^2.
\end{aligned}
$$

Then $f_Y(y) = 2y$ on $(0, 1)$. We used the idea that if the maximum of two numbers is less than y, then they must *both* be less than y.

Intuitively, we think that the distribution of the maximum of two values should be "pushed over to the right," that is, the probability that the maximum of two iid random variables is larger than some value $a \in (0, 1)$ is at least the probability that either random variable is larger than a.

This is easily generalized to a sample of size n. Let $U_1, U_2, \ldots, U_n$ be independent Unif$(0, 1)$, and let $Y = \max(U_1, U_2, \ldots, U_n)$. To find the density function for Y, we again use the CDF method:

$$
\begin{aligned}
F_Y(y) = \mathrm{P}(Y \le y) &= \mathrm{P}(\max(U_1, U_2, \ldots, U_n) \le y) \\
&= \mathrm{P}(U_1 \le y \ \text{ and } \ U_2 \le y \ \text{ and } \ \cdots \ \text{ and } \ U_n \le y) \\
&= \mathrm{P}(U_1 \le y)\mathrm{P}(U_2 \le y) \cdots \mathrm{P}(U_n \le y) \\
&= y^n.
\end{aligned}
$$

Then $f_Y(y) = ny^{n-1}$ on $(0,1)$. As n gets large, the density of the maximum "piles up" near $y = 1$. Here are some sketches of the density for various n:

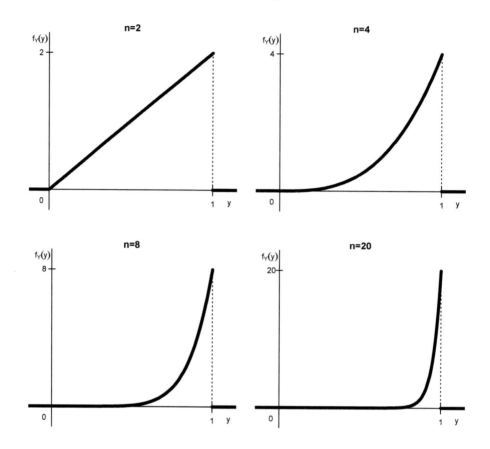

What is the expected value of the maximum Y of n independent $\mathrm{Unif}(0,1)$ random variables? Using the formula for the expected value of a continuous random variable, we have

$$E(Y) = n \int_0^1 y^n dy = \frac{n}{n+1},$$

which shows that as the sample size n gets larger, the expected value of the maximum gets closer to 1.

The variance is also easy to calculate. First we find

$$E(Y^2) = n \int_0^1 y^{n+1} dy = \frac{n}{n+2},$$

then

$$V(Y) = E(Y^2) - E(Y)^2 = \frac{n}{(n+2)(n+1)^2}.$$

The density for the minimum of a sample can also calculated using the CDF method. Again let $U_1, U_2, \ldots, U_n \overset{ind}{\sim} \mathrm{Unif}(0,1)$ and $Y = \min(U_1, U_2, \ldots, U_n)$. This time we use the fact that if a minimum value is greater than y, then *all* of the values must be greater than y. We have

$$
\begin{aligned}
F_Y(y) = \mathrm{P}(Y \le y) &= \mathrm{P}(\min(U_1, U_2, \ldots, U_n) \le y) \\
&= 1 - \mathrm{P}(\min(U_1, U_2, \ldots, U_n) > y) \\
&= 1 - \mathrm{P}(U_1 > y \text{ and } U_2 > y \text{ and } \cdots \text{ and } U_n > y) \\
&= 1 - \mathrm{P}(U_1 > y)\mathrm{P}(U_2 > y) \cdots \mathrm{P}(U_n > y) \\
&= 1 - (1 - y)^n.
\end{aligned}
$$

Then $f_Y(y) = n(1 - y)^{n-1}$ on $(0, 1)$. As n gets large, this density "piles up" near 0, which makes sense intuitively. The density for the minimum of an sample of independent standard uniform random variables is shown for sample sizes $n = 2$, $n = 3$, and $n = 8$:

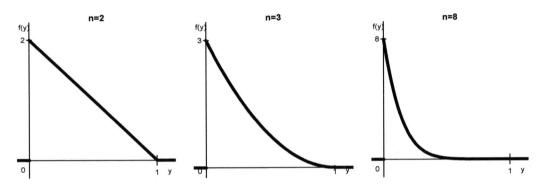

Next we will get a general formula for the maximum of a random sample from a continuous distribution. Suppose $X_1, \ldots, X_n$ are independent random variables with continuous CDF F_X and density f_X, and let $Y = \max(X_1, X_2, \ldots, X_n)$. We again find the density function for Y using the CDF method:

$$
\begin{aligned}
F_Y(y) = \mathrm{P}(Y \le y) &= \mathrm{P}(\max(X_1, X_2, \ldots, X_n) \le y) \\
&= \mathrm{P}(X_1 \le y \text{ and } X_2 \le y \text{ and } \cdots \text{ and } X_n \le y) \\
&= \mathrm{P}(X_1 \le y)\mathrm{P}(X_2 \le y) \cdots \mathrm{P}(X_n \le y) \\
&= F_X(y)^n.
\end{aligned}
$$

Then the density is the derivative of the CDF: $f_Y(y) = nF_X(y)^{n-1}f_X(y)$.

Exercise 33.1 asks you to derive the formula for the density of the minimum of a random sample from a continuous distribution.

The maximum and minimum are special cases of **order statistics**. For a random sample $X_1, X_2, \ldots, X_n \stackrel{iid}{\sim} f_X(x)$, the sorted sample is often written as $X_{(1)}, X_{(2)} \ldots, X_{(n)}$, where $X_{(1)}$ is the minimum and $X_{(n)}$ is the maximum. The random variables $X_{(i)}, i = 1, \ldots, n$, are called order statistics.

Suppose we are interested in the distribution of the order statistic $X_{(k)}$, where k is an integer between 1 and n. To use the CDF method, we start by asking what is $\mathrm{P}(X_{(k)} \le x)$ for some x in the support of f_X. Note that if $X_{(k)} \le x$, then *at least* k

values of the sample are less than or equal to x:

$$P(X_{(k)} \le x) = \sum_{j=k}^{n} P(\text{exactly } j \text{ values of the sample } \le x).$$

The probability that exactly j values of the sample are at most x can be computed using binomial ideas. For any element X_i of the sample, $P(X_i \le x) = F_X(x)$, and so if we think of a "success" as being less than x, the number of the sample less than x is distributed as a binomial random variable with parameters n and $F_X(x)$.

$$P(\text{exactly } j \text{ values of the sample } \le x) = \binom{n}{j} F_X(x)^j [1 - F_X(x)]^{n-j}.$$

Then the CDF for $X_{(k)}$ is

$$F_{(k)}(x) = \sum_{j=k}^{n} \binom{n}{j} F_X(x)^j [1 - F_X(x)]^{n-j}.$$

Instead of taking the derivative in the usual way, we will use the definition of the derivative, plus our ideas about a multinomial distribution. We start with

$$
\begin{aligned}
f_{(k)}(x) &= \frac{d}{dx} F_{(k)}(x) \\
&= \frac{d}{dx} P(X_{(k)} \le x) \\
&= \lim_{\Delta \to 0} \frac{P(X_{(k)} \le x + \Delta) - P(X_{(k)} \le x)}{\Delta} \\
&= \lim_{\Delta \to 0} \frac{P(x < X_{(k)} \le x + \Delta)}{\Delta}.
\end{aligned}
$$

If the kth ordered observation is in $(x, x + \Delta]$, then we have $k - 1$ observations in $(-\infty, x]$ and $n - k - 1$ observations in $(x + \Delta, \infty)$. We can find the probability of this happening using the multinomial, with three groups. Then

$$P(x < X_{(k)} \le x + \Delta)$$
$$= \frac{n!}{(k - 1)!1!(n - k)!} F_X(x)^{k-1} [F_X(x + \Delta) - F_X(x)]^1 [1 - F_X(x + \Delta)]^{n-k}.$$

Plugging this into the numerator of the limit gives

$$
\begin{aligned}
f_{(k)}(x) &= \lim_{\Delta \to 0} \frac{n!}{(k-1)!(n-k)!} F_X(x)^{k-1} \frac{F_X(x+\Delta) - F_X(x)}{\Delta} [1 - F_X(x+\Delta)]^{n-k-1} \\
&= \frac{n!}{(k-1)!(n-k)!} F_X(x)^{k-1} [1 - F_X(x)]^{n-k} \lim_{\Delta \to 0} \frac{F_X(x+\Delta) - F_X(x)}{\Delta} \\
&= \frac{n!}{(k-1)!(n-k)!} F_X(x)^{k-1} [1 - F_X(x)]^{n-k} f(x),
\end{aligned}
$$

where in the last step we used the definition of the derivative of $F_X(x)$.

Example: Suppose we have 9 independent observations from a $\text{Unif}(0, 1)$. What is the distribution for the median value? Using the formula with $k = 5$, we get

$$f_{(5)}(x) = \frac{9!}{4!4!} x^4 (1 - x)^4 = 630 x^4 (1 - x)^4.$$

We recognize that the median $X_{(5)}$ is distributed as Beta$(5, 5)$. Therefore the mean is $1/2$ and the variance is $1/44$.

A quick check using simulations in R verifies our calculations. The following code produces the histogram shown below:

```
n=100000
medu=1:n
for(i in 1:n){
    u=runif(9)
    medu[i]=median(u)
}
hist(medu,freq=FALSE,br=50,main="Histogram of median of 9 uniform random
        variables")
xpl=0:100/100
lines(xpl,dbeta(xpl,5,5),col=2,lwd=2)
```

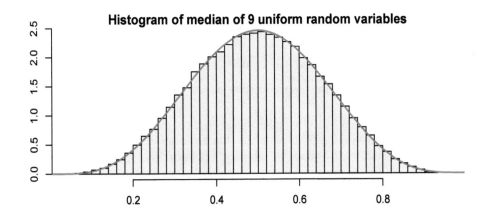

Chapter Highlights

1. We say that random variables are iid if they are independent and identically distributed.

2. A set of iid random variables is called a "random sample" from their common distribution.

3. If $Y = \max(X_1, X_2, \ldots, X_n)$ for $X_1, \ldots, X_n \overset{iid}{\sim} f_X$, the density of the maximum is
$$f_Y(y) = nF_X(y)^{n-1}f_X(y).$$

4. The density for $X_{(k)}$, the kth ordered value of a random sample $X_1, \ldots, X_n \overset{iid}{\sim} f_X$, is
$$f_{(k)}(x) = \frac{n!}{(k-1)!(n-k)!}F_X(x)^{k-1}[1 - F_X(x)]^{n-k}f(x).$$

Exercises

33.1 Suppose the continuous random variables $Y_1, \ldots, Y_n$ are a random sample from $f_Y(y)$. Derive a formula for the density of the minimum using the CDF method. Verify your answer by plugging $k = 1$ into the general formula for the distribution of the $X_{(k)}$ order statistic.

33.2 Using simulations in R, verify that the density for the minimum Y of a random sample of size $n = 5$, from a Unif$(0, 1)$ density, is $f_Y(y) = 5(1 - y)^4$.

33.3 Using simulations in R, verify that the density for the median of a random sample of size $n = 9$, from a Unif$(0, 1)$ density, is that of a Beta$(5, 5)$ random variable.

33.4 Suppose $Y_1, Y_2, \ldots, Y_n$ are independent exponential random variables, all with mean θ. Let $X = \min(Y_1, Y_2, \ldots, Y_n)$, and find the density function for X.

33.5 Suppose $Y_1, \ldots, Y_8 \overset{ind}{\sim} \text{Exp}(1/4)$.

   (a) Find the probability that exactly two of the sample are less than 3.

   (b) Find the probability that at least two of the sample are less than 3.

33.6 Filaments used in optical fibers burn out eventually and no longer work. Suppose the working lifetimes of the filaments are distributed as exponential with mean lifetime 2.4 years. A cable for an optical fiber is composed of twelve filaments, and the cable will function properly as long as at least one filament is still working. What is the probability that the cable lasts for at least ten years? (Assume that the lifetimes of the fibers are independent.)

33.7 Suppose $Y_1, \ldots, Y_n$ are iid continuous random variables with common density $f_Y(y)$ and CDF $F_Y(y)$. Find an expression for the probability that exactly two of the sample values are less than c, where c is a number in the support of the density.

33.8 Suppose we have a random sample $Y_1, \ldots, Y_n$ from Unif$(0, \theta)$ for some $\theta > 0$.

   (a) Find the density function for the random variable $X = \max(Y_i)$.

   (b) Find the expected value of X. What happens to the expected value as n gets large?

   (c) Find the variance of X. What happens to the variance as n gets large?

33.9 Suppose we have a random sample $Y_1, \ldots, Y_n$ from Unif(a, b), where $0 < a < b < \infty$. Find the density function for the random variable $X = \max(Y_i)$.

   (a) Find the density function for the random variable $X = \max(Y_i)$.

   (b) Find the density function for the random variable $X = \min(Y_i)$.

33.10 Suppose we have a random sample $Y_1, \ldots, Y_n$ from the density shown below, where the support is $(0, \theta)$.

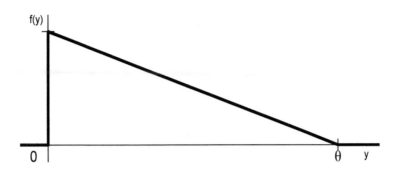

(a) Find the density function for the random variable $X = \max(Y_i)$, and sketch the function for $n = 10$ and $\theta = 2$.

(b) Check your answer to part (a) using simulations.

33.11 Suppose X_1, X_2, X_3, and X_4 are a random sample from the density

$$f(x) = \frac{2x}{\theta} e^{-x^2/\theta} \text{ for } x > 0.$$

Interest is in testing $H_0 : \theta = 2$ versus $H_a : \theta > 2$. Suppose the decision rule is "reject H_0 if the sample maximum is larger than 3."

(a) Find test size α.

(b) Find the power of the test if $\theta = 4$.

33.12 Suppose there are three parking shuttle vans at an airport, operating independently. The round trip from the baggage claim to the lot takes 15 minutes, so for each, the time to arrival at the baggage claim (in minutes) from a randomly selected time is uniformly distributed on $(0, 15)$. Suppose there are three shuttle vans operating, and a person wanting to take the shuttle arrives at the baggage claim at a randomly selected time. The person will get on the first shuttle that arrives.

(a) What is the probability that the person waits less than 5 minutes?

(b) Find an expression for the probability that the person waits for less than m minutes, where $m \in (0, 15)$.

33.13 Let $X_1, \ldots, X_n$ be the lifetimes (in years) of a random sample of components for an electronic device, and suppose that these lifetimes are independent and have an exponential distribution with mean θ. Interest is in testing $H_0 : \theta = 1$ versus $H_a : \theta > 1$.

(a) Statistician A proposes this decision rule: "reject H_0 if the sample maximum is large," where $\alpha = .01$. Compute the critical value for the decision rule for $n = 10$.

(b) Statistician B proposes this decision rule: "reject H_0 if the sample mean is large," where $\alpha = .01$. Compute the critical value for the decision rule for $n = 10$. (*Hint*: Remember that the sample *sum* is a gamma random variable.)

(c) The CEO is a little annoyed that her two statisticians are proposing different tests. She wants to use the one that is *better*. Help her out by computing the powers of the two tests, both at $\alpha = .01$, for a true value of $\theta = 1.5$ and sample size $n = 10$. Which test do you recommend and why?

33.14 Let U_1 and U_2 be independent Unif$(0, 1)$ random variables.

(a) Find the probability that the smaller of the two is less than half of the larger.

(b) Check your answer to part (a) using simulations.

33.15 Let Y_1 and Y_2 be independent exponential random variables, both with mean θ.

(a) Find the probability that the smaller of the two is less than half of the larger.

(b) Check your answer to part (a) using simulations.

33.16 The order statistics considered in this chapter are for random samples from continuous distributions. The ideas are the same for discrete populations. Suppose we have a uniform mass function, with probability $1/10$ for the values $1, \ldots, 10$. That is, $f(i) = 1/10$ for $i = 1, \ldots, 10$. For a random sample of size 5, what is the probability that the largest value is 8?

Chapter 34

Some Distributions Related to Sampling from a Normal Population

When building statistical models to solve real-world problems, our data are often assumed to be realizations from a normal distribution. In this case we have quite a few standard inference methods, many of which (t-tests and F-tests) are presented in introductory statistics classes. In the next few chapters we derive these methods from first principles. We start by defining three new families of continuous random variables. The first one we've already seen as a special branch of the gamma family, important enough to have its own name.

The **chi-squared** random variables are a subfamily of the family of gamma random variables: if $X \sim \text{Gamma}(k/2, 1/2)$, then $X \sim \chi^2(k)$. The density is

$$f_X(x) = \frac{x^{k/2-1}e^{-x/2}}{\Gamma(k/2)2^{k/2}} \text{ for } x > 0.$$

The expected value and variance are k and $2k$, which we get from our expressions for the mean and variance of a gamma random variable.

We have a special name for the chi-squared parameter k: this is called *degrees of freedom*. The rationale for this name comes from linear models theory: the degrees of freedom are equivalent to the dimension of a linear subspace in some important applications. The degrees of freedom for the chi-squared random variables range over the positive integers. For $k = 2$, the distribution is the same as for an exponential random variable with mean 2.

In Chapter 31 we used ideas about the uniqueness of the moment generating function to show that if $Z_1, \ldots, Z_n$ are independent standard normal random variables, then $X = \sum_{i=1}^{n} Z_i^2 \sim \chi^2(n)$. We will use this distribution to make an inference about the variance of a normal population.

The R command dchisq(x,k) will return $f_X(x)$, where f_X is the density for a $\chi^2(k)$ random variable. The command pchisq(y,k) will return the area to the left of y under a $\chi^2(k)$-density. The command qchisq(q,k) will return the value y so that the area to the left of y, under a $\chi^2(k)$-density, is q for $0 \le q < 1$. Finally, rchisq(n,k) will return a vector of n independent random variables from the $\chi^2(k)$ distribution.

Some plots of chi-squared densities are shown, illustrating the increase of both mean and variance as the degrees of freedom get larger:

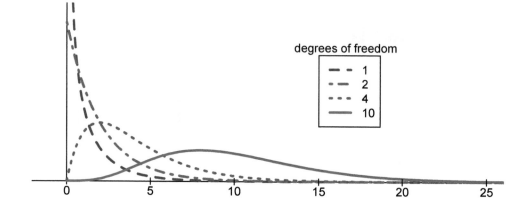

The **Student's t** random variable, often just called a t random variable, is defined as follows: If Z is a standard normal random variable, and $X \sim \chi^2(k)$, and Z and X are independent, then

$$T = \frac{Z}{\sqrt{X/k}} \sim t(k),$$

where $t(k)$ indicates a t distribution with k degrees of freedom. The Student's t random variable is used for test statistics in a variety of applications related to linear models.

Let's derive the density function for the Student's t random variable. It turns out to be easiest to use the Jacobian method for transformation to a joint distribution, so along with the definition of T we let $W = X$. We'll find the joint distribution of T and W, then integrate out W to get the marginal for T. The inverse transformation is defined by

$$z = t\sqrt{w/k} \ \text{ and } \ x = w,$$

and you can confirm that the Jacobian is $\sqrt{w/k}$. The joint density of Z and X is just the product of the marginals by the independence assumption. Therefore, we can write down the joint density for T and W:

$$f_{T,W}(t, w) = \sqrt{\frac{w}{k}} f_Z\left(t\sqrt{w/k}\right) f_X(w).$$

Using the formulas for the densities of the standard normal and the $\chi^2(k)$ gives

$$f_{T,W}(t, w) = \sqrt{\frac{w}{k}} \left[\frac{1}{\sqrt{2\pi}} e^{-\frac{1}{2}t^2 w/k}\right] \left[\frac{1}{\Gamma(k/2)2^{k/2}} w^{k/2-1} e^{-w/2}\right] I\{w > 0\}$$

$$= \frac{1}{\sqrt{k\pi}} \frac{1}{\Gamma(k/2)2^{(k+1)/2}} w^{(k+1)/2-1} e^{-w(1+t^2/k)/2} I\{w > 0\}.$$

Now we want to integrate out w. It's not as bad as it looks! We remember that

$$\int_0^\infty w^{\alpha-1} e^{-\beta w} dw = \frac{\Gamma(\alpha)}{\beta^\alpha},$$

because the gamma density integrates to one. Using $\alpha = (k+1)/2$ and $\beta = (1 + t^2/k)/2$, the marginal for T is

$$f_T(t) = \frac{1}{\sqrt{k\pi}} \frac{1}{\Gamma(k/2)2^{(k+1)/2}} \Gamma((k+1)/2) \left(\frac{2}{1+t^2/k}\right)^{(k+1)/2}$$

$$= \frac{\Gamma((k+1)/2)}{\Gamma(k/2)} \frac{1}{\sqrt{k\pi}} \left(1 + \frac{t^2}{k}\right)^{-(k+1)/2}.$$

There are some things about the T random variable that we can notice right away. First, the density is an even function, so it is symmetric about the vertical axis. Therefore, the mean (if it exists) must be zero. The density is defined on the whole real line, and its value goes to zero as t increases to infinity (or decreases to $-\infty$). A quick calculation of the first derivative of the density function tells us that the density is increasing when $t < 0$, zero when $t = 0$, and decreasing for $t > 0$. A calculation of the second derivative tells us that the density has points of inflection at $t = \pm\sqrt{k/(k+2)}$, which gets closer to one as k gets larger. In fact, the t-density looks a lot like the normal density, only with "heavier tails."

Now we come to a nice result about the limit of the t-density as the degrees of freedom get large—we'll show how to do the limit in pieces.

We don't have a formula to simplify this ratio of gamma functions, but there is the following approximation (using Stirling's formula; see Appendix B.6):

$$\frac{\Gamma\left(r + \frac{1}{2}\right)}{\Gamma(r)} \asymp \sqrt{r},$$

meaning that the ratio of the right- and left-hand sides goes to one as r increases. Therefore, in the expression for the t-density,

$$\frac{\Gamma((k+1)/2)}{\Gamma(k/2)} \frac{1}{\sqrt{k\pi}} \to \frac{1}{\sqrt{2\pi}},$$

as k gets large. Next we need to consider the limit

$$\lim_{k\to\infty} \left(1 + \frac{t^2}{k}\right)^{-(k+1)/2}.$$

Recall that $x^a = e^{a\log(x)}$, so we look at

$$\lim_{k\to\infty} \frac{k+1}{2} \log\left(1 + \frac{t^2}{k}\right) = \lim_{k\to\infty} \frac{k}{2} \log\left(1 + \frac{t^2}{k}\right) + \lim_{k\to\infty} \frac{1}{2} \log\left(1 + \frac{t^2}{k}\right)$$

$$= \frac{1}{2} \lim_{k\to\infty} k \log\left(1 + \frac{t^2}{k}\right)$$

because the second term goes to zero. Now we have a product of two terms, the first of which (k) goes to infinity and the second goes to zero, so we invert the first term and put it in the denominator, to use l'Hôpital's rule:

$$= \frac{1}{2} \lim_{k \to \infty} \frac{\log\left(1 + \frac{t^2}{k}\right)}{\frac{1}{k}}$$

$$= \frac{1}{2} \lim_{k \to \infty} \frac{\frac{-\frac{t^2}{k^2}}{1 + \frac{t^2}{k}}}{-\frac{1}{k^2}} \qquad \text{(taking derivatives)}$$

$$= \frac{1}{2} \lim_{k \to \infty} \frac{t^2}{1 + \frac{t^2}{k}} \qquad \text{(multiplying top and bottom by } k^2)$$

$$= \frac{1}{2} t^2.$$

Therefore,

$$\lim_{k \to \infty} \left(1 + \frac{t^2}{k}\right)^{-(k+1)/2} = e^{-t^2/2}.$$

Putting the two pieces together gives, for any real number t,

$$\lim_{k \to \infty} f_{T_k}(t) = \frac{1}{\sqrt{2\pi}} e^{-t^2/2},$$

which we recognize as the standard normal density.

Because of the symmetry of the Student's t-density, the expected value of a $t(k)$ random variable should be zero. This is the case for t random variables with degrees of freedom greater than one, but the expected value of a $t(1)$ random variable does not exist. To compute the variance, we note that for $T_k \sim t(k)$, $V(T_k) = E(T_k^2) = E(kZ^2/X)$, where $Z \sim N(0,1)$ and $X \sim \chi^2(k)$ are independent random variables. Because of the independence,

$$V(T_k) = kE(Z^2)E(1/X),$$

and in Exercise 34.7 you are asked to show that $E(1/X) = 1/(k-2)$ whenever $k \geq 3$. Therefore, as $E(Z^2) = 1$, we have $V(T_k) = k/(k-2)$, and we note that the variance is undefined for t random variables with one or two degrees of freedom. Further, the variance for a Student's t random variable is always larger than one, but approaches one as the degrees of freedom increase.

The Student's t does not, in general, have a CDF in closed form, nor can we write down its moment generating function. The density does not have an antiderivative, so for $T_k \sim T(k)$ we can't compute $P(T_k > 2)$, for example, unless we use numerical integration. There are tables of t quantiles in most introductory statistics books, and we can also get these using R. The command pt(c,k) will return $P(T_k < c)$, and qt(q,k) will return the value c so that $P(T_k < c) = q$. We can get values of the density function with dt(t,k), which returns $f_{T_k}(t)$, where f_{T_k} is the density for a $t(k)$ random variable. The command rt(n,k) will return a vector of n independent observations from a $t(k)$ distribution.

The F random variable

Suppose $X_1 \sim \chi^2(m)$ and $X_2 \sim \chi^2(n)$ are independent, and define

$$Y_1 = \frac{X_1/m}{X_2/n}$$

to be an F random variable with m numerator and n denominator degrees of freedom, often written $Y_1 \sim F(m, n)$. We again use the Jacobian method of transformation to find the density function, with $Y_2 = X_2$. We can find the joint density of Y_1 and Y_2, then integrate the joint density to get the marginal density of Y_1. If $y_1 = (x_1/m)/(x_2/n)$ and $y_2 = x_2$, the inverse transformation can be written as $x_1 = my_1y_2/n$, $x_2 = y_2$. The Jacobian is my_2/n, and the joint density is

$$
f_Y(y_1, y_2) = \frac{my_2}{n} f_X\left(\frac{m}{n}y_1y_2, y_2\right)
$$

$$
= \frac{my_2}{n} \left[\frac{1}{\Gamma\left(\frac{m}{2}\right)2^{m/2}}\left(\frac{m}{n}y_1y_2\right)^{\frac{m}{2}-1}e^{-\frac{m}{n}y_1y_2/2}\right]
$$

$$
\times \left[\frac{1}{\Gamma\left(\frac{n}{2}\right)2^{n/2}}y_2^{\frac{n}{2}-1}e^{-y_2/2}\right]I\{y_1y_2 > 0, y_2 > 0\}
$$

$$
= \left(\frac{m}{n}\right)^{\frac{m}{2}}\frac{1}{\Gamma\left(\frac{m}{2}\right)\Gamma\left(\frac{n}{2}\right)2^{(m+n)/2}}y_1^{\frac{m}{2}-1}y_2^{\frac{m+n}{2}-1}e^{-\frac{y_2}{2}\left[1+\frac{my_1}{n}\right]}I\{y_1 > 0\}I\{y_2 > 0\},
$$

where we have collected the powers of y_1 and y_2. When we integrate out y_2, we again use the form of a gamma density with $\alpha = (m+n)/2$ and $\beta = (1 + my_1/n)/2$. Then

$$
f_1(y_1) = \left(\frac{m}{n}\right)^{\frac{m}{2}}\frac{1}{\Gamma\left(\frac{m}{2}\right)\Gamma\left(\frac{n}{2}\right)2^{(m+n)/2}}y_1^{\frac{m}{2}-1}\Gamma\left(\frac{m+n}{2}\right)\left(\frac{2}{1+\frac{my_1}{n}}\right)^{\frac{m+n}{2}}I\{y_1 > 0\}
$$

$$
= \frac{\Gamma\left(\frac{m+n}{2}\right)}{\Gamma\left(\frac{m}{2}\right)\Gamma\left(\frac{n}{2}\right)}\left(\frac{m}{n}\right)^{\frac{m}{2}}\frac{y_1^{\frac{m}{2}-1}}{\left(1+\frac{my_1}{n}\right)^{\frac{m+n}{2}}}I\{y_1 > 0\}
$$

is the density function for an $F(m, n)$ random variable.

When we derive the mean and variance for $Y \sim F(m, n)$, we can use the definition rather than the expression for the density. For

$$
Y = \frac{X_1/m}{X_2/n},
$$

where $X_1 \sim \chi^2(m)$ and $X_2 \sim \chi^2(n)$ are independent, we have

$$
E(Y) = \frac{n}{m}E(X_1)E\left(\frac{1}{X_2}\right) = nE\left(\frac{1}{X_2}\right).
$$

It's surprising that the expected value depends only on the denominator degrees of freedom! Again, we use the result of Exercise 34.7 that, for $n \geq 3$,

$$
E\left(\frac{1}{X_2}\right) = \frac{1}{n-2},
$$

so that

$$
E(Y) = \frac{n}{n-2}.
$$

The same exercise asks you to show, in addition, that for $n \geq 5$,

$$
V\left(\frac{1}{X_2}\right) = \frac{2}{(n-2)^2(n-4)}.
$$

Now we use the result of Exercise 23.12 (which works for continuous as well as discrete random variables). Then

$$
\begin{aligned}
V(Y) &= \frac{n^2}{m^2} V\left(X_1 \frac{1}{X_2}\right) = \frac{n^2}{m^2}\left[E(X_1)^2 V(1/X_2) + E(1/X_2)^2 V(X_1) + V(X_1)V(1/X_2)\right] \\
&= \frac{n^2}{m^2}\left[\frac{2m^2}{(n-2)^2(n-4)} + \frac{2m}{(n-2)^2} + \frac{4m}{(n-2)^2(n-4)}\right] \\
&= \frac{2n^2(m+n-2)}{m(n-2)^2(n-4)}.
\end{aligned}
$$

The F distribution does not have a closed form CDF or moment generating function. There are tables of F-density quantiles in statistics textbooks, and we can conveniently find areas under the density and quantiles using R commands: `pf(c,m,n)` returns the area to the left of c under an $F(m,n)$-density. To find c so that $P(Y < c) = q \in (0,1)$, where $Y \sim F(m,n)$, use `qf(q,m,n)`. The command `rf(N,m,n)` returns a vector of N independent observations from an $F(m,n)$ distribution, and to get the value of an $F(m,n)$-density at y, use `df(y,m,n)`.

If the numerator degrees of freedom is $m = 1$, the F-density has an asymptote at zero, and if $m = 2$, the F-density is finite at zero and subsequently decreasing. For $m > 2$, the shape of the F-density is unimodal, that is, increasing then decreasing. Here is a plot showing F-densities with varying degrees of freedom.

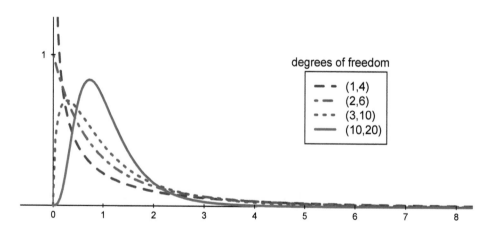

You can make your own plots in R to get an idea of how the shape and scale change with different numerator and denominator degrees of freedom. For example, to plot an $F(4, 80)$-density, use the commands

```
xpl=0:1000/100
plot(xpl,df(xpl,4,80),type="l")
```

In the next chapters, we will see how useful these distributions are for hypothesis tests involving the mean or variance of a population or populations that can be assumed to be normally distributed. We will derive test statistics with chi-squared, t, and F distributions when the null hypothesis is true.

Chapter Highlights

1. The **chi-squared** random variable with k degrees of freedom is a Gamma$(k/2, 1/2)$ random variable. Alternatively, it could be defined as the sum of the squares of k independent standard normal random variables. If X is such a random variable, we write $X \sim \chi^2(k)$.

   (a) The density function for a $\chi^2(k)$ random variable is

   $$f_X(x) = \frac{x^{k/2-1}e^{-x/2}}{\Gamma(k/2)2^{k/2}} \quad \text{for } x > 0.$$

   (b) If $X \sim \chi^2(k)$, then $E(X) = k$ and $V(X) = 2k$.

2. The **Student's t** random variable (also called simply the t random variable) is defined as follows. If $Z \sim N(0,1)$ and $W \sim \chi^2(k)$ are independent, then

   $$T = \frac{Z}{\sqrt{W/k}}$$

   is a t random variable with k degrees of freedom. We write $T \sim t(k)$.

   (a) The density for a t random variable with k degrees of freedom is

   $$f_T(t) = \frac{\Gamma((k+1)/2)}{\Gamma(k/2)} \frac{1}{\sqrt{k\pi}} \left(1 + \frac{t^2}{k}\right)^{-(k+1)/2}.$$

   (b) The mean for any t random variable is zero when the degrees of freedom is greater than 1, since all the densities are symmetric and centered at zero. The mean for a $t(1)$ random variable is undefined.

   (c) The variance for a $t(k)$ random variable is $k/(k-2)$ for $k \geq 3$, and the variance of a $t(1)$ or $t(2)$ random variable is undefined.

3. The **F** random variables are defined as follows. If $X_1 \sim \chi^2(m)$ and $X_2 \sim \chi^2(n)$ are independent, then

   $$F = \frac{X_1/m}{X_2/n}$$

   is an F random variable with m numerator and n denominator degrees of freedom. We write (somewhat confusingly) $F \sim F(m,n)$.

   (a) The density for an F random variable with m numerator and n denominator degrees of freedom is

   $$f(y) = \frac{\Gamma\left(\frac{m+n}{2}\right)}{\Gamma\left(\frac{m}{2}\right)\Gamma\left(\frac{n}{2}\right)} \left(\frac{m}{n}\right)^{\frac{m}{2}} \frac{y^{\frac{m}{2}-1}}{\left(1 + \frac{my}{n}\right)^{\frac{m+n}{2}}} I\{y > 0\}.$$

   (b) The mean of an $F(m,n)$ random variable is $n/(n-2)$.

(c) The variance of an $F(m, n)$ random variable is

$$\frac{2n^2(m + n - 2)}{m(n - 2)^2(n - 4)}.$$

4. The R functions related to these distributions are

- dchisq(x,k) returns $f_X(x)$, where f_X is the density for a $\chi^2(k)$ random variable;

- pchisq(y,k) returns the area to the left of y under a $\chi^2(k)$-density;

- qchisq(q,k) returns the value y so that the area to the left of y, under a $\chi^2(k)$-density, is q for $0 \leq q < 1$, and

- rchisq(n,k) returns a vector of n independent random variables from the $\chi^2(k)$ distribution;

- pt(c,k) returns $P(T_k < c)$, where $T_k \sim t(k)$;

- qt(q,k) returns the value c so that $P(T_k < c) = q$;

- dt(t,k) returns $f_T(t)$, where f_T is the density for a $t(k)$ random variable, and

- rt(n,k) returns a vector of n independent observations from a $t(k)$ distribution;

- pf(c,m,n) returns the area to the left of c under an $F(m, n)$-density;

- qf(q,m,n) returns c so that $P(Y < c) = q \in (0, 1)$, where $Y \sim F(m, n)$;

- rf(N,m,n) returns a vector of N independent observations from an $F(m, n)$ distribution, and

- df(y,m,n) returns the value of an $F(m, n)$-density at y.

Exercises

34.1 Make a nice plot of some t densities, together with the standard normal density, all on the same axes with different line types (solid, dashed, dotted, dot-dash—use lty=3 in the lines command to get dotted, for example). Use 1, 3, 5, and 20 degrees of freedom for the t densities. Be sure to label your densities or provide a legend.

34.2 Use R to fill in the blanks:

(a) The 95th percentile of a $t(12)$-density is _____.

(b) The 10th percentile of an $F(4, 5)$-density is _____.

(c) The probability that an $F(4, 5)$-density is more than 4.6 is _____.

(d) The probability that a $t(10)$-density is more than 2.6 is _____.

(e) The 95th percentile of a $\chi^2(3)$-density is _____.

(f) The probability that a $\chi^2(3)$-density is more than 2.6 is _____.

34.3 Suppose T_1 and T_2 are independent random variables both having $t(12)$ densities. What is the probability that they are both less than 1.2?

34.4 Suppose T_1, T_2, and T_3 are independent random variables having $t(12)$ densities. What is the probability that at least one is greater than 2?

34.5 Suppose W_1 and W_2 are independent random variables both having $\chi^2(12)$ densities.

(a) What is the probability that W_1/W_2 is greater than 1?

(b) What is the probability that W_1/W_2 is greater than 2?

(c) What is the probability that $2W_1/W_2$ is greater than 2?

(d) What is the probability that $W_1 > 2W_2$?

(e) Verify your answer to (d) using simulations.

34.6 Suppose $W_1 \sim \chi^2(3)$ and $W_2 \sim \chi^2(6)$ are independent random variables. What is the probability that W_1/W_2 is greater than 2? Check your answer using simulations in R.

34.7 If $X \sim \chi^2(k)$,

(a) for $k \geq 3$, find the expected value of $1/X$, and verify your answer using simulations in R.

(b) for $k \geq 5$, find the variance of $1/X$, and verify your answer using simulations in R.

34.8 The CDF for the t-density doesn't have a general closed form, but for $k = 1$ it does. Derive it! The $t(1)$-density is also called a *Cauchy* density.

34.9 Suppose $X_1, \ldots, X_n$ are independent $N(\mu, \sigma^2)$. Find

$$\mathrm{E}\left(\frac{\sum_{i=1}^n (X_i - \mu)^2}{n}\right).$$

34.10 Suppose $X_1 \sim N(0, \sigma^2)$ and $X_2 \sim N(0, \sigma^2)$ are independent random variables. Find a so that

$$a\left(\frac{X_1 - X_2}{X_1 + X_2}\right)^2$$

has a distribution from this chapter, and derive the distribution.

34.11 Suppose X_1, X_2, Y_1, and Y_2 are independent random variables, where X_1 and X_2 are distributed as $N(\mu, \sigma^2)$ and Y_1 and Y_2 are distributed as $N(0, 2\sigma^2)$. Find a so that

$$\frac{a(X_1 - X_2)}{\sqrt{Y_1^2 + Y_2^2}}$$

has a distribution from this chapter, and derive the distribution. Verify your answer using simulations in R: In a loop, generate the four normal random variables (for some choice of μ), then compute the given ratio. Then make a probability plot of the sorted values of the ratio against the quantiles of the density that is your answer.

34.12 Suppose $X \sim F(m, n)$. Find the distribution of $Y = 1/X$. Verify your answer using simulations in R using $m = 4$ and $n = 8$.

34.13 Suppose $X \sim t(k)$. Find the distribution of $Y = X^2$. (*Hint:* Use the definition of the t random variable, not its density.)

34.14 Suppose $X \sim N(0, 4)$ and $W \sim \chi^2(6)$. Then

$$P\left(\frac{X^2}{W} > 2\right)$$

is the area to the _____ of _____ under a _____ density.

34.15 Suppose $X_1, X_2, X_3,$ and X_4 are independent random variables, where $X_1 \sim N(\mu, \sigma^2)$ and $X_2, X_3, X_4 \sim N(0, \sigma^2)$. Then

$$P\left(\frac{(X_1 - \mu)^2}{X_2^2 + X_3^2 + X_4^2} > 2\right)$$

is the area to the _____ of _____ under a _____ density.

34.16 Suppose X_1, X_2, X_3 are independent $N(0, \sigma^2)$. Fill in the blanks:

$$P\left(\frac{X_1^2 + X_2^2}{X_3^2} > 6\right)$$

is the area to the _____ of _____ under a _____ density.

34.17 Suppose $X \sim N(\mu, 2\sigma^2)$, and $Y_1, \ldots, Y_{18}$ are independent $N(0, \sigma^2)$. Further, X and Y_i are independent for $i = 1, \ldots, 18$. Fill in the blanks:

$$P\left(\frac{X - \mu}{\sqrt{Y_1^2 + \cdots + Y_{18}^2}} > 1\right)$$

is the area to the _____ of _____ under a _____ density.

34.18 Suppose $X_1, \ldots, X_6$ are independent $N(\mu, \sigma^2)$. Determine the constant a such that the random variable

$$W = \frac{a(X_1 - X_2)}{\sqrt{(X_3 - X_4)^2 + (X_5 - X_6)^2}}$$

has one of the distributions from this chapter, and derive the distribution. Verify your answer using simulations in R with $\mu = 5$ and $\sigma^2 = 4$.

34.19 Suppose $Y_1, \ldots, Y_{10}$ are independent normal random variables, all with unknown mean μ and known variance $\sigma^2 = 4$. Interest is in testing

$$H_0 : \mu = 1 \ \text{ versus } \ H_a : \mu > 1.$$

Derive a decision rule for test size $\alpha = .05$, using the sample mean $\bar{X}$, and state it in terms of "reject H_0 when $\bar{X}$ is _____."

34.20 Suppose $Y_1, \ldots, Y_{10}$ are independent normal random variables, all with known mean μ and unknown variance σ^2. Interest is in testing

$$H_0 : \sigma^2 = 4 \quad \text{versus} \quad H_a : \sigma^2 < 4.$$

Derive a decision rule for test size $\alpha = .05$, using

$$W = \sum_{i=1}^{10} (Y_i - \mu)^2,$$

and state in terms of "reject H_0 when W is _____."

34.21 Suppose $X_1, \ldots, X_n \overset{ind}{\sim} N(0, \sigma^2)$ and $Y_1, \ldots, Y_m \overset{ind}{\sim} N(0, 2\sigma^2)$ are independent random variables. Determine constants $a_1, \ldots, a_n$ and $b_1, \ldots, b_m$ such that the random variable

$$W = \sum_{i=1}^{n} a_i X_i^2 + \sum_{j=1}^{m} b_j Y_j^2$$

has one of the distributions from this chapter, and determine the distribution.

Chapter 35

Hypothesis Tests for a Normal Population Parameter, Part I

Many statistical models involve random variables with normal distributions. For example, suppose that $Y_1, Y_2, \ldots, Y_n$ are independent normal random variables all having mean μ and variance σ^2. That is,

$$Y_1, \ldots, Y_n \stackrel{ind}{\sim} N(\mu, \sigma^2).$$

For many applications, we say that $Y_1, Y_2, \ldots, Y_n$ are a random sample from a *population* that is normally distributed. The sample is also referred to as "the data."

We are now making the transition from probability to statistics. In typical probability problems, you are given the population parameters and asked to compute a probability of an event in a sample; for typical problems in statistics, you are given a sample and some assumptions about the population distribution, but parameters such as the mean and variance are unknown; you are asked to make an *inference* about the parameters. Inference can mean performing a hypothesis test about a parameter, or estimating a parameter and providing a confidence interval.

A **statistic** is a function of the sample that does not involve the unknown parameters. For purposes of inference, the statistic of interest is often a sample mean or a sample variance, or some function of the sample mean and sample variance, but in general a statistic can be any summary of the sample, perhaps a maximum or a median.

In this chapter we will use results and ideas from previous chapters to develop a hypothesis test concerning the population mean μ, when the population variance σ^2 is known, as well as a test concerning σ^2, when μ is known. The former is called a **Z-test**, and the latter does not have a name.

Here is a summary of some useful results we have previously derived concerning random samples from a normal population. These results will be used to derive the testing procedures in the next few chapters, and will first be used for the Z-test.

1. If $Y \sim N(\mu, \sigma^2)$, then for any constants a and b, $aY + b \sim N(a\mu + b, a^2\sigma^2)$. In particular,

$$\text{if } Y \sim N(\mu, \sigma^2), \quad \text{then } Z = \frac{Y - \mu}{\sigma} \sim N(0, 1).$$

2. If $Y_1 \sim N(\mu_1, \sigma_1^2)$ and $Y_2 \sim N(\mu_2, \sigma_2^2)$, and Y_1 and Y_2 are independent, then $Y_1 + Y_2 \sim N(\mu_1 + \mu_2, \sigma_1^2 + \sigma_2^2)$.

3. Therefore, if $Y_1, Y_2 \overset{ind}{\sim} N(\mu, \sigma^2)$, then $\bar{Y} = (Y_1 + Y_2)/2 \sim N(\mu, \sigma^2/2)$.

4. We can generalize the previous result to get

$$\text{if } Y_1, \ldots Y_n \overset{ind}{\sim} N(\mu, \sigma^2), \quad \text{then } \bar{Y} = \frac{1}{n}(Y_1 + \cdots + Y_n) \sim N\left(\mu, \frac{\sigma^2}{n}\right),$$

and hence

$$Z = \frac{\bar{Y} - \mu}{\sqrt{\sigma^2/n}} \sim N(0, 1).$$

Test for a population mean

The last item in the above list is used to test for a population mean μ, when the variance σ^2 is known. This is our Z-test, which is often the first hypothesis testing procedure taught in the introductory statistics classes. We'll derive it as we go through an example.

Example: A machine produces rods for building a device. When the machine is "in spec" the lengths of the rods (in mm) follow a normal distribution with mean $\mu = 120$ and standard deviation $\sigma = 5$. Suppose that to check the machine, we take a random sample of size $n = 4$ rods and measure the lengths. We want to test the null hypothesis that the machine is in spec, i.e., has the correct mean length, versus the alternative that the machine is not in spec. We make the additional (rather unlikely) assumption that the standard deviation is $\sigma = 5$, even when the machine is not in spec. We can state these hypotheses more formally as

$$H_0 : \mu = 120 \quad \text{versus} \quad H_a : \mu \neq 120.$$

For a similar example in Chapter 18, we used the decision rule "reject the null hypothesis unless *all* the rods have lengths within a certain distance D of 120 mm." We will see that a test involving the sample mean has better power with the same sample size and test size.

Assuming the population standard deviation is known ($\sigma = 5$), we can set up a Z-test. Let $\bar{Y}$ be our sample mean; then the statistic

$$Z = \frac{\bar{Y} - 120}{5/\sqrt{4}}$$

has a standard normal distribution if H_0 is true. Otherwise, if H_a is true, the distribution of Z is normal with variance one, but with a nonzero mean.

Let's think about a decision rule, choosing the test size to be $\alpha = .05$. We want to reject H_0 when $\bar{Y}$ is "too big" or "too small," and we want the probability of doing this to be .05 when H_0 is true. Therefore, by the 68-95-99.7 rule, we can reject H_0 when the test statistic Z is bigger than 2 or less than -2, and the test size α will be about .05. An equivalent way to state this decision rule is to reject H_0 when

$$\frac{\bar{Y} - 120}{2.5} > 2 \text{ or when } \frac{\bar{Y} - 120}{2.5} < -2,$$

or when the sample mean $\bar{Y}$ is either greater than 125 or less than 115.

To be concrete, let's imagine that we have a random sample of size $n = 4$ rods, with lengths 115, 120, 113, and 116. The sample mean is readily calculated: $\bar{Y} = 116$. We know that we accept H_0, but let's calculate the p-value. Remember the definition: A p-value is the probability of "our data" or "more extreme" when H_0 is true, where "our data" is the sample values summarized by the test statistic, and "more extreme" means "supports H_a more."

Here, "our data" is $\bar{Y} = 116$. "More extreme" means farther from the H_0 value of 120. So the p-value is $P(\bar{Y} < 116) + P(\bar{Y} > 124)$, calculated under H_0. The command pnorm(116,120,5/2) returns .0548, which is half of the p-value. We accept H_0 because the p-value is larger than the test size $\alpha = .05$.

The density for $\bar{Y}$, assuming that H_0 is true, has mean 120 and standard deviation 2.5, and is shown in the plot on the right. The shaded area represents the p-value of about .11.

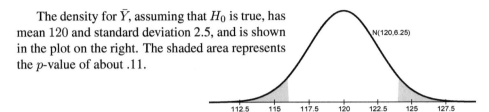

Because of the way we constructed our decision rule, we'll always get a p-value greater than α if $\bar{Y}$ is between 115 and 125, and the p-value will be smaller than α if $\bar{Y}$ is greater than 125 or if $\bar{Y}$ is less than 115.

Let's compute the power for the test when the true mean is $\mu = 125$. The power is the probability of correctly rejecting H_0 when the alternative is true. In this example, we want to compute the probability that $\bar{Y}$ is either greater than 125 or less than 115 when the true mean is 125. We can do this with the 68-95-99.7 rule: The sample mean is greater than 125 with probability .5, and less than 115 with probability almost zero, as 115 is 4 standard deviations away from the mean of 125. Therefore the power is just over 1/2.

The density for $\bar{Y}$, assuming that H_a is true, has mean 125 and standard deviation 2.5, and is shown in the plot on the right. The shaded area represents the power for our test.

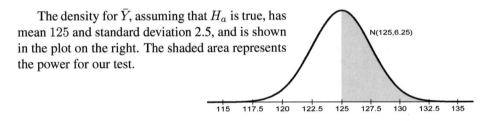

Assessing the Z-test assumptions

When we do a Z-test, we assume that we have a random sample from a normal population. Whether the observations are independent and identically distributed can often be assessed from the context of the problem. However, the context of the problem rarely admits the normality assumption. For example, the lengths of the rods can't technically be normally distributed, because the support of the normal density is the whole real line, and the lengths can't be negative. But for the normal distribution, the tail probabilities "go to zero fast"—if $Y \sim N(120, 25)$, as in the rod length example, the probability that Y is less than zero is *so* small that if we observed a trillion values per second from this distribution, we would likely never see a negative value, even in billions of years.

It's standard practice to assume "approximately normal" for this test. The question of "how approximate is OK" does not have a definitive answer; we can make histograms or probability plots (see Chapter 19) to visually check the assumptions, and there are formal hypothesis tests where the null hypothesis is that the population is normally distributed. Further, we will see in Chapter 41 that the normality assumption can be waived if the sample size is large.

Test for the variance of a population

Next, we'll develop a test for the variance of a population when we assume the mean is known. The results that we need from previous chapters are

1. if $Z \sim N(0, 1)$, then $Z^2 \sim \chi^2(1)$;

2. if $Z_1, \ldots, Z_n \overset{iid}{\sim} N(0, 1)$, then $Z_1^2 + \cdots + Z_n^2 \sim \chi^2(n)$.

Example: Continuing the context of the last example, suppose we have an opportunity to buy a new machine. We'll be interested in the purchase if the new machine makes rods whose lengths have smaller variance than 25, the variance of rods made by the old machine. Suppose for now we can be confident that the mean length of the rods for the new machine is correct. We want to test

$$H_0 : \sigma^2 = 25 \quad \text{versus} \quad H_a : \sigma^2 < 25,$$

where σ^2 is the variance of the rods made by the new machine.

Using our previous results, we see that if $Y_1, \ldots, Y_n \overset{iid}{\sim} N(\mu, \sigma^2)$, then

$$\sum_{i=1}^{n} \left(\frac{Y_i - \mu}{\sigma} \right)^2 \sim \chi^2(n).$$

We can define a statistic X by plugging in the null hypothesis variance and the known mean:

$$X = \frac{\sum_{i=1}^{n}(Y_i - 120)^2}{25}.$$

Then if the null hypothesis is true, $X \sim \chi^2(n)$. If H_a is true, we expect X to be *smaller*, on average. Therefore, small values of X support H_a, and we reject H_0 if the observed value of X is "too small" to be plausible with the null distribution.

Let's find the decision rule, supposing our test size is $\alpha = .05$ and our sample size is $n = 9$. We want the probability of rejecting H_0 to be .05 when H_0 is true, so we need to find the 5th percentile of a $\chi^2(9)$-density. The R command can do this for us: qchisq(.05,9) returns about 3.325. The decision rule is to reject H_0 when the value of the test statistic X, calculated using the sample values, is less than 3.325.

In practice, we often don't bother with a decision rule based on the test statistic values; instead we go right to calculating the p-value. We can always state a decision rule as "reject H_0 when the p-value is less than the test size."

Suppose our nine values of rod lengths are 124.0, 118.5, 119.0, 118.0, 116.0, 118.0, 121.5, 121.5, and 123.5. We calculate the observed value of the test statistic

$$X_{obs} = \frac{\sum_{i=1}^{9}(Y_i - 120)^2}{25} = \frac{60}{25} = 2.4.$$

This is smaller than the value we got for our decision rule, so we know that the p-value will be smaller than .05 (and we reject H_0), but let's compute the p-value anyway.

The p-value is the probability of seeing this value of the test statistic, *or smaller*, when the null hypothesis is true. (Recall that smaller values support H_a but larger values do not.) The null hypothesis distribution of X is $\chi^2(9)$; this density is shown in the plot below. The area to the left of the observed test statistic value is shaded. Using pchisq(2.4,9), we find $p = .0165$. If the new machine has the same variance as the old machine, the probability of seeing this much or more evidence that the new machine has smaller variance is .0165. Because this is less than our test size, we reject H_0 and conclude that the new machine has smaller variance.

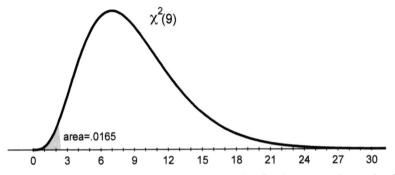

Let's compute the power for this test, given a value for the true variance that is less than the null hypothesis variance. Let's say $\sigma^2 = 16$ and try to calculate the probability of rejecting H_0 under this assumption and the assumption that the mean is $\mu = 120$.

We want

$$
\begin{aligned}
P(X < 3.325) &= P\left(\frac{\sum_{i=1}^{n}(Y_i - 120)^2}{25} < 3.325 \right) \\
&= P\left(\frac{\sum_{i=1}^{n}(Y_i - 120)^2}{16} < 3.325 \times \frac{25}{16} \right) \\
&= P\left(W < 5.195 \right),
\end{aligned}
$$

where $W \sim \chi^2(9)$. Using pchisq(5.195,9), we find the power is .183. Perhaps we would choose a larger sample size, if possible, to get a higher power for this test.

Chapter Highlights

1. To construct a hypothesis test for the mean of a normal population if the variance σ^2 is *known*, we use a Z-test. If $Y_1, \ldots, Y_n \sim N(\mu, \sigma^2)$ and $H_0 : \mu = \mu_0$ is true, then

$$
Z = \frac{\bar{Y} - \mu_0}{\sigma/\sqrt{n}} \sim N(0, 1),
$$

where $\bar{Y}$ is the sample mean.

2. To construct a hypothesis test for the variance of a normal population if the mean μ is *known*: If $Y_1, \ldots, Y_n \sim N(\mu, \sigma^2)$ and $H_0 : \sigma^2 = \sigma_0^2$ is true, then

$$
X = \frac{\sum_{i=1}^{n}(Y_i - \mu)^2}{\sigma_0^2} \sim \chi^2(n).
$$

Exercises

35.1 An employee at the Atlantic Fishing and Tackle Company has a large spool of fishing line that he knows is either from Supplier A or Supplier B. Lengths of Supplier A's fishing line are known to have an average breaking strength of 30 pounds, with a standard deviation of 4 pounds. Lengths of Supplier B's fishing line have an average breaking strength of 25 pounds and a standard deviation of 4 pounds. Assume that both "breaking-strength" distributions are approximately normal.

  (a) Sketch the breaking-strength distributions for fishing line of both suppliers on the same axis.

  (b) Suppose the spool is from Supplier A. The employee cuts a length and tests it. What is the probability that its breaking strength is less than 25 pounds?

  (c) Suppose the spool is from Supplier B. The employee cuts a length and tests it. What is the probability that its breaking strength is less than 25 pounds?

  (d) Now the employee cuts four lengths and tests all of them. Let $\bar{Y}$ be the average breaking strength. Sketch the distribution of $\bar{Y}$ for both suppliers on the same axis.

  (e) If the spool is from Supplier A, what is the probability that $\bar{Y}$ is less than 25?

  (f) If the spool is from Supplier B, what is the probability that $\bar{Y}$ is less than 25?

  (g) The employee is interested in testing

   H_0: Spool is from Supplier A

   against the alternative

   H_a: Spool is from Supplier B,

   with a sample of size $n = 4$. What is the *decision rule* for this test if $\alpha = 0.05$? Write the decision rule in terms of $\bar{Y}$: Reject H_0 if _____ _____.

  (h) What is the *power* for this test?

35.2 A man is shot and killed while hunting in North Forest. It looks like an accident, but police question another hunter who is known to have had a grudge against the victim. They find a pine needle stuck to the suspect's coat. The suspect claims to have been hunting in South Forest on the day of the incident.

North Forest has only Species N pine trees, and South Forest has only Species S pine trees:

  • Species N pine trees have needle lengths that are approximately normally distributed with mean 5.4 cm and standard deviation 0.4 cm.

  • Species S pine trees have needle lengths that are approximately normally distributed with mean 6.4 cm and standard deviation 0.5 cm.

The prosecution hires *you* to be the expert witness statistician. You are to test

   H_0 : needle from Species S   versus H_a : needle from Species N.

(a) The judge says that $\alpha = 0.01$ ("beyond a reasonable doubt"). Determine the decision rule "reject if the length of the pine needle is . . . ."

(b) What is the power for the decision rule?

(c) The pine needle found on the suspect's coat is 5.75 centimeters long. What is the p-value?

(d) Now suppose *four* needles had been found on the suspect's coat. We can use the *average* of the lengths as our test statistic. Determine the decision rule "reject if the average length of the pine needles is . . . ."

(e) What is the power of *this* test?

(f) The lengths are 5.65, 6.12, 5.85, and 5.38. Now what is the p-value and what is the conclusion?

35.3 Suppose X_1, X_2, X_3, X_4, and X_5 are independent normal random variables. All of these random variables have mean μ, X_1 and X_2 have variance 4, and X_3, X_4, and X_5 have variance 9.

(a) Derive a statistic to test $H_0 : \mu = 100$ versus $H_0 : \mu > 100$, and state its distribution when H_0 is true. Make a decision rule for test size $\alpha = .05$.

(b) Apply your test when $X_1 = 101.1$, $X_2 = 99.7$, $X_3 = 103.5$, $X_4 = 102.2$, and $X_5 = 100.6$.

35.4 Suppose $X_1, \ldots, X_{10}$ are a random sample from a normal distribution with mean μ, and we know that the variance of the distribution is $\sigma^2 = 1$. We are interested in testing $H_0 : \mu = 0$ versus $H_a : \mu > 0$.

(a) Find c_1 so that the decision rule "reject H_0 if at least 1 value of the sample is greater than c_1" has test size $\alpha = .01$.

(b) Find c_2 so that the decision rule "reject H_0 if the sample mean $\bar{X} > c_2$" has test size $\alpha = .01$.

(c) Compute the power of the two tests when $\mu = 1$. Which has higher power?

35.5 Suppose $X_1, \ldots, X_{10}$ are a random sample from a normal distribution with known mean $\mu = 0$ and variance σ^2. We are interested in testing $H_0 : \sigma^2 = 1$ versus $H_a : \sigma^2 > 1$.

(a) Find c_1 so that the decision rule "reject H_0 if at least 1 value of the sample is greater than c_1" has test size $\alpha = .01$.

(b) Find c_2 so that the decision rule "reject H_0 if $\sum_{i=1}^{10} X_i^2 > c_2$" has test size $\alpha = .01$.

(c) Compute the power of the two tests when $\sigma^2 = 2$. Which has higher power?

35.6 Suppose X_1, X_2, X_3, and X_4 are independent normal random variables. All of these random variables have mean μ. The random variables X_1 and X_2 have variance 1, but X_3 and X_4 have variance 4. Interest is in testing $H_0 : \mu = 0$ versus $H_a : \mu > 0$.

(a) Statistician A reasons that $X_1 + X_2 + X_3 + X_4$ is normal with mean 4μ and variance 10, so under H_0,

$$Z_A = \frac{X_1 + X_2 + X_3 + X_4}{\sqrt{10}} \sim N(0, 1).$$

Statistician A will reject H_0 when Z_A is greater than 1.645, the 95th percentile of a standard normal density. Determine the power of this test when $\mu = 1$.

(b) Statistician B reasons that X_1, X_2, $X_3/2$, and $X_4/2$ are all standard normal under H_0, so

$$Z_B = \frac{X_1 + X_2 + X_3/2 + X_4/2}{2} \sim N(0, 1).$$

Statistician B will reject H_0 when Z_B is greater than 1.645, the 95th percentile of a standard normal density. Determine the power of this test when $\mu = 1$.

35.7 Suppose $X_1, \ldots, X_8$ are independent normal random variables, all with mean zero, where X_1, X_2, X_3, X_4 have variance σ^2, while X_5, X_6, X_7, X_8 have variance $2\sigma^2$. Derive a statistic to test $H_0 : \sigma^2 = 10$ versus $H_a : \sigma^2 > 10$ and give its distribution under the null hypothesis.

(a) For test size $\alpha = 0.1$, state the decision rule as "reject H_0 if ___(name of your____
___test statistic)___ is ___(greater than, less than)___   ___(value)___."

(b) If $\sigma^2 = 12$, what is the power of your test?

(c) Check your answer to (b) using simulations in R.

35.8 Filaments made at Gloglobe factory are supposed to contain 2.75 mg of chromelite. Because of randomness in the manufacturing process, the amount of chromelite in the filament is actually a random variable. If a filament has more than 2.77 mg or less than 2.73 mg of chromelite, it must be thrown out. Suppose the machines at the factory make filaments with a chromelite distribution which is normally distributed with mean $\mu = 2.75$ mg and standard deviation $\sigma = 0.02$ mg of chromelite. Then about 32% of the filaments must be discarded. The factory boss is considering buying a new machine that is known to make filaments of the same mean $\mu = 2.75$ mg, but might have a smaller variance. The boss is interested in testing $H_0 : \sigma = .02$ versus $H_a : \sigma < .02$. The machines are expensive, so he doesn't want to make a Type I Error; he sets $\alpha = .01$.

(a) Suppose he gets a random sample of size $n = 20$ filaments from the new machine. Construct a test statistic and a decision rule.

(b) Suppose the new machine makes filaments with a standard deviation of .01 mg of chromelite. What is the power for the test?

(c) Repeat (a) and (b) for a sample size of $n = 50$.

35.9 Scientists are interested in measuring the length of a macromolecule that is used in making synthetic fibers. The molecules are hard to measure, and each measurement (in nanometers) consists of some random error that can be modeled as a normal distribution with mean zero and variance $\sigma^2 = 2$. If the true length of the macromolecule is μ nanometers, then a measurement Y of the molecule is normally distributed with mean μ and variance σ^2, given the error introduced by the measuring device.

Suppose we have a random sample $Y_1, \ldots, Y_{10}$ representing length measurements of the macromolecule in nanometers using the measuring device with $\sigma^2 = 2$. Then the scientists have access to a new measuring device, with $\sigma^2 = 1$. Suppose $X_1, \ldots, X_5$ are measurements of the macromolecule using the new device.

(a) Derive a test statistic for $H_0 : \mu = 100$ nanometers versus $H_a : \mu > 100$ nanometers, using all 15 measurements.

(b) If the mean of the first ten measurements is 101.2, and the mean of the second five measurements is 100.9, do we reject H_0 at $\alpha = .05$?

Chapter 36

Hypothesis Tests for a Normal Population Parameter, Part II

Suppose we have a random sample $Y_1, \ldots, Y_n$ from a normal population with mean μ and variance σ^2, and we want to develop tests about μ and about σ^2 that are similar to the tests in the last chapter. In practice, it's unlikely that we can use those tests, because we won't know the other parameter (μ for a test about σ^2, or σ^2 for a test about μ), so we'd like to develop tests where we do not need these parameters.

Test for the mean when the variance is unknown

First, we'll consider a test about the mean μ where we do not know σ^2. We can't use the Z-test we developed in Chapter 35, but we can develop a test using the t distribution defined in Chapter 34. The first order of business is to define the **sample variance** S^2:

$$S^2 = \frac{1}{n-1} \sum_{i=1}^{n} (Y_i - \bar{Y})^2.$$

This is the formula that R uses when implementing the command var. We have already mentioned that the sample variance can be used to estimate the population variance; in this chapter, we will see why S^2 is a good estimator of σ^2. Note that we do not use the population mean μ for the calculation; instead, we use sample mean $\bar{Y}$. If we knew μ, we *could* use

$$\frac{1}{n} \sum_{i=1}^{n} (Y_i - \mu)^2$$

as an estimator for σ^2, because the expected value of this expression is σ^2 (as shown in Exercise 34.9). Comparing this expression with that for S^2, we see that not only did we replace μ with $\bar{Y}$, but we swapped n with $n-1$. We'll figure out why this is the right thing to do, but first we will prove one of the most important results in statistics: For a random sample from a normal population, the sample mean and the sample variance are independent random variables.

Intuitively, we might at first be surprised by the independence of the sample mean and the sample variance. After all, both quantities are calculated using the same numbers. But on reflection, the sample mean indicates a center, and the sample variance indicates a spread, and for the normal distribution, one of these does not "influence" the other (in contrast, for an exponential random variable, the variance grows with the mean). In any case, this independence is an impressively useful result.

For simplicity we start with $\mu = 0$ and $\sigma = 1$, so that we have standard normal random variables. For $Z_1, \ldots, Z_n \overset{ind}{\sim} N(0, 1)$, we will show two important results:

1. The random variables $\bar{Z}$ and $\sum_{i=1}^{n}(Z_i - \bar{Z})^2$ are independent.

2. The distribution of $\sum_{i=1}^{n}(Z_i - \bar{Z})^2$ is $\chi^2(n-1)$.

We've already done most of the work for the $n = 2$ case, so we start with this to give the ideas. Recall this result from Chapter 30: If Z_1 and Z_2 are independent standard normal random variables, then $Z_1 + Z_2$ and $Z_1 - Z_2$ are independent. The sample variance can be written as

$$(Z_1 - \bar{Z})^2 + (Z_2 - \bar{Z})^2 = \left[Z_1 - \frac{Z_1 + Z_2}{2}\right]^2 + \left[Z_2 - \frac{Z_1 + Z_2}{2}\right]^2$$

$$= \left[\frac{Z_1 - Z_2}{2}\right]^2 + \left[\frac{Z_2 - Z_1}{2}\right]^2$$

$$= (Z_1 - Z_2)^2/2.$$

By our previous result, this random variable and $\bar{Z} = (Z_1 + Z_2)/2$ are independent. Because $Z_1 - Z_2 \sim N(0, 2)$, we know $(Z_1 - Z_2)/\sqrt{2} \sim N(0, 1)$, and hence $(Z_1 - Z_2)^2/2 \sim \chi^2(1)$. Hence, the two important results hold for the $n = 2$ case.

How do we tackle $n > 2$? We use a complicated transformation with a multivariate Jacobian; this is an extension of the derivation that we did in Chapter 30, used for the $n = 2$ case. The change of variable is defined by

$$X_1 = \bar{Z} \quad \text{and} \quad X_i = Z_i - \bar{Z}, \quad i = 2, \ldots, n.$$

Because the Z_i random variables can take any value in $\mathbb{R}$, so can the X_i, so the support of each X_i is the whole real line, and the joint support of the X_i random variables is $\mathbb{R}^n$.

For $x_1 = \sum_{i=1}^{n} z_i/n$ and $x_2 = z_2 - x_1, \ldots, x_n = z_n - x_1$, the inverse transformation is

$$z_1 = x_1 - \sum_{i=2}^{n} x_i \quad \text{and} \quad z_i = x_1 + x_i, \quad i = 2, \ldots, n,$$

and the Jacobian is

$$\det \begin{bmatrix} 1 & -1 & -1 & -1 & \cdots & -1 & -1 \\ 1 & 1 & 0 & 0 & \cdots & 0 & 0 \\ 1 & 0 & 1 & 0 & \cdots & 0 & 0 \\ & & & \vdots & & & \\ 1 & 0 & 0 & 0 & \cdots & 1 & 0 \\ 1 & 0 & 0 & 0 & \cdots & 0 & 1 \end{bmatrix} = n.$$

Let $f_Z(z_1, \ldots, z_n)$ be the joint distribution of the independent standard normal random variables. If $f(z) = e^{-z^2/2}/\sqrt{2\pi}$, then

$$f_Z(z_1, \ldots, z_n) = f(z_1)f(z_2) \cdots f(z_n),$$

and the formula for the joint density for the X_i is

$$f_X(x_1, \ldots, x_n) = n f_Z\left(x_1 - \sum_{i=2}^{n} x_i, x_1 + x_2, \ldots, x_1 + x_n\right)$$

$$= n f\left(x_1 - \sum_{i=2}^{n} x_i\right) f(x_1 + x_2) \cdots f(x_1 + x_n),$$

by independence of the Z_i's. Using the formula for f, we get

$$f_X(x_1, \ldots, x_n) = n \left[\frac{1}{\sqrt{2\pi}} e^{-(x_1 - \sum_{i=2}^n x_i)^2/2} \right] \prod_{i=2}^n \left[\frac{1}{\sqrt{2\pi}} e^{-(x_1 + x_i)^2/2} \right]$$

$$= \frac{n}{(2\pi)^{n/2}} \exp \left\{ -\frac{1}{2} \left(x_1 - \sum_{i=2}^n x_i \right)^2 - \frac{1}{2} \sum_{i=2}^n (x_1 + x_i)^2 \right\}$$

$$= \frac{n}{(2\pi)^{n/2}} \exp \left\{ -\frac{1}{2} \left[nx_1^2 + \left(\sum_{i=2}^n x_i \right)^2 + \sum_{i=2}^n x_i^2 \right] \right\}$$

$$= \frac{n}{(2\pi)^{n/2}} e^{-nx_1^2/2} \exp \left\{ -\frac{1}{2} \left[\left(\sum_{i=2}^n x_i \right)^2 + \sum_{i=2}^n x_i^2 \right] \right\}.$$

This shows that $X_1 = \bar{Z}$ is independent of $X_2, \ldots, X_n$, because the term with x_1 can be factored out; in fact, we see that X_1 is normal with mean zero and variance $1/n$ (which we already knew).

The next step is to write $\sum_{i=1}^n (Z_i - \bar{Z})^2$ as a function of $X_2, \ldots, X_n$. First note that $\sum_{i=1}^n (Z_i - \bar{Z}) = 0$, and so $Z_1 - \bar{Z} = -\sum_{i=2}^n (Z_i - \bar{Z})$. Then

$$\sum_{i=1}^n (Z_i - \bar{Z})^2 = (Z_1 - \bar{Z})^2 + \sum_{i=2}^n (Z_i - \bar{Z})^2$$

$$= \left[\sum_{i=2}^n (Z_i - \bar{Z}) \right]^2 + \sum_{i=2}^n (Z_i - \bar{Z})^2 = \left[\sum_{i=2}^n X_i \right]^2 + \sum_{i=2}^n X_i^2.$$

We have shown that $\bar{Z}$ and $S^2 = \sum_{i=1}^n (Z_i - \bar{Z})^2/(n-1)$ are independent.

The next step is to derive the distribution of the sample variance S^2. We know that for $n = 2$, $S^2 \sim \chi^2(1)$, and we'd like to show that for any sample size $n > 1$, $(n-1)S^2 \sim \chi^2(n-1)$. We use the principle of mathematical induction. Suppose the result is true for sample sizes $2, 3, \ldots, n-1$. If $\bar{Z}_n = (Z_1 + \cdots + Z_n)/n$, we can write

$$\bar{Z}_n = \frac{1}{n} \left(\sum_{i=1}^{n-1} Z_i + Z_n \right) = \frac{1}{n} \left[(n-1)\bar{Z}_{n-1} + Z_n \right],$$

where we have written the mean of the sample of size n in terms of the mean of the first $n-1$ elements.

Writing the sample variance for all n elements as

$$S_n^2 = \frac{1}{n-1} \sum_{i=1}^n (Z_i - \bar{Z})^2,$$

our goal is to write this expression in terms of S_{n-1}^2, the sample variance for the first $n-1$ elements. Then we can use the induction hypothesis that $(n-1)S_{n-1}^2 \sim \chi^2(n-1)$.

We have

$$(n-1)S_n^2 = \sum_{i=1}^{n}(Z_i - \bar{Z}_n)^2$$

$$= \sum_{i=1}^{n}\left[Z_i - \frac{n-1}{n}\bar{Z}_{n-1} - \frac{1}{n}Z_n\right]^2$$

$$= \sum_{i=1}^{n}\left[(Z_i - \bar{Z}_{n-1}) + \frac{1}{n}(\bar{Z}_{n-1} - Z_n)\right]^2,$$

where we have broken up the $\bar{Z}_{n-1}$ term into two pieces. Multiplying out the square and bringing the summation sign through, we have

$$(n-1)S_n^2 = \sum_{i=1}^{n}(Z_i - \bar{Z}_{n-1})^2 + \frac{2}{n}\sum_{i=1}^{n}(Z_i - \bar{Z}_{n-1})(\bar{Z}_{n-1} - Z_n) + \frac{1}{n}(\bar{Z}_{n-1} - Z_n)^2$$

$$= \sum_{i=1}^{n}(Z_i - \bar{Z}_{n-1})^2 + \frac{2}{n}(\bar{Z}_{n-1} - Z_n)\sum_{i=1}^{n}(Z_i - \bar{Z}_{n-1}) + \frac{1}{n}(\bar{Z}_{n-1} - Z_n)^2$$

$$= \sum_{i=1}^{n}(Z_i - \bar{Z}_{n-1})^2 + \frac{2}{n}(\bar{Z}_{n-1} - Z_n)\left[\sum_{i=1}^{n-1}(Z_i - \bar{Z}_{n-1}) + (Z_n - \bar{Z}_{n-1})\right]$$

$$+ \frac{1}{n}(\bar{Z}_{n-1} - Z_n)^2,$$

where in the middle term we have written the sum from $i = 1$ to n in two pieces, the first going from $i = 1$ to $n - 1$. Now we can use

$$\sum_{i=1}^{n-1}(Z_i - \bar{Z}_{n-1}) = 0.$$

Continuing with our derivation, we have

$$(n-1)S_n^2 = \sum_{i=1}^{n}(Z_i - \bar{Z}_{n-1})^2 + \frac{2}{n}(\bar{Z}_{n-1} - Z_n)(Z_n - \bar{Z}_{n-1}) + \frac{1}{n}(\bar{Z}_{n-1} - Z_n)^2$$

$$= \sum_{i=1}^{n}(Z_i - \bar{Z}_{n-1})^2 - \frac{2}{n}(\bar{Z}_{n-1} - Z_n)^2 + \frac{1}{n}(\bar{Z}_{n-1} - Z_n)^2$$

$$= \sum_{i=1}^{n}(Z_i - \bar{Z}_{n-1})^2 - \frac{1}{n}(\bar{Z}_{n-1} - Z_n)^2$$

$$= \sum_{i=1}^{n-1}(Z_i - \bar{Z}_{n-1})^2 + (Z_n - \bar{Z}_{n-1})^2 - \frac{1}{n}(\bar{Z}_{n-1} - Z_n)^2$$

$$= \sum_{i=1}^{n-1}(Z_i - \bar{Z}_{n-1})^2 + \frac{n-1}{n}(\bar{Z}_{n-1} - Z_n)^2$$

$$= (n-2)S_{n-1}^2 + \frac{n-1}{n}(\bar{Z}_{n-1} - Z_n)^2.$$

We've written $(n-1)S_n^2$ as a sum of two random variables. Note that S_{n-1}^2 and $\bar{Z}_{n-1}$ are independent by the previous result, and Z_n is independent of both of those. So the two terms are independent. By the induction hypothesis, $(n-2)S_{n-1}^2 \sim \chi^2(n-1)$, so we need to show that the second term is $\chi^2(1)$ and we're done.

We have

$$Z_n - \bar{Z}_{n-1} \sim N\left(0, 1 + \frac{1}{n-1}\right) = N\left(0, \frac{n}{n-1}\right),$$

so

$$\sqrt{\frac{n-1}{n}}(Z_n - \bar{Z}_{n-1}) \sim N(0, 1),$$

and because the square of a standard normal random variable has a $\chi^2(1)$ distribution, we have

$$\frac{n-1}{n}(Z_n - \bar{Z}_{n-1})^2 \sim \chi^2(1).$$

We have written $(n-1)S_n^2$ as a sum of independent χ^2 random variables; we know this sum is also a χ^2 random variable, and the degrees of freedom add. Therefore, $(n-1)S_n^2 \sim \chi^2(n-1)$.

The next step is to extend these results to a random sample from a general normal population. Suppose $Y_1, Y_2, \ldots, Y_n \overset{iid}{\sim} N(\mu, \sigma)^2$, and define S^2 as the variance for this sample:

$$S^2 = \frac{1}{n-1} \sum_{i=1}^{n} (Y_i - \bar{Y})^2.$$

If we make the transformation

$$Z_i = \frac{Y_i - \mu}{\sigma},$$

we get

$$\frac{(n-1)S^2}{\sigma^2} = \frac{1}{\sigma^2} \sum_{i=1}^{n} (Y_i - \bar{Y})^2 = \sum_{i=1}^{n} (Z_i - \bar{Z})^2 \sim \chi^2(n-1).$$

Finally, using the fact that the expected value of a $\chi^2(n-1)$ random variable is $n-1$, we can compute

$$E(S^2) = \frac{\sigma^2}{n-1}(n-1) = \sigma^2,$$

which explains the $n-1$ denominator of the sample variance. We use the sample variance as an estimator of the population variance, because it has the right expected value. If we had divided by n instead of $n-1$, the expectation would be too small.

Now we're ready to derive the **one-sample t-test**, for inference about the mean μ of a normal population when the variance is unknown. Suppose $Y_1, Y_2, \ldots, Y_n \overset{iid}{\sim} N(\mu, \sigma)^2$, and we wish to test

$$H_0 : \mu = \mu_0 \text{ versus } H_a : \mu \neq \mu_0.$$

The first part is the same as for the Z-test: We know $\bar{Y} \sim N(\mu, \sigma^2/n)$, so

$$\frac{\bar{Y} - \mu}{\sigma/\sqrt{n}} \sim N(0, 1).$$

We don't know σ^2, but we know that

$$\frac{(n-1)S^2}{\sigma^2} \sim \chi^2(n-1),$$

and we know that the above two random variables are independent. Now the magic of the T random variable is seen. Using the definition, we know that

$$T = \frac{\frac{\bar{Y}-\mu}{\sigma/\sqrt{n}}}{\sqrt{\frac{(n-1)S^2}{\sigma^2}/(n-1)}} \sim t(n-1).$$

The σ and $n-1$ cancel to give

$$T = \frac{\bar{Y}-\mu}{S/\sqrt{n}} \sim t(n-1).$$

Comparing this with the Z-test, we have substituted S for the unknown σ, which changes the distribution from a standard normal to a $t(n-1)$.

In introductory (noncalculus) statistics courses, we say "if you substitute S for σ in the Z statistic, the distribution changes (magically) from a standard normal to a t." It does seem like magic! Our lengthy derivation led to a nice simple result. On the way, we proved independence of the sample mean and variance, and derived the distribution of the sample variance. These results will be used again in the next chapters to develop more complicated tests.

Given that we showed in Chapter 34 that the t distribution is "close to" a standard normal when n is "large," we infer that for large samples our statistic T is close to our old Z statistic.

Example: NuttyKorn is a snack consisting of caramel popcorn and peanuts. They advertise that the weight of peanuts (in ounces) in a 26-ounce box of snack follows a normal distribution with mean 5.

A consumer group has grave fears that NuttyKorn is skimping on the peanuts. They want to test $H_0 : \mu = 5$ versus $H_a : \mu < 5$, where μ is the *true* average weight of peanuts in a 26-ounce box of snack. They will randomly select 10 boxes and compute the peanut weights for each box. They will reject H_0 if the sample mean is too small. Because σ^2 is unknown, we can't use our Z-test, but we can use our t-test. Suppose they want to use $\alpha = .01$ as a test size. What is the decision rule? We can't state the decision rule in terms of $\bar{Y}$ only; we have to state it in terms of T, which uses both $\bar{Y}$ and S. Under H_0,

$$T = \frac{\bar{Y} - 5}{S/\sqrt{10}} \sim t(9).$$

The decision rule is to reject H_0 when T is "small," and the cutoff for "small" is defined by our chosen α, as shown in the plot. We reject H_0 when $T < -2.821$.

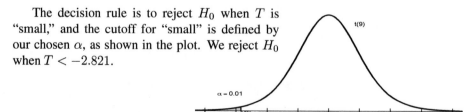

We can't state our decision rule in terms of $\bar{Y}$ as we did with the known-σ case, and it's also more difficult to compute the power for the test. The distribution of T, when $\mu \neq 5$, follows something called a "noncentral" t-density, which we will not cover in this book. However, we can use R to simulate the power.

Suppose the true mean is $\mu = 4.5$, so that the true peanut weight is 10% lower than promised. How likely are we to reject the null hypothesis? It depends on the true population variance σ^2. Typically, the researcher makes some guesses about the various model parameters and computes power for a variety of situations. Let's guess that the true standard deviation of peanut weight is .5 ounces, and find the power for $n = 10$. The following R code simulates samples of size ten from the assumed (alternative) distribution and saves the T statistic for each sample. The proportion of samples for which the null hypothesis is rejected is an approximate power; we get about .64. If we run the code again with $\mu = 4$ (the true weight of peanuts is 20% too low), the power is about .998. We are quite sure of "catching" the skimping if μ is as low as 4.

```
n=10
nloop=100000
tstat=1:nloop
for(iloop in 1:nloop){
    y=rnorm(10,4.5,.5)
    muhat=mean(y)
    sighat=sd(y)
    tstat[iloop]=(muhat-5)/(sighat/sqrt(10))
}
hist(tstat,br=50)
sum(tstat<qt(.01,9))/nloop
```

Test for the variance when the mean is unknown

Next, let's revisit our test of $H_0 : \sigma^2 = \sigma_0^2$, where we have a random sample $Y_1, Y_2, \ldots, Y_n$ from an $N(\mu, \sigma^2)$ population. For the previous test we used a known μ, but in the more common situation where μ is unknown, we can easily use our distribution for the sample variance to construct a test.

First, let's redo the example of the last chapter, where we have the opportunity to buy a new rod-producing machine for our factory. We will be interested in buying the new machine if the variance of the rods it produces is *smaller* than that for the old machine. We are testing $H_0 : \sigma^2 = 25$ versus $H_a : \sigma < 25$, with a sample of $n = 9$ rods. Our sample values are 124.0, 118.5, 119.0, 118.0, 116.0, 118.0, 121.5, 121.5, and 123.5, as before.

This time we don't assume we know the mean, but we use the result that $X = (n-1)S^2/\sigma^2$ is $\chi^2(n-1)$. We calculate $S^2 = 7.5$, and the observed chi-squared test statistic is $X_{obs} = 8 \times 7.5/25 = 2.4$. We reject the null hypothesis if there is evidence that the true variance is *smaller* than 25; hence the p-value is the area to the left of 2.4, under a $\chi^2(8)$-density. Using the R command pchisq(2.4,8) we find $p = .0338$.

Power calculations can be accomplished in the same manner as in the previous chapter, without resorting to simulations. Let's compute the power of the above test when the true population standard deviation is $\sigma = 4$ instead of $\sigma = 5$. To compute the power, we first need a decision rule.

We reject H_0 when the observed X is less than the 5th percentile of a $\chi^2(8)$, or when $X < 2.733$. Then the power is

$$P(X < 2.733) = P\left(\frac{8S^2}{25} < 2.733\right) = P\left(\frac{8S^2}{16} < 2.733 \times \frac{25}{16}\right)$$
$$= P(W < 4.270) = .168,$$

where $W \sim \chi^2(8)$. Comparing this with the power from the last chapter, we find that the power is a bit smaller when the mean is not known.

Two-sided chi-squared tests

Suppose we have a two-sided alternative $H_a : \sigma^2 \neq \sigma_0^2$. Now we want to reject H_0 if the observed sample variance is too large *or* too small. Let's suppose we have a sample of size $n = 10$ and a test size $\alpha = .05$. We know that the test statistic

$$X = \frac{(n-1)S^2}{\sigma_0^2} \sim \chi^2(n-1)$$

if H_0 is true. To find the decision rule, we split the test size in half, so that under the null hypothesis the probability that we reject H_0 because the sample variance is too small is $\alpha/2$ and the probability that we reject H_0 because the sample variance is too big is $\alpha/2$. The rejection region is shown in the plot below. The lower critical value is qchisq(.025,9) (about 2.7) and the upper critical value is qchisq(.975,9) (about 19.0). We reject the null hypothesis if the test statistic is less than 2.7 or greater than 19.0.

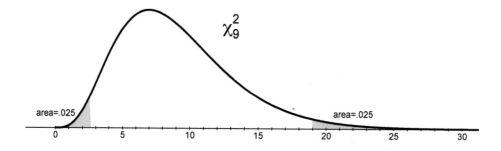

We take our sample and compute the sample mean $\bar{Y}$ and the sample variance S^2; suppose we find $X_{obs} = 4.2$. What is the p-value? We know that values of the test statistic that are less than 4.2 support the alternative more, and we can find that under the null hypothesis, $P(X < 4.2) = .102$. Large values of X also support the alternative more. We can find that $P(X > 14.6) = .102$. We reason that the value 14.6 gives the same amount of evidence against the null hypothesis as our observed value 4.2, and values greater than 14.6 support the alternative more than our observed value 4.2. In short, we find the p-value is $.102 + .102 = .204$. In other words, for the two-sided test, we find the area under the curve (to the left or right) that is smaller, and we double that area to get the p-value. For the observed chi-squared test statistic of 4.2, the two-sided p-value is shown as the shaded area in the plot below:

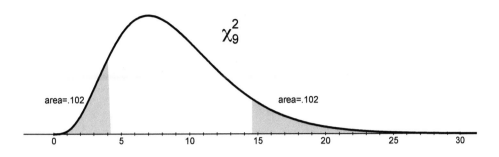

area=.102

area=.102

χ_9^2

Chapter Highlights

1. For a sample $Y_1, \ldots, Y_n$, the sample variance is defined as

$$S^2 = \frac{1}{n-1} \sum_{i=1}^{n} (Y_i - \bar{Y})^2,$$

where $\bar{Y}$ is the sample mean.

2. If $Y_1, \ldots, Y_n \overset{ind}{\sim} N(\mu, \sigma^2)$, then $\bar{Y}$ and S^2 are independent random variables.

3. If $Y_1, \ldots, Y_n \overset{ind}{\sim} N(\mu, \sigma^2)$, the random variable

$$\frac{(n-1)S^2}{\sigma^2}$$

has a $\chi^2(n-1)$ distribution.

4. To construct a hypothesis test for the mean of a normal population if the variance is *unknown*, we use a t-test. If $Y_1, \ldots, Y_n \overset{ind}{\sim} N(\mu, \sigma^2)$ and $H_0 : \mu = \mu_0$ is true, then

$$t = \frac{\bar{Y} - \mu_0}{S/\sqrt{n}} \sim t(n-1),$$

where S^2 is the sample variance.

5. To construct a hypothesis test for the variance of a normal population if the mean is *unknown*, we use a χ^2-test. If $Y_1, \ldots, Y_n \sim N(\mu, \sigma^2)$ and $H_0 : \sigma = \sigma_0$ is true, then

$$X = \frac{(n-1)S^2}{\sigma_0^2} \sim \chi^2(n-1).$$

Exercises

36.1 A company's mixed nuts are sold in cans, and the label says that 25% (by weight) of the contents are cashews. Suspecting this might be an overstatement, an inspector takes a random sample of 36 100-gram cans and measures the weight of the cashews in each can. The mean and standard deviation of these measurements are found to be 23.5 and 3.1, respectively. Do these results constitute strong evidence in support of the inspector's belief? Answer by calculating and interpreting the p-value.

36.2 A factory manager is interested in testing a new machine to see if there is evidence that it makes rods whose lengths have *smaller* variance than those for the old machine, using $\alpha = .01$. It is known that the rod lengths from the old machine have variance $\sigma_0^2 = 225$.

   (a) A sample of size $n = 9$ gives $S^2 = 144$. Compute the p-value and state the decision and conclusion in the context of the problem.

   (b) Suppose that the true variance for the new machine is actually only 100 (standard deviation is 10). Compute the power of the test.

   (c) What sample size would you need to get a power of at least .8?

36.3 High school seniors in a large city have standardized math test scores that follow a normal distribution with mean 580. An alternative school with 20 seniors claims to be more effective at teaching math to high school students. A member of the school board for the city wishes to test this claim, that is, let μ be the average test score at the alternative school, and

$$H_0\colon \mu = 580,$$
$$H_a\colon \mu > 580.$$

Suppose the 20 current seniors can be considered a representative sample of seniors who have been taught with the methods at the alternative school. The average score for seniors at this school is 608, and the sample standard deviation is 47. Give the p-value, the decision at $\alpha = 0.05$, and the conclusion in the context of the problem.

36.4 Lifetimes of a certain brand of switch follow (approximately) a normal distribution with mean 100 hours. Five switches of a new brand are obtained and their lifetimes measured to be 120, 101, 114, 95, and 130. Does this provide strong evidence ($\alpha = .05$) that the new switches have a longer average lifespan?

36.5 The mean travel time between Alewife station and Harvard Square, along the Red Line train, is about 18–20 minutes. The transit authority would like the standard deviation of the time to be not more than three minutes (variance not more than nine). To check for larger variances, she will take a random sample of twenty travel times.

   (a) Suppose the sample variance is 11.2, and that the travel times can be modeled as normally distributed. Test the null hypothesis that the variance is nine against the alternative that the variance is more than nine, at $\alpha = 0.05$. State the conclusion in the context of the problem.

   (b) What is the power of the test in part (a) if the true variance is 11?

36.6 The manager of Fizzypop is concerned about the variance of the bottle-filling procedure. When the filling machine is operating properly, the 12-ounce bottles of Fizzypop have actual fill-weights which are normally distributed with mean 12.05 ounces and standard deviation 0.05 ounces (variance = 0.0025). The manager takes a random sample of five bottles every few hours to check that the variance is not too large. The hypotheses are

$$H_0: \sigma^2 = 0.0025 \quad \text{versus} \quad H_a: \sigma^2 > 0.0025.$$

(a) Why is the test one-sided instead of two-sided?

(b) Suppose the values of the first sample are $\{12.08, 12.13, 12.05, 11.96, 11.95\}$. Test the hypotheses at $\alpha = 0.05$.

36.7 A manufacturing plant produces precision ball bearings. The diameters of the ball bearings are modeled as independent normal random variables with mean μ and variance σ^2. As part of a process control system, ten ball bearings are randomly selected from the production and their diameters measured, in order to test $H_0 : \sigma = 1$ versus $H_a : \sigma > 1$. If the null hypothesis is rejected, the plant is shut down for repairs. Because it is expensive to shut down the plant, the test size $\alpha = .01$ is used. If the true population standard deviation is 1.25 millimeters, what is the probability that the plant is shut down?

36.8 A consumer group testing "low-sodium" claims targets Cheez-Spheres. The label states that there are 200 mg sodium per eight-ounce bag. The test procedure will be to randomly select n bags and measure the average sodium content. We test

$$H_0: \mu = 200 \quad \text{versus} \quad H_a: \mu > 200$$

at $\alpha = 0.01$. We think it's reasonable to assume that the sodium content of Cheez-Spheres follows approximately a normal distribution.

(a) Do you think the Cheez-Spheres company would rather have $\alpha = 0.01$ or $\alpha = 0.05$? Explain.

(b) With $n = 8$, we get $\bar{Y} = 215$, $S = 26$. Perform the test! Do we reject or accept H_0 at $\alpha = 0.01$? State the conclusion in the context of the problem.

(c) The consumer group is wondering if they should have taken a bigger sample. Suppose the true mean is really 215; they would like to be confident of detecting that. They want to know the power of their test with $n = 8$. Do simulations in R with $\mu = 215$ and $\sigma = 26$ to find the power with $n = 8$.

(d) Suppose they want the power to be at least .8. What should their sample size be?

36.9 The average healthy body temperature for a human is usually quoted as $98.6°F$ (or $37.0°C$). Use the data found at

http://www.amstat.org/publications/jse/v4n2/datasets.shoemaker.html

to test the null hypothesis that the average temperature is $98.6°F$ versus the alternative that it is something else. Check the normality assumption by making a probability plot of the data.

36.10 Levels of DDT in kale were measured by laboratories using "pesticide residue measurement" methods. Fifteen values are reported in the data set DDT in the R data library MASS. Suppose there is an EPA threshold of 3 units. Test $H_0 : \mu = 3$ versus $H_a : \mu > 3$, where μ is the average DDT level. Report a p-value and a decision. Make a probability plot to see if there is evidence that the normality assumption is incorrect.

36.11 Let $X_1, \ldots, X_n$ be a random sample from $N(\mu, \sigma^2)$, and let $\bar{X}$ and S^2 be the usual sample mean and sample variance. Define the random variable

$$Y = \frac{c(\bar{X} - \mu)^2}{S^2}.$$

Find c for which Y has one of our "named" distributions, and determine the distribution. Verify your answer using simulations in R using $n = 10$: In a loop, generate the random sample, and compute $\bar{X}$, S^2, and the statistic Y for each sample. Make a probability plot of the sorted Y values against the quantiles from your density. Superimpose a diagonal line.

36.12 Show through simulations that the Z-test of Chapter 35 has better power than the t-test when σ^2 is known, for $n = 12$ and your choice of model parameters.

36.13 Show through simulations that the test for population variance when the population mean in known, derived in Chapter 35, has better power than the test when μ is unknown, for $n = 20$ and your choice of model parameters.

36.14 We showed that if $X_1, \ldots, X_n$ is a random sample from a normal population, and the sample variance is

$$S^2 = \frac{1}{n-1} \sum_{i=1}^{n} (X_i - \bar{X})^2,$$

then

$$\frac{(n-1)S^2}{\sigma^2} \sim \chi^2(n-1),$$

and hence $E(S^2) = \sigma^2$. Show that for a random sample from *any* population with finite variance σ^2, we have $E(S^2) = \sigma^2$.

Chapter 37

Two-Independent-Samples Tests: Normal Populations

In the last two chapters we derived "one-sample tests"—that is, hypothesis testing about the mean or the variance of a single normal population. Often we want to compare means or variances of two normal populations. Let $X_1, \ldots, X_n$ be a random sample from an $N(\mu_x, \sigma_x^2)$ population, and let $Y_1, \ldots, Y_m$ be a random sample from an $N(\mu_y, \sigma_y^2)$ population, and suppose the samples are independent. Let's consider the test of

$$H_0 : \mu_x = \mu_y \quad \text{versus} \quad H_a : \mu_x \neq \mu_y,$$

or possibly a one-sided alternative. Exercise 37.10 asks you to derive a test when the variances σ_x^2 and σ_y^2 are known. Here we make the common assumption that the population variances are unknown but the same, that is, $\sigma_x^2 = \sigma_y^2 = \sigma^2$. We derive the **independent samples t-test**.

Let $\bar{X}$ and $\bar{Y}$ be the sample means; we know the expected values of these statistics are the population means μ_x and μ_y, respectively. The sample variances S_x^2 and S_y^2 are random variables; the expected value of each of these is the common population variance σ^2. From the derivations in the last chapter, we know

$$\frac{(n-1)S_x^2}{\sigma^2} \sim \chi^2(n-1)$$

and

$$\frac{(m-1)S_y^2}{\sigma^2} \sim \chi^2(m-1),$$

and the two terms are independent because the samples are independent. Finally, we recall that the sum of independent chi-squared random variables is also chi-squared, and the degrees of freedom for the sum is the sum of the degrees of freedom. This tells us that

$$\frac{(n-1)S_x^2}{\sigma^2} + \frac{(m-1)S_y^2}{\sigma^2} \sim \chi^2(n+m-2).$$

To construct the test statistic, it's natural to compare $\bar{X}$ and $\bar{Y}$. If these are "far apart," this is evidence for H_a. We know that

$$\bar{X} - \bar{Y} \sim N\left(\mu_x - \mu_y, \frac{\sigma^2}{n} + \frac{\sigma^2}{m}\right),$$

so

$$\frac{(\bar{X} - \bar{Y}) - (\mu_x - \mu_y)}{\sqrt{\frac{\sigma^2}{n} + \frac{\sigma^2}{m}}} \sim N(0, 1).$$

Under H_0,

$$Z = \frac{\bar{X} - \bar{Y}}{\sqrt{\frac{\sigma^2}{n} + \frac{\sigma^2}{m}}} \sim N(0, 1),$$

but we don't know σ^2. However, we know that $\bar{X}$ and S_x are independent (from the last chapter), and $\bar{Y}$ and S_y are independent, and so by the independence of the samples we get that the difference $\bar{X} - \bar{Y}$ is independent of both sample variances. Using the definition of the t random variable, we can construct the random variable:

$$T = \frac{\frac{\bar{X} - \bar{Y}}{\sqrt{\frac{\sigma^2}{n} + \frac{\sigma^2}{m}}}}{\sqrt{\left(\frac{(n-1)S_x^2}{\sigma^2} + \frac{(m-1)S_y^2}{\sigma^2}\right) / (m + n - 2)}},$$

which will have a $t(m + n - 2)$ distribution if H_0 is true. The population variances cancel, to give the simplified test statistic

$$T = \frac{\bar{X} - \bar{Y}}{S_p\sqrt{\frac{1}{n} + \frac{1}{m}}},$$

where

$$S_p^2 = \frac{(n - 1)S_x^2 + (m - 1)S_y^2}{m + n - 2}$$

is called the **pooled sample variance**. If $m = n$, this is simply the average of the two sample variances. When the null hypothesis is true, the random variable T has a $t(m + n - 2)$ distribution, and because the expected value of T is zero, the observed values will be "small" in absolute value. Therefore, values that are "large" (either positive or negative) will support the two-sided alternative. We can compute a p-value for our data, using the areas in the tails of the t-density.

Example: A psychologist explores the impact of background music on memory ability. She has 30 student volunteers who will take a memory test. She randomly assigns 15 students to the treatment group; these will take the test in a room with classical music playing in the background, and the others will take the test in a quiet room. Here are the summary data:

	Music	Quiet
Sample mean	80.7	75.1
Sample std dev	9.7	7.6

Let μ_1 be the true average (population) score for students listening to classical music, and let μ_2 be the true average (population) score for students in quiet rooms. Assume that the scores on the memory test are approximately normally distributed for each population, with equal variances. Is there strong evidence that the population means are different?

The hypotheses are $H_0 : \mu_1 = \mu_2$ and $H_a : \mu_1 \neq \mu_2$. Because the sample sizes are the same, the pooled sample variance is simply the average of the sample variances:

$$S_p^2 = \frac{(9.7)^2 + (7.6)^2}{2} = 75.925.$$

We construct the test statistic

$$T_{obs} = \frac{80.7 - 75.1}{\sqrt{75.925 \left(\frac{1}{15} + \frac{1}{15} \right)}} = 1.760.$$

We find the p-value using R,

```
2*(1-pt(1.76,28))
```

to get $p = .0893$. This is the sum of the shaded areas in the plot of the $t(28)$-density, shown on the right.

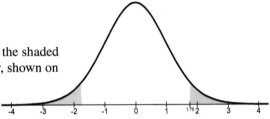

If $\alpha = .05$, we accept H_0. There is not enough evidence to conclude that the background music affects the score on a memory test.

The psychologist is disappointed that the null hypothesis was not rejected and would like to do a power calculation. What if the *population* means were approximately what was found in the samples; what is the power for these sample sizes, and how much would the power increase if the samples were larger? With this information, she can decide if it is worthwhile to redo the study, this time with larger samples.

Power calculations for the independent samples t-test can be accomplished through simulations, where guesses are made for the population parameters. Let's set the expected scores to be 80 and 75 for the scores in the music and quiet rooms, respectively, and compute the power when the common standard deviation of scores is $\sigma = 8$. The following R code will produce the power for the test when $n_1 = n_2 = 15$:

```
mu1=80; mu2=75
sig=8
n1=15;n2=15
nloop=100000
tstat=1:nloop
for(iloop in 1:nloop){
    x=rnorm(n1,mu1,sig)
    y=rnorm(n2,mu2,sig)
    poolvar=((n1-1)*var(x)+(n2-1)*var(y))/(n1+n2-2)
    tstat[iloop]=(mean(x)-mean(y))/sqrt(poolvar*(1/n1+1/n2))
}
(sum(tstat>qt(.975,n1+n2-2))+sum(tstat<qt(.025,n1+n2-2)))/nloop
```

We find that the power is about .38. If we change the code to use $\sigma = 10$ to be more conservative, we get that the power is only about .26. We're not surprised that we could not reject the null hypothesis, as it seems that our original test did not have high power.

Let's double the sample sizes and compute the power again. For $\sigma = 8$, we find the power is about .66, which is not too bad, but for $\sigma = 10$ the power is only about .48. Of course we have no way of knowing for sure what the true standard deviation is; we have only the estimates from the first study. The psychologist can use power calculations from the various guesses, balanced by the higher costs, time, and effort involved with larger samples, to determine a suitable sample size for the second experiment.

The t-statistic is straightforward to calculate, given the summary statistics. An R function saves some steps when starting with the "raw data." Given two vectors containing independent random samples from two populations, the R function `t.test` will do the computations for us and return a p-value. For example,

```
x=c(90, 106, 118, 135, 115, 107,  97, 105)
y=c(116,  81,  88, 100,  79, 102, 105,  74)
t.test(x,y,var.equal=TRUE)
```

will return the results for the two-sided test:

```
Two Sample t-test

data:  x and y
t = 2.2405, df = 14, p-value = 0.04179
alternative hypothesis: true difference in means is not equal to 0
95 percent confidence interval:
  0.6838298 31.3161702
sample estimates:
mean of x mean of y
  109.125    93.125
```

Confidence interval calculation will be covered in Chapter 48. If we use `t.test(x,y)` without the `var.equal=TRUE` option, the equal variances assumption is not made, and a modified t-test is performed that adjusts for nonequal variances.

The F-test for comparing population variances

Next, let's develop a test to compare population variances. We want to test $H_0 : \sigma_x^2 = \sigma_y^2$ versus $H_a : \sigma_x^2 \neq \sigma_y^2$, or possibly a one-sided alternative. A straightforward test uses the sample variances and an F-statistic. Recall from Chapter 34 that if $X_1 \sim \chi^2(\nu_1)$ and $X_2 \sim \chi^2(\nu_2)$ are independent, then

$$F = \frac{X_1/\nu_1}{X_2/\nu_2} \sim F(\nu_1, \nu_2).$$

If two random samples from normal populations are independent, then their sample variances are independent and

$$\frac{(n-1)S_x^2}{\sigma_x^2} \sim \chi^2(n-1) \quad \text{and} \quad \frac{(m-1)S_y^2}{\sigma_y^2} \sim \chi^2(m-1)$$

are independent χ^2 random variables. So,

$$\frac{\frac{(n-1)S_x^2}{\sigma_x^2}/(n-1)}{\frac{(m-1)S_y^2}{\sigma_y^2}/(m-1)} = \frac{S_x^2/\sigma_x^2}{S_y^2/\sigma_y^2} \sim F(n-1, m-1).$$

Under $H_0 : \sigma_x^2 = \sigma_y^2$,

$$F = \frac{S_x^2}{S_y^2} \sim F(n-1, m-1).$$

We can use this to test the equal-variances assumption in the previous example about background music. For those data,

$$F_{obs} = \frac{S_1^2}{S_2^2} = 1.629.$$

It's clear that values greater than 1.629 would support the alternative more, because this would be more evidence that σ_1^2 was bigger than σ_2^2. However, values less than $1/1.629$ would also support the alternative more, because this would be more evidence that σ_2^2 was bigger than σ_1^2.

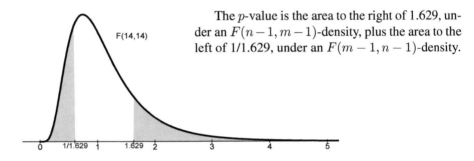

The p-value is the area to the right of 1.629, under an $F(n-1, m-1)$-density, plus the area to the left of $1/1.629$, under an $F(m-1, n-1)$-density.

In Exercise 37.12, you showed that the two areas are the same. Therefore, the p-value can be found as

`2*(1-pf(1.629,14,14)),`

which is .372. We don't have evidence that the population variances are different.

Let's do a power calculation for the variance comparison. If $\sigma_1 = 10$ and $\sigma_2 = 8$, what is the probability that our F-test leads us to reject the null hypothesis of equal variances, when $\alpha = .05$? To find the power for the same sample sizes, we first find the decision rule. The null hypothesis is rejected when the F-statistic is greater than the 97.5th percentile of an $F(14, 14)$-density, or when it is less than the 2.5th percentile. The R command `qf(.975,14,14)` returns 2.979, and `qf(.025,14,14)` returns .336. Then we calculate

$$P\left(F > 2.979\right) = P\left(\frac{S_1^2}{S_2^2} > 2.979\right) = P\left(\frac{S_1^2/100}{S_2^2/64} > 2.979 \times \frac{64}{100}\right),$$

which is the probability that an $F(14, 14)$ random variable is greater than 1.907, which is found using `1-pf(1.907,14,14)`, which returns .120. Similarly,

$$P\left(F < .336\right) = P\left(\frac{S_1^2}{S_2^2} < .336\right) = P\left(\frac{S_1^2/100}{S_2^2/64} < .336 \times \frac{64}{100}\right),$$

which is found using `pf(.215,14,14)`, which gives .003. Then the power is .120+.003, or .123.

The power is typically low for the two-sample test for equal variances, because the sample variances themselves have large variances. If we take samples of size 50 each, with $\sigma_1 = 10$ and $\sigma_2 = 8$, we can go through the same steps: The decision rule is found with `qf(.975,49,49)` and `qf(.025,49,49)`: "reject H_0 when the F-statistic is greater than 1.762 or less than .567." The probability that we see a value of more than 1.762, with the given variance assumptions, is the probability that an $F(49, 49)$

random variable is greater than $1.762 \times 64/100$, or about .338, and the probability that our F-statistic is smaller than .567 (the probability that an $F(49, 49)$ random variable is less than $.567 \times 64/100$) is less than .001, so the power is only about .338. The sample sizes would need to be quite a bit larger to bring the power up to 80% or so.

Checking the normality assumption

For the one-sample problems, we can simply make a histogram or probability plot of the sample observations to assess visually the normality assumption. For two-sample problems, we could make two histograms or two probability plots, but it's better to combine the samples to get an assessment with a larger number of observations. However, if the alternative hypothesis is true, the samples come from different populations. If we could subtract off the population mean for each sample, then we could consider the two samples as from the same population, because of the equal-variances assumption. We don't know the population means, but we can subtract off the sample means, as an approximation. These are called the "centered" observations, and in regression models they are called "residuals."

Let's try this with the data set ex0125, found in the R package Sleuth3. The description is, "The data are the zinc concentrations (in mg/ml) in the blood of rats that received a dietary supplement and rats that did not receive the supplement." We can do a two-independent-samples t-test to see if the zinc levels are different in the two groups. We can check the normality assumption with the centered values using the following code:

```
library(Sleuth3)
zinc=ex0125$Zinc
supp=ex0125$Group
zinc[supp=="A"]=zinc[supp=="A"]-mean(zinc[supp=="A"])
zinc[supp=="B"]=zinc[supp=="B"]-mean(zinc[supp=="B"])
par(mar=c(3,4,1,2))
hist(zinc,main="centered zinc levels")
n=length(zinc)
par(mar=c(4,4,1,1))
plot(qnorm(1:n/(n+1)),sort(zinc),xlab="normal quantiles")
title("probability plot")
```

The code produces the two plots below. The histogram does not look particularly bell-shaped, but with only 39 observations, it's hard to assess normality with a histogram. In the probability plot, the points look roughly linear, so we conclude that we don't see evidence against normality.

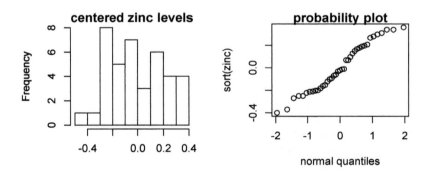

Chapter Highlights

1. If $X_1, \ldots, X_n$ and $Y_1, \ldots, Y_m$ are independent random samples from normal populations with means μ_x and μ_y, respectively, and common population variance σ^2, then under $H_0 : \mu_x = \mu_y$,

$$T = \frac{\bar{X} - \bar{Y}}{\sqrt{S_p^2 \left(\frac{1}{n} + \frac{1}{m}\right)}}$$

has a $t(n + m - 2)$-density, where

$$S_p^2 = \frac{(n-1)S_x^2 + (m-1)S_y^2}{m + n - 2}$$

is the *pooled sample variance*.

2. The two-independent-samples t-test using the pooled sample variance can be accomplished in R using the command `t.test(x,y,var.equal=TRUE)`, which performs a two-sided test. If the option `var.equal=TRUE` is not included, the equal-variances assumption is not made, and an approximate test is performed.

3. If $X_1, \ldots, X_n$ and $Y_1, \ldots, Y_m$ are independent random samples from normal populations with variances σ_x^2 and σ_y^2, respectively, then under $H_0 : \sigma_x^2 = \sigma_y^2$,

$$F = \frac{S_x^2}{S_y^2}$$

has an $F(n - 1, m - 1)$-density, where S_x^2 and S_y^2 are the sample variances.

4. The normality assumption can be assessed using a histogram or probability plot with the $n + m$ centered observations $X_1 - \bar{X}, \ldots, X_n - \bar{X}, Y_1 - \bar{Y}, \ldots, Y_m - \bar{Y}$.

❧❧

Exercises

37.1 A company specializing in determining levels of various pollutants in wetlands has conducted an experiment to determine which of two chelation techniques removes more heavy metals from soil samples. They test the techniques on 16 randomly chosen soil samples and provide the following measurements:

Technique A	26	34	27	33	27	33	29	31
Technique B	22	29	26	28	28	21	22	24

(a) State the assumptions needed for the independent samples t-test to be valid.

(b) Perform the independent samples t-test, report a p-value, and state the conclusion (use $\alpha = .05$) in the context of the problem.

37.2 A study was performed to compare two cholesterol-reducing drugs. Observations of the number of units of cholesterol reduction were recorded for 12 subjects receiving Drug A and 14 subjects receiving Drug B:

	Drug A	Drug B
n	12	14
Mean	5.64	5.03
Standard dev.	1.25	1.82

Researchers are interested in testing $H_0 : \mu_A = \mu_B$ versus $H_a : \mu_A \neq \mu_B$, where μ_A and μ_B are the true expected units of cholesterol reduction for the two drugs.

(a) State the assumptions needed for the independent samples t-test to be valid.

(b) Perform the independent samples t-test, report a p-value, and state the conclusion (use $\alpha = .05$) in the context of the problem.

(c) Test the equal variances assumption, also at $\alpha = .05$.

37.3 Does rye grass grow better (taller) in sandy soil or soil with clay? A botanist plants ten plots of rye grass in sandy soil, and ten in clay soil. After one month, he measures heights of the grass. The summary statistics are shown below:

	Sand	Clay
Sample mean	2.7	3.2
Sample std dev	0.9	0.8

(a) Define parameters and state hypotheses in terms of the parameters.

(b) Perform a two-independent-samples t-test at $\alpha = 0.05$. What is your decision and conclusion in the context of the problem?

(c) State the assumptions for the test in part (b).

37.4 The researchers for the cholesterol drug study of Exercise 37.2 are disappointed that the null hypothesis was accepted. They believe that the drugs really are different, but the sample sizes were too small. Suppose they guess that the difference in population means is really $\mu_A - \mu_B = .6$ units, and the common population variance is $\sigma^2 = 2$. They ask you to compute the power with sample sizes $n_A = 12$ and $n_B = 14$. Notice that you can't do exact calculations by hand. Instead, simulate data sets under the researchers' guesses. For each simulated data set, calculate the t-statistic and determine whether H_0 is rejected at $\alpha = .05$. In a loop, generate 100,000 of these data sets, and report the proportion of the data sets for which H_0 was rejected. This is the power. (You can check your code by simulating data under $H_0 : \mu_A = \mu_B$—the proportion of rejections should be .05.)

The researchers would like to redo the study with a bigger sample size, to have a power of at least .8. If $n_A = n_B$, what is the minimum sample size required?

37.5 The researchers for the rye grass study of Exercise 37.3 are disappointed that the null hypothesis was accepted. They believe that the soils really do result in different heights, but their sample sizes were too small. Suppose they guess that the difference in population mean heights is really .6 units, and the common population variance is $\sigma^2 = 1$. They ask you to compute the power if they had used 16 pots for each type of soil. Compute the power of the test with this larger sample size through simulations.

Now suppose the researchers would like to have a power of at least .8. What should the sample sizes be for the same population means and variances if they use the same number of pots for each type of grass?

37.6 A health researcher is studying the long-term efficacy of radon spa therapy in rheumatoid arthritis. Sixty patients participating in a rehabilitation program including a series of 15 baths were randomly assigned to two groups. One group had an ordinary warm bath, while the other had the radon spa bath treatment. A "functional restriction variable" was measured at baseline for each participant, as well as after treatment, and six months after treatment. The after-treatment measurements showed, on average, decreases in functional restriction, compared to the before-treatment measurements in both groups. A summary of the after- and before-treatment differences in measurements for both groups is below.

	Radon spa group	Ordinary bath group
Sample size	30	30
Mean	18.4	16.5
Std dev	2.90	4.42

(a) Define parameters and state hypotheses in terms of the parameters.

(b) Perform a two-independent-samples t-test at $\alpha = 0.05$. What is your decision and conclusion in the context of the problem?

(c) State the assumptions for the test in part (b).

(d) Test the equal variances assumption at $\alpha = .10$.

37.7 In a longevity study, female mice were randomly assigned to treatment groups to investigate whether restricting dietary intake increases life expectancy. In the table below, we have a randomly selected sample from the complete data set, which can be found in the R package Sleuth2. In the first group, mice ate an unlimited amount of nonpurified, standard diet, and in the second, mice were given a normal diet before weaning and reduced calorie diet (50 kcal/wk) after weaning.

	Lifetimes in months								
Normal diet	35.5	33.8	31.4	30.2	30.0	27.1	24.1	21.5	9.2
Restrict calorie	51.7	48.1	47.2	46.9	46.7	40.9	39.9	38.2	30.9

(a) Read the data (in this table) into R and do a two-independent-samples t-test. Report the p-value and interpret in the context of the problem.

(b) What are the assumptions from part (a)? Make a probability plot to assess the normality assumption.

37.8 Ovenbirds build nests out of clay that are shaped like ovens. Suppose researchers conjecture that ovenbirds near the coasts of Mexico make thicker nests than ovenbirds in central Mexico. After sending graduate students out to find and measure the thickness of nests, they compile the following table:

	Central	Coastal
Sample size	10	12
Sample mean	180 mm	188 mm
Sample std dev	4.4 mm	8.6 mm

Determine if there is evidence for the researchers' conjecture at $\alpha = .05$. State the assumptions you are making in the context of the problem.

37.9 Consider again the data set referenced in Exercise 36.9.

   (a) Determine if there is evidence at $\alpha = .05$ that the average body temperature is different for women and men.

   (b) Do the assumptions for the two-independent-samples t-test seem reasonable? Visually assess the normality assumption using a probability plot.

37.10 Let $X_1, \ldots, X_n$ be a random sample from an $N(\mu_x, \sigma_x^2)$ population, and let $Y_1, \ldots, Y_m$ be a random sample from an $N(\mu_y, \sigma_y^2)$ population, and suppose the samples are independent. Interest is in testing $H_0 : \mu_x = \mu_y$ versus $H_a : \mu_x \neq \mu_y$.

   (a) Derive a test statistic that can be used if σ_x^2 and σ_y^2 are known, and state its distribution if H_0 is true.

   (b) Show through simulations that your test has higher power than the usual independent samples t-test for $n = m = 12$, $\sigma_x^2 = \sigma_y^2 = 4$, and $\mu_x = 10$ and $\mu_y = 9$.

37.11 A device for measuring radiation leakage has a *known* error variance σ^2. That is, if the radiation leakage has a true value μ, then the measurement of the device can be modeled as a random variable with mean μ and variance σ^2. In a physics lab, there are two experimental prototypes for radiation barriers, and interest is in comparing efficacy. For each prototype, 10 independent measurements are taken under identical conditions. The sample mean for prototype A is $\bar{X} = 122.6$ mSv, and the sample mean for prototype B is $\bar{Y} = 124.2$ mSv. If the device has standard deviation $\sigma = 12$ mSv, is there strong evidence that the radiation leakage with prototype A is, on average, different from the radiation leakage for prototype B? (Develop a test statistic for a two-independent-samples test with *known* common variance, and apply it to these data.)

37.12 Let $X_1, \ldots, X_n$ be a random sample from an $N(\mu_x, \sigma_x^2)$ population, and let $Y_1, \ldots, Y_m$ be a random sample from an $N(\mu_y, \sigma_y^2)$ population, and suppose the samples are independent. Interest is in testing $H_0 : \mu_x = \mu_y$ versus $H_a : \mu_x \neq \mu_y$. Suppose it is known that $\sigma_x^2 = 2\sigma_y^2$. Derive a test statistic and state its distribution under the null hypothesis.

37.13 The data set ex0524 in the R library Sleuth2 has observations of levels of zinc (mg/g) in the hair of pregnant women. Some of the women were vegetarians, some were not. Interest is in seeing whether pregnant vegetarians have lower levels of zinc, compared with pregnant nonvegetarians. Suppose you know that the usual assumptions for the two-independent-samples t-test are not met; instead, assume that the variance of zinc levels is twice as high for vegetarians than for nonvegetarians. (Suppose the other assumptions are met.) Perform the test using this assumption, and get a p-value. Interpret the results in the context of the problem.

37.14 Let $X_1, \ldots, X_n$ be a random sample from an $N(\mu_x, \sigma^2)$ population, and let $Y_1, \ldots, Y_m$ be a random sample from an $N(\mu_y, \sigma^2)$ population, and suppose the samples are independent. Interest is in testing $H_0 : \mu_x = \mu_y + 1$ versus $H_0 : \mu_x \neq \mu_y + 1$. Derive a test statistic and state its distribution under H_0.

37.15 A manufacturer is considering Brand A and Brand B switches for his machine. He obtains a (random) sample from each brand and measures their lifetimes, resulting in the data in the table. Brand A switches are cheaper, so he'll only choose Brand B if there is evidence that the average lifetime of Brand B switches is *more than 10 hours greater* than the average lifetime of Brand A switches.

				Lifetimes in hours							Sample mean	Sample variance
Brand A	76	122	111	130	73	94	109	84	82	119	100	430
Brand B	132	109	120	125	122	118	160	118	147	114	127	250

(a) Test the null hypothesis that Brand B switches have lifetimes that are, on average, 10 hours greater than the average lifetime of Brand A switches versus the alternative that the average lifetime of Brand B switches is *more* than 10 hours greater than the average lifetime of Brand A switches. Give the distribution of the test statistic under the null hypothesis, and report your p-value.

(b) What are the assumptions for your test in part (a)?

37.16 If the sample sizes are the same, is the pooled sample standard deviation S_p equal to the average of the sample standard deviations S_1 and S_2?

37.17 Let $X_1, \ldots, X_n$ be a random sample from an $N(0, \sigma_x^2)$ population, and let $Y_1, \ldots, Y_m$ be a random sample from an $N(0, \sigma_y^2)$ population, and suppose the samples are independent.

(a) Derive a test statistic for $H_0 : \sigma_x^2 = \sigma_y^2$ versus $H_a : \sigma_y^2 > \sigma_x^2$, and state its distribution under the null hypothesis.

(b) Find the power of your test when $n = 18$ and $m = 24$, where $\sigma_y = 8$ and $\sigma_x = 4$.

(c) Show through simulations that your answer to (b) is correct.

(d) Show that the power of your test (using zero means) is larger than the standard test for equal variances for which the means are unknown. Use $n = 18$ and $m = 24$, $\sigma_y = 8$, and $\sigma_x = 4$.

Chapter 38

Paired-Samples Test: Normal Populations

A **paired-samples** test to compare population means will sometimes have higher power than an independent-samples test with the same sample sizes. This increase in power happens when some of the variation in the populations can be "controlled for" by pairing. To see what we mean by this, let's start with a motivating example. We'll first do an independent samples "design" for our test, then contrast that with a **paired-samples** design.

Example: Folks at an agricultural experiment station were interested in comparing the yields for two new varieties of corn. Fourteen farmers volunteered for the study; they were randomly assigned to two groups of seven. The first group planted one acre each of variety A corn seeds at the beginning of the season, and the other group planted variety B. The results (in bushels of corn) are listed here, along with the summary statistics:

Variety A	48.2	44.6	49.7	40.5	54.6	47.1	51.4	$\bar{X}_A = 48.01$	$S_A^2 = 21.11$
Variety B	41.5	40.1	44.0	41.2	49.8	41.7	46.8	$\bar{X}_B = 43.50$	$S_B^2 = 12.48$

Using the sample means and variances, we can perform a two-independent-samples t-test and find $p = .066$. If we use $\alpha = .05$, we have to accept H_0, but the results can be considered "borderline." If the sample size was a little bigger, we might have found that the same difference in means was significant. However, the experimenters found it hard to find farmers who are willing to volunteer.

Next, let's consider the paired-samples design, where instead of having fourteen volunteers, we'll use only seven. For this design, each farmer plants *both* varieties of corn, randomly assigned to a different 1-acre plots on his or her farm. For comparison, suppose we get the same data for the new design:

Farm	1	2	3	4	5	6	7
Variety A	48.2	44.6	49.7	40.5	54.6	47.1	51.4
Variety B	41.5	40.1	44.0	41.2	49.8	41.7	46.8

Now the samples are *not* independent, so the assumptions for the independent-samples test of the last chapter are not met. However, we can do a paired-samples test, which is actually easier to set up and perform.

We assume that the *differences* in yields form a random sample from a normal density with mean μ_d and variance σ_d^2, and we test $H_0 : \mu_d = 0$. This is just our one-sample

335

t-test! Under H_0,

$$T = \frac{\bar{D}}{\sqrt{S_d^2/n_d}} \sim t(n_d - 1),$$

where $\bar{D}$ is the sample mean difference in yields, S_d^2 is the sample variance for the differences, and n_d is the number of pairs. For our data we find $\bar{D} = 4.43$ and $S_d^2 = 5.70$, and $n_d = 7$, so the observed t-statistic is 4.91, which gives a (two-sided) $p = .0026$.

Why are the results so dramatically different with the same numbers? The paired-samples test has considerably more power than the independent-samples test if the pairing variable explains some of the variation in the response. For example, suppose there is a lot of variation in yields across farms. Some farms have better soil, in some areas there might be more conducive weather, some farmers might employ better practices, etc. The variation due to farms might "swamp" the variation due to seed variety, so that the sample size for the two-independent-samples test must be very large in order to have a reasonable power. Instead, we control for the variation due to farms by pairing.

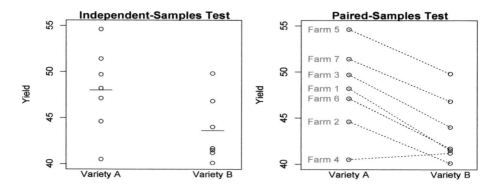

The data for the independent-samples design are shown in the above plot on the left. The mean yields, marked by the horizontal lines, are substantially different, but this difference is not statistically significant (at $\alpha = .05$) because the variation in the samples is large, and the samples are small. In the plot on the right, we see that much of the variation is due to the farms. Farm 5 does well with both varieties, compared to farms 2 and 4, for example, and for 6 of the 7 farms, variety A has higher yield than variety B. The mean of the seven differences is significantly different from zero. We say that the variation due to farm is "controlled for" in the paired-samples design, but not in the independent-samples design.

In summary, the paired-samples design tends to have higher power than the independent samples design if the variable on which the pairing occurs (in the above example this variable was "farm") explains some of the variance in the response ("yield").

Chapter Highlights

1. The paired-sample test is likely to have better power than the corresponding two-independent samples test in the case where there is variation due to the pairing variable.

2. Suppose $D_1, \ldots, D_n$ are differences for the paired observations, and these differences are a random sample from an (approximately) normal distribution. Then under H_0 : the true difference is zero, we have

$$T = \frac{\bar{D}}{\sqrt{S_d^2/n_d}} \sim t(n-1),$$

where $\bar{D}$ is the mean difference, S_d^2 is the sample variance for the differences, and n_d is the number of *pairs*. In short, we use a one-sample t-test, where the sample is from the population of differences.

Exercises

38.1 Two samples were randomly chosen from normally distributed populations 1 and 2 with population means μ_1 and μ_2 and common population variance:

Sample 1	Sample 2
109	106
106	110
91	103
100	115
113	117
86	91
107	108
91	89

(a) Suppose the samples were chosen independently. Test H_0: $\mu_1 = \mu_2$ versus H_a: $\mu_1 \neq \mu_2$ at $\alpha = 0.10$.

(b) Suppose the samples are paired by row. Test the same hypotheses at $\alpha = 0.10$.

38.2 A consumers' group wants to compare the gas mileages of two compact car models to see which is better (more miles per gallon). They hire 6 drivers and have them each drive one car of each model along a fixed route and carefully measure miles per gallon. Here are the data, in miles per gallon:

Driver	1	2	3	4	5	6
Model A	35	34	38	33	39	36
Model B	35	32	34	31	38	33

Perform a test to determine if there is evidence that the population values for the gas mileages of the two models is different, using $\alpha = .10$.

(a) State the hypotheses, your test statistic, and its distribution under H_0.

(b) What are the assumptions for your test in (a)?

38.3 A utility company buries metal pipes in the ground. The pipes must be coated with a corrosion-resistant type of paint. The company wants to compare a new pipe coating (A) with the old one (B) to see if the new one allows less corrosion. They will coat a sample of eight pipes with the new coating, and eight with the old. They have eight boxes of different types of soil, and they bury one pipe with coating A and one pipe with coating B in each box. Then after a period of time, they dig all sixteen pipes up and measure corrosion of each pipe in millimeters. (They choose this design because they think that some soil types are more corrosive than others.) The observed values are shown in the table.

	Soil Type								Sample mean	Sample variance
	1	2	3	4	5	6	7	8		
Coating A	8	7	10	9	13	11	11	11	8.0	3.93
Coating B	11	11	13	12	14	10	17	16	6.625	4.17

(a) Test the null hypothesis that there is no difference in corrosion for the two coatings versus the alternative hypothesis that the new coating results in less corrosion. State the name of the test you perform, give the distribution of the test statistic under the null hypothesis, and report your p-value.

(b) What are the assumptions for your test in part (a)?

38.4 The data set `co.transfer` provided in the R library `boot` has this description:

> Seven smokers with chickenpox had their levels of carbon monoxide transfer measured on entry to a hospital and then again after 1 week. The main question is whether one week of hospitalization has changed the carbon monoxide transfer factor.

(a) Define some parameters and state the hypotheses to be tested.

(b) Explain why a paired-samples test is appropriate, compared to an independent-samples test.

(c) Get a p-value for your test and interpret in the context of the problem.

38.5 A researcher is interested in comparing the effect of an injection of insulin on the blood sugar of men and women. The response variable is the time for the drop in blood sugar after the injection. A two-independent-samples t-test is proposed in which random samples of men and women will be injected, and the average time for men compared with the average time for the women. However, the researchers think that a genetic component is a large source of variation in the time to blood sugar drop. Suggest a paired-samples design that will control for this genetic variation.

38.6 The data set `frets` provided in the R library `boot` has this description:

> The data consist of measurements of the length and breadth of the heads of pairs of adult brothers in 25 randomly sampled families. All measurements are expressed in millimeters.

An older brother makes the claim that older brothers have larger head breadths than younger brothers.

(a) Define some parameters and state the appropriate hypotheses for this claim.

(b) Explain why a paired-samples test is appropriate, compared to an independent-samples test.

(c) Get a p-value for your test and interpret in the context of the problem.

(d) Using your calculated differences, make a probability plot to assess the normality assumption.

38.7 Researchers conjecture that catfish are more likely to absorb and store mercury in their tissues, compared with pike. They go to eight lakes in southern Michigan, and from each lake they sample one catfish and one pike. The concentrations of mercury are measured, and compiled in the following table:

	Lake								Sample mean	Sample variance
	1	2	3	4	5	6	7	8		
Catfish	3	12	8	5	12	7	13	4	8.0	3.93
Pike	1	11	9	3	10	6	11	2	6.625	4.17

Determine if there is evidence for the researchers' conjecture at $\alpha = .05$. What assumptions are you making?

38.8 A national highway safety association is interested in measuring the difference in reaction times for drivers when they have had just one drink compared to when they are completely sober. Reaction times to avoid a simulated crash are measured using a driving simulation machine. There are fifteen volunteers with ages ranging from 21 to 65, with both genders represented. A paired-samples design is chosen. The test is administered twice to each subject, with the testing times a week apart. On one occasion, the subject has had one serving of alcohol administered 30 minutes before the test, and on the other occasion the subject has had no alcohol. A coin is tossed to determine which treatment is administered first.

(a) Define some parameters and state the hypotheses to be tested.

(b) Sketch a design for a study using independent samples.

(c) Explain why a paired-samples design is chosen instead of an independent-samples design.

Here are the reaction times (hundredths of seconds):

drink: 111.3 108.0 113.7 114.6 112.3 111.8 109.8 114.1 111.7
111.9 113.4 113.1 117.1 113.7 119.9

sober: 109.8 112.0 113.7 111.2 110.9 112.0 106.3 113.5 105.9
114.5 111.6 111.8 110.2 108.7 112.4

(d) Suppose the assumptions for a paired-samples t-test are met. Perform the test and report the bounds for the p-value. State the decision at $\alpha = 0.01$ and the conclusion in the context of the problem.

38.9 The data set `darwin` provided in the R library `boot` has this description:

Charles Darwin conducted an experiment to examine the superiority of cross-fertilized plants over self-fertilized plants. 15 pairs of plants were used. Each pair consisted of one cross-fertilized plant and one self-fertilized plant which germinated at the same time and grew in the same pot. The plants were measured at a fixed time after planting and the difference in heights between the cross- and self-fertilized plants are recorded in eighths of an inch.

(a) Define some parameters and state the hypotheses to be tested.

(b) Use a t-test to get a p-value for the test of the null hypothesis that cross- and self-fertilized plants have, on average, the same height versus the two-sided alternative. Interpret your results in the context of the problem.

(c) Discuss the assumptions for this test and whether the test is appropriate in this context.

38.10 The data set anorexia provided in the R library MASS has this description: "The anorexia data frame has 72 rows and 3 columns. Weight change data for young female anorexia patients." One of the variables indicates the treatment: the patients were randomly divided into "Cont" (control), "CBT" (cognitive behavioral treatment), and "FT" (family treatment). The other variables are Prewt (weight of patient before study period, in lbs.) and Postwt (weight of patient after study period, in lbs.).

(a) Suppose we look only at the 29 records for the patients with CBT, and interest is in whether the treatment is effective, that is, do patients weigh, on average, more after the treatment? Estimate the average weight gain for these patients and determine using a t-test if this is significantly different from zero.

(b) Discuss the assumptions for this test and whether the test is appropriate in this context.

Chapter 39

Distribution-Free Test for Percentiles

To apply our one-sample t-test ideas, we assume that the population is normal. Sometimes we don't believe this is true, and we might not have *any* theory about the distribution of the population. The **sign test** is our first example of a "distribution-free" test and is analogous to the one-sample t-test, in that the null hypothesis specifies a value for the "center" of the distribution. For the sign test, the parameter of interest is the population *median*, or 50th percentile, rather than the mean.

The sample mean can be thought of as a balance point if there are weights at each observation on the number line. The median can also be thought of in terms of weight, but there is no "leverage." It's simply the point that divides the weight of the points in two, "middle" point of the sorted sample if there is an odd number of points, or the average of the two middle points if there is an even number of points. If we have a set of numbers, say $\{12, 19, 26, 10, 20, 21\}$, we can plot these on a number line:

The mean, marked with an arrow, is the center of gravity or balance point. The total weight to the right of the median has to be equal to the total weight to the left. (Any value between 19 and 20 can technically be the median, but it is conventional to choose the center value.) The mean is 18 and the median is 19.5.

The median is a more "stable" or "robust" measurement of center, because if, for example, the point at 10 moved to the left, the mean would move as well, but the median would stay put. We say the median is "robust to outliers."

For the sign test, the null hypothesis proposes a median value for the distribution. Like the one-sample t-test, it can be one-sided or two-sided. Under the null hypothesis, the number of observations in the sample that are greater than (alternatively, less than) the median would be approximately 1/2. The p-value is calculated using the binomial probabilities developed in Chapter 8.

Example: Suppose we have a random sample of bass from North Lake. The weights of the fish are 1.2, 4.3, 2.4, 1.4, 0.8, 1.9, 1.5, and 1.1 pounds. Suppose we want to test

$$H_0: m = 2 \text{ pounds} \quad \text{versus} \quad H_a: m < 2 \text{ pounds},$$

where m is the *median* weight of bass in the lake. Because the median is the 50th percentile, under the null hypothesis we would expect about half of the bass in the sample to weigh less than two pounds, and half to weigh more. Sampling from the population can be thought of as tossing a coin, with heads equivalent to "more than two pounds," and tails meaning "less than two pounds." In fact, if H_0 is true, the number of values above 2.0 is a binomial random variable with probability $p = .5$ and $n = 8$.

We compute the p-value as our data (six of eight bass less than two pounds) or more extreme (seven or eight of eight bass less than two pounds), under H_0 (median is two pounds). Remember that "more extreme" means "supports the alternative more." The p-value is the binomial probability of at least six out of eight, when $p = 0.5$. This is

$$\binom{8}{6}(.5)^6(.5)^2 + \binom{8}{7}(.5)^7(.5)^1 + \binom{8}{8}(.5)^8(.5)^0 = 0.1445.$$

There is not enough evidence to reject the null hypothesis at $\alpha = 0.10$. The interpretation of the p-value in terms of repeated samples is: If the median bass weight is two pounds, and we draw many random samples of size eight from the lake, about 14.45% of these samples will contain at least six fish less than two pounds.

Example: Twelve measurements of soil contamination in parts per million of a certain chemical were taken at random coordinates within five miles of a production plant:

6.2,   4.1,   3.5,   5.1,   5.0,   3.6,   4.8,   4.1,   3.6,   4.7,   4.3,   4.2.

If the soil contamination in the region is above 4.0, the plant must take steps to clean the soil and curtail pollution in the future. We want to test

$$H_0: \text{median soil contamination is 4.0 ppm}$$

versus

$$H_a: \text{median soil contamination is more than 4.0 ppm}$$

at $\alpha = .05$. (The production plant operators would like to see a smaller α, and the local grassroots environmental group would like to see a larger α.)

The p-value is the probability of our data (nine of twelve above 4.0) or more extreme (more than nine out of twelve), under the null hypothesis (each of the twelve measurements has probability 0.5 of being more than 4.0). We can use the R command `1-pbinom(8,12,.5)` to get $p = .073$. We accept H_0 at $\alpha = .05$, but we would reject at $\alpha = .10$.

The distribution-free tests typically have lower power than the tests with distributional assumptions, especially when the assumptions are close to true. For example, if we assume that the measurements are from a normal distribution, the one-sample t-test provides a p-value of about .04. Exercise 39.5 asks you to explore the power of the sign test, compared to the t-test, using simulated data. You will find that when the data have "outliers," which is often the case when the true population distribution is "heavy-tailed," the sign test can have better power.

Although the t-test can have nominally better power even when the assumptions are not met, the distribution-free test might still be preferable, because the "true" test size for the t-test might be different from the "target" test size. To demonstrate, suppose we want to test $H_0 : \mu = 4$, and our population has mean $\mu = 4$, but the distribution is skewed like a gamma, shown in the top plot below. If we sample repeatedly from this skewed distribution and calculate the t-statistic, as in the code below, we find that the statistic does not follow a $t(n-1)$-density, as it would if the population were normal. The $t(n-1)$-density is drawn over the histogram of the observed t-statistic; we can see that the values tend to be lower when sampling from this skewed distribution. The proportion of times the null is rejected at $\alpha = .05$ is about .075. Even worse, when $\alpha = .01$, the proportion of times the null is rejected is almost .03. The test size is about 50% larger than it is believed to be when $\alpha = .05$, and almost three times as large when $\alpha = .01$, so the resulting p-values will, on average, be deceptively small.

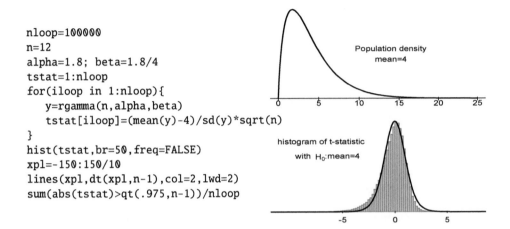

```
nloop=100000
n=12
alpha=1.8; beta=1.8/4
tstat=1:nloop
for(iloop in 1:nloop){
    y=rgamma(n,alpha,beta)
    tstat[iloop]=(mean(y)-4)/sd(y)*sqrt(n)
}
hist(tstat,br=50,freq=FALSE)
xpl=-150:150/10
lines(xpl,dt(xpl,n-1),col=2,lwd=2)
sum(abs(tstat)>qt(.975,n-1))/nloop
```

When the true test size is larger than the target test size, as in the above example, we say that the test size is "inflated" and the test is "anticonservative." This situation is to be avoided because, when we reject H_0, we can't be confident of the amount of evidence on which this decision is based. Doing a distribution-free test, without any distributional assumptions for the population, will guard against anticonservative tests, but at the price of having lower power.

Let's think about doing a power calculation for this sign test. We'd like to determine the probability of rejecting the null hypothesis for some version of the alternative. In this example, the alternative is true when the true median is larger than 4.0 ppm. Instead of trying to compute power when the true median is specified to be some value, we instead identify the alternative scenario by specifying the percentile for the null hypothesis median. We can calculate the probability of rejecting H_0 when the 4.0 ppm is actually the 30th percentile of the distribution of soil contamination. To compute the power for a sample of size 12, we first need to specify a test size and compute a decision rule. Suppose $\alpha = .05$. Recall that we reject H_0 when ten or more out of twelve measurements have soil measurements greater than 4.0 ppm. The power is the probability of rejecting H_0 when 4.0 ppm is the 30th percentile. This is `1-pbinom(9,12,.7)`, or .253.

If we boost the sample size to 24, we can find a new decision rule for $\alpha = .05$: we reject H_0 when 17 or more soil measurements have greater than 4.0 ppm. Then the power is `1-pbinom(16,24,.7)`, or .565.

The sign test concepts can be extended easily to test for other percentiles.

Example: Scientists at East State College are concerned with PCB levels in a local lake. They sample 24 channel catfish, which are bottom feeders, and measure the level of PCBs in micrograms per gram of fish. Here is the sorted sample, which we assume is a random sample of catfish in the lake:

0.00, 0.00, 0.00, 0.00, 0.00, 0.00, 0.01, 0.02, 0.02, 0.14, 0.19, 0.27,

0.31, 0.38, 0.40, 0.59, 0.60, 0.61, 0.76, 0.99, 1.53, 4.04, 5.85, 8.41.

The distributions of pollutants in wildlife are typically skewed, as seen in this sample. Suppose the scientists are concerned with making an inference about the 75th percentile of the distribution of PCB contamination. They want to know, is there evidence at $\alpha = .05$ that the 75th percentile exceeds .5?

Under the null hypothesis that the 75th percentile equals .5, the number of fish in the sample with less than or equal to .5 micrograms of PCBs would follow a binomial distribution, with $n = 24$ and $p = .75$. Our observed value is 15, and smaller values support the alternative more. The p-value for the test is $P(Y \leq 15)$, when $Y \sim \text{Binom}(24, .75)$. The R command pbinom(15,24,.75) returns the p-value .12, so the evidence is not strong that the 75th percentile is above .5.

Chapter Highlights

1. The sign test is used for $H_0 : m = m_0$, where m is the population median. Alternatives can be one- or two-sided. If we have a random sample $Y_1, \ldots, Y_n$ from a population with unknown distribution and median m_0, then the number of sample units that are less than m_0 follows a $\text{Binom}(n, 1/2)$ distribution.

2. The sign test can be extended to tests involving other percentiles. If we have a random sample $Y_1, \ldots, Y_n$ from a population with unknown distribution and qth quantile q_0, then the number of sample units that are less than q_0 follows a $\text{Binom}(n, q)$ distribution.

☙❧

Exercises

39.1 A banker claims that balances for checking accounts tend to be higher now than they were a few years ago. She wants to test

H_0: median balance is $800,

H_a: median balance is more than $800

at $\alpha = 0.05$. A random selection of 12 account balances shows

968, 1220, 1571, 272, 388, 812, 2625, 2183, 992, 1394, 5344, 521.

(a) Do a sign test and report the conclusion in the context of the problem.

(b) Suppose the alternative is true, and the 30th percentile of distributions is $800. What is the power of the test?

39.2 To determine if the economic health in Smalltown has improved over the three years since receiving stimulus funds, the change in net worth was measured for a random sample of residents. A negative measure means that the net worth has decreased. The measurements (in $1000) are

$$-7.1, \quad 240.6, \quad 57.0, \quad 36.8, \quad 2.3, \quad 42.2,$$
$$126.8, \quad 11.4, \quad 184.2, \quad -35.1, \quad -24.2, \quad -26.3.$$

(a) Do a sign test and report the conclusion in the context of the problem.

(b) Do a one-sample t-test of $H_0 : \mu = 0$ versus $H_a : \mu > 0$, where μ is the mean change in net worth of Smalltown residents.

(c) Get a histogram of the data and interpret what you see in the context of the problem. Do you think the answer in (a) or the answer in (b) is more accurate?

39.3 For the data in Exercise 36.9, test the null hypothesis that the median temperature is 98.6°F. Do you think the t-test or the sign test is more appropriate?

39.4 Refer to Exercise 38.9. Use a sign test for the null hypothesis that cross- and self-fertilized plants have the same median height versus the one-sided alternative that the cross-fertilized plants tend to be taller. Compare your p-value to the one-sided p-value from the paired-samples t-test.

39.5 Compare the sign test to the one-sample t-test using simulations in R.

(a) Generate samples of size $n = 13$ from a normal distribution with mean 1 and standard deviation 2, and test the null hypothesis that the center of the distribution is zero. For the one-sample t-test, use $H_0 : \mu = 0$ and $H_a : \mu > 0$, and for the sign test, use $H_0 : m = 0$ and $H_a : m > 0$, where μ is the mean and m is the median. Which test has higher power?

(b) Generate samples of size $n = 13$ with "heavy tailed" distribution with center at one. Specifically, suppose the observations are from a $t(3)$ distribution, with the value 1 added. Test the null hypothesis that the center of the distribution is zero. For the one-sample t-test, use $H_0 : \mu = 0$ and $H_a : \mu > 0$, and for the sign test, use $H_0 : m = 0$ and $H_a : m > 0$, where μ is the mean and m is the median. Which test has higher power? What is the true test size for the t-test?

39.6 The R data set `trees` contains "measurements of the girth, height, and volume of timber in 31 felled black cherry trees." Test the null hypothesis that the 95th percentile to black cherry tree volumes is 50 cubic feet, against the alternative that the 95th percentile is larger than 50 cubic feet.

39.7 Household incomes in Smalltown are known to have a median values of $120,000. We want to test whether the median income in the Belleview neighborhood is higher, using a sample of 12 incomes (in thousands of dollars):

$$173, \quad 122, \quad 115, \quad 119, \quad 98, \quad 92, \quad 132, \quad 198, \quad 193, \quad 225, \quad 138, \quad 200.$$

(a) Perform a sign test and report a p-value and a conclusion at $\alpha = .05$.

(b) Suppose the true distribution is roughly Gamma$(15, .1)$. What is the power of the test?

39.8 Teachers in elementary schools in a city where high levels of lead have been found in the water are concerned that there are lasting effects for children's education. Historically, the 25th percentile in a third grade reading test is 119.5 points. They take a random sample of 20 third grade children; the sorted scores are below:

$$92, \quad 99, \quad 100, \quad 104, \quad 116, \quad 117, \quad 117, \quad 118, \quad 120, \quad 121,$$

$$121, \quad 122, \quad 126, \quad 126, \quad 129, \quad 132, \quad 133, \quad 133, \quad 141, \quad 170.$$

(a) Test the null hypothesis that the 25th percentile score is 119.5 points versus the alternative hypothesis that the 25th percentile score is *lower*. Report a p-value and a conclusion at $\alpha = .10$.

(b) Determine the power of the test if the true distribution of scores is Gamma$(50, .4)$.

Chapter 40

Distribution-Free Two-Sample Tests

We may use the independent and paired sample t-tests of Chapters 37 and 38, under the assumption of normality of the population distributions. When the normality assumption is not met, **distribution-free** tests, which are also called **nonparametric** tests, can be used. The **Wilcoxon tests** are popular examples of distribution-free tests. The **rank-sum test** is used for independent samples, and the **signed rank test** is for paired samples. We'll develop the ideas behind the tests through examples.

Example: A public health group is concerned about workers' exposure to lead in certain procedures in a factory. They take a random sample of eight workers from the factory, and a random sample of eight workers from a factory without the exposure concern. They measure the blood lead levels of all the sampled workers; these measurements are shown in the table.

	Blood lead levels for factory data							
Factory	169.9	161.6	139.0	144.8	152.0	135.0	160.4	155.1
Other	127.8	137.4	156.9	147.0	133.2	145.9	140.8	130.9

We want to test

H_0: blood lead level distribution is the same for the factory workers as for the general population of workers in the industry

versus

H_a: blood lead levels in the factory are on average *higher* than in the general population of workers in the industry.

If we do a two-independent-samples t-test on the population average blood levels, we get $t = 2.25$ and a p-value close to 0.021. But suppose there is evidence that blood lead levels have skewed distributions in the populations, so we cannot assume normality.

The rank-sum test is the distribution-free version of the two-independent-samples test. To motivate this, let's plot the data on one axis. The factory workers are indicated with red dots and the other industry workers with blue dots, as shown in the plot below.

If the null hypothesis were true, we would, for a "typical" data set, see a jumble of red and blue dots, neither group appearing to cluster on one end of the range. If the alternative were true, we would see more red dots on the right side of the range. Let's rank the points by numbering from smallest to largest, disregarding which group the point belongs to, as shown above the circles shown in the plot. Now, sum the ranks of both groups of points.

$$\text{Sum of ranks of factory data} = 4 + 6 + 8 + 11 + 12 + 14 + 15 + 16 = 86$$

and

$$\text{sum of ranks of "other" data} = 1 + 2 + 3 + 5 + 7 + 9 + 10 + 13 = 50.$$

If the null hypothesis were true, we would expect the sums of the rankings to be about the same; if the alternative were true, we would expect the sum of the rankings of the factory group to be bigger. Because the latter happened, this provides some evidence for the alternative hypothesis. To assess the strength of the evidence, we need a p-value, which means we need a distribution for the test statistic (the rank-sum for the factory data) under the null hypothesis.

To get this distribution, we note that under H_0, *any* ordering of the ranks is equally likely. Under H_0, the smallest rank-sum, which is $1 + 2 + 3 + 4 + 5 + 6 + 7 + 8 = 36$, is just as likely as the largest rank-sum, which is $9 + 10 + 11 + 12 + 13 + 14 + 15 + 16 = 100$. These are equally probable with any other, such as $3 + 4 + 6 + 8 + 11 + 12 + 14 + 15 = 73$. The sample space of equally probable events (when H_0 is true) is the listing of all possible rankings.

The p-value is the probability of our data (rank-sum $= 86$), or more extreme (a rank-sum greater than 86 would support the alternative hypothesis more), under the null hypothesis. To get this, we count up all of the possible rankings for which the rank-sum is at least 86, and divide this by the total number of possible rankings. We can enumerate the orderings for very small sample sizes, but this method gets time consuming very quickly. Instead, people have compiled tables for finding p-values, given the sample sizes and the rank-sum. Rather than using tables, we can implement the test in R.

The `wilcox.test` function is structured in the same way as the `t.test` function.

To find the p-value for our test, we input the data and use the function, as shown in the commands after the R prompt ">"; then the results are shown:

```
> y1=c(169.9, 161.6, 139.0, 144.8, 152.0, 135.0, 160.4, 155.1)
> y2=c(127.8, 137.4, 156.9, 147.0, 133.2, 145.9, 140.8, 130.9)

> wilcox.test(y1,y2,alternative="greater")

Wilcoxon rank sum test

data:  y1 and y2
W = 50, p-value = 0.03248
alternative hypothesis: true location shift is greater than 0
```

We see that the one-sided p-value is about .032, so we may reject H_0 at $\alpha = 0.05$. There *is* evidence that the factory workers have higher blood lead levels than the other industry workers.

For the Wilcoxon rank-sum test, we assume that we have independent random samples from the populations. The form of the population densities is unknown, and the null hypothesis is that the two population densities are identical, that is, that the two samples are really from the same population.

Another assumption typically stated for the rank-sum test is this: We assume that the densities are identical, except that one is (perhaps) shifted to the right or left of the other. An example pair of densities where this assumption is true is shown on the left below. However, the test will "work" if the pair is that shown on the right. Because the test size and p-value are computed using the null hypothesis, the assumptions about the population under the alternative are less important. This "shift" assumption guarantees that the test is *unbiased*, which is a nice property of tests to be defined in Chapter 57.

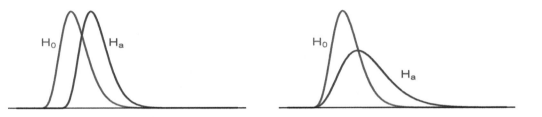

Given this "shift" assumption, the hypotheses could be stated in terms of the means, although it is more common to see the hypotheses in terms of the medians. That is, the null hypothesis for the Wilcoxon rank-sum test is typically $H_0 : m_1 = m_2$, where m_1 and m_2 are the population medians. The two-sided alternative is simply $H_a : m_1 \neq m_2$.

Example: Interest is in comparing gas mileages of two types of compact cars. Twelve drivers are randomly assigned to cars; they each drive the assigned car through a route, and the gas mileages are recorded in the table below.

Model A	35	34	38	33	39	37
Model B	35	32	34	31	38	33

Suppose we have reason to believe that the gas mileage populations are *not* normally distributed, but we are reasonably sure that the shapes of the density functions are the same. We wish to know whether one of the densities is shifted to the right of the other. This is a two-sided hypothesis test. We sort and rank the data:

Mileage	31	32	33	33	34	34	35	35	37	38	38	39
Rank	1	2	3.5	3.5	5.5	5.5	7.5	7.5	9	10.5	10.5	12
Model	B	B	A	B	A	B	A	B	A	A	B	A

Note that for these data, there are duplicate numbers, so that some of the ranks are "ties." Traditionally, the ranks are averaged for the ties, as shown. However, the presence of ties produces a warning in the `wilcox.test` results, because the test is no longer "exact." Theoretically, ties are impossible if the random variables are continuous, but of course in the "real world" there is always rounding.

```
> y1=c(35, 34, 38, 33, 39, 37)
> y2=c(35, 32, 34, 31, 38, 33)
> wilcox.test(y1,y2)

Wilcoxon rank sum test with continuity correction

data:  y1 and y2
W = 27, p-value = 0.1705
alternative hypothesis: true location shift is not equal to 0

Warning message:
In wilcox.test.default(y1, y2) : cannot compute exact p-value with ties
```

There is also a distribution-free version of the paired-samples t-test, called the
Wilcoxon signed rank test. Recall that for small sample sizes, we required the population of differences to be normally distributed in order to do a paired-samples t-test. For
the signed rank test, we require only that the population of differences is symmetric.
If the population of differences is *not* symmetric, we can resort to our sign test of the
previous chapter.

Recall that if a density is symmetric, its mean is the same as its median. The null
hypothesis is that the median (or mean) of the difference distribution is zero, i.e., the
density of differences is symmetric around zero. The alternative may be one- or two-
sided.

To conduct the test, we use paired samples from the two populations, and compute
the differences, just as for the paired-sample t-test. In the table below, we see data from
the crop yield example from Chapter 38.

	Crop yield (bushels/acre)						
Farm	1	2	3	4	5	6	7
Variety A	48.2	44.6	49.7	40.5	54.6	47.1	51.4
Variety B	41.5	40.1	44.0	41.2	49.8	41.7	46.8
Difference	6.7	4.5	5.7	-0.7	4.8	5.4	4.6
Abs. val. of diff.	6.7	4.5	5.7	0.7	4.8	5.4	4.6
Rank	7	2	6	1	4	5	3
Signed rank	7	2	6	-1	4	5	3

We next compute the absolute values of the differences, rank them, and then restore
the signs to the ranks. These steps are shown in the last three rows of the above table.

If the null hypothesis is true, the sum of the positive ranks ought to be, on average,
about the same magnitude as the sum of the negative ranks. If, on the other hand, the
positive differences have systematically larger ranks, this is evidence that the average of
the population differences is positive or (in this example) that the mean of Variety A is
larger.

For such a small sample, we can compute a p-value without using R or tables. Suppose our alternative hypothesis states that the difference in population averages is positive. We see that there is evidence that this alternative is true, since there all but one of
the sample differences are positive, and only the negative difference has low rank.

The p-value is the probability that we have this much evidence for H_a, or more
evidence, when H_0 is the true hypothesis. If H_0 is true, then each of the differences
is positive or negative with probability 0.5. The distribution of the signs is like the
distribution of a series of seven coin tosses. There are $2^7 = 128$ possible sequences of

heads and tails, or positive and negative signs. In other words, if H_0 is true, the sequence of all positive signs has probability $1/128$, and the sequence $+, -, +, -, +, -, +$ also has probability $1/128$. There are 128 of these possible sequences, and they are all equally likely under H_0.

For the data in our small example above, the sum of the positive ranks is 27. Any other sequence of signs with sum of positive ranks *larger than* 27 will support the first alternative hypothesis more, and another sequence with sum equal to 27 will support H_a just as much as our data. However, it is easy to see that there is exactly one possible sequence of positive and negative ranks that provides more evidence for our alternative: if all the differences were positive, the rank sum would be 28.

The p-value is the probability of our result or more extreme under H_0. Because there are two sequences (out of 128 possible sequences) that support H_a at least as much as our data, the (one-sided) p-value is $2/128 \approx 0.016$. The two-sided p-value is twice this, but we still reject H_0 at $\alpha = .05$.

Of course, with larger sample sizes, it is tedious to write out sequences like this, so we use tables or R. The same function is used as for the independent-samples test, but with a `paired=TRUE` option.

```
> y1=c(48.2, 44.6, 49.7, 40.5 ,54.6 ,47.1 ,51.4)
> y2=c(41.5, 40.1, 44.0, 41.2, 49.8, 41.7, 46.8)
> wilcox.test(y1,y2,alternative="greater",paired=TRUE)

Wilcoxon signed rank test

data:  y1 and y2
V = 27, p-value = 0.01563
alternative hypothesis: true location shift is greater than 0
```

Example: A scientist wishes to measure the reaction times of monkeys to a stimulus, when the monkey is calm, compared to when the monkey is distracted. There are twelve monkeys available, and a paired design is used. The reaction time for each monkey is measured under both conditions, with randomized order. The null hypothesis is that the reaction times are the same under both conditions, and the alternative is that the reaction is slower when the monkey is distracted.

	1	2	3	4	5	6	7	8	9	10	11	12
						Monkey						
Distr	2.4	4.3	2.0	4.8	3.4	2.8	2.9	2.6	1.2	3.6	4.1	3.2
Calm	2.2	3.8	2.2	4.4	3.5	2.8	2.6	2.2	1.2	3.3	3.8	2.9
Diff	0.2	0.5	-0.2	0.4	-0.1	0.0	0.3	0.4	0.0	0.3	0.3	0.3

As we can see, there are zero differences, as well as ties in the difference row. If a difference is actually zero, we throw out the observation, so that we really have $n =$ the number of nonzero differences. Using this rule for ties, we have only $n = 10$ readings. Further, there are duplicate nonzero differences, so we have to average the ranks:

	1	2	3	4	5	6	7	8	9	10	11	12
						Monkey						
Diff	0.2	0.5	-0.2	0.4	-0.1	0.0	0.3	0.4	0.0	0.3	0.3	0.3
Abs diff	0.2	0.5	0.2	0.4	0.1	0.0	0.3	0.4	0.0	0.3	0.3	0.3
Rank	2.5	10	2.5	8.5	1	X	5.5	8.5	X	5.5	5.5	5.5
Signed rank	2.5	10	-2.5	8.5	-1	X	5.5	8.5	X	5.5	5.5	5.5

The R function `wilcox.test` makes two complaints: one for the zero differences and one for the tied differences.

```
> y1=c(2.4 , 4.3 , 2.0 , 4.8 , 3.4 ,2.8 , 2.9 , 2.6 , 1.2 , 3.6 , 4.1 , 3.2)
> y2=c(2.2 , 3.8 , 2.2 , 4.4 , 3.5 , 2.8 , 2.6 , 2.2 , 1.2 , 3.3 , 3.8 , 2.9)
> wilcox.test(y1,y2,paired=TRUE)

Wilcoxon signed rank test with continuity correction

data:  y1 and y2
V = 51, p-value = 0.0189
alternative hypothesis: true location shift is not equal to 0

Warning messages:
1: In wilcox.test.default(y1, y2, paired = TRUE) :
  cannot compute exact p-value with ties
2: In wilcox.test.default(y1, y2, paired = TRUE) :
  cannot compute exact p-value with zeroes
```

Chapter Highlights

1. The rank-sum test is used for the two-independent-samples problem. The null hypothesis is that the distributions of two populations are the same versus the alternative that one is shifted so that the mean of the density is higher. No assumptions are made as to the formula for the density of the populations. To apply the test, the observations from both samples are pooled and ranked. The null is rejected if the rank-sum for one sample is significantly higher than the other.

2. The signed rank test is used for the paired-samples problem. The differences are ranked, and "signed" with $+$ if the observation from the first population is larger, and $-$ otherwise. Then the ranks for the $+$ pairs are summed. The null hypothesis that the population of differences is symmetric around zero is rejected if the summed ranks of the $+$ pairs are "too large" or "too small."

❧❧

Exercises

40.1 In a longevity study, female mice were randomly assigned to treatment groups to investigate whether restricting dietary intake increases life expectancy. In the table below, we have a randomly selected sample from the complete data set, which can be found in the R package `Sleuth2`. In the first group, mice ate an unlimited amount of nonpurified, standard diet, and in the second, mice were given a normal diet before weaning and reduced calorie diet (50 kcal/wk) after weaning.

	Lifetimes in months								
Normal diet	35.5	33.8	31.4	30.2	30.0	27.1	24.1	21.5	9.2
Restrict calorie	51.7	48.1	47.2	46.9	46.7	40.9	39.9	38.2	30.9

(a) Read the data into R and do a two-independent-samples t-test for the null hypothesis that the expected lifetime is the same for both groups, against a two-sided alternative. Report the p-value and interpret in the context of the problem.

(b) What are the assumptions from part (a)? Get boxplots or histograms or probability plots of the samples to assess the normality assumption.

(c) Do a Wilcoxon rank-sum test of the null hypothesis that the expected lifetime is the same for both groups, against a two-sided alternative. Report the p-value and interpret in the context of the problem.

40.2 A utility company buries metal pipes in the ground. The pipes must be coated with a corrosion-resistant type of paint. The company wants to compare a new pipe coating (A) with the old coating (B) to see if the new one allows *less* corrosion. They will coat a sample of eight pipes with the new coating, and eight with the old coating, bury the pipes in soil under controlled conditions, then after a period of time dig all sixteen up and measure corrosion of each pipe in millimeters. Let μ_A be the average corrosion in this length of time for coating A, and let μ_B be that for coating B.

Because the type of soil is known to be a source of variation (perhaps more acidic soils cause more corrosion), we could control for type of soil by using the same type for all pipes in the sample. However, since the pipes are to be used in a variety of soil types, we would like to be able to compare coatings for more than one type. We choose a paired-samples test instead *to control for the soil-type variation.*

Suppose there are eight soil types. One of the coating A pipes and one of the coating B pipes are randomly chosen for each soil type. At the end of the experiment, we have eight *differences* in corrosion that we use for the analysis. The hypotheses to be tested are

H_0: corrosion density is the same for coatings A and B (so that the differences density is symmetric about zero)

versus

H_a: coating A tends to allow *less* corrosion than coating B.

We have data:

	\multicolumn{8}{c}{Soil type}
	1	2	3	4	5	6	7	8
Coating A	0.8	0.7	1.0	0.9	1.4	1.1	0.6	1.2
Coating B	1.1	1.1	1.3	1.2	1.4	1.0	1.1	1.1

Compute the test statistic and test the hypotheses at $\alpha = 0.05$.

40.3 Refer to Exercise 38.8. Suppose we cannot assume that the differences in reaction time are normally distributed. Perform the appropriate distribution-free test, and state the decision at $\alpha = 0.01$ and the conclusion in the context of the problem.

40.4 An experiment was conducted to compare the weights of young pigs who had been treated with growth hormones to those untreated. A sample of 14 piglets was randomly divided into two groups. Seven were injected with the hormone and the remainder were left untreated. After a 6-month period, the pigs were weighed:

Hormone treated pigs	132	112	115	106	113	139	118
Untreated pigs	110	113	88	111	92	126	99

Suppose that it is known that the distribution of pig weights is not normal. Test the null hypothesis that there is no difference in the mean weight between treated and untreated pigs versus the alternative hypothesis that there is a higher average weight among the treated pigs. Use $\alpha = 0.05$.

40.5 Refer to Exercise 38.10. If you make a histogram of the differences you might find there is evidence that the distribution is not normal. Do a signed rank test, and compare the results with the results from the previous t-test.

40.6 The data set `cd4` in the R data library `boot` has this description:

> CD4 cells are carried in the blood as part of the human immune system. One of the effects of the HIV virus is that these cells die. The count of CD4 cells is used in determining the onset of full-blown AIDS in a patient. In this study of the effectiveness of a new antiviral drug on HIV, 20 HIV-positive patients had their CD4 counts recorded and then were put on a course of treatment with this drug. After using the drug for one year, their CD4 counts were again recorded. The aim of the experiment was to show that patients taking the drug had increased CD4 counts which is not generally seen in HIV-positive patients.

Do a distribution-free paired-samples test to determine if the patients taking the drug had significantly increased CD4 counts, using $\alpha = .01$.

40.7 The data set `ex0112` in the R data library `Sleuth2` has this description:

> Researchers used 7 red and 7 black playing cards to randomly assign 14 volunteer males with high blood pressure to one of two diets for four weeks: a fish oil diet and a standard oil diet. These data are the reductions in diastolic blood pressure.

Do a distribution-free test to determine if there is strong evidence that fish oil is better at reducing diastolic blood pressure, compared to the "standard" oil, using $\alpha = .01$.

Chapter 41

The Central Limit Theorem

We already think the normal random variable is amazing. Its properties are remarkable: The distribution of the mean of a sample from a normal population is again normal, and the mean of a random sample is independent of the sample variance, allowing us to construct those nice T- and F-statistics for inference. The **central limit theorem**, however, tops all this. It says that if you make a Z statistic using a random sample from *any* distribution (that has finite variance), this Z statistic approximates a standard normal random variable, and the approximation gets better as the sample size increases.

Before we prove this, let's demonstrate it using samples from an exponential population. If $X_1, X_2, \ldots, X_n$ are a random sample from $\text{Exp}(1)$, then we know the mean and variance of the population are both 1, and we construct the Z statistic as

$$Z_n = \frac{\bar{X} - \mu}{\sigma/\sqrt{n}} = \sqrt{n}(\bar{X} - 1) = \sqrt{n}\left(\frac{S}{n} - 1\right) = \frac{S}{\sqrt{n}} - \sqrt{n},$$

where $S = X_1 + \cdots + X_n$. We write Z_n in terms of S, because we know that $S \sim \text{Gamma}(n, 1)$, and Z_n is just a linear transformation of S. Therefore, using our CDF method from Chapter 21, we can write the *true* density for Z_n as

$$f_Z(z) = \sqrt{n} f_S(\sqrt{n}(z + \sqrt{n})),$$

where f_S is the density for a $\text{Gamma}(n, 1)$ random variable.

The expression for the f_S-density does not *look* similar to a standard normal density, but we can make some sketches for various values of n and compare them to the standard normal. In the plots below, we see the population density on the far left. In the next four plots, we see the density for the Z_n-statistic when $n = 2$, 4, 12, and 25. In each plot, the dotted curve is the density for the standard normal. When n is only 25, the densities are very close to each other.

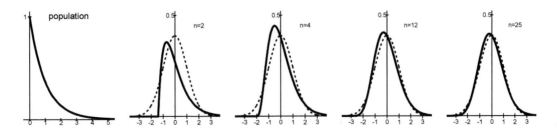

355

The central limit theorem is not difficult to prove, for distributions with a finite *third* moment, and whose moment generating function exists. We'll start with a simpler lemma that we will use to prove the more general case.

Lemma: Suppose $X_1, \ldots, X_n$ is a random sample from a population with mean zero and variance one, with moment generating function (MGF) $M_X(t)$ defined in an open interval containing zero. Then the distribution of

$$Z_n = \sqrt{n}\bar{X} = \frac{1}{\sqrt{n}}(X_1 + \cdots + X_n)$$

approaches that of an $N(0, 1)$ random variable.

To prove this lemma, we use the MGF ideas. In particular, we show that the MGF for Z_n approaches the MGF for the standard normal random variable, as n gets large.

Consider the Taylor's expansion of the MGF about zero:

$$M_X(t) = M_X(0) + M_X'(0)t + \frac{1}{2}M_X''(0)t^2 + \frac{1}{6}M_X'''(\xi)t^3,$$

where $\xi \in [0, t]$. Because $M_X(0) = 1$, $M_X'(0) = 0$, and $M_X''(0) = 1$, we get

$$M_X(t) = 1 + \frac{1}{2}t^2 + ct^3,$$

where $c = M_X'''(\xi)/6$ (c is actually a function of t instead of a constant, but we are assuming it is bounded).

The MGF $M_n(t)$ for Z_n is

$$M_n(t) = \mathrm{E}(e^{tZ_n}) = \mathrm{E}\left(e^{\frac{t}{\sqrt{n}}(X_1 + \cdots X_n)}\right) = \left[\mathrm{E}(e^{\frac{t}{\sqrt{n}}X_1})\right]^n = \left[M_X\left(\frac{t}{\sqrt{n}}\right)\right]^n.$$

Now we plug in the Taylor's expansion to get

$$M_n(t) = \left[1 + \frac{t^2}{2n} + \frac{ct^3}{n^{3/2}}\right]^n.$$

We can take the limit as n goes to infinity using the fact that for any a and $b > 0$, $b^a = e^{a\log(b)}$:

$$\left[1 + \frac{t^2}{2n} + \frac{ct^3}{n^{3/2}}\right]^n = \exp\left[n\log\left(1 + \frac{t^2}{2n} + \frac{ct^3}{n^{3/2}}\right)\right],$$

and we find the limit of the exponent using l'Hôpital's rule:

$$\lim_{n\to\infty} \frac{\log\left(1 + \frac{t^2}{2n} + \frac{ct^3}{n^{3/2}}\right)}{1/n} = \lim_{n\to\infty} \frac{\frac{t^2}{2} + \frac{3t^3}{2n^{1/2}}}{1 + \frac{t^2}{2n} + \frac{3t^2}{n^{3/2}}} = \frac{t^2}{2}.$$

This gives

$$\lim_{n\to\infty} M_n(t) = e^{t^2/2},$$

and we recognize the MGF for the standard normal. Therefore, the "limiting distribution" of Z_n is the standard normal.

This lemma is generalized to the following.

Central limit theorem: Let $Y_1, \ldots, Y_n$ be a random sample from a population with mean μ and variance σ^2. Then the distribution of the random variable

$$Z_n = \sqrt{n}\left(\frac{\bar{Y} - \mu}{\sigma}\right)$$

approaches that of a standard normal as n gets large. You're asked to prove this (starting with the lemma) in Exercise 41.1.

Example: Suppose $Y_1, \ldots, Y_n$ are independent Bernoulli(p), so that $S = \sum_{i=1}^{n} Y_i$ is Binom(n, p), and suppose n is "large." To calculate $P(S \geq k)$, the central limit theorem allows an approximation

$$P(S \geq k) = P\left(\frac{(S - np)}{\sqrt{np(1-p)}} \geq \frac{(k - np)}{\sqrt{np(1-p)}}\right)$$

$$\approx P\left(Z \geq \frac{(k - np)}{\sqrt{np(1-p)}}\right),$$

where $Z \sim N(0, 1)$.

The central limit theorem tell us that our Z-tests are still (approximately) valid if the normal population assumption is violated, as long as the sample size is "large." The number $n = 30$ is often given as large enough, but like all rules of thumb, it depends on how close the population is to normal, and also on how precise you want to be. The normal approximation to the binomial is a standard example of the central limit theorem, but is no longer particularly useful for computing binomial probabilities in practice.

Example: Suppose a population has mean $\mu = 150$ and standard deviation $\sigma = 20$. Estimate the probability that a sample of size $n = 100$ has mean less than 145. We can define

$$Z = \sqrt{100}\left(\frac{\bar{Y} - 150}{20}\right)$$

and estimate

$$P(\bar{Y} \leq 145) = P\left[Z \leq \sqrt{100}\left(\frac{145 - 150}{20}\right)\right] = P(Z \leq -2.5) \approx .0062.$$

Example: Suppose the delivery times for a courier service in a large city follow a distribution that is known to be skewed, with mean 20 minutes and standard deviation 15 minutes. Estimate the number m of minutes such that the time needed for the next 100 deliveries will exceed m with probability .05. In other words, we want to be 95% certain of being able to complete all 100 deliveries in the next m minutes. Let's assume independence of delivery times.

We know that

$$Z = \sqrt{100}\left(\frac{\bar{X} - 20}{15}\right)$$

is approximately standard normal, and we want

$$P\left(\sum_{i=1}^{100} X_i < m\right) = .95.$$

So, we use

$$P\left[10\left(\frac{\bar{X} - 20}{15}\right) \leq 1.645\right] \approx .95$$

and get

$$P\left[\sum_{i=1}^{100} X_i \leq 2247\right] \approx .95,$$

so if we allow for 2247 minutes for 100 deliveries, we'll be (about) 95% certain of being able to complete them in that time.

Chapter Highlights

1. If a population has finite mean μ and finite variance σ^2, and $Y_1, \ldots, Y_n$ is a random sample from the population with sample mean $\bar{Y}$, then

$$Z_n = \sqrt{n}\left(\frac{\bar{Y} - \mu}{\sigma}\right)$$

has approximately a standard normal distribution, and the approximation gets better as n increases.

Exercises

41.1 Prove the central limit theorem, starting with the lemma.

41.2 Get graphical demonstrations of the central limit theorem using simulations.

(a) We know that if we take a sample of size n from a Gamma(2,1/2) density, the sample mean is a random variable with mean 4 and variance $8/n$. We can take a look at the distribution of the sample mean by simulating 10,000 (or more) data sets of size 20 from this density and obtaining 10,000 sample means. Plot the histogram of the sample means using `freq=FALSE` and overlay a normal density with mean 4 and variance .4. Repeat for $n = 100$, and notice that the normal density fits better.

(b) Repeat part (a), where the population is Bernoulli(.25).

(c) Repeat part (a), where the population is Beta(2,2).

41.3 A sample of $n = 40$ bulbs is purchased for a building.

    (a) If the true distribution of lifetimes of the bulbs is exponential with mean 2 years, find the exact probability that the sample average lifetime is greater than 2.5 years.

    (b) Approximate the probability that the sample average lifetime is greater than 2.5 years using the central limit theorem if the mean is 2 years and the variance is 4 years.

41.4 A large cruise ship is preparing for a 30-day voyage with 2500 passengers. From past experience, they know that the average number of passengers visiting the sickbay per day demanding a hangover remedy is 250, with standard deviation 150. Although the ship's statistician believes that the true distribution of passengers per day who demand the remedy is skewed, she still wants to estimate the total number of hangover remedies needed for the voyage, to be 95% sure of having enough for all suffering passengers. Help her out by estimating the number of remedies needed.

41.5 To play a slot machine at a casino, the player drops in a one-dollar coin. With probability 1/20, the machine drops out 10 dollars. Suppose a gambler starts with 100 dollars. Approximate the 99th percentile of the distribution of the amount of money he has after he plays 100 times.

41.6 The lifetime of a sensitive computer component has a mean of 120 hours and a standard deviation of $120\sqrt{3}$ hours. A space shuttle mission requires this component, which can be replaced when it fails.

    (a) Use the central limit theorem to approximate the number N of components the astronauts should bring on the mission, to be 99% sure that they will have enough components to last through a 400-hour mission.

    (b) Your answer to (a) is exact if the distribution is normal. Suppose the true density of component lifetimes (in hours) is

$$f(x) = \frac{1}{80}\left[\frac{x}{240} + 1\right]^{-4} \quad \text{for } x > 0.$$

Using simulations in R, estimate the true probability that the your answer to (a) is adequate. Specifically, simulate N component lifetimes from the density (you can use the method in Chapter 19), and determine if the sum is more than 400. Do this in a loop to determine the true probability that the mission succeeds.

41.7 Accidents at a large mining company happen frequently. The average cost is $1200 and the standard deviation of costs is $1200\sqrt{3}$. The accountants want to have a reserve of money to cover accident costs. Specifically, they want N dollars reserved, so that the probability that the cost of the next 20 accidents is greater than N dollars is only 1%.

    (a) Use the central limit theorem to approximate the number N of dollars to be reserved.

(b) Your answer to (a) is exact if the distribution is normal. Suppose the true density of costs of accidents (in dollars) is

$$f(x) = \frac{1}{800} \left[\frac{x}{2400} + 1 \right]^{-4} \quad \text{for } x > 0.$$

Using simulations in R, estimate the true probability that the your answer to (a) is adequate.

41.8 A weaving machine in a garment factory produces lengths of linen cloth. In a randomly selected yard of cloth, the number of flaws follows this distribution:

# flaws	0	1	2	3
Probability	.6	.2	.15	.05

The manager of the factory has the opportunity to buy a new machine and wants to test the null hypothesis that the distribution of numbers of flaws for the new machine is the same as for the old machine against the alternative that the new machine makes, on average, fewer flaws. The manager will make 20 yards of linen cloth on the new machine. Assume that the numbers of flaws in each yard are independent random variables.

(a) Use the central limit theorem to approximate the distribution of the sample mean, and find the decision rule for $\alpha = .046$.

(b) Because the distribution of the test statistic is not really correct, we don't necessarily reject with probability .046 when the null hypothesis is true. Do simulations in R to find the true test size.

(c) Suppose the true distribution for flaws with the new machine is

# flaws	0	1	2	3
Probability	.75	.15	.1	0

Do simulations to determine the power for the test in (a).

(d) Write code to simulate the *true* distribution of T, the average number of flaws in the 20 yards, when the null hypothesis is true. What is the decision rule that gives a test size as close as possible to .05 (this test size will be .046)?

(e) What is the power for your decision rule in (d) when the true distribution is that in (c)?

Chapter 42

Parameter Estimation

In previous chapters we have considered useful models involving sampling from a normal population or populations. Interest was in estimating the parameters (the means and the variances of the populations) and conducting hypothesis tests about these parameters. Now we look at the more general case of estimating parameters when the model does not (necessarily) involve normal random variables, or not (necessarily) iid random variables.

A statistical model often involves assuming that our data, say $y_1 \ldots, y_n$, are a realization of a collection of random variables $Y_1, \ldots, Y_n$, which has a joint distribution from a certain parametric family. For example, we might assume the Y_i are iid normal with mean μ and variance σ^2. Or, we might think that the Y_i are binomial with parameter p, or iid Gamma(α, β), or log-normal, etc. We assume we know the family, but we want to estimate the parameters, such as μ or α or p, and to interpret the estimates in the context of the problem. Also of interest is the precision or accuracy of our estimator.

The term "estimator" is used for a random variable that is a summary statistic such as a sample average, and the term "estimate" is the realization of this random variable computed from the data. In other words, an estimator is a random variable, and an estimate is a realization of this random variable.

We might have more than one reasonable estimator for a parameter, and we need to choose among them to find one most appropriate to our situation. In this chapter we will look at some properties of estimators, to be used to compare two or more candidate estimators, in our search for a "best" estimator.

Let's start with **bias**. If $\hat{\theta}$ is a random variable used to estimate the parameter θ, then the bias of $\hat{\theta}$ is defined as

$$B(\hat{\theta}) = E(\hat{\theta}) - \theta.$$

If $B(\hat{\theta}) = 0$ for all possible θ, we say the estimator is **unbiased**. This means that the expected value of the estimator is always the thing we're trying to estimate.

Obviously, unbiasedness is a good property for an estimator, but another consideration is the variance of the estimator. If one estimator is unbiased but has large variance, and another estimator has some bias but smaller variance, we might decide to give up unbiasedness for a smaller variance.

A comprehensive measure of the quality of an estimator is the **mean squared error**, or **MSE**. This is defined to be

$$\text{MSE}(\hat{\theta}) = \text{E}[(\hat{\theta} - \theta)^2],$$

or the average squared distance from the estimator to the thing it's estimating. This looks a little like the definition of the variance, and in fact the MSE is the variance if $\text{E}(\hat{\theta}) = \theta$; that is, the MSE and the variance coincide for unbiased estimators. Otherwise, we can write

$$
\begin{aligned}
\text{V}(\hat{\theta}) &= \text{E}\left\{ \left[\hat{\theta} - \text{E}(\hat{\theta})\right]^2 \right\} \\
&= \text{E}\left\{ \left[(\hat{\theta} - \theta) + \left(\theta - \text{E}(\hat{\theta})\right)\right]^2 \right\} \\
&= \text{E}\left\{ (\hat{\theta} - \theta)^2 + 2(\hat{\theta} - \theta)\left(\theta - \text{E}(\hat{\theta})\right) + \left(\theta - \text{E}(\hat{\theta})\right)^2 \right\} \\
&= \text{E}\left\{ (\hat{\theta} - \theta)^2 \right\} + 2\text{E}(\hat{\theta} - \theta)\left(\theta - \text{E}(\hat{\theta})\right) + \left(\theta - \text{E}(\hat{\theta})\right)^2 \\
&= \text{E}\left\{ (\hat{\theta} - \theta)^2 \right\} - \left(\theta - \text{E}(\hat{\theta})\right)^2 \\
&= \text{MSE}(\hat{\theta}) - \text{B}(\hat{\theta})^2.
\end{aligned}
$$

This gives

$$\text{MSE}(\hat{\theta}) = \text{V}(\hat{\theta}) + \text{B}(\hat{\theta})^2.$$

We have decomposed the MSE into "precision" (variance) and "accuracy" (bias). Of course we want both pieces to be small. We'd like to define an estimator as "best" if it has the smallest MSE. In practice, this is impossible, because the MSE depends on the value of θ itself, which is unknown.

Example: Let's consider estimation of a binomial parameter p. Suppose $Y_1, \ldots, Y_n$ are outcomes in a binomial trial, i.e., $\text{P}(Y_i = 1) = p$ and $\text{P}(Y_i = 0) = 1 - p$, and the Y_i are independent. Then the sum of the Y_i is distributed as $\text{Binom}(n, p)$. An obvious way to estimate p given a sample of size n is with

$$\hat{p} = \frac{1}{n} \sum_{i=1}^{n} Y_i,$$

the sample proportion of "successes." Let's compare this to the less-obvious estimator

$$\tilde{p} = \frac{\sum_{i=1}^{n} Y_i + 1}{n + 2}.$$

We'll start by comparing the biases of the estimators:

$$\text{B}(\hat{p}) = \text{E}(\hat{p}) - p = p - p = 0,$$

so $\hat{p}$ is unbiased. But

$$\text{B}(\tilde{p}) = \text{E}(\tilde{p}) - p = \frac{np + 1}{n + 2} - p = \frac{1 - 2p}{n + 2},$$

so $\tilde{p}$ is biased. Note that value of the bias depends on p, and for $p = 1/2$, $\tilde{p}$ has zero bias, but recall that an estimator is called "unbiased" only if it has zero bias for *all* values of the parameter.

We might be inclined to reject $\tilde{p}$ in favor of the unbiased $\hat{p}$, but first let's compute the variance. We find that

$$\text{V}(\hat{p}) = \frac{p(1 - p)}{n}$$

and

$$\text{V}(\tilde{p}) = \frac{np(1 - p)}{(n + 2)^2},$$

so that $\text{V}(\tilde{p})$ is less than $\text{V}(\hat{p})$ for all values of $p \in (0, 1)$ and all sample sizes n. Are we willing to give up unbiasedness for a smaller variance? Let's decide based on the MSE:

$$MSE(\hat{p}) = \text{V}(\hat{p}) = \frac{p(1 - p)}{n}$$

and

$$MSE(\tilde{p}) = \frac{np(1 - p)}{(n + 2)^2} + \left(\frac{1 - 2p}{n + 2}\right)^2 = \frac{1 + (n - 4)p(1 - p)}{(n + 2)^2}.$$

Which is better? Here are some graphical comparisons for values of $p \in (0, 1)$. The solid curve represents the MSE for $\hat{p}$, for each value of p in $(0, 1)$, while the dashed curve is the MSE for $\tilde{p}$. For p in the range $(.16, .84)$, the estimator $\tilde{p}$ has smaller MSE than $\hat{p}$, but $\hat{p}$ is better for p near zero or near one.

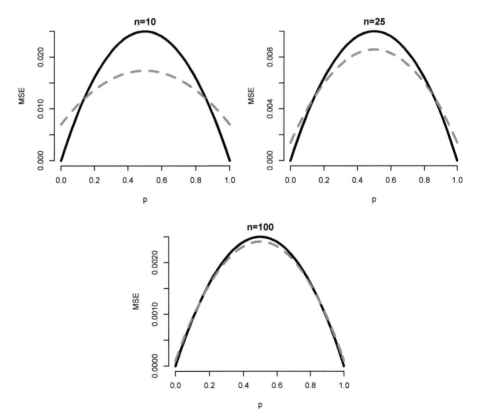

The difference between the MSEs of the estimators disappears as n gets larger, but the difference can be substantial for smaller n. If we have $n = 10$ and the true p is near $1/2$, $\tilde{p}$ can be a much better estimator of p than $\hat{p}$.

In summary, we have three properties of estimators: bias, variance, and MSE, where MSE is the sum of the variance and the squared bias. We'd like to say that the minimum MSE estimator is "best," but we have seen that the MSE can depend on the parameter, and the MSE for one estimator might be better for some parameter values, but worse for others.

It's pretty clear that we can't have a "best MSE" estimator. Suppose we have a random sample from an exponential population with a parameter $\theta > 0$. We can define an estimator $\tilde{\theta} = 2$, which ignores the values in the sample. The variance of such an estimator is zero, so the MSE is simply the square of the bias. Therefore, the MSE for our estimator is zero for $\theta = 2$, although it can be very large for other values of θ. There cannot exist an estimator with a *uniformly* minimum MSE.

If we restrict our attention to unbiased estimators only, we can compare them in terms of **relative efficiency**. Suppose $\hat{\theta}$ and $\tilde{\theta}$ are unbiased estimators of θ. We define the efficiency of $\hat{\theta}$ relative to $\tilde{\theta}$ to be

$$\text{eff}(\hat{\theta}, \tilde{\theta}) = \frac{V(\tilde{\theta})}{V(\hat{\theta})}.$$

So, if the efficiency of $\hat{\theta}$ relative to $\tilde{\theta}$ is larger than one, that means the variance of $\hat{\theta}$ is smaller. In this case we say that $\hat{\theta}$ is "more efficient" than $\tilde{\theta}$.

Example: Suppose $X_1, \ldots, X_n \sim N(\mu, \sigma^2)$ and $Y_1, \ldots, Y_n \sim N(\mu, 4\sigma^2)$ are independent random variables. Let

$$\hat{\mu}_1 = \frac{\bar{X} + \bar{Y}}{2} \quad \text{and} \quad \hat{\mu}_2 = \frac{2\bar{X} + \bar{Y}}{3}.$$

It is easy to check that both estimators are unbiased. What is the efficiency of $\hat{\mu}_2$ relative to $\hat{\mu}_1$?

The variance of $\hat{\mu}_1$ is

$$V(\hat{\mu}_1) = \frac{1}{4}\left[V(\bar{X}) + V(\bar{Y})\right] = \frac{1}{4}\left(\frac{\sigma^2}{n} + \frac{4\sigma^2}{n}\right) = \frac{5\sigma^2}{4n},$$

by independence.

Similarly, the variance of $\hat{\mu}_2$ is

$$V(\hat{\mu}_2) = \frac{1}{9}\left[4V(\bar{X}) + V(\bar{Y})\right] = \frac{1}{9}\left(\frac{4\sigma^2}{n} + \frac{4\sigma^2}{n}\right) = \frac{8\sigma^2}{9n}.$$

Then the relative efficiency of $\hat{\mu}_2$ relative to $\hat{\mu}_1$ is

$$\frac{V(\hat{\mu}_1)}{V(\hat{\mu}_2)} = \frac{45}{32}.$$

Because this is greater than one, $\hat{\mu}_2$ is more efficient (has smaller variance) than $\hat{\mu}_1$.

Example: Suppose $Y_1, \ldots, Y_n \overset{iid}{\sim} \text{Unif}(0, \theta)$, and interest is in estimating θ. Because $E(\bar{Y}) = \theta/2$, the estimator $\tilde{\theta} = 2\bar{Y}$ is unbiased. We have

$$V(\tilde{\theta}) = 4V(Y_1)/n = \frac{\theta^2}{3n}.$$

Another possibility is to construct an unbiased estimator based on the sample maximum. Using the solution to Exercise 33.8, the expected value of the maximum of the sample is $n\theta/(n+1)$, so if we let

$$\hat{\theta} = (n+1)\max(Y_i)/n,$$

we have another unbiased estimator for θ. In the same exercise we found the variance of the sample maximum, so

$$V(\hat{\theta}) = \frac{(n+1)^2}{n^2}V(\max(Y_i)) = \frac{(n+1)^2}{n^2}\frac{n\theta^2}{(n+2)(n+1)^2} = \frac{\theta^2}{n(n+2)}.$$

So, the efficiency of $\hat{\theta}$ relative to $\tilde{\theta}$ is

$$\frac{V(\tilde{\theta})}{V(\hat{\theta})} = \frac{n+2}{3},$$

which is bigger than one for all $n > 1$, and goes to infinity as the sample size grows. The estimator based on the sample maximum is considerably more efficient!

The **asymptotic relative efficiency** is simply the limit of the relative efficiency, as the sample size n increases without bound. For our first example, we found that the relative efficiency did not depend on the sample size, so the asymptotic relative efficiency is also 45/32. For the second example, the asymptotic relative efficiency of the adjusted sample maximum is infinite, relative to twice the sample mean.

Example: Suppose we have a random sample $Y_1, \ldots, Y_n$ from an exponential distribution with mean θ. It seems reasonable to estimate θ using the sample mean, say $\hat{\theta} = \bar{Y}$. The standard deviation of the distribution is also θ, so we can define $\tilde{\theta} = S$, the sample standard deviation. It's possible to compute the MSEs of both estimators analytically, but it's a messy calculation for $\tilde{\theta}$, so let's do simulations instead. The code on the left below takes many samples of size from the exponential population with $\theta = 2$, and computes the two estimates for each sample. The histograms of the estimates are shown on the right.

```
n=10;th=2
nloop=100000
thhat=1:nloop
thtil=1:nloop
for(iloop in 1:nloop){
    y=rexp(n,1/th)
    thhat[iloop]=mean(y)
    thtil[iloop]=sd(y)
}
mean((thhat-1/th)^2)
mean((thtil-1/th)^2)
```

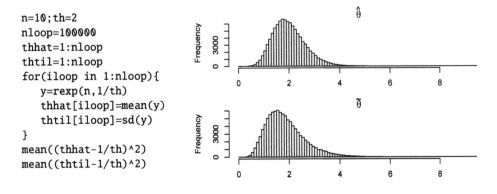

The estimator $\tilde{\theta}$ is biased small; on average the value is about 1.85, so given $\theta = 2$, the bias is about $-.15$. However, the variance is smaller, so $\text{MSE}(\tilde{\theta})$ is about 2.41, compared to about 2.65 for $\text{MSE}(\hat{\theta})$. This is another example of a biased estimator with smaller variance having smaller MSE than an unbiased estimator.

However, another round of simulations with $\theta = 1/2$ shows a smaller MSE for the unbiased estimator. For this example, $\hat{\theta}$ has a smaller MSE for some values of θ, but for other values of θ, $\tilde{\theta}$ has a smaller MSE.

Chapter Highlights

1. The bias of an estimator $\hat{\theta}$ of θ is

$$\text{B}(\hat{\theta}) = \text{E}(\hat{\theta}) - \theta.$$

2. If $\text{B}(\hat{\theta}) = 0$ for all values of θ, then $\hat{\theta}$ is unbiased.

3. The mean squared error of an estimator $\hat{\theta}$ of θ is

$$MSE(\hat{\theta}) = \text{E}[(\hat{\theta} - \theta)^2].$$

4. If $\hat{\theta}$ and $\tilde{\theta}$ are unbiased estimators of θ, the efficiency of $\hat{\theta}$ relative to $\tilde{\theta}$ is

$$\text{eff}(\hat{\theta}, \tilde{\theta}) = \frac{\text{V}(\tilde{\theta})}{\text{V}(\hat{\theta})}.$$

Exercises

42.1 Suppose we have a random sample of size n from a population with mean μ and variance σ^2. The usual estimate for μ is the sample mean $\hat{\mu} = \bar{X}$, but someone proposes

$$\tilde{\mu} = \frac{\sum_{i=1}^{n} X_i}{n + 1}.$$

Compute the bias, variance, and MSE of both estimators. Which has smaller MSE if $n = 10$, $\sigma = 10$, and $\mu = 1$? What if $n = 10$, $\sigma = 1$, and $\mu = 20$?

42.2 Suppose $Y_1, \ldots, Y_n$ are independent exponential random variables all with mean θ. The "usual" estimator for θ is $\hat{\theta} = \bar{Y}$. Define the alternative estimator

$$\tilde{\theta} = \frac{\sum_{i=1}^{n} Y_i + 1}{n + 2}.$$

(a) Compute the bias and variance of $\tilde{\theta}$.

(b) Plot the MSEs of $\hat{\theta}$ and $\tilde{\theta}$ for values of θ in $(0, 1)$ when $n = 10$. For what range of θ does $\tilde{\theta}$ have a smaller MSE?

42.3 Suppose $Y_1, \ldots, Y_n$ are Bernoulli random variables with success probability p, while $X_1, \ldots, X_n$ are Bernoulli random variables with success probability $p/2$. Further, the samples are independent.

   (a) Show that the estimator $\tilde{p} = 2(\bar{Y} + \bar{X})/3$ is an unbiased estimator of p and compute its variance.

   (b) Show that the estimator $\hat{p} = \bar{Y}/2 + \bar{X}$ is an unbiased estimator of p and compute its variance.

   (c) What is the efficiency of $\hat{p}$ relative to $\tilde{p}$?

42.4 Suppose we have a random sample of size n Poisson random variables with mean λ. A natural way to estimate λ is with the sample mean. Remember that the variance of a Poisson random variable is equal to the mean, so the sample variance ought to be another estimator for λ. Write code in R to determine which estimator has a smaller MSE for $\lambda = 2$ and $n = 10$.

42.5 Suppose X_1, X_2, X_3, X_4, and X_5 are independent normal random variables. All of these random variables have mean μ, X_1 and X_2 have variance 4, and X_3, X_4, and X_5 have variance 9. Show that each of the following estimators is unbiased, and determine which has the smallest variance.

   (a) Statistician A says, "$X_1 + X_2 + X_3 + X_4 + X_5$ has mean 5μ, so I'll use the sample mean $\hat{\mu}_A = \bar{X}$ as my estimator."

   (b) Statistician B says, "$X_1/2$, $X_2/2$, $X_3/3$, $X_4/3$, and $X_5/3$ all have the same variance, so I'll use for my estimator:

$$\hat{\mu}_B = \frac{X_1/2 + X_2/2 + X_3/3 + X_4/3 + X_5/3}{2}."$$

   (c) Statistician C uses

$$\hat{\mu}_C = \frac{X_1/4 + X_2/4 + X_3/9 + X_4/9 + X_5/9}{5/6}.$$

42.6 Suppose that $\hat{\theta}_1$ and $\hat{\theta}_2$ are independent random variables, and both are unbiased estimators of the parameter θ. Let $\sigma_1^2 = V(\hat{\theta}_1)$ and $\sigma_2^2 = V(\hat{\theta}_2)$. We want to construct an estimator $\hat{\theta}_3 = a\hat{\theta}_1 + (1 - a)\hat{\theta}_2$ for $a \in [0, 1]$. Show that $\hat{\theta}_3$ is unbiased. What is the value of a that minimizes the variance of $\hat{\theta}_3$?

42.7 Suppose Y_1, Y_2, Y_3, and Y_4 are independent exponential random variables with mean θ. Consider the following estimators for θ:

   - $\hat{\theta}_1 = \bar{Y}$,
   - $\hat{\theta}_2 = Y_2$,
   - $\hat{\theta}_3 = (Y_1 + 2Y_2)/3$,
   - $\hat{\theta}_4 = 4\min(Y_1, Y_2, Y_3, Y_4)$.

   Which estimators are unbiased? Which has the smallest MSE?

42.8 Suppose $Y_1, \ldots, Y_n$ is a random sample from a uniform distribution on $(\theta, \theta + 1)$. Find a function of $\bar{Y}$ that is an unbiased estimator of θ, and compute its variance.

42.9 The random variables Y_1, Y_2, Y_3, and Y_4 are independent exponential random variables, all with mean $\theta > 0$. Consider the statistic

$$T = \frac{1}{8}(Y_1^2 + Y_2^2 + Y_3^2 + Y_4^2).$$

Compute the MSE for T as an estimator of θ^2.

42.10 Suppose X and Y are independent exponential random variables, where the mean of X is θ and the mean of Y is 2θ. Consider estimators of the form $\hat{\theta} = aX + bY$ for constants a and b.

(a) Find an expression for the bias of $\hat{\theta}$ in terms of a and b, and determine a condition on the values of a and b, such that $\hat{\theta}$ is unbiased.

(b) Of all the values of a and b that make the estimator unbiased, find the values of a and b that minimize the variance of the estimator.

42.11 Suppose $X \sim \text{Pois}(\lambda)$ and $Y \sim \text{Pois}(2\lambda)$ are independent random variables. Consider a linear estimator of λ, that is, $\hat{\lambda} = aX + bY$.

(a) Find an expression for the bias of $\hat{\lambda}$ in terms of a and b, and determine a condition on the values of a and b, such that $\hat{\lambda}$ is unbiased.

(b) Of all the values of a and b that make the estimator unbiased, find the values of a and b that minimize the variance of the estimator.

42.12 Suppose $Y_i \sim N(\mu, \sigma_i^2)$ are independent random variables, where σ_i^2 is known, for $i = 1, \ldots, n$. An estimator $\hat{\mu}$ is linear if $\hat{\mu} = \sum_{i=1}^{n} a_i Y_i$.

(a) Find an expression for the bias of $\hat{\mu}$, in terms of the a_i, $i = 1, \ldots, n$, and determine a condition on the values of a_i for which $\hat{\mu}$ is unbiased.

(b) Of all the values of the a_i that make the estimator unbiased, find the values that minimize the variance of the estimator. *Hint:* Use the method of Lagrange multipliers to find the constrained minimum.

42.13 Suppose $Y_1, \ldots, Y_n$ are independent random variables all having the density function

$$f_\theta(y) = \frac{3y^2}{\theta} e^{-y^3/\theta} \quad \text{for } y > 0,$$

where $\theta > 0$ is an unknown parameter. Compute the bias and variance of the estimator

$$\hat{\theta} = \sum_{i=1}^{n} Y_i^3/n.$$

42.14 Let $Y_1, \ldots, Y_n$ be a random sample from an exponential distribution with mean θ, and consider the following two estimators for θ: $\hat{\theta} = \bar{Y}$ and $\tilde{\theta} = n\min(Y_i)$.

(a) Show that both estimators are unbiased. (*Hint:* Recall the solution to Exercise 33.4.)

(b) Compute the variance of both estimators.

(c) What is the relative efficiency of $\hat{\theta}$ with respect to $\tilde{\theta}$? What is the asymptotic relative efficiency?

(d) Simulate 10,000 random samples of size $n = 40$ from an exponential distribution with mean $\theta = 2$, and for each sample compute the two estimators. Make histograms with the same scale on the horizontal axis (same "breaks") to compare the two estimators visually, and find the variances to check your answer to (b).

42.15 Suppose $Y_1, \ldots, Y_n$ is a random sample of a population that is distributed as exponential with mean θ, and interest is in estimating $\alpha = \theta^2$.

(a) Show that the following estimator is unbiased:

$$\hat{\alpha}_1 = \frac{1}{2n} \sum_{i=1}^{n} Y_i^2.$$

(b) Determine the value of a that makes $\hat{\alpha}_2$ unbiased:

$$\hat{\alpha}_2 = a\bar{Y}^2.$$

(c) Find the relative efficiency of $\hat{\alpha}_1$ with respect to $\hat{\alpha}_2$.

42.16 (a) If $Y_1, \ldots, Y_n \overset{iid}{\sim} N(\mu, \sigma^2)$, then both the mean and the median of the sample are unbiased estimators of μ (the latter result uses a symmetry argument). For $n = 20$, $\mu = 100$, and $\sigma^2 = 1$, do simulations to compare the variances of the estimators.

(b) Suppose $\varepsilon_1, \ldots, \varepsilon_n \overset{iid}{\sim} t(3)$, and let $Y_i = \mu + \varepsilon_i$. Then both the mean and the median of $Y_1, \ldots, Y_n$ are unbiased estimators of μ. For $n = 20$ and $\mu = 100$, do simulations to compare the variances of the estimators.

42.17 A "contaminated normal" density is often defined as a "mixture" of normal densities, where the density function can be written as

$$f(x) = (1 - \alpha)f_1(x) + \alpha f_2(x)$$

for $\alpha = (0, 1)$. The densities f_1 and f_2 are both normal densities, and typically the variance for f_1 is small, while the variance for f_2 is large. Suppose that the mean for both f_1 and f_2 is μ, the variances are σ_1^2 and σ_2^2, respectively, and $X_1, \ldots, X_n$ is a random sample from the population with density f.

(a) Show that $\bar{X}$ is an unbiased estimator of μ.

If the population is suspected to be contaminated, some statisticians recommend using a "trimmed sample mean" to estimate the sample mean. Let $\tilde{\mu}$ be the estimator obtained by deleting the maximum and minimum value from the sample, then averaging the remaining values.

(b) If $\sigma_1^2 = \sigma_2^2 = 1$ (i.e., an uncontaminated normal), do simulations to compare the MSE of $\tilde{\mu}$ and $\bar{X}$ for a sample size $n = 40$. (Recall simulation of mixture distributions in Chapter 19.)

(c) If $\sigma_1^2 = 1$ and $\sigma_2^2 = 100$, and $\alpha = .1$, do simulations to compare the MSE of $\tilde{\mu}$ and $\bar{X}$ for a sample size $n = 40$.

Chapter 43

Maximum Likelihood Estimation

Suppose $Y_1, \ldots, Y_n$ have a joint distribution that is known up to a parameter, or set of parameters. For example, we could know that the sample is from a normal population, but we don't know the mean and variance. We might have a random sample from a gamma distribution, but we don't know the shape and scale parameters, or a random sample from a geometric distribution, and we want to estimate p. In this chapter we find the **maximum likelihood estimator** (MLE) for the unknown parameter or parameters.

Example: For a simple example, suppose we have 16 independent binomial trials with probability p of success, and we observe 5 successes. For any $p \in (0, 1)$, we can find probabilities of numbers of successes. Below are three bar charts of distributions of a Binom$(16, p)$ random variable, with our observed value of 5 indicated in red.

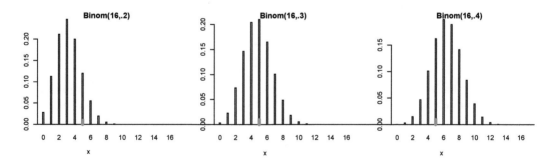

For each p, the height of the bar at 5 (i.e., the probability of observing 5 successes) might be larger or smaller. We want our MLE of p to be that which makes the tallest bar, i.e., the p that maximizes the likelihood of seeing 5 successes.

The **likelihood function** is simply the joint probability distribution, written as a function of the parameters. That is, the likelihood function is the joint density for a continuous population, or the joint mass function for a discrete population. One might say, using GRE-type analogies, that the likelihood is to the joint probability distribution function as statistics is to probability.

For random variables $Y_1, \ldots, Y_n$, suppose their joint density or mass function is $f(y_1, \ldots, y_n)$. If the specification of the joint distribution involves an unknown parameter vector $\boldsymbol{\theta} = (\theta_1, \ldots, \theta_k)$, we will write the density or mass function as

$$f_\theta(y_1, \ldots, y_n)$$

to emphasize that, though the function arguments are $y_1, \ldots, y_n$, the function also depends on $\boldsymbol{\theta}$. We define the likelihood to be

$$L(\boldsymbol{\theta}; y_1, \ldots, y_n) = f_\theta(y_1, \ldots, y_n)$$

so the likelihood is the same as the joint density or mass function, except we think of it as a function of the parameters given the data; that is, the argument of the function is the parameter, and the data are "fixed."

The MLE for $\boldsymbol{\theta}$ is the value of $\boldsymbol{\theta}$ that maximizes the likelihood, given the data.

For our example where $X \sim \text{Binom}(16, p)$, we have

$$P(X = 5) = L(p) = p^5 (1 - p)^{11},$$

and it is straightforward to maximize this expression by taking the derivative with respect to p, setting it equal to zero, and solving for p to get $\hat{p} = 5/16$. The likelihood function is plotted below with the MLE marked by a dotted vertical line.

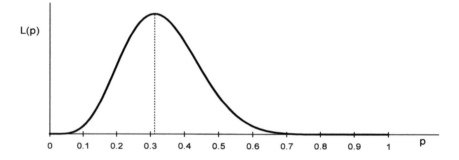

For this simple example, it is not too hard to take the derivative of $L(p)$ using the multiplication rule and the chain rule, but often it's much easier to find the maximizer of the **log-likelihood** function, which is simply the log of the likelihood function. In this case,

$$\ell(p) = 5 \log(p) + 11 \log(1 - p),$$

and it's easier to take the derivative of ℓ with respect to p, set it equal to zero, and solve for p to get $\hat{p} = 5/16$.

More generally, the log-likelihood function of a parameter θ is

$$\ell(\theta; y_1, \ldots, y_n) = \log\left(L(\theta; y_1, \ldots, y_n)\right).$$

Because the log function is strictly increasing, the value of the maximizer of ℓ is the same as the value of the maximizer of L. (You are asked to give a formal proof in Exercise 43.1.) In other words, if $\hat{\theta}$ maximizes the likelihood L, then $\hat{\theta}$ also maximizes the log-likelihood ℓ. Often, but not always, we get our MLE by maximizing the log-likelihood.

Example: Suppose $Y_1, \ldots, Y_n$ is a random sample from an exponential population with mean $\theta > 0$. We want to find the MLE for θ. The joint density is the product of the population densities (because the Y_i are independent), so

$$L(\theta; y_1, \ldots, y_n) = f_\theta(y_1, \ldots, y_n) = \prod_{i=1}^{n} \left[\frac{1}{\theta} e^{-y_i/\theta} \right] = \frac{1}{\theta^n} \exp \left\{ -\sum_{i=1}^{n} y_i/\theta \right\}.$$

The log-likelihood is

$$\ell(\theta; y_1, \ldots, y_n) = -n \log(\theta) - \frac{\sum_{i=1}^{n} y_i}{\theta},$$

and

$$\frac{d}{d\theta} \ell(\theta; y_1, \ldots, y_n) = -\frac{n}{\theta} + \frac{\sum_{i=1}^{n} y_i}{\theta^2},$$

which is zero when $\theta = \bar{y}$. It is easy to check that the second derivative of ℓ, evaluated at $\bar{y}$, is negative, so the likelihood is maximized. Hence, we define the maximum likelihood estimator for θ as

$$\hat{\theta} = \bar{Y},$$

which is a random variable.

Let's compute the bias of the MLE $\hat{\theta}$:

$$\mathrm{E}(\hat{\theta}) = \mathrm{E}(\bar{Y}) = \theta,$$

so $\hat{\theta}$ is unbiased. Therefore, the variance is also the MSE and is

$$\mathrm{V}(\hat{\theta}) = \mathrm{V}(\bar{Y}) = \mathrm{V}(Y_i)/n = \theta^2/n.$$

Example: $Y_1, \ldots, Y_n$ are independent Bernoulli random variables where $P(Y_i = 1) = p$ and $P(Y_i = 0) = 1 - p$ for some $p \in (0, 1)$. The joint mass function turns into the likelihood

$$L(p; y_1, \ldots, n) = \prod_{i=1}^{n} \left[p^{y_i} (1 - p)^{1-y_i} \right],$$

and the log-likelihood is

$$\ell(p; y_1, \ldots, n) = \sum_{i=1}^{n} [y_i \log(p) + (1 - y_i) \log(1 - p)]$$
$$= s \log(p) + (n - s) \log(1 - p),$$

where $s = \sum_{i=1}^{n} y_i$ is the number of ones (or "successes") in the sample. Then

$$\frac{\partial \ell}{\partial p} = \frac{s}{p} - \frac{n - s}{1 - p},$$

which is zero when $p = s/n$. The second derivative of ℓ with respect to p is negative at $p = s/n$, so we have found a maximum, and the MLE is $\hat{p} = S/n$, the sample mean, or sample proportion of successes.

Example: Let $Y_1, \ldots, Y_n$ be a random sample from an $N(\mu, \sigma^2)$ population. Let's find the MLEs for the parameters μ and σ^2. When there is more than one parameter, we maximize the likelihood function over both simultaneously. This often involves two equations and two unknowns. The normal likelihood is

$$L(\mu, \sigma^2; y_1, \ldots, y_n) = \prod_{i=1}^{n} \frac{1}{\sqrt{2\pi\sigma^2}} \exp\left\{ -\frac{(y_i - \mu)^2}{2\sigma^2} \right\}$$

$$= \left(\frac{1}{2\pi\sigma^2} \right)^{n/2} \exp\left\{ -\frac{\sum_{i=1}^{n}(y_i - \mu)^2}{2\sigma^2} \right\}.$$

Computing the log-likelihood function,

$$\ell(\mu, \sigma^2; y_1, \ldots, y_n) = -\frac{n}{2}\log(2\pi) - \frac{n}{2}\log(\sigma^2) - \left\{ \frac{\sum_{i=1}^{n}(y_i - \mu)^2}{2\sigma^2} \right\}.$$

Let's first take the derivative of the log-likelihood function with respect to μ:

$$\frac{\partial \ell}{\partial \mu} = \frac{1}{\sigma^2} \sum_{i=1}^{n}(y_i - \mu),$$

and this is zero when $\mu = \bar{y}$, the sample mean. A quick check of the second derivative shows that the sample mean maximizes the likelihood for any value of σ^2. Therefore, the MLE for μ is $\hat{\mu} = \bar{Y}$.

Next, we take the derivative of the log-likelihood function with respect to σ^2. It's a little confusing that σ^2 is the parameter, and not the square of the parameter. If you like, you can plug in $\theta = \sigma^2$, take the derivative with respect to θ, then switch back to σ^2 notation. We find

$$\frac{\partial \ell}{\partial \sigma^2} = -\frac{n}{2\sigma^2} + \left(\frac{\sum_{i=1}^{n}(y_i - \mu)^2}{2\sigma^4} \right),$$

and this is zero when

$$\sigma^2 = \frac{1}{n} \sum_{i=1}^{n}(y_i - \mu)^2.$$

The MLE for σ^2 is (after plugging in the MLE for μ)

$$\hat{\sigma}^2 = \frac{1}{n} \sum_{i=1}^{n}(Y_i - \bar{Y})^2.$$

The MLE for the population variance is $(n-1)S^2/n$, where S^2 is the sample variance. Therefore, the MLE for σ^2 is biased a little small.

Example: Let $Y_1, \ldots, Y_n$ be a random sample from a $\text{Unif}(0, \theta)$ population where $\theta > 0$ is the unknown parameter. We would like to estimate θ. Using that the joint density of the sample is the product of the individual densities, the likelihood is

$$L(\theta; y_1, \ldots, y_n) = \prod_{i=1}^{n} \left[\frac{1}{\theta} I\{0 \le y_i \le \theta\} \right] = \frac{1}{\theta^n} \prod_{i=1}^{n} I\{0 \le y_i \le \theta\}.$$

Here's a situation where we *don't* want to use the log-likelihood. Instead, we notice that we can write

$$L(\theta; y_1, \ldots, y_n) = \frac{1}{\theta^n} I\{\theta \geq y_{(n)}\},$$

where $y_{(n)} = \max(y_1, \ldots, y_n)$. If we plot the likelihood function, we see that the MLE is $\hat{\theta} = Y_{(n)}$.

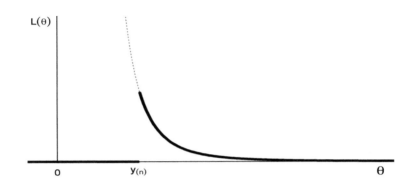

Is the MLE $\hat{\theta}$ unbiased? We compute $\mathrm{E}(\hat{\theta}) = \mathrm{E}(Y_{(n)})$, using the density for the maximum of a sample from a $\mathrm{Unif}(0, \theta)$ density. Recall that this density is

$$f_n(y) = \frac{n}{\theta^n} y^{n-1} \text{ for } y \in (0, \theta),$$

and we have found the mean and variance to be

$$\mathrm{E}(Y_{(n)}) = \frac{n\theta}{n+1} \text{ and } \mathrm{V}(Y_{(n)}) = \frac{n\theta^2}{(n+2)(n+1)^2}.$$

The MLE is biased small:

$$\mathrm{B}(\hat{\theta}) = -\frac{\theta}{n+1},$$

although as n gets larger, the bias goes to zero.

The mean squared error (MSE) of $\hat{\theta}$ is

$$\mathrm{MSE}(\hat{\theta}) = \frac{2\theta^2}{(n+2)(n+1)}.$$

Of course, we can "adjust" the MLE to make an unbiased estimator. The estimator $\tilde{\theta} = (n+1)\hat{\theta}/n$ is a tad bigger than the MLE, and is unbiased. The variance of this adjusted estimator is larger than the variance of the MLE:

$$\mathrm{V}(\tilde{\theta}) = \frac{(n+1)^2}{n^2} \times \mathrm{V}(\hat{\theta}) = \frac{\theta^2}{n(n+1)},$$

but the MSE of the adjusted estimator is smaller whenever $n > 2$.

Invariance property of the MLE

Suppose we want to estimate a function of the parameter, say $t(\theta)$. The invariance property of the MLE says that $t(\hat{\theta})$ is the MLE for $t(\theta)$, where $\hat{\theta}$ is the MLE for θ.

For example, if we have a random sample $Y_1, \ldots, Y_n$ from $\text{Exp}(\theta)$, suppose we want to find the MLE of the variance θ^2. We know that the MLE of θ is the sample mean $\bar{Y}$, so the MLE of θ^2 is $\bar{Y}^2$.

This is fairly clear intuitively, as $t(\theta) = \theta^2$ is monotone increasing, and hence one-to-one, for $\theta > 0$. More formally, let $\eta = t(\theta)$ be an invertible function of θ, and write the likelihood in terms of η:

$$L^*(\eta; y_1, \ldots, y_n) = \prod_{i=1}^n f_{t^{-1}(\eta)}(y_1, \ldots, y_n) = L(t^{-1}(\eta), y_1, \ldots, y_n).$$

The maximum value of L^* has to be the same as the maximum value of L, which occurs at $\hat{\theta}$, so $t^{-1}(\hat{\eta}) = \hat{\theta}$ or $\hat{\eta} = t(\hat{\theta})$.

Chapter Highlights

1. If $Y_1, \ldots, Y_n$ have joint density $f_\theta(y_1, \ldots, y_n)$, where $\boldsymbol{\theta}$ is a vector of parameters to be estimated, then the likelihood is defined to be

$$L(\boldsymbol{\theta}; y_1, \ldots, y_n) = f_\theta(y_1, \ldots, y_n).$$

2. The MLE $\hat{\boldsymbol{\theta}}$ maximizes the likelihood over possible values of $\boldsymbol{\theta}$.

3. The log-likelihood function is $\ell(\boldsymbol{\theta}; y_1, \ldots, y_n) = \log\left[L(\boldsymbol{\theta}; y_1, \ldots, y_n)\right]$. The value of $\boldsymbol{\theta}$ that maximizes ℓ is the value of $\boldsymbol{\theta}$ that maximizes L. For many problems, it is computationally simpler to maximize the log-likelihood to get the MLE.

4. If the likelihood $L(\theta; y_1, \ldots, y_n)$ is not continuous at θ (e.g., if the support depends on θ), then we can't find the MLE by taking the derivative with respect to θ. In this case, you can sketch the likelihood function (it doesn't help to find the log-likelihood) and discover the maximum.

5. Invariance property: The MLE for $t(\boldsymbol{\theta})$ is $t(\hat{\boldsymbol{\theta}})$, where $\hat{\boldsymbol{\theta}}$ is the MLE for $\boldsymbol{\theta}$.

Exercises

43.1 Suppose $\psi(x)$ is a continuous function on the real line, such that for any x, $\psi(x) > 0$, and the first and second derivatives exist. Further, suppose that $\psi(x)$ is maximized at $x = x_0$; specifically, suppose $\psi'(x_0) = 0$ and $\psi''(x_0) < 0$. Prove that $\log(\psi(x))$ is also maximized at $x = x_0$.

43.2 Suppose $Y_1, \ldots, Y_n$ is a random sample from an exponential population with mean $\theta > 0$. Using R, plot the likelihood function when the sample mean is 2.8 and $n = 10$. Indicate the MLE on the plot.

43.3 Suppose $Y_1, \ldots, Y_n \overset{ind}{\sim} \text{N}(0, \sigma^2)$ and find the MLE for σ^2. Is the MLE unbiased?

43.4 Suppose $\theta > 0$ and $Y_1, \ldots, Y_n$ are a random sample from the density

$$f_\theta(y) = \frac{2y}{\theta} e^{-y^2/\theta} \quad \text{for } y > 0.$$

(a) Find the MLE for θ.

(b) Is the MLE unbiased for θ? Explain.

(c) If $n = 8$ and the observations are

$$1.1, \quad 4.2, \quad 5.6, \quad 0.2, \quad 0.4, \quad 6.5, \quad 1.4, \quad 2.9,$$

use R to plot the likelihood function and indicate the MLE on the plot.

43.5 Suppose $Y_1, \ldots, Y_n$ are independent random variables with common density

$$f_\mu(y) = e^{\mu - y} \quad \text{for } y > \mu.$$

(a) Find the MLE for μ.

(b) Find the bias of the estimator in part (a).

(c) Suppose $n = 8$ and we observe the values 7.59, 7.18, 8.39, 6.86, 5.88, 4.99, 4.81, 10.30. Use R to plot the likelihood function and indicate the MLE on the plot.

43.6 Suppose $Y_1, \ldots, Y_n$ are independent random variables with common density

$$f_\theta(y) = \frac{2\theta^2}{y^3} \quad \text{for } y > \theta \text{ and } \theta > 0.$$

Find the MLE for θ.

43.7 Suppose we want to measure a parameter μ, but our measuring device is such that our measurements $Y_1, \ldots, Y_n$ are independent normal random variables with mean μ and variance σ^2. After measuring $Y_1, \ldots, Y_n$, we get a better measuring device and take more measurements: $X_1, \ldots, X_m$ are independent normal random variables with mean μ and variance $\sigma^2/2$. (Assume the X_i measurements and the Y_i measurements are independent.) Derive the MLE for μ (using all the data).

43.8 For some $\theta > 0$, suppose $X_1, \ldots, X_n$ is a random sample from a density

$$f_\theta(x) = \theta^2 x e^{-\theta x} \quad \text{for } x \in [0, \infty).$$

Find the MLE for θ.

43.9 For some $\lambda > 0$, suppose $X_1, \ldots, X_n$ is a random sample from a density

$$f_\lambda(x) = \frac{\lambda}{2\sqrt{x}} e^{-\lambda \sqrt{x}} \quad \text{for } x > 0.$$

Find the MLE for λ.

43.10 Suppose $Y_1, \ldots, Y_n \overset{iid}{\sim} \mathrm{Pois}(\lambda)$.

    (a) Find the MLE for λ.

    (b) If $p_0 = P(Y_1 = 0)$, find the MLE for p_0.

43.11 Devices used in a manufacturing process have a probability p of breaking with each use. Suppose $Y_1, \ldots, Y_n$ are the numbers of successful uses for n devices. Find the MLE for p, assuming independence.

43.12 Suppose $Y_1, \ldots, Y_n$ are a random sample from the density

$$f_\theta(y) = 2\theta^2 y^{-3} \quad \text{for } y > \theta$$

for some $\theta > 0$. Find the MLE for θ.

43.13 Suppose $Y_1, \ldots, Y_n \overset{iid}{\sim} f_\theta(y)$, where

$$f_\theta(y) = \theta y^{\theta - 1} \quad \text{for } y \in (0, 1),$$

and $\theta > 0$. Find the MLE for θ.

43.14 Suppose $Y_1, \ldots, Y_n \overset{iid}{\sim} \mathrm{Unif}\,(\theta, 2\theta)$, where $\theta > 0$. Find the MLE for θ.

43.15 Suppose $Y_1, \ldots, Y_n$ are independent random variables whose density is

$$f_\theta(y) = 3\theta y^2 e^{-\theta y^3} \quad \text{for } y > 0$$

for some $\theta > 0$.

    (a) Find the MLE for θ.

    (b) Write some R code to check your answer, and to estimate the bias and variance when $n = 20$ and $\theta = 2$. You can use the code from Exercise 19.15 to simulate from the distribution.

43.16 Suppose $Y_1, \ldots, Y_n$ are a random sample from the density

$$f_\theta(y) = \frac{\theta}{(y + 1)^{\theta + 1}} \quad \text{for } y > 0$$

for some $\theta > 1$.

    (a) Find the MLE for θ.

    (b) Find the bias and variance of your estimator using simulations R for $n = 20$ and $\theta = 5$. (You can use the answer to Exercise 19.14 to simulate from f_θ.)

    (c) Suppose the observed sample values are

        0.144, 0.912, 0.199, 0.849, 0.329, 0.173, 0.384, 0.360, 0.002, 1.664.

        Using R, plot the likelihood function against $\theta \in (0, 10)$ and mark the MLE on the plot.

43.17 Suppose $Y_1, \ldots, Y_n$ are a random sample from the density

$$f_\lambda(y) = \frac{1}{2\lambda\sqrt{y}} e^{-\sqrt{y}/\lambda} \quad \text{for } y > 0$$

for some $\lambda > 0$.

(a) Find the MLE for λ.

(b) Show that $n\hat{\lambda}$ is a gamma random variable.

(c) Suppose the observed values are

$$3.9, \quad 10.1, \quad 5.8, \quad 2.7, \quad 0.1, \quad 26.2, \quad 1.7, \quad 1.0.$$

Use the distribution in part (b) to test the null hypothesis that $\lambda = 1$ versus the alternative that $\lambda > 1$. Use $\alpha = .01$.

43.18 Suppose that $Y_1, \ldots, Y_n$ are independent "double-exponential" random variables with density function

$$f_\theta(y) = \frac{1}{2\theta} e^{-|y|/\theta} \text{ for } y \in \mathbb{R}.$$

Find the MLE for θ.

43.19 Suppose that $Y_1, \ldots, Y_n$ are independent "double-exponential" random variables with density function

$$f_\mu(y) = \frac{1}{2} e^{-|y-\mu|} \text{ for } y \in \mathbb{R}.$$

Find the MLE for μ.

43.20 Suppose that $Y_1, \ldots, Y_n \stackrel{ind}{\sim} N(\mu_y, \sigma^2)$ and $X_1, \ldots, X_m \stackrel{ind}{\sim} N(\mu_x, \sigma^2)$. Find the MLEs for μ_y, μ_x, and σ^2.

43.21 For $k \geq 2$, suppose that $Y_{ij} \stackrel{ind}{\sim} N(\mu_i, \sigma^2)$ for $i = 1, \ldots, k$ and $j = 1, \ldots, n_i$. Find the MLEs for $\mu_1, \ldots, \mu_k$, and σ^2.

43.22 Suppose $x_1, \ldots, x_n$ are known constants, and $Y_i = \beta x_i + \varepsilon_i$ for $i = 1, \ldots, n$, where $\varepsilon_1, \ldots, \varepsilon_n$ are independent standard normal random variables.

(a) Find the MLE for β.

(b) Find the bias and variance for your estimator in (a).

(c) Suppose $n = 10$, $x_i = i$ for $i = 1, \ldots, 10$, and the observed values of Y are $3.9, 4.6, 5.3, 9.4, 10.0, 12.4, 12.4, 15.2, 17.0, 19.5$. Find your estimate for β. What is the variance of the estimator?

Chapter 44

Estimation Using the Method of Moments

The method of moments estimator can be used when we have a random sample from a population, where the distribution is known up to a parameter or set of parameters. Of the three major types of parameter estimation considered in this book, it could be said that the method of moments estimator is the most straightforward, and (usually) the easiest to obtain.

To apply the method of moments, we set **population moments** equal to the **sample moments** and solve for the parameter(s).

Suppose the population has a distribution in a family of densities (or mass functions) described by $f_\theta(y)$. Recall that the kth population moment is $\mu'_k = E(Y^k)$ for $Y \sim f_\theta$. For a random sample $Y_1, \ldots, Y_n \overset{ind}{\sim} f_\theta(y)$, the **$k$th sample moment** is defined to be

$$m'_k = \frac{1}{n} \sum_{i=1}^{n} Y_i^k.$$

If the parameter is one-dimensional, we can try setting the first population and sample moments to be equal, then solve for the parameter to get the estimator. If the first population moment does not depend on the parameter, then we have to try using a higher moment. If the parameter is two-dimensional, we can try setting the first two population and sample moments to be equal, then solving two equations and two unknowns to get the parameter estimates.

Example: Suppose $Y_1, \ldots, Y_n$ are independent exponential random variables, all with mean θ. Then the first population moment is $E(Y_1) = \theta$, and the first sample moment is simply $\bar{Y}$, so the method of moments estimator for the population mean θ is the sample mean. This is the same estimator as the MLE.

Example: Suppose $Y_1, \ldots, Y_n \overset{ind}{\sim} \text{Unif}(0, \theta)$. Then the first population moment is $E(Y_1) = \theta/2$, and the first sample moment is simply $\bar{Y}$, so this gives that the method of moments estimator for θ is $\hat{\theta} = 2\bar{Y}$. This is a different estimator than the MLE! Recall that the MLE is $\hat{\theta} = Y_{(n)}$, the sample maximum. Which estimator is better? The method of moments estimator is unbiased, because $E(\bar{Y}) = \theta/2$, and we computed the bias of the MLE to be $-\theta/(n+1)$. However, the variance of the MLE was

$$V(\hat{\theta}) = \frac{n\theta^2}{(n+2)(n+1)^2},$$

which is considerably smaller than

$$V(\tilde{\theta}) = 4V(\bar{Y}) = \frac{\theta^2}{3n}.$$

We can compare

$$MSE(\tilde{\theta}) = \frac{\theta^2}{3n} \quad \text{to} \quad MSE(\hat{\theta}) = \frac{2\theta^2}{(n+2)(n+1)}$$

and conclude that the MSE for the MLE is smaller for $n > 2$.

Example: Suppose $Y_1, \ldots, Y_n \overset{ind}{\sim} N(0, \sigma^2)$, and consider using the method of moments to estimate σ^2. The first population moment is zero, so that doesn't help us estimate the parameter. The second population moment is σ^2, so the method of moments estimator for σ^2 is $\tilde{\sigma}^2 = \sum_{i=1}^n Y_i^2/n$. This is the same as the MLE.

Example: Suppose $Y_1, \ldots, Y_n \overset{ind}{\sim} \text{Gamma}(\alpha, \beta)$, and consider using the method of moments to estimate α and β. Here we have two parameters to estimate, so we need two equations. The first equation sets the first sample moment equal to the first population moment:

$$m_1' = \bar{Y} = \tilde{\alpha}/\tilde{\beta}.$$

The second equation sets the second sample moment equal to the second population moment:

$$m_2' = \frac{1}{n}\sum_{i=1}^n Y_i^2 = \tilde{\alpha}/\tilde{\beta}^2 + \tilde{\alpha}^2/\tilde{\beta}^2.$$

We get the estimators

$$\tilde{\alpha} = \frac{(m_1')^2}{m_2' - (m_1')^2} \quad \text{and} \quad \tilde{\beta} = \frac{m_1'}{m_2' - (m_1')^2}.$$

Chapter Highlights

1. The method of moments estimator is computed by setting one or more of the population moments equal to the corresponding sample moment(s). For example, set the population mean to the sample mean, and solve for the parameter.

❧❧

Exercises

44.1 Suppose $Y_1, \ldots, Y_n \overset{ind}{\sim} \text{Pois}(\lambda)$. Find the method of moments estimator for λ.

44.2 Suppose $Y_1, \ldots, Y_n$ are a random sample from the density

$$f_\theta(y; \theta) = \frac{3y^2}{\theta^3} \quad \text{for } 0 < y < \theta.$$

Find the method of moments estimator for θ.

44.3 Suppose $Y_1, \ldots, Y_n$ are independent random variables with common density

$$f_\theta(y) = \frac{2\theta^2}{y^3} \quad \text{for } y > \theta.$$

(a) Find the method of moments estimator for θ.

(b) Find the bias and variance of your estimator in part (a)

(c) Using simulations, compare your estimator in part (a) to the MLE found in Exercise 43.6. (You can simulate from this density using your code from Exercise 19.12.) Use $\theta = 2$ and $n = 50$.

44.4 Let $Y_1, \ldots, Y_n$ be a random sample from a $\text{Beta}(1, \theta)$ population.

(a) Derive the method of moments estimator for θ.

(b) Derive the MLE for θ.

(c) Do simulations to compare the MSE for the estimators in part (a) and (b) for $n = 10$ and $\theta = 4$. That is, generate 100,000 samples of size 10 from a Beta(1,4) density, and for each sample compute both estimators and store them in vectors. Get histograms representing the densities of the two estimators (you can make these look nice by stacking them vertically and using the same breakpoints, for easy comparison). Then for each of the two estimators, find the average squared distance from the true value. Report the MSE for each and turn in your code.

44.5 Let $Y_1, \ldots, Y_n$ be a random sample from a $\text{Beta}(\theta, \theta + 1)$ population. Derive the method of moments estimator for θ.

44.6 Let $Y_1, \ldots, Y_n$ be a random sample from a $\text{Beta}(\theta, \theta)$ population. Derive the method of moments estimator for θ.

44.7 Suppose $Y_1, \ldots, Y_n$ are independent random variables with common density

$$f_\mu(y) = e^{\mu - y} \quad \text{for } y > \mu.$$

Find the method of moments estimator for μ, and compare it to the MLE derived in Exercise 43.5. Which has a smaller MSE?

44.8 For some $\lambda > 0$, suppose $X_1, \ldots, X_n$ is a random sample from a density

$$f(x) = \frac{\lambda}{2\sqrt{x}} e^{-\lambda \sqrt{x}} \quad \text{for } x > 0.$$

(a) Find the method of moments estimator for λ.

(b) Do simulations to compare the MSE for the estimator in (a) with the MSE for the MLE, found in Exercise 43.9. (*Hint:* You can simulate the data using the results of Exercise 19.16.)

44.9 Suppose $Y_1, \ldots, Y_n$ is a random sample from the density

$$f_\theta(y) = \theta y^{\theta-1} \ \text{ for } \ y \in (0,1),$$

where $\theta > 0$.

(a) Derive the method of moments estimator for θ.

(b) Compare the MSE of your answer to part (a), with the MSE of the MLE found in Exercise 43.13, using simulations for $\theta = .5$ and $\theta = 3$, both with $n = 20$.

44.10 Suppose $Y_1, \ldots, Y_n \overset{iid}{\sim} \text{Unif}(\theta, 2\theta)$, where $\theta > 0$.

(a) Find the method of moments estimator for θ.

(b) Compare the method of moments estimator to the MLE found in Exercise 43.14, either by computing the bias and variance of both estimators or through simulations.

44.11 Suppose $Y_1, \ldots, Y_n$ are a random sample from the density

$$f_\theta(y) = \frac{\theta}{(y+1)^{\theta+1}} \ \text{ for } \ y > 0$$

for some $\theta > 1$.

(a) Find the method of moments estimator for θ.

(b) Compare the method of moments estimator to the MLE found in Exercise 43.16, either by computing the bias and variance of both estimators or through simulations.

44.12 Suppose $Y_1, \ldots, Y_n$ are a random sample from the density

$$f_\theta(y) = 2\theta^2 y^{-3} \ \text{ for } \ y > \theta$$

for some $\theta > 0$.

(a) Find the method of moments estimator for θ.

(b) Find the bias and variance of your estimator using simulations R for $n = 20$ and $\theta = 5$. (You can use the answer to Exercise 19.6 to simulate from f_θ.)

(c) For the same $n = 20$ and $\theta = 5$, find the bias and variance of the MLE for θ, found in Exercise 43.12.

(d) Which of the two estimators has a smaller MSE for $n = 20$ and $\theta = 5$?

44.13 Let $Y_1, \ldots, Y_n$ be a random sample; the common density is

$$f_\theta(y) = 3\theta y^2 e^{-\theta y^3} \ \text{ for } \ y > 0$$

for some $\theta > 0$.

(a) Find the method of moments estimator for θ.

(b) Write some R code to check your answer, and to estimate the bias and variance when $n = 20$ and $\theta = 2$. You can use the code from Exercise 19.15 to simulate from the distribution.

(c) Compare your answer to that of Exercise 43.15.

44.14 Suppose that $Y_1, \ldots, Y_n$ are independent "double-exponential" random variables with density function

$$f_\theta(y) = \frac{1}{2\theta} e^{-|y|/\theta} \quad \text{for } y \in \mathbb{R}.$$

(a) Find the method of moments estimator for θ.

(b) Compare this estimator to that of Exercise 43.18, finding the bias and variance of each through simulations in R. Use $\theta = 2$ and $n = 12$.

44.15 Suppose that $Y_1, \ldots, Y_n$ are independent "double-exponential" random variables with density function

$$f_\mu(y) = \frac{1}{2} e^{-|y-\mu|} \quad \text{for } y \in \mathbb{R}.$$

(a) Find the method of moments estimator for μ.

(b) Compare this estimator to that of Exercise 43.19, finding the bias and variance of each through simulations in R. Use $\mu = 2$ and $n = 12$.

44.16 Suppose $Y_1, \ldots, Y_n$ is a random sample from a uniform density on $(-\theta, \theta)$ for some $\theta > 0$.

(a) Find a method of moments estimator.

(b) Find the MLE for θ.

(c) Do simulations to compare the MSE for the estimators in parts (a) and (b) for $n = 10$ and $\theta = 4$. Get histograms representing the densities of the two estimators. Report the MSE for each and turn in your code.

44.17 Suppose $Y_1, \ldots, Y_n$ is a random sample from the density shown below.

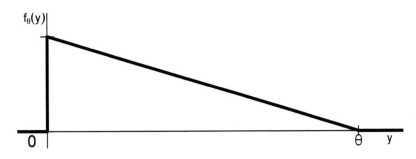

(a) Find the method of moments estimator for θ.

(b) Using simulations in R, compare the bias and variance of the estimator in part (a) with those of the estimator $\hat{\theta} = \max(Y_1, \ldots, Y_n)$ using $\theta = 2$ and $n = 20$. Which estimator has a smaller MSE? (To simulate from the density, you can use the code from Exercise 19.8.)

44.18 Let $X_1, \ldots, X_n$ be a random sample from the density shown below. Find the method of moments estimator for θ.

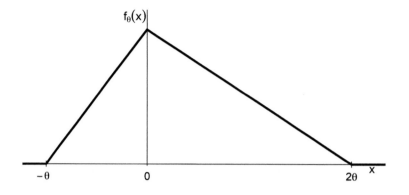

44.19 Let $X_1, \ldots, X_n$ be a random sample from the density shown below. Find the method of moments estimator for θ.

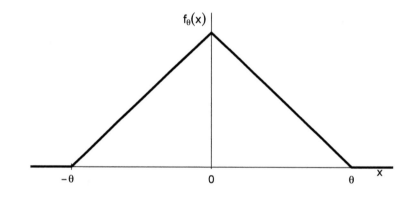

Chapter 45

Bayes Estimation

Bayes estimation is a whole new ball game. Most of the methods in this book are called "frequentist" or "classical" to distinguish them from Bayesian methods. In frequentist statistics, the parameter to be estimated is an unknown but fixed quantity. In the frequentist paradigm, it's incorrect to talk about the probability of a parameter θ being greater than 2, for example, and a parameter cannot have a probability distribution.

For the Bayesian approach, the parameter is treated as another random variable! Under the Bayesian model, the parameter has been generated from a known "prior" distribution and the data are then generated using that parameter. The data are combined with the "prior information," to produce a "posterior distribution" containing all of the information about the parameter. The Bayesian estimator is typically the mean of this posterior distribution.

This method of producing an estimator is optimal if the "true parameter" generating the data is indeed generated from the prior distribution. That is, if we simulate a parameter from a prior density, then simulate data using that parameter, then produce a Bayes estimator with the prior and the data, then the MSE for this estimator will be smaller than that for the MLE, or method of moments estimator, or any other frequentist estimator.

The problem, of course, is that we don't know what the "real prior" is that generated the parameter for our particular data set. In applications of Bayesian ideas, the prior distribution simply summarizes what we know (or believe, or guess) about the parameter, before we take any data.

In this chapter, we will write the joint density or mass function for the data, *given the parameter* θ, as $f_{Y|\theta}(y_1, \ldots, y_n|\theta)$, to emphasize that this is now a conditional density or mass function. The prior distribution is traditionally written as $\pi(\theta)$. There is a joint density for the sample and the parameter; we'll write this as $g_{Y,\theta}(y_1, \ldots, y_n, \theta)$. Of interest is the posterior density, which is $\pi(\theta|y_1, \ldots, y_n)$, i.e., the distribution of the parameter, given the observed data. Using our definition of conditional probability from Chapter 26, the posterior density is

$$\pi(\theta|y_1, \ldots, y_n) = \frac{g_{Y,\theta}(y_1, \ldots, y_n, \theta)}{g_Y(y_1, \ldots, y_n)},$$

where $g_Y(y_1, \ldots, y_n)$ is the marginal joint density for the data, with the parameter "integrated out." (It's important not to confuse the g_Y with $f_{Y|\theta}$. The latter is our "model" for the data, written as f_θ in other chapters, but we don't know the marginal distribution g_Y or the joint distribution $g_{Y,\theta}$, and we don't need to know them.)

Again using the conditional distribution definition in the numerator of our posterior, we have

$$
\begin{aligned}
\pi(\theta|y_1,\ldots,y_n) &= \frac{f_{Y|\theta}(y_1,\ldots,y_n|\theta)\pi(\theta)}{g_Y(y_1,\ldots,y_n)} \\
&= \frac{f_{Y|\theta}(y_1,\ldots,y_n|\theta)\pi(\theta)}{\int_{-\infty}^{\infty} f_{Y|\theta}(y_1,\ldots,y_n|\theta)\pi(\theta)d\theta}.
\end{aligned}
$$

We have written the posterior in terms of the prior and the "model," i.e., the conditional distribution of the data given the parameter.

We often can get away with considering only

$$
\pi(\theta|y_1,\ldots,y_n) \propto f_{Y|\theta}(y_1,\ldots,y_n|\theta)\pi(\theta),
$$

that is, the posterior is *proportional to* the joint density of the data (given the parameter value) times the prior. If we recognize the posterior distribution family, we don't need the proportionality constant. For the examples and exercises in this chapter, we will always recognize the posterior distribution. In the case where the posterior does not belong to a "named" family, numerical methods are typically employed to simulate the posterior distribution of the parameters; these are beyond the scope of this book.

The mean value of the posterior distribution is typically chosen as the Bayes estimator, but the mode or median can also be used. The uncertainty or variance of the estimator is also provided by the posterior distribution—this is a nice bonus that we will explore further in Chapter 49.

Example: Let's go through the Bayes estimation of a binomial proportion. To give the example a context, let's suppose we are interested in the proportion p of voters in Smalltown who support Proposal Q. We obtain a "random sample" of n voters from the population; suppose $X_1,\ldots,X_n$ are independent random variables, where $P(X_i = 1) = p$ and $P(X_i = 0) = 1 - p$ for $i = 1,\ldots,n$.

Let's describe our prior information about the parameter p using a prior density that is Beta(α,β). If we're pretty sure that the true p is somewhere "in the middle," we can use $\alpha > 1$ and $\beta > 1$ to get a "bump" shape to the prior. The prior density is

$$
\pi(p) \propto p^{\alpha-1}(1-p)^{\beta-1},
$$

where again we ignore the proportionality constant. It's strange at first to have a distribution for a parameter. Theoretically, the parameter p is drawn from the prior distribution and, given that parameter, data are collected from the joint mass function $f_{Y|p}$. In practice, the prior density is meant to represent the information the researchers have about the parameter, before any data are collected. This might be information from previous studies, or an "educated guess."

The joint mass function of the data, given p, is

$$
f(x_1,\ldots,x_n|p) = \prod_{i=1}^{n} p^{x_i}(1-p)^{1-x_i},
$$

where x_i are the observed (0–1) values, where $x_i = 1$ is assigned if the ith observed voter supports Proposal Q or, more generally, if the ith observation is a "success." The posterior density for p is proportional to the joint mass function, given p, times the prior for p:

$$\pi(p|x_1,\ldots,x_n) \propto \left[\prod_{i=1}^{n} p^{x_i}(1-p)^{1-x_i}\right]\left[p^{\alpha-1}(1-p)^{\beta-1}\right] = p^{s+\alpha-1}(1-p)^{n-s+\beta-1},$$

where $s = \sum_{i=1}^{n} x_i$ is the number of successes observed. You can recognize a Beta($s + \alpha, n - s + \beta$) density for the posterior distribution. That is, the combined information about the parameter p, from the prior information and from the data, is described by the posterior distribution.

The mean of the posterior distribution is

$$\tilde{p} = \frac{s+\alpha}{n+\alpha+\beta},$$

which is the Bayes estimator. Notice that when n is large, the Bayes estimator is close to the MLE $\hat{p} = s/n$. If the prior is uniform on $(0,1)$, then $\alpha = \beta = 1$ and the Bayes estimator is the $\tilde{p}$ of Chapter 42 on parameter estimators.

In the figure below we see a Beta($2,2$) prior density with two posterior densities. The prior density represents our information before any data are collected about the proportion of voters who support Proposal Q; we think it's likely to be somewhere in the middle of the range, but our information is still "vague." If we have a sample of size $n = 20$ voters, where $s = 13$ support the proposal, the information we have about the parameter is represented by the green density. If $n = 40$ and $s = 26$, the posterior density for the parameter is the red curve; as the sample size gets larger and we get more information, the posterior density gets taller and thinner.

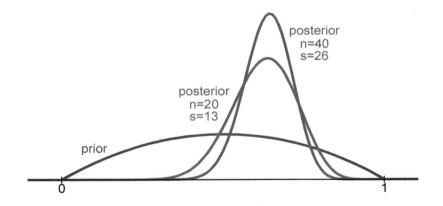

The beta distribution was a fortuitous choice for the family of densities for the prior information, because the posterior is easily recognized as also having a beta distribution. Technically we can use any prior density that describes our information about the parameter before the data are collected, but if we don't recognize the posterior family of densities, it's a lot more work to find the posterior mean.

The parameters α and β in the prior distribution are called **hyperparameters**, indicating that they are not the parameters of interest in the model of the joint density of the data. The hyperparameters describe the prior information about the parameter of interest. Once the family of prior densities is chosen, the hyperparameters are chosen to reflect what is known or guessed about the parameter before the data are collected.

Example: Let's derive a Bayes estimator for the mean μ of a normal population, when the population variance σ^2 is known. To put the example in a particular context, let's imagine that astronomers are interested in the distance of a star cluster from the earth. They will measure the "parallax" of n stars in the cluster, using a satellite telescope as well as a telescope on earth. They know the precision of the measurements: Each measurement is the true distance, plus a "random error," that can be modeled as normal with mean zero and standard deviation σ light-years.

For the prior information, the scientists can make educated guesses based on color and luminosity of the stars in the cluster, and this information can be summarized with a normal distribution with mean θ light-years and standard deviation τ light-years. That is, the hyperparameters θ and τ are chosen by the scientists to represent their guess about the cluster distance, and their uncertainty regarding this guess.

To summarize, $X_1, \ldots, X_n \overset{ind}{\sim} N(\mu, \sigma^2)$, where σ^2 is known and we have a normal prior $N(\theta, \tau^2)$ for μ. Then (leaving off all those tedious constants) the posterior density is proportional to the joint density of the data given μ, times the prior density for μ:

$$\pi(\mu | x_1, \ldots, x_n) \propto \left[\prod_{i=1}^{n} \exp\left(-\frac{(x_i - \mu)^2}{2\sigma^2} \right) \right] \exp\left(-\frac{(\mu - \theta)^2}{2\tau^2} \right)$$

$$= \exp\left(-\frac{1}{2} \left[\frac{\sum_{i=1}^{n}(x_i - \mu)^2}{\sigma^2} + \frac{(\mu - \theta)^2}{\tau^2} \right] \right),$$

which we can recognize as the form of a normal density for μ. The terms in the square brackets can be simplified:

$$\frac{\sum_{i=1}^{n}(x_i - \mu)^2}{\sigma^2} + \frac{(\mu - \theta)^2}{\tau^2} = \mu^2 \left[\frac{n}{\sigma^2} + \frac{1}{\tau^2} \right] - 2\mu \left[\frac{\sum_{i=1}^{n} x_i}{\sigma^2} + \frac{\theta}{\tau^2} \right] + c,$$

where c represents terms not containing μ. If we have a density for μ that is proportional to an expression $\exp\{-[a\mu^2 - 2b\mu + c]/2\}$, the mean of this density is b/a and the variance is $1/a$ (see Exercise 18.6). Therefore, after some further simplification, the posterior for μ is

$$\mu | x_1, \ldots, x_n \sim N\left(\frac{\tau^2 \sum_{i=1}^{n} x_i + \sigma^2 \theta}{\tau^2 n + \sigma^2}, \frac{\sigma^2 \tau^2}{n\tau^2 + \sigma^2} \right),$$

so our Bayes estimate of μ is

$$\tilde{\mu} = \frac{\tau^2 \sum_{i=1}^{n} x_i + \sigma^2 \theta}{n\tau^2 + \sigma^2}.$$

When n or τ is large, this is close to the sample mean. Further, the variance of the posterior is close to the variance of the sample mean, when n is large or τ is large.

We've derived the posterior density for μ using a normal prior; let's go back to our star cluster example, plug in some numbers, and look at these distributions. Let $\sigma = 50$ be the known standard deviation of the measurements. Suppose the scientists' best guess at the distance (before any data were collected) was $\theta = 400$ light-years, and $\tau = 200$ expresses the fact that this guess is not very precise. The prior density is shown below in blue. Now suppose that the average measured distance of 10 stars in the cluster is 582 light-years. Plugging in $n = 10$, $\sum_{i=1}^{10} x_i = 5820$, and the hyperparameter values, we get a posterior with mean 580.9 and standard deviation 15.8; the posterior density is shown in green below; we can see that our information is much more precise after our ten measurements.

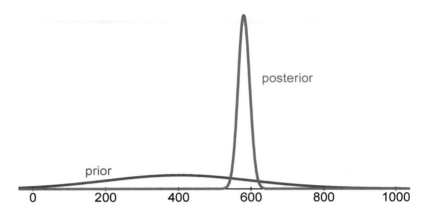

The choice of a prior is an important first step in Bayes estimation. Prior densities are typically chosen to be "vague," or have large variance, so that the information in the data dominates the information in the prior. Also, the choice of the family of priors can make the posterior information easy to plot and to interpret.

Definition: A prior density is called a **conjugate prior** if the resulting posterior is in the same family as the prior.

Note that both of our previous examples of prior densities are conjugate priors. A very common approach to Bayes estimation is to determine the family of conjugate priors, and choose the hyperparameters to correspond to the prior information.

Example: To estimate the parameter for an exponential population, the conjugate prior density is either gamma or inverse gamma, depending on whether the rate or mean parameter is to be modeled, respectively. To illustrate the use of the inverse gamma prior for the mean of an exponential distribution, we'll look at modeling survival times for leukemia patients, using the data set `leuk` in the R data library `MASS`. There are two groups of patients, explained in the data description:

> Survival times are given for 33 patients who died from acute myelogenous leukaemia. Also measured was the patient's white blood cell count at the time of diagnosis. The patients were also factored into 2 groups according to the presence or absence of a morphologic characteristic of white blood cells. Patients termed AG positive were identified by the presence of Auer rods and/or significant granulation of the leukaemic cells in the bone marrow at the time of diagnosis.

Let's model the survival time (in weeks) of patients whose test result is "absent" as exponential with mean θ (we'll ignore the information about the white blood cell count, although that could be incorporated in a more sophisticated survival model). The prior information for θ is summarized in InvGamma(α, β):

$$\pi(\theta) \propto \frac{1}{\theta^{\alpha+1}} e^{-\beta/\theta} \quad \text{for } \theta > 0$$

and the joint density for the (independent) observations, given θ, is

$$f(y_1, \ldots, y_n | \theta) = \prod_{i=1}^{n} \left[\frac{1}{\theta} e^{-y_i/\theta} \right] = \frac{1}{\theta^n} \exp \left\{ -\sum_{i=1}^{n} y_i/\theta \right\}.$$

Therefore the posterior for θ is

$$\pi(\theta | y_1, \ldots, y_n) \propto \frac{1}{\theta^n} \exp \left\{ -\sum_{i=1}^{n} y_i/\theta \right\} \frac{1}{\theta^{\alpha+1}} e^{-\beta/\theta}$$

$$= \frac{1}{\theta^{n+\alpha+1}} \exp \left\{ -\left(\beta + \sum_{i=1}^{n} y_i \right) /\theta \right\},$$

which we recognize is $\text{InvGamma}(n + \alpha, \sum_{i=1}^{n} y_i + \beta)$.

Suppose the oncologists can summarize their prior information with a distribution having mean 20 and standard deviation 40. This is a "vague" prior, because the standard deviation is twice the mean, which indicates that the oncologists are not very confident about their guess of 20 weeks as the average survival time. To find the inverse gamma parameters α and β we solve (see Chapter 16 highlights for the mean and variance of an inverse gamma distribution)

$$\frac{\beta}{\alpha - 1} = 20 \quad \text{and} \quad \frac{\beta^2}{(\alpha - 1)^2 (\alpha - 2)} = 1600.$$

This gives $\alpha = 9/4$ and $\beta = 25$.

The sum of the $n = 16$ survival times in weeks for AG negative patients is found to be 287, using sum(leuk$time[leuk$ag=="absent"]). Therefore, the posterior distribution for θ is $\text{InvGamma}(18.25, 312)$. This is shown in the plot below, along with the prior density. The posterior mean is 18.1, but according to the posterior, a large range of values for θ are plausible, from about 10 to about 30, perhaps.

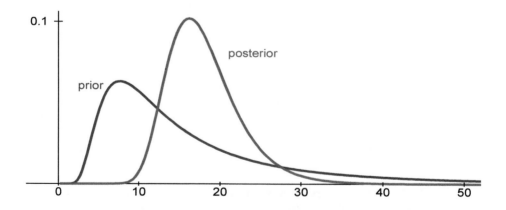

One common objection to Bayes methods, as opposed to frequentist methods, is the fact that the prior is typically "just a guess"—what if the guess is incorrect? What if Expert A has one guess and Expert B has a different guess? These are valid concerns, but in practice the Bayes methods behave pretty well if the prior is either appropriate for the "true" parameter value or simply "vague." For example, let's look again at estimating the binomial parameter, using a Beta$(2, 2)$ prior. It's straightforward to compute the MSE for the sample proportion—this is the variance, $\text{MSE}(\hat{p}) = p(1 - p)/n$. For the Bayes estimator, the variance is smaller: $\text{V}(\tilde{p}) = np(1 - p)/(n + 4)^4$, but the estimator has bias $\text{B}(\tilde{p}) = (2 - 4p)/(n + 4)$. In the three plots below, the MSE is plotted against values of p, where the MSE for the sample proportion is shown as the solid curve, and the MSE for the Bayes estimator is the dashed curve. When the prior guess of "in the middle" is correct, the Bayes estimator "wins," but the MLE performs better when the population proportion p is near zero or near one. The differences in the estimators are largest for smaller sample sizes, and as n grows, the estimates as well as the MSEs get closer together.

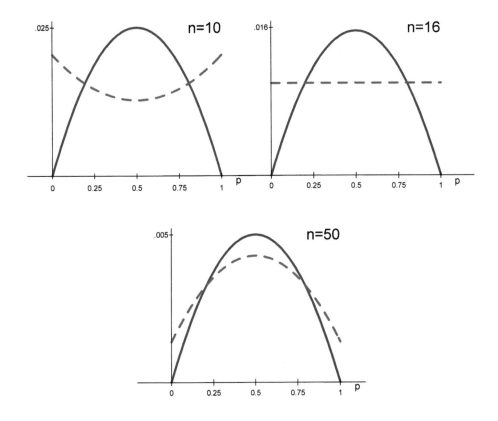

Of course, with the Bayes posterior density, we get more than an estimate of the parameter. We have a distribution of values for the parameter, which also gives some information about the uncertainty of the parameter estimate. We'll revisit these ideas in Chapter 49.

We have considered only models with one unknown parameter, where we recognize our posterior distribution family. However, Bayes ideas are widely applied to large and "messy" models with many parameters, where the joint posterior distribution for the parameters can be simulated with numerical methods. In fact, Bayes methods can be used to tackle problems where traditional statistical methods are intractable. Practitioners also appreciate being able to incorporate prior information, whether from educated guesses or from previous studies.

Chapter Highlights

1. The Bayesian framework for statistical modeling treats the model parameters as random variables.

2. Given a prior distribution $\pi(\theta)$ for the model parameter θ, the posterior distribution for θ is proportional to the joint density (or mass function) for the data, given the parameter, times the prior density. That is,

$$\pi(\theta|y_1,\ldots,y_n) \propto f_{Y|\theta}(y_1,\ldots,y_n|\theta)\pi(\theta).$$

3. If the posterior density for the parameter is from the same family as the prior density, the latter is called a conjugate prior.

4. The Bayes estimator can be defined as the mean of the posterior density.

Exercises

45.1 Let $X_1,\ldots,X_n$ be a random sample from a Poisson(λ) population. Using a Gamma(α,β) prior distribution for λ, find the posterior distribution for λ.

(a) What is the Bayes estimator, i.e., the posterior mean?

(b) Is the prior a conjugate prior? Explain.

(c) Suppose $n = 10$, $\lambda = 4$, $\alpha = 8$, $\beta = 2$, and $\sum_{i=1}^{n} X_i = 48$. Sketch the prior and posterior densities on the same axis.

(d) Suppose $n = 10$, $\lambda = 4$, $\alpha = .8$, $\beta = .2$, and $\sum_{i=1}^{n} X_i = 48$. Sketch the prior and posterior densities on the same axis. Note that the prior mean is the same, but the prior variance is much larger, so that this prior is "vague" compared to the prior in part (b).

(e) Compare the Bayes estimator to the MLE for λ by computing the MSEs for each of the two estimators in the following eight scenarios. Summarize your findings; these will, in general, be representative of most Bayes/MLE comparisons:

(1) $n = 10$, $\lambda = 4$, $\alpha = 8$, $\beta = 2$ (small sample, good prior),
(2) $n = 10$, $\lambda = 4$, $\alpha = .8$, $\beta = .2$ (small sample, vague, good prior),
(3) $n = 10$, $\lambda = 8$, $\alpha = 8$, $\beta = 2$ (small sample, bad prior),
(4) $n = 10$, $\lambda = 8$, $\alpha = .8$, $\beta = .2$ (small sample, vague, bad prior),
(5) $n = 100$, $\lambda = 4$, $\alpha = 8$, $\beta = 2$ (large sample, good prior),
(6) $n = 100$, $\lambda = 4$, $\alpha = .8$, $\beta = .2$ (large sample, vague, good prior),
(7) $n = 100$, $\lambda = 8$, $\alpha = 8$, $\beta = 2$ (large sample, bad prior),
(8) $n = 100$, $\lambda = 8$, $\alpha = .8$, $\beta = .2$ (large sample, vague, bad prior).

45.2 Suppose $Y_1,\ldots,Y_n$ is a random sample from an exponential distribution with mean $1/\beta > 0$, that is, the population density is

$$f(y|\theta) = \beta e^{-\beta y} \quad \text{for} \quad y > 0.$$

(a) Compute the MLE for β.

(b) Use a Gamma(a, b) prior and find the posterior density for β, and use the posterior mean as the Bayes estimator.

(c) Is the prior in (b) a conjugate prior? Explain.

(d) Suppose the $n = 8$ observations are 0.07, 0.28, 0.02, 0.26, 0.94, 1.68, 0.68, 0.10. Compute the MLE and the Bayes estimator, using $a = b = 1$ for the hyperparameters.

(e) Do simulations to compare the MSEs of the ML and Bayes estimators. Suppose the truth is $\beta = 1$, and use $n = 8$, $a = .1$, and $b = .1$. (Note that the mean of the prior is equal to the true value.)

(f) Repeat part (e), using $\beta = 5$. In this case the prior mean is more than one prior standard deviation from the truth.

(g) The idea of a "true parameter" is a frequentist way of thinking! Do simulations where, within the loop, the parameter is generated from the prior distribution, Gamma$(.1, .1)$, and the data with $n = 8$ generated from that parameter. Report the MSE for the Bayes estimator and the MLE. (When the parameter is sampled from the prior, the Bayes estimator will always have the smaller MSE.)

45.3 Let $Y_1, \ldots, Y_n$ be a random sample from a normal population with mean zero and variance σ^2, where interest is in estimating the variance of the population. Define the "precision parameter" $\tau = 1/\sigma^2$, and write the population density as

$$f_{Y|\tau}(y|\tau) = \sqrt{\frac{\tau}{2\pi}} e^{-\tau y^2/2}.$$

Find a Bayes estimate of the precision τ, using a Gamma(α, β) prior. Is the prior conjugate? Explain.

45.4 Suppose $\theta > 0$ and $Y_1, \ldots, Y_n$ is a random sample from a population with density

$$f(y|\theta) = 2y\theta e^{-\theta y^2} \quad \text{for } y > 0.$$

(a) Compute the method of moments for θ.

(b) Compute the MLE for θ.

(c) Find the Bayes estimator (posterior mean) using a gamma prior. Is this prior conjugate?

(d) Do simulations to compare the MSE of your three estimators. (*Hint*: Exercise 19.15 asked you to write code to simulate from a very similar distribution.) Choose a value of θ, and a "moderate" sample size, say $n = 20$, and choose two hyperparameters for your Bayes estimator. The first hyperparameter is such that the prior mean is close to the true population mean for your given θ, and the second is chosen so that the prior mean is far from the true population mean. Compare the four estimators (MLE, method of moments, Bayes with "good" prior, and Bayes with "bad" prior) in terms of MSE.

45.5 For some $\theta \in (0, 1)$, define a density on the interval $(0, 2)$:

$$f(x|\theta) = \begin{cases} \theta & \text{on } (0, 1], \\ 1 - \theta & \text{on } (1, 2). \end{cases}$$

(a) Draw a picture of this density.

Suppose $X_1, \ldots, X_n$ is a random sample from $f(x|\theta)$.

(b) Find the MLE for θ. (*Hint:* Let Y be the number of observed X_i that are less than one.)

(c) Consider a uniform(0,1) prior for θ and determine the Bayes estimator (posterior mean).

45.6 We have a random sample $Y_1, \ldots, Y_n$ from a population with density function

$$f(y|\theta) = \frac{\theta}{2} \left(\frac{y}{2} + 1 \right)^{-(\theta+1)}, \quad y > 0,$$

for some $\theta > 0$ to be estimated. Use a Gamma(α, β) prior for θ, and determine the posterior density for θ. Is the prior conjugate?

45.7 Suppose $\theta > 2$ and $Y_1, \ldots, Y_n$ are a random sample from a population with density function

$$f(y|\theta) = \frac{\theta}{(y+1)^{\theta+1}}, \quad y > 0.$$

Use a Gamma(α, β) prior to find the posterior distribution for θ. What is the posterior mean?

45.8 Suppose we have a random sample $Y_1, \ldots, Y_n$ from a population with density

$$f_{Y|\mu}(y|\mu) = \frac{1}{y\sqrt{2\pi}} \exp \left\{ \frac{[\log(y) - \mu]^2}{2} \right\} \quad \text{for } y > 0,$$

and interest is in estimating μ. Determine a conjugate family of prior densities, and find the posterior density. What is the Bayes estimate (posterior mean) for μ?

45.9 Geologists are studying earthquake activity near an active fault in East Asia, and they are interested in the time between tremors with magnitude 4.0 or higher. They model the time Y between tremors in days as following the density

$$f(y|\theta) = \theta^2 y e^{-2\theta y} \quad \text{for } y > 0.$$

The average time between tremors is then $1/\theta$. Based on historical activity near a similar fault, they use a Gamma$(.25, 6.25)$ prior for θ. This prior has a mean of .04 "inverse days" and a standard deviation of .08, reflecting a "vague" prior with a "guess" of about 25 days between tremors.

(a) Suppose the average of the next $n = 8$ times between tremors with magnitude 4.0 or higher is 45.2 days. What is the Bayes estimate of θ, the inverse of the average time between tremors with magnitude 4.0 or higher?

(b) Suppose the average of the next $n = 40$ times between tremors with magnitude 4.0 or higher is 45.2 days. What is the Bayes estimate of of θ?

(c) Plot the prior density, along with the posteriors from (a) and (b), on the same axes.

45.10 The data set **ex0428**, "Darwin's Data," in the library **Sleuth2** of the R project has this description:

> Plant heights (inches) for 15 pairs of plants of the same age, one of which was grown from a seed from a cross-fertilized flower and the other of which was grown from a seed from a self-fertilized flower.

Suppose we want to model the difference μ in heights of cross- and self-fertilized flowers, using a normal density. We'll also suppose that we know the variance of differences in heights to be $\sigma^2 = 20$, simply because a two-parameter model is beyond the scope of this book. Use a normal prior for μ, with mean zero and large variance, say $\tau^2 = 100$.

(a) Find the Bayes estimator for the difference μ in heights.

(b) Make a plot of the posterior distribution. Would you say there is evidence that the true difference μ is larger than zero?

45.11 Consider again "Darwin's Data" described in Exercise 45.10. This time, interest is in estimating the probability p that the cross-fertilized plant is taller than the self-fertilized plant. Using a uniform prior, find the posterior distribution for p, and sketch it. Would you say there is evidence that p is greater than 1/2?

45.12 The data set **ex0621**, "Failure Times of Bearings," in the data library **Sleuth2** of the R project has this description:

> Data consist of times to fatigue failure (in units of millions of cycles) for 10 high-speed turbine engine bearings made from five different compounds.

Suppose we model the lifetimes of ball bearings made with Compound IV as being exponentially distributed, and interest is in modeling θ, the average lifetime.

(a) An expert engineer guesses that the average lifetime is about ten million cycles, but he is not very confident of this guess. Formulate an inverse gamma prior with mean 10 and variance 100. Use this prior and the data for Compound IV to find a posterior mean for the estimate of the average lifetime.

(b) Another engineer claims to have more expert knowledge about this compound, and her prior guess is eight million cycles for the average lifetime. Formulate an inverse gamma prior with mean 8 and variance 16 to reflect the higher confidence by this engineer. Use this prior and the data for Compound IV to find a posterior mean for the estimate of the average lifetime.

(c) Suppose there are 100 observations instead of 10, but the mean of the observations is the same. Compute the two estimates using the priors of parts (a) and (b). Comment on the difference in the estimates, compared with the difference of the estimates in (a) and (b).

Chapter 46

Consistency of Point Estimators

Conceptually, an estimator is **consistent** if it "converges" to its target parameter as the sample size gets larger; that is, the estimator has to get closer to the thing it's estimating as we get more and more data. Consistency is a "minimum requirement" for any respectable estimator.

To talk about consistency, we have to think of a sequence of random samples of size n, where n is increasing without bound, and a sequence of estimators $\hat{\theta}_n$, calculated from the sample using the formula that defines the estimator. It is a "thought exercise" because of course we never actually have such a sequence of samples that are increasing in size. However, we would like to be assured that *if we did*, our estimator would become increasingly closer to its target.

We say that an estimator $\hat{\theta}_n$ is **weakly consistent** for θ if for any $\delta > 0$,

$$\lim_{n \to \infty} \mathrm{P}\left(|\hat{\theta}_n - \theta| > \delta\right) = 0.$$

In words, if the probability that the estimator is more than a distance δ from its target goes to zero as the sample size increases, for any positive δ, then we say the estimator is weakly consistent.

Example: Show that the maximum of a random sample from a $\mathrm{Unif}(0, \theta)$ density is consistent for $\theta > 0$. Let's start by using the definition. For any $\delta > 0$,

$$\lim_{n \to \infty} \mathrm{P}\left(|Y_{(n)} - \theta| > \delta\right) = \lim_{n \to \infty} \mathrm{P}\left(Y_{(n)} < \theta - \delta\right)$$

$$= \lim_{n \to \infty} \left(\frac{\theta - \delta}{\theta}\right)^n$$

$$= 0.$$

We say an estimator is **consistent in mean squared error (MSE)** if the MSE of the estimator goes to zero as the sample size increases. Because the MSE is the sum of the variance and the squared bias, *both* the variance and the bias have to shrink to zero to get consistency in MSE. This is often easier to check than weak consistency, and is actually stronger, in the sense that consistency in MSE implies weak consistency.

Let's prove this: Suppose $V(\hat{\theta}_n) \to 0$ and $B(\hat{\theta}_n) \to 0$ as $n \to \infty$, and we'll show that $\hat{\theta}_n$ is weakly consistent. Using $|a| + |b| \geq |a + b|$ for any real numbers a and b, we have

$$P\left(|\hat{\theta}_n - \theta| > \delta\right) \leq P\left(|\hat{\theta}_n - E(\hat{\theta}_n)| + |E(\hat{\theta}_n) - \theta| > \delta\right)$$

$$= P\left(\frac{|\hat{\theta}_n - E(\hat{\theta}_n)|}{\sqrt{V(\hat{\theta}_n)}} + \frac{|E(\hat{\theta}_n) - \theta|}{\sqrt{V(\hat{\theta}_n)}} > \frac{\delta}{\sqrt{V(\hat{\theta}_n)}}\right)$$

$$= P\left(\frac{|\hat{\theta}_n - E(\hat{\theta}_n)|}{\sqrt{V(\hat{\theta}_n)}} > \frac{(\delta - |B(\hat{\theta}_n)|)}{\sqrt{V(\hat{\theta}_n)}}\right)$$

$$\leq \frac{V(\hat{\theta}_n)}{(\delta - |B(\hat{\theta}_n)|)^2} \to 0,$$

by hypothesis, where the last line uses Chebyshev's inequality. The numerator goes to zero as n increases, but the denominator goes to δ, so the fraction converges to zero.

There is also a type of "strong" consistency, which implies both consistency in MSE and weak consistency. Strong consistency is beyond the scope of this book, but in practice, the term "consistent" without a modifier refers to (at least) weak consistency.

Example: Let's look again at a random sample from a $\text{Unif}(0, \theta)$ population for some $\theta > 0$. Now we can show that the method of moments estimator $\tilde{\theta} = 2\bar{Y}$ is also consistent for θ. We have already shown that this estimator is unbiased, and

$$V(\tilde{\theta}) = 4V(\bar{Y}) = \frac{4\theta^2}{12n} = \frac{\theta^2}{3n} \to 0 \quad \text{as} \quad n \to \infty.$$

Although we have shown that the maximum likelihood estimator $Y_{(n)}$ has uniformly smaller MSE, compared to $2\bar{Y}$, this method of moments estimator is also consistent.

Example: Suppose $Y_1, \ldots, Y_n$ are independent Bernoulli random variables with probability p of success. The sample mean $\bar{Y} = \sum_{i=1}^{n} Y_i/n$ is easily seen to be consistent in MSE and, hence, weakly consistent. But let's consider estimating the *odds* of success, defined as

$$\theta = \frac{p}{1 - p},$$

that is, the odds of success is the probability of success divided by the probability of failure. It's natural to consider the estimator

$$\hat{\theta} = \frac{\bar{Y}}{1 - \bar{Y}},$$

and we can show that this estimator of the odds is weakly consistent. Beginning with the definition of weak consistency, choose any $\delta > 0$; then

$$P(|\hat{\theta} - \theta| < \delta) = P(\theta - \delta < \hat{\theta} < \theta + \delta)$$

$$= P\left(\frac{\theta - \delta}{1 + \theta - \delta} < \bar{Y} < \frac{\theta + \delta}{1 + \theta + \delta}\right)$$

$$= P\left(\frac{p(1 + \delta) - \delta}{1 - (1 - p)\delta} < \bar{Y} < \frac{p(1 - \delta) + \delta}{1 + (1 - p)\delta}\right)$$

$$\geq P\left(p - \delta < \bar{Y} < p + \delta\right),$$

and the last term goes to one by the consistency of $\bar{Y}$.

It is easy to see, however, that our estimator of odds is *not* consistent in MSE! For any n and p, there is a finite, nonzero probability that $\hat{p} = 1$. As n gets larger, this probability decreases, but because it is nonzero, the expected value and the variance of our estimator for odds is infinity for any finite sample size! There are a few examples of estimators that are weakly consistent but not consistent in MSE—this is one of them.

Chapter Highlights

1. An estimator $\hat{\theta}_n$ is **weakly consistent** for θ if for any $\delta > 0$,

$$\lim_{n \to \infty} P\left(|\hat{\theta}_n - \theta| > \delta\right) = 0.$$

2. If the MSE of an estimator goes to zero as n gets large, then the estimator is **consistent in MSE**.

3. If an estimator is consistent in MSE, then it is also weakly consistent.

4. In practice, we use the term "consistent" to mean "at least weakly consistent."

5. Therefore if the bias and variance both go to zero as n gets large, then the estimator is consistent.

Exercises

46.1 Suppose $Y_1, \ldots, Y_n$ are a random sample from the density

$$f_\beta(y) = \frac{1}{\beta} e^{-(y-2)/\beta} \quad \text{for } y > 2.$$

(a) Find the method of moments estimator for β.

(b) Show that your estimator in part (a) is consistent.

46.2 Suppose $Y_1, \ldots, Y_n$ are a random sample from the density

$$f_\theta(y) = \frac{6}{\theta^3} y(\theta - y) \quad \text{for } y \in (0, \theta).$$

(a) Find a constant a so that $\hat{\theta} = a\bar{Y}$ is an unbiased estimator.

(b) Show that your estimator in part (a) is consistent.

46.3 Suppose $Y_1, \ldots, Y_n$ are independent observations from

$$f_\theta(y) = \frac{3(\theta - y)^2}{\theta^3} \quad \text{for } y \in [0, \theta],$$

and let $Y_{(n)}$ be the maximum of the $Y_1, \ldots, Y_n$. Show that $Y_{(n)}$ is a consistent estimator of θ.

46.4 If $X_1, \ldots, X_n$ is a random sample from a population with mean μ and variance σ^2,

(a) is $\bar{X}^2$ an unbiased estimator of μ^2?

(b) what are the conditions on the population necessary to ensure that $\bar{X}^2$ is consistent for μ^2?

46.5 For a random sample $Y_1, \ldots, Y_n$ from an exponential population with mean θ, suppose we are interested in estimating θ^2, the variance of the population. Statistician A recommends the sample variance, pointing out that this is unbiased for the population variance. Statistician B recommends using $\bar{Y}^2$, pointing out that this is consistent for θ^2.

(a) Show that $\bar{Y}^2$ is consistent for θ^2. Is the estimator unbiased?

(b) Compare the MSEs two estimators through simulations: for $\theta = 4$ and $n = 20$, report the MSEs for the two estimators.

46.6 Suppose $Y_1, \ldots, Y_n$ is a random sample from a Gamma$(2, 1/\theta)$ density. Let $\bar{Y}$ be the sample mean and show that $\hat{\theta} = \bar{Y}/2$ is a consistent estimator for θ.

46.7 Suppose $Y_1, \ldots, Y_n$ are independent exponential random variables with mean $\theta > 0$, and let $\hat{\theta} = n \min(Y_i)$.

(a) Is $\hat{\theta}$ an unbiased estimator of θ? Explain.

(b) Is $\hat{\theta}$ a consistent estimator of θ? Explain.

46.8 Consider a random sample $Y_1, \ldots, Y_n$ from the density shown in the plot below. Show that $3\bar{Y}$ is a consistent estimator of θ.

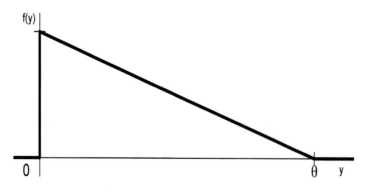

46.9 Show that the method of moments estimator for the density in Exercise 44.18 is consistent for θ.

46.10 Let $X_1, \ldots, X_n$ be a random sample from the density shown below. Find a consistent estimator for θ and show that it is consistent.

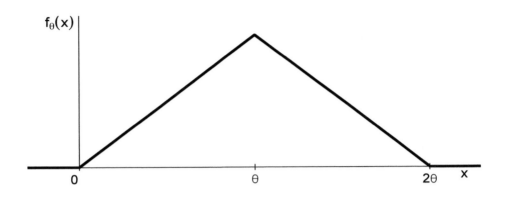

Chapter 47

Modes of Convergence of Random Variables and the Delta Method

In this chapter we consider any sequence of random variables $Y_1, Y_2, \ldots$; this could be a sequence of estimators such as $\hat{\theta}_1, \hat{\theta}_2, \ldots$ for growing sample sizes. There are several ways in which such a sequence of random variables can be said to "converge," and we will look at two of these: **convergence in probability** to a constant value, and **convergence in distribution** to some limiting probability distribution. We have seen these concepts before: A sequence of estimators is consistent if it converges in probability to its target, and the central limit theorem is an example of convergence in distribution. We'll discuss both types of convergence in more generality, in preparation for the **delta method** of estimating variances of random variables. The delta method can be considered an extension of the central limit theorem and is useful for approximating the distribution of an estimator. In a later chapter, we will use the delta method for constructing approximate confidence intervals.

Convergence in probability

We say "Y_n converges in probability to a constant c" if for any $\delta > 0$,

$$\lim_{n \to \infty} \mathrm{P}(|Y_n - c| > \delta) = 0$$

or, equivalently, for any $\delta > 0$,

$$\lim_{n \to \infty} \mathrm{P}(|Y_n - c| < \delta) = 1.$$

We can indicate convergence in probability with more compact notation:

$$Y_n \xrightarrow{p} c \ \text{ as } \ n \to \infty,$$

and this kind of convergence is sometimes called "weak convergence," analogously to the weak consistency of the last chapter. If $\hat{\theta}_n$ is a statistic calculated using a sample of size n from a distribution with parameter θ, then $\hat{\theta}_n \xrightarrow{p} \theta$ as the sample size $n \to \infty$ is another way of saying that $\hat{\theta}_n$ is a consistent estimator of θ.

An important application of this concept is the **law of large numbers**: Let $Y_1, Y_2,$ $\ldots$ be iid random variables with common mean μ and common (finite) variance σ^2.

Define $\bar{Y}_n$ to be the mean of the first n random variables, that is,

$$\bar{Y}_n = \frac{1}{n} \sum_{i=1}^{n} Y_i.$$

Then $\bar{Y}_n \xrightarrow{p} \mu$.

The proof is an application of Markov's inequality from Chapter 20. Recall that if X is any random variable with finite mean and support on the positive reals, then $P(X > 1) \leq E(X)$. Choose any $\delta > 0$. Then

$$\begin{aligned}
P\left(|\bar{Y}_n - \mu| > \delta\right) &= P\left([\bar{Y}_n - \mu]^2 > \delta^2\right) \\
&= P\left(\frac{[\bar{Y}_n - \mu]^2}{\delta^2} > 1\right) \\
&\leq E\left(\frac{[\bar{Y}_n - \mu]^2}{\delta^2}\right) \\
&= \frac{\sigma^2}{\delta^2 n} \to 0 \text{ as } n \to \infty,
\end{aligned}$$

because $E\left([\bar{Y}_n - \mu]^2\right) = \sigma^2/n$, the variance of the sample mean.

This rule is used colloquially, perhaps more than any other result from probability theory. "You win some, you lose some, it averages out" is perhaps an application of the law of large numbers to life in general. The law of large numbers "guarantees" that casinos will make a profit! If the probability of the house winning each game is, say, .52, then to the individual gamblers who play only a few games, this probability might be indistinguishable from 1/2, but over time the house steadily wins.

The law of large numbers is sometimes incorrectly used to predict a particular event. For example, suppose a gambler notices that the marble for the roulette wheel has landed on black for the last eight spins. This is unusual since the probability of black is close to 1/2. He or she might feel that a red is "due" to occur, and might claim this is because of the law of large numbers! However, we know that the spins are independent, and the probability of the next spin being red doesn't depend on the last spins. The law of large numbers refers to an average of many outcomes, not any particular outcome. It is true that over time, the observed proportion of "red" spins will converge to the probability of an individual red spin, but during this time there will be "runs" of consecutive black spins and consecutive red spins.

These results are useful in many applications:

R1. If $X_n \xrightarrow{p} a$ and $Y_n \xrightarrow{p} b$, and $W_n = X_n + Y_n$, then $W_n \xrightarrow{p} a + b$.

R2. If $X_n \xrightarrow{p} a$ and $Y_n \xrightarrow{p} b$, and $W_n = X_n Y_n$, then $W_n \xrightarrow{p} ab$.

R3. If $X_n \xrightarrow{p} a$ and g is a function that is continuous at a, then $g(X_n) \xrightarrow{p} g(a)$.

For example, if $\hat{\theta}_n$ is a consistent estimator for θ, and $X_n \xrightarrow{p} 0$, then $\hat{\theta} + X_n$ is also consistent for θ, by R1. Or, if $\hat{\theta}_n$ is a consistent estimator for θ, and $X_n \xrightarrow{p} 1$, then $\hat{\theta} X_n$ is also consistent for θ, by R2. By R3, if $\hat{\theta}_n$ is a consistent estimator for θ, then $\hat{\theta}_n^2$ is a consistent estimator for θ^2. Also by R3, we find that the sample standard deviation S is a consistent estimator for the population standard deviation σ whenever S^2 is a consistent estimator for the population σ^2.

Proof of R1: Suppose $X_n \xrightarrow{p} a$ and $Y_n \xrightarrow{p} b$. Then

$$
\begin{aligned}
\mathrm{P}\left(|(X_n + Y_n) - (a+b)| > \delta\right) &\leq \mathrm{P}\left(|X_n - a| + |Y_n - b| > \delta\right) \\
&\leq \mathrm{P}\left(|X_n - a| > \delta/2 \ \text{ or } \ |Y_n - b| > \delta/2\right) \\
&\to 0 \ \text{ as } \ n \to \infty,
\end{aligned}
$$

because both $X_n \xrightarrow{p} a$ and $Y_n \xrightarrow{p} b$ as $\to \infty$.

Proof of R2: Suppose $X_n \xrightarrow{p} a$ and $Y_n \xrightarrow{p} b$, and we'll start with the case $a, b > 0$. First choose any $\delta > 0$, then choose any $\xi_1, \xi_2 > 0$, where $(a + \xi_1)(b + \xi_2) = ab + \delta$.

The idea behind the proof can be seen in the plot on the right. The solid curve is $xy = ab + \delta$ and the point $(a + \xi_1, b + \xi_2)$, for positive ξ_1 and ξ_2, is on the curve. Then for any joint probability distribution $f(x, y)$, we must have $\mathrm{P}(XY > ab + \delta) < \mathrm{P}(X > a + \xi_1 \ \text{ or } \ Y > b + \xi_2)$.

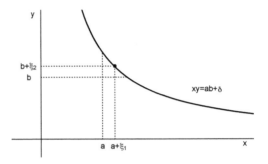

If we set for any $\delta > 0$, $\xi_1 = a + \delta/(2b)$ and $\xi_2 = (ab + \delta)/\xi_1$, we have

$$
\begin{aligned}
\mathrm{P}(X_n Y_n > ab + \delta) &< \mathrm{P}(X_n > a + \xi_1 \ \text{ or } \ Y_n > b + \xi_2) \\
&\leq \mathrm{P}(X_n > a + \xi_1) + \mathrm{P}(Y_n > b + \xi_2) \to 0,
\end{aligned}
$$

by convergence of X_n and Y_n. We can make a similar argument for any $0 < \delta < ab$ by drawing the curve $xy = ab - \delta$. Then for $\xi_1 = a - \delta/(2b)$ and $\xi_2 = (ab - \delta)/\xi_1$,

$$
\begin{aligned}
\mathrm{P}(X_n Y_n < ab - \delta) &< \mathrm{P}(X_n > a - \xi_1 \ \text{ or } \ Y_n > b - \xi_2) \\
&\leq \mathrm{P}(X_n > a - \xi_1) + \mathrm{P}(Y_n > b - \xi_2) \to 0,
\end{aligned}
$$

by convergence of X_n and Y_n. If $a < 0$, we can consider convergence of the sequence $-X_n$ and, similarly, if $b < 0$. The case with $a = 0$ or $b = 0$ is the subject of Exercise 47.1.

Proof of R3: Let's recall the definition of continuity from our calculus class.

A function g in an interval I is continuous at a in the interior of I if, for every $\epsilon > 0$, there is a $\delta > 0$ such that whenever x is within δ units of a, $g(x)$ is within ϵ units of $g(a)$. This idea is shown on the plot to the right.

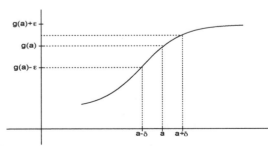

Note that this implies that if $g(x)$ is *more* than ϵ units away from $g(a)$, then x *can't* be within δ units of a; this is the key to proving R3. Start with the definition of convergence in probability, using the random variable $g(X_n)$. Choose any $\epsilon > 0$; then there is a $\delta > 0$ as in the definition of continuity. It follows that

$$P\left(|g(X_n) - g(a)| > \epsilon\right) \le P\left(|X_n - a| > \delta\right)$$
$$\to 0,$$

because of the convergence of X_n.

Convergence in distribution

Let $Y_1, Y_2, \ldots$, be a sequence of random variables where the cumulative distribution function (CDF) for Y_n is $F_n(y)$. Let Y be a random variable with CDF $F(y)$. Then we say the sequence Y_n **converges in distribution** to Y if for all y in the support of F, where F is continuous, we have

$$\lim_{n \to \infty} F_n(y) = F(y).$$

We write $Y_n \overset{\mathcal{D}}{\to} Y$ for convergence in distribution.

We have already seen an example of this! The central limit theorem is about convergence in distribution to a normal random variable. Here, $X_1, X_2, \ldots$ are iid random variables from a population with mean μ and variance σ^2. We defined

$$Z_n = \frac{\bar{X}_n - \mu}{\sigma/\sqrt{n}},$$

where $\bar{X}$ is the sample mean, and we proved (using the moment generating function ideas) that $Z_n \overset{\mathcal{D}}{\to} Z$, where Z is a standard normal random variable. We can alternatively write

$$Z_n \overset{\mathcal{D}}{\to} N(0, 1),$$

that is, we write the distribution on the right, instead of a random variable that has the distribution. Note that this also implies that

$$\sqrt{n}(\bar{X} - \mu) \overset{\mathcal{D}}{\to} N(0, \sigma^2).$$

This form of the central limit theorem is important in the application of the delta method.

The proof of the next important result it not given, as it is more technically difficult. Slutsky's theorem involves two random variables, one converging in distribution and the other converging in probability to a constant.

Slutsky's theorem says that if $Y_n \overset{\mathcal{D}}{\to} Y$ and $X_n \overset{p}{\to} c$, then

S1. $X_n + Y_n \overset{\mathcal{D}}{\to} c + Y$, and

S2. $X_n Y_n \overset{\mathcal{D}}{\to} cY$.

Example: An important application of Slutsky's theorem is the normal approximation with an estimated variance. We known that if the conditions for the central limit theorem hold, then for a random sample from a population with mean μ and finite variance

$\sigma^2 > 0$, the sample mean $\bar{Y}$ is approximately normal. In particular,

$$\frac{\sqrt{n}(\bar{Y}_n - \mu)}{\sigma} \xrightarrow{\mathcal{D}} \mathrm{N}(0, 1).$$

If the population variance σ^2 is unknown, we can estimate it with the sample variance S^2, and we have seen that $S^2 \xrightarrow{p} \sigma^2$. By result R3, we also know that $S \xrightarrow{p} \sigma$ and $1/S \xrightarrow{p} 1/\sigma$. Therefore by Slutsky's theorem,

$$\frac{\sqrt{n}(\bar{Y}_n - \mu)}{S} \xrightarrow{\mathcal{D}} \mathrm{N}(0, 1).$$

This result tells us, for example, that we can use a Z-test when the population variance is unknown, if the sample size is large. (However, a little reflection leads us to conclude that it's still better to use the t-test, because this will be more "conservative" in that the probability of rejecting H_0 when it is true is smaller with the t-test compared to the Z-test.)

The delta method

One of the most important and useful applications of these convergence concepts is the delta method, which extends the central limit theorem to find approximate distributions and variances of estimators that are functions of random variables from populations with finite variances. We will show how to do this using a Taylor's expansion to derive the delta method, but first we'll look at a special case to motivate the ideas.

Example: Suppose interest is in estimating the odds of success from binomial trials; the odds are defined as $g(p) = p/(1-p)$, that is, the probability of a success divided by the probability of a failure. The odds are typically used to compare treatment outcomes in medical studies, where the *odds ratio* is used to compare probabilities of successes under various scenarios.

If $Y_1, \ldots, Y_n$ are independent Bernoulli(p), then $T = \sum_{i=1}^{n} Y_i$ is Binom(n, p), and the maximum likelihood estimator (MLE) for p is $\hat{p} = T/n$. We can find the variance of our estimator for the probability of success,

$$\mathrm{V}(\hat{p}) = \frac{1}{n^2} \mathrm{Var}(T) = \frac{np(1-p)}{n^2} = \frac{p(1-p)}{n}.$$

From the invariance property of maximum likelihood estimation, we know that the MLE for the odds is $\hat{p}/(1 - \hat{p})$. But how do we find the variance of this estimator for the odds? For any finite n and $p \in (0, 1)$, the expected value and variance of the odds estimator is infinity! This is because there is a finite probability that $\hat{p} = 1$, though this probability quickly goes to zero as n gets larger (the last example of Chapter 46 showed that the odds estimator is weakly consistent, but not consistent in MSE).

For practical purposes, we want to use the limiting variance for a moderate-sized sample; this is what we approximate using the delta method. This method will give us a straightforward way to approximate not only the variance of an estimator, but also provide an approximate normal distribution.

More generally, if we know the mean and variance of an estimator $\hat{\theta}$ of a parameter θ, we can often use the central limit theorem to approximate the distribution of $\hat{\theta}$.

Using the delta method, we can generalize the central limit theorem to approximate the distribution of $g(\hat{\theta})$ and find its variance. We'll limit our discussion to a one-dimensional parameter, although the delta method can be derived more generally.

Delta method theorem: Suppose T_n is a sequence of random variables, and suppose that

$$\sqrt{n}(T_n - \mu) \xrightarrow{\mathcal{D}} \mathrm{N}(0, \sigma^2).$$

If $g(t)$ is a function such that $g'(\mu)$ exists and is not zero, and $g''(t)$ is bounded on an open interval containing μ, then

$$\sqrt{n}\left[g(T_n) - g(\mu)\right] \xrightarrow{\mathcal{D}} N\left(0, \sigma^2[g'(\mu)]^2\right).$$

Note that this gives an approximation of the variance of $g(T_n)$ as well as an approximation of the distribution of $g(T_n)$. In a later chapter, we will use this approximate distribution for constructing confidence intervals for functions of parameters.

Proof: We use Taylor's theorem to write

$$g(T_n) = g(\mu) + g'(\mu)(T_n - \mu) + \frac{1}{2}g''(\xi)(T_n - \mu)^2$$

for some ξ between T_n and μ. Then

$$\sqrt{n}[g(T_n) - g(\mu)] = \sqrt{n}\left[g'(\mu)(T_n - \mu) + \frac{1}{2}g''(\xi)(T_n - \mu)^2\right]$$

$$= g'(\mu)\sqrt{n}(T_n - \mu) + \frac{1}{2}g''(\xi)\sqrt{n}(T_n - \mu)^2.$$

The first term converges (by assumption) in distribution to a normal with mean zero and variance $g'(\mu)^2\sigma^2$. For the second term, note that $\sqrt{n}(T_n - \theta)$ converges in distribution to a normal random variable and $T_n - \mu \xrightarrow{P} 0$. So by Slutsky's theorem the second term goes to zero, and that completes the proof.

Let's finish the motivating example for the delta method; that is, approximate the variance of the MLE for the odds, given a random sample from a Bernoulli(p) distribution.

We know that $\hat{p}$, the sample proportion, has mean p and variance $p(1 - p)/n$, so

$$\sqrt{n}(\hat{p} - p) \xrightarrow{\mathcal{D}} \mathrm{N}(0, p(1 - p)).$$

Define $g(p) = p/(1 - p)$ (the odds); then $g(\hat{p})$ is the MLE for the odds. Also, $g'(p) = (1 - p)^{-2}$, so

$$\sqrt{n}[g(\hat{p}) - g(p)] \xrightarrow{\mathcal{D}} \mathrm{N}(0, p(1 - p)^{-3}),$$

and we approximate the variance of $\hat{p}/(1 - \hat{p})$ as $p(1 - p)^{-3}/n$.

How large does n have to be for this to be a "good" approximation? Let's do simulations for $n = 30$ and $p = .2$. The following code determines that in this case the true variance is about .0147, while the variance estimated by the delta method is .0130. The true variance is about 13% larger than the estimated variance, which might not be desirable: We are claiming more precision than we actually have. If we repeat the simulations for $n = 100$ and $p = .2$, we find that the true variance is only about 2% larger than the estimated variance:

```
n=30;p=.2
nloop=100000
ohat=1:nloop
for(iloop in 1:nloop){
   x=rbinom(1,n,p)
   ohat[iloop]=x/n/(1-x/n)
}
var(ohat)
p/(1-p)^3/n
```

It's a bit surprising that the delta method works well here, considering that the true mean and variance of the estimator for the odds are infinite for any finite n. (Recall that this is an example of a sequence of estimators that converges weakly but not in MSE.) We would get in trouble with our simulations if $n = 30$ and $p = .75$, because the probability of "all ones" in this case is about .0002, and we expect to get an infinite value for at least one ohat. With a sample of size $n = 50$, the probability of "all ones" is small enough so that (probably) no infinite values will occur in the simulations, but the estimated variance in this case is only 60% of the true variance!

This example shows that there is no rule of thumb for sample sizes that are large enough to apply these approximations. Common sense, a healthy degree of skepticism, and simulations will help avoid mistakes made through underestimating variances.

Example: Let $Y_1, \ldots, Y_n$ be a random sample from a population with mean $\mu \neq 0$ and variance σ^2. (We don't specify a distribution or family of distributions, or even whether the population is discrete or continuous.) We know from the central limit theorem that

$$\sqrt{n}(\bar{X} - \mu) \xrightarrow{\mathcal{D}} \text{N}(0, \sigma^2).$$

Now suppose the value of μ^2 is important for some application, and suppose we know that μ is not zero. It's reasonable to use $\bar{X}^2$ as an estimator for μ^2 (it may be biased, but at least it's consistent). Of course, in practice we want to find the variance of our estimator. If we knew the family of distributions, we might be able to compute this analytically, but we can always approximate the variance of $\bar{X}^2$ using the delta method.

We have $g(\mu) = \mu^2$, and so $g'(\mu) = 2\mu$. Therefore, just plugging into the statement of the theorem, we get

$$\sqrt{n}[\bar{X}^2 - \mu^2] \xrightarrow{\mathcal{D}} \text{N}(0, 4\mu^2\sigma^2),$$

with the caveat that this won't work if $\mu = 0$. Then the estimated variance of $\bar{X}^2$ is $4\mu^2\sigma^2/n$.

Now suppose instead the researchers need to estimate $1/\mu$ for their application, and they are sure that $\mu > 0$. They need to approximate the variance of $1/\bar{X}$, so we use $g(\mu) = 1/\mu$ and $g'(\mu) = -1/\mu^2$. Then

$$\sqrt{n}[1/\bar{X} - 1/\mu] \xrightarrow{\mathcal{D}} \text{N}(0, \sigma^2/\mu^4),$$

and the estimated variance of $1/\bar{X}$ is $\sigma^2/(n\mu^4)$.

This approximation is (asymptotically) correct when $\mu \neq 0$, but it might not behave very well for μ near zero, until n is very large. For example, if $\mu = .5$ and $\sigma = 1$, the approximation is poor unless n is large enough so that the probability that the sample mean is "away from" zero is close to one. The sample size needs to be well over $n = 100$ to get a good approximation to the variance.

Chapter Highlights

1. Convergence in probability: A sequence of random variables $Y_1, Y_2, \ldots$ converges in probability to a constant c; or $Y_n \overset{p}{\to} c$ as $n \to \infty$ if for any $\delta > 0$,

$$\lim_{n \to \infty} \mathrm{P}(|Y_n - c| > \delta) = 0.$$

2. The definition of convergence in probability leads to these useful results:

   R1. If $X_n \overset{p}{\to} a$ and $Y_n \overset{p}{\to} b$, and $W_n = X_n + Y_n$, then $W_n \overset{p}{\to} a + b$.

   R2. If $X_n \overset{p}{\to} a$ and $Y_n \overset{p}{\to} b$, and $W_n = X_n Y_n$, then $W_n \overset{p}{\to} ab$.

   R3. If $X_n \overset{p}{\to} a$ and g is a function that is continuous at a, then $g(X_n) \overset{p}{\to} g(a)$.

3. Law of large numbers: Let $Y_1, Y_2, \ldots$ be iid random variables with common mean μ and common (finite) variance σ^2. Define $\bar{Y}_n$ to be the mean of the first n random variables, that is,

$$\bar{Y}_n = \frac{1}{n} \sum_{i=1}^{n} Y_i.$$

   Then $\bar{Y}_n \overset{p}{\to} \mu$.

4. Convergence in distribution: Let $Y_1, Y_2, \ldots,$ be a sequence of random variables where the CDF for Y_n is $F_n(y)$. Let Y be a random variable with CDF $F(y)$. Then we say the sequence Y_n converges in distribution to Y if

$$\lim_{n \to \infty} F_n(y) = F(y)$$

   at all points y where F is continuous. We write $Y_n \overset{\mathcal{D}}{\to} Y$.

5. Slutsky's theorem: If $Y_n \overset{\mathcal{D}}{\to} Y$ and $X_n \overset{p}{\to} c$, then

   S1. $X_n + Y_n \overset{\mathcal{D}}{\to} c + Y$, and

   S2. $X_n Y_n \overset{\mathcal{D}}{\to} cY$.

6. Delta method theorem: Suppose T_n is a sequence of random variables and

$$\sqrt{n}(T_n - \theta) \overset{\mathcal{D}}{\to} \mathrm{N}(0, \sigma^2).$$

   If $g(t)$ is a function such that $g'(\theta)$ exists and is not zero, then

$$\sqrt{n}\,[g(T_n) - g(\theta)] \overset{\mathcal{D}}{\to} N\left(0, \sigma^2[g'(\theta)]^2\right).$$

   Then the approximate variance of $g(T_n)$ is

$$\mathrm{V}[g(T_n)] \approx \frac{\sigma^2[g'(\theta)]^2}{n}.$$

Exercises

47.1 Prove result R2 in the case where $a = 0$ or $b = 0$.

47.2 Suppose $Y_1, Y_2, \ldots$ are independent exponential random variables with mean θ. Show that

$$\frac{1}{n} \sum_{i=1}^{n} Y_i^2$$

converges in probability to a constant as n grows, and give the constant.

47.3 Suppose $Y_1, Y_2, \ldots$ are independent random variables with density

$$f_\theta(y) = (\theta + 1)y^\theta \quad \text{for } y \in (0, 1).$$

Show that the sequence

$$\frac{1}{n} \sum_{i=1}^{n} \frac{1}{Y_i}$$

will converge in distribution to a constant under a condition on θ; give the constant and the condition.

47.4 If $X_1, \ldots, X_n$ are independent exponential random variables, all with mean θ, then

(a) approximate the variance of $\bar{X}^2$ using the delta method;

(b) compute the true variance of $\bar{X}^2$ and compare it to your answer in (a).

47.5 Given a $\theta > 0$, the random variables $Y_1, \ldots, Y_n$ are independent with a Gamma$(3, \theta)$ density. Consider the estimator

$$\hat{\theta} = \frac{3}{\bar{Y}}.$$

Approximate the variance of $\hat{\theta}$ using the delta method.

47.6 Consider the example where researchers want to estimate $1/\mu$ using a random sample from N(μ, σ^2). They use $1/\bar{X}$, the inverse of the sample mean, for their estimator, and estimate the variance using the delta method. This is an asymptotic variance, meaning it's approximate, with the approximation getting better when the sample size increases.

(a) Do simulations with $n = 30$, $\mu = 5$ with $\sigma^2 = 1$ to compare the variance of $1/\bar{X}$ with the calculated asymptotic variance.

(b) Use your simulations results from (a) to make density histograms of $\sqrt{n}(1/\bar{X} - 1/\mu)$ and superimpose a mean zero normal density with variance σ^2/μ^4. Is the approximated density close to the true density?

(c) Do simulations with $n = 30$, $\mu = 1/2$ with $\sigma^2 = 1$ to compare the variance of $1/\bar{X}$ with the calculated asymptotic variance.

(d) Repeat (c) with $n = 100$. Is this large enough? What about $n = 1000$?

(e) For $n = 100$ and $n = 1000$, use your simulation results from (b) and (c) to make density histograms of $\sqrt{n}(1/\bar{X} - 1/\mu)$ and superimpose a mean zero normal density with variance σ^2/μ^4.

47.7 Suppose $Y_1, \ldots, Y_n$ are independent $\text{Pois}(\lambda)$ random variables.

    (a) Using the delta method, approximate the variance of $\bar{Y}^2$.

    (b) Check your answer to part (a) by doing simulations to compute the true variance of $\bar{Y}^2$ for $n = 30$ and $\lambda = 2$, and compare this to the result in (a).

47.8 Suppose $Y_1, \ldots, Y_n$ are independent $\text{Beta}(\alpha, 2)$ random variables, and suppose we wish to estimate α using $\hat{\alpha}_n = 2\bar{Y}/(1 - \bar{Y})$. Use the delta method to estimate the variance of $\hat{\alpha}_n$.

47.9 Suppose $Y_1, \ldots, Y_n$ are independent $\text{Beta}(\alpha, 1)$ random variables, and suppose we wish to estimate α using $\hat{\alpha}_n = \bar{Y}/(1 - \bar{Y})$. Use the delta method to estimate the variance of $\hat{\alpha}_n$.

47.10 Let $X_1, \ldots, X_n$ be a random sample from an exponential density with mean θ.

    (a) Find the MLE for $P(X_1 > 1)$.

    (b) Approximate the variance of the estimator in part (a) using the delta method.

47.11 Consider a random sample $Y_1, \ldots, Y_n$ from the density

$$f_\theta(y) = (\theta + 1)y^\theta I\{0 < y < 1\},$$

where $\theta > -1$.

    (a) Find the method of moments estimator for θ.

    (b) Use the delta method to approximate the variance of your estimator in part (a).

47.12 For some $\alpha > 0$, suppose $Y_1, \ldots, Y_n \overset{ind}{\sim} \text{Gamma}(\alpha, (2\alpha)^{-1})$. (Use our usual definition of the gamma distribution parameters, i.e., 2α is the *rate* parameter.)

    (a) Find the method of moments estimator for α.

    (b) Use the delta method to approximate the variance of your estimator in part (a).

    (c) Check your answer to part (b) using simulations with $n = 100$ and $\alpha = 2$. That is, simulate many data sets of size $n = 100$ from $\text{Gamma}(2, 1/4)$, and for each data set, use your estimator to find $\hat{\alpha}$. Compare the variance of your many estimators to your calculated variance.

47.13 Suppose $Y_1, \ldots, Y_n$ are independent random variables with density

$$f_\theta(y) = \frac{\theta}{(y + 1)^{\theta+1}} \quad \text{for } y > 0.$$

    (a) Find the method of moments estimator for θ.

    (b) Use the delta method to approximate the variance of your estimator in part (a).

    (c) Check your answer to part (b) using simulations with $n = 100$ and $\theta = 8$.

47.14 For some $\beta > 0$, suppose $Y_1, \ldots, Y_n \overset{ind}{\sim} \text{Gamma}(2, \beta)$. Using the delta method, approximate the variance of $1/\bar{Y}$.

47.15 Let $Y_1, \ldots, Y_n$ be a random sample from a Geom(p) distribution. Recall from Exercise 43.11 that the MLE for p is

$$\hat{p} = \frac{n}{n + \sum_{i=1}^n Y_i} = \frac{1}{1 + \bar{Y}}.$$

Approximate the variance of the MLE using the delta method.

47.16 The number of whale sightings on a three-hour excursion from Massachusetts is modeled as a Poisson random variable with mean $\lambda > 0$. Interest is in estimating the proportion of excursions with no whale sightings. Suppose $Y_1, \ldots, Y_n$ are independent Poisson(λ) random variables.

(a) Find the MLE for the proportion of excursions with no whale sightings.

(b) Estimate the variance of your estimator in (a) using the delta method.

(c) Using simulations, compare the estimated variance with the true variance when $n = 40$ and $\lambda = 2.4$.

Chapter 48

Quantifying Uncertainty: Standard Error and Confidence Intervals

It's not enough simply to compute an estimate of a parameter; we also want to quantify the uncertainty of our estimator. When an estimate is reported, the next question is sure to be "plus or minus what?" Although we can often derive the variance of the estimator, the expression for the variance almost always involves unknown parameters; in many cases the variance of our parameter estimate can be a function of the parameter we're trying to estimate.

The **standard error** of an estimate is simply the estimate of the standard deviation of the estimator, usually obtained by "plugging in" the parameter estimate into the expression for the variance, then taking the square root. The estimation results are often reported as the estimate and its standard error.

Example: Suppose we have a random sample from an $N(\mu, \sigma^2)$ population, and the sample mean is used to estimate the population mean. We have a measure of uncertainty of our estimate: The variance of the sample mean is σ^2/n, and the standard deviation of the sample mean is $\sigma/\sqrt{n}$. However, in practice this is not helpful, because typically σ is not known. We can estimate the standard deviation of the sample mean using $S/\sqrt{n}$, where S is the sample standard deviation. Therefore, $S/\sqrt{n}$ is the standard error: the estimate of the standard deviation of the estimator.

Example: Suppose $Y_1, \ldots, Y_n$ are independent Bernoulli(p) random variables. We estimate p using $\hat{p} = \bar{Y}$, and we know that the variance of our estimator is $p(1-p)/n$. Because we don't know p, we have to plug in our estimate, and our standard error for our estimate of p is $\sqrt{\hat{p}(1-\hat{p})/n}$.

Confidence intervals are a popular way to express uncertainty in an estimate; these include a *level* of uncertainty about a range of possible values for a parameter or function of a parameter. In this chapter we first cover *exact* confidence intervals, using a **pivotal quantity** with a known distribution. Then we consider *approximate* confidence intervals using an approximate distribution, obtained with the central limit theorem or the delta method. These are both methods in "frequentist" statistics; in the next chapter we'll talk about Bayesian *credible intervals*.

Let's start with an example from introductory statistics. Suppose $X_1, \ldots, X_n$ is a random sample from an $N(\mu, \sigma^2)$ population. We can calculate the sample mean $\bar{X}$ and the sample variance S^2, and we have determined in Chapter 36 that we can construct a statistic that is a t random variable:

$$T = \frac{\bar{X} - \mu}{S/\sqrt{n}} \sim t(n-1).$$

We can find the percentiles of a $t(n-1)$ distribution using tables or R. Let $t_\alpha^{(\nu)}$ be the 100αth percentile of a $t(\nu)$ distribution; for example, the 95th percentile of a t distribution with 5 degrees of freedom is written $t_{.95}^{(5)}$. Then by the definition of percentiles,

$$\mathrm{P}\left(t_{\alpha/2}^{(n-1)} \leq \frac{\bar{X} - \mu}{S/\sqrt{n}} \leq t_{1-\alpha/2}^{(n-1)}\right) = 1 - \alpha,$$

and because of symmetry of the t-density,

$$\mathrm{P}\left(-t_{1-\alpha/2}^{(n-1)} \leq \frac{\bar{X} - \mu}{S/\sqrt{n}} \leq t_{1-\alpha/2}^{(n-1)}\right) = 1 - \alpha.$$

We can rearrange the inequalities to put the parameter in the middle:

$$\mathrm{P}\left(\bar{X} - t_{1-\alpha/2}^{(n-1)}\frac{S}{\sqrt{n}} \leq \mu \leq \bar{X} + t_{1-\alpha/2}^{(n-1)}\frac{S}{\sqrt{n}}\right) = 1 - \alpha,$$

then we say that

$$\left(\bar{X} - t_{1-\alpha/2}^{(n-1)}\frac{S}{\sqrt{n}}, \bar{X} + t_{1-\alpha/2}^{(n-1)}\frac{S}{\sqrt{n}}\right)$$

is a $100(1-\alpha)\%$ confidence interval for μ.

If the observed sample mean is $\bar{X}_{obs} = 45.3$ and the observed sample variance is $S_{obs}^2 = 13.2$ for $n = 10$, we can get a 95% confidence interval for μ using the formula we just derived. The confidence interval can be expressed as

$$\bar{X}_{obs} \pm t_{.975}^{(n-1)}\frac{S_{obs}}{\sqrt{n}}, \quad \text{or} \quad 45.3 \pm 2.262\sqrt{\frac{13.2}{10}} \quad \text{or} \quad 45.3 \pm 2.60,$$

or $(42.7, 47.9)$.

The $100(1-\alpha)\%$ confidence interval can be used to test hypotheses about the population mean μ with test size α. Because we are using the same expression T for both procedures, we can test $H_0 : \mu = \mu_0$ at test size α, by rejecting H_0 if μ_0 is not in the confidence interval.

The above derivation of the confidence interval for the normal population mean is an example of the **pivotal quantity method** for constructing confidence intervals. We can construct confidence intervals for parameters of interest in a wide variety of situations by finding a **pivotal quantity**. A pivotal quantity is not a statistic; it is a function of *both* the data and the parameter of interest. The important characteristic is that **its distribution does not depend on the parameter**. For the above example, we used

$$T = \frac{\bar{X} - \mu}{S/\sqrt{n}}$$

as a pivotal quantity. Because its distribution is $t(n-1)$ for *any* values μ and σ^2, we were able to write down the probability expression and rearrange it to get the confidence interval.

Example: Suppose $X_1, \ldots, X_n \overset{ind}{\sim} N(\mu_X, \sigma^2)$ and $Y_1, \ldots, Y_m \overset{ind}{\sim} N(\mu_Y, \sigma^2)$ are independent random samples. We want to find a confidence interval for $\mu_X - \mu_Y$. Following our derivation for hypothesis testing in Chapter 37, we recall that

$$T = \frac{(\bar{X} - \bar{Y}) - (\mu_X - \mu_Y)}{S_p\sqrt{\frac{1}{m} + \frac{1}{n}}} \sim t(m + n - 2),$$

where the pooled variance is

$$S_p^2 = \frac{(n-1)S_X^2 + (m-1)S_Y^2}{n + m - 2},$$

where S_X^2 is the sample variance for the X values and S_Y^2 is the sample variance for the Y values. Because T contains the parameter of interest (the difference in population means) and its distribution doesn't depend on the parameters, we can use it as a pivotal quantity. Starting with the definition of percentiles and symmetry of the t-density,

$$P\left(-t_{1-\alpha/2}^{(m+n-2)} \le \frac{(\bar{X} - \bar{Y}) - (\mu_X - \mu_Y)}{S_p\sqrt{\frac{1}{m} + \frac{1}{n}}} \le t_{1-\alpha/2}^{(m+n-2)}\right) = 1 - \alpha,$$

and we rearrange: $1 - \alpha$ is equal to the expression

$$P\left((\bar{X} - \bar{Y}) - t_{1-\alpha/2}^{(m+n-2)} S_p\sqrt{\frac{1}{m} + \frac{1}{n}}\right.$$

$$\left. \le \mu_X - \mu_Y \le (\bar{X} - \bar{Y}) + t_{1-\alpha/2}^{(m+n-2)} S_p\sqrt{\frac{1}{m} + \frac{1}{n}}\right).$$

A $100(1 - \alpha)\%$ confidence interval is

$$\left((\bar{X} - \bar{Y}) - t_{1-\alpha/2}^{(m+n-2)} S_p\sqrt{\frac{1}{m} + \frac{1}{n}}, (\bar{X} - \bar{Y}) + t_{1-\alpha/2}^{(m+n-2)} S_p\sqrt{\frac{1}{m} + \frac{1}{n}}\right),$$

or

$$\bar{X} - \bar{Y} \pm t_{1-\alpha/2}^{(m+n-2)} S_p\sqrt{\frac{1}{m} + \frac{1}{n}}.$$

Example: Suppose $X_1, \ldots, X_n$ is a random sample from an $N(\mu, \sigma^2)$ population, and this time we want a confidence interval for the population variance σ^2. We can calculate the sample mean $\bar{X}$ and the sample standard deviation S^2, and we have determined that

$$\frac{(n-1)S^2}{\sigma^2} \sim \chi^2(n-1),$$

so this expression can be used as a pivotal quantity. Therefore if $\chi_\alpha^{(n-1)}$ is the 100αth percentile of a $\chi^2(n-1)$,

$$P\left(\chi_{.025}^{(n-1)} \le \frac{(n-1)S^2}{\sigma^2} \le \chi_{.975}^{(n-1)}\right) = .95.$$

Rearranging the inequalities gives a 95% confidence interval for σ^2:

$$\left(\frac{(n-1)S^2}{\chi_{.975}^{(n-1)}}, \frac{(n-1)S^2}{\chi_{.025}^{(n-1)}} \right).$$

The chi-squared distribution is not symmetric, so we can't write the confidence interval with our "plus or minus" notation.

Example: Suppose $Y_1, \ldots, Y_n$ are independent exponential random variables with mean θ; find a 90% confidence interval for θ. We need to find an expression with the data and the parameter θ, for which the distribution doesn't depend on θ. If Y is an exponential random variable with mean θ, let

$$X = Y/\theta.$$

Then, using our transformation of variables techniques, we can find that

$$\begin{aligned}
F_X(x) = P(X \le x) &= P(Y/\theta \le x) \\
&= P(Y \le \theta x) = F_Y(\theta x) \\
&= 1 - e^{-x},
\end{aligned}$$

because the exponential CDF is $F_Y(y) = 1 - e^{-y/\theta}$. Let $X_i = Y_i/\theta$, and we know that $\sum_{i=1}^{n} X_i \sim \text{Gamma}(n, 1)$, so we can use

$$Q = \sum_{i=1}^{n} \frac{Y_i}{\theta}$$

as our pivotal quantity.

The 5th and 95th percentiles of a Gamma$(n, 1)$ random variable can be found using R; call these a and b, respectively.

For example, if $n = 10$, we can find the 5th and 95th percentiles by

```
qgamma(.05,10,1)
qgamma(.95,10,1)
```

so that the shaded areas in the plot on the right are each .05.

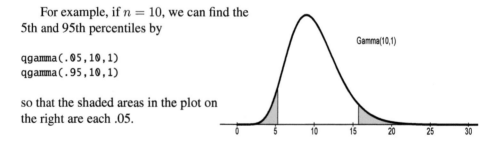

Then, by definition of the percentiles, we know that

$$P\left(a \le \sum_{i=1}^{n} \frac{Y_i}{\theta} \le b \right) = .9;$$

rearranging gives

$$P\left(\sum_{i=1}^{n} \frac{Y_i}{b} \le \theta \le \sum_{i=1}^{n} \frac{Y_i}{a} \right) = .9,$$

and our 90% confidence interval for θ is

$$\left(\sum_{i=1}^{n} \frac{Y_i}{b}, \sum_{i=1}^{n} \frac{Y_i}{a} \right).$$

If $\bar{Y} = 4.3$ and $n = 10$, we find $a = 5.425$, $b = 15.71$, so we compute the confidence interval $(2.74, 7.93)$.

Interpretation: We must not lose sight of the fact that the randomness of the confidence interval is in the data and not the parameter. An *incorrect* but common interpretation is to say that the probability that θ is in $(2.74, 7.93)$ is 90%. Because a parameter is a constant and not random, we cannot correctly make such a probabilistic statement. The correct interpretation is rather long-winded: If we take "many" random samples of size 10 from an $\text{Exp}(\theta)$ population, and for each sample we compute a 90% confidence interval in the same way, then 90% of our samples will contain the true value of θ. In fact, the interpretation of the confidence interval really doesn't address specifically that particular calculated interval!

The above confidence intervals are "exact"—if our assumption about the distribution of the population from which our data are taken is correct, the probability of capturing the true parameter is exactly on target. Suppose instead that we can't assume a distribution for our population—we don't have any theory or prior knowledge to choose a population distribution family. If we have a big enough sample size, approximate confidence intervals can be calculated using the central limit theorem.

Suppose we are interested in a confidence interval for a population mean μ, and we have a "large" random sample from the population. Using the sample mean $\bar{Y}$ and the central limit theorem,

$$\frac{\bar{Y} - \mu}{\sigma/\sqrt{n}} = Z$$

is approximately standard normal, even when the population is not normal. However, we probably don't know the population variance, so we have to substitute the sample standard deviation S for σ. Should we then use the normal table or the t-table? Because the sample size n is "large" the $t(n-1)$ and the normal percentiles are similar, but it's a bit more conservative to use the t-table. That is, the confidence interval you get from using a t approximation will always be a little bigger than that for the normal approximation, and a little more caution is a good thing when the true distribution is unknown.

Because the central limit theorem applies to random samples from discrete as well as continuous populations, we can get approximate confidence intervals for parameters of mass functions.

Example: Confidence intervals for the binomial probability p are of interest in many contexts. The sample proportion $\hat{p}$ is a random variable with mean p and variance $p(1-p)/n$, and if n is large,

$$Z = \frac{\hat{p} - p}{\sqrt{p(1-p)/n}}$$

is approximately $N(0, 1)$. A $100(1 - \alpha)\%$ confidence interval is approximated starting with

$$\mathrm{P}\left(-z_{1-\alpha/2} < \frac{\hat{p} - p}{\sqrt{p(1-p)/n}} < z_{1-\alpha/2}\right) \approx \alpha.$$

If we are willing to use the standard error for $\hat{p}$ in the denominator of Z, we get, for example, the 95% confidence interval

$$\hat{p} \pm 1.96\sqrt{\frac{\hat{p}(1-\hat{p})}{n}}.$$

However, as we will see in Chapter 50, this confidence interval performs poorly for some p, unless n is quite large.

The Wilson confidence interval does *not* substitute the standard error in the denominator of Z. To derive this improved confidence interval, start with the expression from the central limit theorem:

$$\mathrm{P}\left(\hat{p} - z_{1-\alpha/2}\sqrt{p(1-p)/n} < p < \hat{p} + z_{1-\alpha/2}\sqrt{p(1-p)/n}\right) \approx 1 - \alpha.$$

However, we do not plug in $\hat{p}$ for p; instead we solve both inequalities for p. First, solve

$$p = \hat{p} + z_{1-\alpha/2}\sqrt{p(1-p)/n}$$

for p to get the upper bound for the confidence interval, then solve

$$p = \hat{p} - z_{1-\alpha/2}\sqrt{p(1-p)/n}$$

to get the lower bound. Each involves solving a quadratic equation, and the resulting formula gives a Wilson confidence interval:

$$\left(\frac{\hat{p} + \frac{z_{1-\alpha/2}^2}{2n} - z_{1-\alpha/2}\sqrt{\frac{\hat{p}(1-\hat{p})}{n} + \frac{z_{1-\alpha/2}^2}{4n^2}}}{1 + \frac{z_{1-\alpha/2}^2}{n}}, \frac{\hat{p} + \frac{z_{1-\alpha/2}^2}{2n} + z_{1-\alpha/2}\sqrt{\frac{\hat{p}(1-\hat{p})}{n} + \frac{z_{1-\alpha/2}^2}{4n^2}}}{1 + \frac{z_{1-\alpha/2}^2}{n}}\right).$$

This interval is compared with the simpler option in Exercise 50.1.

Example: Suppose next that we want to estimate β using a random sample from a Gamma$(3, \beta)$ population where we assume $\beta > 0$. The population density function is

$$f_\beta(y) = \frac{1}{2}\beta^3 y^2 e^{-\beta y} \text{ for } y > 0,$$

with mean $\mathrm{E}(Y) = 3/\beta$ and variance $3/\beta^2$. The method of moments estimator is then $\tilde{\beta} = 3/\bar{Y}$, where $\bar{Y}$ is the sample mean. We know by the law of large numbers that $\bar{Y} \overset{P}{\to} 3/\beta$, so by R3 of Chapter 47, $\tilde{\beta}$ is consistent for β. To find an approximate confidence interval for β, we can use the delta method, starting with the central limit theorem

$$\sqrt{n}\left(\bar{Y} - \frac{3}{\beta}\right) \overset{D}{\to} N\left(0, 3/\beta^2\right).$$

Writing $\theta = 3/\beta$, the function $g(\theta) = 3/\theta$, we have $g(\bar{Y}) = 3/\bar{Y} = \tilde{\beta}$ and $g(3/\beta) = \beta$, and $g'(\theta)^2 = 9/\theta^4 = \beta^4/9$. Therefore by the delta method,

$$\sqrt{n}\left(\tilde{\beta} - \beta\right) \overset{D}{\to} N\left(0, \beta^2/3\right).$$

Then for large n, we can construct a $100(1 - \alpha)\%$ confidence interval for β starting with

$$P\left(-z_{1-\alpha/2} \leq \frac{\sqrt{n}\left(\tilde{\beta} - \beta\right)}{\sqrt{\beta^2/3}} \leq z_{1-\alpha/2}\right) \approx 1 - \alpha.$$

We could use a further approximation, substituting $\tilde{\beta}$ for the β in the denominator, to get

$$\tilde{\beta} \pm z_{1-\alpha/2}\frac{\tilde{\beta}}{\sqrt{3n}}.$$

However, it's not too much more work to solve

$$\frac{\sqrt{n}\left(\tilde{\beta} - \beta\right)}{\sqrt{\beta^2/3}} = z_{1-\alpha/2}$$

for β to get the upper confidence bound, and to use the other inequality to get the lower confidence bound. The confidence interval for β is then

$$\left(\frac{\tilde{\beta}}{1 + \frac{z_{1-\alpha/2}}{\sqrt{3n}}}, \frac{\tilde{\beta}}{1 - \frac{z_{1-\alpha/2}}{\sqrt{3n}}}\right).$$

If we had a sample of $n = 100$ with $\bar{Y} = 1.25$, our approximate 95% confidence interval for β is $(2.16, 2.71)$.

For our exact intervals, we know that if our assumptions are correct, the probability of "capturing" the parameter with our interval is as intended. Our approximate confidence intervals based on the central limit theorem and the delta method rely on "large" samples and convergence of distributions. In Chapter 50, we will explore the validity of these approximations through simulations.

Chapter Highlights

1. The **standard error** of an estimator is the estimate of the standard deviation of the estimator.

2. A **pivotal quantity** is a function of both the data and the parameter of interest, such that its distribution does not depend on the parameter. It can then be used for constructing confidence intervals.

3. Approximate confidence intervals for a parameter may be constructed using the central limit theorem or the delta method.

❦❦

Exercises

48.1 Suppose we have two drugs, Drug A and Drug B, both designed to lower cholesterol. We conduct a study to compare the drugs, using $n = 40$ patients. We randomly assign treatments, and subsequently determine for each the number of units of cholesterol reduction. We want to test

$$H_0: \mu_A = \mu_B \quad \text{versus} \quad H_a: \mu_A \neq \mu_B,$$

where μ_A is the average of cholesterol reduction using Drug A and μ_B is the average of cholesterol reduction using Drug B.

After we conduct the study, we assume independence and normality so that we can use a pivot with a t distribution to compute a 95% confidence interval for $\mu_A - \mu_B$. This interval is $(1.2, 8.4)$. Answer TRUE, FALSE, or CAN'T TELL for each of the following statements:

(a) The difference in sample averages is in $(1.2, 8.4)$.

(b) The difference in population averages is in $(1.2, 8.4)$

(c) If we did a two-independent-samples t-test, we would reject the above null hypothesis at $\alpha = 0.05$.

(d) If we did a two-independent-samples t-test, we would reject the above null hypothesis at $\alpha = 0.01$.

(e) The probability that $\mu_A - \mu_B$ is in $(1.2, 8.4)$ is .95.

(f) If we repeated our study many times (and the assumptions are correct), the computed confidence interval would capture the true difference $\mu_A - \mu_B$ 95% of the time.

48.2 Suppose $X_1, \ldots, X_n \overset{ind}{\sim} N(\mu_1, \sigma_1^2)$ and $Y_1, \ldots, Y_m \overset{ind}{\sim} N(\mu_2, \sigma_2^2)$, and the samples are independent.

(a) Derive a $100(1 - \alpha)\%$ confidence interval for σ_2^2/σ_1^2. (Do not assume that μ_1 and μ_2 are known.)

(b) Use your formula to get a 95% confidence interval for the ratio of variances for the populations, using the data summary in Exercise 37.6.

48.3 A study was performed to compare two cholesterol-reducing drugs. Observations of the number of units of cholesterol reduction were recorded for 12 subjects receiving Drug A and 14 subjects receiving Drug B:

	Drug A	Drug B
n	12	14
Mean	5.64	5.03
Standard dev.	1.25	1.82

Make the assumptions appropriate for the two-independent-samples t-test derived in Chapter 37.

(a) Calculate a 95% confidence interval for the difference between the mean cholesterol reduction for Drug A and the mean cholesterol reduction for Drug B.

(b) Check to see if it is plausible that the population variances are equal, using a 90% confidence interval and your answer to Exercise 48.2(a).

48.4 Suppose Y_1 and Y_2 are independent draws from a Unif$(0, \theta)$ population; find a 95% confidence interval for θ.

48.5 Astronomers discover an eclipsing binary star in a distant part of the galaxy, and would like to know the period of revolution. Let the time between successive eclipses be θ. If seven astronomers independently decide to point a telescope at the binary star and watch until they see the eclipse, then the times they watch can be considered a random sample of size $n = 7$ from a Unif$(0, \theta)$ distribution. One astronomer gets lucky and sees the eclipse after only 3.2 days, but another astronomer waited 27.1 days before seeing the eclipse. If 27.1 days is the longest time waited, use this to construct a 99% confidence interval for the period of eclipse θ.

48.6 Suppose $X_1, \ldots, X_n, Y_1, \ldots, Y_m$ are independent exponential random variables, where θ is the mean of the X_i, $i = 1, \ldots, n$, and 2θ is the mean of the Y_i, $i = 1, \ldots, m$. Find an exact 95% confidence interval for θ. (*Hint:* You can use the answer to Exercise 21.7 to construct the pivotal quantity.)

48.7 Suppose $X_1, \ldots, X_n \overset{ind}{\sim} N(0, \sigma^2)$. Find a 90% confidence interval for σ^2. Apply your result when your sample has values

$$-0.5, \quad -5.6, \quad 0.1, \quad 0.2, \quad -1.8, \quad -0.2, \quad -1.6, \quad -0.3, \quad -2.2, \quad -1.4.$$

48.8 Lab technicians are using a new type of chromatograph which measures the volume of analyte. They assume that the measurements taken by the device are normally distributed with mean μ, the true volume, and variance σ^2. To estimate σ^2, they take eight measurements using a known $\mu = 10.455$. The measurements are

$$10.480, \quad 10.469, \quad 10.450, \quad 10.488, \quad 10.442, \quad 10.453, \quad 10.430, \quad 10.445.$$

Construct a 90% confidence interval for σ^2.

48.9 Lab technicians have two measuring devices, and they want to know which has a smaller variance. They take n measurements with each of the devices, all with a known mean μ. Let the measurements with the first device be $Y_1, \ldots, Y_n$ and the measurements with the second device be $X_1, \ldots, X_n$. Suppose $X_i \overset{iid}{\sim} N(\mu, \sigma_x^2)$ and $Y_i \overset{iid}{\sim} N(\mu, \sigma_y^2)$, and the samples are independent.

(a) Construct a 98% confidence interval for σ_y^2/σ_x^2.

(b) Suppose $\mu = 12.00$ and the measurements for the first device are

$$12.004, \quad 12.006, \quad 12.001, \quad 12.008, \quad 12.003, \quad 11.989,$$

and for the second device,

$$11.964, \quad 11.992, \quad 11.999, \quad 12.020, \quad 11.983, \quad 12.008.$$

Comment on the evidence that one device is more precise than the other.

48.10 Lab technicians want to determine the quantity μ; they take five measurements with a device known to give an unbiased measurement with known variance σ^2. Then they obtain a better device and take an additional five measurements. The better device is known to give unbiased measurements with known variance $\sigma^2/9$. Assuming all measurements are independent, find a 95% confidence interval for μ.

48.11 Suppose Y_1, Y_2, Y_3, and Y_4 are independent normal random variables, all with variance σ^2. $E(Y_1) = E(Y_2) = 4$ and $E(Y_3) = E(Y_4) = 8$.

   (a) Using a pivotal quantity, find an expression for a 90% confidence interval for σ^2.

   (b) Give the upper and lower limits for your confidence interval in (a) when the observed values are $y_1 = 1.2$, $y_2 = 5.4$, $y_3 = 3.6$, $y_4 = 7.2$.

48.12 Suppose $Y_1, \ldots, Y_n$ are independent normal random variables each with mean μ_Y and variance 1, while $X_1, \ldots, X_n$ are independent normal random variables each with mean μ_X and variance 2. Further, the samples are independent.

   (a) Find a 95% confidence interval for $\mu_Y - \mu_X$.

   (b) Suppose $n = 10$, $\bar{Y} = 5.6$, and $\bar{X} = 3.8$. Compute the bounds of the confidence interval in (a) and comment on the evidence that μ_Y and μ_X are different.

48.13 Suppose Y_1, Y_2, Y_3, and Y_4 are independent normal random variables, all with mean zero. $V(Y_1) = V(Y_2) = \sigma^2$ and $V(Y_3) = V(Y_4) = 4\sigma^2$.

   (a) Using a pivotal quantity, find an expression for a 90% confidence interval for σ^2.

   (b) Give the upper and lower limits when the observed values are $y_1 = -1.2$, $y_2 = 2.4$, $y_3 = 8.6$, $y_4 = -7.2$.

48.14 Suppose Y_1, Y_2, Y_3, and Y_4 are independent normal random variables. The random variables Y_1 and Y_2 have mean μ and variance one, while the random variables Y_3 and Y_4 all have mean 2μ and variance one. Find a pivotal quantity and use it to construct a 95% confidence interval for μ.

48.15 Scientists are measuring times between eruptions of a geyser. If the times between eruptions can be modeled as independent realizations of an exponential random variable with mean θ, find a 95% confidence interval for θ if the measured times are

$$5.70, \quad 13.86, \quad 57.45, \quad 10.79, \quad 25.75, \quad 50.98.$$

48.16 Suppose $Y_1, \ldots, Y_n$ are independent random variables with common density

$$f_\mu(y) = e^{\mu - y} \text{ for } y > \mu.$$

Derive a 95% confidence interval for μ.

48.17 Suppose $Y_1, \ldots, Y_n$ are independent Beta$(\alpha, 2)$ random variables. Using the estimator in Exercise 47.8, derive an approximate 95% confidence interval for α.

48.18 Suppose $Y_1, \ldots, Y_n$ are independent Beta$(\alpha, 1)$ random variables. Using the estimator in Exercise 47.9, derive an approximate 95% confidence interval for α.

48.19 The first example in Chapter 2 involved simulating the probability of getting at least five sixes when a fair six-sided die is rolled ten times. We simulated one million rolls and obtained the estimated probability of .0155. Make a 99% confidence interval for this approximation.

48.20 Let $Y_1, \ldots, Y_n$ be a random sample from a Geom(p) distribution. Recall from Exercise 43.11 that the MLE for p is

$$\hat{p} = \frac{1}{1 + \bar{Y}}.$$

Use the delta method to get an approximate 95% confidence interval for p.

48.21 The time between geyser eruptions is modeled as exponential with mean θ hours. Suppose the last 30 times between eruptions averaged .78 hours. Use the delta method to find an approximate 90% confidence interval for the probability that a randomly selected time between geysers is more than one hour.

48.22 Suppose $Y_1, \ldots, Y_n$ are independent random variables with density

$$f_\theta(y) = \frac{\theta}{(y + 1)^{\theta+1}} \quad \text{for } y > 0.$$

Use the answer to Exercise 47.13 to construct an approximate $100(1 - \alpha)\%$ confidence interval for θ.

Chapter 49

Bayes Credible Intervals

When we perform a Bayesian analysis to estimate a parameter, we obtain a posterior distribution for the parameter. In Chapter 45 we used the mean of this distribution for the Bayes estimator. However, from the posterior distribution we can easily obtain an interval estimate as well.

Going back to an example from Chapter 45, suppose we are estimating p, a binomial proportion. If we have observations $Y_1, \ldots, Y_n \sim$ Bernoulli(p), and we choose a Beta(α, β) prior, we find that the posterior distribution is Beta$(S + \alpha, n - S + \beta)$, where $S = \sum_{i=1}^{n} Y_i$. The posterior density is shown for $n = 20$, $S = 14$, and the uniform prior ($\alpha = \beta = 1$).

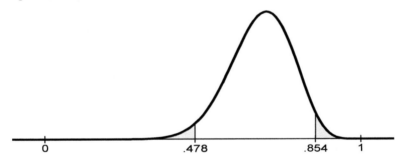

A "Bayesian credible interval" is formed from the posterior in a natural way. For a 95% confidence interval, we can choose the 2.5th and 97.5th percentiles of the Beta$(15, 7)$ distribution (shown as vertical lines in the figure); this is $(.478, .854)$. (Use `qbeta(.025,15,7)` and `qbeta(.975,15,7)`). Isn't that easy?

The above interval is called an "equal-tails" credible interval. The length of the interval is $.854 - .478 = .376$. If we choose instead the 3rd and 98th percentiles, the interval $(.487, .860)$ shifts a bit to the right, but still captures 95% of the area under the posterior. This shifted 95% credible interval has length $.373$, a bit smaller. The smallest possible confidence interval that captures 95% of the posterior is called a "highest posterior density" (HPD) credible interval. The HPD (usually) must be calculated numerically.

A third standard way to choose a credible interval is that for which the posterior mean is in the middle. For our above example, the posterior mean is $\tilde{p} = .682$. We want to find a distance d so that the interval $(.682 - d, .682 + d)$ captures 95% of the posterior

429

density. A bit of trial and error using `pbeta(.682+d,15,7)-pbeta(.682-d,15,7)` finds $d \approx .187$ (using three decimal precision), and $(.495, .869)$ is (approximately) the 95% credible interval centered at the posterior mean.

The interpretation of a Bayes credible interval is satisfying. It *is* permitted to say "the posterior probability that the true parameter is between .478 and .854 is 95%." This would be frowned upon for the frequentist methods, where it is incorrect to make probabilistic statements about a parameter.

Example: Let's revisit the estimation of survival times for leukemia patients from Chapter 45. We modeled the survival times for AG-negative patients as exponential with mean θ, and the prior for θ was InvGamma$(2.25, 25)$, and the posterior, given the 16 observed lifetimes, was InvGamma$(18.25, 312)$. The equal-tailed 95% credible interval for θ can be found with `qigamma(.025,18.25,312)` and `qigamma(.975,18.25,312)`; this is $(11.3, 28.7)$, as shown in the plot below:

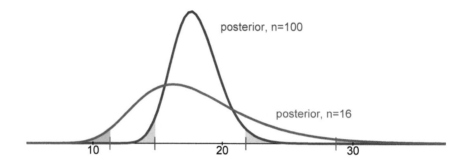

To demonstrate how the credible interval gets narrower as the sample size increases, let's use the same prior, but imagine that we have $n = 100$ patients. Suppose the average lifetime of the 100 patients was the same; in this case, the sum of the survival times would be $287/16 \times 100 \approx 1794$, and the taller, thinner posterior is shown in the above plot, with the equal-tailed 95% credible interval marked.

Chapter Highlights

1. A Bayesian $100(1 - \alpha)\%$ credible interval for a parameter θ is an interval such that the posterior probability that θ falls in that interval is $1 - \alpha$.

2. There are three standard types of $100(1 - \alpha)\%$ credible interval:

   (a) the "equal-tails" interval at the $\alpha/2$ and $1 - \alpha/2$ quantiles of the posterior distribution;

   (b) the "highest probability density" (HPD) interval, which identifies the smallest-length interval capturing $100(1 - \alpha)\%$ of the posterior density; and

   (c) the "mean-centered" interval, which identifies d so that $(\tilde{\theta} - d, \tilde{\theta} + d)$ captures $100(1 - \alpha)\%$ of the posterior density, where $\tilde{\theta}$ is the posterior mean.

Exercises

49.1 Suppose the number of times that an electronic device can be used successfully before failing is modeled as Geom(p), where p is the probability that the device breaks in a single use. Consider a Beta(α, β) prior distribution for p.

(a) The experts determine that the prior should have mean of .2 with a standard deviation of .2. Determine the appropriate hyperparameters for the beta prior.

(b) Suppose $Y_1, \ldots, Y_n \sim$ Geom(p). Using the prior from part (a), find the Bayes posterior density for p.

(c) They test five electronic devices, and the number of successful uses are 12, 4, 20, 8, and 3. What is the Bayes estimator (i.e., the posterior mean)?

(d) For the data in part (c), construct an equal-tailed 90% credible interval for p.

49.2 The number of accidents per month at a large mining company is modeled as Poisson with mean λ. The new management has put in place several new safety measures, and they want to estimate accident rate λ under the new system. They want to use a Bayes estimation, and their prior distribution for λ is that of a gamma random variable, with mean eight and standard deviation four.

(a) Determine appropriate hyperparameters for the gamma prior.

(b) Suppose the numbers of accidents in the first six months are 10, 4, 5, 2, 7, and 4. Find a 95% credible interval for λ.

49.3 Refer to Exercise 45.11. Using a uniform prior, find a 99% credible interval for p, the probability that a cross-fertilized plant is taller than a self-fertilized plant.

49.4 Suppose $X_1, \ldots, X_n \overset{ind}{\sim} N(0, \sigma^2)$, and consider an InvGamma(α, β) prior for σ^2.

(a) Find the posterior distribution for σ^2, and explain how to get an equal-tailed 90% credible interval. (*Note:* The parameter is σ^2, not σ. If this is confusing, change to $\theta = \sigma^2$ and write the joint density for the data, given the parameter, in terms of θ.)

(b) If the researchers want a prior with mean 2 and variance 4, find α and β.

(c) Find a 90% credible interval for σ^2 with the prior in (b) when your sample has values

$$-0.5, \quad -5.6, \quad 0.1, \quad 0.2, \quad -1.8, \quad -0.2, \quad -1.6, \quad -0.3, \quad -2.2, \quad -1.4.$$

Compare your result to the confidence interval in Exercise 48.7.

49.5 Geologists are studying earthquake activity near an active fault in East Asia. They model the rate (in time) of earthquakes with magnitude 5.0 or higher as following an exponential distribution with rate parameter β. They use the conjugate (gamma) prior. Based on historical activity near a similar fault, they think the rate is about every 250 days (that is, the mean of the gamma prior should be $1/250$). To make a "vague" prior, they choose a standard deviation that is twice their prior mean.

(a) What are the hyperparameters of the prior distribution to get the desired prior mean and variance?

(b) Suppose the next eight times between earthquakes are 14, 107, 952, 247, 552, 171, 435, and 56. Find a 90% credible interval for the earthquake rate β.

49.6 The data set `ex0621`, "Failure Times of Bearings," in the data library `Sleuth2` of the R project has this description:

Data consist of times to fatigue failure (in units of millions of cycles) for 10 high-speed turbine engine bearings made from five different compounds.

Suppose we model the lifetimes of ball bearings made with Compound IV as being exponentially distributed, and interest is in modeling θ, the average lifetime.

(a) An expert engineer would guess that the average lifetime is about ten million cycles, but he is not confident of this guess. Formulate an inverse gamma prior with mean 10 and variance 100. Use this prior and the data for Compound IV to construct a 95% credible interval for the mean lifetime.

(b) Another engineer claims to have more expert knowledge about this compound, and her prior guess is eight million cycles for the average lifetime. Formulate an inverse gamma prior with mean 8 and variance 16 to reflect the higher confidence by this engineer. Use this prior and the data for Compound IV to construct a 95% credible interval for the mean lifetime.

(c) Suppose there were 100 observations instead of 10, but the mean of the observations is the same. Construct the credible intervals of (a) and (b), using this (imaginary) larger data set. Comment on the difference between the two intervals compared with those in (a) and (b).

49.7 The time between geyser eruptions is modeled as exponential with mean θ hours. If we use an inverse gamma prior with parameters α and β, we have shown that the posterior for θ is inverse gamma with parameters $\alpha + n$ and $\beta + \sum_{i=1}^{n} x_i$, where $x_1, \ldots, x_n$ are a random sample of waiting times between eruptions. We can use this to get a credible interval for θ, but we can also use the posterior to get a credible interval for the probability p that a randomly selected waiting time is more than one hour. Suppose the last 30 times between eruptions averaged .78 hours. Choose a vague prior for θ, so that the prior mean is one and the prior standard deviation is 2. Find a 95% credible interval for p, and compare it to the confidence interval for p found in Exercise 48.21.

49.8 Suppose $Y_1, \ldots, Y_n$ are independent Beta($\alpha, 1$) random variables. Show that an exponential prior for α leads to a gamma posterior, and find a formula for a 90% credible interval for α.

49.9 Suppose $X_1, \ldots, X_n$ and $Y_1, \ldots, Y_m$ are independent exponential random variables, where the X_i have parameter β (mean $1/\beta$), $i = 1, \ldots, n$, and the Y_i have parameter 2β (mean $1/(2\beta)$), $i = 1, \ldots, m$. Using a Gamma(a, b) prior, find a 90% credible interval for β.

49.10 Suppose $\theta > 2$ and $Y_1, \ldots, Y_n$ are a random sample from a population with density function

$$f(y|\theta) = \frac{\theta}{(y+1)^{\theta+1}}, \quad y > 0.$$

Use a Gamma(α, β) prior to find the posterior distribution for θ. Suppose $\alpha = 1$, $\beta = .1$, $n = 10$, and we have observed values

$$0.10, \quad 0.21, \quad 0.38, \quad 0.04, \quad 0.82, \quad 0.43, \quad 0.08, \quad 2.42, \quad 0.22, \quad 0.88.$$

Give the lower and upper limits for a 90% credible interval for θ.

Chapter 50

Evaluating Confidence Intervals: Length and Coverage Probability

There are two ways to assess confidence intervals. The first is the **coverage probability**, or the probability that the confidence interval captures the parameter. For exact confidence intervals, the coverage probability is equal to the target coverage probability (if the model is correct). For example, the confidence interval for the mean of a normal population, created using a pivotal quantity with a t-density, is exact; therefore, if the population is truly normal, a 95% confidence interval will capture the population mean exactly 95% of the time.

We have also constructed approximate confidence intervals, using the central limit theorem or the delta method. It's of interest to find the "true" coverage probability for these approximate intervals. If the true coverage probability is smaller than the target, this is considered undesirable, because we don't want to claim more precision than we have. If the true coverage probability is larger than the target, this is considered "conservative."

The second assessment is **interval length** (sometimes called width). We'd prefer the confidence or credible interval lengths to be small, reflecting higher precision. If two methods have the same coverage probability, we'd prefer the method that tends to give *smaller* intervals.

For example, suppose $Y_1, \ldots, Y_n$ are iid Poisson(λ), and we want to construct a $100(1 - \alpha)\%$ confidence interval for λ. According to the central limit theorem,

$$\frac{\bar{Y} - \lambda}{\sqrt{\lambda/n}} \xrightarrow{\mathcal{D}} N(0, 1),$$

and we also know that $\sqrt{\bar{Y}/n} \xrightarrow{p} \sqrt{\lambda/n}$, by R3 of Chapter 47. Therefore, by Slutsky's theorem of the same chapter, we know that

$$\frac{\bar{Y} - \lambda}{\sqrt{\bar{Y}/n}} \xrightarrow{\mathcal{D}} N(0, 1),$$

so $\bar{Y} \pm 1.96\sqrt{\bar{Y}/n}$ is an approximate 95% confidence interval for λ. The larger the sample size n, the closer the coverage probability gets to the target, but how large is "large enough"? Questions like this can be answered through simulations.

The following code sets $n = 20$ and $\lambda = 2$, and generates 100,000 samples using the rpois function. For each sample, an approximate 95% confidence interval is constructed using the above formula. The number of times the interval captures the true parameter is recorded.

```
nloop=100000
q=qnorm(.975)
n=20;lambda=2
cov=0
for(iloop in 1:nloop){
    y=rpois(n,lambda)
    ybar=mean(y)
    lower=ybar-qt*sqrt(ybar/n)
    upper=ybar+qt*sqrt(ybar/n)
    if(lower<lambda&upper>lambda){cov=cov+1}
}
cov/nloop
```

We get .942 as the true coverage probability, which is a bit small but not bad. If we want coverage probabilities for this sample size, for a range of λ values, say 1–6, we can embed the above code in a loop through λ values. The following figure shows the true coverage probabilities for 100 values of λ and three sample sizes. We see that for $n = 20$, these probabilities are, on average, too low. The confidence interval performs better for larger sample sizes.

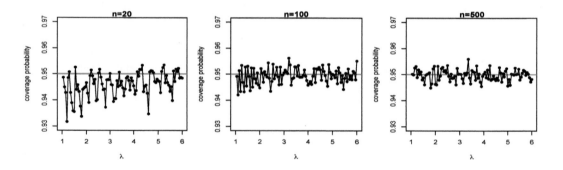

Suppose that instead of using Slutsky's theorem, we do not "plug in" $\bar{Y}$ for λ, and instead solve the two inequalities

$$P\left(-q < \frac{\bar{Y} - \lambda}{\sqrt{\lambda/n}} < q\right) \approx 1 - \alpha$$

for λ, where q is the $1 - \alpha/2$ quantile of the standard normal distribution. Then, after solving a quadratic equation, our confidence interval becomes

$$\bar{Y} + \frac{q^2}{2n} \pm \sqrt{\frac{\bar{Y}q^2}{n} + \frac{q^4}{4n^2}},$$

and because we made only one approximation instead of two, we expect it to perform better. The coverage probabilities for the same sample sizes and λ values are shown in the next figure. They tend to have larger coverage probabilities, on average, but still sometimes too small, especially for the smallest sample size.

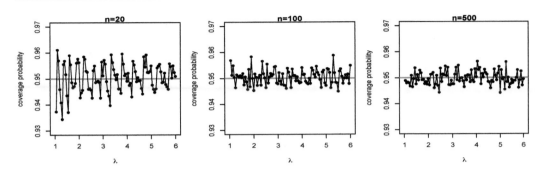

As a third example, let's look at coverage probabilities for a Bayes credible interval, with a vague conjugate prior. If $\alpha = 1/4$ and $\beta = 1/16$, a Gamma(α, β) prior will have mean 4 and standard deviation 8. If $s = \sum_{i=1}^{n} y_i$, the posterior for λ is Gamma$(s + 1/4, n + 1/16)$, and we can use `qgamma(.025,s+1/4,n+1/16)` and `qgamma(.975,s+1/4,n+1/16)` for the lower and upper limits for the credible interval, respectively. The coverage probabilities for the same sample sizes and λ values are shown in the next figure. The performance is about the same.

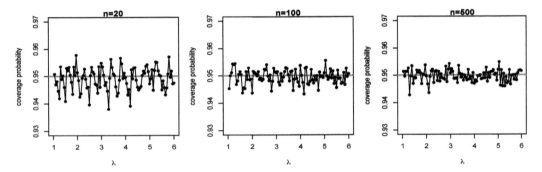

The Bayes credible interval is always "exact" if, instead of fixing λ for each simulated data set, we sample from the prior distribution. For example, suppose $n = 20$, and fix any α and β for the hyperparameters. The following code samples a new λ parameter from the prior before simulating the data set, and in this case the coverage proportion is equal to the target:

```
nloop=100000;n=20
cov=0
for(iloop in 1:nloop){
    lambda=rgamma(1,alpha,beta)
    y=rpois(n,lambda)
    s=sum(y)
    lower=qgamma(.025,s+alpha,n+beta)
    upper=qgamma(.975,s+alpha,n+beta)
    if(lower<lambda&upper>lambda){cov=cov+1}
}
cov/nloop
```

Let's compare our three types of confidence interval for λ, except now we'll compare the length of the confidence or credible intervals. For each loop, we can save the length of the computed interval. In the plot below, the average lengths are plotted against λ for $n = 20$. The average lengths are quite comparable, and are even closer together when the sample size is larger.

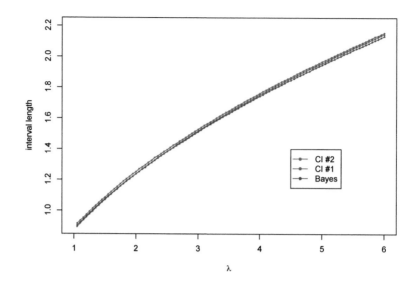

Chapter Highlights

1. The coverage probability is the proportion of confidence intervals, in repeated sampling, that capture the parameter of interest. This is equal to the target coverage probability when the confidence interval is "exact."

2. If two confidence intervals have the same coverage probability, the smaller length interval is preferable.

Exercises

50.1 Let's do some simulations to find coverage probabilities for a binomial proportion. Given a random sample $Y_1, \ldots, Y_n$ from a Bernoulli(p) population, we know that $S = \sum_{i=1}^{n} Y_i \sim$ Binom(n,p).

(a) The approximate 95% confidence interval that we learn in Intro Stat is

$$\hat{p} \pm 1.96 \sqrt{\frac{\hat{p}(1 - \hat{p})}{n}},$$

where $\hat{p} = \bar{Y}$. For $p = .05$ and $n = 59$, find the true coverage probability using simulations.

(b) The Wilson confidence interval was published in 1927. To derive the formula for the $100(1-\alpha)\%$ confidence interval, let z_a be the $100(1-a)$th percentile of a standard normal. Then

$$P\left(\hat{p} - z_{\alpha/2}\sqrt{p(1-p)/n} < p < \hat{p} + z_{\alpha/2}\sqrt{p(1-p)/n}\right) \approx 1 - \alpha.$$

Now we do *not* plug in $\hat{p}$ for p; instead we solve both inequalities for p. This involves solving a quadratic equation. The resulting formula gives the confidence interval

$$\left(\frac{\hat{p} + \frac{z_{\alpha/2}^2}{2n} - z_{\alpha/2}\sqrt{\frac{\hat{p}(1-\hat{p})}{n} + \frac{z_{\alpha/2}^2}{4n^2}}}{1 + \frac{z_{\alpha/2}^2}{n}}, \frac{\hat{p} + \frac{z_{\alpha/2}^2}{2n} + z_{\alpha/2}\sqrt{\frac{\hat{p}(1-\hat{p})}{n} + \frac{z_{\alpha/2}^2}{4n^2}}}{1 + \frac{z_{\alpha/2}^2}{n}} \right).$$

Do simulations to find the coverage probability when $p = .05$ and $n = 59$.

(c) Finally, do simulations to find the coverage probability for the Bayesian credible interval when $p = .05$ and $n = 59$ and the prior is uniform on $(0, 1)$.

50.2 Consider the problem of finding a confidence interval for α, given a random sample $Y_1, \ldots, Y_n$ from a Beta$(\alpha, 1)$ population. Compare the coverage probability and length for the approximate 95% confidence interval computed in Exercise 48.18, with the Bayes credible interval computed in Exercise 49.8, using an exponential prior with mean 1. Use $n = 40$ and three values for α: $\alpha = .1$, $\alpha = 1$, and $\alpha = 10$.

50.3 Suppose $Y_1, \ldots, Y_n$ are independent random variables with common density

$$f_\theta(y) = e^{\theta - y} \text{ for } y > \theta,$$

and interest is in finding a 95% confidence interval for θ.

(a) Find a 95% confidence interval using $\sum_{i=1}^n Y_i - n\theta$ as a pivotal quantity.

(b) Find a 95% confidence interval using $\min(Y_i) - \theta$ as a pivotal quantity.

(c) The intervals in (a) and (b) should have exact coverage probability, but do simulations to compare their average lengths for $n = 10$, $n = 100$, and $\theta = 1$.

50.4 Recall the example in Chapter 48, where we want to estimate β using a random sample from a Gamma$(3, \beta)$ population. We used an approximation to build a confidence interval. Using simulations, find the true coverage probability when the target is 95% when $\beta = 3$. Use $n = 20$ and $n = 200$.

50.5 Recall Exercise 48.10, where $X_1, \ldots, X_5 \sim N(\mu, \sigma^2)$ and $Y_1, \ldots, Y_5 \sim N(\mu, \sigma^2/9)$ for known σ^2. The object is to find a confidence interval for μ.

- Statistician A says, "We know $\bar{X} \sim N(\mu, \sigma^2/5)$ and $\bar{Y} \sim N(\mu, \sigma^2/45)$, so let's use the average

$$\frac{\bar{X} + \bar{Y}}{2} \sim N(\mu, 5\sigma^2/90)$$

as our pivotal quantity."

- Statistician B says, "I want to use

$$\frac{\bar{X} + 3\bar{Y}}{4} \sim N(\mu, \sigma^2/40)$$

as our pivotal quantity."

Both should have exact coverage probability. Determine through simulations which has the smallest length, using $\sigma = 1$ and your choice of μ.

50.6 In Exercise 48.21 we used the delta method to make a 95% confidence interval for the probability that the waiting time for a geyser eruption is more than one hour. In Exercise 49.7 we constructed a 95% credible interval for the same probability and found that it was larger than the confidence interval. Compare the two intervals through simulations using $n = 30$ and $\theta = 1$. What are the coverage probabilities and the average lengths? Repeat for $n = 100$ and $\theta = 1$.

50.7 In Exercise 48.22 we used the delta method to make a 95% confidence interval for θ. In Exercise 49.10 we constructed a 95% credible interval for θ using a Gamma$(1, .1)$ prior. This is a fairly vague prior with mean 10 and standard deviation 10. Compare the coverage probabilities and average lengths of the two interval methods for $n = 30$ and $\theta = 6$. Repeat for $n = 200$ and $\theta = 6$.

50.8 Exercise 48.6 asks you to go to the trouble of finding a pivotal quantity and an exact confidence interval for θ when $X_1, \ldots, X_n, Y_1, \ldots, Y_m$ are independent exponential random variables, where θ is the mean of the X_i, $i = 1, \ldots, n$, and 2θ is the mean of the Y_i, $i = 1, \ldots, m$. Suppose another statistician does not want to do all this work, so instead says, "I know that when n is large, the mean of the X_i is approximately normal with mean θ and variance θ^2/n, and when m is large, the mean of the Y_i is approximately normal with mean 2θ and variance $4\theta^2/n$. Therefore, $\bar{X} + \bar{Y}$ is approximately normal with mean 3θ and variance $5\theta^2/n$. I will construct an approximate confidence interval using

$$Q = \frac{\bar{X} + \bar{Y} - 3\theta}{\sqrt{5\theta^2/n}} \xrightarrow{\mathcal{D}} N(0, 1).$$

My estimate of θ is $\hat{\theta} = (\bar{X} + \bar{Y})/3$, so my approximate 95% confidence interval is

$$\hat{\theta} \pm \frac{2}{3}\sqrt{5\hat{\theta}^2/n}.$$

That should be good enough." Using simulations, compare the coverage probabilities and lengths for the two confidence intervals, exact and approximate. Set $\theta = 4$. If $n_1 = n_2 = 10$, does the approximation work sufficiently well? What about if $n_1 = n_2 = 60$?

Chapter 51

Bootstrap Confidence Intervals

To motivate the bootstrap confidence interval ideas, consider the data shown in the plot below. Suppose this is a sample collected from a population whose distribution is unknown, and interest is in estimating the population mean, or median, or variance, or some other population parameter, and getting a confidence interval to express the uncertainty in our estimate.

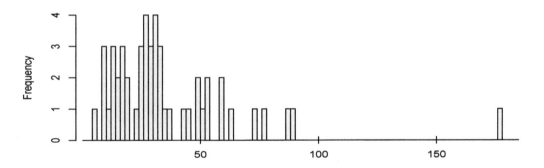

If we knew the family of distributions, we could try to get a pivotal quantity, which would give us an exact confidence interval, for which the coverage probability is equal to the target coverage. Alternatively, we could use Bayesian ideas and an appropriate prior for the population mean to get a Bayes credible interval.

If we don't know the family of distributions, and interest is in a confidence interval for the mean, or a function of the mean, we can use the central limit theorem and/or the delta method and get an approximate confidence interval using a normal limit distribution. This approximate confidence interval is easy to construct for the population mean: It is the sample mean, plus or minus about two standard errors. However, if we wanted a confidence interval for the population *median*, or some other quantile of the population, the central limit theorem will not help us.

The bootstrap is an all-purpose method for constructing approximate confidence intervals for any population quantity, without making assumptions about the population distribution.

The bootstrap idea uses the **empirical distribution function** to approximate the cumulative distribution function (CDF). For values $x_1, \ldots, x_n$ of a random sample from

441

a distribution with CDF F, the empirical distribution function (EDF) for the sample is

$$\hat{F}(x) = \frac{1}{n} \sum_{i=1}^{n} I\{x_i \le x\} \text{ for } x \in \mathbb{R}.$$

In other words, the EDF is the proportion of the sample that is less than or equal to x. For the sample of size $n = 50$ that was shown in the previous figure, we have the following EDF:

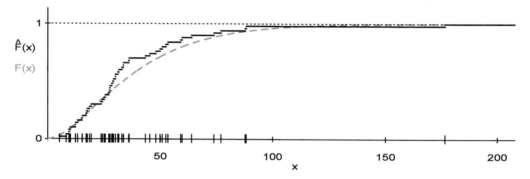

The tick marks for the observations are shown on the horizontal axis, and the EDF jumps upwards a distance $1/n$ at each tick. The CDF for the true population, from which the sample was taken, is shown as the red dashed curve.

Convergence of the EDF to the CDF

If the true CDF F is continuous at x, then $\hat{F}(x) \xrightarrow{p} F(x)$, as the sample size n increases without bound.

To prove this, we simply notice that $n\hat{F}(x)$ is a $\text{Binom}(n, p)$ random variable where n is the sample size and $p = F(x)$. We have already proved that the sample proportion converges in probability to the true proportion; that is, $\hat{F}(x) \xrightarrow{p} p = F(x)$.

If we knew the true CDF, we could sample from it to find the distribution of the sample median, or the distribution of any sample statistic. Instead, we sample from the observed EDF as if it were the mass function for the population. This is easy to do using the `sample` command in R, as we learned in Chapter 2. The fact that the EDF converges to the CDF means that the coverage probabilities of the resulting confidence intervals converge to the target coverage probabilities.

The following code samples from the empirical distribution function, which is the same as sampling from the data vector x, with replacement, and with equal probabilities. This approximates sampling from the true CDF. For each "bootstrap sample," the median is computed and saved in a vector medboot.

```
nboot=100000
medboot=1:nboot
for(iboot in 1:nboot){
    bsamp=sample(x,n,replace=TRUE)
    medboot[iboot]=median(bsamp)
}
hist(medboot,yaxt="n",main="",br=30)
b=sort(medboot)
b[nboot*.025]
b[nboot*.975]
```

The values of the vector `medboot` are shown in the histogram below. The 2.5th and 97.5th percentiles of the 100,000 values are easily found after sorting the values. These percentiles form the lower and upper bounds of a 95% confidence interval for the median and are marked on the histogram with parentheses. The median for the true distribution is about 33.6, which is captured by this confidence interval.

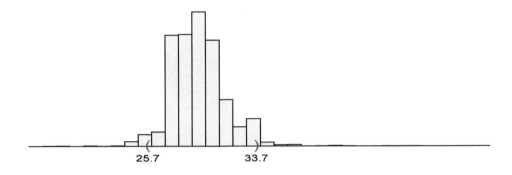

Because the EDF approximates the CDF, the coverage probability should be near the target. To find the true coverage probability for a known distribution and sample size, simulations can be used, with the bootstrap loop inside an outer loop. In the outer loop, we sample from the known distribution. Then the bootstrap interval is computed, whether or not it captures the true parameter. The true coverage probability is the proportion of samples for which the true parameter is captured. For the above example, we can determine that the coverage probability is correct—95%. For a sample size of 30 from the same distribution, the coverage probability is slightly low—about 94%. For different distributions, the bootstrap coverage probability will vary and, of course, will be unknown for a "real-world" problem.

The data set `paulsen` in the `boot` package in R contains measurements of neurotransmission in guinea pig brains, in picoamperes. The histogram of the raw data below shows a skewed shape, possibly with more than one mode. Let's assume that these signals are a random sample from a distribution representative of neurotransmissions in mammals, and suppose scientists are interested in the 90th percentile of this distribution.

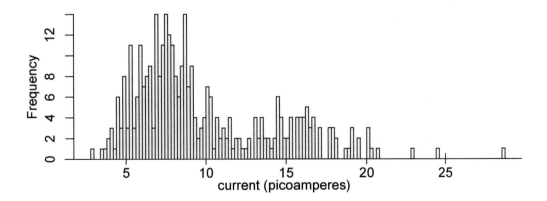

The 90th percentile of the sample is 16.3 picoamperes. This is an estimate of the population's 90th percentile, but the scientists also want a measure of uncertainty for the estimate. Let's find a 95% confidence interval for the 90th percentile of the unknown population distribution using the bootstrap. The following bit of R code finds the 90th percentile for many bootstrap samples of the data vector and determines the middle 95% of these percentiles to create the bootstrap confidence interval. On the right is the histogram of the 90th percentiles of the bootstrap samples, with the confidence interval bounds marked with parentheses.

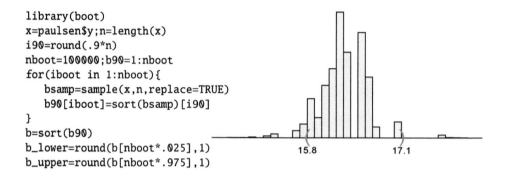

```
library(boot)
x=paulsen$y;n=length(x)
i90=round(.9*n)
nboot=100000;b90=1:nboot
for(iboot in 1:nboot){
    bsamp=sample(x,n,replace=TRUE)
    b90[iboot]=sort(bsamp)[i90]
}
b=sort(b90)
b_lower=round(b[nboot*.025],1)
b_upper=round(b[nboot*.975],1)
```

Chapter Highlights

1. The **empirical distribution function (EDF)** for a random sample $x_1, \ldots, x_n$ from a population with distribution F is defined to be

$$\hat{F}(x) = \frac{1}{n}\sum_{i=1}^{n} I\{x_i \leq x\} \quad \text{for} \ \ x \in \mathbb{R}.$$

2. The EDF $\hat{F}$ can be used to estimate the CDF F, because for each x where $F(x)$ is continuous, $\hat{F}(x) \xrightarrow{P} F(x)$.

3. A bootstrap confidence interval is made by repeatedly taking random samples of size n from the EDF $\hat{F}$ and computing the mean of each "bootstrap sample." These bootstrap sample means approximate the distribution of the sample mean and can be used to find a confidence interval.

4. The steps for obtaining a bootstrap confidence interval are as follows:

   (a) Compute $\hat{\theta} = T(x_1, \ldots, x_n)$ using the sample values.

   (b) For b going from 1 to B:

       i. Take a random sample of size n from the empirical distribution function; i.e., sample with replacement from $\{x_1, \ldots, x_n\}$, with equal probabilities, to get $\{x_{b1}, \ldots, x_{bn}\}$.

       ii. Compute $\hat{\theta}_b = T(x_{b1}, \ldots, x_{bn})$.

   (c) Use the percentiles of $\hat{\theta}_1, \ldots, \hat{\theta}_B$ to obtain the bootstrap confidence interval.

Exercises

51.1 The data set `wind` in the R library `gstat` has this description:

> Daily average wind speeds for 1961–1978 at 12 synoptic meteorological stations in the Republic of Ireland. Wind speeds are in knots (1 knot = 0.5418 m/s) at each of the stations.

Estimate the median wind speed at station ROS using the sample from 1978, and obtain a 95% bootstrap confidence interval.

51.2 The data set `chem` in the R library `MASS` has this description:

> A numeric vector of 24 determinations of copper in wholemeal flour, in parts per million.

Estimate the median amount of copper in wholemeal flour, and obtain a 99% bootstrap confidence interval.

51.3 The data set `shoes` in the R library `MASS` has this description:

> A list of two vectors, giving the wear of shoes of materials A and B for one foot each of ten boys.

Suppose the differences in shoe wear for these ten boys are independent realizations from a "population" of shoe-wear differences for these two materials. Find a 95% bootstrap confidence interval for the median difference. Is there evidence at $\alpha = .05$ that the median difference is not zero, i.e., that the two materials wear differently?

51.4 Consider the problem of constructing a 90% confidence interval for the mean θ of an exponentially distributed population, as we did for an example in Chapter 48. If we use a gamma pivotal quantity, we can get an exact confidence interval for θ.

(a) Do simulations to compare the coverage probability and average length for the exact (gamma) confidence interval with a (distribution-free) bootstrap confidence interval for the mean, using $\theta = 3$ and $n = 100$.

(b) Now suppose your assumption of an exponentially distributed population is incorrect, and the true population is Gamma$(2, 2/3)$. Repeat the simulations in part (a), comparing the (now incorrect) gamma interval with the bootstrap interval.

51.5 The data set `aircondit` in R has the following description:

> Proschan (1963) reported on the times between failures of the air-conditioning equipment in 10 Boeing 720 aircraft. The airconditioning data frame contains the intervals for the ninth aircraft.

Find a 90% bootstrap confidence interval for the median time between failures of the equipment.

51.6 The data set `sunspot.month` in R has the following description:

> Monthly numbers of sunspots, as from the World Data Center, aka SIDC.

Find a 95% bootstrap confidence interval for the 90th percentile of monthly numbers of sunspots.

51.7 The data set `acme` in R package `boot` has the following description:

> The excess returns for the Acme Cleveland Corporation along with those for all stocks listed on the New York and American Stock Exchanges were recorded over a five year period. These excess returns are relative to the return on a riskless investment such as U.S. Treasury bills.

The vector `acme$acme` has 60 values for the excess return for the Acme Cleveland Corporation. The median is negative. Construct a 95% bootstrap confidence interval for the median. Does it contain zero?

Chapter 52

Information and Maximum Likelihood Estimation

In this chapter we develop some results that allow us to make inferences with maximum likelihood estimators (MLEs) using approximate normal distributions when the exact computations for distributions of test statistics or pivotal quantities are too difficult or unwieldy.

Suppose for the random variables $Y_1, \ldots, Y_n$ we know the joint density or joint mass function up to a parameter θ. Here we consider only one unknown parameter, so θ is a scalar rather than a vector. The joint density or mass function is written as $f_\theta(y_1, \ldots, y_n) = f_\theta(\boldsymbol{y})$. Again, we are using capital letters like Y to represent random variables, and the corresponding lowercase y to represent an observed value. We will use bold letters $\boldsymbol{Y} = (Y_1, \ldots, Y_n)$ to represent a **random vector** (which simply means a vector of random variables), and $\boldsymbol{y} = (y_1, \ldots, y_n)$ to represent the observed values.

Recall that the likelihood is simply the joint density for the sample, evaluated at the sample values, and written as a function of the parameter: $L(\theta; \boldsymbol{y}) = f_\theta(\boldsymbol{y})$. The log-likelihood is

$$\ell(\theta; \boldsymbol{y}) = \log\left[L(\theta; \boldsymbol{y})\right].$$

Normal population with known variance

Example: Before we start the technical derivations, let's look at a simple example. Suppose we have a sample of size $n = 10$ from a normal distribution with mean μ and variance $\sigma^2 = 1$. The log-likelihood is a constant plus

$$\ell(\mu; \boldsymbol{y}) = -\frac{1}{2} \sum_{i=1}^{10} (y_i - \mu)^2.$$

This function is quadratic in μ, so it's easy to find the value of μ that maximizes this log-likelihood. This will be $\hat{\mu} = \bar{Y}$, the MLE.

In the following plot on the left, we see the likelihood as a function of μ for three samples of size $n = 10$ from this distribution with $\mu = 2$. Of course, different samples will result in different functions and different estimates of μ. The expression $\ell(\mu; \boldsymbol{Y})$ represents a *random function*, and we see three realizations in the plot. On the right are the functions $\ell(\mu; \boldsymbol{y})$ for three different realizations of $\boldsymbol{Y}$ when $n = 20$.

447

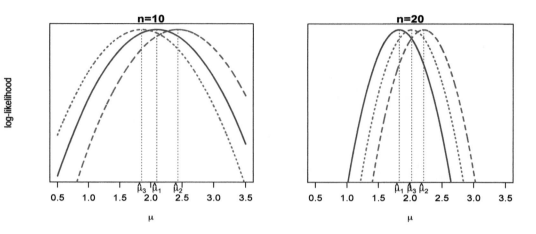

The MLE is found where the derivative is zero, and the second derivative will be negative if we have found a maximum. The purpose of the following derivations is to relate the second derivative of the log-likelihood function to the variance of the estimator.

The second derivative describes the curvature of the function; if this is "large" and negative, then the curve is "pointy," and this "pointiness" is associated with a smaller variance of the estimator. This is quite intuitive, because if ℓ is flatter at $\hat{\mu}$, we think that the true μ is almost as likely to be anywhere in the flat region, while if ℓ is a pointier function (as on the right in the above plot), we feel more confident that $\hat{\mu}$ does not move much when new samples are taken.

We'll derive the results starting with the simple fact that the integral of the joint density, over the arguments, has to equal one. Writing $d\boldsymbol{y} = dy_1 dy_2 \dots dy_n$, we have

$$\int_{\mathbb{R}^n} f_\theta(\boldsymbol{y}) d\boldsymbol{y} = 1,$$

and taking the derivative with respect to θ on both sides, we have

$$\frac{d}{d\theta} \int_{\mathbb{R}^n} f_\theta(y) d\boldsymbol{y} = 0.$$

If the support of the joint density (and hence the limits of integration) don't depend on the parameter θ, we can move the differential operator within the integral to get

$$\int_{\mathbb{R}^n} \left[\frac{d}{d\theta} f_\theta(\boldsymbol{y}) \right] d\boldsymbol{y} = 0.$$

Now we insert $f_\theta(\boldsymbol{y})/f_\theta(\boldsymbol{y})$ inside the integral to get

$$\int_{\mathbb{R}^n} \left[\frac{\frac{d}{d\theta} f_\theta(\boldsymbol{y})}{f_\theta(\boldsymbol{y})} \right] f_\theta(\boldsymbol{y}) d\boldsymbol{y} = 0.$$

We recognize the term in the brackets as the derivative of $\log(f_\theta(\boldsymbol{y}))$ (using the chain rule), so finally

$$\int_{\mathbb{R}^n} \left[\frac{d}{d\theta} \log(f_\theta(\boldsymbol{y})) \right] f_\theta(\boldsymbol{y}) d\boldsymbol{y} = 0. \qquad (\star)$$

Therefore

$$\mathrm{E}\left[\frac{d}{d\theta}\log(f_\theta(\boldsymbol{Y}))\right] = 0,$$

where $\boldsymbol{Y} = (Y_1, \ldots, Y_n)$ is a vector of the random variables. In terms of the log-likelihood function, we have $\mathrm{E}\left[\frac{d}{d\theta}\ell(\theta; \boldsymbol{Y})\right] = 0$.

The function $U(\theta) = \frac{d}{d\theta}\ell(\theta; \boldsymbol{Y})$ is called the **score function**, or the **Fisher's score**, or just the **score**. This is a random function; we use lowercase $u(\theta)$ to represent the observed score function. We have proved that, for the value of θ used to generate the data, $U(\theta)$ is a random variable with mean zero.

In the above plots, the slope of the curves at the true value ($\mu = 2$) is positive or negative depending on whether $\hat{\mu}$ is smaller than the true value or larger than the true value, respectively. The derivation tells us that the *average* slope is zero over many repeated samples.

We take another derivative of ($\star$) with respect to θ:

$$\frac{d}{d\theta}\int_{\mathbb{R}^n}\left[\frac{d}{d\theta}\log(f_\theta(\boldsymbol{y}))\right]f_\theta(\boldsymbol{y})d\boldsymbol{y} = 0,$$

and now we use the product rule after moving the outside differential operator within the integral:

$$
\begin{aligned}
0 &= \int_{\mathbb{R}^n}\left\{\left[\frac{d^2}{d\theta^2}\log(f_\theta(\boldsymbol{y}))\right]f_\theta(\boldsymbol{y}) + \left[\frac{d}{d\theta}f_\theta(\boldsymbol{y})\right]\left[\frac{d}{d\theta}\log(f_\theta(\boldsymbol{y}))\right]\right\}d\boldsymbol{y} \\
&= \int_{\mathbb{R}^n}\left[\frac{d^2}{d\theta^2}\log(f_\theta(\boldsymbol{y}))\right]f_\theta(\boldsymbol{y})d\boldsymbol{y} + \int_{\mathbb{R}^n}\left[\frac{d}{d\theta}\log(f_\theta(\boldsymbol{y}))\right]\left[\frac{\frac{d}{d\theta}(f_\theta(\boldsymbol{y}))}{f_\theta(\boldsymbol{y})}f_\theta(\boldsymbol{y})\right]d\boldsymbol{y} \\
&= \int_{\mathbb{R}^n}\left[\frac{d^2}{d\theta^2}\log(f_\theta(\boldsymbol{y}))\right]f_\theta(\boldsymbol{y})d\boldsymbol{y} + \int_{\mathbb{R}^n}\left[\frac{d}{d\theta}\log(f_\theta(\boldsymbol{y}))\right]^2 f_\theta(\boldsymbol{y})d\boldsymbol{y} \\
&= \mathrm{E}\left[\frac{d^2}{d\theta^2}\log(f_\theta(\boldsymbol{Y}))\right] + \mathrm{E}\left[\left\{\frac{d}{d\theta}\log(f_\theta(\boldsymbol{Y}))\right\}^2\right].
\end{aligned}
$$

Finally, we get the result

$$-\mathrm{E}\left[\frac{d^2}{d\theta^2}\log(f_\theta(\boldsymbol{Y}))\right] = \mathrm{E}\left[\left\{\frac{d}{d\theta}\log(f_\theta(\boldsymbol{Y}))\right\}^2\right].$$

This can be written in terms of the log-likelihood:

$$-\mathrm{E}\left[\frac{d^2}{d\theta^2}\ell(\theta; \boldsymbol{Y})\right] = \mathrm{E}\left[\left\{\frac{d}{d\theta}\ell(\theta; \boldsymbol{Y})\right\}^2\right].$$

How do we interpret this? On the right is the expected squared score function; this is the variance of the score function, because we have already shown that the expected value of the score function is zero. On the left, the second derivative of the log-likelihood function, evaluated at the true parameter, is expected to be negative, and the larger negative values imply more curvature or "pointiness" of the log-likelihood function, and hence more "certainty" for our maximum likelihood estimate.

The term on the left, the negative of the second derivative of the log-likelihood function, has a special name: This is the **information**, or **Fisher's information**, and is written as a function of the parameter:

$$I(\theta) = -\mathrm{E}\left[\frac{d^2\ell(\theta;\mathbf{Y})}{d\theta^2}\right].$$

If the variance of the score function is small, then we expect our estimator to have small variance: In effect, we have "a lot of information" about the parameter. Intuitively, the variance of our estimator is smaller, as the amount of information increases.

It is straightforward to see that if the $Y_1,\ldots,Y_n$ are independent random variables with common density $f_\theta(y)$, then we can write the information as

$$I(\theta) = -n\mathrm{E}\left[\frac{d^2\log(f_\theta(Y))}{d\theta^2}\right],$$

where $Y \sim f_\theta$.

Binomial experiments

Example: For n binomial trials with probability p of success, the log-likelihood function is

$$\ell(p; s) = s\log(p) + (n - s)\log(1 - p),$$

where s is the observed number of successes. Log-likelihood functions for three samples of size $n = 20$ are shown on the left in the plot below, and examples for $n = 60$ are shown on the right.

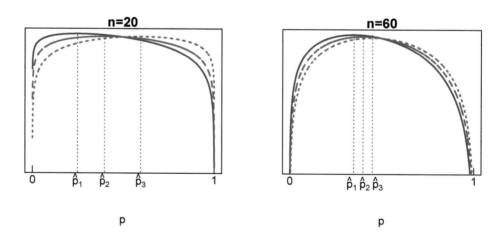

The score function is

$$u(p) = \frac{s}{p} - \frac{n - s}{1 - p},$$

and the information is

$$I(p) = \frac{n}{p(1 - p)}.$$

As the sample size grows, the information grows.

For the binomial case, the information is the inverse of the variance. This is not true for every type of model, but we can use the above derivations to get a nice practical result about *approximating* the variance of an MLE, using the information. We start with a first-order Taylor expansion of the score function about the true parameter value θ. The score function at the MLE can be written as

$$U(\hat{\theta}) \approx U(\theta) + \frac{dU(\theta)}{d\theta}(\hat{\theta} - \theta).$$

The left-hand side is zero by definition of the MLE. Then we write

$$\hat{\theta} \approx \theta - U(\theta)\left[\frac{dU(\theta)}{d\theta}\right]^{-1} \approx \theta + U(\theta)I(\theta)^{-1},$$

where in the second approximation we have replaced the derivative of the score function with its expected value, the negative of the Fisher information.

Recalling that $E(U(\theta)) = 0$, the above approximation tells us that the expected value of $\hat{\theta}$ is approximately θ, and this approximation gets better as n increases—this is the "asymptotic unbiasedness" of the MLE.

Next,

$$V(\hat{\theta}) \approx V(U(\theta))I(\theta)^{-2} = I(\theta)^{-1},$$

because the variance of the score function is the Fisher information. It can be shown that we have "asymptotic normality" of the estimator; in fact, we have the nice result that is both simple and useful:

$$\hat{\theta} \approx N(\theta, I(\theta)^{-1}).$$

Example: Suppose $Y_1, \ldots, Y_n$ is a random sample from $f_\theta(y) = \theta(1 - y)^{\theta-1}$ for $y \in (0, 1)$. Then

$$\ell(\theta; \boldsymbol{Y}) = n \log(\theta) + (\theta - 1)\sum_{i=1}^{n} \log(1 - Y_i),$$

and the score function is

$$U(\theta) = \frac{d\ell(\theta; \boldsymbol{Y})}{d\theta} = \frac{n}{\theta} + \sum_{i=1}^{n} \log(1 - Y_i).$$

This provides the MLE

$$\hat{\theta} = \frac{-n}{\sum_{i=1}^{n} \log(1 - Y_i)}.$$

It might be hard to compute directly the mean and variance of this estimator (although see Exercise 52.9). Instead, we can use the normal approximation, first computing the information:

$$I(\theta) = -E\left(\frac{d}{d\theta}U(\theta)\right) = E\left(\frac{n}{\theta^2}\right) = \frac{n}{\theta^2}.$$

The variance of the MLE is approximately θ^2/n. Of course, because we don't know θ, we need to further approximate using $\hat{\theta}$; the standard error for $\hat{\theta}$ is $\hat{\theta}/\sqrt{n}$.

Using the approximate normal distribution with this mean and variance allows for hypothesis testing about θ and confidence intervals. However, this is quite a bit of approximation, and the test sizes and coverage probabilities will also be approximate, getting better as the sample size increases. An approximate 95% confidence interval for θ is

$$\hat{\theta} \pm 2\hat{\theta}/\sqrt{n}.$$

(The factor of 2 comes from the 68-95-99.7 rule ... there doesn't seem to be much point in using a more precise 1.96 when we have made so many approximations to get this result!)

Let's check the coverage probability using simulations. Suppose $\theta = 3$ and $n = 40$. The following code samples from the population (recognizing that $f_\theta(y)$ is the density for a Beta$(1, \theta)$ random variable).

```
n=40;th=3
nloop=100000
ncap=0
for(iloop in 1:nloop){
    y=rbeta(n,1,th)
    thhat=-n/sum(log(1-y))    ## compute the estimator
    lower=thhat-2*thhat/sqrt(n)   ## lower confidence bound
    upper=thhat+2*thhat/sqrt(n)   ## upper confidence bound
    if(lower<th&upper>th){ncap=ncap+1}   ## did it capture true parameter?
}
ncap/nloop
```

We find that the coverage probability is about .956, which is pretty good. It's not exactly .95, but larger coverage probabilities are "conservative" and do not cause as much concern as coverage probabilities that are too small.

Example: Suppose we have a random sample $Y_1, \ldots, Y_n$ from an Exp(θ) population, and interest is in testing $H_0 : \theta = 4$ versus $H_a : \theta > 4$. The likelihood function is

$$\ell(\theta; \boldsymbol{Y}) = n \log(\theta) - \theta \sum_{i=1}^{n} Y_i,$$

and from the first derivative with respect to θ we get the MLE

$$\hat{\theta} = \frac{n}{\sum_{i=1}^{n} Y_i} = \bar{Y}^{-1}.$$

From the second derivative, we find the information is

$$I(\theta) = \frac{n}{\theta^2},$$

and

$$\hat{\theta} \approx \mathrm{N}(\theta, \theta^2/n).$$

Under H_0, $Z = \sqrt{n}(\hat{\theta}/4 - 1)$ is approximately standard normal, and we can use this result to do hypothesis testing. It is reasonable to ask, how big does n have to be before this is a good approximation of the true distribution of Z?

We can notice that because $\sum_{i=1}^{n} Y_i \sim \text{Gamma}(n, \theta)$, we have $1/[\sum_{i=1}^{n} Y_i] \sim \text{InvGamma}(n, \theta)$, so the *true* density of $\hat{\theta}$ is

$$f_{\hat{\theta}}(t) = \frac{\theta^n}{n!} \left(\frac{t}{n} \right)^{-n-1} e^{-n\theta/t} \quad \text{for } t > 0.$$

In the plot below, both the true and approximate densities for $\hat{\theta}$ are shown for $n = 30$ and $\theta = 4$.

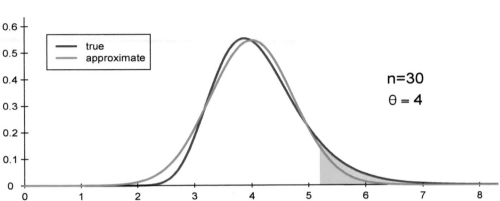

For a target test size of $\alpha = .05$, we reject H_0 when $Z > 1.645$, the 95th percentile of a standard normal. This is equivalent to "reject H_0 is $\hat{\theta} > 5.201$." In the above plot, the pink area is .05. The actual test size is the area to the right of 5.201, under the blue curve. We can see that this test size is inflated. A larger n is necessary to get a true test size that is close to the target, or we could take the trouble to derive the true test statistic distribution!

The approximate test using the information to approximate the variance of the estimator is called a **Wald test** and is a common approach in many data-analysis contexts. Similarly, we can construct **Wald confidence intervals** using the normal approximation to the distribution of the MLE.

Chapter Highlights

Let $Y_1, \ldots, Y_n$ be random variables with joint density or mass function $f_\theta(y_1, \ldots, y_n)$. The likelihood is $L(\theta; y_1, \ldots, y_n) = L(\theta; \boldsymbol{y})$, where $\boldsymbol{y} = (y_1, \ldots y_n)$, and the log-likelihood is $\ell(\theta; \boldsymbol{y})$. The random vector $\boldsymbol{Y}$ is $(Y_1, \ldots, Y_n)$.

1. The **score function** $U(\theta)$ is the derivative of the log-likelihood $\ell(\theta; \boldsymbol{Y})$ with respect to the parameter θ.

   (a) We set the score function to zero, and solve for θ, to get the MLE $\hat{\theta}$.

   (b) The score function is a function of the sample random variables, and so is a random function. The score function evaluated at the MLE, $U(\hat{\theta})$, is a random variable with zero mean, and its variance is the Fisher information.

2. The **Fisher information** is the negative of the expected value of the second derivative of the log-likelihood:

$$I(\theta) = -\mathrm{E}\left[\frac{d^2\ell(\theta; \boldsymbol{Y})}{d\theta^2}\right].$$

If $Y_1, \ldots, Y_n$ are iid with common density $f_\theta(y)$, then

$$I(\theta) = -n\mathrm{E}\left[\frac{d^2\ell(\theta; Y)}{d\theta^2}\right],$$

where $Y \sim f_\theta$.

3. If $\hat{\theta}$ is the MLE for θ, and $I(\theta)$ is the information, then the following approximate distribution holds if some mild conditions are met, with the approximation getting closer to the true density as n increases:

$$\hat{\theta} \sim \mathrm{N}(\theta, 1/I(\theta)).$$

Exercises

52.1 In Exercise 47.15, you were asked to use the delta method to approximate the variance of $\hat{p}$, the MLE for p given a random sample from a Geom(p) distribution. Use the information to approximate the variance of $\hat{p}$, and compare your answer.

52.2 Consider a random sample from a density in the Weibull family; that is, $Y_1, \ldots, Y_n$ are independent random variables with density

$$f_\theta(y) = \frac{2y}{\theta^2}e^{-y^2/\theta^2} \quad \text{for } y > 0.$$

(a) Find the MLE for θ.

(b) Use the Fisher information to approximate the variance of the MLE.

(c) Do simulations with $n = 30$ and $\theta = 2$ to determine the true variance of your estimator (to two or three significant figures), and compare this to the approximate variance found in (b).

52.3 Suppose $Y_1, \ldots, Y_n$ are independent random variables with distribution

$$f_\theta(y) = \theta y^{-(\theta+1)} \quad \text{for } y > 1,$$

where $\theta > 1$.

(a) Compute the maximum likelihood estimator and the Fisher information $I(\theta)$.

(b) Use a Wald test to determine an approximate p-value for $H_0 : \theta = 4$ versus $H_0 : \theta > 4$, using the following random sample from the distribution:

1.53, 1.36, 3.64, 1.18, 1.15, 1.31, 1.59, 1.89, 1.47, 1.25, 1.05, 2.98.

(c) Simulate the null distribution of the MLE from (a), using one million loops. This will give a p-value that is precise to about three decimal places. Compare this to the Wald test p-value, which is based on an approximation of the null distribution. Finally, make a histogram of the true null hypothesis density of your test statistic, and superimpose the approximate normal density to see the reason for the discrepancy of p-values.

52.4 Bufflehead ducks nest in trees along the water. Suppose we can model the number of eggs in a randomly selected clutch (nest) as having a Poisson distribution with mean λ. Suppose it is known that in the Northeastern U.S. habitat, $\lambda = 6.2$, but there is concern that egg clutches are smaller in a region affected by a new industry, so they want to test $H_0 : \lambda = 6.2$ versus $H_a : \lambda < 6.2$. Scientists take a "random sample" of 20 nests and find that the sample average clutch size is 5.4 eggs.

(a) Find the exact p-value using the fact that the total number of eggs in the sample is also a Poisson random variable.

(b) Use the Wald test to approximate the p-value, and compare your answer to that in part (a).

52.5 A device used for drilling in rock has a 5% chance of failing whenever it is used, independently of how many times it has been previously used. A new device claims to have a lower chance of failing. Members of a drilling team wish to test $p = .05$ against $p < .05$, at $\alpha = .01$. They will use $n = 20$ new devices until they fail, and record the average number of successful uses of the new device.

(a) Find the decision rule, using the fact that the total number of times the new devices are used successfully is a negative binomial random variable.

(b) Use the Wald test to approximate the decision rule, using that *each* of the twenty observations of numbers of successful uses is a geometric random variable. Compare your answer to that in part (a).

52.6 The cicklehead fish form mating pairs that take turns guarding the eggs until they hatch. The proportion of time the male spends guarding the eggs can be modeled as a Beta$(1, \theta)$ random variable. Suppose scientists observe 12 mating pairs, and they observe the following proportions:

0.47, 0.03, 0.18, 0.28, 0.04, 0.10, 0.21, 0.10, 0.35, 0.33, 0.25, 0.20.

They want to test $H_0 : \theta = 3$ versus $H_a : \theta \neq 3$.

(a) Use the Wald test to approximate the p-value.

(b) Find the "exact" p-value (to three decimal places) using simulations.

52.7 For some $\lambda > 0$, suppose $X_1, \ldots, X_n$ is a random sample from a density

$$f(x) = \frac{\lambda}{2\sqrt{x}} e^{-\lambda\sqrt{x}} \text{ for } x > 0.$$

(a) Find the MLE $\hat{\lambda}$ and use the information to approximate the variance of $\hat{\lambda}$.

(b) For a sample of size $n = 30$ and $\lambda = 1/2$, use simulations to get a better approximation of the true variance of $\hat{\lambda}$, and compare this to the approximation using the information. (*Hint:* You can use the code from Exercise 19.16 to simulate from the distribution.)

52.8 Suppose the maximum daily wind speed in Ventuvio County is modeled as following the density

$$f_\theta(y) = \frac{\theta}{(y+1)^{\theta+1}} \text{ for } y > 0,$$

and the historical value is known to be $\theta = 8$. Climate researchers wish to test the null hypothesis that θ is still equal to 8 against the alternative that $\theta < 8$ (higher average maximum daily wind speed). They collect 20 maximum daily wind speeds, at random times throughout the year, and treat these measurements as random sample from the new distribution; these are the sorted values (scaled so that 1 represents 100 kph):

0.02,   0.05,   0.07,   0.08,   0.09,   0.10,   0.10,   0.11,   0.13,   0.15,

0.17,   0.19,   0.22,   0.23,   0.24,   0.25,   0.27,   0.34,   0.37,   0.94.

(a) Perform a Wald's test by finding a test statistic that has approximately a standard normal density when H_0 is true, and report an approximate p-value.

(b) Simulate the values of the test statistic in (a) from the null distribution, and get an exact p-value (to 3 significant decimal places).

(c) Get a histogram of the distribution of the Wald test statistic when the null hypothesis is true, and superimpose a standard normal. Comment on whether this sample size is large enough to have a good approximation to the distribution of the test statistic.

(d) Suppose the true value for the new distribution is 6. Use simulations to find the power of the Wald's test when $\alpha = .05$. How big does the sample size need to be in order to have a power of at least .8?

52.9 Suppose $Y_1, \ldots, Y_n$ is a random sample from $f_\theta(y) = \theta(1-y)^{\theta-1}$ for $y \in (0,1)$. Then

$$\ell(\theta; \boldsymbol{Y}) = n \log(\theta) + (\theta - 1) \sum_{i=1}^{n} \log(1 - Y_i),$$

and we have the MLE

$$\hat{\theta} = \frac{-n}{\sum_{i=1}^{n} \log(1 - Y_i)}.$$

The example in this chapter uses an approximation for the variance of this estimator. However, we can compute it using the results of Exercise 21.24. Derive the true mean and variance for $\hat{\theta}$ and compare them to the approximations.

Chapter 53

Sufficient Statistics

At this point we have learned several standard ways of constructing estimators for parameters, given some observations with a density or a mass function that is known up to the parameter. Estimators can be assessed and compared, through bias, variance, and mean squared error. We have learned how to construct exact or approximate confidence intervals, and we can assess and compare different types of confidence intervals in terms of coverage probability and interval length. We also know several ways of constructing hypothesis tests, and we have compared tests with power computations. In this chapter we are interested in identifying a particular type of statistic that helps us choose "good" options for any of these applications.

Conceptually, we say a statistic is **sufficient** if it "contains all the available information about the parameter." We will see that estimators, confidence intervals, and hypothesis tests that are based on sufficient statistics can be shown to have some nice optimality properties. That is, when using these sufficient statistics we can be confident that our estimators have small variance, our tests have good power, and our confidence intervals are narrow. The technical definition follows:

> Let $Y_1, \ldots, Y_n$ be a sample from a probability distribution that is known up to a parameter θ. The statistic $T = \tau(Y_1, \ldots, Y_n)$ is **sufficient for θ** if the conditional distribution of $Y_1, \ldots, Y_n$, given T, does not depend on θ.

A sufficient statistic summarizes the data without losing any information. It's a nice concept, and one of its primary applications is in the construction of "uniformly minimum variance unbiased estimators." In this chapter we will learn how to find a sufficient statistic given a model, but keep in mind that it's motivated by this important result about finding a "best" estimator.

The **Sufficiency principle** states that if $T = \tau(Y_1, \ldots, Y_n)$ is sufficient for θ, and we have two different samples with the same value of T, that is, $\tau(y_1, \ldots, y_n) = \tau(x_1, \ldots, x_n)$, then any inference about θ should be the same whether the sample $y_1, \ldots, y_n$ or $x_1, \ldots, x_n$ is observed.

Example: Suppose we conduct binomial trials, where each of the independent random variables $Y_1, \ldots, Y_n$ is zero or one, with probability p of being one. There are many sequences of zeros and ones that have k ones and lead to the same estimator of p. It's intuitive that the order of the ones and zeros should not matter when we use the data to make an inference about p. We will see that the sample proportion $\bar{Y}$ is sufficient for p, so according to the sufficiency principle, inference about p for all the different samples with $\bar{Y} = .3$ should be exactly the same.

Now let's show formally that the sum $S = \sum_{i=1}^{n} Y_i$ is sufficient for p in the binomial trials example. Using our definition, we must show that the conditional distribution of the data $Y_1, \ldots, Y_n$, given S, doesn't depend on p. Going back to our definition of conditional probabilities, for $k \in \{0, 1, \ldots, n\}$ and $y_1 + \cdots + y_n = k$, we have

$$
\begin{aligned}
\mathrm{P}(Y_1 = y_1, \ldots, Y_n = y_n | S = k) &= \frac{\mathrm{P}(Y_1 = y_1, \ldots, Y_n = y_n \text{ and } S = k)}{\mathrm{P}(S = k)} \\
&= \frac{\mathrm{P}(Y_1 = y_1, \ldots, Y_n = y_n)}{\mathrm{P}(S = k)} \\
&= \frac{p^{y_1}(1-p)^{1-y_1} \cdots p^{y_n}(1-p)^{1-y_n}}{\binom{n}{k} p^k (1-p)^{n-k}} = \frac{1}{\binom{n}{k}}.
\end{aligned}
$$

The final expression for this probability doesn't depend on p, so we have proved sufficiency of the sample sum. It follows that the sample proportion $\bar{Y} = S/n$ is also sufficient. Any one-to-one transformation of a sufficient statistic will again be sufficient, because all the information is preserved in the transformation.

The factorization theorem

Fortunately, there is a nice easy way to determine if a statistic is sufficient, without computing conditional probabilities or conditional distributions. Let $Y_1, \ldots, Y_n$ be random variables having a probability distribution that is known up to a parameter θ. The statistic $T = \tau(Y_1, \ldots, Y_n)$ is sufficient for θ if and only if the joint density or mass function $f_\theta(y_1, \ldots, y_n)$ of the sample can be factored into two functions:

$$
f_\theta(y_1, \ldots, y_n) = g_\theta(t) h(y_1, \ldots, y_n),
$$

where $t = \tau(y_1, \ldots, y_n)$ and the function h doesn't depend on θ.

For the binomial example, we could have simply noted that the joint mass function is

$$
\prod_{i=1}^{n} \left[p^{y_i}(1-p)^{1-y_i} \right] = p^s (1-p)^{n-s},
$$

where s is the sample sum. Here $h(y_1, \ldots, y_n) = 1$ and $g_p(s) = p^s(1-p)^{n-s}$. Let's do another example before we prove the factorization theorem.

Example: Let $U_1, U_2, \ldots,$ be iid $\mathrm{Unif}(0, \theta)$ random variables, and let

$$
U_{(n)} = \max(U_1, \ldots, U_n).
$$

Let's show that $U_{(n)}$ is a sufficient statistic for θ, using the factorization theorem. We can write the joint density of $U_1, \ldots, U_n$ as the product of the marginals:

$$
\begin{aligned}
f_\theta(u_1, \ldots, u_n) &= \prod_{i=1}^{n} \left[\frac{1}{\theta} I\{0 \leq u_i \leq \theta\} \right] \\
&= \frac{1}{\theta^n} \prod_{i=1}^{n} I\{0 \leq u_i \leq \theta\} \\
&= \frac{1}{\theta^n} I\{u_{(n)} \leq \theta\} I\{u_{(1)} \geq 0\}.
\end{aligned}
$$

Therefore if $t = u_{(n)}$, $g_\theta(t) = I\{t \le \theta\}/\theta^n$, and $h(u_1, \ldots, u_n) = I\{u_{(1)} \ge 0\}$, we have shown that $T = U_{(n)}$ is sufficient for θ.

The proof of the factorization theorem will be given for the discrete case, where

$$f_\theta(y_1, \ldots, y_n) = \mathbf{P}_\theta(Y_1 = y_1, \ldots, Y_n = y_n).$$

Suppose there is a statistic $T = \tau(Y_1, \ldots, Y_n)$ so that

$$f_\theta(y_1, \ldots, y_n) = g_\theta(t)h(y_1, \ldots, y_n),$$

where $t = \tau(y_1, \ldots, y_n)$ and h does not depend on θ. The joint support of the random variables, that is, the set of all possible n-tuples $(y_1, \ldots, y_n)$, can be partitioned into sets A_t, where t runs through all of the possible values for T. This is a countable set, because the random variables are discrete. We consider the conditional probability that $Y_1 = y_1, \ldots, Y_n = y_n$, given $T = t$, and show that the expression for this conditional probability does not involve θ. The conditional probability can be written as

$$\mathbf{P}_\theta(Y_1 = y_1, \ldots, Y_n = y_n | T = t) = \frac{\mathbf{P}_\theta(Y_1 = y_1, \ldots, Y_n = y_n \text{ and } T = t)}{\mathbf{P}_\theta(T = t)}.$$

The numerator $\mathbf{P}_\theta(Y_1 = y_1, \ldots, Y_n = y_n \text{ and } T = t) = \mathbf{P}_\theta(Y_1 = y_1, \ldots, Y_n = y_n)$ because T is determined by $Y_1, \ldots, Y_n$. Assume $(y_1, \ldots, y_n) \in A_t$ for some t (otherwise the probability is zero); then the numerator is

$$\mathbf{P}_\theta(Y_1 = y_1, \ldots, Y_n = y_n) = f_\theta(y_1, \ldots, y_n) = g_\theta(t)h(y_1, \ldots, y_n),$$

since by our assumption the joint probability mass function can be factored. The denominator is

$$\begin{aligned} \mathbf{P}_\theta(T = t) &= \sum_{y_1, \ldots, y_n \in A_t} f_\theta(y_1, \ldots, y_n) \\ &= \sum_{y_1, \ldots, y_n \in A_t} g_\theta(t)h(y_1, \ldots, y_n) \\ &= g_\theta(t) \sum_{y_1, \ldots, y_n \in A_t} h(y_1, \ldots, y_n). \end{aligned}$$

Now, putting the fraction back together, we have

$$\begin{aligned} \mathbf{P}_\theta(Y_1 = y_1, \ldots, Y_n = y_n | T = t) &= \frac{g_\theta(t)h(y_1, \ldots, y_n)}{g_\theta(t) \sum_{y_1, \ldots, y_n \in A_t} h(y_1, \ldots, y_n)} \\ &= \frac{h(y_1, \ldots, y_n)}{\sum_{y_1, \ldots, y_n \in A_t} h(y_1, \ldots, y_n)}, \end{aligned}$$

which does not depend on θ. We have shown that the factorization implies that the statistic is sufficient. To show that all sufficient statistics lead to such a factorization (the "only if" part of the theorem), we simply write

$$\begin{aligned} f_\theta(y_1, \ldots, y_n) &= \mathbf{P}_\theta(Y_1 = y_1, \ldots, Y_n = y_n) \\ &= \mathbf{P}_\theta(Y_1 = y_1, \ldots, Y_n = y_n \text{ and } T = t) \\ &= \mathbf{P}_\theta(T = t)\mathbf{P}_\theta(Y_1 = y_1, \ldots, Y_n = y_n | T = t), \end{aligned}$$

and the factorization is accomplished because the second term, by definition of sufficiency, does not depend on θ.

Example: Show that the sample mean is sufficient for the population mean for a random sample from an $N(\mu, \sigma^2)$ distribution, where σ^2 is known.

The joint density is

$$f_\mu(y_1, \ldots, y_n) = \prod_{i=1}^{n} \left[\frac{1}{\sqrt{2\pi\sigma^2}} e^{-(y_i-\mu)^2/(2\sigma^2)} \right]$$

$$= \left(\frac{1}{2\pi\sigma^2} \right)^{n/2} e^{-\sum_{i=1}^{n}(y_i-\mu)^2/(2\sigma^2)}$$

$$= \left(\frac{1}{2\pi\sigma^2} \right)^{n/2} e^{-n\mu^2/(2\sigma^2)} e^{\mu(\sum_{i=1}^{n} y_i)/\sigma^2} e^{-\sum_{i=1}^{n} y_i^2/(2\sigma^2)}.$$

Now if $T = \bar{Y}$, we can define

$$g_\mu(t) = \left(\frac{1}{2\pi\sigma^2} \right)^{n/2} e^{-n\mu^2/(2\sigma^2)} e^{\mu n t/\sigma^2}$$

and

$$h(y_1, \ldots, y_n) = e^{-\sum_{i=1}^{n} y_i^2/(2\sigma^2)}.$$

If we have only one parameter, typically a sufficient statistic is one-dimensional (although contrary examples can be constructed). If we have two parameters, we typically need a two-dimensional sufficient statistic.

Example: Show that the sample mean *and* the sample variance together are sufficient for the population mean and variance for a random sample from an $N(\mu, \sigma^2)$ distribution, where σ is unknown.

Beginning with the joint density again,

$$f_{\mu,\sigma^2}(y_1, \ldots, y_n)$$

$$= \left(\frac{1}{2\pi\sigma^2} \right)^{n/2} \exp\left\{ -\sum_{i=1}^{n}(y_i - \mu)^2/(2\sigma^2) \right\}$$

$$= \left(\frac{1}{2\pi\sigma^2} \right)^{n/2} \exp\left\{ -\sum_{i=1}^{n} \left[(y_i - \bar{y})^2 + (\bar{y} - \mu)^2 + 2(y_i - \bar{y})(\bar{y} - \mu) \right]/(2\sigma^2) \right\}$$

$$= \left(\frac{1}{2\pi\sigma^2} \right)^{n/2} \exp\left\{ -\left[\sum_{i=1}^{n}(y_i - \bar{y})^2 + n(\bar{y} - \mu)^2 \right]/(2\sigma^2) \right\}$$

$$= \left(\frac{1}{2\pi\sigma^2} \right)^{n/2} \exp\left\{ -\left[(n-1)s^2 + n(\bar{y} - \mu)^2 \right]/(2\sigma^2) \right\},$$

where

$$s^2 = \frac{1}{n-1} \sum_{i=1}^{n}(y_i - \bar{y})^2.$$

The joint density can be written in terms of s^2 and $\bar{y}$, so by the factorization theorem, the sample mean $\bar{Y}$ and the sample variance S^2 are jointly sufficient for μ and σ^2.

A sufficient statistic is **minimal sufficient** if it's the "most efficient" summary of the data. Formally, a minimal sufficient statistic is a function of every other sufficient statistic.

Clearly, if we have a sufficient statistic and we add *more* information, we still have a sufficient statistic. For example, we have shown that the sample mean $\bar{Y}$ is sufficient for an exponential population mean; if we add the sample variance S^2, we get a two-dimensional sufficient statistic $(\bar{Y}, S^2)$ that is not minimal. If a sufficient statistic is one-dimensional, it must be minimally sufficient.

If we have a two-dimensional parameter, the minimal sufficient statistic will usually be two-dimensional as well. There are a few exceptions, but these are anomalous cases. For practical purposes, we can look for a sufficient statistic that is the same dimension as the parameter, and conclude that such a sufficient statistic is also minimal sufficient.

Chapter Highlights

1. By definition, a statistic $S = g(Y_1, \ldots, Y_n)$ is sufficient for a parameter θ if the conditional distribution of $Y_1, \ldots, Y_n$, given S, does not depend on θ.

2. The factorization theorem: The statistic $S = g(Y_1, \ldots, Y_n)$ is sufficient for θ if and only if the joint density or mass function $f(y_1, \ldots, y_n; \theta)$ of the sample can be factored into two functions $g(s; \theta)$ and $h(y_1, \ldots, y_n)$, where the function h doesn't depend on θ.

Exercises

53.1 Show that the mean of a random sample from a Poisson distribution is sufficient for the population mean λ.

53.2 Show that the sample mean is sufficient for the population mean θ for a random sample from an exponential distribution.

53.3 Suppose $\theta > 1$ and $Y_1, \ldots, Y_n$ is a random sample from

$$f_\theta(y) = (\theta - 1)y^{-\theta} \text{ for } y \in (1, \infty).$$

Show that $T = \sum_{i=1}^n \log(Y_i)$ is sufficient for θ.

53.4 Suppose $\theta > 1$ and let $Y_1, \ldots, Y_n$ be a random sample from a population with density

$$f_\theta(y) = \frac{\theta}{(y+1)^{\theta+1}} \text{ for } y > 0.$$

(a) Find a statistic that is sufficient for θ.

(b) Find the expected value of your sufficient statistic.

53.5 Let $Y_1, \ldots, Y_n$ be a random sample from a normal population with mean zero and variance σ^2. Find a sufficient statistic for σ^2.

53.6 Suppose $Y_1, \ldots, Y_n$ are iid discrete random variables. We know $P(Y_1 = 0) = \theta$, while $P(Y_1 = -1) = P(Y_1 = 1) = 1 - \theta/2$ for some $\theta \in (0, 1)$. Let S be the number of Y_i that equal zero, and show that S is sufficient for θ.

53.7 Suppose $\theta > 0$ and $Y_1, \ldots, Y_n$ are independent random variables with common density

$$f_\theta(y) = \frac{3y^2}{\theta} e^{-y^3/\theta} \text{ for } y > 0.$$

Find a sufficient statistic for θ.

53.8 Suppose $Y_1 \sim N(\mu, \sigma^2)$ and $Y_2 \sim N(\mu, 4\sigma^2)$ are observed, where σ^2 is known. If Y_1 and Y_2 are independent, find a one-dimensional sufficient statistic for μ.

53.9 Suppose $Y_k \overset{ind}{\sim} N(\mu, k\sigma^2)$, $k = 1, \ldots, n$, are observed, where σ^2 is known. Find a one-dimensional sufficient statistic for μ.

53.10 Suppose $Y_k \overset{ind}{\sim} N(\mu, \sigma_k^2)$, $k = 1, \ldots, n$, are observed, where $\sigma_1^2, \ldots, \sigma_n^2$ are known. Find a one-dimensional sufficient statistic for μ.

53.11 Suppose $Y_1, \ldots, Y_8$ are independent exponential random variables. The random variables Y_1, Y_2, Y_3, and Y_4 all have mean θ, while the random variables Y_5, Y_6, Y_7, and Y_8 all have mean 2θ. Find a sufficient statistic for θ.

53.12 Suppose $Y_1, \ldots, Y_n$ are a random sample from the density

$$f_\theta(y) = 2\theta^2 y^{-3} \text{ for } y > \theta$$

for some $\theta > 0$. Find a sufficient statistic for θ.

53.13 Suppose $Y_1, \ldots, Y_n$ are independent random variables with common density

$$f_\theta(y) = e^{\theta - y} \text{ for } y > \theta.$$

Find a sufficient statistic for θ.

53.14 Suppose $Y_1, \ldots, Y_n$ are independent discrete random variables with common mass function $f_a(-1) = (1 - a)/2$, $f_a(0) = a$, and $f_a(1) = (1 - a)/2$ for some $a \in (0, 1)$. Show that the number of zero observations is sufficient for a.

53.15 Suppose $Y_1, \ldots, Y_n$ are independent random variables with common density

$$f(y) = \frac{\theta}{2} \left(\frac{y}{2} + 1 \right)^{-(\theta+1)} \text{ for } y > 0,$$

where $\theta > 0$.

(a) Find a one-dimensional sufficient statistic for θ.

(b) Find the expected value of your sufficient statistic.

53.16 Given a sample $Y_1, \ldots, Y_n$ of independent Gamma(α, β) random variables, find a two-dimensional statistic sufficient for α and β.

53.17 Consider again Exercise 42.5 and determine which (if any) of the estimators in parts (a), (b), and (c) is a function of a sufficient statistic.

53.18 Consider again Exercise 50.3, where a confidence interval for the parameter θ of a "shifted exponential" distribution was found. Determine if either of the pivotal quantities used in (a) and (b) is based on a sufficient statistic, and if so, which.

53.19 Consider again Exercise 42.14 and determine if $\hat{\theta}$ or $\tilde{\theta}$ is a function of a sufficient statistic.

53.20 Consider again Exercise 42.15 and determine if either of the estimators for α involves a sufficient statistic.

53.21 Let $X_1, \ldots, X_n$ be a random sample from $N(\mu, \sigma^2)$ and let $Y_1, \ldots, Y_m$ be a random sample from $N(2\mu, \sigma^2)$; further assume the samples are independent and σ^2 is known.

    (a) Find a statistic sufficient for μ.

    (b) Use the statistic in part (a) to construct a pivotal quantity for a 95% confidence interval for μ.

    (c) Another statistician reasons $\bar{X}$ and $\bar{Y}/2$ both have mean μ, so

$$(\bar{X} + \bar{Y}/2)/2 \sim N\left(\mu, \frac{\sigma^2}{4}\left(\frac{1}{n} + \frac{1}{4m}\right)\right).$$

I will make a confidence interval using

$$Q = \frac{(\bar{X} + \bar{Y}/2)/2 - \mu}{\frac{\sigma}{2}\sqrt{\frac{1}{n} + \frac{1}{4m}}}$$

as a pivotal quantity. Then Q is standard normal and my confidence interval is

$$(\bar{X} + \bar{Y}/2)/2 \pm 1.96\frac{\sigma}{2}\sqrt{\frac{1}{n} + \frac{1}{4m}}.$$

Show that for all n, m, σ that the confidence interval in part (b) has smaller width.

53.22 Let $Y_1, \ldots, Y_n$ be independent exponential random variables, where X_k has mean $k\theta$.

    (a) Find a statistic that is sufficient for θ.

    (b) Use the statistic in (a) to make a pivotal quantity for a 90% confidence interval for θ.

    (c) Find the bounds for the confidence interval in (b) when the data are (in order)

3.7,   0.3,   2.8,   13.6,   15.9,   31.9,   15.6,   31.7,   26.3,   9.7,   29.9,   29.0.

53.23 In all our examples and exercises so far, we have found a one-dimensional sufficient statistic whenever we have a one-dimensional parameter. Consider, however, the case where $Y_1, \ldots, Y_n$ are independent and uniform on $(\theta, \theta + 1)$, and show that we need a two-dimensional sufficient statistic for θ.

Chapter 54

Uniformly Minimum Variance Unbiased Estimators

We have developed several standard types of estimators, and several ways to assess estimators. We can determine the bias and variance of our estimators to quantify accuracy and precision. The mean squared error (MSE) is a composite measure of the accuracy and precision, that is, the expected squared distance of the estimator from the thing it's trying to estimate.

In this chapter we're looking for a "best" estimator, and it's reasonable to want a minimum-MSE estimator. However, the estimators will have different MSEs for different true values of θ, and it's impossible to find a *uniformly best* estimator, by which we mean an estimator that has smallest MSE for *all* values of θ.

Fortunately it's possible to come up with a *best unbiased* estimator. We can often find an unbiased estimator that has smaller variance than other unbiased estimators for all values of the parameter. We will call such an estimator *best* in the class of unbiased estimators, using the term **uniformly minimum variance unbiased estimator** **(UMVUE)**, where "uniformly" refers to "over all parameter values." (That said, it's important to remember that other (biased) estimators might have smaller MSEs for many parameter values!)

This chapter has three results concerning best unbiased estimation of parameters.

1. The **Cramér–Rao lower bound** tells us what the smallest possible variance is for an unbiased estimator of a given parameter; this can be used to check if an unbiased estimator attains this minimum variance. If it does, it's UMVUE. The proof of this result is a nice application of the Cauchy–Schwarz inequality from advanced calculus, but we will skip the proof and concentrate on the application.

2. The **Rao–Blackwell theorem** is a cute way to take *any* unbiased estimator and a sufficient statistic, and combine them to get a *better* unbiased estimator. The proof is a nice, straightforward application of our conditional expectation results from Chapter 29, so we'll go through it.

3. The **Lehmann–Scheffé theorem** tells us how to construct an UMVUE from a sufficient statistic. This will be our main result (and the most useful of the three). It follows pretty closely from the Rao–Blackwell theorem, but the proof involves definitions of minimal sufficient and "complete" statistics, which we will discuss only briefly.

The Cramér–Rao lower bound

Suppose our random sample $Y_1, \ldots, Y_n$ is from a population with density (or probability mass function) in the family $f_\theta(y)$, and we want to estimate θ. Suppose $\hat\theta$ is any unbiased estimator; then if some mild "regularity conditions" hold,

$$\mathrm{V}(\hat\theta) \geq \frac{1}{I(\theta)},$$

where $I(\theta)$ is the information defined in Chapter 52. For the special case of a joint density of a random sample from a population, we can write the information in terms of the population density or mass function:

$$I(\theta) = -n\mathrm{E}\left[\frac{\partial^2}{\partial\theta^2} \log[f_\theta(Y)]\right].$$

The inequality is also called the "information inequality."

Example: Let's try this for estimation of the mean of an exponential random variable. Suppose $Y_1, \ldots, Y_n$ are iid exponential with mean θ.

$$f_\theta(y) = \frac{1}{\theta}e^{-y/\theta} \ \text{ for } \ y > 0.$$

Then

$$\log[f_\theta(y)] = -\log(\theta) - \frac{y}{\theta}$$

and

$$\frac{\partial^2}{\partial\theta^2} \log[f_\theta(y;)] = \frac{1}{\theta^2} - \frac{2y}{\theta^3},$$

so then

$$\mathrm{E}\left[\frac{\partial^2}{\partial\theta^2} \log[f_\theta(Y)]\right] = \mathrm{E}\left[\frac{1}{\theta^2} - \frac{2Y}{\theta^3}\right] = -\frac{1}{\theta^2}.$$

Finally we have

$$I(\theta) = \frac{n}{\theta^2},$$

and any unbiased estimator of θ will have variance at least as big as θ^2/n. Now we simply note that the sample mean has variance equal to this lower bound, and we conclude that the sample mean is UMVUE. The Cramér–Rao lower bound tells us we can't get a better unbiased estimator, so we can stop looking!

Example: We can also use the Cramér–Rao result for discrete random variables. Suppose $Y_1, \ldots, Y_n$ are independent Bernoulli random variables with $\mathrm{P}(Y = 1) = p$ and $\mathrm{P}(Y = 0) = 1 - p$. We know our sample proportion of "successes" $\bar{Y}$ is an unbiased estimator of p, so let's see if it attains the Cramér–Rao lower bound. Using $f_p(y) = p^y(1-p)^{1-y}$ as our population mass function, we have

$$\log(f_p(y)) = y\log(p) - (1-y)\log(1-p),$$

so

$$\frac{\partial^2 \log(f_p(y))}{\partial p^2} = -\frac{y}{p^2} + \frac{1-y}{(1-p)^2}.$$

Next we find

$$\mathrm{E}\left(\frac{\partial^2 \log(f_p(Y))}{\partial p^2}\right) = -\frac{1}{p} + \frac{1}{1-p} = \frac{-1}{p(1-p)}$$

and

$$I(p) = \frac{n}{p(1-p)},$$

so any unbiased estimator of p will have variance that is at least $p(1-p)/n$. Since this is the variance for $\hat{p} = \bar{Y}$, we know that $\hat{p}$ is UMVUE.

What if we want to estimate a function of our parameter, say $1/\theta$ or p^2, or $p/(1-p)$? **A more general version of the Cramér–Rao result** can be used. Again suppose our random sample $Y_1, \ldots, Y_n$ is from a population with density (or probability mass function) in the family $f_\theta(y)$. If the statistic T is an unbiased estimator of a continuously differentiable function $g(\theta)$ of the parameter, then

$$\mathrm{V}(T) \geq \frac{(g'(\theta))^2}{I(\theta)}.$$

Note that if $g(\theta) = \theta$, then the numerator is 1, and the formula reduces to the simpler form.

Example: Let's look at our exponential example again. The random variables $Y_1, \ldots, Y_n$ are iid $\mathrm{Exp}(\beta)$, so that the mean is $1/\beta$. In Exercise 45.2, we found the MLE for β to be $\hat{\beta} = 1/\bar{Y}$. Let's determine whether this estimator is biased. We recall that $S = Y_1 + \cdots + Y_n \sim \mathrm{Gamma}(n, \beta)$, and the MLE is n/S. Then

$$\mathrm{E}(\hat{\beta}) = n\mathrm{E}\left(\frac{1}{S}\right) = \frac{n\beta^n}{\Gamma(n)}\int_0^\infty \frac{1}{s}s^{n-1}e^{\beta s}ds = \frac{n\beta^n}{\Gamma(n)}\frac{\Gamma(n-1)}{\beta^{n-1}} = \frac{n\beta}{(n-1)}.$$

It looks like the MLE is a little too large, but we can define $T = (n-1)/S$, which is unbiased for $1/\theta$.

To find the variance for our estimator, we can use the solution to Exercise 16.14, to get

$$\mathrm{V}(T) = \frac{\beta^2}{(n-2)}.$$

Let's find out if this is the smallest possible variance. Using the formula for the Cramér–Rao lower bound, we have $g(\theta) = 1/\theta$, so $g'(\theta)^2 = 1/\theta^4$. We use $I(\theta) = n/\theta^2$ as computed previously, and we find that the variance of any unbiased estimator of $1/\theta$ will be at least $1/(n\theta^2) = \beta^2/n$. This is smaller than the variance of our estimator T, so the bound is not achieved.

We know that if our unbiased estimator attains the Cramér–Rao lower bound, our estimator is UMVUE, because we can't find an unbiased estimator that has a lower variance. However, the converse is not true: if our unbiased estimator does not attain the Cramér–Rao lower bound, that doesn't mean it's not UMVUE. Sometimes the lower bound is impossible to attain.

The Rao–Blackwell theorem

The Rao–Blackwell theorem allows us to take any unbiased estimator and a sufficient statistic, and create another unbiased estimator with smaller variance (or at least the variance is not larger). If $\tilde{\theta}$ is an unbiased estimator of θ, and T is a sufficient statistic, define

$$\hat{\theta} = \mathrm{E}(\tilde{\theta}|T).$$

Then the following statements are true:

1. $\hat{\theta}$ is a statistic (i.e., does not depend on θ);

2. $\hat{\theta}$ is unbiased for θ; and

3. $\mathrm{V}(\hat{\theta}) \leq \mathrm{V}(\tilde{\theta})$.

Proof: For the first statement, we use the definition of sufficient statistic. Because T is sufficient, the distribution of the sample, and hence any function of the sample, given T, does not depend on θ. Then $\hat{\theta}$ is a function only of the sample, and hence is a statistic.

For the second statement of the theorem, we use our formula for conditional expectation:

$$\mathrm{E}(\hat{\theta}) = \mathrm{E}\left[\mathrm{E}(\tilde{\theta}|T)\right] = \mathrm{E}(\tilde{\theta}) = \theta.$$

For the third statement, we use the variance formula:

$$\mathrm{V}(\tilde{\theta}) = \mathrm{V}\left[\mathrm{E}(\tilde{\theta}|T)\right] + \mathrm{E}\left[\mathrm{V}(\tilde{\theta}|T)\right] = \mathrm{V}(\hat{\theta}) + \mathrm{E}\left[\mathrm{V}(\tilde{\theta}|T)\right].$$

The second term on the right can't be negative, so we get $\mathrm{V}(\tilde{\theta}) \geq \mathrm{V}(\hat{\theta})$.

Example: Suppose Y_1 and Y_2 are independent exponential random variables, both with mean θ, and we want to find an estimator for θ^2. We know that $\mathrm{E}(Y_1^2) = 2\theta^2$, so $T_1 = Y_1^2/2$ is unbiased for θ. We also know that $S = Y_1 + Y_2$ is sufficient for θ by Exercise 53.2. Further, by Exercise 26.7, we have the distribution of Y_1, given $S = s$, is uniform on $(0, s)$. The expected value of the square of a random variable that is uniform on $(0, s)$ is $s^2/3$, so

$$T_2 = \mathrm{E}(T_1|S) = \frac{S^2}{6}.$$

We see that T_2 is unbiased, because the expected value of S^2, for $S \sim \mathrm{Gamma}(2, 1/\theta)$, is $6\theta^2$. We can also check that we have reduced the variance: $\mathrm{V}(T_1) = \mathrm{V}(Y_1^2)/4 = 5\theta^4$ (see Exercise 16.13), while

$$\mathrm{V}(T_2) = \frac{\mathrm{V}(S^2)}{36} = \frac{(\mathrm{E}(S^4) - \mathrm{E}(S^2)^2)}{36} = \frac{(120 - 36)\theta^4}{36} = \frac{7\theta^4}{3}.$$

The solution to Exercise 16.4 was used to compute the expected values. The variance has been considerably reduced.

The Lehmann–Scheffé theorem

Let's look again at the last line of the Rao–Blackwell proof. We have

$$\mathrm{V}(\tilde{\theta}) = \mathrm{V}(\hat{\theta}) + \mathrm{E}\left[\mathrm{V}(\tilde{\theta}|T)\right].$$

What's to prevent us from conditioning again and getting an estimator with a still smaller variance? When the second term, $\mathrm{E}[\mathrm{V}(\tilde{\theta}|T)]$, is equal to zero, we won't get a better estimator. When does this happen? If the statistic $\tilde{\theta}$ is a function of T, then the variance of $\tilde{\theta}|T$ is zero. (Given T, there is only one possible value of $\tilde{\theta}$.) Then, to get a smaller variance, we would have to condition on a different sufficient statistic.

Recall that a statistic is **minimal sufficient** if it is a function of any other sufficient statistic. Therefore, once we condition on a *minimal* sufficient statistic, we can't use Rao–Blackwell to decrease our variance further. The Lehmann–Scheffé theorem formalizes this idea, and states that if our estimator is a function of a minimal sufficient statistic, it is UMVUE. (Actually, there is another condition: the statistic must also be *complete*. But it is difficult to find an example of a minimal sufficient but not complete statistic, and for practical, real-world problems, we don't need to worry about this distinction.)

In practical terms, we can take any minimal sufficient statistic and find an unbiased estimator that is a function of this statistic, and we know this estimator is UMVUE. We don't have to calculate the conditional expectations of the Rao–Blackwell theorem, or check the Cramér–Rao bound. Easy-peasy!

Example: Let's find an UMVUE for θ when we have $Y_1, \ldots, Y_n \sim \text{Unif}(0, \theta)$. We have already found that our MLE $Y_{(n)}$ always has smaller variance than the method of moments $2\bar{Y}$, but the MLE is biased with $E(Y_{(n)}) = n\theta/(n+1)$. The MLE was shown to be sufficient, however, so if we define

$$\hat{\theta} = \frac{n+1}{n} Y_{(n)}$$

as our unbiased estimator that is a function of a sufficient statistic, then we know that's UMVUE by the Lehmann–Scheffé theorem.

Example: We found that for a random sample $Y_1, \ldots, Y_n$ from an $N(\mu, \sigma^2)$ population with both μ and σ^2 unknown, the two-dimensional statistic $(\bar{Y}, S^2)$, that is, the sample mean and the sample variance, are jointly sufficient for the population mean and variance. They are both unbiased, and hence they are UMVUE.

Example: Going back to the example of estimating $1/\theta$, given a random sample $Y_1, \ldots, Y_n$ from an exponential population with mean θ, we found that our unbiased estimator $T = (n-1)/S$ had a variance slightly larger than the Cramér–Rao lower bound. However, we know that S is sufficient for θ (and hence for $1/\theta$), so by the Lehmann–Scheffé theorem, T is UMVUE. In this case, the Cramér–Rao lower bound cannot be attained.

Confidence intervals

When constructing a confidence interval with the pivotal quantity method, we sometimes have more than one option. Using a sufficient statistic will, in general, produce the shortest-length confidence interval.

For example, suppose that $X_1, \ldots, X_n$ are normal with mean μ and variance one, $Y_1, \ldots, Y_m$ are normal with mean μ and variance four, and all the $m + n$ random variables are pairwise independent. We *could* construct a pivotal quantity using the average of all the random variables. Define

$$\tilde{\mu} = \frac{\sum_{i=1}^{n} X_i + \sum_{i=1}^{m} Y_i}{(n+m)},$$

so that $\tilde{\mu}$ has mean μ and variance $(n + 4m)/(n + m)^2$. Then

$$Z_1 = \frac{\hat{\mu} - \mu}{\sqrt{(n + 4m)/(n + m)^2}}$$

has a standard normal density and can be used as a pivotal quantity, which gives the 95% confidence interval

$$\tilde{\mu} \pm 1.96\sqrt{(n+4m)/(n+m)^2}.$$

A sufficient statistic (see Exercise 54.8) is

$$\sum_{i=1}^{n} X_i + \frac{1}{4}\sum_{i=1}^{m} Y_i,$$

and hence an UMVUE is

$$\hat{\mu} = \frac{\sum_{i=1}^{n} X_i + \frac{1}{4}\sum_{i=1}^{m} Y_i}{n + m/4}$$

(this is also the MLE). It seems intuitive that constructing a pivotal quantity from a sufficient statistic should give a shorter-length confidence interval. The estimator $\hat{\mu}$ has mean μ and variance $1/(n+m/4)$. Then

$$Z_2 = \frac{\hat{\mu} - \mu}{\sqrt{1/(n+m/4)}}$$

is standard normal. The 95% is

$$\hat{\mu} \pm 1.96\frac{1}{\sqrt{n+m/4}}.$$

It is straightforward to see that this interval is, for any $n, m \geq 1$, of shorter length than the interval based on the sample mean.

Chapter Highlights

1. An estimator $\hat{\theta}$ is UMVUE if, for each possible value of θ, $\hat{\theta}$ has the smallest variance in the class of unbiased estimators (and hence the smallest MSE in this class).

2. The Fisher information for a random sample $Y_1, \ldots, Y_n$ from a population with density (or probability mass function) in the family $f_\theta(y)$ is

$$I(\theta) = -n\mathrm{E}\left[\frac{\partial^2}{\partial\theta^2}\log[f_\theta(Y)]\right].$$

3. The Cramér–Rao lower bound for an unbiased estimator $\hat{\theta}$ is

$$\mathrm{V}(\hat{\theta}) \geq \frac{1}{I(\theta)}.$$

If an unbiased estimator $\hat{\theta}$ attains this bound (i.e., has this variance), then it is UMVUE.

4. A more general version of the Cramér–Rao result can be used to bound the variance of an estimator of a function $g(\theta)$ of our parameter θ. If T is unbiased for $g(\theta)$, then

$$V(T) \geq \frac{(g'(\theta))^2}{I(\theta)}.$$

5. Rao–Blackwell: If $\tilde{\theta}$ is an unbiased estimator of θ, and T is a sufficient statistic, define $\hat{\theta} = E(\tilde{\theta}|T)$. Then $\hat{\theta}$ is an unbiased estimator of θ, with $V(\hat{\theta}) \leq V(\tilde{\theta})$. If T is minimal sufficient (and complete), then $\hat{\theta}$ is UMVUE.

6. Lehmann–Scheffé: If an unbiased estimator is a function of a minimally sufficient (and complete) statistic, the estimator is UMVUE.

Exercises

54.1 Given independent random variables $Y_1, \ldots, Y_n$ distributed as Poisson with common mean λ, find the Cramér–Rao lower bound for the variance of unbiased estimators of λ, and compare this lower bound to the variance of the sample mean.

54.2 Suppose $Y_1, \ldots, Y_n \sim N(\mu, \sigma^2)$; find the Cramér–Rao lower bound for the variance of unbiased estimators of μ, showing that the sample mean is UMVUE for the population mean.

54.3 Suppose $Y_1, \ldots, Y_n \sim N(\mu, \sigma^2)$; now we want to estimate μ^2. Find the Cramér–Rao lower bound for the variance of unbiased estimators of μ^2.

54.4 Given independent random variables $Y_1, \ldots, Y_n$ distributed as Poisson with common mean λ, find the Cramér–Rao lower bound for the variance of unbiased estimators with the probability that the count is zero: This is $g(\lambda) = e^{-\lambda}$.

54.5 The Rayleigh density describes the distance from the origin of a point whose coordinates are iid mean zero normal random variables, and hence has useful applications in ecology and target analysis. The one-parameter family of distributions is described by

$$f_\theta(y) = \frac{2y}{\theta} e^{-y^2/\theta} \quad \text{for } y > 0.$$

Suppose $Y_1, \ldots, Y_n$ is a random sample from the Rayleigh density.

(a) Find a sufficient statistic for θ.

(b) Find an UMVUE for θ.

54.6 Suppose $Y_1, \ldots, Y_n$ is a random sample from an exponential population with mean θ.

(a) Find an UMVUE for the population variance θ^2.

(b) Check to see if your UMVUE attains the Cramér–Rao lower bound.

54.7 Suppose $Y_1, \ldots, Y_n$ is a random sample from a normal population with mean zero and variance σ^2.

(a) Find an UMVUE for σ^2.

(b) Show through simulations that your estimator in (a) is better than the unbiased estimator S^2, the sample variance. Use $n = 10$ and $\sigma = 2$.

54.8 Let $X_1, \ldots, X_m$ be a random sample from $N(\mu, \sigma^2)$, and let $Y_1, \ldots, Y_n$ be a random sample from $N(\mu, 4\sigma^2)$; further assume the samples are independent and σ^2 is known.

(a) Find an UMVUE for μ.

(b) Compare the variance of your estimator to that of the unbiased estimator $(\bar{X} + \bar{Y})/2$.

54.9 Let $X_1, \ldots, X_m \overset{ind}{\sim} N(\mu, 1)$, and let $Y_1, \ldots, Y_n \overset{ind}{\sim} N(2\mu, 1)$ be independent random variables.

(a) Find an UMVUE for μ.

(b) Show that your estimator has smaller variance than the unbiased estimator $(\bar{X} + \bar{Y})/3$ for any values of σ^2, m, n.

54.10 Suppose $\theta > 0$ and $Y_1, \ldots, Y_n$ is a random sample from the density

$$f_\theta(y) = \theta y^{\theta-1} \text{ for } y \in (0, 1).$$

It can be shown (using integration by parts) that $E[\log(Y)] = -1/\theta$. Find an UMVUE for $1/\theta$ and show it is UMVUE.

54.11 Suppose $\theta > 0$ and $Y_1, \ldots, Y_n$ is a random sample from the density

$$f(y) = \frac{\theta}{(y+1)^{\theta+1}} \text{ for } y > 0$$

for some $\theta > 0$. Find an UMVUE for $1/\theta$.

54.12 Suppose $Y_1, \ldots, Y_n$ is a random sample of an $N(0, \sigma^2)$ population. Find an UMVUE for σ^2 and show it is UMVUE.

54.13 Suppose $Y_1, \ldots, Y_n$ is a random sample from a population with density

$$f_\mu(y) = e^{\mu-y} I\{y > \mu\}.$$

Find an UMVUE for μ, and show it is UMVUE.

54.14 Suppose $Y_1, \ldots, Y_n$ is a random sample from a population with density

$$f_\theta(y) = \frac{1}{2\theta} e^{-|y|/\theta}$$

with support on the real numbers. Find an UMVUE for θ, and show it is UMVUE.

54.15 Suppose $X_1, X_2, X_3,$ and X_4 are independent exponential random variables where the expected value of X_1, X_2 is θ and the expected value of X_3, X_4 is 2θ, where $\theta > 0$. Find an UMVUE for θ, and show it is UMVUE.

54.16 Suppose $X_1, \ldots, X_n$ are independent Poisson random variables, all with mean λ, and $Y_1, \ldots, Y_m$ are independent Poisson random variables, all with mean 2λ, and each pair X_i, Y_j of random variables is independent. Find an UMVUE for λ.

54.17 Let $X_1, \ldots, X_n \overset{ind}{\sim} N(\theta, 2)$, and interest is in estimating θ^2. Statistician A says, "We have for each $i = 1, \ldots, n$, $E(X_i^2) = \theta^2 + 2$, so let's use $\sum_{i=1}^{n} X_i^2/n - 2$ as our estimator." Statistician B says, "We know that $E(\bar{X}^2) = V(\bar{X}) + E(\bar{X})^2 = 2/n + \theta^2$, so let's use $\bar{X}^2 - 2/n$ for our estimator."

   (a) Determine if either estimator is UMVUE, and if so, which one?

   (b) The other statistician didn't take a graduate-level class in mathematical statistics and needs more convincing. Show through simulations that your choice is better by comparing the MSEs of the two using $n = 20$ and $\theta = .4$.

54.18 There is only one airport shuttle van at the Grand Hotel, and it goes back and forth between the terminal and the hotel all day long. Suppose the time it takes to make one (back and forth) trip is θ minutes. Three guests at the hotel arrived separately at the airport at "random" times, and they reported that they had to wait 2.25, 7.5, and 5.2 minutes for the shuttle. Give an estimate for θ, state your assumptions, and state the properties of your estimator, arguing that your estimator is "optimal" in some sense.

54.19 A silk weaving machine produces bolts of cloth, with occasional flaws occurring. Suppose that the number of flaws in a (randomly selected) yard of cloth follows a Poisson distribution with mean λ. You are given five pieces of cloth produced by the machine. Three pieces are one yard long, and they have zero, two, and three flaws. The other two pieces are two yards long, and they have two and four flaws. Give an estimate for λ, state your assumptions, and state the properties of your estimator, arguing that your estimator is "optimal" in some sense.

Chapter 55

Exponential Families

We've talked a lot about families of distributions: the normal densities, the geometric probability mass functions, etc. A family of distributions is just a collection of densities or mass functions that are the same except for the values of the parameter(s). Most of the named distributions that we have talked about are **exponential families**. The exponential families form a class of families, where the formula for the density or mass function can be written in a certain form. Many nice results hold for all of the families in the class, and exponential families are the starting point for developing a theory about generalized regression models, an important type of statistical data analysis.

We'll start with **one-parameter** exponential families. These can be written as

$$f_\theta(x) = h(x)c(\theta)e^{\omega(\theta)t(x)},$$

where ω and c are functions of θ only (not x), and h and t are functions of x only (not θ). A key feature is this: The part of f_θ that is not in the exponent is factored into a function of x only and a function of the parameter only; this prohibits the support from depending on the parameter. The support for the distribution is expressed as part of the h function.

Example: The family of exponential densities is an exponential family. (This statement is potentially confusing because "exponential" is used to describe *both* a type of density function and a larger family of distributions.) We demonstrate that a particular family is an exponential family by identifying the h, c, ω, and t functions. For the family of exponential densities with mean θ, that is, $f_\theta(x) = e^{-x/\theta}/\theta$ for $x > 0$, we have

$$h(x) = I\{x \geq 0\}; \quad c(\theta) = \frac{1}{\theta}, \quad \omega(\theta) = -\frac{1}{\theta}; \quad \text{and} \quad t(x) = x.$$

This decomposition of f_θ is not quite unique: We could have defined $\omega(\theta) = \frac{1}{\theta}$ and $t(x) = -x$, for example. Either way, we have demonstrated that the family of exponential random variables is an exponential family.

Example: What about our family of binomial mass functions? Here

$$P(Y = y) = \binom{n}{y} p^y (1-p)^{n-y} \quad \text{for} \ \ y = 0, 1, \ldots, n,$$

which certainly doesn't look like an exponential family at first glance. However, we can write

$$h(y) = \binom{n}{y} I\{y = 0, 1, \ldots, n - 1, \text{ or } n\}$$

and

$$p^y(1 - p)^{n-y} = \left(\frac{p}{1 - p}\right)^y (1 - p)^n = (1 - p)^n \exp\left[y \log\left(\frac{p}{1 - p}\right)\right].$$

Then we define

$$c(p) = (1 - p)^n; \quad t(y) = y; \quad \text{and} \quad \omega(p) = \log\left(\frac{p}{1 - p}\right).$$

The ω function is recognized as the *log-odds* of a success. This is important in generalized regression models, where the response is binary.

A **two-parameter** family of distributions is an exponential family if the density or mass function can be written as

$$f_{\theta_1, \theta_2}(x) = h(x)c(\theta_1, \theta_2)e^{\omega_1(\theta_1, \theta_2)t_1(x) + \omega_2(\theta_1, \theta_2)t_2(x)}.$$

The form again prohibits families where the support depends on the parameter. To determine if a particular family is an exponential family, we again try to identify all the components.

Example: The family of normal densities has two parameters, μ and σ^2:

$$f_{\mu, \sigma^2}(x) = \frac{1}{\sqrt{2\pi\sigma^2}} \exp\left[-\frac{1}{2\sigma^2}(x - \mu)^2\right].$$

To get this in the proper form to identify the pieces of the exponential family, we start by multiplying out the exponent:

$$\exp\left[-\frac{1}{2\sigma^2}(x - \mu)^2\right] = \exp\left[-\frac{1}{2\sigma^2}(x^2 - 2\mu x + \mu^2)\right]$$

$$= \exp\left[-\frac{\mu^2}{2\sigma^2}\right] \exp\left[-\frac{x^2}{2\sigma^2} + \frac{\mu x}{\sigma^2}\right].$$

We can write $h(x) = 1$,

$$c(\mu, \sigma^2) = \frac{1}{\sqrt{2\pi\sigma^2}} \exp\left[-\frac{\mu^2}{2\sigma^2}\right],$$

$$t_1(x) = x^2; \quad \omega_1(\mu, \sigma^2) = -\frac{1}{2\sigma^2},$$

and

$$t_2(x) = x; \quad \omega_2(\mu, \sigma^2) = \frac{\mu}{\sigma^2}.$$

This shows that the normal family of densities is a two-parameter exponential family.

Sufficient statistics and exponential families

Suppose we have a random sample $X_1, \ldots, X_n$ from a one-parameter exponential family. Then the joint density of the sample can be written as

$$f_\theta(x_1, \ldots, x_n) = \prod_{i=1}^{n} h(x_i)c(\theta)e^{\omega(\theta)t(x_i)} = \left[\prod_{i=1}^{n} h(x_i)\right] c(\theta)^n e^{\omega(\theta)\sum_{i=1}^{n} t(x_i)},$$

and we can use the factorization theorem to see that

$$T(X_1, \ldots, X_n) = \sum_{i=1}^{n} t(X_i)$$

is a sufficient statistic for the parameter θ. For the exponential and binomial families, we found t to be the identity function ($t(x) = x$), so this result tells us that the sum of the sample is sufficient (and hence the sample mean is sufficient).

Similarly, for a sample $X_1, \ldots, X_n$ from a two-parameter exponential family, it is straightforward that the two statistics

$$T_1(X_1, \ldots, X_n) = \sum_{i=1}^{n} t_1(X_i) \text{ and } T_2(X_1, \ldots, X_n) = \sum_{i=1}^{n} t_2(X_i)$$

are jointly sufficient for the two parameters. If we look back at our normal family example, we see that

$$T_1(X_1, \ldots, X_n) = \sum_{i=1}^{n} X_i \text{ and } T_2(X_1, \ldots, X_n) = \sum_{i=1}^{n} X_i^2$$

are jointly sufficient for μ and σ^2. Because the sample mean and variance are jointly functions of T_1 and T_2, the sample mean and variance are also jointly sufficient for μ and σ^2.

In Chapter 54, we did some hedging about the Lehmann–Scheffé theorem, saying that the required sufficient statistic had to also be "complete" although we did not define completeness. With exponential families, one-dimensional sufficient statistics are *always* complete for one-dimensional parameters, so it's easy to construct UMVUEs. We can also be confident that two-dimensional sufficient statistics are complete for two-dimensional parameters in exponential families. Anomalies occur when there is one parameter, but the minimal sufficient statistic is two-dimensional. A classical example is the normal family where the mean and the variance are θ and θ^2. But in general, it's straightforward to identify UMVUEs in exponential families.

Chapter Highlights

1. A one-parameter exponential family can be written as

$$f(x;\theta) = h(x)c(\theta)e^{\omega(\theta)t(x)},$$

   where the support of the distribution is expressed as part of the h function. (The support may not depend on the parameter.)

2. For a random sample $X_1, \ldots, X_n$ from a one-parameter exponential family,

$$T(X_1, \ldots, X_n) = \sum_{i=1}^{n} t(X_i)$$

   is a sufficient statistic for the parameter θ.

3. A two-parameter exponential family can be written as

$$f(x;\theta_1,\theta_2) = h(x)c(\theta_1,\theta_2)e^{\omega_1(\theta_1,\theta_2)t_1(x)+\omega_2(\theta_1,\theta_2)t_2(x)}.$$

4. For a random sample $X_1, \ldots, X_n$ from a two-parameter exponential family,

$$T_1(X_1, \ldots, X_n) = \sum_{i=1}^{n} t_1(X_i) \text{ and } T_2(X_1, \ldots, X_n) = \sum_{i=1}^{n} t_2(X_i)$$

   are jointly minimal sufficient statistics for the parameters θ_1 and θ_2.

❧❧

Exercises

55.1 Show that the geometric family of mass functions is a one-parameter exponential family. If $Y_1, \ldots, Y_n$ are iid geometric random variables with parameter p, find a sufficient statistic for p.

55.2 Show that the Poisson family of mass functions is a one-parameter exponential family. If $Y_1, \ldots, Y_n$ are iid Poisson random variables with parameter λ, find a sufficient statistic for λ.

55.3 Determine if the Rayleigh family of distributions described by

$$f_\theta(y) = \frac{2y}{\theta}e^{-y^2/\theta} \text{ for } y > 0 \text{ and } \theta > 0$$

is an exponential family. If $Y_1, \ldots, Y_n$ are iid Rayleigh random variables with parameter θ, find a sufficient statistic for θ.

55.4 Determine if the family of distributions described by

$$f_\theta(y) = \frac{6}{\theta^3}y(\theta - y) \text{ for } y \in (0,\theta)$$

is an exponential family, and if so, find a sufficient statistic for θ if $Y_1, \ldots, Y_n$ are a random sample from this density.

55.5 Determine if the family of distributions described by

$$f_\theta(y) = \theta y^{\theta-1} \text{ for } y \in (0,1)$$

is an exponential family, and if so, find a sufficient statistic for θ if $Y_1, \ldots, Y_n$ are a random sample from this density.

55.6 Determine if the family of distributions described by

$$f_\theta(x) = \theta^2 x e^{-\theta x} \text{ for } x \in [0, \infty)$$

is an exponential family, and if so, find a sufficient statistic for θ if $X_1, \ldots, X_n$ are a random sample from this density.

55.7 Determine if the family of distributions described by

$$f_\theta(x) = \frac{\theta}{(x+1)^{\theta+1}} \text{ for } x > 0$$

is an exponential family, and if so, find a sufficient statistic for θ if $X_1, \ldots, X_n$ are a random sample from this density.

55.8 Determine if the family of distributions described by

$$f_\theta(y) = \frac{4\theta^4}{(y+\theta)^5} \text{ for } x > 0$$

is an exponential family, and if so, find a sufficient statistic for θ if $Y_1, \ldots, Y_n$ are a random sample from this density.

55.9 Determine if the Gamma(α, β) family of densities is an exponential family, and if so, find jointly sufficient statistics for α and β if $Y_1, \ldots, Y_n$ are a random sample from this density.

55.10 Determine if the Beta(α, β) family of densities is an exponential family, and if so, find jointly sufficient statistics for α and β if $Y_1, \ldots, Y_n$ are a random sample from this density.

Chapter 56

Evaluating Hypothesis Tests: Test Size and Power

When planning and designing statistical studies, researchers are interested in the power of tests. Once the appropriate statistical model is determined, the power of a test depends on a number of factors, including the test statistic, test size, sample size, and the "effect size," or how different the truth is from the null hypothesis. Other parameters that might be unspecified by the hypotheses, such as the variance when testing the mean of a normal population, can also affect the power. After choosing a test statistic and a desired size α, researchers want to know ahead of time what the power would be for various guesses at effect sizes and values for other parameters, for a range of sample sizes, in order to choose a sample size that is likely to have adequate power.

Preliminary power calculations, with a range of guesses for effect sizes and model parameters for a range of possible sample sizes, can be numerous. The results can be displayed in tables, or in plots called **power curves**. For example, the power can be plotted against the effect size, with several curves shown for different sample sizes, with separate plots for different guesses for the other parameters.

Example: Let's go back to our standard example of a machine making rods that are part of a device. The current machine makes rods with mean length $\mu = 100$ cm and standard deviation $\sigma = 2$. The manager is considering buying a new machine and is interested in the purchase only if the variance of rod lengths is smaller. He will get a sample of $n = 12$ rods made by the new machine and test $H_0 : \sigma = 2$ versus $H_a : \sigma < 2$. He wants to use test size $\alpha = .01$ because of the cost of the new machine.

Let's assume that the machine can be calibrated so that $\mu = 100$ is known. Then if $X_1, \ldots, X_{12}$ are the lengths of the rods made by the new machine, the test statistic is

$$T = \frac{\sum_{i=1}^{12}(X_i - 100)^2}{4},$$

which has a $\chi^2(12)$-density under H_0. If H_a is true, we expect the test statistic to be smaller. We reject H_0 when $T < 3.571$, the 1st percentile of a $\chi^2(12)$ distribution.

The power of this test for can be calculated readily. For $\sigma \in (0, 2)$,

$$P(T < 3.571) = P\left(\frac{\sum_{i=1}^{12}(X_i - 100)^2}{\sigma^2} < \frac{4 \times 3.571}{\sigma^2} \right).$$

Power curves will display the results.

```
sig=1:200/100
n=12
crit=qchisq(.01,n)
pwr=pchisq(4*crit/sig^2,n)
plot(sig,pwr,type="l")
```

The dotted line in the power curve below has height $\alpha = .01$. We can see that if $\sigma = 1.5$, the probability that we will see enough evidence to reject the null hypothesis is low. The true standard deviation has to be smaller than $\sigma = 1$ before we feel confident that the null hypothesis will be (correctly) rejected.

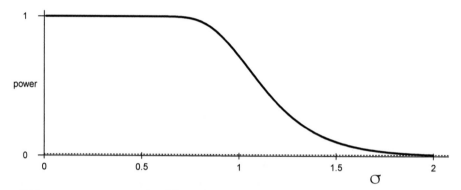

If the manager wants to be able to detect a true standard deviation of $\sigma = 1.5$, while maintaining the small test size, the sample size must be increased. Let's fix $\sigma = 1.5$ and make a power curve for the manager, with increasing sample size.

```
sig=1.5
n=20:100
crit=qchisq(.01,n)
pwr=pchisq(4*crit/sig^2,n)
plot(n,pwr,type="l")
```

The plot below shows the power for 81 values of n. If the manager wants to be 80% sure of detecting a true standard deviation as small as $\sigma = 1.5$, then it looks like a sample of $n = 67$ rods is necessary.

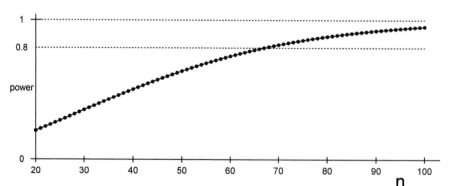

Example: Suppose $Y_1, \ldots, Y_n$ are a random sample from a Beta$(\theta, 1)$ distribution, and we'd like to test $H_0 : \theta = 1$ versus $H_a : \theta > 1$. Under the null hypothesis, the distribution of the population is uniform, and under the alternative, the density is "piled up" to the right, with mean $\theta/(\theta + 1)$, so that we would expect the values to be, on average, *larger* than if the distribution were uniform.

To give an imaginary context for this example, let's model for the proportion of time a male flitterbird spends tending the nest in a randomly selected daylight hour. Suppose flitterbird couples take turns minding the nest while the other forages. The null hypothesis model could be that the proportion of time the male bird is at the nest is uniformly distributed, and the alternative represents the situation in which the male flitterbird tends to spend *more* time tending the nest.

What test statistic should we use? There are a lot of possibilities! The test statistic could simply be the number of observations that are greater than 1/2; then the test size and power could be computed using a binomial distribution. Or, we could estimate θ using various methods and define a test statistic based on the estimator. For example, in Exercise 47.9, we used the delta method to get an approximate distribution of the method of moments estimator for θ. Alternatively, we could approximate the distribution of the maximum likelihood estimator as normal with mean θ using the Fisher information to estimate the variance. How do we determine which of these ideas would be "best?" For now we will investigate these issues through simulations, comparing the power curves, and in the next chapters we will develop some theory that will allow us to find a best option.

Suppose we have $n = 40$ iid observations from the population, and let's start with the binomial idea. (In the context of the application, the researchers could randomly select 40 nests, then randomly select a daylight hour for each nest.) We'll reject H_0 if c or more of the observations are greater than 1/2, and if we try to get the test size α close to .05, we can try qbinom(.95,40,1/2) to get 25. Because the distribution is discrete, we use 1-pbinom(25,40,1/2) to find the actual test size; this returns .0403, telling us that if we reject H_0 when we see 26 or more (out of 40) waiting times more than 1/2 hour, we have $\alpha \approx .04$. This is as close as we can get to the desired test size of .05.

Given this decision rule, we can calculate the power of the test for various values of θ and plot the power against θ. For each value of θ, the probability of the male flitterbird being at the nest for more than 1/2 hour is 1-pbeta(1/2,theta,1), and if we compute this probability for a range of values of θ, we can see how fast the power increases as θ moves away from the null hypothesis value of $\theta = 1$. The following code computes the power for values of θ between 1 and 3 for $n = 40$. The power is saved in a vector and plotted against θ in the following figure. We can change n to get the power curves for the other sample sizes. The power curves are plotted for several sample sizes, and the dashed gray line in the plot indicates the target test size of $\alpha = .05$.

```
theta=0:200/100+1
n=40
crit=qbinom(.95,n,1/2)
pr=1-pbeta(1/2,theta,1)
power=1-pbinom(crit,n,pr)

lines(theta,power)
```

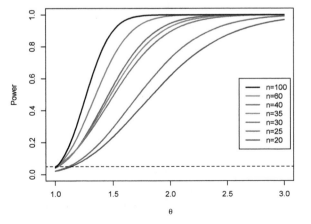

Next, let's look at the method of moments estimator for θ:

$$\tilde{\theta} = \frac{\bar{Y}}{1 - \bar{Y}}.$$

This estimator was first calculated for Exercise 44.9; then in Exercise 47.9, we showed that

$$\sqrt{n}(\tilde{\theta} - \theta) \xrightarrow{\mathcal{D}} \mathrm{N}\left(0, \frac{\theta(\theta+1)^2}{\theta+2}\right).$$

If H_0 is true, the variance of $\tilde{\theta}$ is approximately $4/(3n)$, and

$$Z = \frac{\tilde{\theta} - 1}{\sqrt{4/(3n)}} \approx \mathrm{N}(0,1),$$

with the approximation getting better for larger n. Large values of Z support H_a, so an approximate size .05 test of H_0 versus H_a has the decision rule "reject H_0 when the observed value of Z is greater than 1.645," the 95th percentile of a standard normal density.

There are two important questions about this approximate test statistic: What is the true test size, and how does the power of this test compare to the test with the binomial test statistic? Because we are approximating the distribution of $\tilde{\theta}$, we know that for *large* n, the test size is close to $\alpha = .05$, but how large does n have to be?

The following code samples from a Beta$(\theta, 1)$ density for 201 values of θ between one and three, and for each value, finds the probability that Z is greater than the critical value of 1.645. That is, the power is calculated for each of the values of θ.

```
nloop=10000
n=100
theta=0:200/100+1
crit=qnorm(.95)
power=1:201
for(i in 1:201){
   nrej=0
   for(iloop in 1:nloop){
      y=rbeta(n,theta[i],1)
      thtilde=mean(y)/(1-mean(y))
      z=(thtilde-1)/sqrt(4/3/n)
      if(z>crit){nrej=nrej+1}
   }
   power[i]=nrej/nloop
   print(i+power[i])
}
```

We can run this code for several values of n, and for each we plot the power against the θ values, as shown in the plot below. The power is greater than for the binomial test, but the test size is also larger, especially for the smaller sample sizes. In fact, for $n = 20$, the true test size is more than .09 when the target is only .05.

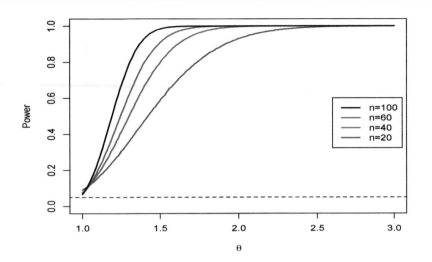

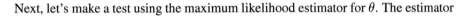

Next, let's make a test using the maximum likelihood estimator for θ. The estimator

$$\hat{\theta} = -\frac{n}{\sum_{i=1}^{n} \log(y_i)}$$

is a function of a statistic that is sufficient for θ. It's a good guess that an estimator using a sufficient statistic will make a better test. To use this estimator, we need its distribution under H_0. We can show that the distribution is related to an inverse gamma random variable (see Exercise 56.8), but let's pretend that's too hard for us to derive.

Instead, let's simply simulate the null distribution of $\hat{\theta}$. That is, we simulate "many" samples of size n from the uniform distribution and calculate the test statistic for each sample. We can make a histogram of test statistic values that is as close to the true null distribution as we would like.

It doesn't take long to get a very precise null distribution for $n = 100$:

```
n=100
nloop=1000000
thetahat=1:nloop
for(iloop in 1:nloop){
   y=runif(n)
   thetahat[iloop]=-n/sum(log(y))
}
hist(thetahat,br=50)
crit=sort(thetahat)[950000]
```

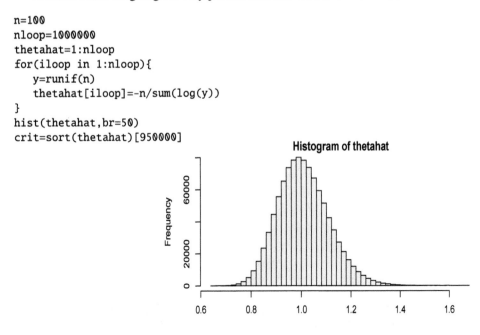

Based on this null distribution, we reject H_0 when $\hat\theta$ is greater than `crit=1.188`, the 95th percentile of the simulated test statistic values. We can use this critical value to compute power curves through simulations as well.

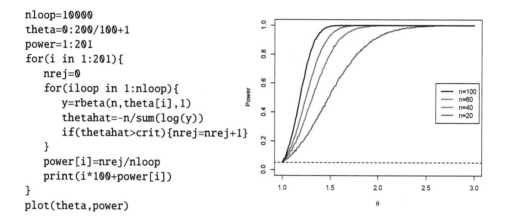

```
nloop=10000
theta=0:200/100+1
power=1:201
for(i in 1:201){
    nrej=0
    for(iloop in 1:nloop){
        y=rbeta(n,theta[i],1)
        thetahat=-n/sum(log(y))
        if(thetahat>crit){nrej=nrej+1}
    }
    power[i]=nrej/nloop
    print(i*100+power[i])
}
plot(theta,power)
```

The power for this test is higher than for the first two, and, better yet, the test size is very close to the target $\alpha = .05$. Although the null distribution was simulated, the test is considered to be "exact" (the true test size equals the target) because we can be as precise as we like with a larger `nloop`.

Given our knowledge to date, we are not surprised that the test using an UMVUE is optimal in terms of power. We'll investigate the concept of *uniformly most powerful tests* in the next couple of chapters.

Chapter Highlights

1. Power curves are a convenient visual display of the power of a test over a range of values of the parameter and the sample size.

2. When the exact distribution of a test statistic is unknown, we can simulate its distribution under the null hypothesis to get a critical value. Using this critical value, we can simulate the test statistic under various alternative scenarios to get power curves.

❧❧

Exercises

56.1 Suppose $X_1, \ldots, X_n$ is a random sample from an exponential distribution with mean θ, and we want to test $H_0 : \theta = 1$ versus $H_a : \theta > 1$. A reasonable test is to reject H_0 when the sample sum is "large"; we can quantify *how* large, given n and α, using a quantile of a gamma distribution. For test size $\alpha = .05$, make nice power curves for θ values between 1 and 4 for sample sizes $n = 10$, 15, 20, and 40.

56.2 Suppose $X_1, \ldots, X_n$ is a random sample from $N(\mu, \sigma^2)$ and we want to test $H_0 :$ $\mu = 100$ versus $H_a : \mu > 100$. We don't know σ^2, so we have to do a t-test instead of a Z-test. To find the power, guesses have to be made for σ^2. Make two sets of power curves for test size $\alpha = .10$: First, fix $\sigma = 20$, and make power curves for values of μ between 100 and 110, for $n = 10$, $n = 20$, and $n = 40$. Next, let $\sigma = 8$ and make power curves for the same values of μ and n. Put all the power curves on the same plot, with a nice legend.

56.3 Let's return to the example about the machine making rods. Suppose we want to test $H_0 : \sigma = 2$ versus $H_a : \sigma < 2$, at $\alpha = .01$, but we don't want to assume that μ is known. Find a test statistic for which we don't need to know μ, and make a plot of power versus sample size (use $n = 10$ through $n = 100$) when the true standard deviation is $\sigma = 1.5$.

56.4 Your clients will be testing Drug A versus Drug B using a two-independent-samples t-test. They are in the design stage and want to know how big a sample they should obtain. They will randomly distribute patients coming into the clinic, so that there are n patients assigned to each drug. They want power curves for effect sizes $\mu_B - \mu_A = 3$ and $\mu_B - \mu_A = 6$, with common population variance $\sigma^2 = 20$ and with $\sigma^2 = 50$. That is, they would like you to plot the power of the test against n for each of four scenarios, to get four curves on the sample plot. Use all values of n in the range of $n = 5$ to $n = 80$.

56.5 The standard two-independent-samples t-test has an assumption that the two population variances are the same. We can do a formal hypothesis test where the null hypothesis is that the variances are the same versus the alternative that they are different. If we do the test and accept the null hypothesis, can we really conclude that the variances are the same? Plot the power of this test against ratios of the variances that are between one and two for equal sample sizes in each group: $n = 20$, $n = 40$, $n = 80$, and $n = 120$.

56.6 For the example of this chapter, we used a method of moments estimator for θ using a random sample from a $Beta(\theta, 1)$ population. We approximated the null distribution using the delta method, and we found that the test size was inflated. Instead of using the normal approximation, simulate the null distribution of $\tilde{\theta}$. Find critical values and a power curve for $n = 40$ and θ ranges from 1 to 2. Compare your power to the power found in the example using the normal approximation, and power found in the example using the MLE.

56.7 Suppose X_1, X_2, X_3, X_4, and X_5 are independent normal random variables. All of these random variables have mean μ, X_1 and X_2 have variance 4, and X_3, X_4, and X_5 have variance 9. Interest is in testing $H_0 : \mu = 100$ versus $H_0 : \mu > 100$ at test size $\alpha = .05$.

(a) Statistician A says, "I know $X_1 + X_2 + X_3 + X_4 + X_5$ has mean 5μ and variance 35, so I'll use

$$Z = \frac{5(\bar{X} - 100)}{\sqrt{35}}$$

as my test statistic and reject when $Z > 1.645$." Make a power curve for this test, for values of μ between 100 and 105.

(b) Statistician B says, "I know $X_1/2$, $X_2/2$, $X_3/3$, $X_4/3$, and $X_5/3$ all have the unit variance, and their sum has mean 2μ, so I'll use

$$Z = \frac{X_1/2 + X_2/2 + X_3/3 + X_4/3 + X_5/3 - 200}{\sqrt{5}}$$

as my test statistic and reject when $Z > 1.645$." Make a power curve for this test for values of μ between 100 and 105, and compare the power to the test in (a).

(c) Statistician C says, "I'm going to make a test based on an UMVUE estimator for μ." Find such a test, and make a power curve for values of μ between 100 and 105. Is the power better than the two previous tests?

56.8 Suppose $Y_1, \ldots, Y_n \sim \text{Beta}(\theta, 1)$.

(a) Show that

$$T = -\frac{1}{\sum_{i=1}^{n} \log(Y_i)}$$

is a sufficient statistic and has an inverse gamma distribution.

(b) Use the distribution in (a) to find a critical value for $H_0 : \theta = 1$ versus $H_a : \theta > 1$, for $n = 60$ and $\alpha = .05$, using the estimator

$$\hat{\theta} = -\frac{n}{\sum_{i=1}^{n} \log(Y_i)}.$$

(c) Make the power curves for values of θ between 1 and 3, and the same values of n as in the example of this chapter. Compare your results to the power curves obtained through the simulated null distributions. Which method do you think is easier?

Chapter 57

The Neyman–Pearson Lemma

A hypothesis is called **simple** if, under the hypothesis, the parameter takes on a single value and the distribution of the data is completely specified. A hypothesis is called **composite** if it places the parameter in a set of values. We have often been looking at a simple null hypothesis and a composite alternative hypothesis. In this chapter we learn how to get the best possible test when *both* hypotheses are simple, and then we extend this to the case where the null hypothesis is simple, and we have a one-sided alternative.

The Neyman–Pearson lemma: Suppose we have a simple null hypothesis and a simple alternative: $H_0 : \theta = \theta_0$ versus $H_a : \theta = \theta_a$. Given data $y_1, \ldots, y_n$ generated from a joint density function $f_\theta(y_1, \ldots, y_n)$, let $L(\theta; y_1, \ldots, y_n) = f_\theta(y_1, \ldots, y_n)$ be the likelihood. Then the most powerful size-α test is given by "reject H_0 if

$$\frac{L(\theta_0; y_1, \ldots, y_n)}{L(\theta_a; y_1, \ldots, y_n)} < c$$

for some number c to be determined." That is, we find c so that if the data are generated from the joint density function with $\theta = \theta_0$, then the probability that this ratio of likelihoods is less than c is α.

Let's do an example before we prove this. Suppose we have a random sample $Y_1, \ldots, Y_n$ from a population with density function

$$f_\theta(y) = \frac{2}{\theta} y e^{-y^2/\theta} \text{ for } y > 0.$$

We want to test $H_0 : \theta = \theta_0$ versus $H_a : \theta = \theta_a$, and suppose that θ_a is greater than θ_0. The likelihood is the product of the densities evaluated at the observations:

$$L(\theta; \boldsymbol{y}) = \left(\frac{2}{\theta}\right)^n \left[\prod_{i=1}^n y_i\right] \exp\left\{-\sum_{i=1}^n y_i^2/\theta\right\}.$$

In the Neyman–Pearson ratio of likelihoods, the product of the y_i terms cancel, and we get the decision rule "reject H_0 if

$$\left(\frac{\theta_a}{\theta_0}\right)^n \exp\left\{-\sum_{i=1}^n y_i^2 \left(\frac{1}{\theta_0} - \frac{1}{\theta_a}\right)\right\} < c."$$

This looks a bit daunting, but remember that θ_0 and θ_a are fixed, and we assumed that $\theta_a > \theta_0$, so $1/\theta_0 - 1/\theta_a > 0$. Therefore, the decision rule is equivalent to "reject H_0 if

$$\sum_{i=1}^{n} y_i^2 > c',\text{"}$$

where c' is determined by α and the distribution of $\sum_{i=1}^{n} y_i^2$. (Notice that if $\theta_a < \theta_0$, we would reject if the sum of the squares is *small*.) It turns out that it is not hard to find the distribution of this sum. We can show using the CDF method that the distribution of Y_i^2 is exponential with mean θ (see Exercise 21.19), so that $\sum_{i=1}^{n} Y_i^2 \sim \text{Gamma}(n, 1/\theta)$.

The Neyman–Pearson lemma tells us that the most powerful test of H_0 versus H_a is to reject H_0 when the sum of the squares of the observations is larger than the $100(1 - \alpha)$th percentile of a $\text{Gamma}(n, 1/\theta_0)$ distribution. We don't have to compare the power of other possible tests, because the lemma says that this power is will be the largest for any given α.

Proof of the Neyman–Pearson lemma: We define a **test function** ϕ_1 as

$$\phi_1(y_1, \ldots, y_n) = \begin{cases} 1 & \text{if } cL(\theta_a; y_1, \ldots, y_n) > L(\theta_0; y_1, \ldots, y_n), \\ 0 & \text{otherwise.} \end{cases}$$

In other words, $\phi_1(y_1, \ldots, y_n) = 1$ if we reject H_0 for these data, and $\phi_1(y_1, \ldots, y_n) = 0$ if we accept H_0. Now let ϕ_2 be the test function for another test with size α; that is, $\phi_2(y_1, \ldots, y_n) = 1$ if the other test rejects H_0, and $\phi_2(y_1, \ldots, y_n) = 0$ otherwise.

Then the product

$$[\phi_1(y_1, \ldots, y_n) - \phi_2(y_1, \ldots, y_n)]\,[cL(\theta_a; y_1, \ldots, y_n) - L(\theta_0; y_1, \ldots, y_n)] \geq 0$$

for all possible values of $y_1, \ldots, y_n$. For, when the term in the second brackets is positive, we have $\phi_1(y_1, \ldots, y_n) = 1$, and the product of two nonnegative terms is nonnegative. On the other hand, if the term in the second brackets is negative, we have $\phi_1(y_1, \ldots, y_n) = 0$, and the product of two nonpositive terms is nonnegative.

Suppose $\mathcal{Y} \subseteq \mathbb{R}^n$ is the domain of f_θ, that is, $\mathcal{Y}$ is the set of all possible values for the data. Then

$$\int_{\mathcal{Y}} [\phi_1(y_1, \ldots, y_n) - \phi_2(y_1, \ldots, y_n)]$$
$$\times [cL(\theta_a; y_1, \ldots, y_n) - L(\theta_0; y_1, \ldots, y_n)]\, dy_1 \ldots, dy_n \geq 0,$$

because the integral of a nonnegative function, over any set, has to be nonnegative. Now we simply break up the integral into four pieces by "multiplying out" the integrand.

Each of the pieces represents a probability:

$$\int_{\mathcal{Y}} \phi_1(y_1, \ldots, y_n) L(\theta_a; y_1, \ldots, y_n) dy_1 \ldots, dy_n$$
$$= \int_{\mathcal{Y}} \phi_1(y_1, \ldots, y_n) f_{\theta_a}(y_1, \ldots, y_n) dy_1 \ldots, dy_n$$
$$= \beta_1,$$

where β_1 is the power of our likelihood ratio test when the alternative is true. Similarly,

$$\int_{\mathcal{Y}} \phi_2(y_1, \ldots, y_n) L(\theta_a; y_1, \ldots, y_n) dy_1 \ldots, dy_n$$
$$= \int_{\mathcal{Y}} \phi_2(y_1, \ldots, y_n) f_{\theta_a}(y_1, \ldots, y_n) dy_1 \ldots, dy_n$$
$$= \beta_2,$$

where β_2 is the power of the other test when the alternative is true. Next,

$$\int_{\mathcal{Y}} \phi_1(y_1, \ldots, y_n) L(\theta_0; y_1, \ldots, y_n) dy_1 \ldots, dy_n = \alpha,$$

and

$$\int_{\mathcal{Y}} \phi_2(y_1, \ldots, y_n) L(\theta_0; y_1, \ldots, y_n) dy_1 \ldots, dy_n = \alpha$$

if the other test also has size α. Putting these pieces together, we have

$$\beta_1 - \beta_2 \geq 0,$$

so the power for our likelihood ratio test in the Neyman–Pearson lemma is at least as large as the power for any other test with size α.

Corollary to the Neyman–Pearson lemma: The power is at least as large as the test size. For, supposing we have a test that does not consider the data, it simply rejects with probability α. Then the power of this funny test is α, and we have $\beta_1 > \alpha$.

Generalization of the Neyman–Pearson lemma to one-sided tests: Now suppose we have a simple null hypothesis and a composite, one-sided alternative hypothesis, so that we want to test $H_0 : \theta = \theta_0$ versus $H_a : \theta > \theta_0$. For any specific $\theta_a > \theta_0$, we can use the Neyman–Pearson lemma to get a most powerful test, out of all tests with size α. The decision rule for this test is calculated using the null hypothesis value of θ, and a consideration of whether the alternative value is smaller or larger than the null value. Therefore, the Neyman–Pearson lemma gives the same most powerful test for all $\theta_a > \theta_0$, and the ratio of likelihoods can be used to construct a most powerful one-sided test.

Going back to the previous example, we see that our decision rule depended only on θ_0, and so we have formulated a *uniformly* most powerful test for $H_0 : \theta = \theta_0$ versus $H_a : \theta > \theta_0$, where "uniformly" means that the test will be most powerful for any $\theta > \theta_0$.

Unfortunately, we can't have a two-sided uniformly most powerful size-α test. Consider a two-sided test such as $H_0 : \theta = \theta_0$ versus $H_a : \theta \neq \theta_0$. For any value of $\theta > \theta_0$, a one-sided uniformly most powerful size-α test will always have higher power than the two-sided test, although this one-sided test will have *lower* power for θ on the other side of θ_0. Practically speaking, the two-sided version of a one-sided uniformly most powerful size-α test is considered "best."

Connection to sufficient statistics: For a sample $Y_1, \ldots, Y_n$ following the joint density $f_\theta(y_1, \ldots, y_n)$, suppose $T = \tau(Y_1, \ldots, Y_n)$ is a sufficient statistic. Then the following test of $H_0 : \theta = \theta_0$ versus $H_a : \theta = \theta_a$, based on T, is most powerful. If $t = \tau(y_1, \ldots, y_n)$, we reject H_0 if

$$\frac{g_{\theta_0}(t)}{g_{\theta_a}(t)} < c,$$

where the joint density is factored as $f_\theta(y_1, \ldots, y_n) = g_\theta(t)h(y_1, \ldots, y_n)$. This result follows easily from the Neyman–Pearson lemma, because in the ratio of likelihoods the h terms cancel. This confirms our intuition that a test based on a sufficient statistic is "best."

The Z, t, χ^2, and F tests we derived for samples from normal populations are all based on sufficient statistics, although we did not approach the construction of the tests from that perspective. We will see in the next chapter that they can be constructed using the ideas of likelihood ratios in a more general context, and we can be confident that these tests have the highest power attainable for their contexts.

Chapter Highlights

1. The Neyman–Pearson lemma: Given data $\boldsymbol{y} = (y_1, \ldots, y_n)$, let $L(\theta; \boldsymbol{y})$ be the likelihood. Then the uniformly most powerful size-α test of $H_0 : \theta = \theta_0$ versus $H_a : \theta = \theta_a$ is given by "reject H_0 if

$$\frac{L(\theta_0; \boldsymbol{y})}{L(\theta_a; \boldsymbol{y})} < c,$$

where the number c is determined by α."

2. For the joint density $f_\theta(y_1, \ldots, y_n)$, suppose $T = \tau(Y_1, \ldots, Y_n)$ is a sufficient statistic. Then the following test of $H_0 : \theta = \theta_0$ versus $H_a : \theta = \theta_a$, based on T, is most powerful: "Reject H_0 if

$$\frac{g_{\theta_0}(t)}{g_{\theta_0}(t)} < c,$$

where the joint density is factored as $f_\theta(y_1, \ldots, y_n) = g_\theta(t)h(y_1, \ldots, y_n)$."

Exercises

57.1 Suppose $X_1, \ldots, X_n$ is a random sample from an exponential population with mean θ.

(a) Find a uniformly most powerful test of $H_0 : \theta = 2$ versus $H_a : \theta = 4$.

(b) Get a p-value for your test when $n = 5$ and the sample observations are 1.1, 2.5, 0.7, 9.8, and 4.7.

(c) What is the power of your test when $n = 5$ and $\alpha = .05$?

57.2 Suppose X_1 and X_2 are independent exponential random variables with $E(X_1) = \theta$ and $E(X_2) = 2\theta$.

    (a) Find a most powerful test of $H_0 : \theta = 2$ versus $H_a : \theta = 4$. Give the decision rule for $\alpha = .05$.

    (b) What is the power of your test in part (a)?

57.3 Suppose $X_1 \sim \text{Pois}(\lambda)$ and $X_2 \sim \text{Pois}(2\lambda)$ are independent random variables. We wish to test $H_0 : \lambda = 4$ versus $H_a : \lambda = 1$.

    (a) Find the test statistic for the most powerful test.

    (b) Suppose we observe $x_1 = 2$ and $x_2 = 0$. What is the p-value?

57.4 Suppose Y_1, Y_2, Y_3, Y_4 are independent Gamma$(2, \beta)$ random variables.

    (a) Find a most powerful test of $H_0 : \beta = 2$ versus $H_a : \beta = 4$ using test size $\alpha = .05$. Use a test statistic for which you can derive the exact distribution under the null hypothesis.

    (b) Get a p-value for your test when the sample observations are 0.469, 0.207, 0.255, and 0.661.

    (c) What is the power of your test when $\alpha = .05$?

57.5 Suppose Y_1, Y_2, Y_3 are independent random variables, where $Y_1 \sim$ Gamma$(2, \beta)$, $Y_2 \sim$ Gamma$(4, \beta)$, and $Y_3 \sim$ Gamma$(6, \beta)$.

    (a) Find a most powerful test of $H_0 : \beta = 2$ versus $H_a : \beta = 4$ using test size $\alpha = .05$. Use a test statistic for which you can derive the exact distribution under the null hypothesis.

    (b) Get a p-value for your test when the sample observations are 0.469, 0.507, and 0.855.

    (c) What is the power of your test when $\alpha = .05$?

57.6 Suppose Y_1, Y_2, Y_3, Y_4, Y_5 are independent Gamma$(\alpha, 1)$ random variables.

    (a) Find a most powerful test of $H_0 : \alpha = 1$ versus $H_a : \alpha = 3$. You don't have to give the distribution of the test statistic under the null hypothesis, but provide a test statistic and state whether to reject when the test statistic is "large" or "small."

    (b) Simulate the distribution of your test statistic when the null hypothesis is true.

    (c) Get a p-value for your test when the sample observations are 1.469, 1.207, 1.255, 2.023, and 0.661.

    (d) What is the power of your test when the test size is .05?

57.7 Suppose Y_1, Y_2, Y_3, and Y_4 are a random sample from the density

$$f_\theta(y) = \frac{2y}{\theta} e^{-y^2/\theta} \text{ for } y > 0.$$

Interest is in testing $H_0 : \theta = 2$ versus $H_a : \theta > 2$. Suppose the decision rule is "reject H_0 if the sample maximum is larger than 3."

(a) Find test size α.

(b) Find c so that the decision rule "reject H_0 if the sample maximum is larger than c" has size $\alpha = .05$.

(c) Find the power of the test in part (b) when $\theta = 4$, and compare this power to that of the uniformly most powerful test discussed in this chapter.

57.8 Let $Y_1, \ldots, Y_n$ be a random sample from an $N(\mu, 1)$-density. What is the most powerful test of $H_0 : \mu = \mu_0$ versus $H_a : \mu = \mu_a$, where $\mu_a > \mu_0$. Is this equivalent to the Z-test of Chapter 35?

57.9 Suppose $Y_1, \ldots, Y_n \sim N(\mu, 10)$ and $X_1, \ldots, X_n \sim N(2\mu, 10)$ are independent random variables.

(a) Find the most powerful test of $H_0 : \mu = 4$ versus $H_a : \mu > 4$.

(b) What is the p-value for your test in (a) if $n = 8$, $\sum_{i=1}^{8} Y_i = 51.8$, and $\sum_{i=1}^{8} X_i = 76.5$?

57.10 Suppose X_1, X_2, X_3, X_4, and X_5 are independent normal random variables. All of these random variables have mean μ, X_1 and X_2 have variance 4, and X_3, X_4, and X_5 have variance 9. Interest is in testing $H_0 : \mu = 100$ versus $H_a : \mu > 100$.

(a) Derive the most powerful test at $\alpha = .05$, and give a decision rule based on a test statistic with known distribution.

(b) Show that the test in (a) uses the UMVUE for μ.

(c) A statistics practitioner who has not taken a class in mathematical statistics says this: "I know that $X_1 + X_2 + X_3 + X_4 + X_5$ has mean 5μ and variance 35, so I'll use

$$Z = \frac{5(\bar{Y} - 100)}{\sqrt{35}}$$

as my test statistic, and reject when Z is larger than the 95th percentile of a standard normal distribution." Show through simulations that your test in (a) has higher power.

57.11 Ecologists are conjecturing that climate change will lead to more frequent flooding of an important river in Southeast Asia. They model the maximum annual height X of the river, as following the density

$$f_\theta(x) = \frac{\theta}{x^{\theta+1}} \quad \text{for } x > 1.$$

Historically, they know that $\theta = 4.2$, and they want to test $H_0 : \theta = 4.2$ versus $H_a : \theta < 4.2$. The mean of the distribution is $\theta/(\theta - 1)$ ($E(X) = 1.3125$ under H_0), so smaller values of θ lead to larger means for X.

(a) Given a random sample $X_1, \ldots, X_n$ from f_θ, show that the most powerful test is to reject H_0 when the observed $\sum_{i=1}^{n} \log(x_i)$ is large.

(b) Find the distribution of the test statistic $T = \sum_{i=1}^{n} \log(X_i)$.

(c) Apply your test to the data from the next four years, where the maximum annual river heights are 2.4, 1.3, 4.0, and 1.9 (assume independence).

Chapter 58

Likelihood Ratio Tests

In practice, the likelihood ratio test (LRT) is popular for its wide applicability and optimality properties. The LRT can be used whenever the likelihood can be expressed. Further, if an exact distribution for an LRT statistic cannot be found, then there is a nice result that will give an approximate distribution in a wide range of situations.

The Neyman–Pearson lemma provides a method for finding a test statistic that provides the highest power, in the case where the null and alternative hypotheses are simple. That statistic is found using a ratio of likelihoods evaluated at the null and alternative values of the parameter. In this chapter we look at a generalization of this idea to the more common case of composite null and alternative hypotheses.

Suppose we collect data to test $H_0 : \boldsymbol{\theta} \in \Omega_0$ versus $H_a : \boldsymbol{\theta} \in \Omega_a$, where $\boldsymbol{\theta}$ might be a single parameter or a vector of parameters. We know (or stipulate) that the data come from a family of densities with likelihood $L(\boldsymbol{\theta}; \boldsymbol{y})$, where $\boldsymbol{y}$ is the data vector. For example, we might have $\boldsymbol{\theta} = (\mu, \sigma^2)$, $\Omega_0 = \{(\mu, \sigma^2) : \mu = 0, \sigma^2 > 0)\}$, and $\Omega_a = \{(\mu, \sigma^2) : \mu \neq 0, \sigma^2 > 0)\}$.

Let $\Omega = \Omega_0 \cup \Omega_a$. The test statistic is a function of the **likelihood ratio**:

$$\lambda = \frac{\max_{\theta \in \Omega_0} L(\boldsymbol{\theta}; \boldsymbol{y})}{\max_{\theta \in \Omega} L(\boldsymbol{\theta}; \boldsymbol{y})}.$$

Note that λ has to be between zero and one. It is greater than zero because the likelihood is the joint density, and has to be positive. It is less than one because the denominator is the maximum over a larger set, and so has to be at least as big as the numerator.

Using λ involves finding two maximum likelihood estimators: for the numerator, we find $\hat{\boldsymbol{\theta}}_0$, maximizing the likelihood over all possible values in Ω_0. Then we evaluate the likelihood function at $\hat{\boldsymbol{\theta}}_0$, so that the numerator of λ is $L(\hat{\boldsymbol{\theta}}_0; \boldsymbol{y})$. (If the null hypothesis is simple, so that there is only one possible value, we simply evaluate the likelihood at that value.) For the denominator, we maximize the likelihood over the larger set Ω, to get $\hat{\boldsymbol{\theta}}$, and evaluate the likelihood function at $\hat{\boldsymbol{\theta}}$: the denominator of λ is $L(\hat{\boldsymbol{\theta}}; \boldsymbol{y})$.

The decision rule for the LRT is to reject H_0 if λ is "small," that is, reject H_0 if $\lambda < k$, where k is chosen so that $P(\lambda < k) = \alpha$ when H_0 is true, for the target test size α. Finding k is the hard part! Generally, an equivalent decision rule is found using a convenient function of λ, as will be illustrated below.

Intuitively, if the value of λ is near one, then allowing the value of $\boldsymbol{\theta}$ to be in the bigger set (containing the alternative hypothesis values) does not improve the likelihood value substantially, so the simpler null model "fits the data" almost as well as the alternative model. If the value of λ is small, then the likelihood value for the alternative model is substantially larger than that for the null, so we might reject the null in favor of the alternative. Of course, the terms "substantially" and "close to" are quantified by the test size α and the distribution of the test statistic.

The likelihood ratio is a class of tests that has some nice optimality properties. We can show that our "sampling from the normal" tests, that is, our Z, t, χ^2, and F tests for means and variances of normal populations, are all LRTs, though a bit disguised. Let's show this explicitly for the simplest case of a one-sample t-test.

Suppose $Y_1, \ldots, Y_n \overset{iid}{\sim} N(\mu, \sigma^2)$ and interest is in testing $H_0 : \mu = \mu_0$ versus $H_a : \mu \neq \mu_0$. Then $\boldsymbol{\theta} = (\mu, \sigma)$, a two-dimensional parameter. The set Ω is two-dimensional, where μ is any real number and $\sigma^2 > 0$. The null set Ω_0 is one-dimensional, where μ is fixed at μ_0 and $\sigma^2 > 0$. The likelihood is familiar:

$$L(\mu, \sigma^2; \boldsymbol{y}) = \left(\frac{1}{2\pi\sigma^2}\right)^{n/2} \exp\left\{-\frac{1}{2\sigma^2} \sum_{i=1}^{n} (y_i - \mu)^2\right\}.$$

To construct the LRT statistic λ, we first maximize the likelihood (separately) over the sets Ω and Ω_0. We do the usual maximization of the log-likelihood function. For the null hypothesis,

$$\ell(\mu_0, \sigma^2; \boldsymbol{y}) = \log L(\mu_0, \sigma^2; \boldsymbol{y}) = (\text{const}) - \frac{n}{2}\log(\sigma^2) - \frac{1}{2\sigma^2}\sum_{i=1}^{n}(y_i - \mu_0)^2;$$

taking the derivative with respect to σ^2, setting it equal to zero, and solving for σ^2 gives

$$\hat{\sigma}_0^2 = \frac{1}{n}\sum_{i=1}^{n}(y_i - \mu_0)^2.$$

For the alternative hypothesis, taking the derivative of ℓ with respect to μ gives $\hat{\mu} = \bar{y}$ for any value of σ^2. Maximizing ℓ with respect to both μ and σ^2 gives

$$\hat{\sigma}^2 = \frac{1}{n}\sum_{i=1}^{n}(y_i - \bar{y})^2.$$

Next, we simplify the expression for λ, noticing that the terms in the exponents conveniently cancel:

$$\lambda = \frac{L(\mu_0, \hat{\sigma}_0^2; \boldsymbol{y})}{L(\hat{\mu}, \hat{\sigma}^2; \boldsymbol{y})} = \frac{\left(\frac{1}{2\pi\hat{\sigma}_0^2}\right)^{n/2} \exp\left\{-\frac{1}{2\hat{\sigma}_0^2}\sum_{i=1}^{n}(y_i - \mu_0)^2\right\}}{\left(\frac{1}{2\pi\hat{\sigma}^2}\right)^{n/2} \exp\left\{-\frac{1}{2\hat{\sigma}^2}\sum_{i=1}^{n}(y_i - \bar{y})^2\right\}}$$

$$= \left(\frac{\hat{\sigma}^2}{\hat{\sigma}_0^2}\right)^{n/2} = \left(\frac{\sum_{i=1}^{n}(y_i - \bar{y})^2}{\sum_{i=1}^{n}(y_i - \mu_0)^2}\right)^{n/2}.$$

Because n is positive, rejecting H_0 when λ is small is the same as rejecting H_0 when

$$\lambda' = \frac{\sum_{i=1}^{n}(y_i - \bar{y})^2}{\sum_{i=1}^{n}(y_i - \mu_0)^2}$$

is small. You might be thinking at this point that the numerator is a function of a χ^2 random variable, and under the null hypothesis, so is the denominator. Do we have an F-statistic? No, because the numerator and denominator are not independent! We need to do some more work.

We can write

$$\sum_{i=1}^{n}(y_i - \mu_0)^2 = \sum_{i=1}^{n}(y_i - \bar{y} + \bar{y} - \mu_0)^2$$

$$= \sum_{i=1}^{n}(y_i - \bar{y})^2 + 2(\bar{y} - \mu_0)\sum_{i=1}^{n}(y_i - \bar{y}) + n(\bar{y} - \mu_0)^2$$

$$= \sum_{i=1}^{n}(y_i - \bar{y})^2 + n(\bar{y} - \mu_0)^2$$

so that

$$\lambda' = \frac{1}{1 + \frac{n(\bar{y}-\mu_0)^2}{\sum_{i=1}^{n}(y_i - \bar{y})^2}}.$$

Then we see that rejecting H_0 when λ' is small is equivalent to rejecting H_0 when

$$\lambda'' = \frac{n(\bar{y} - \mu_0)^2}{\sum_{i=1}^{n}(y_i - \bar{y})^2}$$

is large. We again have a ratio of functions of χ^2 random variables, and this time we can indeed make an F-statistic. We know from Chapter 35 that $\bar{Y}$ and S^2 are independent random variables. We can rewrite the expression to look more familiar:

$$(n-1)\lambda'' = \frac{(\bar{y} - \mu_0)^2}{S^2/n},$$

which is the square of our T statistic:

$$T = \sqrt{(n-1)\lambda''} = \frac{\bar{Y} - \mu_0}{S/\sqrt{n}}.$$

We have shown that the distribution of T under H_0 is $t(n-1)$, and that T^2 is an $F(1, n-1)$ random variable. This is a case where we can get an exact null distribution for a function of our LRT statistic. Hence, our T-test is an LRT.

We can also show that the two-independent-samples t-test, the paired-samples test, the chi-squared test for a normal population variance (see Exercise 58.2), and the F-test to compare population variances (see Exercise 58.4) are all LRTs.

The LRT "automatically" utilizes sufficient statistics. For, if T is a sufficient statistic for the parameter θ, we can factor the likelihood into a function g of the sufficient statistic and the parameters (and no other expression of the data values), and a function h that contains only the data and not the parameters. Therefore, the h function will cancel in the likelihood ratio, and we are left only with the ratio of the g functions. Although we won't directly prove any technical results, our previous experience with sufficient statistics lets us believe that LRTs have some nice optimality properties.

Approximate distribution for LRT

For the previous example, we found that the LRT was equivalent to our old t-test, for which we have an exact distribution under the null hypothesis. It's not always possible to find an exact distribution for the ratio λ or for some equivalent test statistic. However, we have a very nice result that gives us an approximate distribution for $-2\log(\lambda)$ under mild conditions.

To motivate this, let's look at a test for equality of two binomial probabilities. Suppose $S_1 \sim \text{Binom}(n_1, p_1)$ and $S_2 \sim \text{Binom}(n_2, p_2)$. Interest is in testing $H_0 : p_1 = p_2$ versus $H_a : p_1 \neq p_2$.

The likelihood is

$$L(p_1, p_2; s_1, s_2) = p_1^{s_1}(1 - p_1)^{n_1 - s_1} p_2^{s_2}(1 - p_2)^{n_2 - s_2}$$

and the likelihood ratio is

$$\lambda = \left(\frac{\hat{p}_0}{\hat{p}_1}\right)^{s_1} \left(\frac{\hat{p}_0}{\hat{p}_2}\right)^{s_2} \left(\frac{1 - \hat{p}_0}{1 - \hat{p}_1}\right)^{n_1 - s_1} \left(\frac{1 - \hat{p}_0}{1 - \hat{p}_2}\right)^{n_2 - s_2},$$

where $\hat{p}_0 = (s_1 + s_2)/(n_1 + n_2)$, $\hat{p}_1 = s_1/n_1$, and $\hat{p}_2 = s_2/n_2$.

We might be a bit daunted by the idea of trying to come up with an exact null distribution for this ratio! However, we have a very nice result that approximates the null distribution.

Wilk's theorem

Under some mild conditions (including that the support of the joint density does not depend on the parameter value, as well as some differentiability conditions), the distribution under the null hypothesis of

$$-2\log(\lambda)$$

has approximately a $\chi^2(r - r_0)$ distribution, where r is the dimension (number of "free" parameters) under H_a and r_0 is the dimension under H_0.

Let's continue the binomial example before we think about proving this theorem. The test statistic is

$$-2\log(\lambda) = -2\left[s_1 \log\left(\frac{\hat{p}_0}{\hat{p}_1}\right) + s_2 \log\left(\frac{\hat{p}_0}{\hat{p}_2}\right) + (n_1 - s_1)\log\left(\frac{1 - \hat{p}_0}{1 - \hat{p}_1}\right) + (n_2 - s_2)\log\left(\frac{1 - \hat{p}_0}{1 - \hat{p}_2}\right)\right].$$

The expression on the right is undefined if $\hat{p}_1 = 0$, $\hat{p}_2 = 0$, $\hat{p}_1 = 1$, or $\hat{p}_2 = 1$. We can use the result $\lim_{p \to 0} p\log(p) = 0$, however, to get a value for $-2\log(\lambda)$ for any values of the estimators. If $\hat{p}_1 = 0$, for example, we substitute zero for the first term in the sum.

According to Wilk's theorem, the distribution of $-2\log(\lambda)$, when H_0 is true, is approximately $\chi^2(1)$, and the approximation gets better as the sample size grows. Can this really be true? It seems so unlikely! Let's do simulations to convince ourselves that it actually works, and to get an idea of how large n has to be, before the approximation is "reasonable."

For $n_1 = n_2 = 20$ and $p_1 = p_2 = .18$, the code below generates binomial random variables under this particular case of the null hypothesis. The "if" statements take care of the cases where the expression for $\log(\lambda)$ has a zero in the denominator.

```
nloop=1000000
n1=20;n2=20;p1=.18;p2=.18
loglam=1:nloop
for(iloop in 1:nloop){
    s1=rbinom(1,n1,p1)
    s2=rbinom(1,n2,p2)
    p0hat= (s1+s2)/(n1+n2)
    p1hat=s1/n1
    p2hat=s2/n2
    if(s1>0){t1=s1*log(p0hat/p1hat)}else{t1=0}
    if(s2>0){t2=s2*log(p0hat/p2hat)}else{t2=0}
    if(s1<n1){t3=(n1-s1)*log((1-p0hat)/(1-p1hat))}else{t3=0}
    if(s2<n2){t4=(n2-s2)*log((1-p0hat)/(1-p2hat))}else{t4=0}
    loglam[iloop]=-2*(t1+t2+t3+t4)
}
hist(loglam,br=50,freq=FALSE)
xpl=0:200/10
lines(xpl,dchisq(xpl,1))
sum(loglam>qchisq(.95,1))/nloop
```

Histograms of the one million simulated values of $-2\log(\lambda)$ are shown below, with the $\chi^2(1)$-density superimposed. On the left is the histogram with 30 "breaks," and the true distribution seems to follow the approximation pretty well. If we make the histogram with 100 "breaks," we see the discrete nature of the true distribution. We find the true test size for $n_1 = n_2 = 20$ and $p_1 = p_2 = .2$ is about .059, a little inflated. When the sample sizes get larger, the test size gets closer to the target, and the number of possible combinations of values of $\hat{p}_1$ and $\hat{p}_2$ gets larger, and the histogram looks smoother, even with many "breaks."

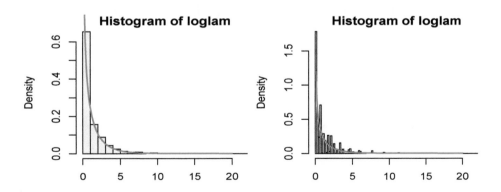

Proof of Wilk's theorem for a one-dimensional parameter

We'll go through the derivation of the asymptotic distribution for $-2\log(\lambda)$ for a one-dimensional parameter vector, where $H_0 : \theta = \theta_0$ versus $H_a : \theta \neq \theta_0$. By definition of the likelihood ratio,

$$\lambda = \frac{L(\theta_0; \boldsymbol{y})}{L(\hat{\theta}; \boldsymbol{y})},$$

where $\hat{\theta}$ is the maximum likelihood estimator. Then

$$-2\log(\lambda) = 2[\log(L(\hat{\theta}; \boldsymbol{y})) - \log(L(\theta_0; \boldsymbol{y}))]$$
$$= 2[\ell(\hat{\theta}; \boldsymbol{y}) - \ell(\theta_0; \boldsymbol{y})].$$

The Taylor expansion of the log-likelihood function about $\hat{\theta}$ gives the following approximation:

$$\ell(\theta; \boldsymbol{y}) \approx \ell(\hat{\theta}; \boldsymbol{y}) + \ell'(\hat{\theta}; \boldsymbol{y})(\theta - \hat{\theta}) + \frac{1}{2}\ell''(\hat{\theta}; \boldsymbol{y})(\theta - \hat{\theta})^2.$$

By definition of the maximum likelihood estimator, $\ell'(\hat{\theta}; \boldsymbol{y}) = 0$. Plugging in θ_0 for θ in the expansion, we get

$$2\left[\ell(\theta_0; \boldsymbol{y}) - \ell(\hat{\theta}; \boldsymbol{y})\right] \approx \ell''(\hat{\theta}; \boldsymbol{y})(\theta_0 - \hat{\theta})^2.$$

We defined the Fisher information

$$I(\theta) = -\mathrm{E}\left[\ell''(\theta; \boldsymbol{Y})\right],$$

and for large n, $I(\theta) \approx -\ell''(\hat{\theta}; \boldsymbol{y})$, because $\hat{\theta}$ is close to θ (and by the law of large numbers). This gives a simple expression for our test statistic in terms of the null hypothesis value of the parameter and the estimated value of the parameter:

$$-2\log(\lambda) \approx I(\theta)(\theta_0 - \hat{\theta})^2.$$

Recall from Chapter 52 that $I(\theta)^{-1}$ approximates the variance of the MLE. Then when H_0 is true,

$$I(\theta)^{1/2}(\theta_0 - \hat{\theta}) \approx \mathrm{N}(0, 1),$$

and finally

$$-2\log(\lambda) \approx \chi^2(1).$$

This approximation "works" under some very mild assumptions. However, we don't expect it to work when the support of the family of densities is a function of the parameter of interest. For example, recall the maximum likelihood estimator for θ, given $Y_1, \ldots, Y_n \sim \mathrm{Unif}(0, \theta)$. To get the maximum likelihood estimator (see Chapter 43), we didn't set the derivative of ℓ to zero—the likelihood function is discontinuous at $\hat{\theta}$. Therefore, the step in the above proof where we use $\ell'(\hat{\theta}; \boldsymbol{y}) = 0$ would be incorrect for this model. One of the assumptions for the approximation is that the support of the family of densities does not depend on the parameters.

This approximation is widely used in generalized regression models and in models with categorical data. The example of comparing two binomial parameters is a special case of the contingency table analysis that we consider in the next chapter.

Chapter Highlights

1. Given a data vector y with likelihood function $L(\theta; y)$, and the test $H_0 : \theta \in \Omega_0$ versus $H_a : \theta \in \Omega_a$, the likelihood ratio is

$$\lambda = \frac{\max_{\theta \in \Omega_0} L(\theta; y)}{\max_{\theta \in \Omega} L(\theta; y)},$$

where $\Omega = \Omega_0 \cup \Omega_a$.

2. The likelihood ratio λ takes values between zero and one, and small values support the alternative hypothesis. If the decision rule for a test is equivalent to "reject H_0 when $\lambda < c$," then the test is called a likelihood ratio test.

3. Under some mild conditions,

$$-2 \log(\lambda)$$

has approximately a $\chi^2(r - r_0)$ distribution if H_0 is true, where r is the number of free parameters under the alternative, and r_0 is the number of free parameters under the null hypothesis.

❧

Exercises

58.1 Suppose $X_1, \ldots, X_n \overset{iid}{\sim} \text{Pois}(\lambda_1)$ and $Y_1, \ldots, Y_n \overset{iid}{\sim} \text{Pois}(\lambda_2)$, and the samples are independent. Interest is in testing $H_0 : \lambda_1 = \lambda_2$ versus $\lambda_1 \neq \lambda_2$.

    (a) Find an LRT statistic λ for this two-sample problem from Poisson populations. Simplify an expression for $-2 \log(\lambda)$ and give its approximate distribution.

    (b) Using simulations in R, find the true test size when $n = 30$ and $\lambda_1 = \lambda_2 = 2.5$, and the target test size is $\alpha = .05$.

58.2 Suppose $X_1, \ldots, X_n \overset{ind}{\sim} \text{N}(0, \sigma^2)$.

    (a) Formulate an LRT for

$$H_0 : \sigma^2 = 1 \quad \text{versus} \quad H_a : \sigma^2 \neq 1.$$

    Simplify the likelihood ratio so that you can find a test statistic with a known distribution under H_0 (not an approximate test).

    (b) Apply your test to the observed values $-0.8, -1.1, 2.2, 0.4, -0.8, -2.2, -1.7, 2.1$.

    (c) Generalize the test in part (a) to

$$H_0 : \sigma^2 = \sigma_0^2 \quad \text{versus} \quad H_a : \sigma^2 \neq \sigma_0^2,$$

    where σ_0^2 is a positive constant. Show that the test derived in Chapter 35 is an LRT.

58.3 Suppose $X_1, \ldots, X_n \overset{ind}{\sim} N(\mu_1, 1)$ and $Y_1, \ldots, Y_m \overset{ind}{\sim} N(\mu_2, 1)$, and the samples are independent. Interest is in testing

$$H_0 : \mu_1^2 = \mu_2^2 \quad \text{versus} \quad H_a : \mu_1^2 \neq \mu_2^2.$$

Derive an LRT and give the (exact) distribution. (*Hint:* With a lot of simplifying, you can get the test from Chapter 35).

58.4 Suppose $X_1, \ldots, X_n \overset{ind}{\sim} N(0, \sigma_1^2)$ and $Y_1, \ldots, Y_{n_2} \overset{ind}{\sim} N(0, \sigma_2^2)$, and the samples are independent. Interest is in testing

$$H_0 : \sigma_1^2 = \sigma_2^2 \quad \text{versus} \quad H_a : \sigma_1^2 \neq \sigma_2^2.$$

Show that the test derived in Exercise 37.17 is an LRT.

58.5 Suppose $X_1, \ldots, X_{n_1} \overset{ind}{\sim} N(0, \sigma_1^2)$, $Y_1, \ldots, Y_{n_2} \overset{ind}{\sim} N(0, \sigma_2^2)$, and $Z_1, \ldots, Z_{n_3} \overset{ind}{\sim} N(0, \sigma_3^2)$ are independent samples from normal populations. Interest is in testing

$$H_0 : \sigma_1^2 = \sigma_2^2 = \sigma_3^2 \quad \text{versus} \quad H_a : \text{at least one variance is different.}$$

(a) Find an LRT; simplify an expression for $-2\log(\lambda)$ and give its approximate distribution.

(b) Apply your test to find out if the variances in tooth growth are different for different dose sizes, using the `ToothGrowth` data set in R, and get an approximate p-value. (Subtract out the mean in each group before applying the test to the data.)

(c) Do simulations to find the true test size when the target test size is $\alpha = .05$, and $n_1 = n_2 = n_3 = 20$, when $\sigma^2 = 10$.

58.6 Suppose $X_1, \ldots, X_n$ is a random sample from an exponential population with mean θ_1, and $Y_1, \ldots, Y_m$ is an independent random sample from an exponential population with mean θ_2. Find the (approximate) LRT for $H_0 : \theta_1 = \theta_2$ versus $H_a : \theta_1 \neq \theta_2$. Apply the test for observed values $\bar{X} = 26.1$, $\bar{Y} = 18.3$, and $n = m = 12$.

58.7 Suppose $X_1, \ldots, X_n$ is a random sample from a Beta($\theta_1, 1$) population, and $Y_1, \ldots, Y_m$ is an independent random sample from a Beta($\theta_2, 1$) population. Find the (approximate) LRT for $H_0 : \theta_1 = \theta_2$ versus $H_a : \theta_1 \neq \theta_2$. Do simulations to check the distribution of your test statistic using a probability plot. Use $\theta_1 = \theta_2 = 2$ and $n_1 = n_2 = 20$.

58.8 Suppose $X_1, \ldots, X_n$ is a random sample from a Gamma($2, 1/\theta_1$) population, and $Y_1, \ldots, Y_m$ is an independent random sample from a Gamma($2, 1/\theta_2$) population. Find the (approximate) LRT for $H_0 : \theta_1 = \theta_2$ versus $H_a : \theta_1 \neq \theta_2$. Do simulations to check the distribution of your test statistic using a probability plot. Use $\theta_1 = \theta_2 = 2$ and $n_1 = n_2 = 20$.

58.9 Lifetimes for a filament used in optical fibers are modeled as Weibull(θ) for some parameter θ. That is, each filament's lifetime is a random variable with density

$$f_\theta(y) = \frac{2y}{\theta} e^{-y^2/\theta} \quad \text{for } y > 0.$$

A team of engineers will choose between Company A and Company B to provide filaments; suppose the filaments from both companies have lifetimes following the above density, with the parameters being θ_A and θ_B, respectively. They obtain m filaments from Company A and n filaments from Company B, and measure the lifetimes.

(a) Find an LRT for $H_0 : \theta_A = \theta_B$ versus $H_a : \theta_A \neq \theta_B$; simplify an expression for $-2\log(\lambda)$ and give its approximate distribution.

(b) Suppose the following are observed lifetimes:

      Company A: 5.30, 3.03, 0.59, 1.43, 5.30, 0.94, 0.91, 2.12, 2.08, 1.25;

      Company B: 1.13, 2.64, 3.89, 1.62, 1.39, 2.54, 0.77, 0.91, 1.78, 1.16.

Find the p-value for this test.

(c) The engineers are worried that the sample sizes are not big enough to use the approximate distribution. Do simulations with $n = m = 10$ and $\theta_A = \theta_B = 6$ to determine the true test size when the target test size is $\alpha = .05$.

58.10 The yearly maximum rainfall in Sepiaville is modeled as

$$f_\theta(x) = \frac{\theta}{x^{\theta+1}} \quad \text{for } x > 1.$$

Climate scientists want to determine if there is evidence that the parameter θ has changed in recent years. They have data from the past: The yearly maximum rainfall in 1946–1965 was

$$1.2, \quad 1.1, \quad 1.0, \quad 1.1, \quad 1.2, \quad 1.1, \quad 1.6, \quad 1.6, \quad 1.5, \quad 1.8, \quad 1.3,$$

$$1.4, \quad 2.3, \quad 1.1, \quad 1.1, \quad 1.1, \quad 1.7, \quad 1.1, \quad 1.0, \quad 1.6$$

and the yearly maximum rainfall for the most recent 20 years was

$$1.7, \quad 1.1, \quad 1.4, \quad 1.1, \quad 3.6, \quad 1.0, \quad 2.0, \quad 1.3, \quad 1.3, \quad 2.2, \quad 1.5,$$

$$1.4, \quad 1.2, \quad 1.4, \quad 1.4, \quad 3.0, \quad 1.9, \quad 1.2, \quad 1.0, \quad 3.9.$$

Suppose we can consider these measurements as random samples from the distribution, with possibly different parameter values.

(a) Give an approximate p-value for the null hypothesis that the distribution of yearly maximum rainfall has not changed (the parameter θ is the same) against the alternative that the θ parameter is now different.

(b) Determine the power of your test if $\theta_1 = 3.7$ and $\theta_2 = 2.2$ with the same sample sizes.

Chapter 59

Chi-Squared Tests for Categorical Data

Three examples of commonly used tests for categorical data are derived from likelihood ratio tests. Goodness-of-fit tests, tests for homogeneity, and tests for independence are all types of inference for "contingency tables" (tables of counts) that use the "Pearson chi-squared statistic." We will show that this test statistic is approximately equivalent to a likelihood ratio, and therefore has the approximate chi-squared distribution.

Goodness-of-fit tests

Suppose we have a sample $Y = (Y_1, \ldots, Y_n)$ from a multinomial distribution with probabilities $p = (p_1, \ldots, p_J)$. That is, for each $i = 1, \ldots, n$, $P(Y_i = j) = p_j$, $j = 1, \ldots, J$, and the Y_i random variables are independent. We consider the null hypothesis $p = p_0$, where p_0 is completely specified.

Define the random variables S_j, $j = 1, \ldots, J$, to be the number of $Y_i = j$, so that $\sum_{j=1}^{J} S_j = n$. The data vector can be summarized by the sufficient statistic $S = (S_1, \ldots, S_J)$; in fact the likelihood is

$$L(p; S) = \prod_{j=1}^{J} p_j^{S_j}$$

and the sample proportions maximize the likelihood under the alternative hypothesis: $\hat{p}_j = S_j/n$. Then the likelihood ratio is

$$\lambda = \frac{\prod_{j=1}^{J} p_{0j}^{S_j}}{\prod_{j=1}^{J} \left(\frac{S_j}{n}\right)^{S_j}} = \prod_{j=1}^{J} \left[\frac{e_j}{S_j}\right]^{S_j},$$

where $e_j = np_{0j}$ is the expected number of observations of the jth level. Now,

$$-2\log(\lambda) = 2\left[\sum_{i=1}^{J} S_j \log\left(\frac{S_j}{e_j}\right)\right],$$

and according to the main result of the last chapter, this statistic is distributed as a $\chi^2(J-1)$ random variable when H_0 is true.

A more common form of this test is a result of a further approximation obtained through a Taylor's expansion. Let $f(x) = x \log(x/c)$, and take the first three terms of a Taylor expansion about c. We have $f'(x) = 1 + \log(x/c)$ and $f''(x) = 1/x$, so that for x near c, we have

$$x \log(x/c) \approx x - c + \frac{1}{2c}(x - c)^2.$$

Then

$$-2 \log(\lambda) \approx 2 \sum_{i=1}^{J} \left[S_j - e_j + \frac{1}{2e_j}(S_j - e_j)^2 \right] = \sum_{i=1}^{J} \frac{(S_j - e_j)^2}{e_j},$$

because $\sum_{i=1}^{J} S_j = \sum_{i=1}^{J} e_j = n$. This is called the Pearson chi-squared test statistic and is often taught in introductory classes (without the likelihood ratio derivation). It is intuitively clear that large values support the alternative, and it is relatively easy to compute by hand with a calculator.

To illustrate, let's take another look at Exercise 7.4: A botanist is studying color patterns of a species of flowering plant. If a certain genetic rule holds, 50% of the plants have pink flowers, 25% have white flowers, and 25% have red flowers. The botanist randomly selects 100 seeds to test the null hypothesis that the genetic rule holds. Recall that in the Chapter 6 exercises, we looked at several ideas for test statistics and compared them through simulations to determine which had the best power. One of those options was the Pearson statistic, and we found this had the best power. We now know that it is an approximation to a likelihood ratio test, so we expect it to be a good test.

Would it be better to use the first form of the test statistic? After all, the first form is not more complex; the second form is preferred mostly because it is more intuitive and easier to explain. Let's compare

$$X_1 = 2 \left[\sum_{i=1}^{J} S_j \log \left(\frac{S_j}{e_j} \right) \right] \quad \text{and} \quad X_2 = \sum_{i=1}^{J} \frac{(S_j - e_j)^2}{e_j}$$

through some simulations. The following code draws samples of size 40 from the multinomial with the null hypothesis from the flower example. One million loops provide three decimal places of accuracy, and we find that using X_1, the first version of the likelihood ratio test statistic, gives a test size of $\alpha = .051$, which is close to the target of .05. However, note that if a cell is empty, i.e., one of the S_j values is zero, the X_1 is undefined; these loops were deleted from the simulated vector of values. (In one million loops, there were about 25 instances of this.) The Pearson approximation can be used in these cases, and we find $\alpha = .053$. This version of the test has a size that is a bit more inflated.

```
k=3;p0=c(.25,.5,.25)
nloop=1000000
x1=1:nloop;x2=1:nloop
n=40;e=p0*n
for(iloop in 1:nloop){
    s=rmultinom(1,n,p0)
    x1[iloop]=2*sum(s*log(s/e))
    x2[iloop]=sum((s-e)^2/e)
}
sum(x1>qchisq(.95,2),na.rm=TRUE)/(nloop-sum(is.na(x1)))
sum(x2>qchisq(.95,2))/nloop
```

The next figure compares the values of X_1 and X_2 for 1000 values simulated under H_0; we see that for small values (which correctly support H_0) the test statistic values are very close to each other. For larger values (more support for H_a), the statistics can be different from each other, but not in a systematic way; sometimes X_1 is larger, but sometimes X_2 is larger. The probability plots against the $\chi^2(2)$ quantiles both look good, with hardly any deviation from the line.

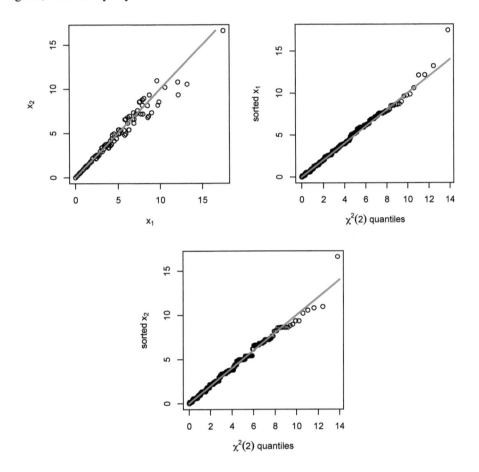

We can change the code slightly to compute power, by sampling from $p = (.2, .6, .2)$, while keeping the same null hypothesis p_0. Surprisingly, the X_1-statistic gives a power of about .186, while the Pearson statistic has a lower power, about .171. However, if the true $p = (1/3, 1/3, 1/3)$, the power for the X_1-statistic is about .462, compared to about .499 for the X_2-statistic.

Both versions of the test statistic, the $-2\log(\lambda)$ and the Pearson statistic, have only approximate χ^2 distributions, with the approximation improving as the sample size grows. How large is large enough? Instead of stipulating a minimum size for n, common advice is that no more than 20% of cells can have expected counts (or estimated expected counts) of less than 5.

We can also test against a family of distributions, rather than against a completely specified null distribution. Here the probabilities for the levels are estimated under both the null and alternative distributions.

Example: A statistician would like to model the number Y of times a flowering plant blooms in a season as $Y \sim \text{Poisson}(\lambda)$. She finds that of a random sample of $n = 20$ plants, nine flowered once, five flowered twice, three flowered more than twice, and three didn't flower at all. Is there evidence that the distribution is not Poisson? Here the null hypothesis is that the probabilities for each of the four levels (zero, one, two, and more than two) can be determined by a single parameter λ, and given lambda these probabilities are

$$p_0 = \left(e^{-\lambda}, \lambda e^{-\lambda}, \frac{1}{2}\lambda^2 e^{-\lambda}, 1 - \left(e^{-\lambda} + \lambda e^{-\lambda} + \frac{1}{2}\lambda^2 e^{-\lambda} \right) \right).$$

We can estimate λ as the sample mean (we showed this is the MLE in Exercise 43.10). Suppose there were two plants that flowered three times and one plant that flowered five times, so $\hat{\lambda} = 1.5$, and $\hat{p}_0 = (.223, .335, .251, .191)$. Because we estimated one parameter, $r_0 = 1$.

The alternative is that the probabilities of the levels follow some other distribution with $p = (p_0, p_1, p_2, p_3)$. Given our sample, our alternative hypothesis estimate is $\hat{p} = (.1, .45, .25, .15)$. Under this general alternative, there are $r = 3$ free parameters.

Our vector of estimated expected values under H_0 is $e = (4.46, 6.70, 5.02, 3.82)$, and our observed counts are $S = (3, 9, 5, 3)$. The observed value of the Pearson chi-squared test statistic is

$$\sum_{i=1}^{4} \frac{(S_j - e_j)^2}{e_j} = 1.44,$$

which, if H_0 is true, is a realization from a χ^2 distribution with $r - r_0 = 2$ degrees of freedom. The p-value is the area to the right of this observed value, under a $\chi^2(2)$-density, or about .49. There is not enough evidence to reject the null hypothesis that the distribution of the flowering frequency is Poisson.

The sample size for this test is not large, and in fact two of the four cells had estimated expected count less than five. We are not sure about the validity of the test, and we could not expect good power for such a small sample size. Next, let's assume we have the same sample mean, and the same sample proportions for the four levels, but a bigger sample size. Suppose for a sample of $n = 60$, we saw 27 plants that flowered once, 15 that flowered twice, nine that flowered more than twice, and nine that did not flower at all, and the sample mean number of flowerings is 1.5. Now the observed value of the Pearson chi-squared test statistic is 4.33, and we have $p = .11$. Goodness-of-fit tests generally have low power unless the sample size is large, so with low sample sizes, we have to be careful not to conclude that there is evidence *in favor of* the null distribution, when it is simply "not rejected."

Tests for homogeneity

Suppose we have two discrete random variables with the same support, and we want to test the null hypothesis that the mass functions are equal, compared to the general alternative that they are not equal. We are not specifying the mass function for the null hypothesis, so that has to be estimated as well. Suppose there are J levels, and we take a sample of size n_1 from the first distribution and a sample of size n_2 from the second distribution. Let $S_{1,j}$ be the number of observations from the first sample that are level j, and let $S_{2,j}$ be the number of observations from the second sample that are level j for $j = 1, \ldots, J$. Then $\sum_{j=1}^{J} S_{1,j} = n_1$ and $\sum_{j=1}^{J} S_{2,j} = n_2$.

Let $p_{1,j}$, $j = 1, \ldots, J$, be the probabilities given by the mass function for the first random variable, and let $p_{2,j}$, $j = 1, \ldots, J$, be the probabilities given by the mass function for the second random variable.

The likelihood function is

$$\prod_{\ell=1}^{2} \prod_{j=1}^{J} p_{\ell,j}^{S_{\ell,j}}.$$

Under H_0, $p_{1,j} = p_{2,j}$ for $j = 1, \ldots, J$, and the estimates are

$$\hat{p}_{0j} = \frac{S_{1,j} + S_{2,j}}{n_1 + n_2} \quad \text{for } j = 1, \ldots, J,$$

and under H_a,

$$\hat{p}_{\ell,j} = \frac{S_{\ell,j}}{n_\ell} \quad \text{for } \ell = 1, 2 \text{ and } j = 1, \ldots, J.$$

Then the likelihood ratio is

$$\lambda = \prod_{j=1}^{J} \left[\frac{\hat{p}_{0j}^{S_{1j}+S_{2j}}}{\hat{p}_{1j}^{S_{1j}} \hat{p}_{2j}^{S_{2j}}} \right] = \prod_{j=1}^{J} \left[\frac{S_{1j}+S_{2j}}{S_{1j}} \frac{n_1}{n_1+n_2} \right]^{S_{1,j}} \left[\frac{S_{1j}+S_{2j}}{S_{2j}} \frac{n_2}{n_1+n_2} \right]^{S_{2,j}},$$

and writing

$$e_{1,j} = (S_{1,j} + S_{2,j}) \frac{n_1}{n_1+n_2} \quad \text{and} \quad e_{2,j} = (S_{1,j} + S_{2,j}) \frac{n_2}{n_1+n_2}$$

to represent the estimate of the expected count when H_0 is true, we have

$$\lambda = \prod_{j=1}^{J} \left[\frac{e_{1,j}}{S_{1,j}} \right]^{S_{1,j}} \left[\frac{e_{2,j}}{S_{2,j}} \right]^{S_{2,j}}.$$

Then the test statistic is

$$-2\log(\lambda) = 2 \sum_{j=1}^{J} \left[S_{1,j} \log \left(\frac{S_{1,j}}{e_{1,j}} \right) + S_{2,j} \log \left(\frac{S_{2,j}}{e_{2,j}} \right) \right].$$

Using the same approximation as the goodness-of-fit test, we get that the following test statistic has an approximate chi-squared distribution under the null hypothesis:

$$\sum_{\ell=1}^{2} \sum_{j=1}^{J} \frac{(S_{\ell,j} - e_{\ell,j})^2}{e_{\ell,j}}.$$

The degrees of freedom: Under H_a there are $r = 2(J - 1)$ unspecified parameters (because $\sum_{i=1}^{J} p_{\ell,j} = 1$ for $\ell = 1$ and $\ell = 2$). Under H_0 we estimated only $r_0 = J - 1$ parameters, so there the degrees of freedom is $r - r_0 = J - 1$.

Let's extend the flowering plant example, where now the botanists have two species of flowers, both of which produce either white, pink, or red flowers when mature. The null hypothesis is that the probabilities of the colors are the same for both species. Suppose we have 50 seeds of each, considered to be random samples from the color distributions. If the counts are 10, 30, and 10 for the first species, and 15, 20, and 15, is this strong evidence that the population color distributions are different? Our null hypothesis estimate is $\hat{p}_0 = (.25, .5, .25)$ and $e = (25, 50, 25)$. There are two degrees of freedom, because under H_0 there are two parameters to estimate, and under H_a there are four. The observed value of the Pearson statistic is 4, and the p-value is `1-pchisq(4,2)`, about .135. The evidence that the distributions are different is not strong.

We can easily extend the test for homogeneity to more than two distributions. Samples of size n_ℓ for $\ell = 1, \ldots, L$ are taken from the L populations, and the null hypothesis is that all L populations have the same probability mass function for the same support of J values. The likelihood function is

$$\prod_{\ell=1}^{L} \prod_{j=1}^{J} p_{\ell,j}^{S_{\ell,j}},$$

and the derivation of the Pearson statistic

$$\sum_{\ell=1}^{L} \sum_{j=1}^{J} \frac{(S_{\ell,j} - e_{\ell,j})^2}{e_{\ell,j}}$$

is similar to the case for $L = 2$. For the null hypothesis, there are $r_0 = J - 1$ parameters to be estimated, and for the alternative we have $r = L(J - 1)$ parameters. The distribution of the Pearson statistic when H_0 is true is approximate chi-squared with $r - r_0 = (L - 1)(J - 1)$ degrees of freedom.

Tests for independence

Suppose we have jointly distributed discrete random variables Y_1 and Y_2, where Y_1 has J levels and Y_2 has L levels. There are $r = JL - 1$ probabilities $p_{l,j}$ to be specified in the joint distribution. Let's consider the null hypothesis that the random variables are independent; in this case there are only $r_0 = (J-1)(L-1)$ probabilities to be specified: If the marginal distribution of Y_1 is $P(Y_1 = j) = p_{01j}$, $j = 1, \ldots, J$, and the marginal distribution of Y_2 is $P(Y_2 = \ell) = p_{02\ell}$, $\ell = 1 \ldots, L$, then $P(Y_1 = j \text{ and } Y_2 = \ell) = p_{01j}p_{02\ell}$ under H_0 for all $j = 1, \ldots, J$ and $\ell = 1, \ldots, L$.

The likelihood function is

$$\prod_{\ell=1}^{L} \prod_{j=1}^{J} p_{\ell,j}^{S_{\ell,j}},$$

where $S_{\ell,j}$ is the observed count of $Y_1 = j$ and $Y_2 = \ell$. Under the alternative hypothesis, the estimated probabilities are simply the sample proportions $\hat{p}_{j,\ell}$ for each combination of levels. Under the null hypothesis, we compute the estimates of the two marginal distributions using sample proportions $\hat{p}_{01,j}$ and $\hat{p}_{02,\ell}$. Then we compute the probability for each combination of levels as products of the marginal distribution probabilities, and the estimate of the expected count for a combination of levels is $e_{l,j} = n\hat{p}_{01,j}\hat{p}_{02,\ell}$.

The likelihood ratio is

$$\lambda = \prod_{j=1}^{J} \prod_{\ell=1}^{L} \left[\frac{\hat{p}_{01,j}\hat{p}_{02,\ell}}{\hat{p}_{j,\ell}} \right]^{S_{j,\ell}} = \prod_{j=1}^{J} \prod_{\ell=1}^{L} \left[\frac{e_{j,\ell}}{S_{j,l}} \right]^{S_{j,\ell}}.$$

Then

$$-2\log(\lambda) = 2 \sum_{j=1}^{J} \sum_{\ell=1}^{L} \left[S_{j,\ell} \log \left(\frac{S_{j,\ell}}{e_{j,\ell}} \right) \right].$$

The derivation of the Pearson statistic

$$\sum_{\ell=1}^{L} \sum_{j=1}^{J} \frac{(S_{\ell,j} - e_{\ell,j})^2}{e_{\ell,j}}$$

follows that of the test for homogeneity. In fact, the test of homogeneity and the test for independence are often confused, and although the assumptions are different, the confusion is not a cause of error, because the tests are, in fact, performed in exactly the same way!

In practice, for the test of independence, we take a sample of size n from a single population and observe the values of two categorical random variables. For the test of homogeneity, we take samples of size $n_1, \ldots, n_L$ from L populations and observe the values of one categorical random variable for each sample.

For example, suppose we take a sample of 100 voters and determine their political party (for simplicity we will consider only two: conservative and liberal) and gender (again for simplicity we consider only two: male and female). The null hypothesis is that the party and gender are independent. We can construct a two-by-two table, and for each of the four "cells" (combinations of levels) we can enter both the observed $S_{j,\ell}$ and the estimated expected count under the null hypothesis $e_{j,\ell}$. The latter can be calculated as row total (for the row containing the cell) times the column total (for the column containing the cell) divided by the sample size n.

If we had instead sampled a predetermined number of male voters, and a predetermined number of female voters, and observed the party affiliations of each, this would be a test for homogeneity. If we constructed a two-by-two table, we can enter in each cell both the observed $S_{j,\ell}$ and the estimated expected count under the null hypothesis $e_{j,\ell}$. Although the reasoning and the assumptions for the test of homogeneity differ substantially from those of the test for independence, the $e_{j,\ell}$ can be calculated as row total (for the row containing the cell) times the column total (for the column containing the cell) divided by the sample size n.

The tests for homogeneity and for independence have the same procedure, so that if a practitioner mistakes one design for the other, for a given table of counts, the correct p-value is obtained. The practical difference in the designs can be seen in the power of the tests.

Example: Let's look again at the flower-color example, and suppose we are designing a test to see if the color distribution is the same for both species. Suppose that instead of choosing 50 seeds from each species, we choose 100 seeds "at random" from a large bin of seeds that contains both types. After the 100 plants are grown, we determine both the species and the observed color distributions. This test is a test for independence. Once the data are collected and the table of counts is made, we can proceed as if the data were collected under either design.

If there are equal proportions of seeds from each species in the bin, then the power of the two tests is approximately the same. However, if there are more of one species, so that the sample size for Species 1, say, tends to be larger than the sample size for Species 2, then the test for homogeneity (with equal sample sizes) will have higher power. You are asked to show this through simulations in Exercise 59.9.

Chapter Highlights

1. The likelihood ratio test applied to categorical data results in several important types of chi-squared tests. For each, the Pearson tests can be derived from a Taylor's expansion. These all have the same form: determine the expected counts, or estimates of expected counts, under the null hypothesis, then use the test statistic

$$X = \sum_{\text{all cells}} \frac{(O_i - E_i)^2}{E_i}.$$

2. For the goodness-of-fit test, we have a sample from a distribution that, under the null hypothesis, is specified exactly. We compare our observed values to those expected if H_0 were true.

3. For the test for homogeneity, we have samples from two distributions, and the null hypothesis is that the distributions are the same. We estimate the common distribution under the null hypothesis, and the test statistic measures how far the actual values are from the estimated H_0 values.

4. For the test for independence, we have a single sample, but for each unit, two measurements are made. The null hypothesis is that the probability that one of the measurements has any given value is independent of the value of the other measurement. We estimate the distribution under the null hypothesis, and the test statistic measures how far the observed values are from the null hypothesis estimate.

Exercises

59.1 In the past, Fizzypop has held 40% of the market share of colas, Yippee has held 30%, Gassaqua has held 20%, and other brands had the remaining 10% of the market. After massive advertising campaigns by the competitors, a survey is taken of 200 consumers. They state their cola preferences as follows:

Observed counts

Fizzypop	Yippee	Gassaqua	others
70	69	43	18

Is there evidence at $\alpha = 0.10$ that the market distribution has changed? (Assume that consumer's stated preference is a good measure of market share.)

59.2 Suppose we know the distribution of car colors in a certain state has the following mass function:

Mass function

white	black	blue	red	green	other
0.22	0.18	0.37	0.10	0.05	0.08

A random sample of 100 moving violations citations in 1998 reveals the following distribution of car colors:

	Observed counts				
white	black	blue	red	green	other
17	12	32	23	6	10

Is there evidence at $\alpha = 0.05$ that the distribution of car colors of cars that were involved in moving violations in 1998 was different from the distribution of car colors in the general population in this state?

59.3 A climatologist would like to model the yearly count of tropical storms in the Atlantic Basin as following a Poisson distribution. The data set `ex1028` in the R package `Sleuth2` has counts of hurricanes for the years 1950–1997. If these counts can be considered a random sample of counts from a Poisson(λ) distribution, find the MLE for λ. Next, test the null hypothesis that the probabilities for the hurricane counts are indeed Poisson. Use a goodness-of-fit test with five levels: 0–3, 4, 5–6, 7–8, 9+.

59.4 Researchers at the Carston Clinic set out to determine the validity of the old belief that a glass of orange juice in the morning will help prevent the common cold.

(a) One hundred volunteers at the clinic were divided randomly into two groups of 50. One group drank a glass of orange juice at breakfast every day, the other group drank a glass of plain water with breakfast. After three months, it was found that 20% of the orange juice group has at least one episode of the common cold, while 30% of the other group got colds. Determine if this difference is statistically significant at $\alpha = .05$.

(b) Repeat part (a), except that four hundred volunteers at the clinic were divided randomly into two groups of 200. The proportions of colds in each group are the same as in (a). Determine if this difference is statistically significant at $\alpha = .05$.

59.5 Suppose we want to compare the duration of nausea after using a new types of anesthetic versus the standard anesthetic used in oral surgery. Eighty patients are randomized into two treatment groups, and the outcomes categorized as follows: Symptoms less than one hour, symptoms between one and five hours, between five and twelve hours, and more than twelve hours. The data are compiled in the table.

Observed counts of nausea duration categories				
	< 1 hour	1–5 hours	5–12 hours	> 12 hours
New anesthesia	15	35	25	5
Standard anesthesia	7	28	34	11

The null hypothesis is that the distributions for the nausea duration levels is the same for both anesthetics, and the alternative is that they are not the same. Perform the test at $\alpha = .05$. Is this a test for homogeneity or a test for independence?

59.6 A gardener wishes to compare an organic pesticide with a standard version for tomato plants. He has 100 plants at the beginning of the season, and randomly assigns fifty of each for treatments. The response levels refer to damage of the leaves: no damage, some damage, and severe damage. At the end of the season he collects the following counts:

Observed counts of leaf damage categories

	none	some	severe
Organic	10	28	12
Standard	16	30	4

Test the homogeneity hypothesis at $\alpha = 0.05$. State the conclusion in the context of the problem.

59.7 The abstract for the article "Mediterranean Alpha-Linoleic Acid-Rich Diet in Secondary Prevention of Coronary Heart Disease" in the journal *Lancet* (1995) reads

> In a prospective, randomised single-blinded secondary prevention trial we compared the effect of a Mediterranean alpha-linolenic acid-rich diet to the usual post-infarct prudent diet. After a first myocardial infarction, patients were randomly assigned to the experimental (n = 302) or control group (n = 303). Patients were seen again 8 weeks after randomization, and each year for 5 years ... there were 16 cardiac deaths in the control and 3 in the experimental group

Determine if there is evidence at $\alpha = 0.01$, that the probability of cardiac death in the treatment group is different from the probability of cardiac death in the control group.

59.8 An article in the *Annals of Internal Medicine* reported on a study of "Gastrointestinal symptoms in volunteers ingesting snack foods containing olestra or triglycerides." In the 1990s, food researchers came up with a "fake fat" called Olestra, which was supposed to taste like fat but have zero calories. It never became very popular, due to (among other things) the notion that it caused digestive problems. Here is a summary of the study: Out of 3181 volunteers, 1620 people consumed snack foods made with the fat substitute Olestra, and 1561 consumed snack foods made with "regular" oil. All volunteers believed they were consuming products made with Olestra. Of the Olestra group, 619 reported "at least one gastrointestinal symptom," as did 576 of the control group. (This is an example of the "nocebo effect." The "placebo effect" is the tendency for patients to improve when they think they are getting treatment, but the nocebo effect is the tendency for people to invent symptoms when they think they are being made ill.) Determine if there is evidence at $\alpha = 0.05$ that the probability of symptoms in the Olestra group is different from the probability of symptoms in the control group.

59.9 Botanists are interested in flower-color distributions of Species 1 and Species 2 plants. Plants from both species can be pink, white, or red flowering. Botanist A will carefully identify 50 seeds of Species 1, grow each into plants, and count how many plants have pink, white, and red flowers. Botanist B will take 100 seeds from a bin containing 25% of Species 1 seeds and 75% of Species 2 seeds. She will then grow each into plants, determine the species of each plant, and count how many plants of each species have pink, white, and red flowers. Suppose that the color distribution for Species 1 is 20% white, 60% pink, and 20% red, while the color distribution for Species 2 is 30% white, 40% pink, and 30% red. Using simulations, find the power for the test against the null hypothesis that the color distributions are the same for the design of each botanist. Who has the better design, and is the difference in power substantial? If the sample sizes are doubled, so that $n_1 = n_2 = 100$ for Botanist A's design, and $n = 200$ for Botanist B's, again compare the difference in power.

Chapter 60

One-Way Analysis of Variance

The one-way analysis of variance (ANOVA) model can be thought of as a generalization of the two-independent-samples t-test, where there are more than two independent samples. Interest is in comparing the means $\mu_1, \ldots, \mu_k$ of $k \geq 2$ populations or groups. The standard null hypothesis is that all the means are equal, while the alternative is that at least one of the means is different.

We have observations from each of the groups to make the decision. If we assume normal distributions with equal variances, we can develop a likelihood ratio test statistic that has an (exact) F distribution under the null hypothesis. Formally, we assume

$$Y_{ij} \overset{ind}{\sim} \mathrm{N}(\mu_i, \sigma^2) \ \text{ for } \ i = 1, \ldots, k \ \text{ and } \ j = 1, \ldots, n_i.$$

That is, for each group $i = 1, \ldots, k$, we have a random sample of size n_i that is normally distributed with mean μ_i and variance σ^2. Further, the samples are independent.

Example: A clinic wants to compare the effects of three cholesterol-lowering drugs. They recruit 30 patients with high cholesterol and assign 10 to each drug. In the course of the study, they measure the amount of cholesterol reduction in all 30 patients. The value y_{ij} is the amount of cholesterol reduction seen in the jth patient given the ith drug, while μ_i is the expected cholesterol reduction for a patient given drug i. The null hypothesis is that all drugs are equally effective, and the alternative is that (at least) one of drugs has a different effect than another. We have $k = 3$ and $n_1 = n_2 = n_3 = 10$, so j runs from 1 to 10, while i can be 1, 2, or 3.

The following plot shows an example data set, with the sample means marked with horizontal lines. The mean for Treatment C is higher than the means for the other two treatments, but how strong is the evidence against the null hypothesis that all the population means are the same? We want to quantify the evidence with a p-value, and we turn to our likelihood ratio ideas to construct a test.

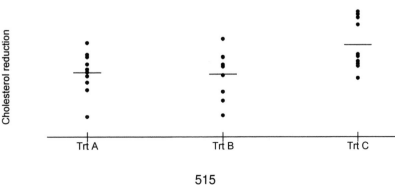

To begin constructing the likelihood ratio test for the general case, we find the MLEs under both the null and alternative hypotheses. Under the null hypothesis, the random variables Y_{ij} are iid normal with common mean $\mu_0 = \mu_1 = \cdots = \mu_k$ and variance σ^2. Therefore the MLE for μ is the mean of all of the Y_{ij} values. That is,

$$\hat{\mu}_0 = \frac{1}{n} \sum_{i=1}^{k} \sum_{j=1}^{n_i} Y_{ij},$$

where $n = n_1 + \cdots + n_k$ is the total sample size. The MLE for σ^2 under the null hypothesis model is

$$\hat{\sigma}_0^2 = \frac{1}{n} \sum_{i=1}^{k} \sum_{j=1}^{n_i} (Y_{ij} - \hat{\mu}_0)^2.$$

Under the alternative hypothesis, the MLEs for the μ_i parameters are the sample means:

$$\hat{\mu}_i = \frac{1}{n_i} \sum_{j=1}^{n_i} Y_{ij} = \bar{Y}_i \text{ for } i = 1, \ldots, k,$$

and the alternative-hypothesis MLE for σ^2 is

$$\hat{\sigma}_a^2 = \frac{1}{n} \sum_{i=1}^{k} \sum_{j=1}^{n_i} (Y_{ij} - \hat{\mu}_i)^2.$$

(Recall Exercise 43.21.) Now that we have the MLEs for each hypothesis, we can construct the likelihood ratio test.

Plugging in the null hypothesis MLEs for the parameters, we have the numerator of the likelihood ratio:

$$\left(\frac{1}{2\pi\hat{\sigma}_0^2} \right)^{n/2} \exp\left\{ -\frac{1}{2\hat{\sigma}_0^2} \sum_{i=1}^{k} \sum_{j=1}^{n_i} (y_{ij} - \hat{\mu}_0)^2 \right\} = \left(\frac{1}{2\pi\hat{\sigma}_0^2} \right)^{n/2} e^{-n/2}.$$

For the denominator, we plug in the alternative hypothesis MLEs to get

$$\left(\frac{1}{2\pi\hat{\sigma}_a^2} \right)^{n/2} \exp\left\{ -\frac{1}{2\hat{\sigma}_a^2} \sum_{i=1}^{k} \sum_{j=1}^{n_i} (y_{ij} - \hat{\mu}_i)^2 \right\} = \left(\frac{1}{2\pi\hat{\sigma}_a^2} \right)^{n/2} e^{-n/2}.$$

Therefore the likelihood ratio is

$$\lambda = \left(\frac{\hat{\sigma}_a^2}{\hat{\sigma}_0^2} \right)^{n/2}.$$

Small values of $\hat{\sigma}_a^2 / \hat{\sigma}_0^2$ support the alternative: This makes sense because if the true means are different, then the $\hat{\sigma}_a^2$ will be the correct estimator of σ^2, but $\hat{\sigma}_0^2$ will be too large. Equivalently, large values of the reciprocal $\hat{\sigma}_0^2 / \hat{\sigma}_a^2$ support the alternative.

Let's try to find a distribution for the random variable

$$\frac{\hat{\sigma}_0^2}{\hat{\sigma}_a^2} = \frac{\sum_{i=1}^{k} \sum_{j=1}^{n_i} (Y_{ij} - \hat{\mu}_0)^2}{\sum_{i=1}^{k} \sum_{j=1}^{n_i} (Y_{ij} - \hat{\mu}_i)^2}.$$

Both the numerator and denominator (if divided by σ^2) have chi-squared distributions, which looks promising for an F-statistic. However, the numerator and denominator are not independent, so we have to do some more work. The numerator can be factored:

$$
\begin{aligned}
\sum_{i=1}^{k}\sum_{j=1}^{n_i}(Y_{ij} - \hat{\mu}_0)^2 &= \sum_{i=1}^{k}\sum_{j=1}^{n_i}(Y_{ij} - \hat{\mu}_i - \hat{\mu}_i - \hat{\mu}_0)^2 \\
&= \sum_{i=1}^{k}\sum_{j=1}^{n_i}[(Y_{ij} - \hat{\mu}_i)^2 + 2(Y_{ij} - \hat{\mu}_i)(\hat{\mu}_i - \hat{\mu}_0) + (\hat{\mu}_i - \hat{\mu}_0)^2] \\
&= \sum_{i=1}^{k}\sum_{j=1}^{n_i}(Y_{ij} - \hat{\mu}_i)^2 + 2\sum_{i=1}^{k}\left[(\hat{\mu}_i - \hat{\mu}_0)\sum_{j=1}^{n_i}(Y_{ij} - \hat{\mu}_i)\right] \\
&\quad + \sum_{i=1}^{k} n_i(\hat{\mu}_i - \hat{\mu}_0)^2 \\
&= \sum_{i=1}^{k}\sum_{j=1}^{n_i}(Y_{ij} - \hat{\mu}_i)^2 + \sum_{i=1}^{k} n_i(\hat{\mu}_i - \hat{\mu}_0)^2,
\end{aligned}
$$

where the middle term of the next-to-last line is zero because for all i, $\sum_{j=1}^{n_i}(Y_{ij} - \hat{\mu}_i) = 0$. Now the random variable of interest becomes

$$
\frac{\hat{\sigma}_0^2}{\hat{\sigma}_a^2} = \frac{\sum_{i=1}^{k}\sum_{j=1}^{n_i}(Y_{ij} - \hat{\mu}_i)^2 + \sum_{i=1}^{k} n_i(\hat{\mu}_i - \hat{\mu}_0)^2}{\sum_{i=1}^{k}\sum_{j=1}^{n_i}(Y_{ij} - \hat{\mu}_i)^2} = 1 + \frac{\sum_{i=1}^{k} n_i(\hat{\mu}_i - \hat{\mu}_0)^2}{\sum_{i=1}^{k}\sum_{j=1}^{n_i}(Y_{ij} - \hat{\mu}_i)^2}.
$$

In Chapter 36, we went through some trouble to show that if random variables $X_1, \ldots, X_n$ are iid $N(\mu, \sigma^2)$ random variables, then the sample mean $\bar{X}$ is independent of the sample variation $\sum_{i=1}^{n}(X_i - \bar{X})^2$. We use this result here for each of the groups i to assert that the $\hat{\mu}_i$ are independent of $\sum_{i=1}^{k}\sum_{j=1}^{n_i}(Y_{ij} - \hat{\mu}_i)^2$, and because $\hat{\mu}_0$ is a function of the $\hat{\mu}_i$, we have that the numerator of the above test statistic is independent of the denominator.

Next, let's show that the random variables in the numerator and the denominator have chi-squared distributions (when divided by σ^2) when H_0 is true. Subsequently, we can find a multiple of the ratio that will have an F distribution, and this is what we will finally use for our test statistic.

For the numerator, we note that under H_0, there is a common mean μ_0, so each $\hat{\mu}_i$ has mean μ_0 and variance σ^2/n_i. Therefore for each $i = 1, \ldots, k$,

$$
\frac{\hat{\mu}_i - \mu_0}{\sigma/\sqrt{n_i}} \sim N(0, 1) \quad \text{and} \quad \frac{n_i(\hat{\mu}_i - \mu_0)^2}{\sigma^2} \sim \chi^2(1).
$$

Because the samples are independent,

$$
\frac{1}{\sigma^2}\sum_{i=1}^{k} n_i(\hat{\mu}_i - \mu_0)^2 \sim \chi^2(k).
$$

Now we write $\hat{\mu}_i - \mu_0 = \hat{\mu}_i - \hat{\mu}_0 + \hat{\mu}_0 - \mu_0$ and expand the square as

$$
\begin{aligned}
\sum_{i=1}^{k} n_i(\hat{\mu}_i - \mu_0)^2 &= \sum_{i=1}^{k} n_i[(\hat{\mu}_i - \hat{\mu}_0)^2 + 2(\hat{\mu}_i - \hat{\mu}_0)(\hat{\mu}_0 - \mu_0) + (\hat{\mu}_0 - \mu_0)^2 \\
&= \sum_{i=1}^{k} n_i(\hat{\mu}_i - \hat{\mu}_0)^2 + 2(\hat{\mu}_0 - \mu_0)\sum_{i=1}^{k} n_i(\hat{\mu}_i - \hat{\mu}_0) + n(\hat{\mu}_0 - \mu_0)^2 \\
&= \sum_{i=1}^{k} n_i(\hat{\mu}_i - \hat{\mu}_0)^2 + n(\hat{\mu}_0 - \mu_0)^2.
\end{aligned}
$$

Now, $\hat{\mu}_0$ is a random variable with mean μ_0 and variance σ^2/n, so

$$
\frac{\hat{\mu}_0 - \mu_0}{\sigma/\sqrt{n}} \sim \mathrm{N}(0,1) \quad \text{and} \quad \frac{n(\hat{\mu}_0 - \mu_0)^2}{\sigma^2} \sim \chi^2(1).
$$

Now we use the result that the sum of independent chi-squared random variables is itself a chi-squared random variable, and the degrees of freedom add. We have

$$
\frac{1}{\sigma^2}\sum_{i=1}^{k} n_i(\hat{\mu}_i - \mu_0)^2 = \frac{1}{\sigma^2}\sum_{i=1}^{k} n_i(\hat{\mu}_i - \hat{\mu}_0)^2 + \frac{n}{\sigma^2}(\hat{\mu}_0 - \mu_0)^2,
$$

and we have shown that the left-hand side is $\chi^2(k)$ and the second term on the right-hand side is $\chi^2(1)$. The two terms on the right are independent random variables, and we conclude that the first term on the right is distributed as $\chi^2(k-1)$. This is the numerator distribution for our ratio.

Now let's tackle the denominator! That's easier, because we know from the Chapter 36 material that for each i,

$$
\sum_{j=1}^{n_i}(Y_{ij} - \hat{\mu}_i)^2/\sigma^2 \sim \chi^2(n_i - 1),
$$

and, hence, by independence we have

$$
\sum_{i=1}^{k}\sum_{j=1}^{n_i}(Y_{ij} - \hat{\mu}_i)^2 \sim \chi^2(n-k),
$$

because $\sum_{i=1}^{k}(n_i - 1) = n - k$. This denominator has a $\chi^2(n-k)$ whether or not the null hypothesis is true. The numerator, however, has a $\chi^2(k-1)$ only if the null is true.

Finally, we can construct our F-statistic. Define

$$
F = \frac{\sum_{i=1}^{k} n_i(\hat{\mu}_i - \hat{\mu}_0)^2/(k-1)}{\sum_{i=1}^{k}\sum_{j=1}^{n_i}(Y_{ij} - \hat{\mu}_i)^2/(n-k)},
$$

then if H_0 is true, it has an $F(k-1, n-k)$ distribution. We know that large values of F support the alternative.

Our test proceeds as follows: Compute the null and alternative estimates of the sample means, and use these to compute the sums of squares in the F-statistic. Then the p-value is the area under the $F(k-1, n-k)$-density, to the right of the observed test statistic. (It's a little confusing that the test statistic and its distribution have the same name, but this is traditional.)

There is a standard language for the sums of squares in the numerator and in the denominator of the F-statistic. We say that

$$SSA = \sum_{i=1}^{k} n_i(\hat{\mu}_i - \hat{\mu}_0)^2$$

is the sum of squares **among** groups. It is a measure of variation among the k sample means. The sum of squares in the denominator is

$$SSW = \sum_{i=1}^{k}\sum_{j=1}^{n_i}(Y_{ij} - \hat{\mu}_i)^2$$

and is the sum of squares **within** groups. We add the variations within the samples.

We showed (in the factorization of the numerator above) that $SSA + SSW = SST$, where

$$SST = \sum_{i=1}^{k}\sum_{j=1}^{n_i}(Y_{ij} - \hat{\mu}_0)^2$$

is called the **total variation** in the sample. The SSA is often called the "explained" variation and the SSW the "unexplained" variation. The proportion of total variation that is explained by the groups is called the **coefficient of determination**; the nickname is the "R-squared" or

$$R^2 = \frac{SSA}{SST}.$$

The **ANOVA table** is often used to display the sums of squares and the steps to compute the F-statistic. In the table below the column labeled df contains the degrees of freedom of the chi-squared distribution associated with the sum of squares. The column labeled MS contains the **mean squares**; these are the sums of squares divided by the degrees of freedom. The ratio of the mean squares among (MSA) to the mean squares within (MSW) is the F-statistic:

Source	SS	df	MS	F_{obs}	p-value
Among	SSA	$k-1$	$SSA/(k-1)$	MSA/MSW	$P(F(k-1, n-k) > F_{obs})$
Within	SSW	$n-k$	$SSW/(n-k)$		
Total	SST	$n-1$			

For the cholesterol example data, the ANOVA table is shown below. Only two of the sums of squares are needed; the rest of the table can be filled out starting with the two numbers. The small p-value indicates that there is evidence against the hypothesis that all the means are equal. For ANOVA tables generated by statistical software packages or functions in R, the "among" row is usually labeled with the variable or group name;

in this example the groups are treatments. The "within" row is often label "residual" or sometimes "error."

Source	SS	df	MS	F_{obs}	p-value
Treatment	33.31	2	16.65	5.20	.012
Residual	86.41	27	3.20		
Total	119.72	30			

We make a lot of assumptions when we use an F-statistic. The assumptions about the independence of the observations can be assessed given the context of the problem. The normality assumption can be checked with a probability plot. We assume that each $Y_{ij} - \mu_i$ is a mean-zero normal with variance σ^2. If we do a probability plot of the n "residuals" $y_{ij} - \bar{y}_i$, the n points will lie near a straight line if the normality assumption holds. The plot of the sorted residuals for the cholesterol example, against the normal quantiles, is shown below, and does not contradict the assumption.

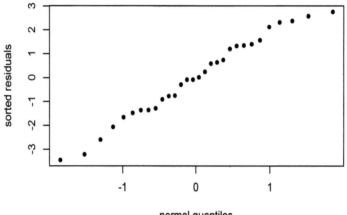

For the two-independent-samples t-test, we could check the assumption of equal variances using an F-test (see Chapter 37). The **Bartlett** test or the **Levene** test can be used for checking the equal-variances assumption for three or more groups. The Bartlett test uses a test statistic whose distribution has been shown to be *approximately* that of a chi-squared random variable, while the Levene test uses a test statistic whose distribution has been shown to be *approximately* that of an F random variable. For each, the null hypothesis is that the variances are equal across the groups. The p-value summarizes the evidence for the alternative that at least one group has a different population variance.

To demonstrate how to perform a one-way ANOVA in R, we use the `ChickWeight` data set. This has information about weights of chicks that have been put on one of four diets. There are measurements for several points in time for each chick, but we look at only the weights at week 21:

```
y=ChickWeight$weight[ChickWeight$Time==21]
diet=ChickWeight$Diet[ChickWeight$Time==21]
```

The weights at 21 weeks are shown for each diet in the plot below. There are $n_1 = 16$ chicks on Diet 1, $n_2 = 10$ chicks on Diet 2, $n_3 = 10$, and $n_4 = 9$, for a total of $n = 45$ chicks.

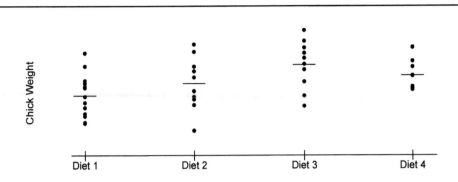

To get an ANOVA table, we use the command

```
anova(lm(y~diet))
```

which produces the following table:

```
Analysis of Variance Table

Response: y
          Df Sum Sq Mean Sq F value   Pr(>F)
diet       3  57164 19054.7  4.6547 0.006858 **
Residuals 41 167839  4093.6
```

We see that $SSA = 57{,}164$ and $SSW = 167{,}839$; the degrees of freedom for the among sum of squares is one less than the number of diets. The sums of squares are used to construct the mean squares, which are used to construct the F-statistic. The p-value is the area to the right of the observed F-statistic value under an $F(3, 41)$-density.

The residuals are in the vector `resid(lm(y diet))` and a probability plot can be constructed to check the normality assumption as follows:

```
plot(qnorm(1:45/46),sort(resid(lm(y~diet))),xlab="Normal quantiles",
                ylab="sorted residuals")
```

The points in the normal probability plot look close to a straight line, so we conclude that the normality assumption is not violated.

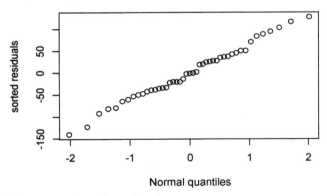

To check the assumption of equal variances:

```
bartlett.test(y~diet)
```

which produces the following output:

```
Bartlett test of homogeneity of variances

data:  y by diet
Bartlett's K-squared = 3.0524, df = 3, p-value = 0.3836
```

The large p-value tells us that we do not have evidence that the equal-variances assumption is violated. If any of the assumptions are violated, the model needs to be changed or the data transformed until the assumptions are (or seem to be) valid. These procedures are beyond the scope of this book.

The phrase "one-way" means that there is only one predictor variable. More complicated ANOVA models consist of more than one predictor; the development of these can be found in linear models textbooks.

Chapter Highlights

1. Because the one-way ANOVA is a generalization of the two-independent-samples t-test to more than two samples, the assumptions for one-way ANOVA are very similar.

   (a) The random variables Y_{ij} are normally distributed with mean μ_i and variance σ^2.

   (b) The Y_{ij} are independent random variables.

2. The null hypothesis is

   $$H_0 : \text{all the means are the same, i.e., } \mu_1 = \mu_2 = \cdots = \mu_k,$$

   and the alternative is

   $$H_a : \text{at least one mean is different from another.}$$

3. We define the sum of squares among groups as

   $$SSA = \sum_{i=1}^{k} n_i(\hat{\mu}_i - \hat{\mu}_0)^2$$

   and the sum of squares **within** groups as

   $$SSW = \sum_{i=1}^{k}\sum_{j=1}^{n_i}(Y_{ij} - \hat{\mu}_i)^2,$$

   while the total sum of squares is

   $$SST = \sum_{i=1}^{k}\sum_{j=1}^{n_i}(Y_{ij} - \hat{\mu}_0)^2.$$

   Under the null hypothesis,

   $$F = \frac{SSA/(k-1)}{SSW/(n-k)} \sim F(k-1, n-k),$$

   and large values support the alternative.

4. $SST = SSA + SSW.$

Exercises

60.1 A pharmaceutical company wants to compare the efficacy of cholesterol-reducing drugs, which are Drug A, Drug B, Drug C, and a placebo. They conducted randomized clinical trials with six subjects in each treatment (24 subjects total). Here is a partial ANOVA table for the results:

Source	SS	df	MS	F-stat	p-value bounds
Treatment (among)	127.0				
Residual (within)					
Total	293.5				

(a) Finish filling in the ANOVA table, given two of the sums of squares.

(b) State the decision at $\alpha = 0.05$ and the conclusion in the context of the problem.

(c) What is the MLE for the common population variance σ^2?

60.2 A psychiatrist explores the impact of background music on memory ability. She has 30 student volunteers who will take a memory test; she randomly assigns them to three groups of 10 students each. Group 1 will take the test in a quiet room, Group 2 in a room with classical music playing in the background, and Group 3 in a room with jazz music in the background. Here are the summary data:

	Quiet	Classical	Jazz
Sample mean	89.5	89.7	79.4
Sample std dev	8.91	10.25	7.06

Let μ_1 be the true average (population) score for students in quiet rooms, let μ_2 be the true average score for students listening to classical music, and let μ_3 be the true average score for students listening to jazz. Assume that the scores on the memory test are approximately normally distributed for each population, with equal variances.

(a) Given SST=2802.8, fill in an ANOVA table.

(b) Test

$$H_0: \mu_1 = \mu_2 = \mu_3 \quad \text{versus} \quad H_a: \text{at least one mean different}$$

at $\alpha = 0.05$. State the decision and the conclusion in the context of the problem.

(c) Give the R^2 and interpret in the context of the problem.

(d) What is the MLE for the common population variance σ^2?

60.3 An experiment was conducted to compare the starch content of tomato plants grown in sandy soil supplemented by one of four different nutrient packages, A, B, C, and D.

Nutrient package			
A	B	C	D
45	39	60	56
47	39	42	61
67	43	55	60
61	46	44	47
56	59	30	69
72	27	18	42

    (a) Enter the data into R and get the ANOVA table.

    (b) Report the p-value and interpret in the context of the problem.

    (c) What is the R^2?

60.4 Twenty-four expert typists are hired by a computer company to compare the ease of use for three keyboard designs. Eight are randomly assigned to each design, and they will each type the same document on the keyboard. The response variable is the time to type the document; smaller times will indicate better designs. The collected data are shown below.

Design	Typing times							
A	364	366	394	386	379	398	371	370
B	355	359	374	342	378	355	376	358
C	360	345	374	390	386	373	393	366

    (a) Enter the data into R to obtain an ANOVA table.

    (b) State the decision at $\alpha = 0.05$ and the conclusion in the context of the problem.

    (c) What is the MLE for the common population variance?

60.5 The R data set `chickwts` has this description: "An experiment was conducted to measure and compare the effectiveness of various feed supplements on the growth rate of chickens." Determine if there are significant differences in the chick weights by type of feed supplements; report a p-value. Check the assumptions for the ANOVA model.

60.6 The R dataset `PlantGrowth` has this description: "Results from an experiment to compare yields (as measured by dried weight of plants) obtained under a control and two different treatment conditions." Determine if there are significant differences in the dried plant weights by type of treatment; report a p-value. Check the assumptions for the ANOVA model.

60.7 Determine the expected value of the MLE for the common population variance σ^2. From this, determine an unbiased estimator for σ^2. Notice that this is one of the entries of the ANOVA table!

60.8 Suppose we actually know the common variance σ^2. Develop a likelihood ratio test for the same hypotheses.

Appendix A

Answers to Exercises

1.1 The pink area represents $A \cap B$, and the blue area represents $A^c \cap B$. The colored areas together are the union, which simplifies to B.

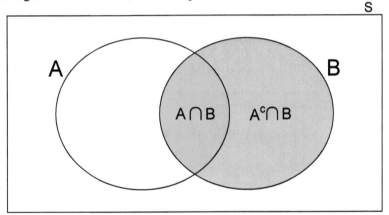

1.3 We'll show $(A \cup B)^c = A^c \cap B^c$; the demonstration of $(A \cap B)^c = A^c \cup B^c$ is similar. For the plot on the left, the horizontal lines shade A^c, and the vertical lines shade B^c. Then $A^c \cap B^c$ is shaded by the cross-hatches; this area coincides with that shown in the plot on the right, where $(A \cup B)^c$ is shaded. The set $A^c \cup B^c$ is the areas in the plot on the right that have *any* shading in either direction. This is the complement of $A \cap B$.

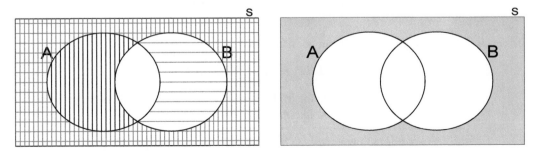

1.5 To prove rule 4, $P(\emptyset) = 0$, we define $A_1 = S$, and $A_i = \emptyset$ for $i = 2, 3, 4, \ldots$. Then by axiom 3, we must have

$$\sum_{i=1}^{\infty} P(A_i) = 1 + \sum_{i=2}^{\infty} P(A_i) \le 1$$

by axiom 2. This implies that $\sum_{i=2}^{\infty} P(A_i) = 0$. Because axiom 1 forces all the terms in the sum to be nonnegative, we have $P(\emptyset) = 0$.

To prove rule 5, we can define $A_i = \emptyset$ for $i = n+1, n+2, \ldots$. Then

$$\sum_{i=1}^{\infty} P(A_i) = \sum_{i=1}^{n} P(A_i) + \sum_{i=n+1}^{\infty} P(A_i) = \sum_{i=1}^{n} P(A_i).$$

Also,

$$\cup_{i=1}^{\infty} A_i = \left[\cup_{i=1}^{n} A_i\right] \cup \left[\cup_{i=n+1}^{\infty} A_i\right] = \cup_{i=1}^{n} A_i.$$

Putting these two parts together gives the result.

For rule 6, it is sufficient to prove $P(A) + P(A^c) = 1$. Define $A_1 = A$ and $A_2 = A^c$. Then A_1 and A_2 are disjoint, so rule 5 holds, and $P(A) + P(A^c) = P(A \cup A^c) = P(S) = 1$ by axiom 1.

For rule 7, define $A_1 = A$ and $A_2 = B \cap A^c$. Then A_1 and A_2 are disjoint with $A_1 \cup A_2 = B$. By rule 5, $P(B) = P(A_1 \cup A_2) = P(A_1) + P(A_2)$, and by axiom 1, $P(A_2) \ge 0$. Therefore $P(A) \le P(B)$.

For rule 8, define $A_1 = A \cap B^c$, $A_2 = A \cap B$, and $A_3 = A^c \cap B$. Then A_1, A_2, and A_3 are disjoint sets and $A_1 \cup A_2 \cup A_3 = A \cup B$. By rule 5,

$$\begin{aligned}
P(A_1 \cup A_2 \cup A_3) &= P(A_1) + P(A_2) + P(A_3) \\
&= [P(A_1) + P(A_2)] + [P(A_2) + P(A_3)] - P(A_2) \\
&= [P(A_1 \cup A_2)] + [P(A_2 \cup A_3)] - P(A_2).
\end{aligned}$$

Now we recognize that $A_1 \cup A_2 \cup A_3 = A \cup B$, $A_1 \cup A_2 = A$ and $A_2 \cup A_3 = B$.

1.7 Let's use the notation RB to mean you get red and your friend gets blue. Then the sample space can be written as

$$\left\{ \begin{array}{ccc} RR & RB & RG \\ BR & BB & BG \\ GR & GB & GG \end{array} \right\}.$$

It is reasonable to assume that each outcome is equally likely, so we assign probability 1/9 to each. The event "same color" has three outcomes, so the probability is 3/9 or 1/3.

1.9 The sample space can be written as

$$\{HHH, HHT, HTH, THH, THT, TTH, HTT, TTT\},$$

where the first letter in the triplet is your toss, and the two others are your friends' tosses. If the coins are fair, these eight outcomes are equally likely.

(a) For the outcomes HTT and THH you have to pay for everyone; this has probability 1/4.

(b) For outcome TTT and HHH the bill is shared; this has probability 1/4.

(c) For all other outcomes you get a free lunch; this has probability 1/2.

1.11 If A is the event that the selected household has at least one dog, and B is the event that the selected household has at least one cat, then we are looking for $P(A \cap B)$. We can compute $P(A \cup B) = .8$, and so $P(A \cap B) = P(A) + P(B) - P(A \cup B) = .2$.

1.13 (a)

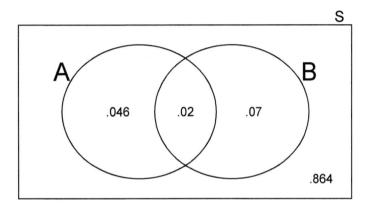

(b) The proportion of U.S. residents who reside in New York and are immigrants.

(c) The proportion of U.S. residents who either reside in New York or have immigrated.

(d) .136.

1.15 Let A be the event that a randomly selected elderly person has diabetes, and let B be the event that a randomly selected elderly person is living below the poverty level.

(a) $P(A \cap B) = P(A) + P(B) - P(A \cup B) = .05$.

(b) $P(A^c \cap B^c) = P((A \cup B)^c) = .65$.

1.17 Using the Venn diagram below we see that the proportion of bees with both phenotypes is .2.

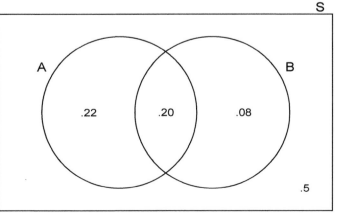

1.19   (a) The outcome is in $A \cap B$ if the fruit fly has *both* the Cy and e genetic
           markers.

           We can organize the information in a Venn diagram to make answering the
           next questions easy:

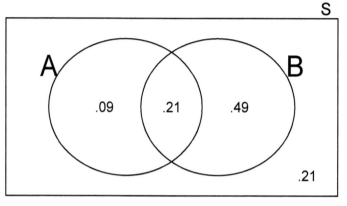

       (b) The outcome is in $B \cap A^c$ if the fruit fly has the e genetic marker but does
           not have the Cy genetic marker. From the Venn diagram, we see that $P(B \cap A^c) = .49$.

       (c) From the Venn diagram, we see that $P(B \cup A^c) = .91$.

       (d) No, the intersection of A and B is not zero. It is possible for fruit flies to
           have both genetic markers, which would result in a black body and black,
           slightly curled wings.

2.1 The following code gives a proportion of .0155.

```
nloop=1000000
numall=0
for(iloop in 1:nloop){
   roll=sample(1:6,6,replace=TRUE)
   sroll=sort(roll)
   if(all(sroll==c(1,2,3,4,5,6))){numall=numall+1}
}
numall/nloop
```

2.3 The probability of a large straight is about .031. Some code:

```
nlstr=0
nrep=1000000
for(irep in 1:nrep){
   dice=sample(1:6,5,replace=TRUE)
   sdice=sort(dice)
   if(all(sdice==c(1,2,3,4,5))|all(sdice==c(2,3,4,5,6))){
      nlstr=nlstr+1
   }
}
nlstr/nrep
```

2.5 The following code gives a proportion of about 23%, so that if one of the treatments seems to do twice as well as another, this is really not strong evidence that it's better!

```
nloop=100000
pa=1:nloop
pb=1:nloop
for(iloop in 1:nloop){
    trt=sample(1:3,40,replace=TRUE)
    cure=sample(1:2,40,replace=TRUE,prob=c(.2,.8))
    pa[iloop]=sum(trt==1&cure==1)/sum(trt==1)
    pb[iloop]=sum(trt==2&cure==1)/sum(trt==2)
}
sum(pa>=2*pb)/nloop
```

2.7 The following code gives a probability of 3.0%.

```
nloop=1000000
sumdiff=0
for(iloop in 1:nloop){
    x=sample(1:4,4,replace=TRUE,prob=c(.48,.28,.19,.05))
    if(all(sort(x)==c(1,2,3,4))){sumdiff=sumdiff+1}
}
sumdiff/nloop
```

2.9 (a) The following code provides a probability of about .013.

```
nloop=100000
nexact=0
for(iloop in 1:nloop){
    kids=sample(1:90,50,replace=TRUE)
    skids=sort(kids)
    if(all(skids[1:4]==c(1,2,3,4))){nexact=nexact+1}
}
nexact/nloop
```

(b) The following code provides a probability of about .030.

```
nloop=100000
npr=0
for(iloop in 1:nloop){
    k4=1:4<0
    kids=sample(1:90,50,replace=TRUE)
    for(i in 1:4){
        if(sum(kids==i)>0){k4[i]=TRUE}
    }
    if(all(k4)){npr=npr+1}
}
npr/nloop
```

2.11 In the following piece of R code, we number the executives 1–26, with 1–12 being those from Tampa. The variable cnt0 gets incremented if the random sample of three from the numbers 1–26 has no numbers less than 13. We find that the proportion of samples with no numbers less than 13 (no executives from Tampa) is about .14.

```
cnt0=0
nloop=100000
for(iloop in 1:nloop){
    retreat=sample(1:26,3)
    if(sum(retreat<13)==0){cnt0=cnt0+1}
}
cnt0/nloop
```

2.13 The following code gives a probability of about .75 that there is no pair among Sam's four socks.

```
nloop=100000
nno=0
socks=trunc(0:23/2)+1
for(iloop in 1:nloop){
    sam=sample(socks,4)
    if(length(unique(sam))==4){nno=nno+1}
}
nno/nloop
```

3.1 The order of the numbers *is* important, so there are $26^2 \times 10^3 = 676,000$ different IDs.

3.3 The order of the numbers *is* important, so there are $9 \times 10^6 = 9$ million different numbers.

3.5 The number of ways to choose three executives from all executives in the corporation is $\binom{26}{3}$, and the number of ways to choose three executives *not* from Tampa is $\binom{14}{3}$. Therefore, the probability is the ratio: .14.

3.7 If the order of the scoops is not important, the number of sundaes is

$$\binom{12}{3} = 220.$$

If the order of the scoops *is* important, the number of sundaes is

$$12 \times 11 \times 10 = 1320.$$

3.9 The number of possible ordered draws of three balls from twenty is $20*19*18 = 6840$, so the probability of winning is $1/6840 \approx .000146$. We can simulate draws using the following code.

```
nloop=1000000
cnt=0
for(iloop in 1:nloop){
draw=sample(1:20,3)
if(all(draw==c(1,2,3))){cnt=cnt+1}
}
cnt/nloop
```

3.11 Choose the value for three of a kind, choose the value for two of a kind, choose the three cards, choose the two cards:

$$\frac{\binom{13}{1}\binom{12}{1}\binom{4}{3}\binom{4}{2}}{\binom{52}{5}} \approx .00144.$$

3.13 (a) If we line up the candies, then the job of splitting the candies into three piles can be looked at as choosing where to put two dividers. There are 9 places to put the dividers, so this number is $\binom{9}{2} = 36$.

(b) Now we need to place $m - 1$ dividers, and there are $n - 1$ places to put them. So the answer is $\binom{n-1}{m-1}$.

3.15 There are 12! ways to arrange blocks that are all different. However, in this case we can rearrange the blues without changing the color arrangement, so we have to divide by 6!, the number of arrangements of the blue blocks. We also have to divide by 4!, the number of ways to rearrange the red blocks. (We could also divide by 1! and 1! for the yellow and green blocks, if we wanted to be really thorough.) So the answer is

$$\frac{12!}{6!4!} = 27{,}720.$$

3.17 If you pick up the 5 and roll one die, the probability of getting a 1 or a 4 is $1/3$. Alternatively, if you pick up the two 1's and 5 and roll three dice, the only way *not* to get at least three of a kind is if none of the three is a 4. The probability of this is $(5/6)^3 = .579$, so the probability of at least one four is .421, which is higher than $1/3$.

3.19 Using the formula,

$$(x + 2y)^5 = x^5 + \binom{5}{4}x^4(2y) + \binom{5}{3}x^3(2y)^2 + \binom{5}{2}x^2(2y)^3$$

$$+ \binom{5}{1}x(2y)^4 + (2y)^5$$

$$= x^5 + 10x^4y + 40x^3y^2 + 80x^2y^3 + 80xy^4 + 32y^5.$$

4.1 Let A be the event that a randomly selected student from State University has been to Europe, and let B be the event that a randomly selected student from State University has been to Africa. If we fill out the Venn diagram (below), we see that $P(A \cap B) = .04$, so using the definition of conditional probability, we see that eighty percent of students who have been to South America have also been to Europe.

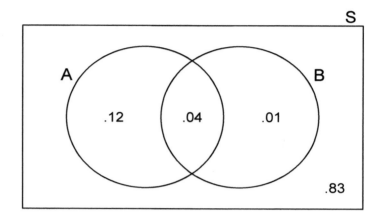

4.3 In Exercise 1.11, we defined A to be the event that the selected household has at least one dog, and B is the event that the selected household has at least one cat; then we computed $P(A \cap B) = .2$. Therefore, $P(B|A) = .2/.4 = 1/2$.

4.5 If A is the event that a randomly chosen employee is a woman, and B is the event that a randomly chosen employee is promoted, we are given that $P(A) = .7$ and $P(B) = .2$. Further, we are given that $P((A \cup B)^c) = P(A^c \cap B^c) = .24$; this allows us to fill up the Venn diagram as below. This gives $P(A \cap B) = .14$, which is $P(A)(B)$, so the events are independent.

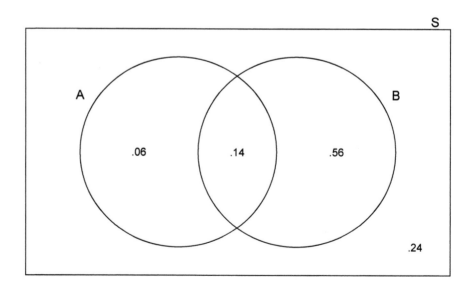

4.7 Filling in the Venn diagram as shown, we find the desired probability is .73.

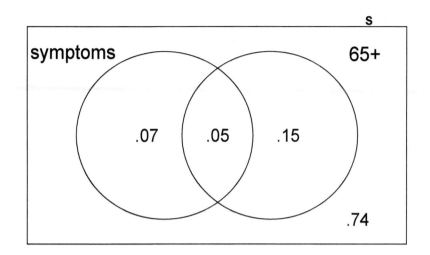

4.9 Filling in the Venn diagram as shown, we find the desired probability is $.24/(.24 + .31) = .436$.

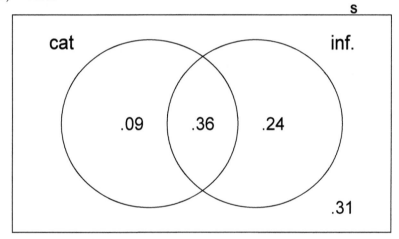

4.11 Let A be the event that the three numbers are all greater than 1/2, and let B be the event that the sum is less than 2. Then we can simulate $P(A)$, $P(B)$, and $P(A \cap B)$ with the following code. Then $P(A|B) = P(A \cap B)/P(B) \approx .025$.

```
nloop=1000000;na=0;nb=0;nab=0
for(iloop in 1:nloop){
x=runif(3)
eventA=all(x>.5)
if(eventA){na=na+1}
eventB=sum(x)<2
if(eventB){nb=nb+1}
if(eventA&eventB){nab=nab+1}
}
na/nloop
nb/nloop
nab/nloop
nab/nb
```

4.13 The sample space is $S = \{GB, BG, BBG, GGB, GGG, BBB\}$. They are *not* equally likely. The probabilities (in order) are

$$1/4, 1/4, 1/8, 1/8, 1/8, 1/8.$$

4.15 We can make a tree diagram, or reason as follows: If the first toss goes to Charley, then he wins, and this happens 1/2 of the time. If the first toss goes to Sam, then he will win unless both of the next two go to Charley. The sequence Sam, Charley, Charley of winners happens 1/8 of the time, so all together Charley Whiney wins 5/8 of the time.

4.17 Let A be the event that the randomly selected person has shingles, and let B be the event that the person had been vaccinated. (a) We easily get $P(A \cup B) = .75$, which leads to $P(A \cap B) = .05$, and we can fill out the Venn diagram below. Then we can compute $P(A|B) = P(A \cap B)/P(B) = .05/.20 = .25$.

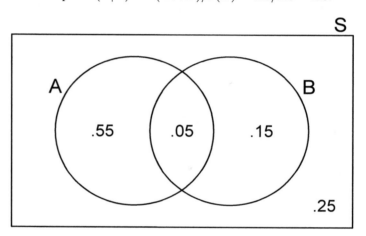

(b) No, $P(A|B) \neq P(A)$.

4.19 Using the definition of conditional probability twice, we have

$$P(A|B) = \frac{P(A \cap B)}{P(B)}$$
$$= \frac{P(B|A)P(A)}{P(B)}$$
$$= \frac{P(B)P(A)}{P(B)}$$
$$= P(A),$$

where the assumption $P(B|A) = P(B)$ is used in the third step.

4.21 For this exercise, we are given proportions rather than conditional probabilities, so we make a Venn diagram. Let A be the event that a randomly selected person over 65 has condition A, and similarly for B. Then because the events are independent, we know that 12% have both conditions, and hence 42% have neither.

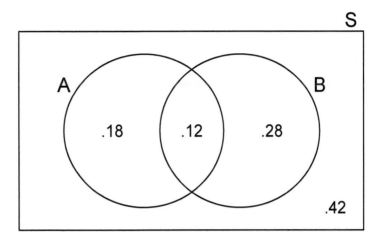

4.23 We make the tree diagram with the same reasoning as for Exercise 4.22, but this time we keep going if we do not get a pair.

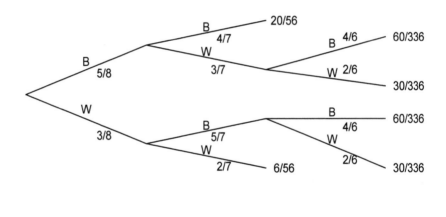

| Draw 1 | Draw 2 | Draw 3 |

The probability that the pair is blue is $20/56 + 60/336 + 60/336 = .7143$.

4.25 Let A be the event that a randomly selected vase was made by Susan, and let B be the event that a randomly selected vase has no flaws.

(a) We want to find $P(A|B)$, and we are given $P(B|A)$. Therefore we use Bayes' theorem to get

$$P(A|B) = \frac{P(B|A)P(A)}{P(B|A)P(A) + P(B|A^c)P(A^c)} = \frac{(.7)(.3)}{(.7)(.3) + (.3)(.2)} = \frac{.21}{.39}$$

$$\approx .538.$$

(b) Let C be the event that a randomly selected vase has two flaws. Then using the law of total probability, we get

$$P(B) = P(B|A)P(A) + P(B|A^c)P(A^c) = (.1)(.3) + (.2)(.7) = .17.$$

4.27 (a) Using the tree diagram below, we can calculate the probability of an explosion to be .0018 + .0499 = .0517.

(b) Then the probability that the circuit breaker was flawed, given that there was an explosion, is .0018/.0517 = .0348.

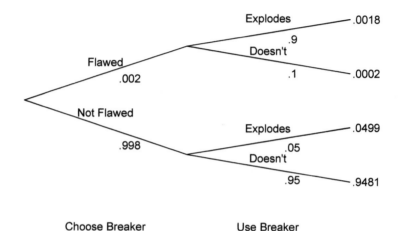

Choose Breaker   Use Breaker

4.29 Let's use the law of total probability. We select a person at random from Smalltown. Let F be the event the person gets the flu, and let V be the event that the person gets the vaccination. Then

$$P(F) = P(F|V)P(V) + P(F|V^c)P(V^c) = .02 \times .8 + .05 \times .2 = .026.$$

If instead we decide to make a tree diagram we see again that 2.6% of Smalltown residents will get the flu:

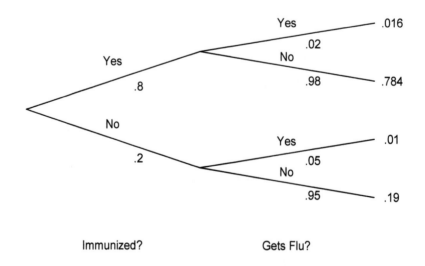

Immunized?   Gets Flu?

4.31 Filling out the tree diagram, we find that Machine A produces 5/6 of the defective boards. That is,

$$P(\text{Machine A}|\text{defective}) = \frac{.05}{.05 + .01} = \frac{5}{6}.$$

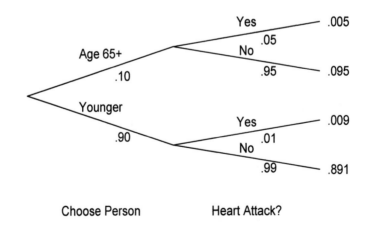

Choose Machine   Choose Board

4.33 Adding the two "yes" probabilities in the tree diagram below, we find that 1.4% of the population will have heart attacks.

Choose Person   Heart Attack?

4.35 After filling out the tree diagram below, it is clear that the proportion of jailed people that are Bacmenian is $.0045/(.0045 + .0035) = 56.25\%$.

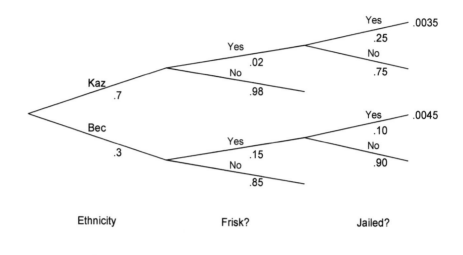

5.1 (a) The probability that both are women is $(.7)^2 = .49$, and the probability that both are men is $(.3)^2 = .09$. Therefore, the probability of exactly one woman is $1 - .49 - .09 = .42$. That is $P(Y = 0) = .09$, $P(Y = 1) = .42$, and $P(Y = 2) = .49$.

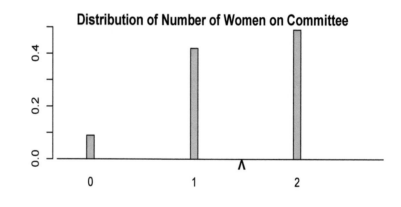

(b) $E(Y) = (0)(.09) + (1)(.42) + 2(.49) = 1.4$.

5.3 (a)

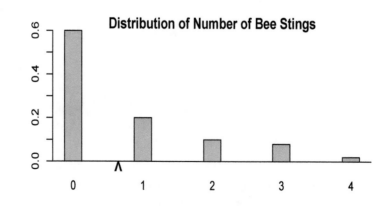

(b) $1 - .6 = .4$.

(c) $(500)(.08) = 40$.

(d) $E(Y) = (0)(.6) + (1)(.2) + (2)(.1) + (3)(.08) + (4)(.02) = .72$.

(e) This code can be used to check your answer to (d):

```
x=sample(0:4,1000000,prob=c(.6,.2,.1,.08,.02),replace=TRUE)
mean(x)
```

(f) $E(Y^2) = (0)(.6) + (1)(.2) + (4)(.1) + (9)(.08) + (16)(.02) = 1.64$, so $V(Y) = 1.64 - .72^2 = 1.1216$, and the standard deviation is about 1.059.

(g) This code can be used to check your answer to (f):

```
x=sample(0:4,1000000,prob=c(.6,.2,.1,.08,.02),replace=TRUE)
sqrt(var(x))
```

(The function **sd** also gives the standard deviation.)

5.5 (a)

y = # candies	1	2	3	4
Proportion $f(y)$	.2	.16	.128	.512

(b) 2.952.

5.7 By symmetry, we know $E(Y) = 0$, so

$$V(Y) = E(Y^2) = (0)^2(1/2) + (-a)^2(1/4) + (a)^2(1/4) = a^2/2.$$

5.9 Here

$$E(X) = (0)(1 - p/2) + (1)(p/3) + (2)(p/6) = 2p/3, \quad \text{and}$$

$$E(X^2) = (0^2)(1 - p/2) + (1^2)(p/3) + (2^2)(p/6) = p,$$

so

$$V(X) = p - (2p/3)^2 = p(1 - 4p/9).$$

5.11 To get the probability mass function we can make a tree diagram as shown (later we will be able to use ideas about a *hypergeometric* distribution).

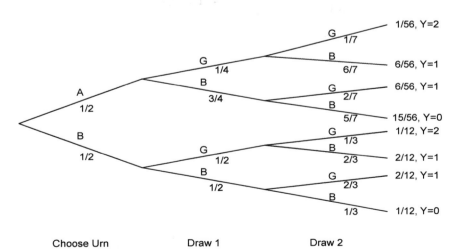

Gathering values of Y, we compile the following table:

y = #gold balls drawn	0	1	2
$P(Y = y)$	.3512	.5476	.1012

The expected amount of money won is $W = 1000Y$, so $E(W) = 1000E(Y) = \$750$.

5.13 (a) The probability mass function is

y = #phenotypes	0	1	2
$P(Y = y)$	.5	.3	.2

(b) $E(Y) = .7$.

5.15 Starting with the formula, we have

$$E(Y) = \sum_{y=0}^{\infty} yP(Y = y)$$

$$= \sum_{y=1}^{\infty} y\left(\frac{e^{-1/2}}{y!2^y}\right) \qquad \text{(start sum at 1)}$$

$$= \sum_{y=1}^{\infty} \left(\frac{e^{-1/2}}{(y-1)!2^y}\right) \qquad \text{(cancel } y\text{)}$$

$$= \sum_{z=0}^{\infty} \left(\frac{e^{-1/2}}{z!2^{z+1}}\right) \qquad \text{(let } z = y - 1\text{)}$$

$$= \frac{1}{2}\sum_{z=0}^{\infty} \left(\frac{e^{-1/2}}{z!2^z}\right) \qquad \text{(factor out } 1/2\text{)}$$

$$= \frac{1}{2},$$

using the fact that the probabilities must sum to one.

5.17 Once we set up a tree diagram, we can readily calculate the probabilities of getting gold and brown:

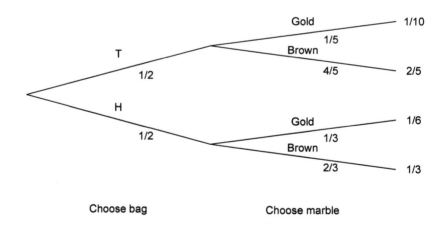

Then the probability of winning \$100 is $1/10 + 1/6$ and the probability of losing \$20 is $2/5 + 1/3$. The expected winnings are

$$(1/10 + 1/6) \times \$100 - (2/5 + 1/3) \times \$20 = \$12.$$

6.1 (a) $H_0 : \mu_G = \mu_B$ versus $H_a : \mu_G > \mu_B$.

  (b) If we repeated the same study many times, and the null hypothesis is true, then about 7% of the studies would provide as much or more evidence for the alternative hypothesis.

6.3 (a) Let p_0 be the proportion of tumors found using the old machine, and let p_1 be the proportion of tumors found using the new machine. Then the hypotheses are

$$H_0 : p_1 = p_0 \quad \text{versus} \quad H_a : p_1 > p_0.$$

  (b) A Type I Error occurs when we decide that the new machine is better when it really is not better. A Type II Error occurs when we decide that the new machine is not better, when really it is better.

  (c) The consequence of a Type I Error might be to buy new machines that will not perform better. The consequence of a Type II Error would be to miss a proportion of tumors that might have been caught if the new machines had been purchased.

  (d) If the test size is high, it is "easier" to reject H_0, making a Type I Error more likely, which might result in spending a lot of money on new machines that are not any better than the old machines. However, if the test size is small, then a Type II Error would be more likely, and in this case the hospital might fail to buy better equipment.

  (e) The company that sells the machines would probably prefer to have a large test size, because the probability of concluding that the new machines are better would be greater.

6.5 (a) The test size is the probability of all heads or all tails when the coin is fair: $\alpha = 1/2^8 + 1/2^8 = .0078$.

  (b) The power is the probability of all heads or all tails when $p = .8$: $(.8)^8 + (.2)^8 = .168$.

6.7 The p-value is the probability of "our data" (no wins) or "more extreme" (there is no outcome that would support the alternative more) when H_0 is true (proportion of winning tickets is .05) is $(1 - .05)^{50} = .0769$.

6.9 (a) The probability of observing at least two bee stings for both children (i.e., rejecting H_0) is $.2^2 = .04$ if H_0 is true. Therefore $\alpha = .04$.

  (b) For the alternative distribution, the probability of observing at least two bee stings for both children (i.e., rejecting H_0) is $.4^2 = .16$, which is the power of this test.

7.1 For part (a), we want to simulate from the null distribution.

```
nloop=1000000
n=30
smean=1:nloop
for(iloop in 1:nloop){
    u=runif(n)
    p0=sum(u<..6)/n
    p1=sum(u>=.6&u<.8)/n
    p2=sum(u>=.8&u<.9)/n
    p3=sum(u>=.9&u<.98)/n
    p4=sum(u>=.98)/n
    smean[iloop]=p1+2*p2+3*p3+4*p4
}
sum(smean>1)/nloop
hist(smean,breaks=100)
```

When we make the histogram of the distribution of sample means, we can also find the proportion of sample means that are greater than 1.

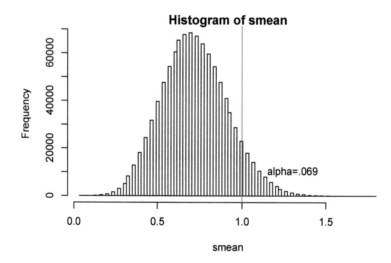

(a) The test size for rejecting when the sample mean is greater than one is .069.

(b) If we want a test size of .01, we find the value of the sample mean, for which 1% of the distribution is above this value. We find that if our decision rule is to reject H_0 when the sample mean is greater than 1.2, we get a test size of about .01.

(c) To find the power of the new test with $\alpha = .01$, we simulate from the alternative distribution (code below). We find that the power, i.e., the probability that our sample mean is greater than 1.2, is about .815.

```
nloop=1000000
n=30
smean=1:nloop
for(iloop in 1:nloop){
    u=runif(n)
    p0=sum(u<..2)/n
    p1=sum(u>=.2&u<.6)/n
```

```
        p2=sum(u>=.6&u<.86)/n
        p3=sum(u>=.86&u<.96)/n
        p4=sum(u>=.96)/n
        smean[iloop]=p1+2*p2+3*p3+4*p4
    }
    sum(smean>1.2)/nloop
```

7.3 (a) To simulate from the null distribution:

```
nloop=1000000
sampmean=1:nloop
for(iloop in 1:nloop){
    x=sample(0:4,20,replace=TRUE,prob=c(.3,.3,.2,.1,.1))
    sampmean[iloop]= mean(x)
}
hist(sampmean,br=200,xaxt="n",freq=FALSE,col='beige')
sum(sampmean>=2)/nloop
```

We find the test size is about .026.

(b) The p-value is returned by the command sum(sampmean>=2)/nloop; it is about .073.

(c) Changing the probability vector in the simulations to prob=c(.15,.3,.25, .2,.1), we find the power is about .258.

7.5 (a) Given that H_0 is true, we can simulate values of T_1 using the following code.

```
nloop=1000000
teststat=1:nloop
for(iloop in 1:nloop){
    x=sample(1:4,100,replace=TRUE,prob=c(.48,.28,.19,.05))
    xo=sum(x==1)
    xa=sum(x==2)
    xb=sum(x==3)
    xab=sum(x==4)
    teststat[iloop]=abs(xo-48)+abs(xa-28)+abs(xb-19)+abs(xab-5)
}
hist(teststat)
```

If we reject H_0 when $T_1 \geq 23$, then the test size is $\alpha = .0491$. Because the test statistic has integer values, we can't formulate a test whose size is exactly $\alpha = .05$.

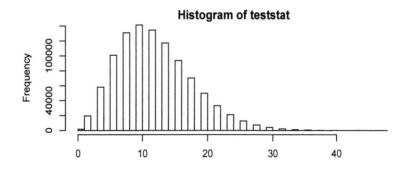

Histogram of teststat

(b) We get that the observed test statistic value is 22 and we accept H_0.

7.7 We can compute the power for the tests, under the given alternative, with the following code. The power for T_1 in this case is .778, while the power for T_2 is .792. We see that T_2 is preferred, although the powers are similar.

```
nloop=1000000
nrej1=0
nrej2=0
for(iloop in 1:nloop){
    x=sample(1:4,100,replace=TRUE,prob=c(.40,.40,.12,.08))
    xo=sum(x==1)
    xa=sum(x==2)
    xb=sum(x==3)
    xab=sum(x==4)
    teststat1=abs(xo-48)+abs(xa-28)+abs(xb-19)+abs(xab-5)
    teststat2=(xo-48)^2/48+(xa-28)^2/28+(xb-19)^2/19+(xab-5)^2/5
    if(teststat1>22){nrej1=nrej1+1}
    if(teststat2>7.817){nrej2=nrej2+1}
}
nrej1/nloop
nrej2/nloop
```

8.1 (a) The following bit of R code and outputs demonstrates how to verify the mean and variance of a binomial random variable.

```
> y=rbinom(100000,20,.2)
> mean(y)
[1] 4.00655
> var(y)
[1] 3.172979
```

(b) We can approximate $E(Y^3)$ with

```
> mean(y^3)
[1] 104.3726
```

8.3 The p-value is the probability that she gets *at least* five out of eight correct guesses. If the null hypothesis is true, this is $P(Y \geq 5)$, where $Y \sim \text{Binom}(8, .25)$. Then $p = .0273$.

8.5 The probability of at least nine of ten shots is

$$(.8)^{10} + 10(.8)^9(.2) = .376.$$

We have to assume that the probability is the same for each shot and the trials are independent. This would not be true if, for example, she gets nervous after missing a shot and is more likely to miss the next shot.

8.7 There are six spins, and it makes sense to assume the outcomes are independent. So, the probability of at least one pizza is one minus the probability of zero pizzas, and the probability of zero pizzas is $(.9)^6$. The answer is about .4686.

8.9 (a) Under H_0, the number of switches that fail within the year is $Y \sim \text{Binom}(16, .2)$. The test size is

$$\alpha = P(Y \leq 1) = (.8)^{16} + 16(.8)^{15}(.2) = .141.$$

(b) The power is $P(Y \leq 1)$, but this time $Y \sim \text{Binom}(16, .1)$. The power is

$$(.9)^{16} + 16(.9)^{15}(.1) = .515.$$

(c) The owner of brand B likes a nice big test size, because rejecting H_0 means ordering his or her switches. A bigger test size means that the null hypothesis is more likely to be rejected, whether or not it is true!

8.11 Let Y be the number of patients in the sample that have high blood pressure.

(a) The test size is $P(Y \geq 8)$, when $\theta = .42$ can be found using the command `1-pbinom(7,12,.42)` to get $\alpha = .076$.

(b) The power is $P(Y \geq 8)$ when $\theta = .54$, or `1-pbinom(7,12,.54)` tells us that the power is .280.

8.13 The test size is the probability that four or more sixes appear, with 10 rolls of the die, when the probability of a six is 1/6 for each roll. The rolls are independent. So,

$$\alpha = 1 - \left(\frac{5}{6}\right)^{10} - 10\left(\frac{5}{6}\right)^9\left(\frac{1}{6}\right) - \binom{10}{2}\left(\frac{5}{6}\right)^8\left(\frac{1}{6}\right)^2 - \binom{10}{3}\left(\frac{5}{6}\right)^7\left(\frac{1}{6}\right)^3$$
$$\approx .0698.$$

8.15 (a) The command `pbinom(3,40,1/6)` gives the test size .0811.

(b) The command `pbinom(3,40,.08)` gives .601.

(c) The command `pbinom(2,40,1/6)` gives .0274, which is the largest attainable test size that does not exceed .03. Therefore, the decision rule is to reject H_0 when there are 2 or fewer ones out of 40 tosses.

(d) The command `pbinom(2,40,.08)` gives a power of .369, not very high.

(e) If $n = 112$, we can reject H_0 when there are 11 or fewer ones to get a test size of $\alpha = .0288$ and a power of .815.

8.17 For Company A chips, the probability of at most one defective out of 12 chips is `pbinom(1,12,.2)` or .275; for Company B chips, the probability of at most one defective out of 12 chips is `pbinom(1,12,.05)` or .882. We can fill in a tree diagram:

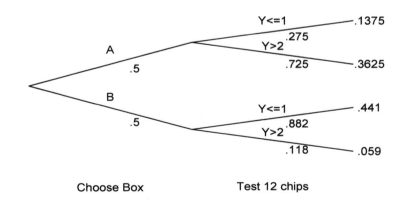

We find that the probability that the box is from Company A, given that there is at most one defective, is

$$\frac{.1375}{.1375 + .441} \approx .238.$$

8.19 (a) Let Y be the number of sixes, and let X be the amount you win in dollars. Then $X = 100Y - 25(5 - Y) = 125Y - 125$, and we can calculate $E(X) = 125E(Y) - 125 = 125(5/6) - 125 = -20.83$ dollars. Don't play!

(b) Let c be the amount you have to pay for each nonsix; then the amount you win is $X = 100Y - c(5 - Y) = (100 + c)Y - 5c$. The expected value of X is $(100 + c)(5/6) - 5c$, which is zero when $c = 20$. So, the game is fair if you have to pay \$20 for each nonsix, and you get \$100 for each six.

9.1 (a) This bit of R code and output can be used to verify the mean and variance of $Y \sim \text{Geom}(p)$.

```
> y=rgeom(100000,.4)
> mean(y)
[1] 1.49648
> var(y)
[1] 3.761545
```

(b) Using the above y vector,

```
> mean(y^3)
[1] 35.5376
```

approximates the expected value of Y^3.

9.3 Using the answer to Exercise 9.2, we have

$$P(Y \geq k_2 | Y \geq k_1) = \frac{P(Y \geq k_2 \text{ and } Y \geq k_1)}{P(Y \geq k_1)} = \frac{P(Y \geq k_2)}{P(Y \geq k_1)} = \frac{(1 - p)^{k_2 - 1}}{(1 - p)^{k_1 - 1}}$$
$$= (1 - p)^{k_2 - k_1} = P(Y \geq k_2 - k_1).$$

9.5 (a) If Y is the number of successful surgeries, then $Y \sim \text{Binom}(7, .8)$. $P(Y \geq 5) = P(Y = 5) + P(Y = 6) + P(Y = 7) = .275 + .367 + .210 = .852$.

(b) The expected number of successful surgeries before a failure occurs is $(1 - p)/p = .8/.2 = 4$.

9.7 (a) Cindy will draw at most three balls. $P(X = 1) = 2/3$, $P(X = 2) = 4/15$, $P(X = 3) = 1/15$, and $E(X) = 21/15 = 7/5$.

(b) Now $X - 1$ is geometric with $p = 2/3$, and $E(X) = (1 - p)/p + 1 = 3/2$.

9.9 The R code is

```
nloop=1000000
nuse=1:nloop
for(iloop in 1:nloop){
x=rgeom(3,.1)
```

```
nuse[iloop]=sum(x)
}
hist(nuse)
sum(nuse>=20)/nloop
```

and we get the power is about .62.

9.11 The following R code gives a probability of just over one third.

```
x=rgeom(1000000,.01)
y=rgeom(1000000,.02)
sum(y>=x)/1000000
```

9.13 The following code simulates the rolls for you and Eddie, and counts up the times you "win" when the number of rolls is greater than 12. (Note that the simulated geometric random variables give the number of rolls *before* the six, so we have to add one roll for each player.) Then for each win you get $10, but for each loss (number of rolls less than or equal to 12), you have to pay $10. Dividing by nloop gives the expected profit, which is about −$2.38. Don't play!

```
nloop=1000000
win=0
for(iloop in 1:nloop){
    rolls=rgeom(2,1/6)
    if(sum(rolls)+2>12){win=win+1}
}
(win*10-(nloop-win)*10)/nloop
```

10.1 Using the formula,

$$\frac{P(Y = k + 1)}{P(Y = k)} = \frac{\frac{e^{-\lambda}\lambda^{k+1}}{(k+1)!}}{\frac{e^{-\lambda}\lambda^{k}}{k!}} = \lambda\frac{k!}{(k + 1)!} = \frac{\lambda}{k + 1}.$$

10.3 Using the answers to Exercises 10.1 and 10.2, we see that for $\lambda > 1$, the sequence of probabilities is increasing initially, because $P(Y = 0) = e^{\lambda}$ and $P(Y = 1) = \lambda e^{\lambda}$. The probabilities increase until the ratio $\lambda/(k + 1)$ becomes negative, so that the largest probability is $P(Y = k)$, where $k \leq \lambda < k + 1$.

10.5 Let Y be the number of times the geyser erupts. Then

$$P(Y \geq 3) = 1 - P(Y = 0) - P(Y = 1) - P(Y = 2) = .456.$$

10.7 The probability that for a randomly selected day there were no accidents is $e^{-.3} = .741$, so the probability that there were no accidents on both days is $e^{-.6} = .549$.

10.9 (a) The probability that a randomly selected camper has at least one bee sting is

$$P(X \geq 1) = 1 - P(X = 0) = 1 - e^{-.8} = .551,$$

so the probability that all five have had at least one bee sting is $(.551)^5 = .051$.

(b) If $\lambda = 1.2$, the probability that a randomly selected camper has at least one bee sting is

$$P(X \geq 1) = 1 - P(X = 0) = 1 - e^{-1.2} = .699,$$

so the probability that all five have had at least one bee sting is $(.699)^5 = .167$. The power is not very high!

10.11 (a) When $\lambda = 3.7$, the probability of two or fewer injuries in a randomly chosen week is .2854 using the formula for Poisson probabilities and adding the values for zero, one, and two. Then the probability of two or fewer injuries in the next few weeks is $(.2854)^3 = .0233$, which is the test size.

(b) When $\lambda = 2.1$, the probability of two or fewer injuries in a randomly chosen week is .6496 using the formula for Poisson probabilities and adding the values for zero, one, and two. Then the probability of two or fewer injuries in the next few weeks is $(.6496)^3 = .274$, which is the power.

10.13 Calculate the probabilities for the numbers of flaws using the Poisson formula. Then make a tree diagram:

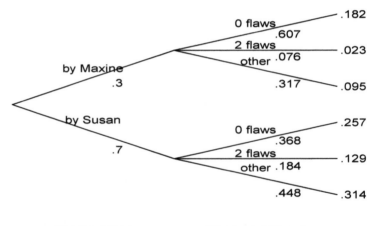

(a) For Susan, the expected number of flaws is $\lambda = 1$.

(b) Use the formula for conditional probability to get the probability that Susan made a randomly selected vase, given that it has no flaws: This is $.257/(.257 + .182) = .585$.

(c) The proportion of vases with two flaws is $.023 + .129 = .152$.

10.15 The probability of 20 or more tropical storms, when the average number is 12.1, is 1-ppois(19,12) or about .023. Therefore, this is an unusually high number of storms. If we were testing $H_0 : \lambda = 12.1$ versus $H_a : \lambda > 12.1$, the p-value for the test would be about .023, and we'd reject at $\alpha = 05$. However, we expect this many storms (or more) in about 2.3% of the years, when H_0 is true, so this could simply be an unusual year.

10.17 The probability that all four have at least two blossoms is the probability that a single stalk has at least two blossoms, raised to the fourth power.

(a) Under H_0, this is

$$\alpha = [1 - e^{-1.5} - 1.5e^{-1.5}]^4 = [.442]^4 = .038.$$

(b) If $\lambda = 3$, the power is

$$[1 - e^{-3} - 3e^{-3}]^4 = .411.$$

(c) We calculated in (a) the probability under H_0 that a single stalk has at least two blossoms is .442. Then for this decision rule,

$$\alpha = \binom{7}{6}(.442)^6(.558) + (.442)^7 = .032.$$

The test size is a bit smaller than for the experiment and decision rule in (a)

(d) If $\lambda = 3$, the probability that a single stalk has at least two blossoms is .801, so the power is

$$\binom{7}{6}(.801)^6(.199) + (.801)^7 = .579.$$

10.19 Use the conditional probability definition and the tree diagram to get that the probability that a randomly selected typist went to School A, given there were no typos, is .71.

10.21 (a) If H_0 is true, the probability of at least two homicides in a randomly selected week is
$$1 - e^{-1.23} - 1.23e^{-1.23} = .348.$$

Therefore, the probability that at least two homicides will occur in the next three weeks (assuming independence) is $(.348)^3 = .042$, which is the test size.

(b) Using $\lambda = 1.85$, the probability of at least two homicides in a randomly selected week is
$$1 - e^{-1.85} - 1.85e^{-1.85} = .552.$$

Therefore, the probability that at least two homicides will occur in the next three weeks (assuming independence) is $(.552)^3 = .168$, which is the power.

11.1 The p-value is the probability that we see as much or more evidence for H_a, given that H_0 is true. Here, this is the probability that the magician turns over three or four black cards. This is

$$p = \frac{\binom{4}{3}\binom{10}{1}}{\binom{14}{4}} + \frac{\binom{4}{4}\binom{10}{0}}{\binom{14}{4}} = .041.$$

11.3 (a) This is a hypergeometric probability. If X is the number of ears without worms, then $P(X = 6)$ is the probability of choosing six from 36 and two from 4, and

$$P(X \geq 6) = P(X = 6) + P(X = 7) + P(X = 8)$$

$$= \frac{\binom{36}{6}\binom{4}{2}}{\binom{40}{8}} + \frac{\binom{36}{7}\binom{4}{1}}{\binom{40}{8}} + \frac{\binom{36}{8}\binom{4}{0}}{\binom{40}{8}}$$

$$= .9796.$$

(b) This is a binomial probability. If X is the number of ears without worms, then $X \sim \text{Binom}(8, .1)$. Then

$$P(X \geq 6) = P(X = 6) + P(X = 7) + P(X = 8)$$

$$= \binom{8}{6}(.9)^6(.1)^2 + 8(.9)^7(.1) + (.9)^8$$

$$= .9619.$$

11.5 Let A be the event that the box is from Manufacturer A, and let D be the event that at least two of the three parts are defective. We can use Bayes' rule to calculate $P(A|D)$. We know $P(D|A)$ is a hypergeometric probability. If X is the number of defective parts chosen in a sample of size 3 from box A, then

$$P(X \geq 2) = P(X = 2) + P(X = 3) = \frac{\binom{10}{2}\binom{10}{1}}{\binom{20}{3}} + \frac{\binom{10}{3}\binom{10}{0}}{\binom{20}{3}} = .50.$$

So $P(D|A) = .50$. If Y is the number of defective parts chosen in a sample of size 3 from box B (where there are only two defectives), then

$$P(Y \geq 2) = P(Y = 2) = \frac{\binom{2}{2}\binom{18}{1}}{\binom{20}{3}} = .0158.$$

Then $P(D|A^c) = .0158$. Now using Bayes' formula,

$$P(A|D) = \frac{P(D|A)P(A)}{P(D|A)P(A) + P(D|A^c)P(A^c)} = .9694.$$

11.7 The R command

```
1-phyper(9,14,17,15)
```

gives that the p-value is about .024.

11.9 For a box with only one defective, the probability that it does not get rejected is

$$p_1 = \frac{\binom{9}{3}}{\binom{10}{3}} = 7/10,$$

and for a box with four defectives, the probability that it does not get rejected is

$$p_1 = \frac{\binom{6}{3}}{\binom{10}{3}} = 1/6.$$

Now we can make a tree diagram:

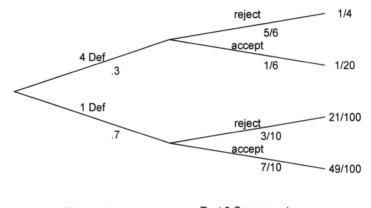

Choose Lot   Test 3 Components

The proportion of rejected lots is $.25 + .41 = .46$.

11.11 Under the null hypothesis, each ordering of the twelve rats is equally likely. It's like choosing six rats "at random" for the "fastest" group. The probability that, just by chance, *all* six omega-3 rats are chosen for the fastest group is

$$\frac{\binom{6}{6}}{\binom{12}{6}} = \frac{1}{924}.$$

The probability that exactly five out of six omega-3 rats are chosen for the fastest group is

$$\frac{\binom{6}{5}\binom{6}{1}}{\binom{12}{6}} = \frac{36}{924},$$

so

$$\alpha = \frac{37}{924} \approx .04.$$

11.13 (a) The probability that Eddie gets two golds is

$$\frac{\binom{4}{2}}{\binom{12}{2}} = \frac{1}{11},$$

while the probability that you get two golds is

$$\frac{\binom{3}{2}}{\binom{9}{2}} = \frac{1}{12}.$$

(b) The simulated probability is about .093. The code is

```
nloop=1000000
cnt=0
for(iloop in 1:nloop){
    eddie=rhyper(1,4,8,2)
    you=rhyper(1,3,6,2)
    if(eddie+you>2){cnt=cnt+1}
}
cnt/nloop
```

11.15 You and your partner have 26 cards between you and the opponents have 26 cards. Your 26 cards can be thought of as a sample without replacement from the deck of 52 cards, in which there are 13 spades.

(a) The probability that you and your partner have exactly 10 spades is

$$\frac{\binom{39}{16}\binom{13}{10}}{\binom{52}{26}} = .0217.$$

(b) The probability that you and your partner have at least 10 spades is

$$\frac{\binom{39}{16}\binom{13}{10}}{\binom{52}{26}} + \frac{\binom{39}{15}\binom{13}{11}}{\binom{52}{26}} + \frac{\binom{39}{14}\binom{13}{12}}{\binom{52}{26}} + \frac{\binom{39}{13}\binom{13}{13}}{\binom{52}{26}} = .0261.$$

12.1 For each toss, the probability of getting "doubles" is 1/6, and the tosses are independent.

(a) If Y is the number of nondoubles that are rolled, then $Y \sim \text{NB}(1/6, 2)$. Let $W = 100 - 8(Y + 2)$ be the amount you win. Then $\text{E}(W) = 100 - 8(\text{E}(Y) + 2) = 4$ because $\text{E}(Y) = 10$. Your expected winnings is positive: \$4!

(b) You make a profit if W is positive, that is, if $100 > 8(Y + 2)$, or if $Y \leq 11$ (Y has to be an integer). The probability that $Y \leq 11$ can be found using pnbinom(11,2,1/6), and we find that the probability of making a profit is .664.

12.3 If Y is the number of the patient who is the third patient to have the disease, and we can assume independent trials, then $X = Y - 3 \sim \text{NB}(3, .25)$, and the probability that the third patient is found in the fourth hour is

$$\text{P}(Y = 7) + \text{P}(Y = 8) = \text{P}(X = 4) + \text{P}(X = 5) \approx .152.$$

12.5 (a) Let Y be the number of failed attempts. Then $Y \sim NB(1, .25)$, or rather, $Y \sim Geom(.25)$. The charge is $2(Y + 1)$, so the expected charge is $2E(Y) + 2 = 8$.

(b) Let Y be the number of failed attempts. Then $Y \sim NB(2, .25)$. The charge is $X = 2(Y + 2)$, so

$$P(X > 20) = P(Y > 8) = 1 - P(Y \le 8).$$

This can be found with the R command `1-pnbinom(8,2,.25)`, and we get that the probability that we spend more than \$20 is .244.

12.7 Yes, we can use the negative binomial distribution to solve this problem. If Y is the number of nonsixes in both sequences of rolls, then $Y \sim NB(2, 1/6)$. Eddie pays you \$10 if $Y > 10$ (total number of rolls is greater than 12). The probability for this is found using `1-pnbinom(12,2,1/6)` and is about .3813. Then your expected profit is

$$10(.3813) - 10(1 - .3813) \approx -2.37.$$

13.1 (a) The triangle and rectangle have the same area a, so $a = 1/2$; (b) 1/8; (c) 1.

(d) Now we need the density function:

$$f_Y(y) = \begin{cases} 1/2 & \text{for } a \in (0, 1], \\ (3 - y)/4 & \text{for } a \in (1, 3), \end{cases}$$

so

$$E(Y) = \frac{1}{2} \int_0^1 y\,dy + \frac{1}{4} \int_1^3 y(3 - y)\,dy = 13/12.$$

(e) The probability that two independent waiting times are greater than two minutes is $(1/8)(1/8) = 1/64$.

13.3 (a) It's easiest to find $P(Y \le 1)$ and subtract this from one. The height of the density is 2/3, so the height at $y = 1$ is 1/3. Using the formula for area of a triangle, $P(Y \le 1) = 1/6$, so $P(Y > 1) = 5/6$.

(b) The area under the density to the right of b is the area of a triangle with base b and height $b/3$ (using similar triangles), so we have

$$\frac{1}{2} = \frac{1}{2}(b)\left(\frac{b}{3}\right),$$

or $b = \sqrt{3}$.

(c) The density is

$$f_Y(y) = \begin{cases} y/3 & \text{for } a \in (0, 2], \\ 2(3 - y)/3 & \text{for } a \in (2, 3), \end{cases}$$

so

$$E(Y) = \frac{1}{3} \int_0^2 y^2\,dy + \frac{2}{3} \int_2^3 y(3 - y)\,dy = 5/3.$$

(d) Suppose x is the number of minutes; then the area of the triangle representing the probability of waiting more than x minutes has base $3 - x$ and height $2 * (3 - x)/3$. Therefore,

$$.05 = \left(\frac{1}{2}\right) \left(\frac{2}{3}\right) (3 - x)^2,$$

so $(3 - x)^2 = .15$ and $x = 2.613$ minutes.

13.5 The expected value of $1/Y$ is

$$E(1/Y) = \frac{1}{b - a} \int_a^b \frac{1}{y} dy = \frac{\log(b) - \log(a)}{b - a}.$$

To confirm this answer, we can use

```
y=runif(100000,1,4)
mean(1/y)
```

which returns a number very close to $\log(4)/3 \approx .462$.

13.7 (a) We need

$$1 = c \int_a^b e^{-y} dy,$$

so

$$c = \frac{1}{e^{-a} - e^{-b}}.$$

(b) For $a = 1$ and $b = 3$, the density looks like

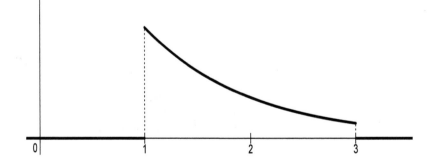

(c) Based on the idea of a balance point, we conclude that the mean is between a and $(a + b)/2$.

13.9 (a) To integrate the density, do a change of variable $u = y^2$ to get $c = 2$.

(b) With the same change of variable,

$$P(Y > 2) = 2 \int_2^\infty y e^{-y^2} dy$$

$$= \int_4^\infty e^{-u} du = e^{-4} \approx .0183.$$

(c) Again with the same change of variable,

$$E(Y) = 2 \int_0^\infty y^2 e^{-y^2} dy$$

$$= \int_0^\infty \sqrt{u} e^{-u} du = \Gamma(3/2) = \sqrt{\pi}/2 \approx .8862.$$

(d) To find $E(Y^2)$,

$$E(Y^2) = 2 \int_0^\infty y^3 e^{-y^2} dy$$

$$= \int_0^\infty u e^{-u} du = \Gamma(2) = 1.$$

Now $V(Y) = 1 - .8862^2 = .2146$.

13.11 Using the definition of expected value and making the substitution $u = y + 1$,

$$E(Y) = 3 \int_0^\infty \frac{y}{(y+1)^4} dy$$

$$= 3 \int_1^\infty \frac{u-1}{u^4} du = \frac{3}{2} - 1 = \frac{1}{2}.$$

To compute the variance,

$$E(Y^2) = 3 \int_0^\infty \frac{y^2}{(y+1)^4} dy$$

$$= 3 \int_1^\infty \frac{(u-1)^2}{u^4} du = 1.$$

Then $V(Y) = 3/4$.

13.13 (a)

$$\int_0^\infty f(y) dy = a \int_0^\infty (y+1)^{-(a+1)} dy = -(y+1)^{-a} \Big|_0^\infty = 1.$$

(b) To compute the integral below, we use a substitution $u = y + 1$:

$$E(Y) = a \int_0^\infty y(y+1)^{-a-1} dy = a \int_1^\infty (u-1)u^{-a-1} du = \frac{1}{a-1}.$$

(c) To compute $E(Y^2)$, we use the same substitution:

$$E(Y^2) = a \int_0^\infty y^2 (y+1)^{-a-1} dy = a \int_1^\infty (u-1)^2 u^{-a-1} du = \frac{2}{(a-1)(a-2)}.$$

Then

$$V(Y) = \frac{2}{(a-1)(a-2)} - \left(\frac{1}{a-1}\right)^2 = \frac{a}{(a-1)(a-1)^2}.$$

(d)

$$E((Y+1)^2) = \int_0^\infty (y+1)^2 f(y) dy = a \int_0^\infty (y+1)^{-a+1} dy$$

$$= -\frac{a+1}{a-1}(y+1)^{-a+1} \Big|_0^\infty = \frac{a}{a-2}.$$

(e)

$$E(\log(Y+1)) = \int_0^\infty \log(y+1)f(y)dy = a\int_0^\infty \log(y+1)(y+1)^{-a-1}dy$$

$$= -(y+1)^{-a}\log(y+1)|_0^\infty + \int_0^\infty (y+1)^{-a-1}dy = 1/a.$$

We used integration by parts with $u = \log(y+1)$ and $dv = a(y+1)^{-a-1}dy$, as well as l'Hôpital's rule, to get that the first term on the last line is zero.

13.15   (a) First, notice that the density function is even, so that

$$\frac{1}{\pi}\int_{-\infty}^\infty \frac{dx}{1+x^2} = \frac{2}{\pi}\int_0^\infty \frac{dx}{1+x^2}.$$

Solving the integral is easy if you know the right trick.

Using the substitution

$$\sin(\theta) = \frac{1}{\sqrt{1+x^2}}$$

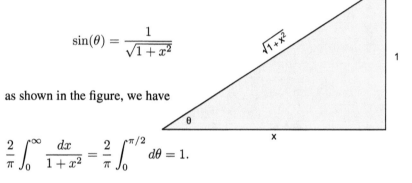

as shown in the figure, we have

$$\frac{2}{\pi}\int_0^\infty \frac{dx}{1+x^2} = \frac{2}{\pi}\int_0^{\pi/2} d\theta = 1.$$

(b) Using the definition of expected value,

$$E(Y) = \frac{1}{\pi}\int_{-\infty}^\infty \frac{xdx}{1+x^2}.$$

Let's look at

$$\frac{1}{\pi}\int_0^\infty \frac{xdx}{1+x^2} \le \frac{1}{\pi}\int_1^\infty \frac{xdx}{1+x^2} \le \frac{1}{\pi}\int_1^\infty \frac{xdx}{2x^2} = \frac{1}{2\pi}\int_1^\infty \frac{dx}{x},$$

which is not bounded. Therefore the expected value is not defined.

13.17   (a) We integrate the density over the support:

$$c\int_0^1 y^{\theta-1}dy = \frac{c}{\theta},$$

so $c = \theta$.

(b) By the definition of expected value,

$$E(Y) = \theta \int_0^1 y^\theta dy = \frac{\theta}{\theta + 1}.$$

(c) By the definition of expected value,

$$E(\log(Y)) = \theta \int_0^1 \log(y) y^{\theta-1} dy = y^\theta \log(y) \Big|_0^1 - \int_0^1 y^{\theta-1} dy,$$

using integration by parts with $u = \log(y)$ and $dv = \theta y^{\theta-1} dy$. For the first term, we use l'Hôpital's rule to find $\lim_{y \to 0} [y^\theta \log(y)] = 0$, and the second term gives $E(\log(Y)) = -1/\theta$.

13.19 (a)
$$P(Y < .2) = 3 \int_0^{.2} (1 - y)^2 dy = -(1 - y)^3 \Big|_0^{.2} = .488.$$

(b)
$$E(Y) = 3 \int_0^1 (1 - y)^2 dy = .25.$$

13.21 (a) The density function needs to integrate to one:

$$c\theta^4 \int_0^\infty (y + \theta)^{-5} dy = -\frac{c\theta^4}{4}(y + \theta)^{-4} \Big|_0^\infty = \frac{c}{4} = 1,$$

so $c = 4$.

(b)
$$E(Y + \theta) = 4\theta^4 \int_0^\infty (y + \theta)^{-4} dy = -\frac{4\theta^4}{3}(y + \theta)^{-3} \Big|_0^\infty = \frac{4\theta}{3}.$$

(c) Making a substitution $u = y + \theta$ we get

$$E(Y) = 4\theta^4 \int_0^\infty y(y + \theta)^{-5} dy$$
$$= 4\theta^4 \int_\theta^\infty (u - \theta) u^{-5} du$$
$$= 4\theta^4 \left[\int_\theta^\infty u^{-4} du - \theta \int_\theta^\infty u^{-5} du \right] = \theta/3.$$

13.23 (a) The probability that all four tomatoes have greater than 1500 IU of vitamin A, given that the distribution is the same as that for conventional tomatoes, is $(1/2)^4 = 1/16 = .0625$.

(b) The height of the alternative density is 1/200, so the area under the density, to the right of 1500 IU, is 3/4. Therefore the power for the test is $(3/4)^4 = .3164$.

(c) Suppose the decision rule is "reject H_0 if all four tomatoes have greater than c IU units of vitamin A." Then we want the area to the right of c to be $.02^{1/4}$. We know that c is larger than 1500, because $.02^{1/4} < 1/2$. Suppose the height of the density at c is h. Then we need $(1600 - c)h/2 = .02^{1/4}$.

The height for the null density is 1/100, so we can use similar triangles to get $(1600 - c)/h = 10,000$. Then

$$.02^{1/4} = (1600 - c)h/2 = (1600 - c)^2/20,000,$$

which gives $1600 - c = 97.25$, approximately, or $c = 1513.275$.

(d) The area to the right of 1513.275, under the alternative density, is .6851, so the power is about .220.

14.1 Let $Y \sim \text{Pois}(\lambda)$; then

$$M_Y(t) = E(e^{tY}) = \sum_{y=0}^{\infty} e^{ty} \frac{e^{-\lambda} \lambda^y}{y!}$$

$$= \sum_{y=0}^{\infty} \frac{e^{-\lambda}(\lambda e^t)^y}{y!}$$

$$= \frac{e^{-\lambda}}{e^{-\lambda e^t}} \sum_{y=0}^{\infty} \frac{e^{-\lambda e^t}(\lambda e^t)^y}{y!}$$

$$= e^{\lambda(e^t - 1)}.$$

14.3 Using the definition of the moment generating function,

$$M_Y(t) = E\left(e^{tY}\right)$$

$$= \frac{1}{\theta} \int_0^\infty e^{ty} e^{-y/\theta} dy$$

$$= \frac{1}{\theta} \int_0^\infty e^{-y(1/\theta - t)} dy$$

$$= \frac{1/\theta}{1/\theta - t} = \frac{1}{1 - t\theta}.$$

14.5 Starting with the definition,

$$M_Y(t) = E\left(e^{tY}\right)$$

$$= \frac{1}{2} \int_0^2 e^{ty} y \, dy,$$

then

$$\frac{1}{2} \int_0^2 e^{ty} y \, dy = \frac{e^{2t}(2t - 1) + 1}{2t^2}$$

after doing integration by parts with $u = y$ and $dv = e^{ty} dy$.

14.7 Using the definition of the MGF,

$$M_Y(t) = E\left(e^{tY}\right)$$

$$= e^a \int_a^\infty e^{-y(1-t)} dy$$

$$= \frac{e^a}{1 - t} e^{-a(1-t)} = \frac{e^{at}}{1 - t}.$$

15.1 To find the CDF,

$$F(y) = P(Y \le y) = \sum_{k=1}^{y} \left[p(1-p)^{k-1}\right].$$

If

$$S = \sum_{k=1}^{y}(1-p)^{k-1} = 1 + (1-p) + (1-p)^2 + \cdots + (1-p)^{y-1},$$

then

$$(1-p)S = (1-p) + (1-p)^2 + \cdots + (1-p)^y,$$

and subtracting the two equations gives

$$S = \frac{1-(1-p)^y}{p}.$$

Then

$$F(y) = p\sum_{k=1}^{y}\left[(1-p)^{k-1}\right] = pS$$
$$= 1 - (1-p)^y.$$

For $p = 1/3$, the sketch is shown:

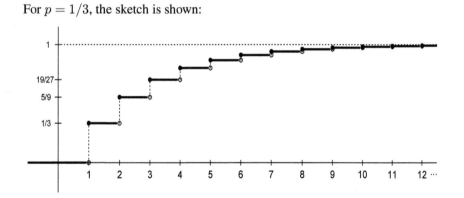

15.3 (a) The height of the density must be 2/3, so we can write the density as

$$f_Y(y) = \begin{cases} \frac{y}{3} & \text{for } y \in (0,2), \\ 2 - \frac{2y}{3} & \text{for } y \in [2,3), \\ 0 & \text{otherwise.} \end{cases}$$

Then for $y \in (0,2)$,

$$F_Y(y) = \int_0^y \frac{s}{3}ds = \frac{1}{6}y^2,$$

and for $y \in (2,3)$,

$$F_Y(y) = \frac{2}{3} + \int_2^y \left(2 - \frac{2s}{3}\right)ds = 2y - \frac{y^2}{3} - 2.$$

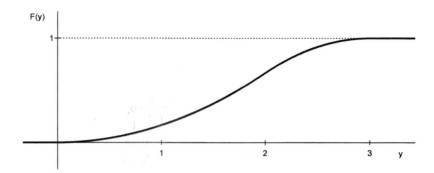

(b) Solve $y^2/6 = .1$ to get that $y = .36$ is the 10th percentile of the distribution.

15.5 We know $F_Y(\theta) = 0$, $F_Y(2\theta) = 1$, and F_Y is linear on $(\theta, 2\theta)$. Therefore, $F_Y(y) = (y - \theta)/\theta$, on $(0, \theta)$. The sketch is on the right.

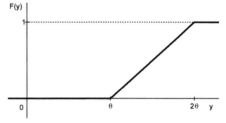

15.7 The height of the triangle is one, so the density is

$$f_Y(y) = \begin{cases} 1 + y & \text{for } y \in (-1, 0), \\ 1 - y & \text{for } y \in [0, 1), \\ 0 & \text{otherwise.} \end{cases}$$

Therefore

$$F_Y(y) = \begin{cases} 0 & \text{for } y \le -1, \\ \frac{1}{2}(1 + y)^2 & \text{for } y \in (-1, 0), \\ 1 - \frac{1}{2}(1 - y)^2 & \text{for } y \in [0, 1), \\ 1 & \text{for } y > 1. \end{cases}$$

(a) The sketch is

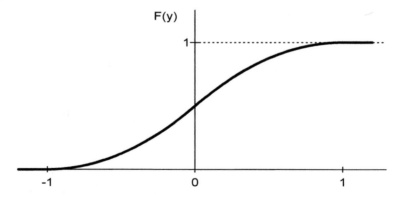

15.9 (a) The CDF is

$$F_Y(y) = 2\int_0^y e^{-2t} dt = 1 - e^{-2y},$$

and the sketches are

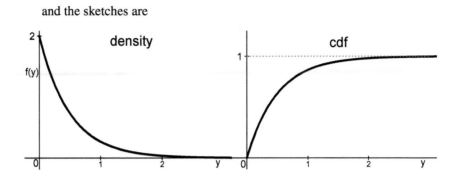

(b) To find the 90th percentile y, we solve

$$.9 = 1 - e^{-2y}$$

and get $y = 1.15$.

15.11 (a) The CDF is

$$F_Y(y) = \frac{1}{1 - e^{-3}} \int_0^y e^{-w} dw = \frac{1 - e^{-y}}{1 - e^{-3}}.$$

(b) Solving

$$\frac{1}{2} = \frac{1 - e^{-y}}{1 - e^{-3}}$$

gives $y \approx .645$ as the median.

15.13 The CDF is

$$F_Y(y) = a \int_0^y (t + 1)^{-(a+1)} dt = -(t + 1)^{-a} \big|_0^y = 1 - \frac{1}{(y + 1)^a}$$

for $y > 0$. You can take the derivative of $F_Y(y)$ to ensure that you get the density function. Another quick check is to plug in $y = 0$ to get $F_Y(0) = 0$, and as y gets large, $F_Y(y)$ approaches 1.

16.1 An exponential random variable Y with mean θ has CDF

$$F_Y(y) = 1 - e^{-y/\theta},$$

so the median is found by solving

$$.5 = 1 - e^{-y/\theta},$$

and we get

$$y = \theta \log(2).$$

16.3 Using the definition of expected value,

$$E(Y^n) = \frac{1}{\theta} \int_0^\infty y^n e^{-y/\theta} = \int_0^\infty (\theta u)^n e^{-u} du = n! \theta^n,$$

where the transformation $u = y/\theta$ and the definition of the gamma function were used. For $n = 4$ and $\theta = 2$, this is 384, and the following R code returns a close value.

```
y=rexp(100000,1/2)
mean(y^4)
```

16.5 Four gamma densities, with the expected value marked:

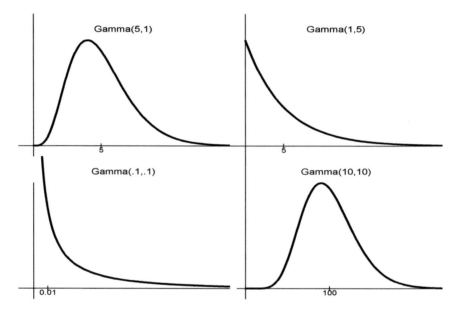

16.7 Because the integrand is part of a gamma density with $\alpha = 9$ and $\beta = 4$, we know right away that

$$\int_0^\infty x^8 e^{-4x} dx = \frac{\Gamma(\alpha)}{\beta^\alpha} = \frac{8!}{4^9} \approx .1538.$$

16.9 To find the 95th percentile of an exponential random variable with mean 56, we solve

$$.95 = 1 - e^{y/56}$$

to get "reject H_0 if the lifetime is more than 167.8 days." If the true mean is 72, the probability of rejecting is $e^{-167.8/72} = .097$, which is a low value for the power.

16.11 The moment generating function for a chi-squared random variable Y with ν degrees of freedom is

$$M_Y(t) = \frac{1}{(1 - 2t)^{\nu/2}}.$$

Then

$$M_Y'(t) = \frac{\nu}{(1 - 2t)^{\nu/2+1}} \quad \text{and} \quad M_Y'(0) = \nu = E(Y).$$

Continuing,

$$M_Y''(t) = \frac{2\nu(\nu/2 + 1)}{(1 - 2t)^{\nu/2+2}} \quad \text{and} \quad M_Y''(0) = 2\nu(\nu/2 + 1) = \nu^2 + 2\nu = E(Y^2).$$

Then

$$V(Y) = E(Y^2) - E(Y)^2 = 2\nu.$$

16.13 Using the definition of variance, we have

$$V(Y^2) = E(Y^4) - E(Y^2)^2$$
$$= 4!\theta^4 - (2!\theta^2)^2 = 20\theta^4$$

using the result of Exercise 16.3.

16.15 Let $Y \sim \text{InvGamma}(\alpha, \beta)$. Starting with the mean, and using the change of variable $x = -\beta/y$, we get

$$\begin{aligned}
E(Y) &= \frac{\beta^\alpha}{\Gamma(\alpha)} \int_0^\infty y^{-\alpha} e^{-\beta/y} dy \\
&= \frac{\beta^\alpha}{\Gamma(\alpha)} \int_\infty^0 \left(\frac{x}{\beta}\right)^\alpha e^{-x} \left(-\frac{\beta}{x^2}\right) dx \\
&= \frac{\beta}{\Gamma(\alpha)} \int_0^\infty x^{\alpha-2} e^{-x} dx \\
&= \frac{\beta}{\Gamma(\alpha)} \Gamma(\alpha - 1) = \frac{\beta}{\alpha - 1}.
\end{aligned}$$

For the variance we first compute $E(Y^2)$ and use the same transformation.

$$\begin{aligned}
E(Y^2) &= \frac{\beta^\alpha}{\Gamma(\alpha)} \int_0^\infty y^{-\alpha+1} e^{-\beta/y} dy \\
&= \frac{\beta^\alpha}{\Gamma(\alpha)} \int_0^\infty \left(\frac{x}{\beta}\right)^{\alpha-1} e^{-x} \left(\frac{\beta}{x^2}\right) dx \\
&= \frac{\beta^2}{\Gamma(\alpha)} \int_0^\infty x^{\alpha-3} e^{-x} dx \\
&= \frac{\beta^2}{\Gamma(\alpha)} \Gamma(\alpha - 2) = \frac{\beta^2}{(\alpha - 1)(\alpha - 2)}.
\end{aligned}$$

17.1 Using the formula for the beta density, we have

$$\begin{aligned}
E(Y(1-Y)^2) &= \frac{\Gamma(\alpha + \beta)}{\Gamma(\alpha)\Gamma(\beta)} \int_0^1 y^\alpha (1-y)^{\beta+1} \\
&= \frac{\Gamma(\alpha + \beta)}{\Gamma(\alpha)\Gamma(\beta)} \frac{\Gamma(\alpha + 1)\Gamma(\beta + 2)}{\Gamma(\alpha + \beta + 3)} \\
&= \frac{\Gamma(\alpha + \beta)}{\Gamma(\alpha + \beta + 3)} \frac{\Gamma(\beta + 2)}{\Gamma(\beta)} \frac{\Gamma(\alpha + 1)}{\Gamma(\alpha)} \\
&= \frac{\alpha\beta(\beta + 1)}{(\alpha + \beta)(\alpha + \beta + 1)(\alpha + \beta + 2)}.
\end{aligned}$$

For $\alpha = 2$ and $\beta = 1$, this is $1/15$, which can be confirmed with the R code:

```
y=rbeta(100000,2,1)
mean(y*(1-y)^2)
```

17.3 We recognize that the integrand is part of a beta density with $\alpha = 4$ and $\beta = 5$, so

$$\int_0^1 x^3(1-x)^4 dx = \frac{\Gamma(4)\Gamma(5)}{\Gamma(9)} = 1/280.$$

17.5 Using the definition of expected value, we must find

$$6\int_0^1 y^3(1-y)^5 dy.$$

Because we know that

$$\int_0^1 y^{\alpha-1}(1-y)^{\beta-1} = \frac{\Gamma(\alpha)\Gamma(\beta)}{\Gamma(\alpha+\beta)},$$

we have

$$E(Y^2(1-Y)^4) = \frac{(6)(3!)(5!)}{9!} = .0119.$$

17.7 The density function is $f(y) = \alpha y^{\alpha-1}$ for $y \in (0,1)$, so using the formula for expected value, we get

$$E(\log(Y)) = \alpha \int_0^1 \log(y)y^{\alpha-1} dy$$
$$= y^\alpha \log(y)\Big|_0^1 - \int_0^1 y^\alpha \frac{1}{y} dy$$
$$= 0 - \frac{1}{\alpha}y^\alpha\Big|_0^1 = -\frac{1}{\alpha}$$

using integration by parts and $u = \log(y)$ and $dv = \alpha y^{\alpha-1}$.

17.9 (a) When H_0 is true, the probability that a single particle is "in the middle" (between 1/4 and 3/4) is 1/2. The number of particles X that are in the middle is a $\text{Binom}(40, 1/2)$ random variable, so qbinom(.90,40,1/2) returns 24. The decision rule is "reject H_0 when the observed X is greater than 25." The actual test size is 1-pbinom(24,40,1/2) or .106, but we can't get closer than this to .10.

(b) If $\theta = 2$, we can find the probability of rejecting H_0 by first finding the probability that a single particle is "in the middle" (between 1/4 and 3/4) when $\theta = 2$. This is pbeta(.75,2,2)-pbeta(.25,2,2) or .6875. Then 1-pbinom(24,40,.6875) returns .847, the power of the test when $\theta = 2$.

18.1 (a) .926. (b) .802.

18.3 By the 68-95-99.7 rule, this is $.5 + .16 = .66$.

18.5 (a)

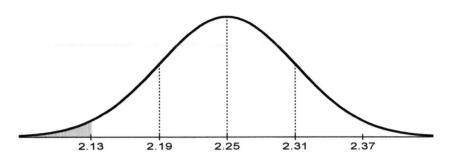

(b) By the 68-95-99.7 rule, this proportion is about .025.

18.7 We have

$$\int_{-\infty}^{\infty} \exp\left\{-\frac{1}{2}(ay^2 - 2by + c)\right\} dy = \int_{-\infty}^{\infty} \exp\left\{-\frac{a}{2}\left(y^2 - \frac{2b}{a}y + \frac{c}{a}\right)\right\} dy$$

$$= \exp\left\{\frac{1}{2}\left(\frac{b^2}{a} - c\right)\right\} \int_{-\infty}^{\infty} \exp\left\{-\frac{a}{2}\left(y^2 - \frac{2b}{a}y + \frac{b^2}{a^2}\right)\right\} dy$$

$$= \exp\left\{\frac{1}{2}\left(\frac{b^2}{a} - c\right)\right\} \int_{-\infty}^{\infty} \exp\left\{-\frac{a}{2}\left(y - \frac{b}{a}\right)^2\right\} dy$$

$$= \exp\left\{\frac{1}{2}\left(\frac{b^2}{a} - c\right)\right\} \sqrt{2\pi/a}.$$

18.9 By the 68-95-99.7 rule, this proportion is about .025.

18.11 We can fill out the following tree diagram (recall Chapter 4):

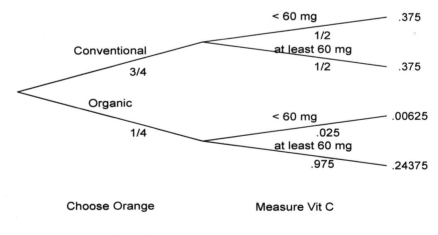

Then, using the 68-95-99.7 rule,

$$P(\text{organic}| < 60\text{mg}) = \frac{.0025}{.00625 + .375} = .01639.$$

18.13 The two distributions of tensile strengths are shown below.

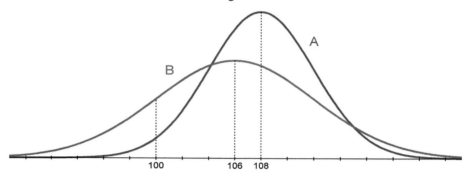

Using the 68-95-99.7 rule, we find that the probability of a randomly selected filament from Supplier B having a tensile strength less than 100 is about .16, while the probability of a randomly selected filament from Supplier A having a tensile strength less than 100 is only about .025. Therefore, if we fill out a tree diagram as shown below, we find that

$$P(B|C) = \frac{P(B \cap C)}{P(C)} = \frac{.032}{.032 + .02} = .615,$$

where B is the event that the chosen filament is from Supplier B, and C is the event that the filament breaks when used in the device.

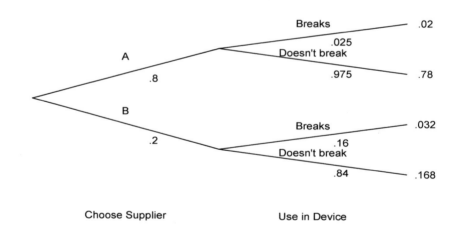

18.15   (a) $\alpha = .025$ by the 68-95-99.7 rule, and the probability of a Type II Error is about .309.

   (b) If we make a Type I Error, we decide that the child has the condition when really s/he doesn't. This could be a bad thing if medication is given, but maybe not so bad if the child is just watched or sent for another test. If we make a Type II Error, we decide that the child does not have the condition when really s/he does. If the condition is dangerous, this could be a bad thing.

   (c) $\alpha = .091$ and the probability of a Type II Error is about .16 (by the 68-95-99.7 rule).

(d) The Type I Error probability is larger but the Type II Error probability is smaller for the second decision rule. I'd prefer the second one if the condition is dangerous.

18.17 (a) We want to find D so that the area between $120 - D$ and $120 + D$, under an N(120, 5)-density, is $(.95)^{1/8} = .9936$. Then the small area to the left of $120 - D$ is .00320, and qnorm(.00320,120,5) returns $120 - D \approx 106.4$. Then $D = 13.6$.

(b) The area under an N(125, 5) curve, in the interval $(120 - D, 120 + D) = (106.4, 133.6)$, is given by pnorm(133.6,125,5)-pnorm(106.4,125,5) and is .9572. The probability that all eight rods are in this interval when the true mean is 125 mm is $(.9572)^8 = .705$. Then the power is $1 - .705 = .295$. It's just a little bit bigger than for $n = 4$.

19.1 The probability plot is obtained by

```
plot(qnorm(1:20/21),sort(case1602$Baseline),xlab="Normal Quantiles",
      ylab="sorted Baseline measurements")
```

and looks like

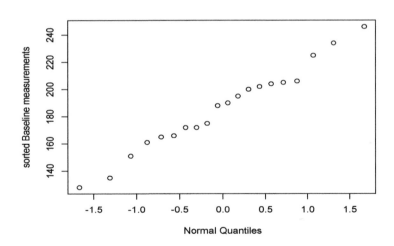

The plot indicates that the measurements are approximately normal, or at least does not give evidence otherwise.

19.3 First we generate 1000 values from a standard normal using x=rnorm(1000). Then we can plot the sorted, squared values of x against the $\chi^2(1)$ quantiles, as in the following plot, with the diagonal line superimposed, to verify that the distribution of the square of a standard normal is chi-squared with one degree of freedom.

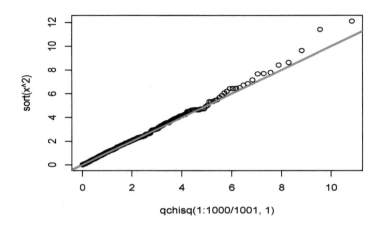

19.5 The distribution is "heavy-tailed," so in order to make a "nice" histogram, we plot only the values that are less than 10, which is about 99% of the values.

```
u=runif(10000)
x=(1-u)^(-1/2)-1
hist(x[x<10],br=50,freq=FALSE)
xpl=0:1000/100
lines(xpl,2/(xpl+1)^3,col=2,lwd=2)
```

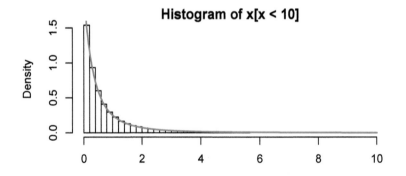

19.7 The CDF is

$$F(x) = 1 - \left(1 + \frac{x}{240}\right)^{-3},$$

so setting $F(x) = u$ and solving for x gives

$$X = 240 \left[\left(\frac{1}{1-U}\right)^{1/3} - 1\right],$$

where U is uniform on $(0, 1)$. The histogram of 10,000 realizations of X, using the transformation from the uniform, is shown:

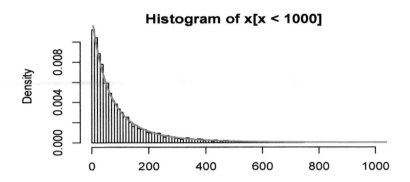

Histogram of x[x < 1000]

19.9 (a) A similar CDF was found in Exercise 15.7. Using the same ideas, we get

$$F_Y(y) = \begin{cases} 0 & \text{for } y \le 0, \\ \frac{1}{2}y^2 & \text{for } y \in (0,1), \\ 1 - \frac{1}{2}(2-y)^2 & \text{for } y \in [1,2), \\ 1 & \text{for } y \ge 2. \end{cases}$$

Therefore we simulate U from a standard uniform, and if U is less than 1/2, we solve

$$u = \frac{1}{2}y^2$$

to get $Y = \sqrt{2U}$, and if U is greater than 1/2, we solve

$$u = 1 - \frac{1}{2}(2-y)^2$$

to get $Y = 2 - \sqrt{2(1-U)}$. The code is

```
u=runif(1000000)
y=1:1000000
y[u<=1/2]=sqrt(2*u[u<=1/2])
y[u>1/2]=2-sqrt(2*(1-u[u>1/2]))
hist(y,freq=FALSE,main="")
lines(c(0,1),c(0,1),lwd=3,col=2)
lines(c(1,2),c(1,0),lwd=3,col=2)
```

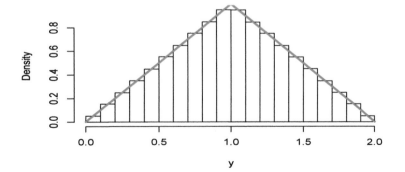

(b) For the smaller sample, it's hard to tell whether the points are following a straight line, but for the larger sample, the "S" shape is noticeable.

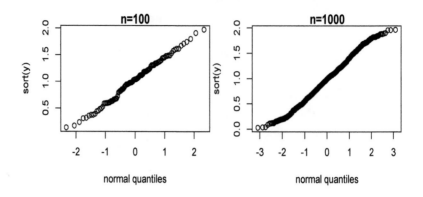

19.11   (a) The CDF is $F(x) = 1 - e^{-x^2/6}$. The sketches are

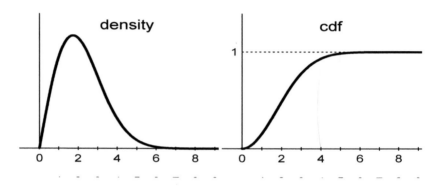

(b) To find the mean, we make the substitution $u = x^2/6$, so $x = \sqrt{6u}$:

$$E(X) = \int_0^\infty \frac{x^2}{3} e^{-x^2/6} dx$$
$$= \int_0^\infty \sqrt{6u}\, e^{-u} du$$
$$= \sqrt{\frac{3\pi}{2}} \approx 2.17$$

using $\Gamma(1/2) = \sqrt{\pi}$.

To find the median, we solve $1/2 = 1 - e^{-x^2/6}$ and get $\sqrt{6\log(2)} \approx 2.04$.

(c) To simulate using replicates from the uniform density, we use the code

```
u=runif(100000)
x=sqrt(-6*log(1-u))
hist(x,freq=FALSE,br=45)
xpl=0:900/100
fpl=xpl/3*exp(-xpl^2/6)
lines(xpl,fpl,lwd=3,col=2)
```

and the histogram is

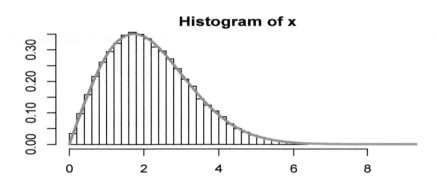

(d) The plot uses a sample of only 1000 replicates:

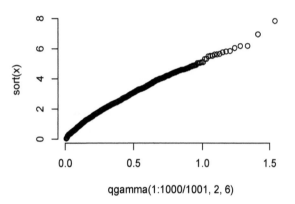

The sample values are larger than the Gamma(2,6) values, and the shape of the density "rises faster" than the Gamma(2,6) density—we conclude this from the concave shape of the points in the plot.

19.13 (a) Setting $F_Y(y) = u$, we can find $F_Y^{-1}(u)$:

$$u = \frac{1}{1 + e^{-y}} \Rightarrow y = -\log(1/u - 1).$$

The code for generation is

```
u=runif(10000)
y=-log(1/u-1)
hist(y,freq=FALSE, br=30)
ypl=-200:200/10
lines(ypl,exp(-ypl)/(1+exp(-ypl))^2,col=2,lwd=2)
```

and the histogram is

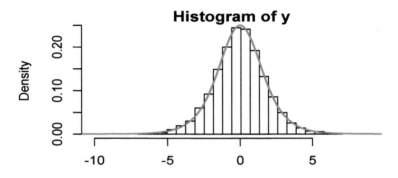

(b) The QQ plot with 100 replicates, on the left, looks like it could be from the normal density, but when there are 1000 replicates, we start to see that the density is heavier-tailed than the normal.

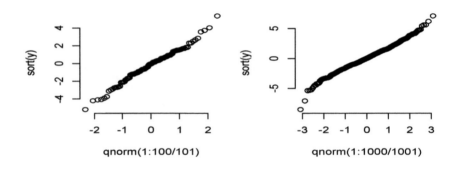

19.15 The CDF is

$$F(y) = 3\theta \int_0^y t^2 e^{-\theta t^3} dt = \int_0^{\theta y^3} e^{-u} du = 1 - e^{-\theta y^3} \quad \text{for } y > 0.$$

Code to sample from the distribution and make a histogram is

```
theta=2;u=runif(100000)
y=(log(1-u)/-theta)^(1/3)
hist(y,br=50,freq=FALSE,main="")
ypl=0:2000/1000
lines(ypl,3*theta*ypl^2*exp(-theta*ypl^3),lwd=2,col=2)
```

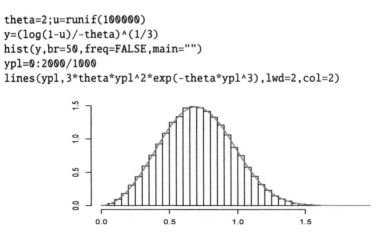

19.17 The density function for the vitamin A content of conventional tomatoes is

$$
f_Y(y) = \begin{cases} \frac{y-1400}{10000} & \text{for } y \in (1400, 1500], \\ \frac{1600-y}{10000} & \text{for } y \in (1500, 1600), \\ 0 & \text{otherwise.} \end{cases}
$$

We use this to sample from the null hypothesis density. The code for one sample of size four is

```
u=runif(4);y=1:4
y[u<1/2]=1400+sqrt(20000*u[u<1/2])
y[u>=1/2]=1600-sqrt(20000*(u[u>=1/2]-1/2))
```

Now we can sample many times and compute the means of the samples to get a distribution of sample means:

```
nloop=10000
mean4=1:nloop
for(iloop in 1:nloop){
    u=runif(4);y=1:4
    y[u<1/2]=1400+sqrt(20000*u[u<1/2])
    y[u>=1/2]=1600-sqrt(20000*(u[u>=1/2]-1/2))
    mean4[iloop]=mean(y)
}
hist(mean4)
```

To get the 95th percentile of averages, we use

```
sort(mean4)[9500]
```

The decision rule is "reject H_0 when the average vitamin A content of the four tomatoes is greater than 1534 units." Then the test size is close to $\alpha = .05$.

To find the power, we have to simulate samples of size four under the alternative distribution. The density function for the alternative is

$$
f_Y(y) = \begin{cases} \frac{y-1400}{20000} & \text{for } y \in (1400, 1500], \\ \frac{1800-y}{60000} & \text{for } y \in (1500, 1800), \\ 0 & \text{otherwise.} \end{cases}
$$

The code for sampling is

```
nloop=10000
mean4=1:nloop
for(iloop in 1:nloop){
u=runif(4);y=1:4
y[u<1/4]=1400+sqrt(40000*u[u<1/4])
y[u>=1/4]=1800-sqrt(120000*(1-u[u>=1/4]))
mean4[iloop]=mean(y)
}
hist(mean4)
```

and to find the power,

```
sum(mean4>1534)
```

returns about .76. This is considerably better than the previous test!

20.1 (a) The probability that Y is within 3 standard deviations of its mean is **greater** than **8/9**.

      (b) The probability that Y is more than 10 standard deviations from the mean is less than .01, so the probability that Y is more than 10 standard deviations *above* the mean is also **less** than **.01**

      (c) If Y is between 0 and 17, then Y is within 7/3 standard deviations from the mean (0 is 10/3 below the mean, and 17 is 7/3 above the mean). Therefore, the probability that Y is between 0 and 17 is **greater** than $40/49$. (We can't get a tighter bound than this using Chebyshev.)

      (d) If $Y^2 > 10$, then Y is more than $\sqrt{10}/3$ standard deviations aways from its mean; the probability of this is **less** than **.9**.

20.3 (a) Using Markov's inequality, the probability of a flood in a randomly selected year does not exceed 1/2:

$$P(Y > 24) \leq \frac{E(Y)}{24} = 1/2.$$

      (b) Using Chebyshev's inequality, the probability of a flood in a randomly selected year does not exceed 1/9:

$$\begin{aligned} P(Y > 24) &= P(Y - 12 > 12) \\ &\leq P(|Y - 12| > 12) \\ &= P(|Y - 12| > 3\sigma) \leq 1/9, \end{aligned}$$

because the standard deviation is $\sigma = 4$.

      (c) Using Chebyshev's inequality, the probability of a flood in a randomly selected year does not exceed 1/36.

20.5 (a) We want to find c so that $P(X > c) = .1$, where X is the number of calls in a randomly selected hour. We know $P(X > c) \leq E(X)/c$, so setting $E(X)/c = .1$ gives a bound of 3400 on the 90th percentile of this distribution.

      (b) Starting with Chebyshev's inequality,

$$P\left(|X - 340| > \sqrt{10}\sqrt{20000}\right) \leq .1,$$

so $P(X > 340 + 447) \leq .1$, and the 90th percentile of the distribution is at most 787.

21.1 The support for Y is $(0,1)$, so for $y \in (0,1)$,

$$F_Y(y) = \text{P}(Y \le y) = \text{P}(X^n \le y) = \text{P}(X \le y^{1/n}) = y^{1/n}.$$

Taking the derivative with respect to y gives

$$f_Y(y) = \frac{1}{n} y^{1/n-1} \text{ for } y \in (0,1).$$

The R code and the histogram to confirm are

```
u=runif(100000)
y=u^3
hist(y,freq=FALSE,br=50,col='beige')
ypl=0:1000/1000
lines(ypl,ypl^(-2/3)/3,col=2,lwd=3)
```

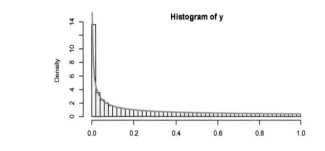

21.3 The support for Y is $(0,a)$. Using the CDF method, for $y \in (0,a)$,

$$\begin{aligned}
F_Y(y) = \text{P}(Y \le y) &= \text{P}(|X| \le y) \\
&= \text{P}(-y \le X \le y) \\
&= F_X(y) - F_X(-y).
\end{aligned}$$

Now we take the derivative with respect to y to get $f_X(y) + f_X(-y)$, and by symmetry of f_X, the density for Y is

$$f_Y(y) = 2f_X(y) \text{ for } y \in (0,a).$$

21.5 The support of X is $(0,1)$, so for $x \in (0,1)$,

$$\begin{aligned}
F_X(x) = \text{P}(X \le x) &= \text{P}(1 - Y \le x) \\
&= \text{P}(Y \ge 1 - x) \\
&= 1 - F_Y(1 - x).
\end{aligned}$$

Taking the derivative with respect to x, we have

$$f_X(x) = f_y(1-x) = \frac{\Gamma(\alpha+\beta)}{\Gamma(\alpha)\Gamma(\beta)}(1-x)^{\alpha-1}x^{\beta-1},$$

so $X \sim \text{Beta}(\beta, \alpha)$.

21.7 Using the CDF method,

$$F_X(x) = P(X \le x)$$
$$= P(\beta Y \le x) = F_Y(x/\beta),$$

so

$$f_X(x) = f_Y(x/\beta) = \frac{1}{\Gamma(\alpha)} x^{\alpha-1} e^{-x},$$

which we recognize as Gamma$(\alpha, 1)$.

21.9 The support for X is the positive reals. For $x > 0$,

$$F_X(x) = P(X \le x) = P(\sqrt{Y} \le x) = P(Y \le x^2).$$

We know $F_Y(y) = 1 = e^{-\theta x}$, so

$$F_X(x) = 1 - e^{-\theta x^2},$$

and taking the derivative, we get the density

$$f_X(x) = 2\theta x e^{-\theta x^2} \text{ for } x \in (0, \infty).$$

The following R code will produce the plot below, and confirm this is correct.

```
theta=2
y=rexp(100000,theta)
x=sqrt(y)
hist(x,freq=FALSE,main=expression(paste("X=",sqrt(Y))),br=40)
xpl=0:300/100
lines(xpl,2*theta*xpl*exp(-xpl^2*theta),col=2,lwd=2)
```

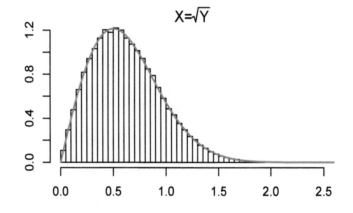

21.11 The new random variable Y takes values in $(0, 2)$. For $y \in (0, 2)$,

$$F_Y(y) = P(Y \le y)$$
$$= P\left(\sqrt{U} \le y\right)$$
$$= P\left(U \le y^2\right)$$
$$= \frac{1}{4}y^2.$$

Then $f_Y(y) = y/2$, $y \in (0, 2)$. Some code to check your answer:

```
u=runif(100000)*4
y=sqrt(u)
hist(y,freq=FALSE,br=50)
xpl=0:2
lines(xpl,xpl/2,lwd=3,col=4)
```

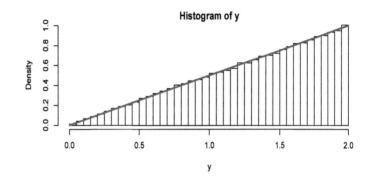

Histogram of y

21.13 (a) The density is $f_U(u) = 1/(b-a)$ on (a, b), so for $u \in (a, b)$, the CDF is

$$F_U(u) = \frac{1}{b-a} \int_a^u du = \frac{u-a}{b-a}$$

for $u \in (a, b)$.

(b) Using the CDF method,

$$
\begin{aligned}
F_Y(y) &= P(Y \le y) \\
&= P\left(\frac{1}{U} \le y\right) \\
&= P\left(U \ge \frac{1}{y}\right) \\
&= 1 - F_U(1/y) \\
&= 1 - \frac{1/y - a}{b-a}.
\end{aligned}
$$

Then the density is

$$f_Y(y) = F_Y'(y) = \frac{1}{y^2(b-a)}.$$

To check, here is some R code:

```
a=1;b=4
u=runif(100000)*(b-a)+a
y=1/u
hist(y,br=50,freq=FALSE)
ypl=250:1000/1000
lines(ypl,1/ypl^2/(b-a),col=2,lwd=3)
```

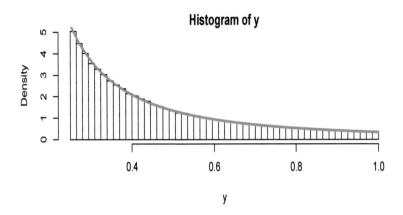

21.15 The support for Y is the positive real numbers. For $y > 0$, the CDF for Y is

$$
\begin{aligned}
F_Y(y) &= \mathrm{P}(Y \le y) \\
&= \mathrm{P}\left(-2\log(1-U) \le y\right) \\
&= \mathrm{P}(\log(1-U) \ge -y/2) \\
&= \mathrm{P}(1 - U \ge e^{-y/2}) \\
&= \mathrm{P}(U \le 1 - e^{-y/2}) \\
&= 1 - e^{-y/2}.
\end{aligned}
$$

We recognize that Y is an exponential random variable with mean 2.

To check, here is some R code:

```
u=runif(10000)
y=-2*log(1-u)
hist(y,freq=FALSE,br=60,col='beige')
ypl=0:2000/100
lines(ypl,dexp(ypl,1/2),col=2,lwd=3)
```

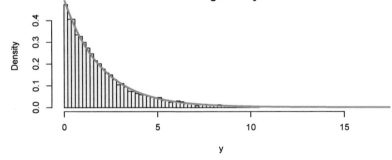

21.17   (a) The density for X is $f_X(x) = 4xe^{-2x}$ for $x > 0$. We start by finding the CDF for Y:

$$
\begin{aligned}
F_Y(y) &= \mathrm{P}(Y \le y) \\
&= \mathrm{P}(\sqrt{X} \le y) \\
&= \mathrm{P}(X \le y^2) \\
&= F_X(y^2).
\end{aligned}
$$

Then, taking the derivative and using the chain rule, we have

$$f_Y(y) = 2yf_X(y^2)$$
$$= 8y^3e^{-2y^2}$$

for $y > 0$.

(b) The code on the left produces the plot on the right:

```
x=rgamma(100000,2,2)
y=sqrt(x)
hist(y,br=50,freq=FALSE)
ypl=0:300/100
lines(ypl,8*ypl^3*exp(-2*ypl^2))
```

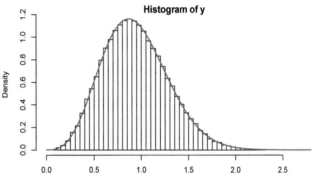

21.19 (a) The CDF for Y is

$$F_Y(y) = \frac{2}{\theta}\int_0^y we^{-w^2/\theta}dw = 1 - e^{-y^2/\theta} \quad \text{for } y > 0.$$

Then

$$F_X(x) = \mathrm{P}(X \le x) = \mathrm{P}(Y \le \sqrt{x}) = 1 - e^{-x/\theta} \quad \text{for } x > 0,$$

which we recognize as the CDF for an exponential random variable with mean θ.

(b) The code on the left produces the plot on the right:

```
x=rexp(10000,1/2) # theta=2
y=sqrt(x)
hist(y,freq=FALSE,br=50)
ypl=0:500/100
lines(ypl,ypl*exp(-ypl^2/2),col=2)
```

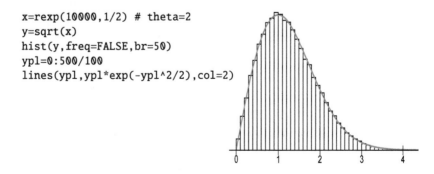

21.21 Set $a = b = 2$: $Y = 2X + 2$ has support on $(2, 4)$ and the density is

$$f_Y(y) = \frac{1}{2}f_X\left(\frac{y-2}{2}\right) = \frac{\theta}{2^\theta}(y-2)^{\theta-1} \quad \text{for } y \in (2, 4).$$

21.23 The random variable Y takes values in $(0, 1)$. For $y \in (0, 1)$,

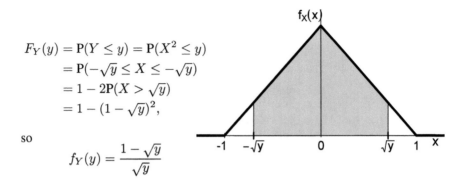

$$F_Y(y) = \mathrm{P}(Y \le y) = \mathrm{P}(X^2 \le y)$$
$$= \mathrm{P}(-\sqrt{y} \le X \le -\sqrt{y})$$
$$= 1 - 2\mathrm{P}(X > \sqrt{y})$$
$$= 1 - (1 - \sqrt{y})^2,$$

so

$$f_Y(y) = \frac{1 - \sqrt{y}}{\sqrt{y}}$$

The following code makes the histograms shown beneath.

```
u=runif(10000)
u1=runif(10000)
x=u[u<u1]
u2=runif(length(x))
x[u2<1/2]=-x[u2<1/2]
hist(x,freq=FALSE,br=50,col='beige')
lines(c(-1,0),c(0,1),lwd=3,col=2)
lines(c(0,1),c(1,0),lwd=3,col=2)

y=x^2
hist(y,freq=FALSE,br=50,col='beige')
ypl=0:1000/1000
lines(ypl,(1-sqrt(ypl))/sqrt(ypl),lwd=3,col=2)
```

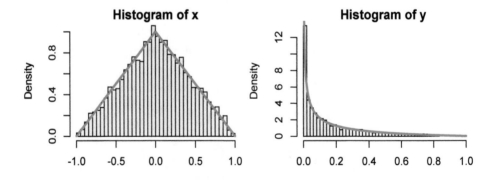

22.1 Using the change of variable $u = (\log(y) - \mu)/\sigma$, we have

$$\mathrm{E}(Y^2) = \frac{1}{\sqrt{2\pi\sigma^2}} \int_0^\infty \frac{1}{y} \exp\left(-\frac{(\log(y) - \mu)^2}{2\sigma^2}\right) dy$$
$$= \frac{1}{\sqrt{2\pi}} \int_{-\infty}^\infty e^{2\mu + 2u\sigma} e^{-u^2/2} du$$
$$= e^{2\mu + 2\sigma^2}.$$

Then

$$V(Y) = E(Y^2) - E(Y)^2 = e^{2\mu+2\sigma^2} - e^{2\mu+\sigma^2} = e^{2\mu+\sigma^2}(e^{\sigma^2} - 1).$$

22.3 If $Y \sim$LN$(1,2)$, then $\log(Y) \sim$N$(1,2)$, and P$(Y > 20) = $ P$(\log(Y) > \log(20)$, which is found with `1-pnorm(log(20),1,2)`, returning about .159.

22.5 (a) `qweibull(.95,2,4)` returns 6.92.

(b) `1-pweibull(8,2,4)` returns about .018.

22.7 The following code will, for any specified n, approximate the probability that the sum of n independent Weibull$(2, 20)$ random variables is at least 110. We find that $n = 11$ is large enough to give a probability of .999.

```
n=11;nloop=100000;cumlife=1:nloop
for(iloop in 1:nloop){
    x=rweibull(n,2,20)
    cumlife[iloop]=sum(x)
}
sum(cumlife>=110)/nloop
```

22.9 Let $X \sim$Pareto$(3, 4)$; then E$(X) = 2$ and V$(X) = 12$. By Chebyshev's inequality,
$$P(X > 42) = P(|X - 2| > 40) \leq 12/40^2 = .0075,$$
because 40 is $40/\sqrt{12}$ standard deviations from the mean. The true probability is found with `1-ppareto(42,3,4)`, which returns .00066.

22.11 If $X \sim \chi^2(2)$, the density for X is $f_X(x) = e^{-x/2}/2$. For $Y = \sqrt{X}$,
$$F_Y(y) = P(Y \leq y) = P(X \leq y^2) = F_X(y^2),$$
so the density for Y is
$$f_Y(y) = 2yf_X(y^2) = ye^{-y^2/2}.$$
This is a Weibull$(2, \sqrt{2})$ or a Rayleigh(2) random variable.

23.1 By the definition of conditional probability,
$$P(Y_1 = y_1 | Y_2 = y_2) = \frac{P(Y_1 = y_1, Y_2 = y_2)}{P(Y_2 = y_2)},$$
but because Y_1 and Y_2 are independent, we have
$$P(Y_1 = y_1 | Y_2 = y_2) = P(Y_1 = y_1).$$
Therefore,
$$P(Y_1 = y_1) = \frac{P(Y_1 = y_1, Y_2 = y_2)}{P(Y_2 = y_2)},$$
so
$$P(Y_1 = y_1, Y_2 = y_2) = P(Y_1 = y_1)P(Y_2 = y_2).$$

23.3   (a) The joint distribution is

		Y_1 =# red balls	
		0	1
	0	1/3	1/6
Y_2 = # green balls	1	1/6	1/3

(b) The marginal for Y_2 is

y	0	1
$P(Y_2 = y)$	1/2	1/2

(c) $P(Y_1 = 0|Y_2 = 0) = 2/3$, $P(Y_1 = 10|Y_2 = 0) = 1/3$.

(d) No; we have $P(Y_1 = 0|Y_2 = 0) = 1/3$, but $P(Y_1 = 0) = 1/2$.

23.5   (a) The joint distribution is

		Y_1					
		1	2	3	4	5	6
	1	1/36	0	0	0	0	0
	2	1/18	1/36	0	0	0	0
Y_2	3	1/18	1/18	1/36	0	0	0
	4	1/18	1/18	1/18	1/36	0	0
	5	1/18	1/18	1/18	1/18	1/36	0
	6	1/18	1/18	1/18	1/18	1/18	1/36

(b) The marginal for Y_1 is

y	1	2	3	4	5	6
$P(Y_1 = y)$	11/36	9/36	7/36	5/36	3/36	1/36

(c) We compute $P(Y_2 = y|Y_1 = 4) = 0$ for $y = 1, 2, 3$. $P(Y_2 = 4|Y_1 = 4) = 1/5$, $P(Y_2 = 5|Y_1 = 4) = P(Y_2 = 6|Y_1 = 4) = 2/5$.

(d) No, Y_1 and Y_2 are not independent. If $Y_1 = 6$, Y_2 has to be 6 as well.

23.7   (a) Using the marginal for Y_1,

$$
\begin{aligned}
E(Y_1) &= p \sum_{y_1=1}^{\infty} y_1 (1 - p)^{y_1 - 1} \\
&= p \sum_{k=0}^{\infty} (k + 1)(1 - p)^k \\
&= p \left[\sum_{k=0}^{\infty} k(1 - p)^k + \sum_{k=0}^{\infty} (1 - p)^k \right] \\
&= p \left[\frac{1 - p}{p^2} + \frac{1}{p} \right] = \frac{1}{p}.
\end{aligned}
$$

(b) Using the marginal for Y_2,

$$E(Y_2) = p^2 \sum_{y_2=2}^{\infty} y_2(y_2 - 1)(1 - p)^{y_2 - 2}$$

$$= p^2 \sum_{k=0}^{\infty} (k + 2)(k + 1)(1 - p)^k$$

$$= p^2 \left[\sum_{k=0}^{\infty} k^2(1 - p)^k + 3 \sum_{k=0}^{\infty} k(1 - p)^k + 2 \sum_{k=0}^{\infty} (1 - p)^k \right]$$

$$= p^2 \left[\frac{(1 - p)(2 - p)}{p^3} + \frac{3(1 - p)}{p^2} + \frac{2}{p} \right] = \frac{2}{p}.$$

(c) Using the joint mass function for Y_1 and Y_2, we have

$$E(Y_1 Y_2) = p^2 \sum_{y_2=2}^{\infty} \sum_{y_1=1}^{y_2 - 1} y_1 y_2 (1-p)^{y_2-2} = p^2 \sum_{y_2=2}^{\infty} \left[\frac{(y_2-1)y_2}{2} \right] y_2(1-p)^{y_2-2}$$

using the fact that the sum of the first n positive integers is $n(n + 1)/2$. Then making the change of variable $k = y_2 - 1$, we have

$$E(Y_1 Y_2) = \frac{p^2}{2} \sum_{k=1}^{\infty} (k + 1)^2 k (1 - p)^{k-1}$$

$$= \frac{p^2}{2(1 - p)} \sum_{k=1}^{\infty} \left[k^3 + 2k^2 + k \right] (1 - p)^k$$

$$= \frac{p^2}{2(1 - p)} \left[\frac{(1 - p)^3 + 4(1 - p)^2 + (1 - p)}{p^4} \right.$$

$$\left. + \frac{2(2 - p)(1 - p)}{p^3} + \frac{1 - p}{p^2} \right]$$

$$= \frac{3 - p}{p^2}$$

using the last formula in Appendix B.2.

23.9 (a) The marginal for Y_1 is obtained by totaling the rows: $P(Y_1 = 0) = .10$, $P(Y_1 = 1) = .38$, and $P(Y_1 = 2) = .52$.

(b) $P(Y_2 = 0 | Y_1 = 2) = P(Y_2 = 0, Y_1 = 2)/P(Y_1 = 2) = .14/.52 = .269$. Similarly, $P(Y_2 = 1 | Y_1 = 2) = .3/.52 = .577$ and $P(Y_2 = 2 | Y_1 = 2) = .08/.52 = .154$.

(c) No. We can see, for instance, that $P(Y_2 = 1 | Y_1 = 0) = 0$, while $P(Y_2 = 1) = .56$.

(d) The diagonal element of the table, with probabilities .10, .26, and .08, represent the same number of blossoms as fruit. One more blossom than fruit (i.e., one blossom that does not form fruit) happens when there is 1 blossom and no fruit (probability .12) or 2 blossoms and 1 fruit (probability .30). The only instance of two blossoms that do not form fruit has probability .14, so the expected number of blossoms that do not form fruit is $(1)(.42) + (2)(.14) = .7$.

(e) The following code simulates many pairs of values for the jointly distributed random variables.

```
n=100000
y1=1:n;y2=1:n
for(i in 1:n){
    u=runif(1)
    if(u<.1){
        y1[i]=0;y2[i]=0
    }else if(u<.22){
        y1[i]=1;y2[i]=0
    }else if(u<.48){
        y1[i]=1;y2[i]=1
    }else if(u<.62){
        y1[i]=2;y2[i]=0
    }else if(u<.92){
        y1[i]=2;y2[i]=1
    }else{
        y1[i]=2;y2[i]=2
    }
}
```

To check (a):

```
sum(y1==0)/n
sum(y1==1)/n
sum(y1==2)/n
```

To check (b):

```
sum(y2==0&y1==2)/sum(y1==2)
sum(y2==1&y1==2)/sum(y1==2)
sum(y2==2&y1==2)/sum(y1==2)
```

23.11 (a) We need only notice that $e^{-6} = e^{-2}e^{-4}$; then

$$\sum_{y_1=0}^{\infty} \sum_{y_2=0}^{\infty} e^{-6} \frac{4^{y_1} 2^{y_2}}{y_1! y_2!} = \left[\sum_{y_1=0}^{\infty} e^{-4} \frac{4^{y_1}}{y_1!} \right] \left[\sum_{y_2=0}^{\infty} e^{-2} \frac{2^{y_2}}{y_2!} \right],$$

and we recognize Poisson probabilities in each of the brackets, which must sum to 1.

(b) We simply plug in $y_1 = y_2 = 0$ to get $P(Y_1 =, Y_2 = 0) = .0025$. There are not many months with no machine failures.

(c) As in the solution to (a), we can write the probabilities as

$$P(Y_1 = y_1, Y_2 = y_2) = \left[e^{-4} \frac{4^{y_1}}{y_1!} \right] \left[e^{-2} \frac{2^{y_2}}{y_2!} \right],$$

and to find the marginal probabilities $P(Y_1 = y_1)$, we sum over Y_2 values:

$$P(Y_1 = y_1) = \sum_{y_2=0}^{\infty} \left[e^{-4} \frac{4^{y_1}}{y_1!} \right] \left[e^{-2} \frac{2^{y_2}}{y_2!} \right] = \left[e^{-4} \frac{4^{y_1}}{y_1!} \right] \sum_{y_2=0}^{\infty} \left[e^{-2} \frac{2^{y_2}}{y_2!} \right]$$

$$= e^{-4} \frac{4^{y_1}}{y_1!}.$$

Similarly, we can find the marginal probabilities $P(Y_2 = y_2)$ to be

$$P(Y_2 = y_2) = e^{-2}\frac{2^{y_2}}{y_2!}.$$

Then

$$P(Y_2 = y_2|Y_1 = y_1) = \frac{P(Y_2 = y_2, Y_1 = y_1)}{P(Y_1 = y_1)} = e^{-2}\frac{2^{y_2}}{y_2!} = P(Y_2 = y_2),$$

and we conclude that Y_1 and Y_2 are independent.

(d) $E(Y_1 + Y_2) = E(Y_1) + E(Y_2)$, and we notice that the marginals are both Poisson, with parameters 4 and 2. Therefore, the expected number of failures is 6.

(e) The expected monthly repair cost is $E(200Y_1 + 150Y_2) = 200E(Y_1) + 150E(Y_2) = 1100$ dollars.

24.1 In the plot below, the joint density is integrated over the beige area to get $F(b_1, b_2)$. To get $F(a_1, b_2)$, we integrate the joint density over the area that has horizontal shading, and to get $F(b_1, a_2)$, we integrate the joint density over the area that has vertical shading. The integral of the joint density over the cross-hatched shading is $F(a_1, a_2)$. If we subtract $F(a_1, b_2)$ and $F(b_1, a_2)$ from $F(b_1, b_2)$, we have subtracted the integral over the cross-hatched area twice, so we need to add it back in. We're left with the integral of the joint density over the beige area with no shading, which is $P(a_1 < Y_1 < b_1, a_2 < Y_2 < b_2)$.

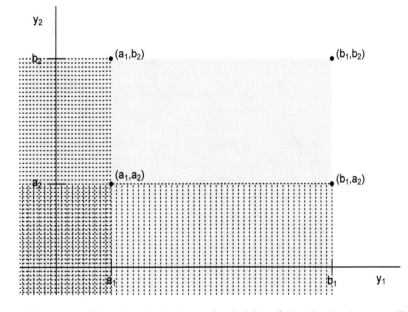

24.3 (a) The area of the triangle is 1, so the height of the density is one. Then $P(Y_2 > 1/2)$, or the volume under the density over the triangle shaded on the left in the plot below, is the same as the area of the triangle, or $P(Y_2 > 1/2) = 1/4$.

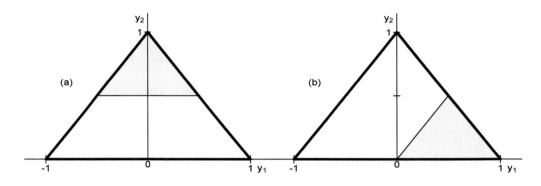

(b) Following the same reasoning, $P(Y_1 > Y_2)$ is the area of the shaded triangle in the plot on the right, or $P(Y_1 > Y_2) = 1/4$.

(c) From symmetry ideas, we can guess that $ev(Y_1) = 0$, but let's do the calculation anyway. Using the equations for the lines forming the support, we have

$$E(Y_1) = \int_0^1 \int_{y_2-1}^{1-y_2} y_1 dy_1 dy_2$$

$$= \frac{1}{2} \int_0^1 \left[(1 - y_2)^2 - (y_2 - 1)^2 \right] dy_2 = 0.$$

(d) We guess that $ev(Y_2)$ is less than 1/2, because for most of the support $y_2 < 1/2$. We have

$$E(Y_2) = \int_0^1 \int_{y_2-1}^{1-y_2} y_2 dy_1 dy_2$$

$$= \int_0^1 y_2 \left[(1 - y_2) - (y_2 - 1) \right] dy_2$$

$$= \int_0^1 y_2 \left[2 - 2y_2 \right] dy_2$$

$$= 1 - 2/3 = 1/3.$$

(e) We can use the answers to (c) and (d): $E(Y_2 - 4Y_1) = E(Y_2) - 4E(Y_1) = 1/3$.

(f) The following R code generates a sample from the joint density and approximates the answers to the above.

```
y1=runif(10000)*2-1
y2=runif(10000)
keep=y2<y1+1&y2<1-y1
y1=y1[keep];y2=y2[keep]
n=sum(keep)
#a:
sum(y2>1/2)/n
#b:
sum(y1>y2)/n
#c:
mean(y1)
```

```
#d:
mean(y2)
#e:
mean(y2-4*y1)
```

24.5 (a) To find $c = 3/4$,

$$1 = c \int_0^2 \int_{(y_2-2)/2}^{(2-y_2)/2} y_2 dy_1 dy_2 = c \int_0^2 y_2(2-y_2)dy_2 = \frac{4c}{3}.$$

(b) The lines $y_1 = y_2$ and $y_2 = 2 - 2y_1$ intersect at $(2/3, 2/3)$, so we integrate the triangle shown in the figure:

$$P(Y_1 > Y_2) = \frac{3}{4} \int_0^{2/3} \int_{y_2}^{1-y_2/2} y_2 dy_1 dy_2$$

$$= \frac{3}{4} \int_0^{2/3} y_2(1 - 3y_2/2)dy_2$$

$$= 1/18.$$

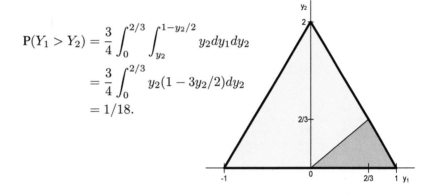

(c) By symmetry, $P(Y_1 > 0) = 0$.

(d) By symmetry, $E(Y_1) = 0$.

(e) It is straightforward to compute

$$\frac{3}{4} \int_0^2 \int_{(y_2-2)/2}^{(2-y_2)/2} y_2^2 dy_1 dy_2 = 1.$$

(f) The following code generates a sample from the joint density and approximates the answers to the above.

```
y1=runif(40000)*2-1
y2=runif(40000)*2
keep1=y2<y1+1&y2<1-y1
y1=y1[keep1];y2=y2[keep1]
u=runif(n)*3/2
keep2=u<3*y2/4
y1=y1[keep2];y2=y2[keep2]
n=sum(keep2)
#b:
sum(y1>0)/n
#c:
sum(y1>y2)/n
#d:
mean(y1)
#e:
mean(y2)
```

24.7   (a) We solve the integral

$$1 = c \int_0^\infty \int_0^\infty e^{-y_1/2} e^{-y_2/4} dy_1 dy_2 = 8c,$$

so $c = 1/8$. The support and density are shown. We see that the density "goes down faster" in the y_1 direction, compared to the y_2 direction.

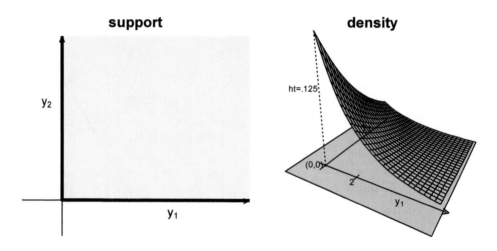

(b) We integrate the density over the subset of the support that corresponds to $y_1 > 2$:

$$P(Y_1 > 2) = \frac{1}{8} \int_2^\infty \int_0^\infty e^{-y_1/2} e^{-y_2/4} dy_2 dy_1$$

$$= \frac{1}{2} \int_2^\infty e^{-y_1/2} dy_1 = e^{-1} \approx .368.$$

(c) To find the expected value, we integrate

$$E(Y_1^2) = \frac{1}{8} \int_0^\infty \int_0^\infty y_1^2 e^{-y_1/2} e^{-y_2/4} dy_2 dy_1$$

$$= \frac{1}{2} \int_0^\infty y_1^2 e^{-y_1/2} dy_1 = 8.$$

24.9   (a) The total volume under the density function, over the support, is

$$\int_0^1 \int_0^{y_2} c(1 - y_2) dy_1 dy_2 = c \int_0^1 y_2(1 - y_2) dy_2 = c \left(\frac{1}{2} - \frac{1}{3} \right) = \frac{c}{6},$$

so $c = 6$. The support and density are shown.

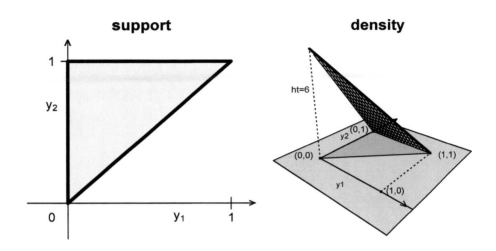

For parts (b)–(d) refer to these shadings of the support:

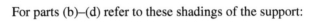

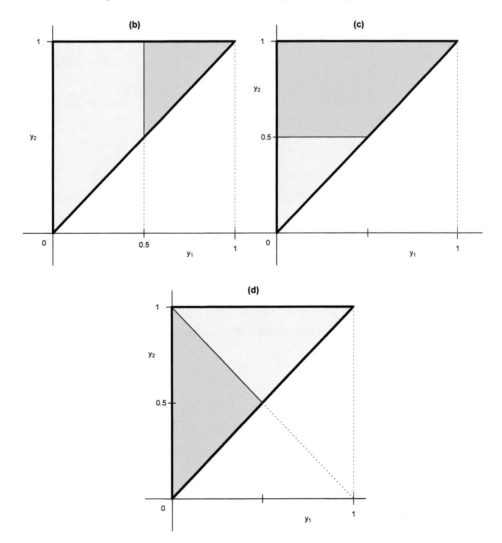

(b)

$$P(Y_1 > 1/2) = 6 \int_{1/2}^{1} \int_{y_1}^{1} (1 - y_2) dy_2 dy_1$$

$$= 3 \int_{1/2}^{1} (1 - y_1)^2 dy_1$$

$$= (1 - 1/2)^3 = 1/8.$$

(c)

$$P(Y_2 > 1/2) = 6 \int_{1/2}^{1} \int_{0}^{y_2} (1 - y_2) dy_1 dy_2$$

$$= 6 \int_{1/2}^{1} (1 - y_2) y_2 dy_2$$

$$= 3(1 - 1/4) - 2(1 - 1/8) = 1/2.$$

(d)

$$P(Y_1 + Y_2 < 1) = 6 \int_{0}^{1/2} \int_{y_1}^{1-y_1} (1 - y_2) dy_2 dy_1$$

$$= 3 \int_{0}^{1/2} [(1 - y_1)^2 - y_1^2] dy_1$$

$$= 3 \int_{0}^{1/2} (1 - 2y_1) dy_1$$

$$= 3(1/2 - 1/4) = 3/4.$$

(e)

$$E(Y_1 Y_2) = 6 \int_{0}^{1/2} \int_{0}^{y_2} y_1 y_2 (1 - y_2) dy_1 dy_2$$

$$= 3 \int_{0}^{1} y_2^3 (1 - y_2) dy_2$$

$$= 3(1/4 - 1/5) = 3/20.$$

(f) The following code will simulate from the joint density.

```
n=50000
y1=runif(n)
y2=runif(n)
use=y1<=y2
y1=y1[use]
y2=y2[use]

new.n=sum(use)
u=runif(new.n,0,6)
keep=u<6*(1-y2)

y1=y1[keep]
y2=y2[keep]
```

Then we can check the answers. For (b):

```
sum(y1>1/2)/sum(keep)
```

For (c):

```
sum(y2>1/2)/sum(keep)
```

For (d):

```
sum(y1+y2<1)/sum(keep)
```

For (e):

```
mean(y1*y2)
```

24.11 (a) Integrating over the positive orthant, we have

$$1 = c \int_0^\infty \int_0^\infty e^{-y_1/8} e^{-y_2/12} dy_1 dy_2 = 96c,$$

so $c = 1/96$.

(b) To get $P(Y_1 + Y_2 < 10)$ we need to integrate over the triangle with vertices $(0, 0)$, $(0, 10)$, and $(10, 0)$:

$$P(Y_1 + Y_2 < 10) = \frac{1}{96} \int_0^{10} \int_0^{10-y_2} e^{-y_1/8} e^{-y_2/12} dy_1 dy_2$$

$$= -\frac{1}{12} \int_0^{10} e^{-y_2/12} e^{-y_1/8} |_0^{10-y_2} dy_2$$

$$= \frac{1}{12} \int_0^{10} e^{-y_2/12} [1 - e^{-(10-y_2)/8}] dy_2$$

$$= 1 - 3e^{-5/6} + 2e^{-5/4}.$$

We wanted $P(Y_1 + Y_2 > 10)$, so we subtract this from one to get

$$P(Y_1 + Y_2 > 10) = 3e^{-5/6} - 2e^{-5/4} \approx .731.$$

It's likely that the EPA will be concerned.

(c) It's not hard to compute $E(Y_1)$:

$$E(Y_1) = \frac{1}{96} \int_0^\infty \int_0^\infty y_1 e^{-y_1/8} e^{-y_2/12} dy_1 dy_2$$

$$= \frac{1}{8} \int_0^\infty y_1 e^{-y_1/8} dy_1 = 8,$$

and similarly $E(Y_2) = 12$. Then $E(Y_1 + Y_2) = 20$.

24.13 (a) To determine c, we integrate

$$1 = c \int_0^\infty \int_{y_1}^\infty e^{.9y_1} e^{-y_2} dy_2 dy_1 = c \int_0^\infty e^{-.1y_1} dy_1 = \frac{c}{.1},$$

so $c = 1/10$.

The support and the joint density function are pictured:

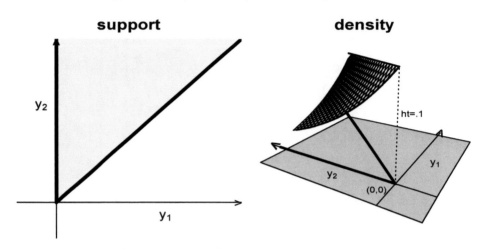

support   **density**

y_2

y_1

(b) To find the expected life of the power cell,

$$E(Y_2) = \frac{1}{10} \int_0^\infty \int_0^{y_2} y_2 e^{-y_2} e^{.9y_1} dy_1 dy_2$$

$$= \frac{1}{10} \int_0^\infty y_2 e^{-y_2} \left[\frac{1}{.9} (e^{.9y_2} - 1) \right] dy_2$$

$$= \frac{1}{9} \left[\int_0^\infty y_2 e^{-.1y_2} dy_2 - \int_0^\infty y_2 e^{-y_2} dy_2 \right]$$

$$= \frac{1}{9} \left[\frac{1}{.1^2} - 1 \right] = 11.$$

(c) We already know $E(Y_2)$, so let's find $E(Y_1)$:

$$E(Y_1) = \frac{1}{10} \int_0^\infty \int_{y_1}^\infty y_1 e^{.9y_1} e^{-y_2} dy_2 dy_1$$

$$= \frac{1}{10} \int_0^\infty y_1 e^{.9y_1} \left[e^{-y_1} \right] dy_1$$

$$= \frac{1}{10} \int_0^\infty y_1 e^{-.1y_1} dy_1 = \frac{1}{10} \frac{1}{.1^2} = 10.$$

Now, $E(Y_2 - Y_1) = 11 - 10 = 1$ is the expected amount of humming time before the machine fails.

24.15 Let Y_1 and Y_2 be the times for the two deliveries; then

$$P(Y_1 + Y_2 < 3) = \int_0^3 \int_0^{3-y_1} e^{-(y_1+y_2)} dy_2 dy_1 \approx .80.$$

25.1   (a) The marginal for Y_1 is

$$f_1(y_1) = \begin{cases} 1 + y_1 & \text{for } y_1 \in (-1, 0], \\ 1 - y_1 & \text{for } y_1 \in (0, 1). \end{cases}$$

(b) The marginal for Y_2 is $f_2(y_2) = 2 - 2y_2$, on $(0, 1)$.

(c) The following code simulates from the joint density and makes the histograms shown below.

```
n=100000
y1=runif(n,-1,1)
y2=runif(n)
keep=y1+y2<1&y2<y1+1
y1=y1[keep]
y2=y2[keep]
hist(y1,freq=FALSE,br=50,main="marginal for y1")
lines(c(-1,0),c(0,1),lwd=2,col=2)
lines(c(0,1),c(1,0),lwd=2,col=2)
hist(y2,freq=FALSE,br=50,main="marginal for y2")
lines(c(0,1),c(2,0),lwd=2,col=2)
```

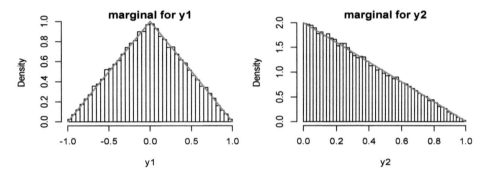

25.3 (a) Integrating over the values of y_2 for a fixed y_1, we get the marginal

$$f_1(y_1) = \frac{9}{5} \int_0^1 (1 - \sqrt{y_1 y_2})\, dy_2 = \frac{9}{5} - \frac{6}{5}\sqrt{y_1} \text{ for } y_1 \in (0, 1).$$

(b) By symmetry we have

$$f_2(y_2) = \frac{9}{5} - \frac{6}{5}\sqrt{y_2} \text{ for } y_2 \in (0, 1).$$

(c) The code to simulate from the joint density and make a histogram of the y_2 values is shown on the left, and the histogram is shown on the right, with the density function superimposed to verify our calculations.

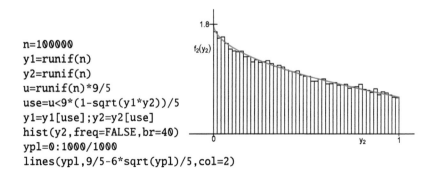

```
n=100000
y1=runif(n)
y2=runif(n)
u=runif(n)*9/5
use=u<9*(1-sqrt(y1*y2))/5
y1=y1[use];y2=y2[use]
hist(y2,freq=FALSE,br=40)
ypl=0:1000/1000
lines(ypl,9/5-6*sqrt(ypl)/5,col=2)
```

25.5 (a) Given $y_1 \in (0,1)$, we "integrate out" y_2:

$$f_1(y_1) = 24 \int_{y_1}^1 y_1(1-y_2)dy_2 = 12y_1(1-y_1)^2$$

on $y_1 \in (0,1)$.

(b) Given $y_2 \in (0,1)$, we "integrate out" y_1:

$$f_2(y_2) = 24 \int_0^{y_2} y_1(1-y_2)dy_1 = 12y_2^2(1-y_2)$$

on $y_2 \in (0,1)$.

(c) Code to sample from the joint density and make histograms of the marginals:

```
n=100000
y1=runif(n)
y2=runif(n)
use=y1<=y2
y1=y1[use]
y2=y2[use]
new.n=sum(use)
u=runif(new.n,0,24)
keep=u<24*y1*(1-y2)
y1=y1[keep]
y2=y2[keep]

hist(y1,br=50,freq=FALSE,main="marginal for y1")
ypl=0:100/100
lines(ypl,12*ypl*(1-ypl)^2,col=2,lwd=2)
hist(y2,br=50,freq=FALSE,main="marginal for y2")
lines(ypl,12*ypl^2*(1-ypl),col=2,lwd=2)
```

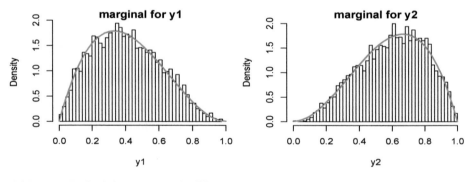

25.7 (a) To find the marginal for Y_1:

$$f_1(y_1) = \frac{3}{8}e^{-y_1/2}\int_0^{y_1} e^{-y_2/4}dy_2 = \frac{3}{2}e^{-y_1/2}(1-e^{-y_1/4}),$$

which is not one of our named densities.

(b) To find the marginal for Y_2:

$$f_2(y_2) = \frac{3}{8}e^{-y_2/4}\int_{y_2}^{\infty} e^{-y_1/2}dy_2 = \frac{3}{4}e^{-3y_2/4};$$

we recognize this as the exponential distribution with mean $4/3$.

25.9 We get the marginal for Y_1 by integrating over the range of Y_2 for a fixed value of y_1:

$$f_1(y_1) = 6 \int_0^{1-y_1} (1 - y_1 - y_2)dy_2 = 3(1-y_1)^2 \text{ for } y_1 \in (0,1).$$

This is the density for the proportion of time, in a randomly selected hour, that worker 1 is digging. Similarly, the density for the proportion of time, in a randomly selected hour, that worker 2 is digging is $f_2(y_2) = 3(1-y_2)^2$ for $y_2 \in (0,1)$.

26.1 (a) The marginal for Y_1 is

$$f_1(y_1) = 2e^{-y_1}(1 - e^{-y_1}) \text{ for } y_1 > 0,$$

so the conditional distribution for $y_1 > 0$ is

$$f(y_2|y_1) = \frac{\frac{2}{a}e^{-y_1}e^{-y_2/a}I\{0 \le y_2 \le ay_1\}}{2e^{-y_1}(1 - e^{-y_1})}$$

$$= \frac{e^{-y_2/a}I\{0 \le y_2 \le ay_1\}}{a(1 - e^{-y_1})}.$$

This is a truncated exponential. Here is the plot for $a = 2$ and $y_1 = 3$:

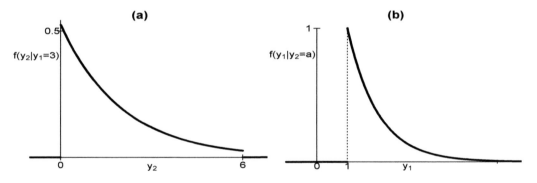

(b) The marginal for Y_2 is

$$f_2(y_2) = \frac{2}{a}e^{-2y_2/a} \text{ for } y_2 > 0,$$

so the conditional distribution for Y_1, given $Y_2 = y_2$, is

$$f(y_1|y_2) = \frac{\frac{2}{a}e^{-y_1}e^{-y_2/a}I\{y_2 \le ay_1 < \infty\}}{\frac{2}{a}e^{-2y_2/a}}$$

$$= e^{-y_1+1} \text{ for } y_1 > 1.$$

26.3 (a) We compute the probability that they both have lifetimes longer than one day, then subtract this from one:

$$P(Y_1 > 1 \text{ and } Y_2 > 1) = 2 \int_1^\infty \int_1^\infty e^{-y_1}e^{-2y_2} dy_1 dy_2 = e^{-3},$$

so the probability that at least one fails within one day is $1 - e^{-3} \approx .950$.

(b) Integrating out the second lifetime gives $f_1(y_1) = e^{-y_1}$ on $y > 0$.

(c) Yes, we can factor the joint densities into the marginals $f_1(y_1)$ and $f_2(y_2) = 2e^{-2y_2}$, both with support on the positive reals.

26.5 We computed the marginal for Y_2, in Exercise 25.9, to be

$$f_2(y_2) = 3(1 - y_2)^2 \text{ for } y_2 \in (0, 1),$$

so the conditional distribution for Y_1, given $Y_2 = y_2$, is

$$f(y_1|y_2) = \frac{6(1 - y_1 - y_2)I\{0 < y_1 < 1 - y_2\}}{3(y_2 - 1)^2}.$$

Plugging in $1/2$ for y_2, we get

$$f(y_1|y_2 = 1/2) = \frac{6(1/2 - y_1)I\{0 < y_1 < 1/2\}}{3/4}$$

$$= 8(1/2 - y_1) \text{ for } y_1 \in (0, 1/2).$$

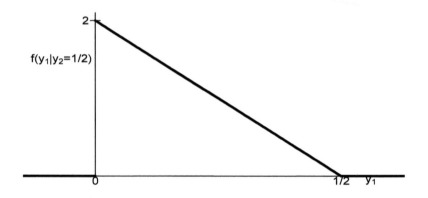

26.7 To find the joint distribution of Y_1 and $S = Y_1 + Y_2$, we first find the joint CDF, then we take derivatives. We have for $0 < y < s < \infty$, $F_{Y_1,S}(y, s) = P(Y_1 \le y, S \le s)$. We integrate the joint density for Y_1 and Y_2 over the shaded region:

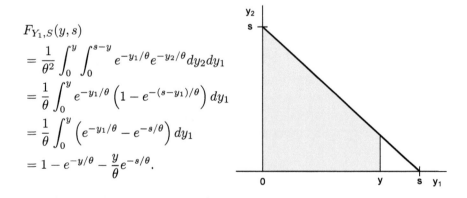

$$F_{Y_1,S}(y, s)$$

$$= \frac{1}{\theta^2} \int_0^y \int_0^{s-y} e^{-y_1/\theta} e^{-y_2/\theta} dy_2 dy_1$$

$$= \frac{1}{\theta} \int_0^y e^{-y_1/\theta} \left(1 - e^{-(s-y_1)/\theta}\right) dy_1$$

$$= \frac{1}{\theta} \int_0^y \left(e^{-y_1/\theta} - e^{-s/\theta}\right) dy_1$$

$$= 1 - e^{-y/\theta} - \frac{y}{\theta} e^{-s/\theta}.$$

Now

$$f_{Y_1,S}(y,s) = \frac{\partial F^2_{Y_1,S}}{\partial y \partial s} = \frac{1}{\theta^2} e^{-s/\theta} I\{0 < y < s < \infty\}.$$

The marginal for S is Gamma$(2, 1/\theta)$, so the conditional distribution is

$$f(y_1|s) = \frac{\frac{1}{\theta^2} e^{-s/\theta} I\{0 < y < s < \infty\}}{\frac{1}{\theta^2 \Gamma(2)} s e^{-s/\theta}} = \frac{1}{s} I\{0 < y < s < \infty\}.$$

In other words, given the sum $Y_1 + Y_2 = s$, the distribution of Y_1 is uniform on $(0, s)$.

26.9 Using the definitions of variance and independence, we have

$$\begin{aligned}
V(XY) &= E(X^2 Y^2) - E(XY)^2 \\
&= E(X^2) E(Y^2) - \mu_X^2 \mu_Y^2 \\
&= (\mu_X^2 + \sigma_X^2)(\mu_Y^2 + \sigma_Y^2) - \mu_X^2 \mu_Y^2 \\
&= \sigma_X^2 \sigma_Y^2 + \mu_X^2 \sigma_Y^2 + \mu_Y^2 \sigma_X^2.
\end{aligned}$$

27.1 (a) By definition,

$$\begin{aligned}
\mathrm{cov}(X_1, X_2) &= \mathrm{cov}(a_1 Y_1 + b_1, a_2 Y_2 + b_2) \\
&= E[(a_1 Y_1 + b_1)(a_2 Y_2 + b_2)] - E(a_1 Y_1 + b_1) - E(a_2 Y_2 + b_2) \\
&= E[a_1 a_2 Y_1 Y_2 + a_1 b_2 Y_1 + a_2 b_1 Y_2 + b_1 b_2] \\
&\quad - (a_1 \mu_1 + b_1)(a_2 \mu_2 + b_2) \\
&= a_1 a_2 E(Y_1 Y_2) + a_1 b_2 \mu_1 + a_2 b_1 \mu_2 + b_1 b_2 \\
&\quad - a_1 a_2 \mu_1 \mu_2 - a_1 b_2 \mu_1 - a_2 b_1 \mu_2 - b_1 b_2 \\
&= a_1 a_2 E(Y_1 Y_2) - a_1 a_2 \mu_1 \mu_2 \\
&= a_1 a_2 \sigma_{12}.
\end{aligned}$$

(b) We know the variance of X_1 is $a_1^2 \sigma_1^2$, and the variance of X_2 is $a_2^2 \sigma_2^2$, so if a_1 and a_2 are both positive, the correlation between X_1 and X_2 is

$$\frac{a_1 a_2 \sigma_{12}}{a_1 \sigma_1 a_2 \sigma_2} = \rho_{12},$$

the correlation between Y_1 and Y_2.

(c) Similarly, if a_1 and a_2 are both negative,

$$\frac{a_1 a_2 \sigma_{12}}{|a_1| \sigma_1 |a_2| \sigma_2} = \rho_{12},$$

the correlation between Y_1 and Y_2 (because a negative times a negative is a positive).

(d) If a_1 is positive and a_2 is negative,

$$\frac{a_1 a_2 \sigma_{12}}{a_1 \sigma_1 |a_2| \sigma_2} = -\rho_{12},$$

the negative of the correlation between Y_1 and Y_2.

27.3   (a) The integral of the density over the support is $3c/4$; therefore $c = 4/3$. To compute the covariance we first compute

$$E(Y_1 Y_2) = \frac{4}{3} \int_0^1 \int_0^1 y_1 y_2 (1 - y_1 y_2) dy_1 dy_2 = 5/27.$$

Second,

$$E(Y_2) = \frac{4}{3} \int_0^1 \int_0^1 y_2 (1 - y_1 y_2) dy_1 dy_2 = 4/9,$$

and by a symmetry argument $E(Y_1) = E(Y_2)$. So, the covariance is $-1/81$.

(b) No, the covariance is not zero, so the random variables cannot be independent.

(c) The following code generates 10,000 values of the jointly distributed random variables and confirms that the sample covariance is close to $-1/81$.

```
n=10000
y1=runif(n)
y2=runif(n)
u=runif(n,0,4/3)
keep=4/3*(1-y1*y2)>u
y1=y1[keep]
y2=y2[keep]
cov(y1,y2)
```

27.5   (a) $\mathrm{cov}(Y_1, Y_2) = E(Y_1 Y_2) - E(Y_1) E(Y_2) = 1/3 - (1/4)(1/4) = 1/12$.

(b) No; if they were independent, the covariance would have to be zero.

27.7   (a) To start, we calculate

$$E(Y_1 Y_2) = (1 - a)(1 - ab) \sum_{y_2=0}^{\infty} \sum_{y_1=y_2}^{\infty} y_1 y_2 a^{y_1} b^{y_1 - y_2}$$

and use the change of variable $u = y_1 - y_2$ and $v = y_2$. Then

$$E(Y_1 Y_2) = (1 - a)(1 - ab) \sum_{v=0}^{\infty} \sum_{u=0}^{\infty} (u + v) v a^{u+v} b^u$$

$$= (1 - a)(1 - ab) \sum_{v=0}^{\infty} \sum_{u=0}^{\infty} u v a^v (ab)^u$$

$$+ (1 - a)(1 - ab) \sum_{v=0}^{\infty} \sum_{u=0}^{\infty} u v^2 a^v (ab)^u$$

$$= \frac{1}{(1 - a)(1 - ab)} + \frac{1 - a}{(1 - ab)} \sum_{v=0}^{\infty} v^2 a^v$$

$$= \frac{1}{(1 - a)(1 - ab)} + \frac{a^2 + a - 1}{(1 - a)^3}$$

$$= \frac{a^2}{(1 - a)^2 (1 - ab)}.$$

(b) No; if they were independent, the covariance would have to be zero.

27.9 We answer (b) first.

(b) We can factor the joint density into marginals

$$f_1(y_1) = \frac{1}{8}e^{-y_1/8} \quad \text{for } y_1 > 0$$

and

$$f_2(y_2) = \frac{1}{12}e^{-y_1/12} \quad \text{for } y_2 > 0$$

and hence the random variables are independent. In fact the concentrations follow exponential distributions.

(a) The covariance must be zero.

27.11 (a) The integral of the density over the support is $c/24$; therefore, $c = 24$.

(b) For any $y_2 \in (0,1)$,

$$f_2(y_2) = 24 \int_0^{1-y_2} y_1 y_2 dy_1 = 12y_2(1-y_2)^2,$$

which is a $\text{Beta}(2,3)$ density. This describes the proportion of the hour that the female spends in nest building.

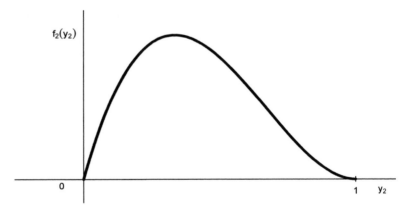

(c) By symmetry, this must be the same as in part (b).

(d) For any $y_1 \in (0,1)$, the formula gives

$$f(y_2|y_1) = \frac{24y_1y_2 I\{0 < y_2 < 1 - y_1\}}{12y_1(1-y_1)^2},$$

so plugging in $y_1 = 1/4$ gives

$$f(y_2|y_1 = 1/4) = \frac{32y_2}{9} I\{0 < y_2 < 3/4\},$$

which is the proportion of the time that the male spends building the nest, given that the female spends 1/4 hour.

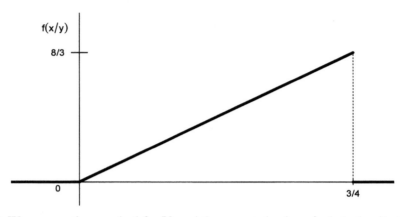

(e) We can use the marginal for Y_2 and the expected value of a beta density to get $E(Y_2) = 2/5$.

(f) First,

$$E(Y_1 Y_2) = 24 \int_0^1 \int_0^{1-y_2} y_1^2 y_2^2 \, dy_1 dy_2 = 8 \int_0^1 (1 - y_2)^3 y_2^2 dy_2$$

$$= \frac{8\Gamma(4)\Gamma(3)}{\Gamma(7)} = \frac{2}{15}.$$

Then $\operatorname{cov}(Y_1, Y_2) = -.02\overline{6}$. When the male spends more time on the nest, the female tends to spend less time on the nest.

(g) The code on the left generates a large sample from the joint density, makes a histogram of the marginal for Y_2, and checks the above answers.

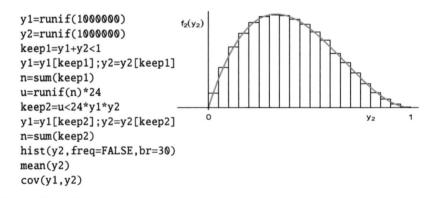

```
y1=runif(1000000)
y2=runif(1000000)
keep1=y1+y2<1
y1=y1[keep1];y2=y2[keep1]
n=sum(keep1)
u=runif(n)*24
keep2=u<24*y1*y2
y1=y1[keep2];y2=y2[keep2]
n=sum(keep2)
hist(y2,freq=FALSE,br=30)
mean(y2)
cov(y1,y2)
```

27.13 The covariance is

$$\operatorname{cov}\left(Y_1 + \cdots + Y_n, Y_1 x_1 + \cdots + Y_n x_n\right)$$

$$= \sum_{i=1}^n \operatorname{cov}(Y_i, x_i Y_i) + \sum_{i=1}^n \sum_{j \neq i} \operatorname{cov}(Y_i, x_j Y_j)$$

$$= \sum_{i=1}^n x_i \sigma^2 + 0 \text{ (by independence)}$$

$$= \sigma^2 \sum_{i=1}^n x_i = 0.$$

28.1 Let Y_1 be the number that lasts less than 20 days, let Y_2 be the number that lasts between 20 and 30 days, and let Y_3 be the number that lasts more than 30 days. Then the triplet (Y_1, Y_2, Y_3) follows a multinomial distribution with $n = 5$ and $p_1 = 1 - e^{-20/26} = .537$, $p_2 = e^{-20/26} - e^{-30/26} = .148$, and $p_3 = e^{-30/26} = .315$:

$$P(Y_1 = 2, Y_2 = 1, Y_3 = 2) = \frac{5!}{2!1!2!}(.537)^2(.148)^1(.315)^2 = .127.$$

28.3 (a) If Y_1 is the number with no flaws, Y_2 is the number with one flaw, and Y_3 is the number with no flaws, then we can calculate $p_1 = e^{-.2} \approx .8187$, $p_2 = .2e^{-.2} \approx .1637$, and $p_3 = .0176$. Then the required probability is

$$P(Y_1 = 5, Y_2 = 4, Y_3 = 1) = \frac{10!}{5!4!1!}(.8187)^5(.1637)^4(.0176) \approx .00588.$$

(b) The expected profit is $5(.8187) + .1637 - 2 \approx 2.26$.

28.5 (a) Using the formula, the probability is

$$\frac{8!}{2!2!4!}(.25)^2(.25)^2(.5)^4 \approx .410.$$

(b) It's easier to use binomial ideas: The probability of a single plant having red or pink flowers is .75, so the probability that all eight have red or pink is $(.75)^8 \approx .100$.

(c) The following R code generates multinomial random vector values representing white, pink, and red flowers, and increments whenever there are more white flowers than pink. The probability is about .204.

```
nloop=100000
flower=rmultinom(nloop,3,c(.25,.5,.25))
cnt=0
for(i in 1:nloop){
if(flower[1,i]>flower[2,i]){cnt=cnt+1}
}
cnt/nloop
```

28.7 (a) The probability that all ten have no flaws is $.6^{10} = .0060$, and the probability that nine have no flaws and one has one flaw (and none have more than one flaw) is

$$\frac{10!}{9! \times 1! \times 0!} = 10 \times .6^9 \times .2 = .0202,$$

so α is about .026.

(b) For the power we do the same calculation but with the new distribution: The probability of rejecting H_0 is $.85^{10} + 10 \times .85^9 \times .1 = .232$.

29.1 We start with the definition of expectation:

$$E(Y) = \sum_{all\ x} \sum_{all\ y} yP(X = x, Y = y)$$

$$= \sum_{all\ x} \sum_{all\ y} yP(Y = y|X = x)P(X = x)$$

$$= \sum_{all\ x} \left[\sum_{all\ y} yP(Y = y|X = x) \right] P(X = x)$$

$$= \sum_{all\ x} E(Y|X = x)P(X = x)$$

$$= E\left[E(Y|X)\right].$$

29.3 Let X be the roll of the dice. Then

$$E(Y) = E\left[E(Y|X)\right]$$
$$= E[.1X] = .1E(X) = .7.$$

On average, you lose 30 cents every time you play!

$$V(Y) = V\left[E(Y|X)\right] + E\left[V(Y|X)\right]$$
$$= V[.1X] + E[X(.1)(.9)]$$
$$= .01V(X) + .09E(X).$$

It's kind of a pain to compute the variance of X, but you can use this code:

```
x=2:12
for(i in 1:11){px[i]=min(x[i]-1,13-x[i])/36}
sum(x^2*px)-49
```

We get $V(X) = 35/6$, so $V(Y) \approx .688$

29.5 (a) Let Y be the tensile strength of a randomly selected rod. Then

$$E(Y) = E\left[E(Y|N)\right]$$
$$= E[100 - N] = 95$$

and

$$V(Y) = V\left[E(Y|N)\right] + E\left[V(Y|N)\right]$$
$$= V[100 - N] + E[6]$$
$$= 5 + 6 = 11.$$

(b) The median is about 95 (the histogram looks symmetric!), and the 5th percentile is about 89.4.

```
nloop=100000
str=1:nloop
for(iloop in 1:nloop){
    flaws=rpois(1,5)
```

```
        str[iloop]=rnorm(1,100-flaws,sqrt(6))
}
sstr=sort(str)
sstr[5000]   ## 5th percentile
sstr[50000]  ## median
sstr[95000]  ## 95th percentile
```

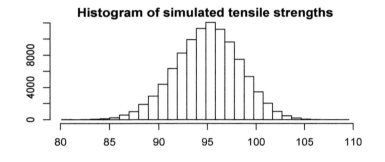

29.7   (a) We are given $X \sim \text{Beta}(3,6)$ and $Y|X = x \sim \text{Geom}(x)$. Then

$$
\begin{aligned}
\mathrm{E}(Y) &= \mathrm{E}[\mathrm{E}(Y|X)] \\
&= \mathrm{E}[(1-X)/X] \\
&= 168 \int_0^1 \frac{1-x}{x} x^2 (1-x)^5 dx \\
&= 168 \int_0^1 x(1-x)^6 dx \\
&= 168 \times \frac{1}{56} = 3.
\end{aligned}
$$

(b)

$$
\begin{aligned}
\mathrm{V}(Y) &= \mathrm{E}[\mathrm{V}(Y|X)] + \mathrm{V}[\mathrm{E}(Y|X)] \\
&= \mathrm{E}[(1-X)/X^2] + \mathrm{V}[(1-X)/X] \\
&= \mathrm{E}[(1-X)/X^2] + \mathrm{E}[(1-X)^2/X^2] - 9 \\
&= 24 + 21 - 9 = 36.
\end{aligned}
$$

(c) We have to do many simulations to get close to the computed variance (the variance of the variance is large!).

```
nsim=1000000
x=rbeta(nsim,3,6)
y=1:nsim
for(isim in 1:nsim){
y[isim]=rgeom(1,x[isim])
}
mean(y)
var(y)
```

29.9 Let X be the mercury concentration for a randomly selected water sample, and let Y be the number of eggs laid by a randomly selected dragonfly. Then

$$
\mathrm{E}(Y) = \mathrm{E}\left[\mathrm{E}(Y|X)\right] = \mathrm{E}\left(\frac{200}{X^2}\right)
$$

$$
= \int_0^\infty \frac{200}{x^2} \frac{1}{2^3\Gamma(3)} x^2 e^{-x/2} dx
$$

$$
= \frac{200}{8 \times 2} \int_0^\infty e^{-x/2} dx
$$

$$
= \frac{200}{8} = 25.
$$

29.11 Let X be the number of gold balls that you draw from the urn. Then $X|Y = y$ is hypergeometric with $M = y$, $N = 20 - y$, and $n = 3$. Using the formula for the expected value of such a hypergeometric, we have $\mathrm{E}(X|Y = y) = 3y/20$. Then

$$
\mathrm{E}(X) = \mathrm{E}[\mathrm{E}(Y|X)] = \mathrm{E}(3Y/20) = 6/20 = .3.
$$

Some code to check the answer:

```
nloop=100000
x=1:nloop
for(iloop in 1:nloop){
    y=rpois(1,2)
    if(y<21){
        urn=c(rep(1,y),rep(2,20-y))
        draw=sample(urn,3)
        x[iloop]=sum(draw==1)
    }else{
        x[iloop]=-99
    }
}
mean(x[x>=0])
```

30.1 To find the marginal for X_2, we have to do two cases: For $x_2 > 0$,

$$
f_2(x_2) = \frac{1}{2\theta^2} \int_{x_2}^\infty e^{-x_1/\theta} dx_1 = \frac{1}{2\theta} e^{-x_2/\theta},
$$

and for $x_2 < 0$,

$$
f_2(x_2) = \frac{1}{2\theta^2} \int_{-x_2}^\infty e^{-x_1/\theta} dx_1 = \frac{1}{2\theta} e^{x_2/\theta}.
$$

The two expressions can be combined:

$$
f_2(x_2) = \frac{1}{2\theta} e^{-|x_2|/\theta};
$$

this is the density for a *double exponential* random variable. The following R code makes the figure below.

```
y1=rexp(10000)
y2=rexp(10000)
x=y2-y1
hist(x,freq=FALSE,br=80,col='beige',main="")
xpl=-100:120/10
lines(xpl,exp(-abs(xpl))/2,col=2,lwd=3)
text(7,.35,expression(X==Y[2]-Y[1]),cex=2)
```

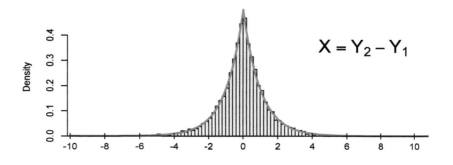

For $y \in (\theta, 2\theta)$, the total shaded area is formed by the support and the line $y_1 + y_2 = x$. We calculate $P(X \leq x)$ as the volume of the density over the total shaded area. This is

$$F_X(x) = P(X \leq x) = 1 - P(X > x) = 1 - \frac{1}{2\theta^2}(2\theta - x)^2.$$

Therefore, the density for X is

$$f_X(x) = \begin{cases} \frac{x}{\theta^2} & \text{for } x \in (0, \theta), \\ \frac{1}{\theta^2}(2\theta - x) & \text{for } x \in [\theta, 2\theta), \end{cases}$$

shown below:

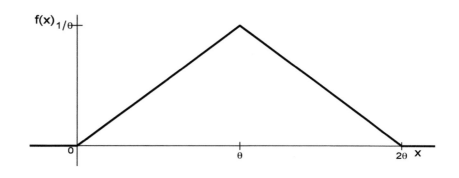

30.3 This is a slightly more complicated version of the previous problem. The support for $X = Y_1 + Y_2$ is $(0, 3\theta)$, and the density for X has three parts to the function, as indicated in the sketch of the support of (Y_1, Y_2) below.

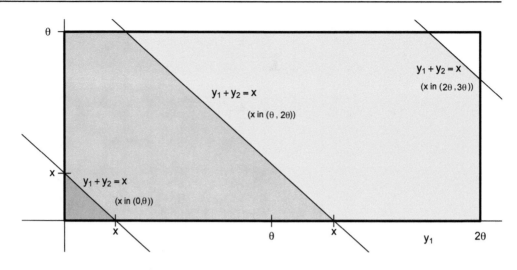

First, for $x \in (0, \theta)$,

$$F_X(x) = P(X \le x) = \frac{x^2}{4\theta^2}.$$

This is the darkest shaded area. (The height of the density is $1/(2\theta^2)$.) Second we consider $x \in (\theta, 2\theta)$,

$$F_X(x) = P(X \le x) = \frac{x - \theta}{2\theta} + \frac{1}{4}.$$

This is the area in the support, to the left of the middle diagonal line, divided by $2\theta^2$. We have added the areas of a rectangle and a triangle. Finally, for $x \in (2\theta, 3\theta)$,

$$F_X(x) = P(X \le x) = 1 - P(X > x) = 1 - \frac{(3\theta - x)^2}{4\theta^2},$$

where we have subtracted the volume over the white triangle from one. Combining these parts of the CDF for X, and taking the derivative with respect to x, we get

$$f_X(x) = \begin{cases} \frac{x}{2\theta^2} & \text{for } x \in (0, \theta], \\ \frac{1}{2\theta^2} & \text{for } x \in (\theta, 2\theta], \\ \frac{3\theta - x}{2\theta^2} & \text{for } x \in (2\theta, 3\theta]. \end{cases}$$

The sketch is shown below:

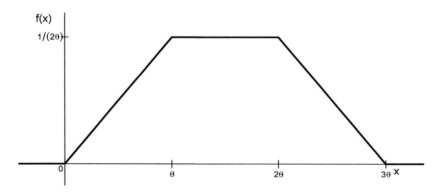

Finally, we check our answer using the following R code for $\theta = 2$.

```
y1=runif(100000,0,4)
y2=runif(100000,0,2)
x=y1+y2
hist(x,freq=FALSE,col='beige',br=100)
lines(c(0,2),c(0,1/4),col=2,lwd=2)
lines(c(2,4),c(1/4,1/4),col=2,lwd=2)
lines(c(4,6),c(1/4,0),col=2,lwd=2)
```

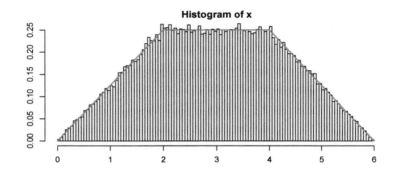

30.5 The random variable Y takes values in $(0,1)$. Using the CDF method, for $y \in (0,1)$,

$$F_Y(y) = \mathrm{P}(Y \le y) = \mathrm{P}\left(\frac{U_1}{U_1 + U_2} \le y\right)$$

$$= \mathrm{P}\left(U_2 \ge \left(\frac{1}{y} - 1\right)U_1\right).$$

The CDF can be calculated as an area determined by a line through the support, but these areas look different for $y < 1/2$ and $y > 1/2$, as shown in the plot on the right.

When $y < 1/2$, the probability that $U_2 \ge (1/y - 1)U_1$ is the darker shaded triangular area, but when $y > 1/2$, this probability corresponds to the total shaded area.

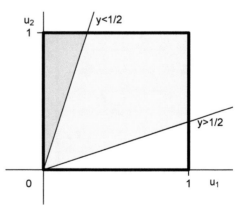

We have

$$F_Y(y) = \begin{cases} \frac{1}{2(1-y)} & \text{for } 0 < y \le 1/2, \\ \frac{3}{2} - \frac{1}{2y} & \text{for } 1/2 < y < 1, \end{cases}$$

so the density is

$$f_Y(y) = \begin{cases} \frac{1}{2(1-y)^2} & \text{for } 0 < y \le 1/2, \\ \frac{1}{2y^2} & \text{for } 1/2 < y < 1. \end{cases}$$

Finally, we check our calculations using simulations in R. The code on the left produces the histogram on the right.

```
u1=runif(100000)
u2=runif(100000)
y=u1/(u1+u2)
hist(y,freq=FALSE,br=50)
ypl=0:100/100
fy=ypl
fy[ypl<=1/2]=1/2/(1-ypl[ypl<=1/2])^2
fy[ypl>1/2]=1/2/ypl[ypl>1/2]^2
lines(ypl,fy,col=2,lwd=2)
```

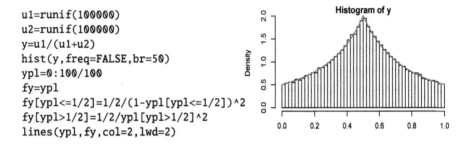

30.7 We notice that X takes values in $(0, \infty)$, and the height of the density function is 2 over the support and zero elsewhere.

Using the CDF method,

$$P(X \le x) = P(Y_2 \le xY_1) = \frac{x}{1+x},$$

where this probability is the shaded area of the support as shown below.

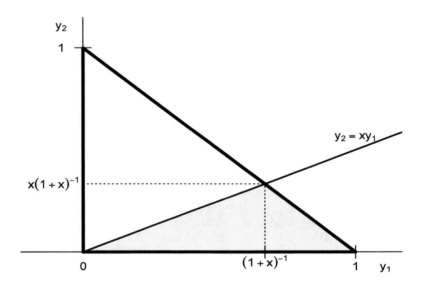

Therefore,

$$f_X(x) = \frac{d}{dx} \frac{x}{1+x} = \frac{1}{(1+x)^2}$$

for $x > 0$.

The code on the left produces the histogram on the right. The density is very "heavy-tailed," so we truncate the values in the histogram.

```
u1=runif(100000)
u2=runif(100000)
y1=u1[u1+u2<1]
y2=u2[u1+u2<1]
x=y2/y1
hist(x[x<10],freq=FALSE,br=50)
xpl=0:1000/100
lines(xpl,1/(1+xpl)^2,col=4,lwd=4)
```

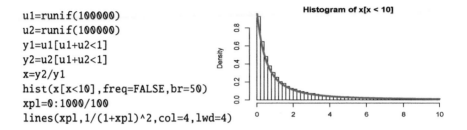

30.9 (a)

$$\frac{1}{8}\int_1^\infty\int_1^\infty y_1 e^{-(y_1+y_2)/2}dy_1dy_2 = \frac{1}{8}\int_1^\infty y_1 e^{-y_1/2}dy_1\int_1^\infty e^{-y_2/2}dy_2$$

$$= \frac{1}{8}\left[6e^{-1/2}\right]\left[2e^{-1/2}\right]$$

$$= \frac{12}{8}e^{-1} \approx .552.$$

(b) Yes, it factors into

$$f_1(y_1) = \frac{1}{4}y_1 e^{-y_1/2}$$

and

$$f_2(y_2) = \frac{1}{2}e^{-y_2/2}.$$

(c) From (b) we get $Y_1 \sim \text{Gamma}(2,2)$ and $Y_2 \sim \text{Exp}(2)$, and the sketch for $f_1(y_1)$ is

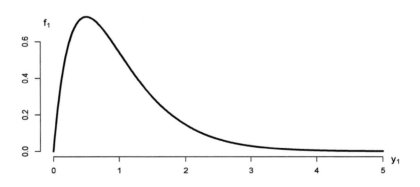

(d) For $X = Y_2/Y_1$, we note that the support for X is the positive reals. We can find the CDF for X:

$$F_X(x) = \text{P}(X \le x) = \text{P}(Y_2/Y_1 \le x) = \text{P}(Y_2 \le xY_1).$$

Now we draw the area of the support for (Y_1, Y_2) over which we integrate:

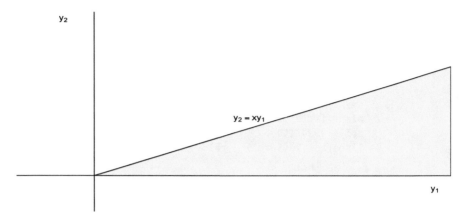

Then

$$P(Y_2 \leq xY_1) = \frac{1}{8} \int_0^\infty \int_0^{xy_1} y_1 e^{-y_1/2} e^{-y_2/2} dy_2 dy_1$$

$$= \frac{1}{4} \int_0^\infty y_1 e^{-y_1/2} \left[1 - e^{-xy_1/2}\right] dy_1$$

$$= 1 - \frac{1}{4} \int_0^\infty y_1 e^{-y_1/\left[\frac{2}{x+1}\right]} dy_1$$

$$= 1 - \frac{1}{4}\Gamma(2)\left(\frac{2}{x+1}\right)^2 \left\{ \frac{1}{\Gamma(2)\left(\frac{2}{x+1}\right)^2} \int_0^\infty y_1 e^{-y_1/\left[\frac{2}{x+1}\right]} dy_1 \right\}$$

$$= 1 - \frac{1}{(1+x)^2}$$

for $x \in (0, \infty)$. (We used the form of the gamma density with parameters 2 and $2/(x + 1)$ to get that the term in the curly brackets is one.) Then we take the derivative with respect to x to get

$$f_X(x) = \frac{2}{(1 + x)^3} \quad \text{for } x > 0.$$

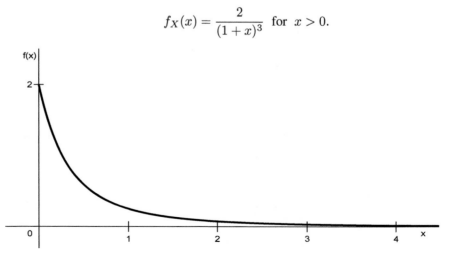

(e) The R code is

```
y1=rgamma(1000000,2,1/2)
y2=rexp(1000000,1/2)
x=y2/y1
hist(x[x<4],freq=FALSE,br=80)
xpl=0:400/100
lines(xpl,2/(xpl+1)^3,col=2,lwd=3)
```

and the histogram with density function is

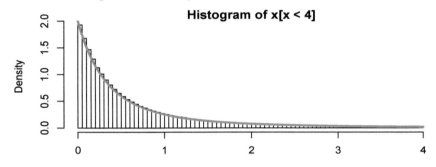

Histogram of x[x < 4]

30.11 (a) The inverse transformations are $y_1 = (x_1 + x_2)/2$ and $y_2 = (x_2 - x_1)/2$. The Jacobian is $J = 1/2$, so the joint density for X_1 and X_2 is

$$f_X(x_1,x_2)=e^{-(x_1+x_2)/2}e^{-(x_2-x_1)/2}I\{0\leq(x_2-x_1)/2\leq(x_2+x_1)/2<\infty\}$$
$$=e^{-x_2}I\{0\leq x_1\leq x_2<\infty\}.$$

(b) If we "integrate out" x_2, we get

$$f_{X_1}(x_1) = \int_{x_1}^{\infty} e^{-x_2}dx_2 = e^{-x_1}$$

for any $x_1 > 0$. So $X_1 \sim \text{Exp}(1)$.

(c) If we "integrate out" x_1, we get

$$f_{X_2}(x_2) = e^{-x_2}\int_0^{x_2} dx_1 = x_2e^{-x_2}$$

for any $x_2 > 0$. So $X_2 \sim \text{Gamma}(2,1)$.

31.1 The following code generates, in a loop, four independent normal variables, squares them, and adds the squares. The values of these sums are plotted in the following histogram, with a $\chi^2(4)$-density curve superimposed:

```
nloop=100000
y=1:nloop
for(iloop in 1:nloop){
    z=rnorm(4)
    y[iloop]=sum(z^2)
}
hist(y,freq=FALSE,breaks=50,main=expression(Y=Z[1]^2+Z[2]^2+Z[3]^2
                                            +Z[4]^2))
ypl=0:100/100*max(y)
lines(ypl,dchisq(ypl,4),col=2,lwd=2)
```

The plot is

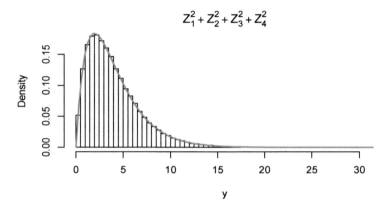

$$Z_1^2 + Z_2^2 + Z_3^2 + Z_4^2$$

31.3 We can find the MGF for the sum, using independence:

$$M_X(t) = \mathrm{E}(e^{t(Y_1+Y_2)}) = \mathrm{E}(e^{tY_1})\mathrm{E}(e^{tY_2}) = \left[e^{\lambda_1(e^t-1)}\right]\left[e^{\lambda_2(e^t-1)}\right]$$
$$= e^{(\lambda_1+\lambda_2)(e^t-1)},$$

which is the MGF for a $\mathrm{Pois}(\lambda_1 + \lambda_2)$ random variable.

31.5 If Y_1 and Y_2 are independent, then

$$M_X(t) = \mathrm{E}(e^{tX}) = \mathrm{E}(e^{t(Y_1+Y_2)}) = \mathrm{E}(e^{tY_1})\mathrm{E}(e^{tY_2}) = M_{Y_1}(t)M_{Y_2}(t).$$

We recall the MGF for $Y \sim \mathrm{Binom}(n,p)$ to be $M_Y(t) = (pe^t + 1 - p)^n$, so

$$M_X(t) = (pe^t + 1 - p)^{n_1}(pe^t + 1 - p)^{n_2} = (pe^t + 1 - p)^{n_1+n_2}.$$

We recognize this as a binomial MGF $X \sim \mathrm{Binom}(n_1 + n_2, p)$.

31.7 If $Y \sim \mathrm{Unif}(0,1)$ and $X = 1 - Y$,

$$M_X(t) = \mathrm{E}(e^{tX}) = \mathrm{E}(e^{t(1-Y)}) = e^t\mathrm{E}(e^{-tY}) = e^t M_Y(-t) = e^t\frac{e^{-t}-1}{-t} = \frac{e^t-1}{t},$$

which is the MGF for $\mathrm{Unif}(0,1)$.

31.9 If X_1, X_2, X_3, X_4, X_5 are independent exponential random variables with mean θ, then their sum $S = X_1 + X_2 + X_3 + X_4 + X_5$ has a $\mathrm{Gamma}(5, 1/\theta)$ density. The probability that the mean is greater than 100 is the probability that the sum is greater than 500, so the power of the test is found with `1-pgamma(500,5,1/72)`, which returns .178.

32.1 (a) From the exponent in the density function we see that

$$\Sigma^{-1} = \begin{pmatrix} 1 & 1/2 \\ 1/2 & 1 \end{pmatrix},$$

so

$$\Sigma = \frac{4}{3}\begin{pmatrix} 1 & -1/2 \\ -1/2 & 1 \end{pmatrix}.$$

We can read off $\sigma_1^2 = \sigma_2^2 = 4/3$ and $\sigma_{12} = -2/3$. Therefore $\rho = (-2/3)/\sqrt{(4/3)(4/3)} = -1/2$.

(b) To simulate we use the Cholesky decomposition of Σ:

```
amat=matrix(c(1,-.5,-.5,1),nrow=2)
amat=amat*4/3
umat=chol(amat)
z1=rnorm(1000)
z2=rnorm(1000)
y=cbind(z1,z2)%*%umat
y1=y[,1]
y2=y[,2]
plot(y1,y2)
cor(y1,y2)
```

32.3 For $\mu = \begin{pmatrix} 2 \\ 1 \end{pmatrix}$ and $\Sigma = \begin{pmatrix} 9 & 2 \\ 2 & 4 \end{pmatrix}$, we calculate $\Sigma^{-1} = \frac{1}{32}\begin{pmatrix} 4 & -2 \\ -2 & 9 \end{pmatrix}$ and

$$\begin{pmatrix} x-2 & y-1 \end{pmatrix}\begin{pmatrix} 9 & 2 \\ 2 & 4 \end{pmatrix}^{-1}\begin{pmatrix} x-2 \\ y-1 \end{pmatrix} = \frac{1}{8}(x-2)^2 - \frac{1}{8}(x-2)(y-1) + \frac{9}{32}(y-1)^2$$

and

$$\det\begin{pmatrix} 9 & 2 \\ 2 & 4 \end{pmatrix} = 32.$$

Then the density is

$$\frac{1}{8\pi\sqrt{2}}\exp\left\{-\left[\frac{1}{16}(x-2)^2 - \frac{1}{16}(x-2)(y-1) + \frac{9}{64}(y-1)^2\right]\right\}$$

for $(x,y) \in \mathbb{R}^2$.

33.1 If $Y_1, \ldots, Y_n \overset{ind}{\sim} f_Y(y)$, then

$$P(T_{(1)} \leq y) = 1 - P(\text{all } Y_i > y)$$
$$= 1 - [1 - F_Y(y)]^n.$$

Then
$$f_Y(y) = n[1 - F_Y(y)]^{n-1} f_Y(y).$$

This corresponds to the formula for the density of $Y_{(k)}$ when $k = 1$.

33.3 The following code simulates uniform random variables and finds the median of samples of size 9.

```
nloop=200000
y=1:nloop
for(iloop in 1:nloop){
    u=runif(9)
    y[iloop]=median(u)
}
hist(y,freq=FALSE,breaks=50,main="Y=median(U)",xlim=c(0,1))
ypl=0:100/100
lines(ypl,dbeta(ypl,5,5),col=2,lwd=2)
```

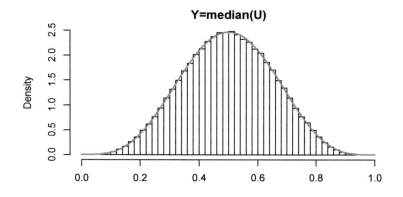

33.5 The probability that $Y_i < 3$ is $1 - e^{-3/4} = .528$ for $i = 1, \ldots, 8$. Then the probability that exactly two are less than 3 is

$$\binom{8}{2} (.528)^2 (1 - .528)^6 = .086.$$

33.7 The probability that $Y_i < c$ is $F_Y(c)$, so the probability that exactly two of the sample are less than c is

$$\binom{n}{2} [F_y(c)]^2 [1 - F_Y(c)]^{n-2}.$$

33.9 The CDF for the uniform density on (a, b) is

$$F_Y(y) = \frac{y - a}{b - a} \quad \text{for } y \in (a, b).$$

Then

$$F_X(x) = P(\max(Y_i) \le x) = \left(\frac{x - a}{b - a} \right)^n$$

and

$$f_X(x) = \frac{n(x - a)^{n-1}}{(b - a)^n} \quad \text{for } x \in (a, b).$$

33.11 We need the CDF of the random variable $Y = \max(X_1, X_2, X_3, X_4)$:

$$F_Y(y) = P(\max(X_1, X_2, X_3, X_4) \le y) = P(\text{all } X_1, X_2, X_3, X_4 \le y) = F_X(y)^4.$$

We can compute

$$F_X(x) = \int_0^x \frac{2t}{\theta} e^{-t^2/\theta} dt = \int_0^{x^2/\theta} e^{-u} du = 1 - e^{-x^2/\theta} \quad \text{for } x > 0,$$

using the substitution $u = x^2/\theta$. So,

$$F_Y(y) = \left[1 - e^{-y^2/\theta} \right]^4.$$

(a) The test size is $P(Y > 3)$ when $\theta = 2$; this is $1 - F_Y(3) = \left[1 - e^{-3^2/2}\right]^4 = .0437$.

(b) The power is $P(Y > 3)$ when $\theta = 4$; this is $\left[1 - e^{-3^2/4}\right]^4 = .3595$.

33.13 (a) For test A, the critical value is the 99th percentile of the distribution of the sample maximum. The CDF for the sample maximum is

$$F(y) = \left[1 - e^{-y/\theta}\right]^{10}.$$

Under $H_0 : \theta = 1$, we see that $F(6.90) = .99$, so we reject H_0 when the sample maximum is greater than 6.90.

(b) For test B, the critical value is the 99th percentile of the distribution of the sample mean. The distribution of the sample *sum* is Gamma$(10, \theta)$, and under $H_0 : \theta = 1$, the 99th percentile is 18.8. So, we reject H_0 if the sample mean is greater than 1.88.

(c) If $\theta = 1.5$, the probability that the sample maximum is greater than 6.9 is .096. The probability that the sample sum is greater than 18.8 is .199. We explain to the CEO that if she uses statistician A's test, the power is only .096; that is, if the true mean lifetime is 1.5 instead of 1, the test proposed by statistician A has only a 9.6% chance of detecting this. The power for statistician B's test is much larger (about 20%) but still rather small. Maybe a larger sample size should be used!

33.15 (a) This happens when $Y_1 < Y_2/2$ or when $Y_2 < Y_1/2$. The lines $Y_1 = 2Y_2$ and $Y_2 = 2Y_1$ are shown in the picture; we have to integrate the joint density over the shaded areas:

$$P(Y_1 > 2Y_2 \text{ or } Y_2 > 2Y_1)$$
$$= 2\theta^2 \int_0^\infty \int_{2y_2}^\infty e^{-\theta y_1} e^{-\theta y_2} dy_1\, dy_2$$
$$= 2\theta \int_0^\infty e^{-3\theta} y_2\, dy_2 = 2/3.$$

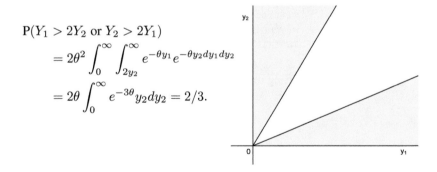

(b) The following code confirms this proportion.

```
y1=rexp(10000,1/2)
y2=rexp(10000,1/2)
s1=sum(y1>2*y2)
s2=sum(y2>2*y1)
(s1+s2)/10000
```

34.1 The plot shows how the t gets close to the normal as the degrees of freedom get
large.

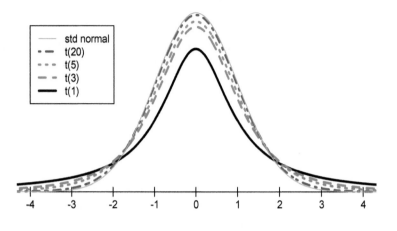

34.3 `pt(1.2,12)`$\wedge$`2` gives .7627.

34.5 (a) By symmetry, the answer must be 1/2, because it's equally likely that W_1 is
larger or that W_2 is larger.

(b) $P(W_1/W_2 > 2) = P(Y > 2)$ if $Y \sim F(12, 12)$. `1-pf(2,12,12)` gives .122.

(c) Same as the answer to (a)!

(d) Same as the answer to (b)!

(e) The following code estimates the probability.

```
w1=rchisq(100000,12)
w2=rchisq(100000,12)
sum(w1>2*w2)/100000
```

34.7 (a) Using the expression for the chi-squared density,

$$
\begin{aligned}
\mathrm{E}(1/X) &= \frac{1}{\Gamma(\nu/2)2^{\nu/2}} \int_0^\infty (1/x)x^{\nu/2-1}e^{-x/2}dx \\
&= \frac{1}{\Gamma(\nu/2)2^{\nu/2}} \int_0^\infty x^{\nu/2-2}e^{-x/2}dx \\
&= \frac{1}{\Gamma(\nu/2)2^{\nu/2}}\Gamma(\nu/2-1)2^{\nu/2-1} \\
&= \frac{1}{\nu-2}.
\end{aligned}
$$

A bit of R code,

```
x=rchisq(100000,4)
mean(1/x)
```

confirms that $\mathrm{E}(1/X) = 1/2$ when $X \sim \chi^2(4)$.

(b) We can find

$$E(1/X^2) = \frac{1}{\Gamma(\nu/2)2^{\nu/2}} \int_0^\infty (1/x^2)x^{\nu/2-1}e^{-x/2}dx$$

$$= \frac{1}{\Gamma(\nu/2)2^{\nu/2}} \int_0^\infty x^{\nu/2-3}e^{-x/2}dx$$

$$= \frac{1}{\Gamma(\nu/2)2^{\nu/2}}\Gamma(\nu/2-2)2^{\nu/2-2}$$

$$= \frac{1}{2^2(\nu/2-2)(\nu/2-1)} = \frac{1}{(\nu-2)(\nu-4)}.$$

Then

$$V(1/X) = \frac{1}{(\nu-2)(\nu-4)} - \frac{1}{(\nu-2)^2} = \frac{2}{(\nu-2)^2(\nu-4)}.$$

A bit of R code,

```
k=6
y=rchisq(100000,k)
var(1/y)
2/(k-2)^2/(k-4)
```

confirms that $V(1/X) = 1/16$ when $X \sim \chi^2(6)$.

34.9 We know that

$$\frac{X_i - \mu}{\sigma} \sim N(0,1), \quad \text{hence} \quad \left(\frac{X_i - \mu}{\sigma}\right)^2 \sim \chi^2(1).$$

By independence, we have

$$\sum_{i=1}^n \left(\frac{X_i - \mu}{\sigma}\right)^2 \sim \chi^2(n), \quad \text{hence} \quad E\left[\frac{\sum_{i=1}^n (X_i - \mu)^2}{\sigma^2}\right] = n,$$

and from this we have

$$E\left[\frac{\sum_{i=1}^n (X_i - \mu)^2}{n}\right] = \sigma^2.$$

34.11 We know that

$$\frac{X_1 - X_2}{\sqrt{2\sigma^2}} \sim N(0,1)$$

and that

$$\frac{Y_1^2}{2\sigma^2} + \frac{Y_2^2}{2\sigma^2} \sim \chi^2(2).$$

Therefore,

$$\frac{(X_1 - X_2)/\sqrt{2\sigma^2}}{\sqrt{\left(\frac{Y_1^2 + Y_2^2}{2\sigma^2}\right)/2}} \sim t(2),$$

and $a = \sqrt{2}$. The following code produces the probability plot.

```
sig=2;mu=3
nloop=1000
td=1:nloop
for(iloop in 1:nloop){
    x1=rnorm(1,mu,sig)
    x2=rnorm(1,mu,sig)
    y1=rnorm(1,0,sqrt(2)*sig)
    y2=rnorm(1,0,sqrt(2)*sig)
    td[iloop]=(x1-x2)/sqrt(y1^2+y2^2)
}
td=td*sqrt(2)
plot(qt(1:nloop/(nloop+1),2),sort(td),xlab="t(2) quantiles",
     ylab="sorted sample")
lines(c(-40,40),c(-40,40),col=2,lwd=2)
```

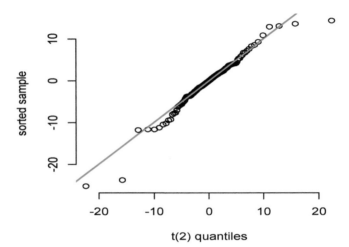

34.13 Let

$$X = \frac{Z}{W/\sqrt{k}},$$

where $Z \sim \mathrm{N}(0,1)$ and $W \sim \chi^2(k)$ are independent. Then

$$X^2 = \frac{Z^2}{W/k},$$

where $Z^2 \sim \chi^2(1)$ and $W \sim \chi^2(k)$ are independent. Then $X^2 \sim F(1,k)$ by definition of the F random variable.

34.15 We know that

$$\left[\frac{X_1 - \mu}{\sigma}\right]^2 \sim \chi^2(1)$$

and

$$\left(\frac{X_2}{\sigma}\right)^2 + \left(\frac{X_3}{\sigma}\right)^2 + \left(\frac{X_3}{\sigma}\right)^2 \sim \chi^2(3),$$

so

$$\frac{3(X_1 - \mu)^2}{X_2^2 + X_3^2 + X_4^2} \sim F(1,3).$$

Finally,

$$P\left(\frac{(X_1 - \mu)^2}{X_2^2 + X_3^2 + X_4^2} > 2\right) = P\left(\frac{3(X_1 - \mu)^2}{X_2^2 + X_3^2 + X_4^2} > 6\right),$$

so the probability is the area to the **right** of **6** under an **$F(1,3)$**-density.

34.17 We know that

$$\frac{X - \mu}{\sqrt{2}\sigma} \sim N(0, 1)$$

and

$$\left(\frac{Y_1}{\sigma}\right)^2 + \cdots + \left(\frac{Y_{18}}{\sigma}\right)^2 \sim \chi^2(18),$$

so

$$\frac{(X - \mu)/(\sqrt{2}\sigma)}{\sqrt{\left[\left(\frac{Y_1}{\sigma}\right)^2 + \cdots + \left(\frac{Y_{18}}{\sigma}\right)^2\right]/18}} \sim t(18),$$

so

$$\frac{3(X - \mu)}{\sqrt{Y_1^2 + \cdots + Y_{18}^2}} \sim t(18),$$

so

$$P\left(\frac{X - \mu}{\sqrt{Y_1^2 + \cdots + Y_{18}^2}} > 1\right) = P\left(\frac{3(X - \mu)}{\sqrt{Y_1^2 + \cdots + Y_{18}^2}} > 3\right)$$

is the area to the **right** of **3** under a **$t(18)$** density.

34.19 We know that $\bar{X}$ is normal with mean μ and variance $4/10$, so

$$Z = \frac{\bar{X} - \mu}{4/\sqrt{10}} \sim N(0, 1).$$

We reject H_0 if Z is larger than 1.645, the 95th percentile of a standard normal random variable. This is equivalent to "reject H_0 when $\bar{X} > 2.04$."

34.21 If $a = 1/\sigma^2$ and $b = 1/(2\sigma^2)$, the sum is a $\chi^2(m + n)$ random variable.

35.1 (a)

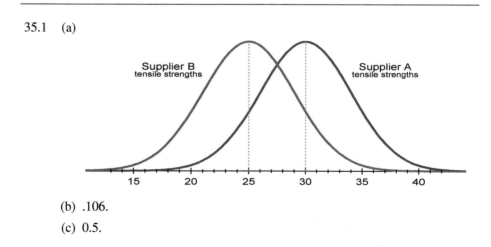

(b) .106.

(c) 0.5.

(d) The distributions of the sample means:

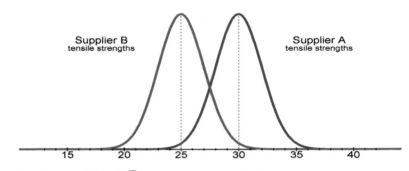

(e) $Z = (25 - 30)/(4/\sqrt{4}) = -2.5$, so the probability is 0.006.

(f) 0.5.

(g) For $z = -1.645$, the area to the left is 0.05. Solve $z = (x - \mu)/(\sigma/\sqrt{n})$ to get $x = 26.71$. So, we reject H_0 if the average of the 4 tensile strengths is less than 26.71.

(h) The power for the test is the probability of getting an average tensile strength of less than 26.71 (rejecting H_0) when the spool is from Supplier B (H_a is true). This is the area to the left of 26.71 under a normal density with mean 25 and standard deviation 2. We get power $= 0.80$.

35.3 These random variables all have mean zero and standard deviation 1:

$$\frac{X_1 - \mu}{2}, \frac{X_2 - \mu}{2}, \frac{X_3 - \mu}{3}, \frac{X_4 - \mu}{3}, \quad \text{and} \quad \frac{X_5 - \mu}{3}.$$

Therefore,

$$\frac{1}{\sqrt{5}} \left[\frac{X_1 - \mu}{2} + \frac{X_2 - \mu}{2} + \frac{X_3 - \mu}{3} + \frac{X_4 - \mu}{3} + \frac{X_5 - \mu}{3} \right] \sim N(0, 1).$$

(a) So, the test statistic is

$$Z = \frac{1}{\sqrt{5}} \left[\frac{X_1 - 100}{2} + \frac{X_2 - 100}{2} + \frac{X_3 - 100}{3} + \frac{X_4 - 100}{3} + \frac{X_5 - 100}{3} \right]$$

and we reject when $Z > 1.645$, the 95th percentile of a standard normal density.

(b) The test statistic value is 1.12, so we accept H_0.

35.5 (a) We need the probability that all 10 values are less than c_1 to be .99 when H_0 is true, so $F(c_1)^{10} = .99$, or $F(c_1) = .99^{1/10} = .999$, where F is the CDF for the standard normal random variable.. Using R, qnorm(.999) returns $c_1 = 3.09$.

(b) Under H_0, $\sum_{i=1}^{10} X_i^2$ has a $\chi^2(10)$ distribution, so we find c_2 with qchisq(.99,10), which returns $c_2 = 23.2$.

(c) For the test in part (a), the power is the probability that at least one value of the sample is greater than $c_1 = 3.09$, when the distribution is $N(0, 2)$. The probability that one is less than 3.09 is found with pnorm(3.09,0, sqrt(2)), which returns .9856. Then the probability that all ten are less than 3.09 is $(.9856)^{10} = .6850$, so the power is about .135.

For the test in part (b), the power is the probability that $\sum_{i=1}^{10} X_I^2$ is greater than $c_2 = 23.2$ when the distribution is $N(0, 2)$. This is

$$P\left(\sum_{i=1}^{10} X_i^2 > 23.2\right) = P\left(\sum_{i=1}^{10} \left[\frac{X_i}{\sqrt{2}}\right]^2 > \frac{23.2}{2}\right).$$

Each $X_i/\sqrt{2}$ is standard normal under the alternative, so the sum of the squares is $\chi^2(10)$. The power is found with 1-$\text{pchisq}(23.2/2,10)$, which returns $.313$. The test with the sum of squared values has much greater power!

35.7 We know that X_i^2/σ^2 are $\chi^2(1)$ for $i = 1,\ldots,4$ and $X_i^2/(2\sigma^2)$ are $\chi^2(1)$ for $i = 5,\ldots,8$, and all of these chi-squared random variables are independent. Therefore,

$$W = \frac{1}{\sigma^2}\left[\sum_{i=1}^{4} X_i^2 + \frac{1}{2}\sum_{i=5}^{8} X_i^2\right] \sim \chi^2(8),$$

and when H_0 is true,

$$W = \frac{1}{10}\left[\sum_{i=1}^{4} X_i^2 + \frac{1}{2}\sum_{i=5}^{8} X_i^2\right] \sim \chi^2(8).$$

(a) Reject H_0 if W is greater than 13.36, the 90th percentile of $\chi^{(}8)$.

(b) If the true σ^2 is 12, then $10W/12 \sim \chi^2(8)$, so the power is

$$P(W > 13.36)P(10W/12 > 11.13) = .194.$$

(c) The following code confirms the power calculation.

```
nloop=100000
nrej=0
x=1:8
sig=sqrt(12)
for(iloop in 1:nloop){
    x[1:4]=rnorm(4,0,sig)
    x[5:8]=rnorm(4,0,sqrt(2)*sig)
    w=(sum(x[1:4]^2)+sum(x[5:8]^2/2))/10
    if(w>qchisq(.9,8)){nrej=nrej+1}
}
nrej/nloop
```

35.9 (a) If $X_1,\ldots,X_5 \sim N(\mu,1)$ and $Y_1,\ldots,Y_{10} \sim N(\mu,2)$, then

$$X_1 + \cdots + X_5 + Y_1 + \cdots + Y_{10} \sim N(15\mu, 25),$$

and so

$$\frac{X_1 + \cdots + X_5 + Y_1 + \cdots + Y_{10} - 15\mu}{\sqrt{25}} = \frac{5\bar{X} + 10\bar{Y} - 15\mu}{5}$$
$$= \bar{X} + 2\bar{Y} - 3\mu \sim N(0,1).$$

Let $W = \bar{X} + 2\bar{Y} - 300$; then W can be used as a test statistic, and to get a test with size α, we reject H_0 when W is greater than the 100αth percentile of a standard normal density.

(b) The observed test statistic is $W_{obs} = 100.9 + 2(202.1) - 300 = 3.3$. The p-value is the area to the right of 3.3 under a standard normal density; this is less than .0015 by the 68-95-99.7 rule, so we reject H_0 and conclude that the macromolecule length is greater than 100 nanometers.

36.1 We have $H_0 : \mu = 9$ versus $H_a : \mu < 9$, where μ is the true mean weight of the cashews in the cans. We can build a t-statistic:

$$T_{obs} = \frac{23.5 - 25}{3.1/\sqrt{36}} = -2.90.$$

If H_0 is true, this is a realization from a $t(35)$ distribution, and small values support H_a. So, the p-value is the area to the left of -2.90 under a $t(35)$, or $p = .0032$. If the company's statement is correct, then the probability of seeing such a small (or smaller) sample mean is only .0032. We'd reject H_0 at $\alpha = .01$, and conclude there is strong evidence for the alternative.

36.3 We can build a t-statistic:

$$T_{obs} = \frac{608 - 580}{47/\sqrt{20}} = 2.66.$$

If H_0 is true, this is a realization from a $t(19)$ distribution, and large values support H_a. So, the p-value is the area to the right of 2.66 under a $t(19)$, or $p = .0077$. We would reject H_0 at $\alpha = .05$ (and even at $\alpha = .01$), and conclude that the mean score for the alternative students is higher than 580.

36.5 (a) We use that, for a random sample of size n from a normal density with variance σ^2,

$$W = \frac{(n-1)S^2}{\sigma^2} \sim \chi^2(n-1),$$

the observed test statistic is $W = 23.6$. The p-value is the area to the right of 23.6, under a chi-squared density with 19 degrees of freedom. Using R, we find $p = .21$, so we accept H_0.

(b) In part (a), we reject H_0 if $W > 30.14$, the 95th percentile of a $\chi^2(19)$ random variable. Or, we reject if the sample variance S^2 is greater than 14.28. If $\sigma^2 = 11$, the probability of rejecting is

$$P(W > 30.14) = P\left(\frac{19S^2}{9} > 30.14\right)$$

$$= P\left(\frac{19S^2}{11} > \frac{9}{11} 30.14\right)$$

$$= P\left(\frac{19S^2}{11} > 24.66\right),$$

which is the area to the right of 24.66, under a chi-squared density with 19 degrees of freedom—the power is .172. Perhaps the transit authority ought to use a bigger sample size.

36.7 We know that for a random sample of size n from an $N(\mu, \sigma^2)$ population,

$$\frac{(n-1)S^2}{\sigma^2} \sim \chi^2(n-1),$$

where S^2 is the sample variance. So, we plug in $n = 10$ and $\sigma^2 = 1$ to find the critical value of `qchisq(.99,9)` is 21.67. We will reject H_0 if $9S^2$ is greater than 21.67. To find the power when $\sigma = 1.25$, we compute

$$P(9S^2 > 21.67) = P\left(\frac{9S^2}{1.25^2} > \frac{21.67}{1.25^2}\right)$$
$$= P(X > 13.87),$$

where $X \sim \chi^2(9)$. This is .127, found with `1-pchisq(13.87,9)`.

36.9 The mean of the $n = 130$ temperatures is 98.249 and the standard deviation is .7332, so the observed test statistic value is

$$\frac{98.249 - 98.6}{.7332/\sqrt{130}} = -5.46.$$

The p-value is `2*pt(-5.46,129)` or 2.35×10^{-7}. If this is a random sample of people, we can decisively reject the null hypothesis that the mean temperature, for the population of people, is 98.6°F.

A histogram and a probability plot of the data are shown below. There is a bit of a curve in the probability plot, suggesting a skew in the distribution. (We will see in Exercise 37.9 that the mean body temperature is significantly different for men and women, so this skew might be due to the fact that the sample is actually from *two* populations.)

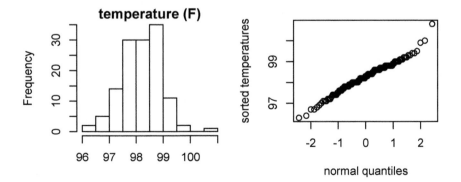

36.11 We know that

$$\frac{X-\mu}{\sigma/\sqrt{n}} \sim N(0,1), \quad \text{so} \quad \left(\frac{X-\mu}{\sigma/\sqrt{n}}\right)^2 \sim \chi^2(1),$$

and $(n-1)S^2/\sigma^2 \sim \chi^2(n-1)$. We also know that $\bar{X}$ and S^2 are independent. The σ^2 cancels out of the ratio, so if $c = n$, we have an $F(1, n-1)$-density. The code on the left makes the probability plot on the right.

```
n=10
nloop=1000
ff=1:nloop
for(iloop in 1:nloop){
    x=rnorm(n)
    xbar=mean(x)
    s2=var(x)
    ff[iloop]=xbar^2/s2
}
ff=ff*n
plot(qf(1:nloop/(nloop+1),1,n-1),sort(ff))
lines(c(0,40),c(0,40),col=2,lwd=2)
```

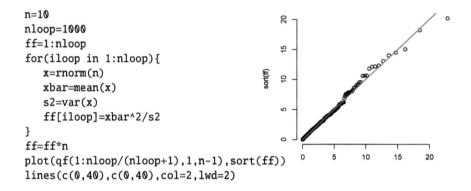

36.13 The following code generates samples of size $n = 20$ from a normal distribution with mean $\mu = 100$ and standard deviation σ. Then for $H_0 : \sigma = 4$ versus $H_a : \sigma \neq 4$, both the W-statistic (with known mean) and the X-statistic (using the sample standard deviation) are computed. This is done many times in a loop, and the proportion of samples for which the null hypothesis is rejected is calculated. We find that the power for the chi-squared test with known mean is about .767, when the true standard deviation is 6, and the chi-squared test with unknown mean has a power of about .749.

```
mu=100
sig0=4
n=20
nloop=100000
wstat=1:nloop
xstat=1:nloop
sig=6
for(iloop in 1:nloop){
    y=rnorm(n,mu,sig)
    wstat[iloop]=sum((y-mu)^2)/sig0^2
    cstat[iloop]=sum((y-mean(y))^2)/sig0^2
}
(sum(wstat>qchisq(.975,20)) + sum(wstat<qchisq(.025,20)))/nloop
(sum(xstat>qchisq(.975,19)) + sum(xstat<qchisq(.025,19)))/nloop
```

37.1 (a) If the populations are normally distributed with equal variances, and the samples are selected independently, and the two samples themselves are random samples, then all the assumptions for the two-independent-samples t-test are met.

   (b) The sample means for Techniques A and B are 30 and 25, respectively, and the sample variances are both 10. Therefore the pooled sample variance is 10, and the t-statistic is

$$T = \frac{5}{\sqrt{10(1/8 + 1/8)}} = \sqrt{10},$$

and the (two-sided) p-value is about .007. The conclusion is that Technique A removes more heavy metals than Technique B.

37.3 (a) Let μ_R be the average height of rye grass grown in sandy soil under the conditions of the experiment, and let μ_C be the average height of rye grass grown in clay soil under the conditions of the experiment. Then $H_0 : \mu_R = \mu_C$ and $H_a : \mu_R \neq \mu_C$.

(b) For a two-independent-samples t-test, we plug into the formula to get $T = 1.314$, and the two-sided p-value is larger than 0.05. We accept the null hypothesis and conclude that rye grows equally tall in both soils.

(c) We assume that the populations of grass heights are (approximately) normally distributed in both types of soil, that the variances are equal, and that the measurements of all the heights are independent.

37.5 The R code is

```
n1=16;n2=n1
nloop=100000
tstat=1:nloop
for(iloop in 1:nloop){
   y1=rnorm(n1,2.7,1)
   y2=rnorm(n1,3.3,1)
   psv=((n1-1)*var(y1)+(n2-1)*var(y2))/(n1+n2-2)
   tstat[iloop]=(mean(y1)-mean(y2))/sqrt(psv*(1/n1+1/n2))
}
sum(abs(tstat)>qt(.975,n1+n2-2))/nloop
```

and the power is about .378.

They would need about 45 pots of each type of grass to get a power of .8.

37.7 (a) The code below returns a p-value of about .0002, so we reject the null hypothesis that the average lifetime of the mice on the restricted diet is the same as that of mice on the unrestricted diet, and it looks like, alas, the mice on the restricted diet live longer.

```
y1=c(35.5, 33.8, 31.4, 30.2, 30.0, 27.1, 24.1, 21.5, 9.2)
y2=c(51.7, 48.1, 47.2, 46.9, 46.7, 40.9, 39.9, 38.2, 30.9)
t.test(y1,y2,var.equal=TRUE)
```

(b) We assume that the populations of lifetimes are normally distributed, and have a common variance. We assume that we have random samples from the populations, and the samples are independent. The following code makes plots to assess visually the normality assumption. The plots show that the distribution may be skewed to the left.

```
y=1:18;y[1:9]=y1-mean(y1);y[10:18]=y2-mean(y2)
hist(y)
plot(qnorm(1:18/19),sort(y),xlab="normal quantiles")
title("probability plot")
```

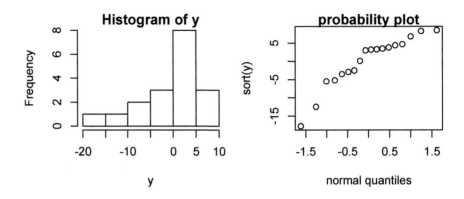

37.9 The vector `degf` holds the $n = 130$ temperatures, and the vector `gen` has the gender. Then the R code does a two-independent-samples t-test, where the null is that the average body temperature for men is the same as the average body temperature for women:

```
t.test(degf[gen==1],degf[gen==2],var.equal=TRUE)
```

We find that the two-sided p-value is about .024, which provides some evidence that the body temperatures are different, with women's temperatures being, on average, higher.

The independent samples assumption seems reasonable, but the assumption that these subjects represent random samples from the population is probably not true; these represent a "convenience sample." However, if the sample is what is called "representative," standard practice assumes that this assumption can be considered to be met. The code below produces the following plots to assess the normality assumption. There is no visual evidence against the normality assumption; in fact, the probability plot looks surprisingly linear for "real" data.

```
res=degf
res[gen==1]=degf[gen==1]-mean(degf[gen==1])
res[gen==2]=degf[gen==2]-mean(degf[gen==2])
hist(res)
n=length(res)
plot(qnorm(1:n/(n+1)),sort(res),xlab="normal quantiles")
title("probability plot")
```

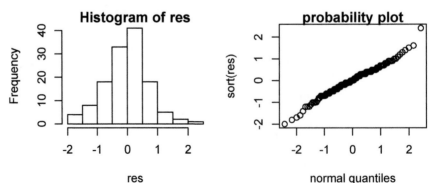

37.11 Using the test statistic developed for Exercise 37.10, we compute

$$Z = \frac{(\bar{X} - \bar{Y}) - (\mu_x - \mu_y)}{\sqrt{\frac{\sigma_x^2}{n} + \frac{\sigma_y^2}{m}}} = \frac{122.6 - 124.2}{\sqrt{\frac{12^2}{10} + \frac{12^2}{10}}} = -.298.$$

There is not evidence that the population means are different.

37.13 The mean observed zinc level for the nonvegetarian pregnant women is 178.0 for vegetarian pregnant women, and the sample mean is 177.1. The sample standard deviations are 14.48 and 20.87, respectively. The t-test derived in Exercise 37.11 can be accomplished: The pooled sample variance (using the formula appropriate for the assumptions about the population variances) is $S_p^2 = 215.3$, which is an estimate of the population variance for zinc levels for pregnant nonvegetarians. Then $T = .108$, so there is no evidence that the population mean zinc levels are different.

37.15 (a) If H_0 is true, then

$$T = \frac{\bar{Y}_B - \bar{Y}_A - 10}{S_p\sqrt{1/10 + 1/10}} \sim t(18).$$

The observed test statistic is 2.06, so the (one-sided) p-value is .027.

(b) We assume that the populations of lifetimes are normally distributed with equal variances, and the observations are independent random samples from the populations.

37.17 (a) Because the population means are zero, we know that

$$\frac{\sum_{i=1}^n X_i^2}{\sigma_x^2} \sim \chi^2(n) \quad \text{and} \quad \frac{\sum_{i=1}^m Y_i^2}{\sigma_y^2} \sim \chi^2(m).$$

We can construct an F random variable using the definition:

$$\frac{\left(\sum_{i=1}^n X_i^2/\sigma_x^2\right)/n}{\left(\sum_{i=1}^m Y_i^2/\sigma_y^2\right)/m} \sim F(n, m).$$

Then when the null hypothesis is true,

$$F = \frac{\sum_{i=1}^n X_i^2/n}{\sum_{i=1}^m Y_i^2/m}$$

has an F distribution with n numerator and m denominator degrees of freedom.

(b) The following R code compares the power for the test in (a) with the F-test that does not use the known means, for $\sigma_x = 8$, $\sigma_y = 4$, $n = 18$, and $m = 24$. The power for the test using known means is about .871, compared to about .854 for the "usual" test.

```
nloop=100000
n=18;m=24
```

```
sigx=8;sigy=4
fstat1=1:nloop
fstat2=1:nloop
for(iloop in 1:nloop){
   x=rnorm(n,0,sigx)
   y=rnorm(m,0,sigy)
   fstat1[iloop]=var(x)/var(y)
   fstat2[iloop]=(sum(x^2)/n)/(sum(y^2)/m)
}
sum(fstat1>qf(.975,n-1,m-1)|fstat1<qf(.025,n-1,m-1))/nloop
sum(fstat2>qf(.975,n,m)|fstat2<qf(.025,n,m))/nloop
```

38.1 (a) The sample means are $\bar{Y}_1 = 100.375$ and $\bar{Y}_2 = 104.875$, and the pooled sample variance is $S_p^2 = 101.911$. The t-statistic for the two-sample test follows a $t(14)$ distribution if the population means are equal. The statistic is $-.892$, and the two-sided p-value is .387.

(b) The mean of the 8 differences is -4.5, and the t-statistic follows a $t(7)$ if the true mean difference is zero. The observed value of the test statistic is -2.02, and so the p-value for the two-sided test is .083.

38.3 (a) We need to do a paired-samples test, because the measurements are paired by soil type. We compute the differences: For coating B minus coating A corrosion measurements, we have 3, 4, 3, 3, 1, -1, 6, 5. We construct a t-statistic that has a $t(7)$ distribution when H_0 is true. The observed value is 3, and the one-sided p-value is about .0031. There is strong evidence that the new coating is better.

(b) We assumed that the population of differences is normally distributed, and that we have a random sample of differences from this population.

38.5 If the researchers could recruit brother and sister pairs, they could do a paired-samples test. For each pair, both brother and sister would be injected with insulin, and the time for the drop in blood sugar would be measured. Because brother and sister are similar genetically, much of this variation would be "controlled for" by the pairing. It is reasonable to hope that the paired test would have greater power.

38.7 This is a paired-samples design, so we will not use the sample means and variances given. The lakes seem to have different levels of mercury, so both fish from Lake 5, say, are high in mercury, but both fish from Lake 1 are low. We compute the differences: 2, 1, -1, 2, 2, 1, 2, 2, and the sample mean and variance are $\bar{D} = 1.375$ and $S_d^2 = 1.125$. Then

$$T_{obs} = \frac{1.375}{1.125/\sqrt{8}} = 3.457,$$

and we compare this to a $t(7)$ distribution. The p-value is .0053, found with `1-pt(3.457,7)`.

38.9 (a) Let μ_D be the difference in cross-fertilized plant heights and self-fertilized plant heights in the population of this kind of plant. The null hypothesis is $H_0 : \mu_d = 0$ versus $H_a : \mu_d > 0$, where the alternative reflects that the cross-fertilized plants are taller.

(b) The two-sided p-value is .0497. If the null hypothesis of equal average plant heights is true, and we repeated the experiment many times, then we'd see this much or more evidence for a difference about 5% of the time.

(c) The probability plot shows two low "outliers" but with a sample this small it is hard to make a conclusion.

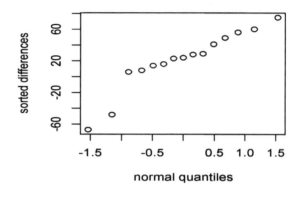

39.1 (a) There are nine balances greater than 800, so the p-value is the probability of nine or more greater than 800, when the median balance is 800. This is `1-pbinom(8,12,1/2)` or .073. At $\alpha = .05$, we do not have enough evidence that the median balance is greater than 800.

(b) If there were ten (or more) balances greater than 800, we would reject H_0, because the p-value for ten is .019. The probability of ten or more balances being greater than 800, when 800 is the 30th percentile, is `1-pbinom(9, 12,.7)` or .25. The power is not very high—maybe a larger sample size should be used!

39.3 There are 39 temperatures above 98.6°F, out of 130 temperatures. The probability of 39 or fewer above 98.6°F, given that 98.6°F is the median, is very small: about 1.2×10^{-6}. The histogram of the data is shown below, with a probability plot of the sorted data against the normal quantiles. Neither looks alarmingly different from what you might expect with a sample from a normal population, so the t-test is OK, but a sign test can be performed to be extra conservative.

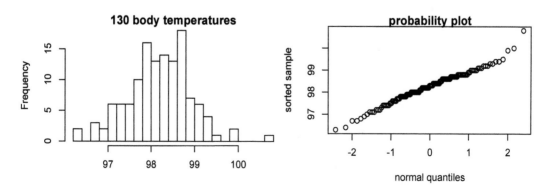

39.5 The following piece of R code generates samples of size $n = 12$ from a normal density with mean 1 and standard deviation 2, then for each sample computes both the t-statistic for $H_0 : \mu = 0$ and the number of positive units in the sample. The power for the t-test is the proportion of samples for which the t-statistic is greater than the 95th percentile of $t(11)$, and the power for the sign test is the proportion of samples for which the number of positive units is at least 9—the p-value will be less than $\alpha = .05$ for these samples. We find that the power for the t-test is about .52, while the power for the sign test is about .39.

```
n=13
nloop=10000
tstat=1:nloop
sstat=1:nloop
sn=sqrt(n)
for(iloop in 1:nloop){
    y=rnorm(n,1,2)
    tstat[iloop]=mean(y)/sd(y)*sn
    sstat[iloop]=sum(y>0)
}
sum(sstat>=10)
sum(tstat>qt(.95,n-1))
```

The next piece of R code generates samples of size $n = 12$ from a $t(3)$ and adds 1 to each unit. The statistics are calculated, and the power is .74 for the t-test and .76 for the sign test.

```
n=13
nloop=10000
tstat=1:nloop
sstat=1:nloop
sn=sqrt(n)
for(iloop in 1:nloop){
    y=rt(n,3)+1
    tstat[iloop]=mean(y)/sd(y)*sn
    sstat[iloop]=sum(y>0)
}
sum(sstat>=10)
sum(tstat>qt(.95,n-1))
```

39.7 (a) There are eight incomes greater than 120K—the p-value is the probability of seeing 8 or more out of 12 when the probability is 1/2. This is .193. We accept the null hypothesis and conclude there is not enough evidence that the median income is higher in Belleview.

   (b) We reject H_0 if we see 10 or more incomes (out of 12) that are more than 120K. The value 120 is the 22.8 percentile of a Gamma(15,.1). The probability of seeing 10 or more greater than 120 is 1-pbinom(9,12,1=.228), or .538.

40.1 (a) The two-sample t-test provides a (two-sided) p-value of .0002. There is strong evidence that the restricted diet group has higher longevity.

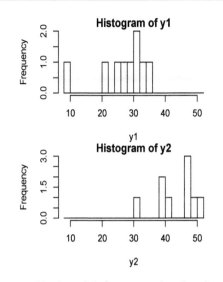

(b) Because the sample sizes are so small, it's almost impossible to see whether the data "look normal." In the absence of something like a couple of definite clusters, we conclude that the populations might be normal, or then again they might not be. We simply don't have the information to make a determination.

(c) The rank-sum test provides a (two-sided) p-value of .0003. Because both tests give the same conclusion, our answer to (a) is reinforced.

(A nice trick for comparing data in two histograms is to stack them vertically and use the same scale for both.)

40.3 The two-sided p-value for the signed rank test is about .026, so the one-sided p-value is about .013, and there is evidence that the reaction time is quicker for sober people.

40.5 The following code produces the histogram.

```
y1=anorexia$Prewt[anorexia$Treat=="CBT"]
y2=anorexia$Postwt[anorexia$Treat=="CBT"]
gain=y2-y1
hist(gain)
```

The sample size $n = 29$ is too small to say definitively whether or not the distribution is normal, but the histogram suggests that there might be two subgroups of girls: one for which the treatment works (the bump on the right), and a larger subgroup for which the treatment fails. The statistician might suggest to the researchers that the subgroups should be examined to see if there is a variable, such as socioeconomic status, that explains the groups. In that case, a more complicated analysis should be done.

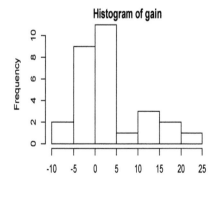

In any case, the p-value for the one-sided test of zero median gain against positive median gain is about .0496, a smidgen under the $\alpha = .05$. We reject H_0; this is a "borderline" case.

40.7 The command in R and the results follow; we accept H_0 at $\alpha = .01$, but we would reject at $\alpha = .05$.

```
> wilcox.test(ex0112$BP[ex0112$Diet=="fishoil"],
          ex0112$BP[ex0112$Diet=="regularoil"])

Wilcoxon rank sum test with continuity correction

data:  ex0112$BP[ex0112$Diet == "fishoil"] and
                     ex0112$BP[ex0112$Diet == "regularoil"]
W = 41, p-value = 0.03915
alternative hypothesis: true location shift is not equal to 0
```

41.1 Suppose $Y_1, \ldots, Y_n$ are a random sample from $N(\mu, \sigma^2)$, and let

$$Z_n = \sqrt{n} \left(\frac{\bar{Y} - \mu}{\sigma} \right).$$

Now let $X_i = (Y_i - \mu)/\sigma$, so that the X_i have mean zero and variance one, and note that

$$Z_n = \sqrt{n} \bar{X}.$$

By the lemma, the distribution of Z_n approaches that of a standard normal random variable as n increases.

41.3 (a) This is the probability that the sum of the lifetimes is greater than $(2.5)(40)$ $= 100$. The sum is $S \sim \text{Gamma}(40, 2)$, and $P(S > 100) = .0646$.

(b) Using the central limit theorem,

$$Z = \frac{\bar{X} - 2}{\sqrt{4/n}}$$

is approximately standard normal. The probability that the sample mean is greater than 2.5 is *approximately* the area to the right of $.5/\sqrt{4/40}$ under a standard normal. This is $P(Z > 1.581) = .0569$.

41.5 Let X_i be the amount of money (in dollars) won on the ith play. Then $P(X_i = -1) = 19/20$ and $P(X_i = 9) = 1/20$, and

$$E(X_i) = 9 \times \frac{1}{20} + (-1) \times \frac{19}{20} = -\frac{1}{2}$$

and

$$E(X_i^2) = 81 \times \frac{1}{20} + \frac{19}{20} = 5, \text{ so } V(X_i) = 4.5.$$

Using the approximation

$$\frac{\bar{X} + 1/2}{\sqrt{4.5/100}} \approx N(0, 1),$$

and that 1.645 is the 95th percentile of the standard normal, we get that the 95th percentile of $\sum_{i=1}^{n} X_i$ is

$$(1.645\sqrt{.045} - 1/2) * 100 = -15.1 \text{ dollars.}$$

41.7 (a) Let $S = X_1 + \cdots + X_{20} < N$; so S has mean 24,000 and standard deviation $120\sqrt{60}$. We want to find N so that $P(S > N) = .01$, which is

$$P\left(\frac{S - 24000}{120\sqrt{60}} > \frac{N - 24000}{120\sqrt{60}}\right) \approx P\left(Z > \frac{N - 24000}{120\sqrt{60}}\right),$$

where Z is a standard normal random variable. The 99th percentile of a standard normal distribution is 2.326, so we solve

$$\frac{N - 24000}{1200\sqrt{60}} = 2.326,$$

giving $N = 45{,}620$ dollars.

(b) The following code calculates that the probability that the cost of the 20 accidents exceeds 45,620 is about 2.4%.

```
nloop=1000000
n=20
cost=1:nloop
for(iloop in 1:nloop){
    u=runif(20)
    y=2400*((1-u)^(-1/3)-1)
    cost[iloop]=sum(y)
}
sum(cost>45620)/nloop
```

42.1 The bias of $\hat{\mu}$ is $E(\hat{\mu}) - \mu = 0$, and the variance is σ^2/n, so the MSE is σ^2/n. The bias of $\tilde{\mu}$ is

$$E(\tilde{\mu}) - \mu = \frac{n\mu}{n+1} - \mu = -\frac{\mu}{n+1}.$$

The variance of $\tilde{\mu}$ is

$$V\left(\frac{\sum_{i=1}^{n} Y_i}{n+1}\right) = \frac{n\sigma^2}{(n+1)^2},$$

so the MSE is

$$\frac{n\sigma^2}{(n+1)^2} + \frac{\mu^2}{(n+1)^2} = \frac{n\sigma^2 + \mu^2}{(n+1)^2}.$$

If $n = 10$, $\sigma^2 = 100$, and $\mu = 1$, then the MSE for $\hat{\mu}$ is 10, and the MSE for $\tilde{\mu}$ is only about 8.27. On the other hand, if $n = 10$, $\sigma = 1$, and $\mu = 20$, then the MSE for $\hat{\mu}$ is .10, and the MSE for $\tilde{\mu}$ is a whopping 3.31. If we expect μ to be small compared with σ, then the person's suggestion might be a good one, but otherwise $\tilde{\mu}$ is much worse than $\hat{\mu}$, even though the variance is smaller.

42.3 (a) The expected value of $\tilde{p}$ is

$$E(\tilde{p}) = \frac{2}{3}E(\bar{X} + \bar{Y}) = \frac{2}{3}\left(\frac{3p}{2}\right) = p,$$

so $\tilde{p}$ is unbiased. The variance of $\tilde{p}$ is

$$V(\tilde{p}) = \frac{4}{9}[V(\bar{X}) + V(\bar{Y})] = \frac{4}{9}\left[\frac{p(1-p)}{n} + \frac{\frac{p}{2}\left(1 - \frac{p}{2}\right)}{n}\right] = \frac{4p}{9n}\left[\frac{3}{2} - \frac{5p}{4}\right].$$

(b) The expected value of $\hat{p}$ is

$$\mathrm{E}(\hat{p}) = \frac{\mathrm{E}(\bar{Y})}{2} + \mathrm{E}(\bar{X}) = \frac{p}{2} + \frac{p}{2} = p,$$

so $\hat{p}$ is unbiased. The variance of $\hat{p}$ is

$$\mathrm{V}(\hat{p}) = \frac{1}{4}\mathrm{V}(\bar{Y}) + \mathrm{V}(\bar{X}) = \frac{1}{4}\frac{p(1-p)}{n} + \frac{\frac{p}{2}\left(1-\frac{p}{2}\right)}{n} = \frac{p}{4n}(3-2p).$$

(c) The efficiency of $\hat{p}$ relative to $\tilde{p}$ is

$$\frac{\mathrm{V}(\tilde{p})}{\mathrm{V}(\hat{p})} = \frac{4}{9}\left(\frac{6-5p}{3-2p}\right) < 1,$$

and the variance of $\tilde{p}$ is always smaller for $p \in (0,1)$.

42.5   (a) The estimator has mean μ and variance

$$\frac{1}{25}(4+4+9+9+9) = \frac{7}{5}.$$

(b) The estimator has mean μ and variance

$$\frac{1}{4}(1+1+1+1+1) = \frac{5}{4}.$$

(c) The estimator has mean μ and variance

$$\frac{36}{25}\left(\frac{1}{4}+\frac{1}{4}+\frac{1}{9}+\frac{1}{9}+\frac{1}{9}\right) = \frac{6}{5},$$

and Statistician C is the winner!

42.7 All four estimators are unbiased, so we just need to compute the variances. These are

- $\mathrm{V}(\bar{Y}) = \theta^2/4$.
- $\mathrm{V}(Y_2) = \theta^2$.
- $\mathrm{V}\left(\dfrac{Y_1 + 2Y_2}{3}\right) = \dfrac{\theta^2 + 4\theta^2}{9} = \dfrac{5\theta^2}{9}$.
- The random variable $Y = \min(Y_1, Y_2, Y_3, Y_4)$ is distributed as $\mathrm{Exp}(\theta/4)$, so the estimator $\hat{\theta}_4 = 4Y$ has density

$$f_{\hat{\theta}_4}(y) = \frac{1}{\theta}e^{-y/\theta}.$$

Then the variance is computed to be θ^2.

The winner is the sample mean!

42.9 To start, we will consider $Y \sim \mathrm{Exp}(1/\theta)$ and compute the mean and variance of Y^2:

$$\mathrm{E}(Y^2) = \frac{1}{\theta}\int_0^\infty y^2 e^{-y/\theta}dy = \frac{1}{\theta}\frac{\Gamma(3)}{(1/\theta)^3} = 2\theta^2,$$

remembering that we can use the gamma density formula in this integration (see Chapter Highlight 5 in Chapter 16). Similarly,

$$E(Y^4) = \frac{1}{\theta} \int_0^\infty y^4 e^{-y/\theta} dy = \frac{1}{\theta} \frac{\Gamma(5)}{(1/\theta)^5} = 24\theta^4,$$

so $V(Y^2) = 20\theta^4$.

Now let's look at the estimator. The expected value of T is θ^2, so T is unbiased. Its variance is

$$V(T) = \frac{1}{64} \times 4 \times 20\theta^4 = \frac{5}{4}\theta^4.$$

Because the estimator is unbiased, the variance is the MSE.

42.11 (a) We can compute the expected value

$$E(\hat{\lambda}) = aE(X) + bE(Y) = a\lambda + 2b\lambda,$$

so $\hat{\lambda}$ is unbiased if $a + 2b = 1$.

(b) Using independence, we can compute the variance

$$V(\hat{\lambda}) = a^2 V(X) + b^2 V(Y) = a^2\lambda + 2b^2\lambda.$$

Plugging in $a = 1 - 2b$, we have

$$V(\hat{\lambda}) = (1 - 2b)^2 V(X) + b^2 V(Y) = \lambda(1 - 4b + 6b^2),$$

which is minimized when $b = 1/3$. Then a is also $1/3$, and the linear unbiased estimator that minimizes the variance is $\hat{\lambda} = (X + Y)/3$.

42.13 To compute $E(\hat{\theta})$, we first compute $E(Y^3)$ for $Y \sim f_\theta$. This is

$$E(Y^3) = \frac{3}{\theta} \int_0^\infty y^5 e^{-y^3/\theta} dy = \theta \int_0^\infty u e^{-u} du = \theta\Gamma(2) = \theta,$$

where we used the substitution $u = y^3/\theta$, $du = 3y^2/\theta dy$. Therefore, $E(\hat{\theta}) = \theta$ and the estimator is unbiased.

For the variance, we use

$$V(\hat{\theta}) = \frac{1}{n^2} \sum_{i=1}^n V(Y_i^3)$$

and

$$V(Y_i^3) = E(Y_i^6) - E(Y_i^3)^2.$$

For the first term,

$$E(Y_i^6) = \frac{3}{\theta} \int_0^\infty y^8 e^{-y^3/\theta} dy = \theta^2 \int_0^\infty u^2 e^{-u} du = \theta^2\Gamma(3) = 2\theta^2,$$

so $V(Y_i^3) = 2\theta^2 - \theta^2 = \theta^2$. Then $V(\hat{\theta}) = \theta^2/n$. (Another way to do this is to show first that $X_i = Y_i^3$ is exponential with mean θ.)

42.15 (a)

$$E(\hat{\alpha}_1) = \frac{1}{2n} \sum_{i=1}^{n} E(Y_i^2) = \frac{1}{2n} \sum_{i=1}^{n} (2\theta^2) = \theta^2,$$

so $\hat{\alpha}_1$ is unbiased.

(b) The following calculation gives $a = n/(n+1)$:

$$E(\hat{\alpha}_2) = aE \left[\frac{1}{n} \sum_{i=1}^{n} Y_i \right]^2 = \frac{a}{n^2} E \left[\sum_{i=1}^{n} Y_i^2 + \sum_{i \neq j} Y_i Y_j \right]$$

$$= \frac{a}{n^2} E \left[2n\theta^2 + n(n-1)\theta^2 \right] = a\theta^2(1 + 1/n).$$

(c) The variance of $\hat{\alpha}_1$ is

$$V(\hat{\alpha}_1) = \frac{1}{4n^2} \sum_{i=1}^{n} V(Y_i^2) = \frac{1}{4n} V(Y_1^2) = \frac{1}{4n} \left[E(Y_1^4) - E(Y_1)^2)^2 \right].$$

Using the result from Exercise 16.3, we have

$$V(\hat{\alpha}_1) = \frac{1}{4n} \left[24\theta^4 - (2\theta^2)^2 \right] = \frac{5\theta^4}{n}.$$

Next,

$$V(\hat{\alpha}_2) = \frac{n^2}{(n+1)^2} V(\bar{Y}^2).$$

If we let $X = n\bar{Y}$, then $X \sim \text{Gamma}(n, \theta)$ and

$$V(\hat{\alpha}_2) = \frac{n^2}{(n+1)^2} V\left(\frac{X^2}{n^2} \right) = \frac{1}{n^2(n+1)^2} V\left(X^2 \right)$$

$$= \frac{1}{n^2(n+1)^2} \left[E(X^4) - E(X^2)^2 \right].$$

Now using the result from Exercise 16.4, this is

$$V(\hat{\alpha}_2) = \frac{1}{n^2(n+1)^2} \left[\theta^4(n+3)(n+2)(n+1)n - (\theta^2(n+1)n)^2 \right]$$

$$= \frac{2\theta^4(2n+3)}{n(n+1)}.$$

Finally, the relative efficiency of $\hat{\alpha}_1$ with respect to $\hat{\alpha}_2$ is

$$\left(\frac{2\theta^4(2n+3)}{n(n+1)} \right) \Big/ \left(\frac{5\theta^4}{n} \right) = \frac{2(2n+3)}{5(n+1)}.$$

This is always smaller than one (for $n > 1$), so $\hat{\alpha}_2$ has a smaller variance for all n.

42.17 (a) Using the definition of expected value,

$$E(\bar{X}) = \int_{-\infty}^{\infty} x f(x) dx = (1-\alpha) \int_{-\infty}^{\infty} x f_1(x) dx + \alpha \int_{-\infty}^{\infty} x f_2(x) dx$$

$$= (1-\alpha)\mu + \alpha\mu = \mu.$$

(b) We can sample from a mixture using the algorithm in Chapter 19. The following code shows that for the uncontaminated normal, the variance of sample mean is about .0250, while the variance for the trimmed mean is about .0253. However, for the contaminated normal (in the code below, delete the # character to un-comment the lines), the variance of the sample mean is about .269, while the variance of the trimmed mean is .137, a big improvement.

```
nloop=100000
muhat1=1:nloop
muhat2=1:nloop
mu=100;n=40
alpha=.1
obs=1:n
for(iloop in 1:nloop){
    y=rnorm(n)+mu
#   u=runif(n)
#   nu=sum(u<alpha)
#   y[u<alpha]=rnorm(nu,100,10)
    muhat1[iloop]=mean(y)
    use=1:n>0
    use[y==min(y)]=FALSE
    use[y==max(y)]=FALSE
    muhat2[iloop]=mean(y[use])
}
var(muhat1)
var(muhat2)
```

43.1 Let $\tilde{\psi}(x) = \log(\psi(x))$; then

$$\tilde{\psi}'(x) = \frac{\psi'(x)}{\psi(x)},$$

so clearly $\tilde{\psi}'(x_0) = 0$. Also,

$$\tilde{\psi}''(x) = \frac{\psi(x)\psi(x)''(x) - \psi'(x)^2}{\psi(x)^2},$$

which is negative at x_0, because $\psi''(x_0) < 0$.

43.3 The likelihood is

$$L = \prod_{i=1}^{n}\left[\frac{1}{\sqrt{2\pi\sigma^2}}\exp\left\{-\frac{y_i^2}{2\sigma^2}\right\}\right] = \left(\frac{1}{2\pi\sigma^2}\right)^{n/2}\exp\left\{-\frac{\sum_{i=1}^{n}y_i^2}{2\sigma^2}\right\}$$

and the log-likelihood is

$$\ell = -\frac{n}{2}\log(2\pi) - \frac{n}{2}\log(\sigma^2) - \frac{\sum_{i=1}^{n}y_i^2}{2\sigma^2}.$$

Then taking the derivative,

$$\frac{\partial\ell}{\partial\sigma^2} = -\frac{n}{2\sigma^2} + \frac{\sum_{i=1}^{n}y_i^2}{2\sigma^4},$$

which is zero when $\sigma^2 = \sum_{i=1}^n y_i^2/n$. The second derivative is negative at this value of σ^2, so the MLE is

$$\hat{\sigma}^2 = \sum_{i=1}^n Y_i^2/n.$$

The estimator is unbiased.

43.5   (a) The likelihood is

$$L(\mu; y_1, \ldots, y_n) = e^{n\mu} e^{-\sum_{i=1}^n y_i} \prod_{i=1}^n I\{y_i > \mu\}$$

$$= e^{n\mu} e^{-\sum_{i=1}^n y_i} I\{\mu < \min(y_i)\}.$$

The likelihood is increasing in μ, so the MLE is $\hat{\mu} = \min(Y_i)$.

(b) To find the expected value of the minimum, we first find the distribution, using the CDF method. The CDF for Y_i is $F(y) = 1 - e^{\mu - y}$, and for $x > \mu$,

$$P(\min(Y_i) \leq x) = 1 - P(\min(Y_i) > x)$$
$$= 1 - P(\text{all}(Y_i) > x) = 1 - e^{n(\mu - x)},$$

so the density for $X = \min(Y_i)$ is $ne^{n(\mu - x)}$ for $x > \mu$. To find the expected value,

$$E(X) = n \int_\mu^\infty x e^{n(\mu - x)} dx = \int_0^\infty \left(\frac{u}{n} + \mu\right) e^{-u} du = \frac{1}{n} + \mu,$$

where we have made the substitution $u = n(x - \mu)$. The bias of the MLE is $1/n$.

(c) For these data, the likelihood is plotted below:

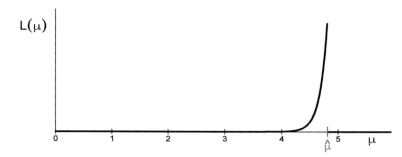

43.7 The likelihood is $L(\mu, \sigma^2; y_1, \ldots, y_n, x_1, \ldots, x_n)$:

$$L = \prod_{i=1}^n \left[\frac{1}{\sqrt{2\pi\sigma^2}} \exp\left\{-\frac{(y_i - \mu)^2}{2\sigma^2}\right\}\right] \times \prod_{i=1}^m \left[\frac{1}{\sqrt{\pi\sigma^2}} \exp\left\{-\frac{(x_i - \mu)^2}{\sigma^2}\right\}\right]$$

$$= A\left(\frac{1}{\sigma^2}\right)^{m/2 + n/2} \exp\left\{-\frac{1}{2\sigma^2}\left[\sum_{i=1}^n (y_i - \mu)^2 + 2\sum_{i=1}^n (x_i - \mu)^2\right]\right\},$$

where A is some constant that we don't need to bother with. The log-likelihood is

$$\ell = \log(A) - \left(\frac{m}{2} + \frac{n}{2}\right)\log(\sigma^2) - \frac{1}{2\sigma^2}\left[\sum_{i=1}^n (y_i - \mu)^2 + 2\sum_{i=1}^n (x_i - \mu)^2\right].$$

Taking the derivative with respect to μ,

$$\frac{\partial \ell}{\partial \mu} = \frac{1}{2\sigma^2} \left[2 \sum_{i=1}^{n} (y_i - \mu) + 4 \sum_{i=1}^{n} (x_i - \mu) \right],$$

which is zero when $n\mu + 2m\mu = \sum_{i=1}^{n} y_i + 2 \sum_{i=1}^{m} x_i$, so

$$\hat{\mu} = \frac{\sum_{i=1}^{n} y_i + 2 \sum_{i=1}^{m} x_i}{n + 2m}.$$

43.9 The likelihood is

$$L(\lambda; x_1, \ldots, x_n) = \frac{\lambda^n}{2^n \prod_{i=1}^{n} x_i^{1/2}} e^{-\lambda \sum_{i=1}^{n} x_i^{1/2}},$$

and the log-likelihood is

$$\ell(\lambda; x_1, \ldots, x_n) = n \log(\lambda) - \lambda \sum_{i=1}^{n} x_i^{1/2} + \text{(constant)}.$$

Then

$$\frac{\partial \ell}{\partial \lambda} = \frac{n}{\lambda} - \sum_{i=1}^{n} x_i^{1/2}$$

and

$$\hat{\lambda} = \frac{n}{\sum_{i=1}^{n} x_i^{1/2}}.$$

43.11 If Y_i is the number of successful uses for the ith device, then $Y_1, \ldots, Y_n \overset{ind}{\sim}$ Geom(p). (Here we are calling a "break" a "success.") The likelihood is the joint mass function

$$L(p; y_1, \ldots, y_n) = \prod_{i=1}^{n} [p(1-p)^{y_i}] = p^n (1-p)^{\sum_{i=1}^{n} y_i}.$$

The log-likelihood is

$$\ell(p; y_1, \ldots, y_n) = n \log(p) + \sum_{i=1}^{n} y_i \log(1-p)$$

and

$$\frac{\partial \ell}{\partial p} = \frac{n}{p} - \frac{\sum_{i=1}^{n} y_i}{1-p},$$

which is zero when $p = n/(n + \sum_{i=1}^{n} y_i)$. The second derivative is negative at this value of p, so the MLE for p is

$$\hat{p} = \frac{n}{n + \sum_{i=1}^{n} Y_i}.$$

43.13 The likelihood function is the joint density

$$L(\theta; y_1, \ldots, y_n) = \theta^n \prod_{i=1}^{n} y_i^{\theta-1},$$

and the log-likelihood is

$$\ell(\theta; y_1, \ldots, y_n) = n \log(\theta) + (\theta - 1) \sum_{i=1}^{n} \log(y_i).$$

Then

$$\frac{\partial \ell}{\partial \theta} = \frac{n}{\theta} + \sum_{i=1}^{n} \log(y_i),$$

and the second derivative is always negative. Then

$$\hat{\theta} = -\frac{n}{\sum_{i=1}^{n} \log(y_i)}.$$

43.15 The likelihood function is the joint density

$$L(\theta; y_1, \ldots, y_n) = \prod_{i=1}^{n} \left[3\theta y_i^2 e^{-\theta y_i^3} \right],$$

so the log-likelihood is a constant plus

$$\ell(\theta; y_1, \ldots, y_n) = n \log(\theta) - \theta \sum_{i=1}^{n} y_i^3.$$

Taking the derivative with respect to θ, we get

$$\frac{\partial \ell}{\partial \theta} = \frac{n}{\theta} - \sum_{i=1}^{n} y^3,$$

so the MLE is

$$\hat{\theta} = \frac{n}{\sum_{i=1}^{n} y_i^3}.$$

Here is code to check the answer and find the mean and variance of $\hat{\theta}$ when $\theta = 2$ and $n = 20$:

```
n=20;theta=2
nloop=100000
thhat=1:nloop
c=gamma(4/3)^3
for(iloop in 1:nloop){
    u=runif(n)
    y=(log(1-u)/-theta)^(1/3)
    thhat[iloop]=n/sum(y^3)
}
mean(thhat)
var(thhat)
```

The mean is about 2.11, and the variance is about .25.

43.17 (a) The likelihood function is the joint density

$$L(\lambda; y_1, \ldots, y_n) = \frac{1}{2^n \lambda^n \prod_{i=1}^{n} \sqrt{y_i}} e^{-\sum_{i=1}^{n} \sqrt{y_i}/\lambda} \quad \text{for } y_i > 0, \ i = 1, \ldots, n.$$

The log-likelihood is

$$\ell(\lambda; y_1, \ldots, y_n) = -n\log(\lambda) - \sum_{i=1}^{n} \sqrt{y_i}/\lambda + (\text{constant}),$$

and

$$\frac{\partial \ell}{\partial \lambda} = \frac{-n}{\lambda} + \frac{\sum_{i=1}^{n} \sqrt{y_i}}{\lambda^2},$$

and setting to zero gives $\hat{\lambda} = \sum_{i=1}^{n} \sqrt{y_i}/n$.

(b) Let's use the CDF method to find the distribution of $X_i = \sqrt{Y_i}$. For $x > 0$,

$$F_X(x) = P(X_i \le x) = P(Y_i \le x^2)$$
$$= F_Y(x^2).$$

Taking the derivative with respect to x (and using the chain rule), we get

$$f_X(x) = f_Y(x^2)2x = \frac{1}{\lambda}e^{-x/\lambda},$$

which we recognize as an exponential density with mean λ. Then $n\hat{\lambda}$ is the sum of the X_i's, and we know that the sum of independent exponential random variables with mean λ is a Gamma$(n, 1/\lambda)$ random variable.

(c) For these data, $n = 8$ and the distribution of $n\lambda$, when H_0 is true, is Gamma$(8, 1)$. When $\lambda > 1$, we expect that the test statistic is larger. Our observed test statistic, the sum of the square roots of the values, is 16.9. The 99th percentile of a Gamma$(8, 1)$ random variable is only 16.0, so we reject H_0 at $\alpha = .01$. The p-value, or the area to the right of 16.9 under a Gamma$(8, 1)$ density, is about .006.

43.19 The likelihood function is

$$L(\mu; \boldsymbol{y}) = \prod_{i=1}^{n} \left[\frac{1}{2} e^{-|y_i - \mu|} \right] = \frac{1}{2^n} e^{-\sum_{i=1}^{n} |y_i - \mu|}$$

and the log-likelihood is a constant plus

$$\ell(\mu; \boldsymbol{y}) = -\sum_{i=1}^{n} |y_i - \mu|.$$

The derivative of each term in the sum, with respect to μ, is -1 if $y_i > \mu$ and $+1$ if $y_i < \mu$; so the MLE is any number μ such that half the y_i are greater and half are less than μ; traditionally we use the median.

43.21 The likelihood is

$$\prod_{i=1}^{k} \prod_{j=1}^{n_i} \left[\frac{1}{\sqrt{2\pi\sigma^2}} \exp\left\{ \frac{(y_{ij} - \mu_i)^2}{2\sigma^2} \right\} \right]$$

and the log-likelihood is

$$\ell(\mu_1, \ldots, \mu_k, \sigma^2 \boldsymbol{y}_1, \ldots, \boldsymbol{y}_k) = -\frac{n}{2}\log(2\pi\sigma^2) - \frac{1}{2\sigma^2}\sum_{i=1}^{k}\sum_{j=1}^{n_i}(y_{ij} - \mu_i)^2$$

where $n = n_1 + \cdots + n_k$, the total number of observed values. For each $i = 1, \ldots, k$,

$$\frac{\partial \ell}{\partial \mu_i} = \frac{1}{\sigma^2} \sum_{j=1}^{n_i} (y_{ij} - \mu_i),$$

and $\hat{\mu}_i = \bar{y}_i$, the mean of the observed $y_{i1}, \ldots, y_{in_i}$. The derivative of ℓ with respect to σ^2 (not σ!) is

$$\frac{\partial \ell}{\partial \sigma^2} = \frac{n}{2\sigma^2} - \frac{1}{2\sigma^4} \sum_{i=1}^{k} \sum_{j=1}^{n_i} (y_{ij} - \mu_i)^2,$$

which is zero for $\sigma^2 = \frac{1}{n} \sum_{i=1}^{k} \sum_{j=1}^{n_i} (y_{ij} - \mu_i)^2$. We plug in the MLEs for the μ_i to get

$$\hat{\sigma}^2 = \frac{1}{n} \sum_{i=1}^{k} \sum_{j=1}^{n_i} (y_{ij} - \bar{y}_i)^2.$$

44.1 The sample mean is $\bar{Y}$, the population mean is λ, so the method of moments estimator is $\tilde{\lambda} = \bar{Y}$.

44.3 (a) We find the first moment,

$$E(Y) = 2\theta^2 \int_{\theta}^{\infty} y^{-2} dy = 2\theta,$$

so $\tilde{\theta} = \bar{Y}/2$ is the method of moments estimator for θ.

(b) By definition, the estimator is unbiased. To find the variance of this estimator,

$$V(\tilde{\theta}) = \frac{V(Y)}{4n},$$

where $Y \sim f_\theta$. But when we try to find $E(Y^2)$, we find that the second moment is undefined. The estimator has infinite variance!

(c) The following code finds the mean and variance of the estimators:

```
nloop=100000;n=50;theta=2
thhat=1:nloop;thtil=thhat
for(iloop in 1:nloop){
    u=runif(n)
    y=theta/sqrt(1-u)
    thhat[iloop]=min(y)
    thtil[iloop]=mean(y)/2
}
mean(thhat)
mean(thtil)
var(thhat)
var(thtil)
```

We find the mean for the MLE is about 2.002, or biased slightly large. The variance for the MLE is about .0004. The variance for the method of moments estimator is much larger and changes, sometimes dramatically, when we repeat the simulations. This is because the variance is theoretically undefined, though the variance of a finite number of estimators can be computed.

44.5 Setting the first population moment equal to the first sample moment gives

$$\bar{Y} = \frac{\tilde{\theta}}{2\tilde{\theta} + 1}$$

or

$$\tilde{\theta} = \frac{\bar{Y}}{1 - 2\bar{Y}}.$$

Notice that this estimator doesn't have to be between zero and one, but as n gets larger, the probability that it's between zero and one goes to one.

44.7 The population expected value is $\mu + 1$, and so the method of moments is $\tilde{\mu} = \bar{Y} - 1$. The method of moments is unbiased, and the variance is $1/n$. The MLE, $\min(Y_i)$, has an exponential distribution with mean $\mu + 1/n$ (see Exercise 33.4). Therefore, the bias for the MLE is $1/n$, and the variance is $1/n^2$. Therefore, the MSE for the MLE is $2/n^2$, and this is smaller than the MSE $1/n$ for the method of moments whenever $n > 2$.

44.9 (a) The population density has mean $\theta/(1 + \theta)$, so the method of moments is $\tilde{\theta} = \bar{Y}/(1 - \bar{Y})$.

(b) Using the MLE $\hat{\theta} = -n/\sum_{i=1}^{n} \log(y_i)$, the following code compares the MSE of the two estimators. For $\theta = .5$, the MSE for $\hat{\theta}$ is about .16, compared with .26 for $\tilde{\theta}$. For $\theta = 3$, the MSEs for the estimators are closer together, with about .58 for $\hat{\theta}$ and about .60 for $\tilde{\theta}$.

```
nloop=10000
n=20
theta=.5
thhat=1:nloop;thtil=1:nloop
for(iloop in 1:nloop){
    u=runif(n)
    y=u^(1/theta)
    ybar=mean(y)
    thhat[iloop]=-n/sum(log(y))
    thtil[iloop]=ybar/(1-ybar)
}
mean((thhat-theta)^2)
mean((thtil-theta)^2)
```

44.11 (a) The first step is to find $E(Y)$ for $Y \sim f_\theta(y)$. This is

$$\begin{aligned}
E(Y) &= \theta \int_0^\infty \frac{y}{(y+1)^{\theta+1}} \, dy \\
&= \theta \int_1^\infty \frac{u-1}{u^{\theta+1}} \, du \\
&= \theta \left[\int_1^\infty u^{-\theta} \, du + \int_1^\infty u^{-\theta-1} \, du \right] \\
&= \frac{1}{\theta - 1}.
\end{aligned}$$

Then the method of moments estimator is found by solving

$$\bar{Y} = \frac{1}{\tilde{\theta} - 1}, \quad \text{to get } \tilde{\theta} = \frac{1}{\bar{Y}} + 1.$$

(b) It would be hard to analytically compare this estimator with the MLE, so we use simulations. The following code incorporates sampling from the distribution f_θ, which was the subject of Exercise 19.14.

```
nloop=10000
n=10
thhat=1:nloop
thtil=1:nloop
th=4
for(iloop in 1:nloop){
   u=runif(n)
   y=(1-u)^(-1/th)-1
   thhat[iloop]=n/sum(log(y+1))
   thtil[iloop]=1/mean(y)+1
}
mean(thhat)
mean(thtil)
var(thhat)
var(thtil)
mean((thhat-th)^2)
mean((thtil-th)^2)
```

For the given values of $n = 10$ and $\theta = 4$, we find that the MLE has a smaller variance and bias, and smaller MSE. Other values also show that the winner is the MLE, although sometimes $\tilde{\theta}$ has a smaller variance, but bigger bias.

44.13 (a) First we find the expected value of a random variable having the given distribution:

$$\mathrm{E}(Y) = 3\theta \int_0^\infty y^3 e^{-\theta y^3} dy = \int_0^\infty \left(\frac{u}{\theta}\right)^{1/3} e^u du = \Gamma\left(\frac{4}{3}\right)/\theta^{1/3}.$$

(b) The following code will find the mean and variance of the estimator.

```
n=20;theta=2
nloop=100000
thhat=1:nloop;thtil=thhat
c=gamma(4/3)^3
for(iloop in 1:nloop){
   u=runif(n)
   y=(log(1-u)/-theta)^(1/3)
   thtil[iloop]=c/mean(y)^3
}
mean(thtil)
var(thtil)
```

We find that the mean is about 2.08, and the variance is about .285.

(c) The bias is slightly smaller than the bias of the MLE, but the variance is larger. The MSE for the method of moments estimator is about .29, while the MSE for the MLE is about .26.

44.15 (a) The method of moments is easy: The first population moment is μ, so $\tilde{\mu} = \bar{Y}$.

(b) The following code shows that for $n = 12$ and $\mu = 2$, the median is a better estimate than the mean: The MSE for the MLE is .118, while the MSE for the method of moments is .167.

```
nloop=100000
n=12;mu=2
muhat=1:nloop
mutil=1:nloop
for(iloop in 1:nloop){
    x=rexp(n,1)
    u=runif(n)
    x[u<1/2]=-x[u<1/2]
    x=x+mu
    muhat[iloop]=median(x)
    mutil[iloop]=mean(x)
}
mean((muhat-mu)^2)
mean((mutil-mu)^2)
```

44.17 (a) The density function is

$$f_\theta(y) = \frac{2(\theta - y)}{\theta^2},$$

and the expected value is

$$\frac{2}{\theta^2} \int_0^\theta y(\theta - y)dy = \theta/3.$$

Therefore the method of moments is $\tilde{\theta} = 3\bar{Y}$.

(b) The following code simulates many samples of size $n = 20$ from the triangular density with $\theta = 2$, and for each sample computes and saves the two estimators. For the method of moments estimator, the bias is zero but the variance is about .10. For the maximum, the variance is much less (about .039), but the bias is about $-.39$. The MSE for the method of moments is .10 and for the maximum, the MSE is about .19, so the method of moments wins for this sample size and $\theta = 2$.

44.19 We can find the density function:

$$f_\theta(x) = \begin{cases} \frac{1}{\theta} + \frac{x}{\theta^2} & \text{for } x \in (-\theta, 0], \\ \frac{1}{\theta} - \frac{x}{\theta^2} & \text{for } x \in (0, \theta), \\ 0 & \text{elsewhere.} \end{cases}$$

It is no good setting the first population moment equal to the first sample moment, because the first population moment does not involve the parameter (it is zero). The second population moment is

$$E(X^2) = \int_{-\theta}^0 \left[\frac{x^2}{\theta} + \frac{x^3}{\theta^2} \right] dx + \int_0^\theta \left[\frac{x^2}{\theta} - \frac{x}{\theta^3} \right] dx = \frac{\theta^2}{6}.$$

To find the method of moments estimator, solve

$$\frac{\tilde{\theta}^2}{6} = \frac{1}{n} \sum_{i=1}^n X_i^2, \quad \text{to get } \tilde{\theta} = \sqrt{\frac{6}{n} \sum_{i=1}^n X_i^2}.$$

45.1 The likelihood is

$$\prod_{i=1}^{n} \left[\frac{e^{-\lambda}\lambda_i^x}{x_i!} \right],$$

and the prior is $\pi(\lambda) \propto \lambda^{\alpha-1}e^{-\beta\lambda}$. Therefore, the posterior is the product

$$\pi(\lambda|x_1,\ldots,x_n) \propto \prod_{i=1}^{n} \left[\frac{e^{-\lambda}\lambda_i^x}{x_i!} \right] \times \lambda^{\alpha-1}e^{-\beta\lambda} = \lambda^{\sum_{i=1}^{n} x_i+\alpha-1}e^{-\lambda(n+\beta)},$$

so the posterior distribution for λ is Gamma$(\sum_{i=1}^{n} x_i + \alpha, n + \beta)$.

(a) The Bayes estimator, i.e., the posterior mean, is

$$\tilde{\lambda} = \frac{\sum_{i=1}^{n} x_i + \alpha}{n + \beta}.$$

(b) The prior is conjugate, because the family for the posterior (gamma) is the same as for the prior.

(c) The sketch is on the left below:

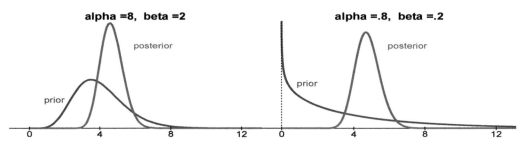

(d) The sketch is on the right above. The posterior looks similar to that in part (b), but has a larger variance.

(e) The table below contains the MSE results for the eight scenarios. It seems that for small sample sizes, the Bayes estimator can be a substantial improvement if you have good prior information; otherwise a vague prior is preferable. When the sample size is larger, the same patterns hold, but the difference between the MSEs is quite small.

Scenario	Sample size	λ	Prior	Bayes MSE	MLE MSE
(1)	small	4	good, not vague	.278	.40
(2)	small	4	good, vague	.384	.40
(3)	small	8	bad, not vague	1.00	.80
(4)	small	8	bad, vague	.775	.80
(5)	large	4	good, not vague	.0384	.040
(6)	large	4	good, vague	.0398	.040
(7)	large	8	bad, not vague	.0830	.080
(8)	large	8	bad, vague	.0797	.080

45.3 The posterior is proportional to the joint density for the data, given τ, times the prior for τ:

$$\pi(\tau|y_1,\ldots,y_n) \propto \tau^{n/2} \exp\left\{-\frac{\tau\sum_{i=1}^{n} y_i^2}{2}\right\} \times \tau^{\alpha-1} \exp\{-\beta\tau\}$$

$$\propto \tau^{n/2+\alpha-1} \exp\left\{-\tau\left[\frac{\sum_{i=1}^{n} y_i^2}{2} + \beta\right]\right\}.$$

We recognize another gamma density, so the prior is conjugate. The posterior mean is

$$\tilde{\tau} = \frac{n/2+\alpha}{\frac{\sum_{i=1}^{n} y_i^2}{2}+\beta} = \frac{n+2\alpha}{\sum_{i=1}^{n} y_i^2 + 2\beta}.$$

45.5 (a) A sketch of the density for an example θ is

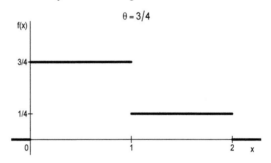

(b) Let Y be the number of X_i that are less than 1. Then the likelihood for observed $Y = y$ is

$$L(\theta; y) = \theta^y(1-\theta)^{n-y},$$

so if we maximize the log-likelihood over θ, we get $\hat{\theta} = Y/n$.

(c) The Bayes posterior density is distributed as Beta$(y+1, n-y+1)$, so the posterior mean is

$$\tilde{\theta} = \frac{y+1}{n+2}.$$

45.7 The prior is

$$\pi(\theta) \propto \theta^{\alpha-1}e^{-\beta\theta},$$

and the joint density given the parameter value is

$$f(y_1,\ldots,y_n|\theta) = \frac{\theta^n}{\prod_{i=1}^{n}(y_i+1)^{\theta+1}} = \theta^n \exp\left\{-(\theta+1)\sum_{i=1}^{n} \log(y_i+1)\right\}.$$

The posterior is

$$\pi(\theta|y_1,\ldots,y_n) \propto \theta^{n+\alpha-1} \exp\left\{-\theta\left[\beta+\sum_{i=1}^{n} \log(y_i+1)\right]\right\},$$

which we recognize as Gamma$(n+\alpha, \beta+\sum_{i=1}^{n} \log(y_i+1))$. Therefore the Bayes estimator is

$$\tilde{\theta} = \frac{n+\alpha}{\beta+\sum_{i=1}^{n} \log(y_i+1)}.$$

45.9 The joint density for the sample of size n times between tremors (assuming independence) is

$$f(y_1, \ldots, y_n|\theta) \propto \theta^{2n} e^{-2\theta \sum_{i=1}^{n} y_i} \quad \text{for } y_i > 0, \quad i = 1, \ldots, n,$$

and the prior is

$$\pi(\theta) \propto \theta^{\alpha-1} e^{-\beta\theta} \quad \text{for } \theta > 0.$$

Therefore the posterior for θ is

$$\pi(\theta|y_1, \ldots, y_n) \propto \theta^{2n+\alpha-1} e^{\theta(2s+\beta)} \quad \text{for } \theta > 0,$$

where $s = \sum_{i=1}^{n} y_i$.

(a) The posterior mean is $(2n+\alpha)/(2s+\beta)$, so if the average of eight observations is 45.2, the posterior mean for θ is about .0223, and the Bayes estimate for the mean time between tremors is about 44.9 days.

(b) If the average of 40 observations is 45.2, the posterior mean for θ is about .0222, and the Bayes estimate for the mean time between tremors is about 45.1 days.

(c) The prior and the two posteriors are shown together below. It might not look like the prior has mean .04, but the distribution is quite spread out, with standard deviation .08. As the sample size gets larger, the posterior gets more concentrated about its mean.

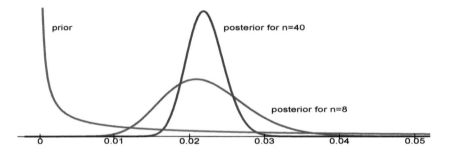

46.1 (a) The population mean is $\beta + 2$ so the method of moments is $\tilde{\beta} = \bar{Y} - 2$.

(b) The population variance is β^2, so the variance of $\bar{Y} - 2$ is β^2/n. The estimator is unbiased, and the variance goes to zero, hence the MSE goes to zero and the estimator is consistent.

46.3 First show that the CDF for Y is

$$F_Y(y) = 1 - \left(\frac{\theta - y}{\theta}\right)^3,$$

so the CDF for $Y_{(n)}$ is

$$F_{Y_{(n)}}(y) = \left[1 - \left(\frac{\theta - y}{\theta}\right)^3\right]^n.$$

Then use this to show

$$P(|Y_{(n)} - \theta| > \delta) = P(Y_{(n)} < \theta - \delta)$$

$$= \left[1 - \left(\frac{\theta - (\theta - \delta)}{\theta} \right)^3 \right]^n$$

$$= \left[1 - \left(\frac{\delta}{\theta} \right)^3 \right]^n$$

$$\to 0$$

as $n \to \infty$.

46.5 (a) Let $S = Y_1 + \cdots + Y_n$; then $S \sim \text{Gamma}(n, 1/\theta)$ and $\bar{Y}^2 = S^2/n^2$. Then

$$E(\bar{Y}^2) = \frac{1}{n^2} E(S^2) = \frac{1}{n^2} \left[V(S) + E(S)^2 \right] = \frac{1}{n^2} (n\theta^2 + n^2\theta^2) = \frac{(n+1)\theta^2}{n}.$$

The estimator is biased a bit large, but the bias goes to zero as n increases without bound. We can find

$$V(S^2) = E(S^4) - E(S^2)^2 = n(n+1)(n+2)(n+3)\theta^4 - [n(n+1)\theta^2]^2$$
$$= n(n+1)\theta^4[(n+2)(n+3) - n(n+1)] = 2n(n+1)(2n+3)\theta^4$$

using the result of Exercise 16.4. Then

$$V(\bar{Y}^2) = \frac{1}{n^4} V(S^2) = \frac{2n(n+1)(2n+3)\theta^4}{n^4} \to 0 \text{ as } n \to \infty.$$

Both the bias and variance go to zero as n increases, so the estimator is consistent.

(b) The MSE for the sample variance is about 104, and for the squared sample mean the MSE is about 58.5. Statistician B is the winner! Here is the code:

```
nloop=1000000
tha=1:nloop
thb=1:nloop
n=20
for(iloop in 1:nloop){
    x=rexp(20,1/4)
    tha[iloop]=var(x)
    thb[iloop]=mean(x)^2
}
mean((tha-16)^2)
mean((thb-16)^2)
```

(c) We showed in Chapter 53 that $\bar{X}$ is (minimally) sufficient for θ, so the estimator

$$T = \frac{n}{n+1} \bar{X}^2$$

must be UMVUE for θ^2, because the result in (a) gives that T is unbiased.

46.7 (a) From Exercise 33.4, we know that the minimum of a random sample from an exponential density with mean θ also has an exponential distribution, with mean θ/n. Then $\hat{\theta} = n \min(Y_i)$ is unbiased for θ.

(b) The estimator $\hat{\theta}$ is *not* consistent. For any $\delta > 0$, we have

$$P(|\hat{\theta} - \theta| > \delta) > P(\hat{\theta} - \theta > \delta)$$
$$= P(\hat{\theta} > \theta + \delta)$$
$$= P\left(\min(Y_i) > \frac{\theta + \delta}{n}\right)$$
$$= P\left(\text{all } Y_i > \frac{\theta + \delta}{n}\right)$$
$$= P\left(Y_1 > \frac{\theta + \delta}{n}\right)^n$$
$$= \left(e^{-(\theta+\delta)/n}\right)^n = e^{-(\theta+\delta)},$$

which does not go to zero as n increases.

46.9 The method of moments estimator was found by setting the population mean equal to the sample mean:

$$\bar{X} = \theta/3, \quad \text{therefore} \quad \hat{\theta} = 3\bar{X}.$$

This is unbiased (by construction!).

Now we need to show that the variance goes to zero. Instead of schlepping through the calculation of the variance, let's be clever. The variance is $E[(X - \theta/3)^2]$, but we know that $|X - \theta/3| < 2\theta$, so if we let σ^2 be the variance of this, we know $\sigma^2 < 4\theta$. Now,

$$V(\hat{\theta}) = V(3\bar{X}) = \frac{9\sigma^2}{n} \le \frac{36\theta}{n} \to 0 \quad \text{as} \quad n \to \infty.$$

We know that if an estimator is unbiased and the variance goes to zero as the sample size grows, it is consistent.

47.1 If both $a = 0$ and $b = 0$, then it is easy to show $X_n Y_n \xrightarrow{p} 0$:

$$P(|X_n Y_n| > \delta) \le P(|X_n| > \sqrt{\delta} \text{ or } |Y_n| > \sqrt{\delta})$$
$$\le P(|X_n| > \sqrt{\delta}) + P(|Y_n| > \sqrt{\delta})$$
$$\to 0.$$

Now if $a = 0$ but $b \neq 0$,

$$P(|X_n Y_n| > \delta) = P(X_n Y_n - X_n b + X_n b| > \delta)$$
$$\le P(|X_n(Y_n - b)| > \sqrt{\delta} \text{ or } |X_n b| > \sqrt{\delta})$$
$$\to 0,$$

using the $a = 0$ and $b = 0$ case for the last step, because $Y_n - b \xrightarrow{p} 0$ and $X_n b \xrightarrow{p} 0$.

47.3 It is straightforward to show that

$$E(1/Y_i) = \frac{\theta + 1}{\theta} \quad \text{and} \quad E(1/Y_i^2) = \frac{\theta + 1}{\theta - 1}.$$

Therefore we require $\theta > 1$ and the sequence converges to $(\theta + 1)/\theta$.

47.5 From the central limit theorem, we know that

$$\sqrt{n}(\bar{Y} - 3/\theta) \xrightarrow{\mathcal{D}} N(0, 3/\theta^2).$$

Letting $\mu = 3/\theta$, we define $g(\mu) = 3/\mu$ and find $g'(\mu) = -3/\mu^2$. The delta method then gives

$$\sqrt{n}(3/\bar{Y} - \theta) \xrightarrow{\mathcal{D}} N(0, \theta^2/3),$$

because $g'(\mu)^2 = 9/\mu^4 = \theta^4/9$.

47.7 (a) From the central limit theorem, we know

$$\sqrt{n}(\bar{Y} - \lambda) \xrightarrow{\mathcal{D}} N(0, \lambda).$$

Defining $g(\lambda) = \lambda^2$, we have $g'(\lambda)^2 = 4\lambda^2$, so

$$\sqrt{n}(\bar{Y}^2 - \lambda^2) \xrightarrow{\mathcal{D}} N(0, 4\lambda^3).$$

Therefore, $V(\bar{Y}^2) \approx 4\lambda^3/n$.

(b) The following code returns the true variance (to two decimal places) for $\lambda = 2$ and $n = 30$: 1.09. The approximation from the delta method gives 1.067.

```
nloop=1000000
x2=1:nloop
n=30;lambda=2
for(iloop in 1:nloop){
    x=rpois(n,lambda)
    x2[iloop]=mean(x)^2
}
var(x2)
4*lambda^3/n
```

47.9 It might be easiest to define

$$\theta = \frac{\alpha}{\alpha + 1} \quad \text{and} \quad \sigma^2 = \frac{\alpha}{(\alpha + 1)^2(\alpha + 2)}$$

to be the mean and variance of our Beta$(\alpha, 1)$ population. Then we know from the central limit theorem that

$$\sqrt{n}(\bar{Y} - \theta) \xrightarrow{\mathcal{D}} N(0, \sigma^2).$$

Let

$$g(\theta) = \frac{\theta}{1 - \theta}, \quad \text{so } g(\bar{Y}) = \frac{\bar{Y}}{1 - \bar{Y}} \quad \text{and} \quad g(\theta) = \alpha.$$

Then

$$g'(\theta) = \frac{1}{(1 - \theta)^2}$$

and

$$\sqrt{n}\left(\frac{\bar{Y}}{1 - \bar{Y}} - \frac{\theta}{1 - \theta}\right) \xrightarrow{\mathcal{D}} N\left(0, \frac{\sigma^2}{(1 - \theta)^4}\right).$$

Notice that $\theta/(1 - \theta) = \alpha$, and plugging in the definition of σ^2 we get the expression in terms of α:

$$\sqrt{n}\,(\hat{\alpha}_n - \alpha) \xrightarrow{\mathcal{D}} N\left(0, \frac{\alpha(\alpha + 1)^2}{\alpha + 2}\right).$$

Then the variance of $\hat{\alpha}_n$ is approximately

$$\frac{\alpha(\alpha + 1)^2}{n(\alpha + 2)}.$$

47.11   (a) It is straightforward to find

$$E(Y) = \frac{\theta + 1}{\theta + 2}.$$

Setting this equal to the sample mean and solving for θ gives the method of moments estimator:

$$\tilde{\theta} = \frac{2\bar{Y} - 1}{1 - \bar{Y}}.$$

Also,

$$E(Y^2) = \frac{\theta + 1}{\theta + 3} \quad \text{and} \quad V(Y) = \frac{\theta + 1}{(\theta + 3)(\theta + 2)^2}.$$

We know

$$\sqrt{n}\left(\bar{Y} - \frac{\theta + 1}{\theta + 2}\right) \xrightarrow{\mathcal{D}} N\left(0, \frac{\theta + 1}{(\theta + 3)(\theta + 2)^2}\right).$$

Let $\mu = (\theta + 1)/(\theta + 2)$ and define $g(\mu) = (2\mu - 1)/(1 - \mu)$; then $g'(\mu) = (1 - \mu)^{-2}$. Finally, the delta method gives

$$\sqrt{n}\left(\tilde{\theta} - \theta\right) \xrightarrow{\mathcal{D}} N\left(0, \frac{(\theta + 1)(\theta + 2)^2}{(\theta + 3)}\right)$$

using $(1 - \mu)^{-2} = (\theta + 2)^2$. The estimated variance of the estimator is

$$V(\tilde{\theta}) \approx \frac{(\theta + 1)(\theta + 2)^2}{(\theta + 3)n}.$$

47.13   (a) In Exercise 13.13, we found that the expected value and the variance of a random variable having this density is $(\theta - 1)^{-1}$. Setting the population mean equal to the sample mean and solving for the parameter gives the estimator

$$\tilde{\theta} = \frac{1}{\bar{Y}} + 1.$$

(b) The central limit theorem tells us that

$$\sqrt{n}\left(\bar{Y} - \frac{1}{\theta + 1}\right) \xrightarrow{\mathcal{D}} N\left(0, \frac{\theta}{(\theta - 2)(\theta - 1)^2}\right).$$

Letting $\mu = 1/(\theta + 1)$, and defining $g(\mu) = 1/\mu + 1$, we have $g'(\mu) = -1/\mu^2$. Using $1/\mu^4 = (\theta - 1)^4$, the delta method theorem gives

$$\sqrt{n}\left(\frac{1}{\bar{Y}} + 1 - \theta\right) \xrightarrow{\mathcal{D}} N\left(0, \frac{\theta(\theta - 1)^2}{\theta - 2}\right).$$

Therefore,

$$V(\tilde{\theta}) \approx \frac{\theta(\theta-1)^2}{(\theta-2)n}.$$

(c) We simulated from this density in Exercise 19.14. The code to check the variance of the estimator is below.

```
th=10
n=100;th=8
nloop=100000
thtil=1:nloop
for(iloop in 1:nloop){
    u=runif(n)
    y=(1-u)^(-1/th)-1
    thtil[iloop]=1/mean(y)+1
}
mean(thtil)
var(thtil)
th*(th-1)^2/(th-2)/n
```

47.15 The CLT tells us that

$$\sqrt{n}\left(\bar{Y}-\frac{1-p}{p}\right) \xrightarrow{\mathcal{D}} N\left(0,\frac{1-p}{p^2}\right).$$

If we define $\theta = (1-p)/p$ and $g(\theta) = 1/(1+\theta)$, then $g'(\theta)^2 = (1+\theta)^{-4}$. The delta method gives

$$\sqrt{n}\left(\frac{1}{1+\bar{Y}}-p\right) \xrightarrow{\mathcal{D}} N\left(0,(1-p)p^2\right).$$

Then $V(\hat{p}) \approx (1-p)p^2/n$.

48.1 (a) TRUE. (b) CAN'T TELL. (c) TRUE. (d) CAN'T TELL. (e) FALSE. (f) TRUE.

48.3 (a) The pivotal quantity is obtained from the derivation of the test statistic for the two-independent-samples t-test. If $X_1,\ldots,X_n \overset{iid}{\sim} N(\mu_A,\sigma^2)$ and $Y_1,\ldots,Y_m \overset{iid}{\sim} N(\mu_B,\sigma^2)$ and the samples are independent, we have

$$T = \frac{(\bar{X}-\bar{Y})-(\mu_A-\mu_B)}{S_p\sqrt{\frac{1}{n}+\frac{1}{m}}} \sim t(m+n-2).$$

Then

$$P(t_{.025}^{(m+n-2)} \leq T \leq t_{.975}^{(m+n-2)}) = .95.$$

Rearranging, we get the confidence interval

$$\left(\bar{X}-\bar{Y}+t_{.025}^{(m+n-2)}S_p\sqrt{\frac{1}{n}+\frac{1}{m}}, \bar{X}-\bar{Y}+t_{.975}^{(m+n-2)}S_p\sqrt{\frac{1}{n}+\frac{1}{m}}\right)$$

for $\mu_A - \mu_B$. Plugging in the data, we get $(-.676, 1.90)$.

(b) We know that $(n_A - 1)S_A^2/\sigma_A^2 \sim \chi^2(n_A - 1)$, and the same for B, so

$$\frac{S_A^2/\sigma_A^2}{S_B^2/\sigma_B^2} \sim F(n_A - 1, n_B - 1).$$

Then

$$P\left(F_{.05}^{(n_A-1, n_B-1)} \leq \frac{S_A^2/\sigma_A^2}{S_B^2/\sigma_B^2} \leq F_{.95}^{(n_A-1, n_B-1)}\right) = .90,$$

and rearranging we get

$$P\left(F_{.05}^{(n_A-1, n_B-1)}\frac{S_B^2}{S_A^2} \leq \frac{\sigma_B^2}{\sigma_A^2} \leq \frac{S_B^2}{S_A^2}F_{.95}^{(n_A-1, n_B-1)}\right).$$

Then a 90% confidence interval for the ratio σ_B^2/σ_A^2 is $(.767, 5.58)$. Because this confidence interval contains one (indicating equal variances) the assumption is plausible.

48.5 Let Y be the maximum of the seven observations; then we can use $Q = Y/\theta$ as a pivotal quantity. The CDF for Q is $F_Q(q) = q^7$ on $(0, 1)$, so we can construct an equal-tailed confidence interval as follows. The .5% percentile for Q is found by solving $.005 = q^7$ and, similarly, solve $.995 = q^7$. Then

$$P(.4691 \leq Q \leq .9993) = .99,$$

which leads to

$$P\left(\frac{Y}{.9993} \leq \theta \leq \frac{Y}{.4691}\right) = .99.$$

Our observed confidence interval is $(27.1, 57.8)$.

48.7 We know that

$$\frac{\sum_{i=1}^n X_i^2}{\sigma^2} \sim \chi^2(n),$$

so we can use this expression as a pivotal quantity. Then

$$P\left(\chi_{.05}^{2(n)} \leq \frac{\sum_{i=1}^n X_i^2}{\sigma^2} \leq \chi_{.95}^{2(n)}\right) = .90,$$

so

$$\left(\frac{\sum_{i=1}^n X_i^2}{\chi_{.95}^{2(n)}}, \frac{\sum_{i=1}^n X_i^2}{\chi_{.05}^{2(n)}}\right)$$

is a 90% confidence interval for σ^2. When $n = 10$ we have $\chi_{.05}^{2(n)} = 3.94$ and $\chi_{.95}^{2(n)} = 18.31$, and for our sample, $\sum_{i=1}^n x_i^2 = 44.39$. Therefore, the 90% confidence interval is about $(2.43, 11.27)$.

48.9 (a) We know that

$$\frac{\sum_{i=1}^n (y_i - \mu)^2}{\sigma_y^2} \sim \chi^2(n) \quad \text{and} \quad \frac{\sum_{i=1}^n (x_i - \mu)^2}{\sigma_x^2} \sim \chi^2(n),$$

and because the samples are independent, we have

$$\frac{\sum_{i=1}^{n}(x_i - \mu)^2/\sigma_x^2}{\sum_{i=1}^{n}(y_i - \mu)^2/\sigma_y^2} \sim F(n, n).$$

Let a be the 1st percentile of $F(n, n)$, and let b be the 99th percentile of $F(n, n)$; then we have

$$P\left(a \le \frac{\sum_{i=1}^{n}(x_i - \mu)^2/\sigma_x^2}{\sum_{i=1}^{n}(y_i - \mu)^2/\sigma_y^2} \le b\right) = .98.$$

Doing some rearranging we have

$$P\left(a\frac{\sum_{i=1}^{n}(y_i - \mu)^2}{\sum_{i=1}^{n}(x_i - \mu)^2} \le \frac{\sigma_y^2}{\sigma_x^2} \le b\frac{\sum_{i=1}^{n}(y_i - \mu)^2}{\sum_{i=1}^{n}(x_i - \mu)^2}\right) = .98,$$

so a 98% confidence interval is

$$\left(a\frac{\sum_{i=1}^{n}(y_i - \mu)^2}{\sum_{i=1}^{n}(x_i - \mu)^2}, b\frac{\sum_{i=1}^{n}(y_i - \mu)^2}{\sum_{i=1}^{n}(x_i - \mu)^2}\right).$$

(b) For the given data we have $\sum_{i=1}^{n}(x_i - 12)^2 = .002114$ and $\sum_{i=1}^{n}(y_i - 12)^2 = 0.000247$, and $n = 6$, so $a = .118$ and $b = 8.466$. Then our confidence interval is $(.0138, .989)$. The confidence interval does not contain one, so there is strong evidence that the first device is more precise.

48.11 (a) Let $S^2 = (X_1 - 4)^2 + (X_2 - 4)^2 + (X_3 - 8)^2 + (X_4 - 8)^2$. Then

$$\frac{S^2}{\sigma^2} \sim \chi^2(4)$$

and

$$P\left(.711 \le \frac{S^2}{\sigma^2} \le 9.49\right) = .95,$$

where $.711$ and 9.49 are the 5th and 95th percentiles of a $\chi^2(4)$ density. The observed confidence interval is

$$\left(\frac{S^2}{9.49}, \frac{S^2}{.711}\right).$$

(b) For the given data, $S^2 = 29.8$, and the confidence interval is $(3.14, 41.9)$.

48.13 (a) We know that

$$\frac{Y_1^2}{\sigma^2} + \frac{Y_2^2}{\sigma^2} + \frac{Y_3^2}{4\sigma^2} + \frac{Y_4^2}{4\sigma^2} \sim \chi^2(4),$$

so, looking up the 5th and 95th percentiles of a $\chi^2(4)$ random variable, we find

$$P\left(.711 \le \frac{Y_1^2}{\sigma^2} + \frac{Y_2^2}{\sigma^2} + \frac{Y_3^2}{4\sigma^2} + \frac{Y_4^2}{4\sigma^2} \le 9.488\right) = .9$$

and the confidence interval is

$$\left(\frac{Y_1^2 + Y_2^2 + Y_3^2/4 + Y_4^2/4}{9.488}, \frac{Y_1^2 + Y_2^2 + Y_3^2/4 + Y_4^2/4}{.711}\right).$$

(b) For the given values, the 90% confidence interval is $(4.07, 54.36)$.

48.15 Using the pivotal quantity $Q = \sum_{i=1}^{6} Y_i/\theta$, where $Y_1, \ldots, Y_6$ is a random sample from an $\text{Exp}(\theta)$ population, we know that $Q \sim \text{Gamma}(6, 1)$. The 2.5th percentile of this density is 2.20, and the 97.5th percentile is 11.67, so the 95% confidence interval is

$$(164.53/11.67, 164.53/2.20) = (14.10, 74.79).$$

48.17 In Exercise 47.8, we used $\hat{\alpha} = 2\bar{Y}/(1 - \bar{Y})$ and determined that

$$\sqrt{n}(\hat{\alpha} - \alpha) \xrightarrow{\mathcal{D}} N\left(0, \frac{(\alpha + 2)^2}{2(\alpha + 3)}\right).$$

Therefore,

$$P\left(-1.96 \leq \frac{\sqrt{n}(\hat{\alpha} - \alpha)}{\sqrt{\frac{(\alpha+2)^2}{2(\alpha+3)}}} \leq 1.96\right) \approx .95.$$

Using a further approximation, we have that

$$\hat{\alpha} \pm 1.96 \sqrt{\frac{(\hat{\alpha} + 2)^2}{2n(\hat{\alpha} + 3)}}$$

is an approximate 95% confidence interval.

48.19 The simulation can be thought of as Bernoulli trials with n equal to one million, and p is the true probability of getting at least five sixes. The (almost exact!) confidence interval is

$$\hat{p} \pm 2\sqrt{\hat{p}(1 - \hat{p})/n}, \quad \text{or} \quad .0155 \pm .00025.$$

48.21 Let $X_1, \ldots, X_n$ be independent exponential random variables with mean θ, representing the wait between eruptions. The probability that the wait is more than one hour is $p = e^{-1/\theta}$; let $\hat{p} = e^{-1/\bar{X}}$, the MLE for this probability. The CLT tells us that $\sqrt{n}(\bar{X} - \theta)$ is approximately normal with mean zero and variance θ^2. Let $g(\theta) = e^{-1/\theta}$. By the delta method theorem,

$$\sqrt{n}(e^{-1/\bar{X}} - e^{-1/\theta}) \xrightarrow{\mathcal{D}} N(0, e^{-2/\theta}\theta^2),$$

and an approximate 95% confidence interval for p is

$$\hat{p} \pm 2\sqrt{e^{-2/\bar{X}}/\bar{X}^2/n},$$

where we have substituted $\bar{X}$ for the unknown θ in the expression for the variance. For the observed data, $n = 30$ and $\bar{x} = .78$, we have that the confidence interval is $.277 \pm .130$, or $(.147, .407)$.

49.1 (a) We solve

$$\frac{\alpha}{\alpha + \beta} = .2 \quad \text{and} \quad \frac{\alpha\beta}{(\alpha + \beta)^2(\alpha + \beta + 1)} = .04,$$

to get $\alpha = 3/5$ and $\beta = 12/5$.

(b) The joint mass function for the data, given p, is

$$f(y_1, \ldots, y_n | p) = \prod_{i=1}^{n} [p(1-p)_i^y] = p^n (1-p)^{\sum_{i=1}^{n} y_i}.$$

The prior is $\pi(p) \propto p^{\alpha-1}(1-p)^{\beta-1}$, so the posterior is

$$\pi(p | y_1, \ldots, y_n) = p^{n+\alpha-1}(1-p)^{\sum_{i=1}^{n} y_i + \beta - 1},$$

that is, $p | y_1, \ldots, y_n \sim \text{Beta}(n + \alpha, \sum_{i=1}^{n} y_i + \beta)$.

(c) The posterior mean is

$$\tilde{p} = \frac{n + \alpha}{n + \sum_{i=1}^{n} y_i + \alpha + \beta}.$$

The posterior for p, using the five observations with $\sum_{i=1}^{5} y_i = 47$, is Beta$(5.6, 49.4)$. The Bayes estimator (posterior mean) is about .102.

(d) The Bayes 90% credible interval is $(.044, 176)$, which is found using qbeta(.05,5.6,49.4) and qbeta(.95,5.6,49.4).

49.3 There are 13 out of 15 instances of the cross-fertilized plant being taller, so the posterior for p is Beta$(14, 3)$. The 99% credible interval is $(.537, .978)$, found using qbeta(.005,14,3) and qbeta(.995,14,3). So we are confident that the probability of the cross-fertilized plant being taller is greater than 1/2.

49.5 (a) We solve

$$\frac{\alpha}{\beta} = \frac{1}{250} \quad \text{and} \quad \frac{\alpha}{\beta^2} = \frac{2}{250},$$

to get $\alpha = 1/500$ and $\beta = 1/2$.

(b) Then the posterior is Gamma$(n + \alpha, \sum_{i=1}^{n} y_i + \beta)$. The given numbers sum to 2534, so we use qgamma(.05,8+1/500,2534+1/2) and qgamma(.95, 8+1/500,2534+1/2) to get $(.00157, .00519)$ as the 90% credible interval for the rate of earthquakes.

49.7 To get the hyperparameters we solve, simultaneously,

$$\frac{\beta}{\alpha - 1} = 1 \quad \text{and} \quad \frac{\beta^2}{(\alpha - 1)^2 (\alpha - 2)},$$

to get $\alpha = 9/4$ and $\beta = 5/4$. The posterior is an inverse gamma with parameters 32.25 and 24.65, and we can find the 2.5th and 97.5th percentiles of this to get a credible interval for θ: Using the pscl library, qigamma(.025,32.25,24.65) returns .557 and qigamma(.975,32.25,24.65) returns 1.116. Therefore a 95% credible interval for $e^{-1/\theta}$ is $(e^{-1/.557}, e^{-1/1.116}) = (.166, .408)$. This is a little smaller than the confidence interval found with the delta method, which was $(.147, .407)$.

49.9 The joint density for the data, given β, is

$$\left[\prod_{i=1}^{n} \beta e^{-\beta x_i} \right] \left[\prod_{i=1}^{m} 2\beta e^{-2\beta y_i} \right] \propto \beta^{n+m} \exp\left\{ -\beta \left(\sum_{i=1}^{n} x_i + 2 \sum_{i=1}^{n} y_i \right) \right\}.$$

The prior for β is

$$\pi(\beta) \propto \beta^{a-1} e^{-b\beta},$$

so the posterior density for β is

$$\pi(\beta|\boldsymbol{x}, \boldsymbol{y}) \propto \beta^{n+m+a-1} \exp\left\{-\beta\left(\sum_{i=1}^{n} x_i + 2\sum_{i=1}^{n} y_i + b\right)\right\},$$

which we recognize as Gamma$(n + m + a, \sum_{i=1}^{n} x_i + 2\sum_{i=1}^{n} y_i + b)$. The 90% credible interval is made with the 5th and 95th percentiles of this posterior.

50.1 The code for the Intro Stat interval is

```
nloop=1000000
n=59;p=.05
cov=0
covw=0
covu=0
z=qnorm(.975)
len=0;lenw=0;lenu=0
for(iloop in 1:nloop){
    s=rbinom(1,n,p)
    phat=s/n
    pm=z*sqrt(phat*(1-phat)/n)
    if(phat-pm<p&phat+pm>p){cov=cov+1}
    len=len+2*pm
    pm=z*sqrt(phat*(1-phat)/n+z^2/4/n^2)
    p1=phat+z^2/2/n
    den=1+z^2/n
    lower=(p1-pm)/den
    upper=(p1+pm)/den
    lenw=lenw+upper-lower
    if(upper>p&lower<p){covw=covw+1}
    lb=qbeta(.025,s+1,n-s+1)
    ub=qbeta(.975,s+1,n-s+1)
    if(ub>p&lb<p){covu=covu+1}
    lenu=lenu+ub-lb
}
cov/nloop
covw/nloop
covu/nloop
len/nloop
lenw/nloop
lenu/nloop
```

For the Intro Stat interval, we get a paltry .799 for the coverage probability, and average length .104. For the Wilson interval, we get 97.3% coverage and length .118.

Finally, for the Bayes interval with uniform prior, we also get 97.3% coverage and a length .117.

50.3 (a) Because $Y_i - \theta \stackrel{ind}{\sim} \text{Exp}(1)$, $i = 1, \ldots, n$, $\sum_{i=1}^{n} Y_i - n\theta \sim \text{Gamma}(n, 1)$ and

$$\left(\frac{\sum_{i=1}^{n} Y_i - b}{n}, \frac{\sum_{i=1}^{n} Y_i - a}{n} \right)$$

is a 95% confidence interval for θ, where a is the 2.5th percentile and b is the 97.5th percentile of $\text{Gamma}(n, 1)$.

(b) The minimum of $Y_i - \theta$ has an exponential distribution with mean $1/n$. Then if d is the 2.5th and c is the 97.5th percentile of $\text{Exp}(n)$,

$$(\min(Y_i) - c, \min(Y_i) - d)$$

is a 95% confidence interval for θ.

(c) Clearly, we don't need to do simulations to find the lengths: For the confidence intervals in (a), each has length $(b - a)/n$ and in (b), the lengths are each $c - d$.

For $n = 10$, we have (qgamma(.975,10,1)-qgamma(.025,10,1))/10 or about 1.23 for the (a) confidence interval lengths, and we have qexp(.975,10)-qexp(.025,n) = .366 for the (b) confidence interval length.

For $n = 100$, the lengths are .392 and .0366. The confidence interval in (b) is much better!

50.5 The following code determines that Statistician B is the winner!

```
sig=1;za=qnorm(.975)
nloop=100000;lenA=0;lenB=0
capA=0;capB=0
for(iloop in 1:nloop){
    x=rnorm(5,12,sig)
    y=rnorm(5,12,sig/3)

    lowA=(mean(x)+mean(y))/2-za*sig*sqrt(5/90)
    uppA=(mean(x)+mean(y))/2+za*sig*sqrt(5/90)
    lenA=lenA+uppA-lowA
    if(lowA<=12&uppA>=12){capA=capA+1}

    lowB=(mean(x)+3*mean(y))/4-za*sig/sqrt(40)
    uppB=(mean(x)+3*mean(y))/4+za*sig/sqrt(40)
    lenB=lenB+uppB-lowB
    if(lowB<=12&uppB>=12){capB=capB+1}
}
capA/nloop
capB/nloop
lenA/nloop
lenB/nloop
```

50.7 The following code finds that the coverage probability for the frequentist confidence interval is about 96.1%, while the Bayes credible interval has 95.0% coverage probability. The average length for the confidence interval is 4.66, while the average length for the credible interval is *smaller*, about 4.41.

For $n = 100$, the coverage probability for the frequentist confidence interval is about 95.7%, while the Bayes credible interval has 95.1% coverage probability.

The average length for the confidence interval is 2.48, while the average length for the credible interval is smaller, about 2.37.

```
n=100;th=6
nloop=1000000
capf=0;capb=0
lenf=0;lenb=0
for(iloop in 1:nloop){
    u=runif(n)
    y=(1-u)^(-1/th)-1
    thtil=1/mean(y)+1
    lowerf=thtil - 2*sqrt(thtil*(thtil-1)^2/(thtil-2)/n)
    upperf=thtil + 2*sqrt(thtil*(thtil-1)^2/(thtil-2)/n)
    lenf=lenf+upperf-lowerf
    if(upperf>th&lowerf<th){capf=capf+1}
    a=n+1;b=sum(log(y+1))+.1
    lowerb=qgamma(.025,a,b)
    upperb=qgamma(.975,a,b)
    lenb=lenb+upperb-lowerb
    if(upperb>th&lowerb<th){capb=capb+1}
}
capb/nloop
capf/nloop
lenb/nloop
lenf/nloop
```

51.1 The code below gives $(9.92, 11.0)$ as the 95% bootstrap confidence interval.

```
nboot=100000
medboot=1:nboot
x=wind$ROS[wind$year==78]
n=length(x)
    for(iboot in 1:nboot){
        draw=sample(x,n,replace=TRUE)
        medboot[iboot]=median(draw)
    }
hist(medboot)
b=sort(medboot)
b[nboot*.025]
b[nboot*.975]
```

51.3 The 100,000 bootstrap samples do not take long to obtain; the 95% confidence interval made by the following code gives $(-.65, -.1)$ for the median difference in wear, suggesting that material B has more wear, and hence material A is better.

```
nboot=100000
medboot=1:nboot
x=shoes$A-shoes$B
n=length(x)
for(iboot in 1:nboot){
    draw=sample(x,n,replace=TRUE)
```

```
    medboot[iboot]=median(draw)
}
hist(medboot)
b=sort(medboot)
b[nboot*.025]
b[nboot*.975]
```

51.5 The following code gives an approximate 90% confidence interval for the median time between failures: $(18, 110.5)$.

```
y=aircondit$hours;n=length(y)
bmed=1:nboot
for(iboot in 1:nboot){
    yboot=sample(y,n,replace=TRUE)
    bmed[iboot]=median(yboot)
}
bsort=sort(bmed)
lower=bsort[250]
upper=bsort[9750]
```

51.7 The following code gives an approximate 95% confidence interval for the median excess returns: $(-.114, -.071)$. It does not contain zero.

```
x=acme$acme
n=length(x)
medhat=median(x)
nboot=100000
bmed=1:nboot
for(iboot in 1:nboot){
    xboot=sample(x,n,replace=TRUE)
    bmed[iboot]=median(xboot)
}
sort(bmed)[nboot*.025]
sort(bmed)[nboot*.975]
```

52.1 The likelihood is the joint mass function

$$L(p; S) = p^n (1-p)^S,$$

where S is the sum of the observations, i.e., the total number of failures. The log-likelihood is

$$\ell(p; S) = n \log(p) + S \log(1-p),$$

and the second derivative is

$$\frac{d^2 \ell(p; S)}{dp^2} = -\frac{n}{p^2} - \frac{S}{(1-p)^2}.$$

The information is then

$$-\mathrm{E}\left(\frac{d^2 \ell(p; S)}{dp^2}\right) = \frac{n}{p^2(1-p)},$$

and the estimate of the variance is the reciprocal of this: $\mathrm{V}(\hat{p}) \approx p^2(1-p)/n$. This is the same as the approximation using the delta method.

52.3 (a) The log-likelihood function is

$$\ell(\theta; \boldsymbol{y}) = n\log(\theta) = (\theta+1)\sum_{i=1}^{n}\log(y_i),$$

so

$$\frac{d\ell}{d\theta} = \frac{n}{\theta} - \frac{1}{\sum_{i=1}^{n}\log(y_i)},$$

and the MLE is $\hat{\theta} = n / \sum_{i=1}^{n}\log(y_i)$. Taking another derivative,

$$\frac{d^2\ell}{d\theta^2} = -\frac{n}{\theta^2},$$

so the information is $I(\theta) = \theta^2/n$.

(b) For the Wald test, we use

$$Z = \frac{\hat{\theta} - \theta_0}{\theta_0/\sqrt{n}} = \frac{2.20 - 4}{4/\sqrt{12}} = -1.56.$$

The approximate p-value is .06, the area to the left of -1.56 under a standard normal density.

(c) We can simulate from the density using the "CDF trick"—we find that if U is uniform on $(0, 1)$, then $Y = (1 - U)(1/\theta)$ has the given density. The following code produces the true distribution of the test statistic.

```
th=4;n=12
nloop=1000000
for(iloop in 1:nloop){
   u=runif(n)
   y=(1-u)^{-1/th}
   thhat[iloop]=n/sum(log(y))
}
hist(thhat,br=100)
ydat=c(1.53, 1.36, 3.64, 1.18, 1.15, 1.31, 1.59, 1.89, 1.47,
       1.25, 1.05, 2.98)
thobs=n/sum(log(ydat))
sum(thhat<thobs)/nloop
```

We get a p-value of about .008! Now we can reject the null hypothesis with decision. Why is the result so different? Let's make a histogram of the true density of our test statistic and compare it to the normal approximation:

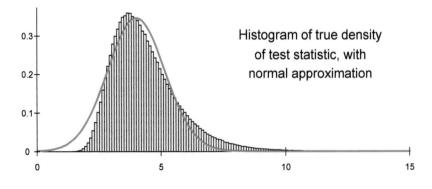

Histogram of true density of test statistic, with normal approximation

Looking at the left tail, we see the reason for the big difference in the p-values. A sample size of 12 is really not large enough for the Wald approximation to hold.

52.5 (a) Suppose S is the number of successful uses. If p is smaller, we expect S to be *larger*. Then $S \sim \mathrm{NB}(20, p)$, and the 99th percentile of this distribution is 611, found with qnbinom(.99,20,.05). The exact decision rule is "reject H_0 if there are more than 611 successful uses of the 20 new devices."

(b) The joint mass function for $X_1, \ldots, X_{20}$ (the numbers of successful uses of the device) is

$$f_p(\boldsymbol{x}) = p^n (1-p)^{\sum_{i=1}^n x_i}.$$

The log-likelihood is

$$\ell(p; \boldsymbol{x}) = n \log(p) + \left(\sum_{i=1}^n x_i \right) \log(1-p).$$

The score function is

$$u(p) = \frac{n}{p} - \frac{\sum_{i=1}^n x_i}{1-p},$$

and

$$\hat{p} = \frac{n}{\sum_{i=1}^n x_i + n}.$$

The information is

$$I(p) = \frac{n}{p^2(1-p)}.$$

Then

$$Z = \frac{\hat{p} - p}{\sqrt{p^2(1-p)/n}}$$

is approximately normal when n is "large." We reject H_0 when $Z > 2.326$; plugging in the H_0 value for p, this decision rule is "reject H_0 if there are more than 793 successful uses of the new devices." (This number of uses corresponds to $\hat{p} = .0246$.) This is quite different from the solution to (a). However, the sample size is quite small, and we can't expect the Wald test to be great.

52.7 (a) The log-likelihood function is a constant plus

$$n \log(\lambda) - \lambda \sum_{i=1}^n x_i^{1/2},$$

and the score function is

$$\frac{n}{\lambda} - \sum_{i=1}^n x_i^{1/2},$$

and the MLE is $\hat{\lambda} = n / \sum_{i=1}^n x_i^{1/2}$.

Taking the derivative of the score function with respect to λ gives the information $I(\lambda) = n/\lambda^2$, and $\mathrm{var}(\hat{\lambda}) \approx \lambda^2/n$. For $n = 30$ and $\lambda = 1/2$, $\mathrm{var}(\hat{\lambda}) \approx 1/120 = .008\overline{3}$.

(b) The following code produces a true variance of about .0095, so the approximation using the Fisher information will underestimate the variance.

```
n=30;lam=.5
nloop=100000
lhat=1:nloop
for(iloop in 1:nloop){
    u=runif(n)
    x=(log(1-u)/lam)^2
    lhat[iloop]=n/sum(sqrt(x))
}
var(lhat)
```

52.9 We know that $-\log(1-Y)$ has an $\mathrm{Exp}(\theta)$-density (that is, the mean is $1/\theta$). Therefore $-\sum_{i=1}^{n}\log(1-Y) \sim \mathrm{Gamma}(n,\theta)$, and

$$\frac{-1}{\sum_{i=1}^{n}\log(1-Y)} \sim \mathrm{InvGamma}((n,\theta).$$

The mean of an $\mathrm{InvGamma}(n,\theta)$ random variable is $\theta/(n-1)$ and the variance is $\theta^2/[(n-1)^2(n-2)]$. Finally,

$$\mathrm{E}(\hat{\theta}) = \frac{n\theta}{n-1} \quad \text{and} \quad \mathrm{V}(\hat{\theta}) = \frac{n^2\theta^2}{(n-1)^2(n-2)}.$$

These are both a bit larger than the approximated mean and variance, but of course the approximation gets closer to the truth as n increases.

53.1 The joint mass function is

$$f_\lambda(y_1,\ldots,y_n) = \prod_{i=1}^{n}\left[\frac{e^{-\lambda}\lambda^{y_i}}{y_i!}\right] = \frac{e^{-n\lambda}\lambda^{\sum_{i=1}^{n}y_i}}{\prod_{i=1}^{n}y_i!} = \frac{e^{-n\lambda}\lambda^{s}}{\prod_{i=1}^{n}y_i!},$$

where $s = \sum_{i=1}^{n}y_i$. Then the sum of observations is sufficient, which implies the mean is sufficient. We have

$$g_\lambda(s) = e^{-n\lambda}\lambda^{s} \quad \text{and} \quad h(y_1,\ldots,y_n) = \prod_{i=1}^{n}\frac{1}{y_i!}.$$

53.3 The joint density of the sample is

$$f_\theta(y_1,\ldots,y_n) = (\theta-1)^n \left(\prod_{i=1}^{n}y_i\right)^{-\theta} = (\theta-1)^n e^{-\theta\sum_{i=1}^{n}\log(y_i)}.$$

So if $S = \sum_{i=1}^{n}\log(Y_i)$, we can write

$$f_\theta(y_1,\ldots,y_n) = g(s,\theta)h(y_1,\ldots,y_n),$$

where $h(y_1,\ldots,y_n) = 1$ and

$$g_\theta(s) = (\theta-1)^n e^{-\theta s}.$$

53.5 The joint density for the sample is

$$f_{\sigma^2}(y_1, \ldots, y_n) = \prod_{i=1}^{n} \left(\frac{1}{\sqrt{2\pi\sigma^2}} \exp\left\{ -\frac{y_i^2}{2\sigma^2} \right\} \right)$$

$$= \left(\frac{1}{2\pi\sigma^2} \right)^{n/2} \exp\left\{ -\frac{\sum_{i=1}^{n} y_i^2}{2\sigma^2} \right\},$$

so $T = \sum_{i=1}^{n} Y_i^2$ is sufficient for σ^2.

53.7 The joint density is

$$f_\theta(y_1, \ldots, y_n) = \frac{3^n}{\theta^n} \left(\prod_{i=1}^{n} y_i^2 \right) e^{-\sum_{i=1}^{n} y_i^3/\theta} \quad \text{for } y_1 > 0, \ldots, y_n > 0,$$

so $T = \sum_{i=1}^{n} Y_i^3$ is sufficient for θ by the factorization theorem.

53.9 The joint density function is

$$f_\mu(y_1, \ldots, y_n)$$

$$= \prod_{k=1}^{n} \left(\frac{1}{\sqrt{2\pi k\sigma^2}} \exp\left\{ -\frac{(y_k - \mu)^2}{2k\sigma^2} \right\} \right)$$

$$= \left(\frac{1}{2\pi\sigma^2} \right)^{n/2} \left(\prod_{k=1}^{n} \frac{1}{\sqrt{k}} \right) \exp\left\{ -\frac{1}{2\sigma^2} \sum_{k=1}^{n} \frac{(y_k - \mu)^2}{k} \right\}$$

$$= \left(\frac{1}{2\pi\sigma^2} \right)^{n/2} \left(\prod_{k=1}^{n} \frac{1}{\sqrt{k}} \right) \exp\left\{ -\frac{1}{2\sigma^2} \left[\sum_{k=1}^{n} \frac{y_k^2}{k} - 2\mu \sum_{k=1}^{n} \frac{y_k}{k} + \frac{n\mu^2}{k} \right] \right\}$$

$$= g_\mu(s) h(y_1, \ldots, y_n),$$

where

$$g_\mu(s) = \exp\left\{ -\frac{1}{2\sigma^2} \left[\frac{n\mu^2}{k} - 2\mu s \right] \right\}, \quad \text{where} \quad s = \sum_{k=1}^{n} \frac{y_k}{k},$$

and

$$h(y_1, \ldots, y_n) = \left(\frac{1}{2\pi\sigma^2} \right)^{n/2} \left(\prod_{k=1}^{n} \frac{1}{\sqrt{k}} \right) \exp\left\{ -\frac{1}{2\sigma^2} \sum_{k=1}^{n} \frac{y_k^2}{k} \right\}.$$

Then

$$S = \sum_{k=1}^{n} \frac{Y_k}{k}$$

is sufficient for μ by the factorization theorem.

53.11 The joint density function is

$$\left[\frac{1}{\theta} e^{-y_1/\theta} \right] \left[\frac{1}{\theta} e^{-y_2/\theta} \right] \left[\frac{1}{\theta} e^{-y_3/\theta} \right] \left[\frac{1}{\theta} e^{-y_4/\theta} \right] \left[\frac{1}{2\theta} e^{-y_5/(2\theta)} \right]$$

$$\times \left[\frac{1}{2\theta} e^{-y_6/(2\theta)} \right] \left[\frac{1}{2\theta} e^{-y_7/(2\theta)} \right] \left[\frac{1}{2\theta} e^{-y_8/(2\theta)} \right]$$

which simplifies to

$$\frac{1}{16\theta^4} \exp\left(\left[\sum_{i=1}^{4} y_i + \frac{1}{2}\sum_{i=5}^{8} y_i\right] / \theta\right),$$

so

$$S = \sum_{i=1}^{4} Y_i + \frac{1}{2}\sum_{i=5}^{8} Y_i$$

is sufficient for θ.

53.13 The joint density for the sample must be written with the indicators for the support:

$$f_\theta(y_1, \ldots, y_n) = \prod_{i=1}^{n} \left[e^{\theta - y_i} I\{y_i > \theta\}\right]$$

$$= e^{-\sum_{i=1}^{n} y_i} \times e^{n\theta} I\{\theta < \min(y_i)\}.$$

The first term in the product is $h(y_1, \ldots, y_n)$, and $g_\theta(s) = e^{n\theta} I\{\theta < s\}$; therefore $S = \min(Y_i)$ is sufficient for θ.

53.15 (a) The joint density for the sample is

$$f_\theta(y_1, \ldots, y_n) = \left(\frac{\theta}{2}\right)^n \prod_{i=1}^{n} \left(\frac{y_i}{2} + 1\right)^{-(\theta+1)}$$

$$\propto \theta^2 \exp\left[-(\theta+1)\sum_{i=1}^{n} \log(y_i/2 + 1)\right].$$

Therefore, $T = \sum_{i=1}^{n} \log(Y_i/2 + 1)$ is sufficient for θ.

(b) We will find $E[\log(Y/2 + 1)]$, where $Y \sim f_\theta$, then multiply this by n:

$$E[\log(Y/2 + 1)] = \frac{\theta}{2} \int_0^\infty \log(y/2 + 1)(y/2 + 1)^{-(\theta+1)} dy$$

$$= \theta \int_1^\infty \log(x) x^{-(\theta+1)} dx.$$

Now we use integration by parts with $u = \log(x)$ and $dv = x^{(\theta+1)} dx$, to get

$$E[\log(Y/2 + 1)] = \int_1^\infty x^{-(\theta+1)} dx = 1/\theta.$$

Therefore $E(T) = n/\theta$.

53.17 The joint density can be written as

$$f_\mu(x_1, x_2, x_3, x_4, x_5) = \frac{1}{36\sqrt{\pi}} \exp\left\{-\frac{1}{2}\left[\frac{x_1^2 - 2x_1\mu + \mu^2}{4} + \frac{x_2^2 - 2x_2\mu + \mu^2}{4}\right.\right.$$

$$\left.\left.+ \frac{x_3^2 - 2x_3\mu + \mu^2}{9} + \frac{x_4^2 - 2x_4\mu + \mu^2}{9} + \frac{x_5^2 - 2x_5\mu + \mu^2}{9}\right]\right\}.$$

If we define

$$h(x_1, x_2, x_3, x_4, x_5) = \frac{1}{36\sqrt{\pi}} \exp\left\{-\frac{1}{2}\left[\frac{x_1^2}{4} + \frac{x_2^2}{4} + \frac{x_3^2}{9} + \frac{x_4^2}{9} + \frac{x_5^2}{9}\right]\right\}$$

and

$$t = \frac{x_1}{4} + \frac{x_2}{4} + \frac{x_3}{9} + \frac{x_4}{9} + \frac{x_5}{9},$$

then

$$g_\mu(t) = \exp\left\{-\frac{1}{2}\left[\frac{5\mu^2}{6} - 2t\mu\right]\right\}$$

and we have successfully factored the joint density. Therefore part (c) has the estimator which uses the sufficient statistic, and is also the smallest variance estimator.

53.19 The joint density function is

$$f(y_1, \ldots, y_n; \theta) = \frac{1}{\theta^n} \exp\left\{-\sum_{i=1}^{n} y_i/\theta\right\},$$

so $\sum_{i=1}^{n} Y_i$ is a sufficient statistic. Therefore, $\hat{\theta} = \bar{Y}$ is the estimator that is based on the sufficient statistic.

53.21 (a) The joint density of the observations $f_\mu(\boldsymbol{x}, \boldsymbol{y})$ is

$$\left[\left(\frac{1}{2\pi\sigma^2}\right)^{n/2} \exp\left\{-\frac{1}{2\sigma^2}\sum_{i=1}^{n}(x_i - \mu)^2\right\}\right]$$
$$\times \left[\left(\frac{1}{2\pi\sigma^2}\right)^{m/2} \exp\left\{-\frac{1}{2\sigma^2}\sum_{i=1}^{m}(y_i - 2\mu)^2\right\}\right].$$

Simplifying, we have

$$f_\mu(\boldsymbol{x}, \boldsymbol{y}) = \left(\frac{1}{2\pi\sigma^2}\right)^{(n+m)/2} \exp\left\{-\frac{1}{2\sigma^2}\left[\sum_{i=1}^{n}(x_i - \mu)^2 + \sum_{i=1}^{m}(y_i - 2\mu)^2\right]\right\}$$
$$= \left(\frac{1}{2\pi\sigma^2}\right)^{(n+m)/2} \exp\left\{-\frac{1}{2\sigma^2}\left[\sum_{i=1}^{n}x_i^2 + \sum_{i=1}^{m}y_i^2 - 2\mu\left(\sum_{i=1}^{m}x_i + 2\sum_{i=1}^{m}y_i\right)\right.\right.$$
$$\left.\left. + n\mu^2 + 4m\mu^2\right]\right\}.$$

Let $t = \sum_{i=1}^{m} x_i + 2\sum_{i=1}^{m} y_i$ and defining

$$h(\boldsymbol{x}, \boldsymbol{y}) = \exp\left\{-\frac{1}{2\sigma^2}\left[\sum_{i=1}^{n}x_i^2 + \sum_{i=1}^{m}y_i^2\right]\right\}$$

and

$$g(t, \mu) = \left(\frac{1}{2\pi\sigma^2}\right)^{(n+m)/2} \exp\left\{-\frac{1}{2\sigma^2}\left[-2\mu t + n\mu^2 + 4m\mu^2\right]\right\},$$

we have factored the joint density to show that $T = \sum_{i=1}^{m} X_i + 2\sum_{i=1}^{m} Y_i$ is sufficient for μ.

(b) We see that T is a normal random variable with mean $(n + 4m)\mu$ and variance $(n + 4m)\sigma^2$, so $\hat{\mu} = T/(n + 4m)$ is an unbiased estimator of μ with variance $\sigma^2/(n + 4m)$. Then

$$Q = \frac{\hat{\mu} - \mu}{\sigma/\sqrt{n + 4m}} \sim \mathrm{N}(0, 1)$$

and the confidence interval is

$$\hat{\mu} \pm \frac{1.96\sigma}{\sqrt{n + 4m}}.$$

(c) We want to show that

$$\frac{1}{2}\sqrt{\frac{1}{n} + \frac{1}{4m}} \geq \frac{1}{\sqrt{n + 4m}}$$

or (squaring both sides)

$$\frac{n + 4m}{4mn} \geq \frac{4}{n + 4m},$$

which is equivalent to $16mn \leq (n + 4m)^2 = n^2 + 8nm + 16m^2$. This inequality holds because $0 \leq n^2 - 8nm + 16m^2 = (n - 4m)^2$.

In the case where $n = 4m$, the lengths of the confidence intervals are the same; otherwise the confidence interval made with the sufficient statistic has a smaller length.

53.23 The joint density is

$$f(y_1, \ldots, y_n) = \prod_{i=1}^{n} I\{\theta < y_i < \theta + 1\} = I\{\theta < \min(y)\}I\{\theta > \max(y) - 1\}.$$

The sufficient statistic is $T = (\min(y), \max(y))$.

54.1 The probability mass function for the population is

$$f_\lambda(y) = \frac{e^{-\lambda}\lambda^y}{y!},$$

so

$$\log(f_\lambda) = -\lambda + y\log(\lambda) - \log(y!)$$

and

$$\frac{\partial^2}{\partial \lambda^2}\log[f_\lambda] = -\frac{y}{\lambda^2},$$

and finally

$$I(\lambda) = \frac{n}{\lambda}.$$

Any unbiased estimator of λ has variance at least λ/n, which is the variance of the sample mean. Therefore, the sample mean is UMVUE.

54.3 The information was calculated in Exercise 54.2; we just have a different numerator. Using $g(\mu) = \mu^2$, the inequality tells us that all unbiased estimators of μ^2 have variance at least $4\mu^2\sigma^2/n$.

54.5 (a) The joint density for the sample is

$$f_\theta(y_1, \ldots, y_n) = \left(\frac{2}{\theta}\right)^n \left(\prod_{i=1}^n y_i\right) e^{-\sum_{i=1}^n y_i^2/\theta},$$

so if $s = \sum_{i=1}^n y_i^2$ we can factor the joint density as

$$f_\theta(y_1, \ldots, y_n) = h(y_1, \ldots, y_n) g_\theta(s),$$

where $h(y_1, \ldots, y_n) = \prod_{i=1}^n y_i$ and

$$g_\theta(s) = \left(\frac{2}{\theta}\right)^n e^{-s/\theta}.$$

Therefore

$$S = \sum_{i=1}^n Y_i^2$$

is sufficient for θ.

(b) To find an UMVUE for θ, we need to find an unbiased estimator that is a function of S. If $Y \sim f_\theta$, then

$$E(Y^2) = \frac{2}{\theta} \int_0^\infty y^3 e^{-y^2/\theta} dy = \theta \int_0^\infty u e^{-u} du = \theta,$$

after making the substitution $u = y^2/\theta$. Therefore, S/n is UMVUE for θ.

54.7 (a) We can write the joint density as

$$f(y_1, \ldots, y_n) = \left(\frac{1}{\sqrt{2\pi\sigma^2}}\right)^n \exp\left\{-\frac{\sum_{i=1}^n y_i^2}{2\sigma^2}\right\},$$

so clearly $W = \sum_{i=1}^n y_i^2$ is sufficient for σ^2. Because the y_i/σ are independent standard normal, we know that

$$\frac{W}{\sigma^2} \sim \chi^2(n),$$

and this gives that W/n is unbiased for σ^2. Hence W/n is UMVUE.

(b) The following code shows that when $n = 10$ and $\sigma = 2$, the variance of W/n is about 3.20 and the variance of S^2 is about 3.58.

```
nloop=10000
shat1=1:nloop
shat2=1:nloop
truesig=2;n=10
for(iloop in 1:nloop){
    x=rnorm(n,0,truesig)
    shat1[iloop]=sum(x^2)/n
    shat2[iloop]=var(x)
}
var(shat1)
var(shat2)
```

54.9 (a) Writing out the joint density for the observations, we can factor the exponent to get that

$$T = \sum_{i=1}^{m} X_i + 2 \sum_{i=1}^{n} Y_i$$

is sufficient for μ. Therefore

$$\hat{\mu} = \frac{1}{m + 4n} \left[\sum_{i=1}^{m} X_i + 2 \sum_{i=1}^{n} Y_i \right]$$

is UMVUE.

(b) We want to show that the variance of our estimator is smaller than that of the alternative unbiased estimator. That is, we want to show

$$\frac{\sigma^2}{m + 4n} \leq \frac{\sigma^2/m + \sigma^2/n}{9},$$

or

$$\frac{1}{m + 4n} \leq \frac{m + n}{9mn}.$$

We can show

$$9mn \leq (m + 4n)(m + n) = m^2 + 5mn + 4n^2$$

because $m^2 - 4mn + 4n^2 = (m - 2n)^2 \geq 0$. We notice that if $m = 2n$, the variances are equal!

54.11 The joint density for the sample is

$$f_\theta(y_1, \ldots, y_n) = \theta^n \exp \left\{ (\theta + 1) \sum_{i=1}^{n} \log(y_i + 1) \right\},$$

so $S = \sum_{i=1}^{n} \log(Y_i + 1)$ is sufficient for θ. From Exercise 13.13 we know that $E(\log(Y_i + 1)) = 1/\theta$, so $E(S) = n/\theta$.

54.13 The joint density for the sample is

$$f_\mu(y_1, \ldots, y_n) = e^{n\mu} e^{-\sum_{i=1}^{n} y_i} I\{y_1 > \mu, \ldots, y_n > \mu\}$$
$$= e^{n\mu} e^{-\sum_{i=1}^{n} y_i} I\{\mu < \min(y_i)\}.$$

Therefore, by the factorization theorem, $\min(Y_i)$ is sufficient for μ, and because $\hat{\mu} = \min(Y_i) - 1/n$ is unbiased, it is UMVUE.

54.15 The joint density is

$$f_\theta(y_1, \ldots, y_n) = \frac{1}{4\theta^4} \exp \left\{ -\frac{2y_1 + 2y_2 + y_3 + y_4}{2\theta} \right\},$$

so $2Y_1 + 2Y_2 + Y_3 + Y_4$ is sufficient, and $T = (2Y_1 + 2Y_2 + Y_3 + Y_4)/6$ is UMVUE.

54.17 We have already shown that $\bar{Y}$ is sufficient for the mean θ, so it is also sufficient for θ^2. Therefore, Statistician B has the better estimator. The following code shows that the MSE for estimator A is about .46, compared to about .08 for estimator B.

```
n=20;theta=.4
nloop=100000
estA=1:nloop
estB=1:nloop
for(iloop in 1:nloop){
    y=rnorm(n,theta,sqrt(2))
    estA[iloop]=mean(y^2)-2
    estB[iloop]=mean(y)^2-2/n
}
mean((estA-theta^2)^2)
mean((estB-theta^2)^2)
```

54.19 Given three one-yard pieces of cloth, the numbers of flaws in each piece, say Y_1, Y_2, and Y_3, are each Poisson(λ). For the two-yard pieces of cloth, the numbers of flaws Y_4 and Y_5 are each Poisson(2λ). Let's assume independence, and hope that's reasonable, because otherwise we can't solve the problem. Then the joint mass function is

$$f_\lambda(y_1, y_2, y_3, y_4, y_5) = \prod_{i=1}^{3}\left[\frac{e^{-\lambda}\lambda^{y_i}}{y_i!}\right]\prod_{i=4}^{5}\left[\frac{e^{-2\lambda}(2\lambda)^{y_i}}{y_i!}\right] = \frac{e^{-7\lambda}2^{y_4+y_5}\lambda^{\sum_{i=1}^{5}y_i}}{\prod_{i=1}^{5}y_i!}.$$

Then $S = \sum_{i=1}^{5}y_i$ is sufficient by the factorization theorem, and $E(S) = 7\lambda$. Therefore, $S/7$ is UMVUE for λ.

55.1 If $Y \sim$ Geom(p), the probability mass function is

$$f_p(y) = P(Y = y) = p(1-p)^{y-1} = pe^{(y-1)\log(1-p)} \quad \text{for} \quad y = 0, 1, 2, \ldots.$$

Then $h(y) = 1/eI\{y = 0, 1, 2, \ldots\}$, $c(p) = p$, $t(y) = y$, and $\omega(p) = \log(1-p)$. Given a random sample from a geometric distribution, a sufficient statistic is the sample mean.

55.3 It's straightforward to see that

$$h(y) = yI\{y > 0\}, \quad c(\theta) = \omega(\theta) = \frac{1}{\theta}, \quad \text{and} \quad t(y) = y^2.$$

Therefore, given a random sample $Y_1, \ldots, Y_n$ from this density, $T(Y_1, \ldots, Y_n) = \sum_{i=1}^{n}Y_i^2$ is sufficient for θ.

55.5 The density can be written as

$$f_\theta(y) = \theta e^{(\theta-1)\log(y)} \quad \text{for} \quad y > 0,$$

so $h(y) = I\{y > 0\}$, $c(\theta) = \theta$, $\omega(\theta) = \theta - 1$, and $t(y) = \log(y)$. Therefore, given a random sample $Y_1, \ldots, Y_n$ from this density, $T(Y_1, \ldots, Y_n) = \sum_{i=1}^{n}\log(Y_i)$ is sufficient for θ.

55.7 We can write

$$f_\theta(x) = \theta e^{-(\theta+1)\log(x+1)}I\{x > 0\},$$

so $h(y) = I\{y > 0\}$, $c(\theta) = \theta$, $\omega(\theta) = \theta + 1$, and $t(x) = \log(x + 1)$. Therefore, given a random sample $X_1, \ldots, X_n$ from this density, $T(X_1, \ldots, X_n) = \sum_{i=1}^{n}\log(X_i + 1)$ is sufficient for θ.

55.9 The two-parameter density can be written as

$$f_{\alpha,\beta}(y) = \frac{\beta^\alpha}{\Gamma(\alpha)}e^{(\alpha-1)\log(y)-\beta y} \text{ for } y > 0,$$

so

$$h(y) = I\{y > 0\}, \quad c(\alpha,\beta) = \frac{\beta^\alpha}{\Gamma(\alpha)}, \quad \omega_1(\alpha,\beta) = \alpha - 1, \quad \omega_2(\alpha,\beta) = -\beta,$$

$$t_1(y) = \log(y), \text{ and } t_2(y) = y.$$

Therefore, if we have a random sample $Y_1, \ldots, Y_n$ from this density,

$$T_1(y_1, \ldots, y_n) = \sum_{i=1}^n \log(y_i) \text{ and } T_2(y_1, \ldots, y_n) = \sum_{i=1}^n y_i$$

are jointly sufficient for α and β.

56.1 For example, if $n = 10$ and $\alpha = .05$, we reject H_0 if the sample sum is bigger
than the 95th percentile of a Gamma$(10, 1)$ distribution, which is 15.7. If $\theta = 2$,
the probability that the sample sum is bigger than 15.7 is .735; this is the power
of the test when $n = 10$ and $\theta = 2$. We can compute the power for $n = 10$ and θ
values between 1 and 4 to get the solid curve in the figure below.

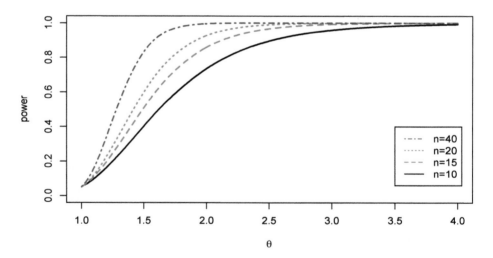

If $n = 15$, we reject H_0 when the sample sum is bigger than the 95th percentile
of a Gamma$(15, 1)$ distribution, which is 21.9. The probabilities of rejection are
depicted for $n = 15$ and values of θ that range from 1–4 by the dashed curve
in the figure. Similarly, we can find the power curves for $n = 20$ and $n = 40$.
Note that the power ranges from α, when the null hypothesis is true, to one, when
the value of θ increases away from the null hypothesis value. Depending on our
guess as to what θ is, we can choose a sample size that gives a sufficiently high
power. For example, if we believe that $\theta = 1.5$, we'd get a power of about .8 if
$n = 40$, but the power would be less than .5 if $n = 10$. On the other hand if we
think $\theta = 2$, we'd get power over 80% when $n = 20$.

56.3 We reject H_0 when $S^2/4$ is smaller than c_n, the 1st percentile of a $\chi^2(n-1)$ random variable. This is

$$P(S^2/4 < c_n) = P(S^2/2.25 < 4c_n/2.25) = P(X < 4c_n/2.25),$$

where $W \sim \chi^2(n-1)$. Here is the code for the power curve shown below:

```
n=10:100
pwr=1:91
for(i in 1:91){
    crit=qchisq(.01,n[i]-1)
    pwr[i]=pchisq(4/2.25*crit,n[i]-1)
}
plot(c(10,100),c(0,1),pch="",xlab="sample size",ylab="power")
lines(n,pwr,lwd=1.5)
points(n,pwr,pch=20,cex=1.5)
```

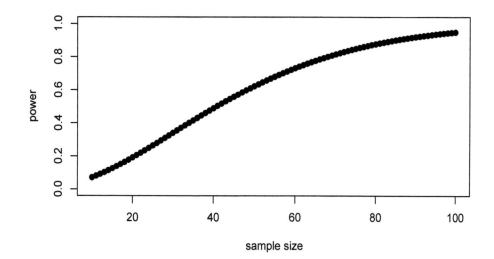

56.5 Let c_1 and c_2 be the 2.5th and 97.5th percentiles of $F(n-1, n-1)$, respectively. Then we will reject H_0 if S_1^2/S_2^2 is larger than c_2 or smaller than c_1. Suppose the ratio of variances is $\sigma_2^2/\sigma_1^2 = r$. Then the power is

$$P(S_1^2/S_2^2 < c_1) + P(S_1^2/S_2^2 > c_2) = P\left(\frac{S_1^2/\sigma_1^2}{S_2^2/\sigma_2^2} < c_1 r\right) + P\left(\frac{S_1^2/\sigma_1^2}{S_2^2/\sigma_2^2} > c_2 r\right)$$
$$= P(W < c_1 r) + P(W > c_2 r),$$

where $W \sim F(n-1, n-1)$.

The following code produces one of the curves on the plot below. We find that the sample size has to be large to provide a reasonable amount of power, even when one of the population variances is twice the other!

```
n=20
rvec=20:40/20
pwr=20:40*0
crit1=qf(.025,n-1,n-1)
crit2=qf(.975,n-1,n-1)
```

```
for(r in 1:21){
    pwr[r]=1-pf(crit2*rvec[r],n-1,n-1)+pf(crit1*rvec[r],n-1,n-1)
}
plot(c(1,2.1),c(0,1),pch="",xlab="ratio of variances",ylab="power")
lines(rvec,pwr,lwd=3)
text(2.051,pwr[21],"n=20")
```

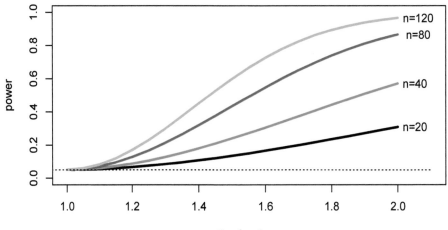

56.7 (a) For Statistician A, the power of the test is

$$P\left(\frac{5(\bar{X} - 100)}{\sqrt{35}} > c\right) = P\left(Z > c - (\mu_a - 100)\sqrt{\frac{5}{7}}\right).$$

(b) For Statistician B, the power of the test is

$$P\left(\frac{X_1/2 + X_2/2 + X_3/3 + X_4/3 + X_5/3 - 200}{\sqrt{5}} > c\right)$$
$$= P\left(Z > c - \frac{2(\mu_a - 100)}{\sqrt{5}}\right)$$

The power curves are shown below; Statistician B has a higher-power test than Statistician A over the range of μ values.

(c) To find an UMVUE estimator, we first find a sufficient statistic. The joint density is $f_\mu(x_1, x_2, x_3, x_4, x_5) =$

$$\frac{1}{(2\pi)^{5/2}} \exp\left\{-\frac{1}{2}\left[\frac{(x_1 - \mu)^2}{4} + \frac{(x_2 - \mu)^2}{4} + \frac{(x_3 - \mu)^2}{9}\right.\right.$$
$$\left.\left. + \frac{(x_4 - \mu)^2}{9} + \frac{(x_5 - \mu)^2}{9}\right]\right\},$$

which becomes

$$\frac{1}{(2\pi)^{5/2}} \exp\left\{-\frac{1}{2}\left[\frac{x_1^2 + x_2^2}{4} + \frac{x_3^2 + x_4^2 + x_5^2}{9}\right.\right.$$
$$\left.\left. - 2\mu\left(\frac{x_1 + x_2}{4} + \frac{x_3 + x_4 + x_5}{9}\right) + \frac{5\mu^2}{6}\right]\right\},$$

so we see by the factorization theorem that

$$T = \frac{X_1 + X_2}{4} + \frac{X_3 + X_4 + X_5}{9}$$

is sufficient for μ. The statistic T has a normal distribution with mean $5\mu/6$ and variance $5/6$, so under H_0,

$$Z = \frac{T - 500/6}{\sqrt{5/6}} \sim N(0, 1).$$

The following plot shows that the statistic in (c) has the best power, with the power for the test in (b) coming in a close second.

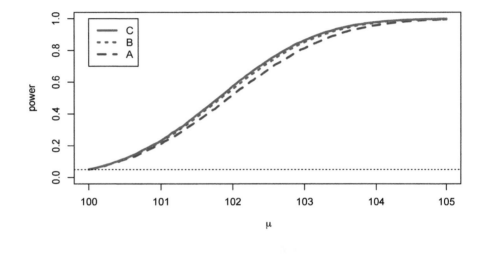

57.1 (a) The ratio of likelihoods is proportional to $e^{\sum_{i=1}^{n} y_i/4}$; the most powerful test rejects H_0 when this is large. This is equivalent to rejecting H_0 when $\sum_{i=1}^{n} y_i$ is large. The statistic $S = \sum_{i=1}^{n} Y_i$ has a Gamma$(n, 1/2)$ distribution when H_0 is true, so we reject H_0 when S is larger than the $100(1 - \alpha)$th percentile of a Gamma$(n, 1/2)$ distribution. By the Neyman–Pearson lemma, this is uniformly the most powerful of all size-α tests.

(b) The sum of the observed values is 18.8, and we compare that to a Gamma$(5, 2)$ distribution to get a p-value: `1-pgamma(18.8,5,1/2)` returns .043, so we would reject H_0 if $\alpha = .05$, but accept if $\alpha = .01$.

(c) For $n = 5$, we reject H_0 when $S > 18.3$, and when H_a is true, S is Gamma$(5, 1/4)$ and the power is P$(S > 18.3) = .52$.

57.3 (a) The likelihood function is

$$L(\lambda; x_1, x_2) = \frac{e^{-3\lambda} \lambda^{x_1 + x_2} 2^{x_2}}{x_1! x_2!},$$

and the ratio to construct the test statistic is

$$\frac{L(4; x_1, x_2)}{L(1; x_1, x_2)} = e^{-9} 4^{x_1 + x_2}.$$

The most powerful test rejects H_0 when $T = X_1 + X_2$ is small.

(b) The test statistic in (a) is distributed as Pois(12) under H_0. The p-value is the probability that $T \leq 2$ when H_0 is true:

$$P(T \leq 2) = e^{-12} + 12e^{-12} + \frac{1}{2}12^2 e^{-12} = .0005.$$

57.5 (a) The joint likelihood is

$$L(\beta; y_1, y_2, y_3) = \frac{\beta^2}{\Gamma(2)} y_1 e^{-\beta y_1} \times \frac{\beta^4}{\Gamma(4)} y_1^3 e^{-\beta y_2} \times \frac{\beta^6}{\Gamma(6)} y_1^5 e^{-\beta y_3}$$
$$\propto \beta^{12} e^{-\beta(y_1+y_2+y+3)}.$$

The ratio of likelihoods is

$$\frac{L(2; y_1, y_2, y_3)}{L(4; y_1, y_2, y_3)} \propto e^{2(y_1+y_2+y_3)},$$

so we reject H_0 when $T = Y_1 + Y_2 + Y_3$ is small. Under the null hypothesis, T has a Gamma(12, 2) density, so we reject when T is smaller than the αth quantile of Gamma(12, 2).

(b) The observed value of T is 1.831, so `pgamma(1.831,12,2)` returns a p-value: .0004.

(c) When $\alpha = .05$ the critical value is found with `qgamma(.05,12,2)`, which is 3.462, and the power is found with `pgamma(3.462,12,4)`, which is about .727.

57.7 Let X be the sample maximum; then using the method from Chapter 33, we can find that the CDF for X is

$$F_X(x) = \left[1 - e^{-x^2/\theta} \right]^4.$$

(a) Under H_0, $P(X > 3) = 1 - \left[1 - e^{-9/2} \right]^4 = .0437$.

(b) We solve $P(X > c) = 1 - \left[1 - e^{-c^2/2} \right]^4 = .05$ to get $c = 2.954$.

(c) The power if $\theta = 4$ is $1 - \left[1 - e^{-c^2/4} \right]^4 = .381$. The UMP derived in the text will have the decision rule "reject when $S = Y_1^2 + Y_2^2 + Y_3^2 + Y_4^2$ is greater than the 95th percentile of a Gamma(4, 1/2) density, or reject when $S > 15.2$." The power if $\theta = 4$ is the probability that a Gamma(4, 1/4) random variable is larger than 15.5, which is .458. The power for the test using the maximum is smaller, as expected.

57.9 (a) The likelihood is

$$L(\mu) = \prod_{i=1}^{n} \left[\frac{1}{\sqrt{20\pi}} \exp\left\{ -\frac{1}{20}(y_i - \mu)^2 \right\} \right] \prod_{i=1}^{n} \left[\frac{1}{\sqrt{20\pi}} \exp\left\{ -\frac{1}{20}(x_i - 2\mu)^2 \right\} \right]$$
$$= \left(\frac{1}{20\pi} \right)^n \exp\left\{ -\frac{1}{20} \left[\sum_{i=1}^{n}(y_i - \mu)^2 + \sum_{i=1}^{n}(x_i - 2\mu)^2 \right] \right\}.$$

We will reject H_0 when $L(4)/L(\mu_a)$ is "small," where we know $\mu_a > 4$. This ratio is

$$\exp\left\{-\frac{1}{20}\left[\sum_{i=1}^{n}(y_i - 4)^2 + \sum_{i=1}^{n}(x_i - 8)^2\right]\right.$$
$$\left. + \frac{1}{20}\left[\sum_{i=1}^{n}(y_i - \mu_a)^2 + \sum_{i=1}^{n}(x_i - 2\mu_a)^2\right]\right\}$$
$$= \exp\left\{-\frac{1}{20}\left[(8 - 2\mu_a)\sum_{i=1}^{n}y_i + (16 - 4\mu_a)\sum_{i=1}^{n}x_i - 80n + 5\mu_a^2\right]\right\}.$$

Noting that $\mu_a > 4$, we see that rejecting when this ratio is small is equivalent to rejecting when $\sum_{i=1}^{n}y_i + 2\sum_{i=1}^{n}x_i$ is small. Therefore let

$$T = \sum_{i=1}^{n}Y_i + 2\sum_{i=1}^{n}X_i$$

be our test statistic. This is normal with mean $5n\mu$ and variance $50n$; given a test size α, we reject H_0 when our observed T is larger than the $100(1-\alpha)$th percentile of this distribution when $\mu = 4$.

(b) For the given data, our test statistic is $T_{obs} = 204.8$, and the null distribution of T has mean 160 and variance 400. The p-value is shown below:

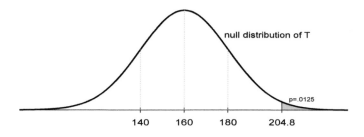

57.11 (a) Given θ_0 and $\theta_a < \theta_0$, the ratio of likelihoods is

$$\frac{\theta_a^n / \left(\prod_{i=1}^{n}x_i\right)^{\theta_a - 1}}{\theta_0^n / \left(\prod_{i=1}^{n}x_i\right)^{\theta_0 - 1}} = \left(\frac{\theta_a}{\theta_0}\right)^n \left[\prod_{i=1}^{n}x_i\right]^{\theta_0 - \theta_a}.$$

We reject when this ratio is *large*, which is equivalent to rejecting H_0 when $\prod_{i=1}^{n}x_i$ is large (because $\theta_0 > \theta_a$), which in turn is equivalent to rejecting when $\sum_{i=1}^{n}\log(x_i)$ is large.

(b) Suppose $X \sim f_\theta$, and note that the CDF for X is $F_X(x) = 1 - x^{-\theta}$. Let $Y = \log(X)$; we can find the distribution of Y using the CDF method:

$$F_Y(y) = P(Y \leq y) = P(\log(X) \leq y) = P(X \leq e^y) = 1 - e^{-\theta y}.$$

This is the CDF for an exponential random variable with rate θ or mean $1/\theta$. Therefore,

$$\sum_{i=1}^{n}\log(X_i) \sim \text{Gamma}(n, \theta).$$

(c) Our sample size is $n = 4$, and we observe $\sum_{i=1}^{n} \log(x_i) = 3.166$. We compare this to the 95th percentile of a Gamma$(4, 4.2)$ density, found using qgamma(.95,4,4.2), which is only 1.846, so we would reject H_0 at $\alpha = .05$. The p-value is found with 1-pgamma(3.166,4,4.2), which is about $.0008$. There is strong evidence that the true θ is less than 4.2, and the maximum river heights have, on average, increased.

58.1   (a) We can write the likelihood

$$L(\lambda_1, \lambda_2; \boldsymbol{x}, \boldsymbol{y}) = \left[e^{-n\lambda_1} \frac{\lambda_1^{\sum_{i=1}^{n} x_i}}{\prod_{i=1}^{n} x_i!} \right] \left[e^{-n\lambda_2} \frac{\lambda_2^{\sum_{i=1}^{n} y_i}}{\prod_{i=1}^{n} y_i!} \right].$$

It is not hard to show that the MLEs under the alternative hypothesis are $\hat{\lambda}_1 = \bar{X}$ and $\hat{\lambda}_2 = \bar{Y}$. Under H_0, the data are really just a sample of size $2n$ from a single Poisson population, so $\hat{\lambda}_0 = (\bar{X} + \bar{Y})/2$. Then (with apologies for too many lambdas)

$$\lambda = \frac{L(\hat{\lambda}_0, \hat{\lambda}_0; \boldsymbol{x}, \boldsymbol{y})}{L(\hat{\lambda}_1, \hat{\lambda}_2; \boldsymbol{x}, \boldsymbol{y})} = \frac{\hat{\lambda}_0^{\sum_{i=1}^{n} x_i + \sum_{i=1}^{n} y_i}}{\hat{\lambda}_1^{\sum_{i=1}^{n} x_i} \hat{\lambda}_2^{\sum_{i=1}^{n} y_i}},$$

after some canceling, and

$$-2\log(\lambda) = 2 \left[\left(\sum_{i=1}^{n} x_i \right) \log(\hat{\lambda}_1) + \left(\sum_{i=1}^{n} y_i \right) \log(\hat{\lambda}_2) \right.$$
$$\left. - \left(\sum_{i=1}^{n} x_i + \sum_{i=1}^{n} y_i \right) \log(\hat{\lambda}_0) \right].$$

When n is "large" the distribution of this test statistic is close to $\chi^2(1)$ ($r = 2$ and $r_0 = 1$). If H_a is true, we expect the values of $-2\log(\lambda)$ to be larger, so we reject H_0 when our observed value of $-2\log(\lambda)$ is larger than the $100(1 - \alpha)$th percentile of a $\chi^2(1)$-density.

(b) The code below will sample from two Poisson populations with mean $\lambda_1 = \lambda_2 = 2.5$, with sample sizes $n_1 = n_2 = 30$. The true test size is very close to the target—we get .050.

```
n=30;nloop=1000000
truelam1=2.5;truelam2=2.5
tst=1:nloop
for(iloop in 1:nloop){
    x=rpois(n,truelam1)
    y=rpois(n,truelam2)t
    lam1=mean(x)
    lam2=mean(y)
    lam0=(lam1+lam2)/2
    tst[iloop]=(sum(x)*log(lam1)+sum(y)*log(lam2)
            -(sum(x)+sum(y))*log(lam0))
}
tst=2*tst
sum(tst>qchisq(.95,1))/nloop
```

58.3 The likelihood is

$$L(\mu_1, \mu_2; \boldsymbol{x}, \boldsymbol{y}) = (\text{constant}) \exp\left\{ -\frac{1}{2}\left[\sum_{i=1}^{n}(x_i - \mu_1)^2 + \sum_{i=1}^{m}(y_i - \mu_2)^2 \right] \right\}.$$

Here $\Omega_0 = \{(\mu_1, \mu_2) : \mu_1 = \mu_2\}$, $\Omega_a = \{(\mu_1, \mu_2) : \mu_1 \neq \mu_2\}$, and $\Omega = \{(\mu_1, \mu_2) \in \mathbb{R}^2\}$. It is straightforward to maximize the likelihood over Ω: $\hat{\mu}_1 = \bar{X}$ and $\hat{\mu}_2 = \bar{Y}$. Maximizing the likelihood over Ω_0 gives

$$\hat{\mu}_0 = \frac{\sum_{i=1}^{n} X_i + \sum_{i=1}^{m} Y_i}{m+n},$$

because under H_0, the two samples can be treated as one sample of size $m + n$. Then

$$\lambda = \exp\left\{ -\frac{1}{2}\left[\sum_{i=1}^{n}(x_i - \hat{\mu}_0)^2 + \sum_{i=1}^{m}(y_i - \hat{\mu}_0)^2 - \sum_{i=1}^{n}(x_i - \hat{\mu}_1)^2 \right.\right.$$
$$\left.\left. - \sum_{i=1}^{m}(y_i - \hat{\mu}_2)^2 \right] \right\}$$

$$= \exp\left\{ -\frac{1}{2}\left[-2\hat{\mu}_0\left(\sum_{i=1}^{n}x_i + \sum_{i=1}^{m}y_i \right) + (m+n)\hat{\mu}_0^2 \right.\right.$$
$$\left.\left. + 2\hat{\mu}_1\sum_{i=1}^{n}x_i + 2\hat{\mu}_2\sum_{i=1}^{m}y_i - n\hat{\mu}_1^2 - m\hat{\mu}_2^2 \right] \right\}$$

$$= \exp\left\{ -\frac{1}{2}\left[-(m+n)\hat{\mu}_0^2 + n\hat{\mu}_1^2 + m\hat{\mu}_2^2 \right] \right\}.$$

Rejecting H_0 when λ is small is equivalent to rejecting H_0 when

$$W = n\hat{\mu}_1^2 + m\hat{\mu}_2^2 - (m+n)\hat{\mu}_0^2$$

is large. Now we use

$$\hat{\mu}_0 = \frac{n\hat{\mu}_1 + m\hat{\mu}_2}{n+m},$$

so

$$W = n\hat{\mu}_1^2 + m\hat{\mu}_2^2 - \frac{(n\hat{\mu}_1 + m\hat{\mu}_2)^2}{n+m}$$
$$= n\hat{\mu}_1^2 + m\hat{\mu}_2^2 - \frac{n^2\hat{\mu}_1^2 + m^2\hat{\mu}_2^2 + 2nm\hat{\mu}_1\hat{\mu}_2}{n+m}$$
$$= \hat{\mu}_1^2\left(n - \frac{n^2}{n+m} \right) + \hat{\mu}_2^2\left(m - \frac{m^2}{n+m} \right) - \frac{2nm\hat{\mu}_1\hat{\mu}_2}{n+m}$$
$$= \frac{mn}{n+m}(\hat{\mu}_1^2 + \hat{\mu}_2^2 - 2\hat{\mu}_1\hat{\mu}_2) = \frac{mn}{n+m}(\hat{\mu}_1 - \hat{\mu}_2)^2.$$

Finally, we get what we always expected: Reject H_0 when $|\hat{\mu}_1 - \hat{\mu}_2|$ is large. We have, when H_0 is true,

$$\frac{\hat{\mu}_1 - \hat{\mu}_2}{\sqrt{\frac{1}{n} + \frac{1}{m}}} \sim \mathrm{N}(0, 1).$$

58.5   (a) The likelihood is

$$L(\sigma_1^2, \sigma_2^2, \sigma_3^2; data) \propto \left(\frac{1}{\sigma_1^2}\right)^{\frac{n_1}{2}} \left(\frac{1}{\sigma_2^2}\right)^{\frac{n_2}{2}} \left(\frac{1}{\sigma_3^2}\right)^{\frac{n_3}{2}}$$

$$\times e^{-\frac{1}{2\sigma_1^2}\sum_{i=1}^{n_1} x_i^2 - \frac{1}{2\sigma_2^2}\sum_{i=1}^{n_2} y_i^2 - \frac{1}{2\sigma_3^2}\sum_{i=1}^{n_3} z_i^2}.$$

This is maximized for $\hat{\sigma}_1^2 = \sum_{i=1}^{n_1} x_i^2/n_1$, $\hat{\sigma}_2^2 = \sum_{i=1}^{n_2} y_i^2/n_2$, and $\hat{\sigma}_3^2 = \sum_{i=1}^{n_3} z_i^2/n_3$.

When $\sigma_1 = \sigma_2 = \sigma_3 = \sigma_0$, the likelihood is maximized at

$$\hat{\sigma}_0^2 = \frac{\sum_{i=1}^{n_1} x_i^2 + \sum_{i=1}^{n_2} y_i^2 + \sum_{i=1}^{n_3} z_i^2}{n_1 + n_2 + n_3}.$$

Therefore the logarithm of the likelihood ratio is

$$\log(\lambda) = \left[\frac{n_1}{2}\log(\hat{\sigma}_1^2) + \frac{n_2}{2}\log(\hat{\sigma}_2^2) + \frac{n_3}{2}\log(\hat{\sigma}_3^2)\right] - \left[\frac{n_1+n_2+n_3}{2}\log(\hat{\sigma}_0^2)\right]$$

and

$$-2\log(\lambda) \approx \chi^2(2).$$

Therefore the decision rule for the LRT is "reject H_0 if our observed value of $-2\log(\lambda)$ is greater than the $100(1 - \alpha)$th percentile of a $\chi^2(2)$ distribution."

(b) For the `tooth` data, $p = .699$.

(c) For $n_1 = n_2 = n_3 = 20$ and $\sigma^2 = 10$, we can do simulations to find that the true test size is a little inflated. For the target test size of $\alpha = .05$, the likelihood ratio test rejects H_0 about 5.33% of the time when H_0 is true. The code is

```
n=20
nloop=1000000
nrej=0
lrt=1:nloop
for(iloop in 1:nloop){
    x=rnorm(n,0,sqrt(10));y=rnorm(n,0,sqrt(10));z=rnorm(n,0,sqrt(10))
    sx=sum(x^2);sy=sum(y^2);sz=sum(z^2);
    sig0=(sx+sy+sz)/3/n
    l1=3*n*log(sig0)
    l2=n*log(sx/n)+n*log(sy/n)+n*log(sz/n)
    lrt[iloop]=l1-l2
}
sum(lrt>qchisq(.95,2))
```

58.7 The Beta$(\theta, 1)$ density is $f_\theta(y) = \theta y^{\theta-1}$ for $y \in (0, 1)$, so the likelihood function for the two independent samples is

$$L(\theta_1, \theta_2; \boldsymbol{y}, \boldsymbol{x}) = \theta_1^n \left[\prod_{i=1}^n y_i\right]^{\theta_1-1} \theta_2^m \left[\prod_{i=1}^m x_i\right]^{\theta_2-1}$$

$$= \theta_1^n e^{(\theta_1-1)\sum_{i=1}^n \log(y_i)} \theta_2^m e^{(\theta_2-1)\sum_{i=1}^n \log(x_i)}.$$

Maximizing the likelihood under H_a gives

$$\hat{\theta}_1 = -\frac{n}{\sum_{i=1}^{n} \log(y_i)} \quad \text{and} \quad \hat{\theta}_2 = -\frac{m}{\sum_{i=1}^{m} \log(x_i)},$$

and under H_0 we have

$$\hat{\theta}_0 = -\frac{n + m}{\sum_{i=1}^{n} \log(y_i) + \sum_{i=1}^{m} \log(x_i)}.$$

When we compute the likelihood ratio, we get a lot of cancellations (including everything in the exponents): The ratio is

$$\lambda = \frac{\hat{\theta}_0^{n+m}}{\hat{\theta}_1^n \hat{\theta}_2^m}.$$

Then the test statistic $-2\log(\lambda)$ has approximately a $\chi^2(1)$ distribution.

```
nloop=10000
th0=2;n=20;m=20
llhrat=1:nloop
for(iloop in 1:nloop){
   x=rbeta(n,th0,1)
   y=rbeta(m,th0,1)
   th0hat=-(n+m)/(sum(log(x))+sum(log(y)))
   th1hat=-n/sum(log(x))
   th2hat=-m/sum(log(y))
   llhrat[iloop]=th0hat^(n+m)/th1hat^n/th2hat^m
}
cstat=-2*log(llhrat)
plot(qchisq(1:nloop/(nloop+1),1),sort(cstat),xlab="Chi(1) quantiles")
lines(c(0,20),c(0,20),col=2,lwd=3)
```

The probability plot of the last lines of the code is

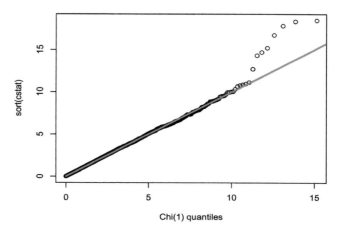

58.9 (a) Let $X_1, \ldots, X_m$ be the filament lifetimes for the sample from Company A, and let $Y_1, \ldots, Y_n$ be the filament lifetimes for the sample from Company B. The likelihood is

$$L(\theta_A, \theta_B; \boldsymbol{x}, \boldsymbol{y}) = \frac{2^{m+n} \left[\prod_{i=1}^{m} x_i\right] \left[\prod_{i=1}^{n} y_i\right]}{\theta_A^m \theta_B^n} e^{-\sum_{i=1}^{n} x_i^2 - \sum_{i=1}^{n} y_i^2}.$$

Finding the MLEs under the alternative hypothesis results in $\hat{\theta}_A = \sum_{i=1}^{m} x_i^2$ and $\hat{\theta}_B = \sum_{i=1}^{n} y_i^2$; under the null hypothesis the common MLE for θ_A and θ_B is $\hat{\theta}_0 = (\sum_{i=1}^{m} x_i^2 + \sum_{i=1}^{n} y_i^2)/(m+n)$. Plugging these into the likelihood ratio gives

$$\lambda = \frac{\hat{\theta}_A^m \hat{\theta}_B^n}{\hat{\theta}_0^{m+n}},$$

canceling the products of the observations and the terms in the exponents. Now

$$-2\log(\lambda) = 2(m+n)\log(\hat{\theta}_0) - 2m\log(\hat{\theta}_A) - 2n\log(\hat{\theta}_B)$$

is approximately $\chi^2(1)$ if the sample size is "large."

(b) The data give $\hat{\theta}_A = 4.03$, $\hat{\theta}_A = 7.98$, and $\hat{\theta}_0 = 6.01$, so that $-2\log(\lambda) = 2.29$ and the p-value is .13.

(c) The following code shows that for $n = m = 10$ and $\theta = 6$, the true test size is .053 when the target is .05. (The CDF trick was used to generate data from the Weibull density.)

```
m=10;n=10
th=6
nloop=1000000
tstat=1:nloop
for(iloop in 1:nloop){
    u=runif(m)
    x=sqrt(-th*log(1-u))
    u=runif(n)
    y=sqrt(-th*log(1-u))
    th0=(sum(x^2)+sum(y^2))/(m+n)
    th1=sum(x^2)/m
    th2=sum(y^2)/n
    tstat[iloop]=2*(m+n)*log(th0)-2*m*log(th1)-2*n*log(th2)
}
sum(tstat>qchisq(.95,1))
```

59.1 The expected values are

Fizzypop	Yippee	Gassaqua	others
80	60	40	20

and the chi-squared statistic is 3.025. Because the 90th percentile of a chi-squared distribution with 3 degrees of freedom is 6.25, we cannot reject the null hypothesis at $\alpha = 0.10$.

59.3 The code gives a p-value of about .78, indicating that there is *no* evidence in this data set that the counts of hurricanes do not follow a Poisson distribution. The expected Poisson counts $(8.40, 6.96, 15.67, 10.83, 6.15)$ with the estimated mean are actually quite close to the observed counts $(9, 9, 13, 11, 6)$.

```
library(Sleuth2)
x=ex1028$Hurricanes
n=length(x)
s=c(sum(x<4),sum(x==4),sum(x==5)+sum(x==6),sum(x==7)+sum(x==8),sum(x>8))
lam=mean(x)
pr=1:13
for(i in 1:12){pr[i]=exp(-lam)*lam^(i-1)/gamma(i)}
pr[13]=1-sum(pr[1:12])
p=c(sum(pr[1:4]),pr[5],sum(pr[6:7]),sum(pr[8:9]),sum(pr[10:13]))
e=p*n
xobs=sum((e-s)^2/e)
1-pchisq(xobs,3)
```

59.5 The table of expected counts:

Expected counts of nausea duration categories

	< 1 hour	1–5 hours	5–12 hours	> 12 hours
New anesthesia	11	31.5	29.5	8
Standard anesthesia	11	31.5	29.5	8

The chi-squared statistic is computed by summing over all eight cells:

$$\chi^2_{obs} = \frac{(15-11)^2}{11} + \frac{(35-31.5)^2}{31.5} + \frac{(25-31.5)^2}{31.5} + \frac{(5-8)^2}{8}$$
$$+ \frac{(7-11)^2}{11} + \frac{(28-31.5)^2}{31.5} + \frac{(34-31.5)^2}{31.5} + \frac{(11-8)^2}{8} = 7.48.$$

If we test the hypotheses at $\alpha = 0.10$, we compare our observed statistic 7.48 with the 90th percentile of a chi-squared density with $(2-1) \times (4-1) = 3$ degrees of freedom, and reject H_0 at $\alpha = .10$. However, we would not reject H_0 at $\alpha = .05$.

59.7 The observed χ^2-statistic is about 9.1, so we reject H_0 at $\alpha = 0.01$ and conclude that the subjects on the Mediterranean diet have a significantly lower proportion of cardiac deaths.

60.1 (a) Using $SSA + SSW = SST$, the definitions for the mean squares, etc., we have the ANOVA table

Source	SS	df	MS	F-stat	p-value bounds
Treatment (among)	127.0	3	42.333	5.085	.0215
Residual (within)	166.5	20	8.325		
Total	293.5	23			

(b) At $\alpha = .05$, we reject the null hypothesis and conclude that at least one of the four treatments has a different mean effect.

(c) The MLE for σ^2 is the SSW divided by the sample size: $166.5/24 = 6.94$.

60.3 (a) We can enter the data into R and obtain the ANOVA table with these commands:

```
y=c(45,39,60,56,47,39,42,61,67,43,55,60,61,46,44,47,56,59,30,69,
   72,27,18,42)
trt=rep(1:4,6)
anova(lm(y~factor(trt)))

Analysis of Variance Table

Response: y
            Df Sum Sq Mean Sq F value  Pr(>F)
factor(trt)  3 1380.5  460.15  3.2506 0.04338
Residuals   20 2831.2  141.56
```

(b) The p-value is .04338. This means that if the null hypothesis that there is no difference in the nutrient packages is true, and we repeat this experiment many times, then about 4.3% of the time we will have this much or more evidence for the alternative hypothesis that at least one nutrient package results in a different average starch content than another.

(c) The R^2 is $SSA/SST = SSA/(SSA + SSW) = .328$. About 32.8% of the variation in starch content of tomatoes is explained by the nutrient package.

60.5 The ANOVA table is obtained:

```
anova(lm(chickwts$weight~chickwts$feed))

Response: chickwts$weight
               Df Sum Sq Mean Sq F value    Pr(>F)
chickwts$feed   5 231129   46226  15.365 5.936e-10 ***
Residuals      65 195556    3009
```

There is strong evidence that there is a difference in chick weight depending on the type of feed; the p-value is about 6×10^{-10}. The normal probability plot is obtained by the commands

```
res=resid(lm(chickwts$weight~chickwts$feed))
plot(qnorm(1:71/72),sort(res),xlab="normal quantiles",
              ylab="sorted residuals")
```

and, as shown below, there is no evidence against the normal errors assumption.

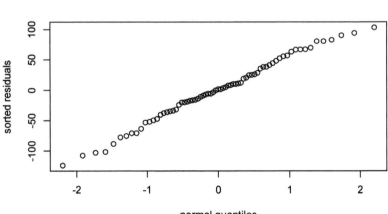

60.7 We know that (under either hypothesis)

$$\frac{SSW}{\sigma^2} \sim \chi^2(n - k).$$

Therefore (because the expected value of a chi-squared random variable is the degrees of freedom),

$$\mathrm{E}\left(\frac{SSW}{n}\right) = \sigma^2 \frac{n - k}{n}.$$

An unbiased estimate of σ^2 is called S^2 (as with the t-tests) and is

$$S^2 = \frac{SSW}{n - k}.$$

This is the mean squares within (MSW) of the table!

Appendix B
Useful Mathematics

Readers are expected to have learned calculus through multiple integration. This review focuses on the pieces of the math background that are most needed for concepts in this book.

B.1 ▪ Countable and Uncountable Infinity

It might seem odd that there can be more than one "size" of infinity. How can one infinity be larger than another? How can we measure and compare one infinite set to another?

Here is an analogy: Suppose we live in prehistoric times and do not know how to count very well—there are words for one, two, three, and many. If there are two heaps of stones that are both "many," how can we tell which heap has more if we can't count them? One way to compare without counting is with a "one-to-one" mapping. If we match up every stone in one pile with exactly one stone in the other pile, and we have no stones left over, then we conclude that the piles have the same number of stones.

This is how we compare sizes of infinity. Here is the rule: If we can make a one-to-one mapping from one set to another with no leftovers in either group, the sets must be the same size. This works for both finite and infinite sets.

The set of counting numbers $\{1, 2, 3, \ldots\}$ is the smallest infinity; we say this set is "countable" or "enumerable." Any set for which there is a one-to-one and onto mapping with the positive integers is a countable set.

For example, the set of even positive integers is countably infinite. You might think this set is "smaller" than the set of *all* the positive integers, but the mapping $k \to 2k$ maps all the positive integers onto the even integers, with no leftovers in either set. Hence, the two sets are the same size, both countably infinite.

Here is where infinite sets can behave differently from finite sets. If we can match up all elements in two finite sets (A and B, say) with no leftovers, then there can't be *another* mapping that uses up all elements in set A with leftovers in set B. However, with infinite sets we *can* have two such mappings. For the above example where set A is the positive integers and set B is the even positive integers, we have the given one-to-one and onto mapping, but there are also one-to-one mappings *with* leftovers. We can map $k \rightarrow k$ and have all the odd numbers left over in set A, or we can map k in A to $4k$ in B, thus using up all of A but having leftovers in B.

It's important to remember that two infinite sets are the same size if *there exists* a one-to-one and onto mapping.

Suppose set A is countable and set B is finite. The union of the two sets, $C = A \cup B$, is countable. For, suppose B has n units, where n is a positive integer. Then define $C_i = B_i$ for $i = 1, \ldots, n$, and define $C_i = A_{n-i}$, for $i > n$. Then we have enumerated the set C.

Next, suppose sets A and B are both countably infinite. The union of the two sets, $C = A \cup B$, is countable: Let A be enumerated as $A = \{x_1, x_2, x_3, \ldots\}$, and let B be enumerated as $B = \{y_1, y_2, y_3, \ldots\}$. Then we can write $C_i = y_{(i+1)/2}$ if i is odd and $C_i = x_{i/2}$ if i is even. Thus, $A \cup B$ is countable. The enumeration of $A \cup B$ can alternatively be shown by a picture:

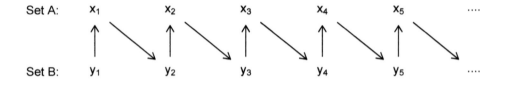

The arrows denote the order in which the set is counted. Now we can put the previous two results together to see that the set of all integers (positive, negative, and zero) is countable.

It's easy to generalize the last result to three or more countable sets: If $A_1, A_2, \ldots, A_k$ are countable, then we can count the union as shown:

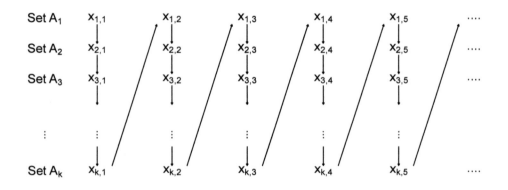

With some care a formula can be given, but we're only after the concepts.

A countable union of countable sets is also countable. This time we do a "zig-zag" enumeration as shown in the following plot. If we count along the diagonals, starting from the upper left corner, we eventually count all the elements of all the sets.

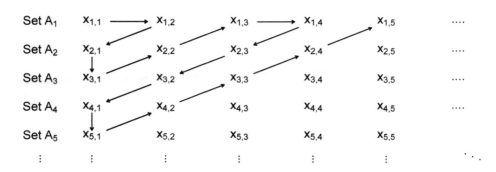

In this way, we can show that the rational numbers are countable. If we let $x_{i,j} = i/j$, then all positive rational numbers are counted using the scheme in the figure above. The set of negative rational numbers can be counted in the same way, and the set of all rational numbers is the union of these two countable sets, plus zero, and so is countable.

The next logical question is, what about the irrational numbers? An irrational number is simply a real number that is not rational, so that the union of the rational numbers and irrational numbers is the set of real numbers. In fact we will show that the irrational numbers are uncountable by showing that the real numbers are uncountable. Actually we will show that we can't even count the real numbers in $(0, 1)$.

Showing that a set is *not* countable is more difficult because we have to show that a scheme for counting *does not exist*, not just that we can't think of one. We use proof by contradiction. Suppose that a counting of the reals in $(0, 1)$ *does* exist; then we can enumerate them and line up the decimal expansions as shown:

$$.372631862....$$
$$.009271194....$$
$$.994712843....$$
$$\vdots$$

Now, we can make a new real number as follows: For the first decimal place, choose something other than 3. For the second, choose something other than 0, and for the third, choose any number except 4. For the kth decimal place, choose something other than the number in the kth decimal place of the kth real number in the enumeration. If we do this for all of the (countably infinite) decimal places, we make a real number that is not in the list. Therefore, the real numbers have *not* been enumerated, and because we can do this for any enumeration, we have shown that the reals are uncountable. It follows that the irrational numbers are also uncountable.

The size of a set is called its "cardinality." This can be a finite number like 7; for sizes of infinite sets it is traditional to use a symbol from the Hebrew alphabet. The cardinality of the countable sets is written as $\aleph_0$ ("aleph-naught"), and the cardinality of the reals is $\aleph_1$ ("aleph-one"). And yes, there are infinitely many sizes of infinity! We have $\aleph_2$, $\aleph_3$, etc. For our purposes, we will need only $\aleph_0$ and $\aleph_1$.

The *continuum hypothesis* states that there is no size of infinity that is greater than α_0 and less than α_1.

B.2 ▪ Sums and Series

We'll start with the sum of the first n integers. Suppose we want to add up the first 100 positive integers: $1 + 2 + \cdots + 99 + 100$. We can notice that the sum of the first and the last number in the list is $1 + 100 = 101$. Also, the second and second-to-last number: $2 + 99 = 101$. The same for the third and third-to-last, and in fact we can do this 50 times, so that $1 + 2 + \cdots + 99 + 100 = 50 \times 101$. You can show that for *both* odd and even n, we have the formula

$$\sum_{k=1}^{n} k = 1 + 2 + \cdots + (n-1) + n = n(n+1)/2.$$

If we let n get larger, of course the sum gets larger; it *diverges* as n increases without bound.

Let's look next at some series that converge. This means we can add a (countably) infinite number of terms and get a finite number. Of course, the terms must get small as we get further out in the summation.

Intuitively, it's easy to add up the following infinite series of numbers:

$$\sum_{k=1}^{\infty} \frac{1}{2^k} = \frac{1}{2} + \frac{1}{4} + \frac{1}{8} + \frac{1}{16} + \frac{1}{32} + \cdots.$$

We can think about traversing the unit interval, starting at zero and taking steps that are the lengths of the terms in the sum. For the first step, we go 1/2 of a unit, which is halfway to one. The next step is 1/4 of a unit, which is half of the remaining distance. The next step of 1/8 of a unit is again half of the remaining distance, and so on. You never get to 1 in any finite number of steps, but you get closer and closer, in fact "arbitrarily close," which means that if you want to get within any small ϵ of 1, there is a finite number of terms you can sum to get there. We say that the series adds to one, or *converges to one*:

$$\sum_{k=1}^{\infty} \frac{1}{2^k} = \lim_{n \to \infty} \sum_{k=1}^{n} \frac{1}{2^k} = 1.$$

There is a trick to adding any series that looks like

$$\sum_{k=0}^{\infty} a^k,$$

as long as $0 < a < 1$.

Let

$$S = 1 + a + a^2 + a^3 + a^4 + \cdots$$

and notice that

$$aS = a + a^2 + a^3 + a^4 + a^5 + \cdots.$$

If we subtract the second equation from the first, most of the terms cancel, and we get

$$S - aS = 1,$$

so that

$$\sum_{k=0}^{\infty} a^k = \frac{1}{1-a}.$$

Sometimes we will need to start with $k = 1$:

$$\sum_{k=1}^{\infty} a^k = \frac{1}{1-a} - 1 = \frac{a}{1-a}.$$

Next, let's compute the sum

$$S = \sum_{k=1}^{\infty} k a^{k-1}$$

using the same ideas. Write

$$S = 1 + 2a + 3a^2 + 4a^3 + 5a^4 + \cdots,$$

then

$$aS = a + 2a^2 + 3a^3 + 4a^4 + \cdots,$$

and now when we subtract we get

$$(1-a)S = 1 + a + a^2 + a^3 + a^4 + \cdots = \frac{1}{1-a},$$

so

$$S = \frac{1}{(1-a)^2}.$$

Here is another way to sum $S = \sum_{k=1}^{\infty} k a^{k-1}$ that is more generalizable. We start by noticing

$$\frac{d}{da} a^k = k a^{k-1},$$

and so

$$\sum_{k=1}^{\infty} k a^{k-1} = \sum_{k=1}^{\infty} \frac{d}{da} a^k = \frac{d}{da} \left[\sum_{k=1}^{\infty} a^k \right].$$

The swapping of the order of the sum and the derivative is allowed if the sum is defined, i.e., if $0 < a < 1$. Using the first summation formula (note that the sum starts at $k = 1$ instead of $k = 0$),

$$\sum_{k=1}^{\infty} k a^{k-1} = \frac{d}{da} \left[\frac{a}{1-a} \right] = \frac{1}{(1-a)^2}.$$

We will also want to find the sum:

$$\sum_{k=2}^{\infty} k(k-1) a^{k-1}.$$

The same trick works with the second derivative:

$$\frac{d^2}{da^2} a^k = k(k-1) a^{k-2},$$

and so

$$\sum_{k=2}^{\infty} k(k-1) a^{k-2} = \sum_{k=2}^{\infty} \frac{d^2}{da^2} a^k = \frac{d^2}{da^2} \left[\sum_{k=2}^{\infty} a^k \right] = \frac{d^2}{da^2} \left[\frac{a^2}{1-a} \right] = \frac{2}{(1-a)^3}.$$

By combining the two previous results, we get

$$\sum_{k=1}^{\infty} k^2 a^{k-1} = \frac{a+1}{(1-a)^3}.$$

With one more iteration (exercise for student!) we get

$$\sum_{k=1}^{\infty} k^3 a^{k-1} = \frac{a^3 + 4a^2 + a}{(1-a)^4}.$$

B.3 ▪ The Taylor Expansion

An important theorem in calculus states that for any function f that is continuous on $[a, x]$, $-\infty < a < x < \infty$, there exists a $c \in (a, x)$ such that

$$\int_a^x f(t)dt = f(c)(x-a).$$

This is illustrated in the figure below, where the shaded area is the integral on the left, while the (equal) striped rectangular area is the expression on the right.

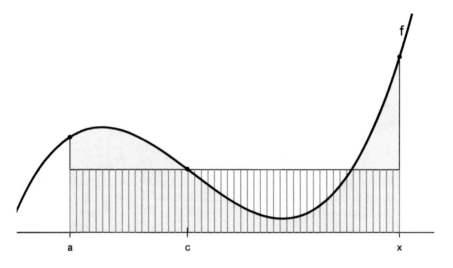

The simplest form of Taylor's theorem is easily derived from this result: If f has a continuous derivative on $[a, x]$, then there is a $c \in (a, x)$ such that

$$f(x) - f(a) = \int_a^x f'(t)dt = f'(c)(x-a).$$

Rewriting, we have for some $c \in (a, x)$

$$f(x) = f(a) + f'(c)(x-a).$$

If f has two continuous derivatives on $[a, x]$, and $x_0 < x$, then for some $c \in (a, x_0)$,

$$f'(x_0) - f'(a) = \int_a^{x_0} f''(t)dt = f''(c)(x_0 - a).$$

Now we integrate again,

$$\int_a^x [f'(x_0) - f'(a)]dx_0 = \int_a^x f''(c)(x_0 - a)dx_0,$$

to get

$$f(x) - f(a) - f'(a)(x - a) = \frac{1}{2}f''(c)(x - a)^2,$$

which is usually written as

$$f(x) = f(a) + f'(a)(x - a) + \frac{1}{2}f''(c)(x - a)^2.$$

We can keep going: If f has three derivatives on $[a, x]$, then for some $c \in (a, x)$,

$$f(x) = f(a) + f'(a)(x - a) + \frac{1}{2}f''(a)(x - a)^2 + \frac{1}{6}f'''(c)(x - a)^3.$$

Generally, if f has k derivatives on $[a, x]$, then for some $c \in (a, x)$,

$$f(x) = f(a) + f'(a)(x - a) + \frac{1}{2}f''(a)(x - a)^2 + \cdots + \frac{1}{k!}f^{(k)}(c)(x - a)^k.$$

The last term is called the "remainder." Even if f has to be "infinitely differentiable" on $[a, b]$, it is not always the case that this remainder goes to zero for $x \in [a, b]$, as k gets larger. If we can show that the remainder *does* go to zero, we have

$$f(x) = \sum_{k=0}^{\infty} \frac{1}{k!}f^{(k)}(a)(x - a)^k.$$

The last expression with $f(x) = e^x$ and $a = 0$ gives us the nifty summation

$$e^x = 1 + x + \frac{x^2}{2} + \frac{x^3}{6} + \cdots + \frac{x^k}{k!} + \cdots.$$

The sequence on the right in the equation above converges to the left-hand side value no matter how far from zero x is. But here is an interesting case where the function f has a k derivative for all $k = 1, 2, \ldots$ on the positive reals, but the Taylor expansion only "works" for a and x "near" each other. Suppose $f(x) = (1 + x)^{1/2}$, and we want to expand f about zero. It is easy to see that for $k = 1, 2, \ldots$

$$f^{(k)}(0) = \prod_{j=1}^{k} \left(\frac{3}{2} - j\right).$$

If we want to approximate $f(2) = \sqrt{3}$ based on this, the kth term of the Taylor series is

$$\frac{2^k}{k!} \prod_{j=1}^{k} \left(\frac{3}{2} - j\right),$$

which increases without bound. Although the derivatives of f exist at zero for all k, the magnitude increases without bound. This is hard to "picture" as the function looks so well-behaved. However, we can use the Taylor expansion about zero to approximate $f(x)$ only for $|x| < 1$.

Taylor's theorem is often used to approximate functions in an interval. For a linear approximation, we say that for x *near* a,

$$f(x) \approx f(a) + f'(a)(x - a).$$

For a quadratic approximation, we can use another term:

$$f(x) \approx f(a) + f'(a)(x - a) + \frac{1}{2}f''(a)(x - a)^2,$$

etc. Here is an illustration of how a curve may be approximated to varying degrees. The black curve is the function f to be approximated, and the red curve is the approximation. The point is at $(a, f(a))$. We sometimes refer to the red curve as the Taylor expansion of f about a.

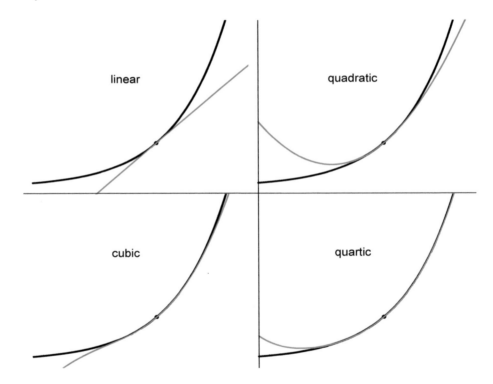

B.4 ▪ L'Hôpital's rule

Suppose we want to find

$$\lim_{x \to 0} \frac{\cos(x) + 2x - 1}{3x}.$$

If we plug in $x = 0$, we get $0/0$, which is undefined. If we plug in increasingly small values of x, we find that the ratio has values getting close to 2/3, so it seems like the answer might be 2/3, but how do we show this? The answer is l'Hôpital's rule.

Suppose that the functions f and g are twice differentiable in (a, b), and $a < c < b$. If $f(c)/g(c) = 0/0$ or $f(c)/g(c) = \infty/\infty$, and if $g'(x) \neq 0$ for $x \in (a, b)$ but $x \neq c$, then

$$\lim_{x \to c} \frac{f(x)}{g(x)} = \lim_{x \to c} \frac{f'(x)}{g'(x)}.$$

Using this with our example, we find

$$\lim_{x \to 0} \frac{\cos(x) + 2x - 1}{3x} = \lim_{x \to 0} \frac{-\sin(x) + 2}{3} = \frac{2}{3}.$$

Graphing the function in R gives the picture below, where the dotted line is at 2/3.

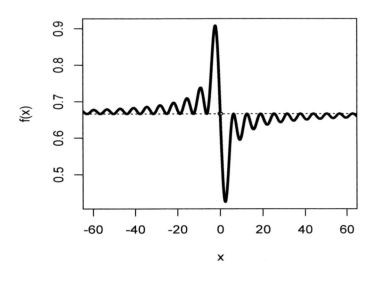

L'Hôpital's rule is used to prove this useful result:

$$\lim_{x \to \infty} \left(1 + \frac{1}{x}\right)^x = e.$$

To do this, we write

$$\left(1 + \frac{1}{x}\right)^x = \exp\left[\log\left\{\left(1 + \frac{1}{x}\right)^x\right\}\right] = \exp\left[x \log\left(1 + \frac{1}{x}\right)\right].$$

Now we find

$$\lim_{x \to \infty} x \log\left(1 + \frac{1}{x}\right) = \lim_{x \to \infty} \frac{\log\left(1 + \frac{1}{x}\right)}{\frac{1}{x}}.$$

Applying l'Hôpital's rule,

$$\lim_{x \to \infty} \frac{\log\left(1 + \frac{1}{x}\right)}{\frac{1}{x}} = \lim_{x \to \infty} \frac{\left(\frac{1}{1 + 1/x}\right)\left(\frac{-1}{x^2}\right)}{-1/x^2} = \lim_{x \to \infty} \frac{1}{1 + 1/x} = 1.$$

Now take $\exp(1)$ to get e.

To see why l'Hôpital's rule works when $f(c) = g(c) = 0$, we consider the Taylor expansions of f and g about c. For $x \in (a, b)$,

$$f(x) = f(c) + f'(c)(x - c) + f''(\xi)(x - c)^2 = f'(c)(x - c) + f''(\xi)(x - c)^2$$

for some $\xi \in (a, b)$, and for some $\zeta \in (a, b)$, we have

$$g(x) = g(c) + g'(c)(x - c) + g''(\zeta)(x - c)^2 = g'(c)(x - c) + g''(\zeta)(x - c)^2.$$

Then

$$\frac{f(x)}{g(x)} = \frac{f'(c)(x - c) + f''(\xi)(x - c)^2}{g'(c)(x - c) + g''(\zeta)(x - c)^2} = \frac{f'(c) + f''(\xi)(x - c)}{g'(c) + g''(\zeta)(x - c)}$$

and taking the limit $x \to c$ gives the result.

B.5 ▪ The Gamma Function

For $a > 0$, the gamma function is defined as

$$\Gamma(a) = \int_0^\infty x^{a-1} e^{-x} dx.$$

Let's find $\Gamma(2)$ using our integration-by-parts method:

$$\Gamma(2) = \int_0^\infty x e^{-x} dx.$$

Letting $u = x$ and $dv = e^{-x} dx$, so that $du = dx$ and $v = -e^{-x}$, we get

$$\int_0^\infty x e^{-x} dx = -x e^{-x} \Big|_0^\infty + \int_0^\infty e^{-x} dx.$$

How do we evaluate $x e^{-x}$ at infinity? We can use l'Hôpital's rule:

$$\lim_{x \to \infty} x e^{-x} = \lim_{x \to \infty} \frac{x}{e^x} = \lim_{x \to \infty} \frac{1}{e^x} = 0,$$

and this gives $\Gamma(2) = 1$.

If we want to find $\Gamma(3)$, we have to use integration by parts twice:

$$\Gamma(3) = \int_0^\infty x^2 e^{-x} dx,$$

so let $u = x^2$ and $dv = e^{-x} dx$. Then $du = 2x dx$ and $v = -e^{-x}$, and

$$\int_0^\infty x^2 e^{-x} dx = -x^2 e^{-x} \Big|_0^\infty + 2 \int_0^\infty x e^{-x} dx.$$

We can use l'Hôpital's rule again for the first term, to get zero (try it), and for the second term, we repeat the steps for $\Gamma(2)$. Finally, we get $\Gamma(3) = 2$.

You might be able to see now that finding $\Gamma(4)$ would involve three applications of integration by parts. Instead, let's use our insight to get a nice formula. Suppose $a > 0$, and find

$$\Gamma(a+1) = \int_0^\infty x^a e^{-x} dx$$

by letting $u = x^a$ and $dv = e^{-x} dx$, so that $du = a x^{a-1} dx$ and $v = -e^{-x}$, and

$$\int_0^\infty x^a e^{-x} dx = -x^a e^{-x} \Big|_0^\infty + a \int_0^\infty x^{a-1} e^{-x} dx.$$

We have proved

$$\Gamma(a+1) = a\Gamma(a)$$

for $a > 0$.

Now it is easy to see that if a is a positive integer, $\Gamma(a) = (a-1)!$ (where the exclamation point means factorial). To be consistent, we must define $0! = 1$.

The gamma function has values on the whole positive real half-line. Let's find $\Gamma(1/2)$:

$$\Gamma(1/2) = \int_0^\infty x^{-1/2} e^{-x} dx.$$

We can make the substitution $u = x^{1/2}$, so that $du = x^{-1/2}/2\,dx$, and noting that the limits stay the same, we get

$$\Gamma(1/2) = 2\int_0^\infty e^{-u^2} du = \int_{-\infty}^\infty e^{-u^2} du$$

by symmetry.

From here, we use a neat trick with polar coordinates. Let

$$I = \int_{-\infty}^\infty e^{-u^2} du,$$

so that

$$I^2 = \int_{-\infty}^\infty \int_{-\infty}^\infty e^{-x^2} e^{-y^2} dx dy = \int_{-\infty}^\infty \int_{-\infty}^\infty e^{-(x^2+y^2)} dx dy.$$

Letting $r^2 = x^2 + y^2$ and $\tan(\theta) = y/x$, the transformation to polar coordinates (recall $dx dy = r dr d\theta$) is

$$I^2 = \int_0^{2\pi} \int_0^\infty e^{-r^2} r dr d\theta.$$

It's easy to integrate with respect to θ to get

$$I^2 = 2\pi \int_0^\infty e^{-r^2} r dr,$$

and substituting $u = r^2$, $du = 2r dr$ results in $I^2 = \pi$. Therefore,

$$\Gamma\left(\frac{1}{2}\right) = \sqrt{\pi}.$$

This gives a whole bunch more values of the gamma function, as, for example,

$$\Gamma\left(\frac{3}{2}\right) = \frac{1}{2}\Gamma\left(\frac{1}{2}\right) = \frac{\sqrt{\pi}}{2},$$

etc.

Here is the picture of our gamma function, with the factorial points marked:

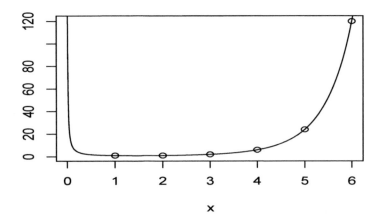

B.6 · The Laplace Method and Stirling Approximation

The Laplace method is a useful tool for a statistician's box and is often used in Bayesian methods to approximate posterior distributions. Here we use it to derive Stirling's formula, which can be used to approximate the factorial function. Consider the problem of approximating

$$\int_a^b e^{Mf(x)} dx$$

in the special case where M is "large" and $f(x)$ has a global maximum at x_0, which is in the interior of the interval of integration: $x_0 \in (a, b)$. The function f is smooth at x_0, so that the first and second derivatives exist at x_0, with $f'(x_0) = 0$ and $f''(x_0) < 0$.

Then, using a quadratic Taylor approximation near x_0,

$$f(x) \approx f(x_0) - \frac{1}{2}|f''(x_0)|(x - x_0)^2,$$

and we approximate the integral as

$$e^{Mf(x_0)} \int_a^b \exp\left\{-\frac{M|f''(x_0)|}{2}(x - x_0)^2\right\} dx.$$

Now we use the assumptions that M is "large" and x_0 is in the interior of (a, b). We recognize the integrand as looking like a normal density with mean x_0 and variance $[M|f''(x_0)|]^{-1}$. This is a small variance, so that we can consider that the function is close to zero except when x is very close to x_0; hence, we can integrate over $(-\infty, \infty)$ without changing the value much. Using the expression for a normal density, we obtain Laplace's approximation:

$$\int_a^b e^{Mf(x)} dx \approx e^{Mf(x_0)}\sqrt{\frac{2\pi}{M|f''(x_0)|}}.$$

Let's use this formula to get the Stirling approximation of $M!$ for a large integer M. Recall the gamma function

$$\Gamma(M + 1) = \int_0^\infty x^M e^{-x} dx = \int_0^\infty e^{M[\log(x) - x/M]} dx.$$

Using Laplace's method with $f(x) = \log(x) - x/M$, we have $x_0 = M$, and

$$\Gamma(M + 1) \approx e^{M(\log(M)-1)}\sqrt{2\pi M} = e^{-M}M^M\sqrt{2\pi M} = \left(\frac{M}{e}\right)^M \sqrt{2\pi M}.$$

When M is an integer, this is a nifty approximation of $M!$.

The application of Stirling's formula in Chapter 34 is for a large integer n:

$$\frac{\Gamma\left(n + \frac{1}{2}\right)}{\Gamma(n)} \approx \frac{e^{-(n+1/2)}(n + 1/2)^{n+1/2}\sqrt{2\pi(n + 1/2)}}{e^{-n}n^n\sqrt{2\pi n}}.$$

This simplifies to

$$e^{-1/2}\left(1 + \frac{1}{2n}\right)^n \sqrt{n + 1/2}\sqrt{\frac{n + 1/2}{n}} \approx \sqrt{n}.$$

The first two terms cancel (approximately) because

$$\lim_{n\to\infty}\left(1 + \frac{1}{2n}\right)^n = e^{1/2}$$

by l'Hôpital's rule.

B.7 ▪ Multiple Integration

Once you have mastered integration over a single variable, multiple integration is easy if the area of integration is a rectangle. Consider a function $f(x, y) : \mathbb{R}^2 \to \mathbb{R}$, as shown:

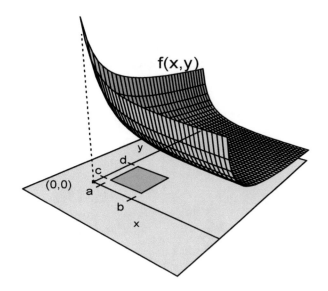

Then the area under $f(x, y)$, over the rectangle $[a, b] \times [c, d]$, is

$$\int_c^d \int_a^b f(x, y)\, dx\, dy.$$

The double integral can be performed as two consecutive single-variable integrals.

Now consider double integration where the area of integration is not a rectangle. The figure below shows a triangular area of integration.

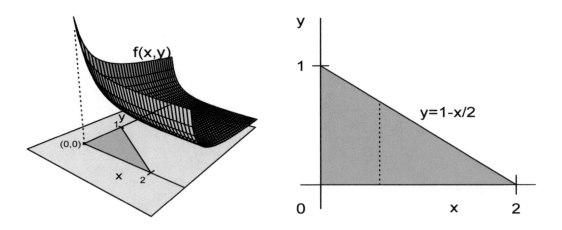

To set up the integral, we first choose the order: Should x or y be the "outside" variable for the integral? Let's choose x so that the infinitesimal of integration is $dy\, dx$. We look at the area of integration to find the range for x. In our example, x goes from 0 to 2. Now we ask, if we fix $x \in [0, 2]$, what values can y take? The answer is, y ranges from 0 to $1 - x/2$. Therefore, the integral over the triangle is

$$\int_0^2 \int_0^{1-x/2} f(x,y)dydx.$$

If we use $dxdy$ as the infinitesimal of integration, then y is "on the outside" and we observe that for our triangular area of integration, y ranges from 0 to 1. For a fixed y, x ranges from 0 to $2(1 - y)$ (looking at the above figure, we can picture the dotted line "sweeping through" the area of integration). Therefore, the integral over the triangle is

$$\int_0^1 \int_0^{2(1-y)} f(x,y)dxdy.$$

To solve the integral, we work from the inside. If $f(x,y) = xy$, we have

$$\int_0^1 \int_0^{2(1-y)} xydxdy = \int_0^1 \left[\int_0^{2(1-y)} xdx \right] ydy$$

$$= \int_0^1 \left[\frac{1}{2}x^2 \Big|_0^{2(1-y)} \right] ydy$$

$$= 2 \int_0^1 (1-y)^2 ydy$$

$$= 2 \int_0^1 (y - 2y^2 + y^3)dy$$

$$= 2(1/2 - 2/3 + 1/4) = 1/6.$$

B.8 ▪ Matrix Notation and Arithmetic

An $m \times n$ matrix A in $\mathbb{R}^m \times \mathbb{R}^n$ is an array with mn numbers arranged in m rows and n columns, in parentheses or brackets. For example, suppose $m = 2$ and $n = 3$ and

$$A = \begin{pmatrix} 4 & 0 & 1 \\ 2 & -2 & 1 \end{pmatrix}.$$

We use subscripts to indicate elements of the matrix; for example, $A_{22} = -2$ and $A_{13} = 1$.

We can add matrices elementwise if they have the same dimensions. If

$$B = \begin{pmatrix} 1 & 2 & -1 \\ 0 & 2 & 1 \end{pmatrix},$$

then

$$A + B = \begin{pmatrix} 5 & 2 & 0 \\ 2 & 0 & 2 \end{pmatrix}.$$

We can multiply our $m \times n$ matrix A by an $n \times k$ matrix C. The i,jth element of the $m \times k$ **product** AC is

$$[AC]_{ij} = \sum_{\ell=1}^n A_{i\ell} C_{\ell j}.$$

For example, if A is the 2×3 matrix given above and

$$C = \begin{pmatrix} 1 & 2 & 0 & 2 \\ 1 & 0 & 2 & -1 \\ 0 & 0 & -1 & 1 \end{pmatrix},$$

then the 2×4 product matrix is

$$AC = \begin{pmatrix} 4 & 8 & -1 & 9 \\ 0 & 4 & -5 & 7 \end{pmatrix}.$$

The **transpose** $A^\top$ of an $m \times n$ matrix A is the $n \times m$ matrix defined by

$$A_{ij}^\top = A_{ji}$$

for $i = 1, \ldots, n$ and $j = 1, \ldots, m$.

A **vector** in $\mathbb{R}^n$ is an $n \times 1$ matrix or an n-tuple written in a column. For $n = 3$, for example, we can define the vector $x \in \mathbb{R}^n$ (by convention vectors are written in boldface) as

$$x = \begin{pmatrix} 3 \\ -1 \\ 2 \end{pmatrix}.$$

We could also write $x = (3, -1, 2)^\top$.

The **determinant** of a square matrix is related to the volume of a "parallelotope" generated by the columns of the matrix. Consider

$$A = \begin{pmatrix} 2 & 1 \\ 1 & 3 \end{pmatrix}.$$

Consider the columns of A, letting $a_1 = \begin{pmatrix} 2 \\ 1 \end{pmatrix}$ and $a_2 = \begin{pmatrix} 1 \\ 3 \end{pmatrix}$. The determinant of A is the area "swept out by" these vectors, as shown as the yellow area below.

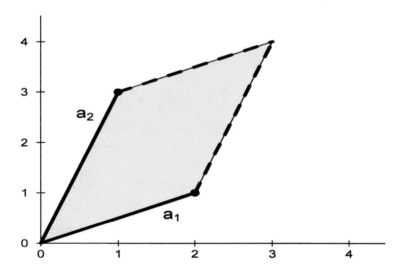

For 3×3 matrices, the determinant is a volume, and for larger square matrices, the determinant is a "hypervolume." (The determinant can be negative, in which case its absolute value is the area or volume.)

The formula for the determinant of a 2×2 matrix is

$$\det \begin{pmatrix} a & b \\ c & d \end{pmatrix} = ad - bc.$$

B.9 · Jacobian Matrix and Transformation of Variables

The Jacobian is often introduced in the third semester of calculus, as an integration technique involving a change of variables. Suppose we have the double integral over a region R in the plane:

$$\iint\limits_{R} f(x, y) dx dy.$$

We define $w = g_1(x, y)$ and $z = g_2(x, y)$, and we would like to perform the integral over w and z instead of x and y. We can break this task down into three pieces:

1. Find the region S so that $(x, y) \in R$ if and only if $(w, z) \in S$.

2. Find expressions for x and y, in terms of the new variables, to plug into $f(x, y)$ to get a function of w and z. Suppose $x = h_1(w, z)$ and $y = h_2(w, z)$ are the inverse transformations.

3. This step is not so intuitive. We have to transform $dx dy$ into an expression with $dw dz$.

For a very simple example, suppose $w = 2x$ and $z = 3y$, and $R = (0, 1) \times (0, 1)$. The area S must be $(0, 2) \times (0, 3)$, and we have the "inverse transformations" $x = w/2$ and $y = z/3$. It is clear that $6 dx dy = dz dw$, so our integral is transformed as follows:

$$\int_0^1 \int_0^1 f(x, y) dx dy = \frac{1}{6} \int_0^2 \int_0^3 f(w/2, z/3) dz dw.$$

To complete the example, suppose $f(x, y) = xy$. It is easy to compute

$$\int_0^1 \int_0^1 xy \, dx dy = 1/4$$

and

$$\frac{1}{6} \int_0^2 \int_0^3 \left(\frac{w}{2}\right) \left(\frac{z}{3}\right) dz dw = 1/4.$$

In the general case, suppose we have found the region S and the inverse transformations $x = h_1(w, z)$ and $y = h_2(w, z)$. The **Jacobian of the transformation** is

$$J = \det \begin{pmatrix} \frac{\partial h_1}{\partial w} & \frac{\partial h_1}{\partial z} \\ \frac{\partial h_2}{\partial w} & \frac{\partial h_2}{\partial z} \end{pmatrix}$$

and the infinitesimal of integration is transformed as follows:

$$dx dy = |J| \, dw dz.$$

Finally, we have

$$\iint\limits_{R} f(x, y) dx dy = \iint\limits_{S} f(h_1(w, z), h_2(w, z) |J| dw dz.$$

Let's use this method to compute

$$\iint\limits_{\mathbb{R}^2} e^{-(x^2+y^2)}\,dx\,dy.$$

We don't have an antiderivative for e^{-x^2}, but we can transform the problem in **polar coordinates**. Suppose $z^2 = x^2 + y^2$ and $\tan(w) = y/x$. The inverse transformation is $x = z\cos(w)$ and $y = z\sin(w)$. If we let z range from zero to infinity while w ranges from zero to 2π, then (w, z) takes all values in $\mathbb{R}^2$.

The Jacobian is

$$J = \det\begin{pmatrix} -z\sin(w) & \cos(w) \\ z\cos(w) & \sin(w) \end{pmatrix} = -z,$$

so $dx\,dy = z\,dz\,dw$. Continuing with the integration, we have

$$\iint\limits_{\mathbb{R}^2} e^{-(x^2+y^2)}\,dx\,dy = \int_0^\infty \int_0^{2\pi} e^{-z^2} z\,dw\,dz = 2\pi \int_0^\infty e^{-z^2} z\,dz.$$

Defining $u = z^2$ and $du = 2z\,dz$, we find

$$2\pi \int_0^\infty e^{-z^2} z\,dz = \pi \int_0^\infty e^{-u}\,du = \pi.$$

Index